£8-00

D1187723

The Psychology of
Animal Learning

The Psychology of Animal Learning

N. J. MACKINTOSH

Laboratory of Experimental Psychology
University of Sussex, Brighton, Sussex, England

1974

ACADEMIC PRESS: London · New York · San Francisco

A Subsidiary of Harcourt Brace Jovanovich, Publishers

ACADEMIC PRESS INC. (LONDON) LTD.
24–28 Oval Road,
London NW1

United States Edition published by
ACADEMIC PRESS INC.
111 Fifth Avenue
New York, New York 10003

Copyright © 1974 by
ACADEMIC PRESS INC. (LONDON) LTD.

Second printing 1975

Third printing 1976

Library of Congress Catalog Card Number: 73 19009
ISBN: 0 12 464650 6

Printed in Great Britain by
The Garden City Press Limited
Letchworth, Hertfordshire SG6 1JS

PREFACE

My aim in this book has been to provide a reasonably detailed and comprehensive treatment of the main areas of research that have developed from the pioneering work of Pavlov and Thorndike. This is, therefore, a book largely concerned with the topics of classical and instrumental conditioning, and it does not cover everything that could be said about learning in animals. I have been selective partly because it was necessary to keep the length of the book to manageable proportions. This is the best excuse I can offer for the exclusion of most comparative, physiological, and pharmacological work. In other cases, a strict interpretation of my subject matter has led to the exclusion of questions concerning motivation and the defining characteristics of reinforcers, and to a concentration on work involving animals other than man.

It is possible that some of these restrictions reflect little more than my own prejudices. It is certain that some of the judgements and conclusions reached during the course of the book will be seen by others in this light. This is inevitable if a single author is to cover, in its entirety, even this restricted aspect of research on animal learning. The book is not intended to develop a particular theory of learning. But it does attempt to bring out the parallels between areas of research often considered in isolation, and it does contain numerous theoretical judgements. Indeed, I have deliberately and systematically sought to establish connections between problems in one area and those in another, and to relate empirical findings to theoretical issues. Facts that seemed to me to be without great theoretical significance have received scant attention. I have, of course, tried to judge theoretical issues impartially, have always tried to present the evidence for my more explicit judgements, and have not resolved issues where I thought the evidence insufficient. I recognize, nevertheless, that I may have been occasionally somewhat brisk in my dismissal of views I did not favour. I do not suppose that many of my judgements will command universal assent, nor even that I shall myself agree with all of them in five years' time.

Although I have assumed as little knowledge as possible on the part of the reader, I cannot pretend that the book will always make easy reading. Terms and procedures have generally been defined whenever they are first introduced, and the subject index contains explicit references to these definitions. But I have not attempted to write an introductory text for the beginning student. Where there was a choice between ignoring or sliding over some problem, or digging deeper in order to reach a more securely based conclusion, I have usually opted for the latter course. This undoubtedly means that the book makes substantial demands on the reader. I can only hope that the effort will seem worthwhile.

CONTENTS

ACKNOWLEDGEMENTS

This book was planned, and its first draft written, while I held a Killam Research Professorship at Dalhousie University. It is no exaggeration to say that without the opportunity provided by this appointment I should never have embarked on such a project. My first debt of gratitude, therefore, is to the Killam trustees, and to my colleagues in the psychology department at Dalhousie, for providing me with such congenial surroundings for so many years. The final draft was revised during the course of a year spent at the University of Hawaii, and I am grateful for the opportunity and facilities provided there.

I am profoundly indebted to Bruce R. Moore and N. S. Sutherland for carefully reading the entire manuscript. They have removed numerous inelegant expressions, saved me from an embarrassing number of inaccuracies and errors, and have helped to clarify the argument and organization of many sections. The errors that remain are, of course, my own. I am also grateful to Caroline Madsen, Michael Peacock, and Roger Thompson of the University of Hawaii, who also read the manuscript, and provided a large number of helpful suggestions.

Margaret Ross typed the first draft, organized the bibliography, and kept me writing to schedule. Janet Lord and Nancy McLearn typed the final draft, Virginia Vaughan drew the figures, and Valerie Leroy assisted in the preparation of the author and subject indices. I am very grateful to them all.

Some of the costs of preparing the book were borne by Grant A259 from the National Research Council of Canada. It is a pleasure to acknowledge the support I have received from that body.

Although all the figures have been freshly drawn for the book, I thank the following publishers and authors for permission to adapt copyright material. (Numbers in parentheses following each reference refer to figure numbers in the present book).

Academic Press Inc., New York

Bolles, R. C., Moot, S. A., and Grossen, N. E. (1971). *Learning and Motivation* **2,** 324–33. (6.10)

Church, R. M., Wooten, C. L., and Matthews, T. J. (1970). *Learning and Motivation,* **1,** 1–17. (6.1)

Clayton, K. N. (1969). *In Reinforcement and Behavior* (J. T. Tapp, Ed.), pp. 96–119. (4.3a)

Flaherty, C. F., Riley, E. P., and Spear, N. E. (1973). *Learning and Motivation*, **4,** 163–75. (7.9)

Kimble, G. A. (1964). *In Categories of Human Learning* (A. W. Melton, Ed.), pp. 32–45. (3.3d).

Warren, J. M. (1965). *In Behavior of Nonhuman Primates*, Vol. 1, (A. M. Schrier, H. F. Harlow, and F. Stollnitz, Eds.), pp. 249–81. (10.11)

xii ACKNOWLEDGEMENTS

American Association for the Advancement of Science, Washington, D.C.
Hearst, E. (1968). *Science, N.Y.*, **162**, 1303–6. (9.11)
Schneiderman, N., Fuentes, I., and Gormezano, I. (1962). *Science, N.Y.* **136**, 650–2. (2.1b; 2.2b)

The American Psychological Association, Washington, D.C.
Annau, Z., and Kamin, L. J. (1961). *J. Comp. Physiol. Psychol.* **54**, 428–32. (2.1c; 2.2c)
Bower, G. H. (1961). *J. Exp. Psychol.* **62**, 196–9. (7.7)
Bower, G. H., and Trapold, M. A. (1959). *J. Comp. Physiol. Psychol.* **52**, 727–9. (7.6a)
Brown, R. T., and Wagner, A. R. (1964). *J. Exp. Psychol.* **68**, 503–7. (8.6)
Colavita, F. B. (1965). *J. Comp. Physiol. Psychol.* **60**, 218–22. (3.2)
Donahoe, J. W., McCroskery, J. H., and Richardson, W. K. (1970). *J. Exp. Psychol.* **84**, 58–63. (9.5)
Estes, W. K. (1944). *Psychol. Monogr.* **57**, (3, Whole No. 263). (6.3)
Franchina, J. J., and Brown, T. S. (1971). *J. Comp. Physiol. Psychol.* **76**, 365–70. (7.8)
Grice, G. R. (1948). *J. Exp. Psychol.* **38**, 1–16. (4.4)
Grice, G. R. and Saltz, E. (1950). *J. Exp. Psychol.* **40**, 702–8. (9.1c)
Guttman, N., and Kalish, H. I. (1956). *J. Exp. Psychol.* **51**, 79–88. (9.1a)
Hanson, H. M. (1959). *J. Exp. Psychol.* **58**, 321–4. (9.12)
Hearst, E., and Koresko, M. B. (1968). *J. Comp. Physiol. Psychol.* **66**, 133–8. (9.2c; 9.2d)
Hull, C. L. (1947). *J. Exp. Psychol.* **37**, 118–35. (4.8)
Hilgard, E. R. (1936). *Psychol. Rev.* **43**, 366–85. (3.3b)
Hilgard, E. R. and Campbell, A. A. (1936). *J. Exp. Psychol.* **19**, 227–47. (3.3)
Honig, W. K., Boneau, C. A., Burstein, K. R., and Pennypacker, H. S. (1963). *J. Comp. Physiol. Psychol.* **56**, 111–16. (9.10b)
Ison, J. R. (1962). *J. Exp. Psychol.* **64**, 314–17. (8.2)
Jenkins, H. M., and Harrison, R. H. (1960). *J. Exp. Psychol.* **59**, 246–53. (9.10a)
Kalish, H. I., and Haber, A. (1965). *J. Comp. Physiol. Psychol.* **60**, 125–8. (9.8b)
Kamin, L. J. (1957). *J. Comp. Physiol. Psychol.* **50**, 450–6. (6.4)
Kremer, E. F. (1971). *J. Comp. Physiol. Psychol.* **76**, 441–8. (2.8)
Leonard, D. W. (1969). *J. Comp. Physiol. Psychol.* **67**, 204–11. (8.4; 8.5)
Lubow, R. E., Markman, R. E., and Allen, J. (1968). *J. Comp. Physiol. Psychol.* **66**, 688–94. (2.7)
MacKinnon, J. R. (1967). *J. Exp. Psychol.* **75**, 329–38. (7.6b)
Marsh, G. (1967). *J. Comp. Physiol. Psychol.* **64**, 284–9. (9.4)
Mellgren, R. L. and Ost, J. W. P. (1969). *J. Comp. Physiol. Psychol.* **67**, 390–4. (5.1)
Newman, J. R. and Grice, G. R. (1965). *J. Exp. Psychol.* **69**, 357–62. (9.8a)
Prokasy, W. F. (1956). *J. Comp. Physiol. Psychol.* **49**, 131–4. (5.9)
Rescorla, R. A. (1968). *J. Comp. Physiol. Psychol.* **66**, 1–5. (2.5)
Smith, M. C. (1968). *J. Comp. Physiol. Psychol.* **66**, 679–87. (2.13)
Smith, M. C., Coleman, S. R., and Gormezano, I. (1969). *J. Comp. Physiol. Psychol.* **69**, 226–31. (2.13)
Sprow, A. J. (1947). *J. Exp. Psychol.* **37**, 197–213. (4.8)
Sutterer, J. R., and Obrist, P. A. (1972). *J. Comp. Physiol. Psychol.* **80**, 314–26. (3.4)
Thomas, D. R., and Switalski, R. W. (1966). *J. Exp. Psychol.* **71**, 236–40. (9.2a; 9.2b)

Wagner, A. R. (1959). *J. Exp. Psychol.* **57**, 130–6. (5.11)
Wagner, A. R., Siegel, S., Thomas, E., and Ellison, G. D. *J. Comp. Physiol. Psychol.* **58**, 354–8. (2.2a)
Weisman, R. G., and Litner, J. S. (1969). *J. Comp. Physiol. Psychol.* **68**, 597–603. (2.6; 6.6)
Wickens, D. D., Born, D. G., and Wickens, C. D. (1963). *J. Comp. Physiol. Psychol.* **56**, 727–31. (2.10)

Canadian Psychological Association, Montreal
Kamin, L. J. (1957). *Can. J. Psychol.* **11**, 48–56. Used by permission. (6.5)
Mackintosh, N. J., McGonigle, B., Holgate, V., and Vanderver, V. (1968). *Can. J. Psychol.* **22**, 85–95. Used by permission. (10.10)

McGraw-Hill Book Company, New York
Estes, W. K. (1959). *Psychology: A Study of a Science*, Vol. 2 (S. Koch, Ed.), pp. 380–491. (4.2)

Polish Scientific Publications, Warsaw
Konorski, J., and Szwejkowska, G. (1956). *Acta Biol. Exp.* **17**, 141–65. (2.4)

Prentice-Hall Inc., Englewood Cliffs, New Jersey
Ferster, C. B., and Skinner, B. F. (1957). *Schedules of Reinforcement*, New York: Appleton-Century-Crofts (4.5; 4.6)
Gormezano, I. (1965). *In Classical Conditioning: A Symposium* (W. F. Prokasy, Ed.), New York: Appleton-Century-Crofts, pp. 48–70. (3.7)
Hoffman, H. S. (1969). *In Punishment and Aversive Behavior* (B. A. Campbell and R. M. Church, Eds.), New York: Appleton-Century-Crofts, pp. 185–234. (9.10c)
Kamin, L. J. (1965). *In Classical Conditioning: A Symposium* (W. F. Prokasy, Ed.), New York: Appleton-Century-Crofts, pp. 118–47. (2.9)
Moore, J. W. (1972). *In Classical Conditioning II: Current Research and Theory* (A. H. Black and W. F. Prokasy, Eds.), New York: Appleton-Century-Crofts, pp. 206–30. (9.1)
Sidman, M. (1966). *In Operant Behavior* (W. K. Honig, Ed.), New York: Appleton-Century-Crofts, pp. 448–98. (6.8)
Terrace, H. S. (1966). *In Operant Behavior: Areas of Research and Application* W. K. Honig, Ed.), New York: Appleton-Century-Crofts, pp. 271–344. (7.4)
Wahlsten, D. L., and Cole, M. (1972). *In Classical Conditioning II: Current Research and Theory* (A. H. Black and W. K. Prokasy, Eds.), New York: Appleton-Century-Crofts, pp. 379–408. (3.5; 3.6)

The Psychonomic Society Inc., Austin, Texas
Coulson, G., Coulson, V., and Garner, L. (1970). *Psychon. Sci.* **18**, 309–10 (6.2)
Garcia, J., Ervin, F. R., and Koelling, R. A. (1966). *Psychon. Sci.* **5**, 121–2. (2.14)
Garcia, J., and Koelling, R. A. (1966). *Psychon. Sci.* **4**, 123–4. (2.11)
Gormezano, I., and Hiller, G. W. (1972) *Psychon. Sci.* **29**, 276–8. (3.7)
Rescorla, R. A. (1966) *Psychon. Sci.* **4**, 383–4. (2.6)

The Ronald Press Company, New York
Miller, N. E. (1944). *In Personality and the Behaviour Disorders* (J. McV. Hunt, Ed.), pp. 431–65. (9.9)

St. Claire-Smith, R. Blocking of punishment. Paper presented at Eastern Psychological Association, Atlantic City, 1970. (5.5)

Society for the Experimental Analysis of Behavior Inc., Bloomington, Indiana

Hearst, E., Koresko, M. B., and Poppen, R. (1964) *J. Exp. Anal. Behav.* **7**, 369–380. (9.7)

Herrnstein, R. J. (1970). *J. Exp. Anal. Behav.* **13**, 243–66. (4.9)

Herrnstein, R. J., and Morse, W. H. (1958). *J. Exp. Anal. Behav.* **1**, 15–24. (4.7)

Kelleher, R. T. (1966). *J. Exp. Anal. Behav.* **9**, 475–85. (5.6)

Nevin, J. A., and Shettleworth, S. J. (1966). *J. Exp. Anal. Behav.* **9**, 305–15. (7.2)

Reynolds, G. S. (1961) *J. Exp. Anal. Behav.* **4**, 57–71. (7.4)

Sidman, M. (1962). *J. Exp. Anal. Behav.* **5**, 97–104. (6.9)

University of California Press, and the Regents of the University of California, Berkley, California

Elliott, M. H. (1928). *University of California Publications in Psychology*, **4**, 19–30. (5.3)

Tolman, E. C., and Honzik, C. H. (1930). *University of California Publications in Psychology*, **4**, 257–75. (5.2)

University of Illinois Press, Urbana, Illinois

Brogden, W. J., Lipman, E. A., and Culler, E. (1938) *Am. J. Psychol.* **51**, 109–17. (3.5)

Jenkins, W. O. (1950) *Am. J. Psychol.* **63**, 237–43. (5.8)

CHAPTER 1

Introduction

I. HISTORICAL BACKGROUND

The study of learning in animals has a short history. It is, however, possible to discern throughout that history two distinct and complementary sets of interests. On the one hand there have been investigators who sought evidence of learning in a variety of animals for the light such evidence might throw on the evolution and adaptive specialization of capacities for behavioural modification in different animal groups. This comparative approach may be contrasted with a more analytic study of the associative processes underlying learned modifications of behaviour, and the experimental investigation of the functional relations between changes in behaviour and changes in antecedent and consequent conditions.

The comparative study of animal learning is usually traced back to Lloyd Morgan and Romanes (Warden, 1927; Bitterman, 1967). For Romanes (1882), the aim of animal psychology was to find evidence of mind in animals other than man, in order to demonstrate the psychological continuity of man and other animals. The study of learning was important because the occurrence of adaptive behaviour, if it could be shown to be a consequence of individual experience rather than innate endowment, provided the single most important criterion for the attribution of mind. Morgan (1894) concurred with Romanes in his estimation of the major aim of animal psychology and of the importance of evidence of learning, but while Romanes was happy to infer rational understanding and insight, logical inference or conceptual thought from his accumulation of instances of individual modification of behaviour, Morgan argued that the apparent complexity of overt behaviour was not a reliable guide to the complexity of the underlying processes involved. Just as the adaptiveness of much instinctive behaviour implied nothing about the animal's insight into the long-term adaptive significance of such behaviour, so the adaptiveness of learned behaviour might well be attributable to simpler associative processes.

In this attitude, Morgan was followed by Thorndike (1898, 1911).

Thorndike was certainly interested in questions concerning the evolution of mental capacities in animals, although, on the basis of his own research, reaching rather conservative conclusions as to the nature of that evolution. Thorndike's major importance in the history of the study of animal learning is as one of the originators of the second tradition in that study. He saw that answers to questions concerning the continuity of mental capacities left unanswered questions concerning the nature of the associative processes underlying those capacities. He also saw that the answers to these questions required the development of controlled, experimental techniques for the analysis of learning.

Thorndike, therefore, shares with Pavlov the honour of initiating the controlled, experimental analysis of animal learning. As a physiologist, Pavlov's scientific antecedents differed markedly from those of Thorndike. Unlike Thorndike, Pavlov had little or no interest in the comparative tradition of Romanes and Morgan; significantly enough, it was the European mechanists, relatively uninfluenced by the growing Darwinian tradition, such as Loeb (1900) and Beer, Bethe and Von Uexküll (1899), as well as the physiological reflexologists such as Sechenov (1863) and Sherrington (1906), whom Pavlov acknowledged as his intellectual precursors (Pavlov, 1927, pp. 4–5). Pavlov regarded his studies of conditioning, therefore, as an extension of the physiological analysis of the mechanisms of bodily functions and behaviour.

The comparative approach to the study of animal learning, although historically older, has for the past 50 years taken second place to the experimental, analytic approach. After an initial phase, during which a wide variety of animals was studied in the different types of experimental apparatus then being developed, the range of animals sampled by psychologists narrowed drastically, with the laboratory rat, of course, becoming by far the most popular (Beach, 1950). It remains true that the psychology of animal learning is based very largely on experiments with dogs, rats, and pigeons, animals selected almost entirely for reasons of convenience. Only rarely has attention been paid to the possibility that there might be substantive differences between the processes of learning in different vertebrate groups, and only very recently have psychologists considered the possibility that these processes might reflect specialized adaptations to particular selection pressures (Hinde and Tinbergen, 1958).

It can hardly be doubted that these are important possibilities, requiring extended and careful study. Nevertheless, there is little force to the argument, recently advanced by a number of psychologists (e.g., Lockard, 1971), that analysis of the processes of learning is vain and doomed to frustration, unless conducted within an explicitly phylogenetic or ecological frame of reference. Although there may be important differences in the processes of behavioural modification to be found in different animals, the study of those processes, independent of any particular concern for their specialization in different groups, remains an important and valid branch of science. Just as the sciences of genetics, embryology, and neurophysiology were initially advanced by the search for general, fundamental principles, undertaken with subjects chosen solely for reasons of convenience, so it is reasonable to suppose that some principles will have emerged from the analysis of associative learning in dogs, rats, and pigeons. It is, at any rate, this tradition, started by Thorndike in America and, more systematically, by Pavlov in Russia, that forms the subject matter of this book.

II. METHODS, PROCEDURES, AND DEFINITIONS

The influence of Thorndike and Pavlov on the science of animal learning has been more profound than that of many historical figures dutifully acknowledged as the founders of other branches of science. The experimental procedures they developed, and their interpretation of those procedures, have had a decisive effect on the study of learning in animals, and on some of its basic theoretical assumptions. Thus questions concerning the definition of learning, and the procedures to be followed in studying learning, may be most appropriately answered by reference to their practice.

A. Classical and Instrumental Conditioning

The experimental situations devised by Thorndike and Pavlov for studying the modification of behaviour may be regarded as arrangements of particular contingencies between events. Thorndike placed a hungry subject in a box, with food visible outside, and required the subject to learn to press a catch or pull a lever in order to escape from the box and obtain the food. The experimenter has arranged a particular contingency between the subject's behaviour and a given outcome, and the experiment is thus a prototypical study of instrumental learning or operant conditioning. Pavlov placed a hungry dog on a stand and, regardless of the subject's behaviour, presented food when a particular stimulus, such as a light or a bell, was momentarily turned on. The experimenter has arranged a particular contingency between a stimulus and an outcome, and the experiment is a prototypical study of classical conditioning.

Although classical and instrumental experiments have not always been defined in this way, and although for a long time the difference between their experimental operations was not recognized (Solomon and Brush, 1956), these two experimental paradigms have determined the course of most subsequent research on animal learning conducted in psychological laboratories. Experiments on classical conditioning may be defined as those in which a contingency is arranged between a stimulus and an outcome; experiments on instrumental learning may be defined as those in which a contingency is arranged between a response and an outcome. These definitions, it should be stressed, are not definitions of learning, but of experimental arrangements for the study of learning; the fact that the operations performed by the experimenter in an experiment on classical conditioning differ from those performed in an experiment on instrumental learning, does not imply that different processes of learning are engaged in the two classes of experiment.

An experimenter does not observe learning directly. He exposes subjects to particular contingencies, and observes changes in their behaviour; if these changes in behaviour can be attributed to the exposure to these contingencies, the experimenter may infer that learning has occurred. In experiments on instrumental learning, he may hope to observe a change in the probability of the response upon whose occurrence the outcome has been made dependent. In experiments on classical conditioning he may hope to observe changes in behaviour whose nature is somehow determined by the nature of the outcome used. But these changes in behaviour are not themselves to be identified with learning, and it follows, therefore, that learning may occur in the absence of current changes in behaviour. Such learning may be detected if a subsequent test reveals changes in behaviour that may be attributed to the subject's earlier exposure to a particular set of contingencies.

B. Reinforcement

In a typical experiment on classical conditioning, the presentation of food to a hungry dog (contingent on the presentation of a particular stimulus) increases the probability that the dog will salivate whenever that stimulus is again presented. In a typical experiment on instrumental learning, the presentation of food to a hungry rat, contingent upon the depression of a lever, increases the probability that the rat will press the lever again. The presentation of food may be said to reinforce the tendency to salivate or press the lever. The experimental operation of arranging outcomes contingent upon events, whose effect is to increase the probability of certain classes of behaviour, is referred to as reinforcement. By a natural extension, the term reinforcer has come to be used to refer to those outcomes whose presentation does increase the probability of a particular response.

In this sense, reinforcers are events which have a particular effect on behaviour. It does not follow that the presentation of a reinforcer is necessary for the formation of associations: as noted above, learning may occur in the absence of any current change in behaviour. Strictly speaking, moreover, reinforcement is an operation which *increases* the probability of certain classes of behaviour. This leaves out of consideration cases where learning might involve a decrease in the probability of certain classes of behaviour. It is, in fact, difficult to disentangle some of the confusions which have arisen in discussions of the concept of reinforcement. If the term reinforcement is used to refer to any operation which increases the probability of a designated response, then in a classical conditioning experiment, not only may the presentation of food to a hungry dog reinforce the response of salivating but the presentation of shock to the paw may reinforce the response of flexing the leg. Both food and shock, therefore, are reinforcers. The problem arises when these terms are applied to instrumental experiments: although the presentation of food contingent on a lever press may increase the probability of that response, it is more natural to expect that the presentation of shock contingent on a lever press would decrease the probability of that response. The presentation of a certain class of reinforcer may suppress rather than reinforce responses on which it is contingent. Similarly, it may be found that a sufficient condition for the reinforcement of an instrumental response is that the *omission* of a certain class of reinforcer be contingent on the occurrence of that response: if a rat can avoid a severe electric shock by running in a shuttle box, it is reasonable to expect it to learn to do so, but the contingency specified by the experimenter in this instrumental experiment is between the occurrence of a response and the omission of a particular outcome.

It seems reasonable to distinguish between those reinforcers whose effect is to increase the probability of instrumental responses, and those whose effect is to decrease the probability of such responses. Thorndike (1911) referred to the former as satisfiers and to the latter as annoyers. Here they will be referred to as appetitive reinforcers or rewards, and aversive reinforcers or punishing stimuli. The distinction should not, however, be tied too closely to the outcome of any particular instrumental experiment: it is probably just as valid to draw a distinction between appetitive and aversive reinforcers in experiments on classical conditioning, and not all instrumental responses are invariably suppressed by the contingent presentation of events which one would normally be inclined to regard as aversive reinforcers (see Chapter 6).

C. Associative and Non-associative Learning

The arrangement of contingencies between stimuli and outcomes in classical conditioning, and between responses and outcomes in instrumental experiments, may be taken as defining procedures for the study of associative learning. Classical and instrumental contingencies do not, of course, necessarily exhaust the range of possibilities that an experimenter may manipulate; nor is associative learning the only possible means for individual modification of behaviour. In experiments on habituation (Thompson and Spencer, 1966; Hinde, 1970), the experimenter does no more than repeatedly present a particular stimulus to the subject. This procedure may result in an orderly decline in the probability of the response initially elicited by that stimulus, but although this change in behaviour may reasonably be subsumed under the category of learning, it is questionable whether it is useful to regard it as an instance of associative learning. Habituation, therefore, is discussed only very cursorily in the following pages. Similarly, studies of perceptual learning, as exemplified by imprinting (Lorenz, 1935; Hinde, 1970) and song learning in birds (Thorpe, 1961; Marler, 1970) are completely ignored, as are studies of the learning of motor skills. This should not, of course, be taken as any reflection on the importance of these phenomena.

III. THEORETICAL ANALYSIS

Thorndike and Pavlov not only developed paradigms for the study of associative learning in animals, each believed his experimental preparation succeeded in revealing the fundamental processes involved in such learning. Each may be said to have propounded a theory of associative learning: more exactly, each proposed a theory of the nature of reinforcement in conditioning. For Thorndike, the presentation of a reinforcer served to strengthen a connection between a preceding response and the situation in which it had occurred. For Pavlov, a reinforcer elicited a pattern of behaviour, which, by association, would then come to be elicited by stimuli preceding the reinforcer. These two principles of reinforcement, Thorndike's law of effect and Pavlov's principle of stimulus substitution, remain the most important analyses of reinforcement to this day. Theories of learning since the time of Thorndike and Pavlov have usually been understood to mean theories about the elements of association rather than theories of reinforcement, and such theories of reinforcement as have been proposed have usually been attempts to specify the characteristics of events which endows them with the ability to act as reinforcers. This question is not considered here.

An experimenter may infer that associative learning has occurred by observing changes in the behaviour of a subject exposed to particular contingencies between stimuli, responses, and reinforcers. But the observation of such changes does not carry any implications about the nature of the associations so formed. Two classes of theory have been proposed, each having a certain face validity. According to the first, animals may associate any pair of events that occur contiguously. It is reasonable to suppose that if the experimenter arranges a contingency between a response and a reinforcer, or between a stimulus and a reinforcer, then these are the associations that are formed. This theory of learning is usually identified with Tolman (1932, 1949). Alternatively, since what the experimenter observes in a typical classical or instrumental experiment is an increase in the probability of a particular response, it is reasonable to suppose that what is learned is an association between that response and some set of antecedent stimuli. This S–R (or stimulus–response) theory of learning, is obviously similar to that proposed by Thorndike; Guthrie (1935) and Hull (1943) are usually taken as the major proponents of such a view.

The importance of this particular theoretical issue should not be exaggerated: it is arguably of less interest than the question whether Thorndike's or Pavlov's is the more general principle of reinforcement. At any rate, it is clear enough that the attention it commands has declined in the past 20 years, as psychologists have applied themselves to the analysis of more detailed, specific questions and theorists have developed less global, more precise theories, designed to accommodate a more restricted range of data or to explore the implications of a particular theoretical principle. There is no gainsaying the benefits that have accrued from this strategy. One consequence, however, has been a severe failure of communication between workers in closely related areas. Thus the analysis of reinforcement in classical conditioning has had little impact on theories of instrumental learning; the analysis of instrumental learning maintained by appetitive reinforcers has proceeded in almost complete disregard of the analysis of avoidance and punishment; and the study of contrast effects has been taken up by experimentalists of various theoretical persuasions, with no concern for the effects observed by other experimenters using different paradigms. It is certainly past time that some attempt was made to draw together some of these diverse approaches.

CHAPTER 2

Classical Conditioning: Basic Operations

I. BASIC PHENOMENA OF CLASSICAL CONDITIONING

The operations involved in classical conditioning are straightforward enough. In a typical experiment, the experimenter presents two stimuli, a conditional stimulus (CS) followed by an unconditional stimulus (UCS)*, and records changes in the subject's behaviour to the CS. It looks as if the situation is as simple as possible, and that the experimenter can exercise a considerable degree of control over the experimental events: the subject is often restrained, the stimuli are salient, the delivery of CS and UCS is timed by the experimenter and does not depend upon the subject's behaviour. It is not surprising, therefore, that conditioning experiments should have been thought to open the way to the fundamental study of associative learning, or, even worse, to reveal the basic units of which all habits are formed. Such hopes have done little to advance the study of conditioning itself.

In Pavlov's terminology, the presentation of the UCS following the CS constituted reinforcement of the conditional reflex. The term reinforcement has by now come to mean a variety of things; as used by Pavlov, however, a reinforcer was that which increased the probability of a CR occurring when the CS was presented. For the time being, it should be possible to preserve this relatively neutral terminology.

A. Acquisition

Repeated reinforcement with a UCS results in the acquisition of a CR. Pavlov himself provided essentially no data on the course of acquisition of

---

* The unconditional stimulus was so named by Pavlov because it unconditionally elicited an unconditional response (UCR). The conditional stimulus elicited a conditional response (CR) only after appropriate training (i.e., conditionally on that training). The translation of these terms as conditioned and unconditioned is, perhaps, unfortunate, although the term "conditioned" has, of course, now acquired a technical sense, as when a response is said to be conditioned to a stimulus.

a CR, but American psychologists, especially those interested in the con-
struction of formal models of conditioning, have devoted more attention
to the question. Fig. 2.1 shows acquisition curves for three types of conditioning.

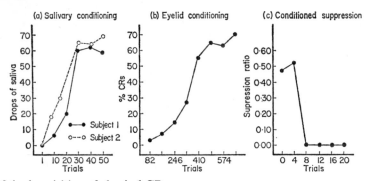

FIG. 2.1. Acquisition of classical CRs.
(a) Salivary CRs in two dogs. (*After* Anrep, 1920.) (b) Eyelid CRs in rabbits.
(*After* Schneiderman *et al.*, 1962.) (c) Conditioned suppression in rats. (*After*
Annau and Kamin, 1961.)

1. *Salivary Conditioning in Dogs*

The dog undergoes a minor operation to the parotid gland so that salivary
flow is directed to a fistula outside the mouth where it may be recorded
(Gormezano, 1966). During training the animal is usually restrained. A
wide range of CSs was used in Pavlov's laboratory (tones, metronomes,
bells, lights, "whirligigs", tactile stimulation). The UCS may be either
food or acid, and the food may be delivered directly to the animal's mouth,
or to a food bowl: the former procedure produces notably stronger salivary
CRs (Kierylowicz *et al.*, 1968). The data shown in Fig. 2.1a are from two
dogs (Anrep, 1920); the CS was a 5 sec tone, the UCS was meat powder
in a tray; CRs were recorded on occasional trials when the CS was presented alone for 30 sec.

2. *Eyelid Conditioning in Rabbits*

The rabbit is placed in a restraining stock, with recording attachments
fixed to the eyelid. The CS is usually auditory (a tone), and the UCS may
be either a puff of air to the cornea or an electric shock to the orbital
region. The CR consists of closure of the eyelid, but a related response,
closure of the nictitating membrane or third eyelid has been more usually
recorded in recent studies (see Gormezano, 1966, for details). In the data
shown in Fig. 2.1b (Schneiderman, *et al.*, 1962), the CS was a 600-msec
tone, the UCS a 100 msec puff of air delivered during the final 100-msec

period of the CS, and the CR, recorded on every trial, was a closure of the eyelid occurring before onset of the UCS.

3. *Conditioned Suppression in Rats*

In this procedure, developed by Estes and Skinner (1941), a freely moving rat is initially trained to press a lever for occasional presentations of food. Once a stable rate of lever pressing has been established, a CS (usually an overhead light, or a tone or white noise) is presented (for up to three minutes), its termination coinciding with the delivery of a brief electric shock (the UCS) to the grid floor of the apparatus. The CR measured is a reduction in the rate of lever pressing during presentation of the CS; it is usually measured in the form of a ratio, $a : (a + b)$, where $a =$ the number of lever responses during the CS, and $b =$ the number of responses in an equivalent period immediately preceding the CS. Such a ratio has a value of 0·50 when the CS has no effect on responding, and 0·00 when the CS completely suppresses responding (hence Fig. 2.1c shows an acquisition curve which declines to zero). Similar experiments have been undertaken with pigeons pecking a lit key for food (e.g., Hoffman and Barrett, 1971), and with rats licking at a tube for water (Leaf and Muller, 1965). In another closely related procedure, Sidman *et al.* (1957) trained monkeys to press a lever, and Rescorla and LoLordo (1965) trained dogs to jump over a hurdle, in order to avoid occasionally presented un-signalled shocks, and found that a CS paired with shock reliably increased the rate of avoidance response. In the data shown in Fig. 2.1c (Annau and Kamin, 1961), the rats were pressing a lever for food on a variable interval schedule which delivered food on average every 2·5 min; the CS was a 3 min white noise signal; and the UCS was a 0·5 sec 0·85 ma shock.

4. *Analysis of Acquisition Data*

Several points are already worth noting in these examples. The measure of conditioning is quite different in each case. In the first example, CRs are measured on unreinforced test trials; in the remaining two cases, CRs are measured on every trial. Although there is occasional need to adopt the test trial procedure (as noted by Bitterman, 1965, this is particularly true in studies of the role of CS–UCS interval; see below p. 64, the most common procedure is to measure anticipatory CRs on every trial. But there is a further, more important difference in measurement. The measure used by Anrep (and Pavlov) for salivary conditioning is an amplitude measure—the amount of salivation occurring during the CS. The measure of eyelid conditioning is a probability measure—the probability that a response occurred on each trial. The measure of conditioned suppression is a comparison between rate of responding during the CS and during an equivalent interval immediately preceding the CS. The difference between these measures should not be exaggerated. They are obviously related: salivation is usually recorded in drops, and the number of drops in a given interval of time is a rate measure, which in many recent experiments has been compared to the pre-CS rate of salivation (e.g., Fitzgerald, 1963; Wagner et al., 1964). An eyelid closure has to be of an amplitude greater than the experimenter's criterion before it is counted as a CR, and this probability measure simply dichotomizes responses into those greater and those less than a particular amplitude. Nevertheless, we shall see that the differences between these measures do have some important consequences.

A second difference between the three acquisition curves of Fig. 2.1 is the striking variation in rate of learning. Conditioned suppression was complete in less than 10 trials, while the probability of eyelid CRs was still increasing after 400 trials. Some of this variety may reflect no more than the difference between measures and the type of UCS used; a relatively strong shock is commonly used in studies of conditioned suppression, while use of an electric shock as the UCS for eyelid conditioning produces much more rapid acquisition (Brelsford and Theios, 1965). It is relatively clear, however, that different response systems do condition at different rates (p. 22). Why this should be so is something of a puzzle.

The final point worth noting is that all three acquisition curves in Fig. 2.1 are of relatively similar shape—a period of positive acceleration is followed by one of negative acceleration. Formal models which postulate such a function for the growth of associative strength (e.g., Spence, 1936) can take satisfaction from this. Those (the large majority) that postulate a simple negatively accelerated growth function (Bush and Mosteller, 1951; Estes, 1950; Hull, 1943; Spence, 1956; Rescorla and Wagner, 1972), must find some way of reconciling data and theory. One formal solution, adopted, for example, by Hull and by Spence in his later theorizing, is to assume a threshold below which associative strength is not translated into performance: early acquisition trials, therefore, although producing large increments in associative strength, result in little increase in the probability of a CR since the associative strength is still below threshold. Another formal solution has been suggested by Norman (1964). He assumed that increments to response probability might not begin until after a variable number of acquisition trials. This assumption is easily represented in the models proposed by Bush and Mosteller or Estes by the addition of a new parameter, k, which corresponds to the number of initial trials that pass before conditioning begins.

One advantage of Norman's idea is that it encourages the experimenter to look for reasons for this absence of learning on early conditioning trials. Several possible reasons can be suggested. The subject might require a certain amount of time to adapt to the apparatus, and this process of adaptation might be incomplete at the outset of conditioning. More interestingly, as we shall note below, it is possible that during early trials CRs are conditioned not only to the nominal CS, but also to the background stimuli provided by the apparatus and experimental situation. Part of what the subject must learn, then, is to discriminate between the experimental situation without the CS (when no UCS occurs) and the situation with the CS (when the UCS does occur). Until this discrimination is formed, it might be impossible to detect conditioning to the CS as such. Finally, a standard procedure in most conditioning experiments is to give the subject a number of nonreinforced trials with the CS alone before conditioning starts. The ostensible purpose of this procedure is to habituate any UCRs to the CS which might complicate the recording of CRs; but it is well established that one consequence of such habituation trials is a significant retardation of subsequent conditioning. The result is referred to as latent inhibition (p. 36). In one study which explicitly examined the acquisition of a CR as a function of prior exposure to the CS, Kremer (1971) found that the acquisition curve was negatively accelerated in a group given no nonreinforced exposure to the CS, but had an initial phase of positive acceleration in a group given prior habituation training (see Fig. 2.8, p. 40).

A rather different class of theory, which handles some of the data very successfully, is the type of Markov model proposed by Theios and Brelsford (1966b). A Markov model assumes that a subject, on any trial, is in one of several possible states, with transitions between these discrete states occurring according to predetermined probabilities. Theios and Brelsford suggested that a three-state model was sufficient to account for the acquisition of a conditioned eyelid response in rabbits. According to this model, the subject starts in an initial, unconditioned state, moves first to an intermediate state, and finally to a terminal, conditioned state. The model has five parameters, corresponding to the probability of a CR in each of the states and the probability of a transition from initial to intermediate, and from intermediate to terminal, states.

It is hardly to be expected that formal models explicitly designed to predict acquisition data should be unable to do so. Yet there is, perhaps, something discouraging about the ease with which such diverse models can be made to succeed. Is there no way of discriminating between these different approaches? The main strategy of recent mathematical theorists has been to examine the fine grain of acquisition data, for example the conditional probability of a CR given the outcome of the preceding one, two or three trials. Theios and Brelsford report such an analysis for an experiment on eyelid conditioning in rabbits. The results were striking:

the Markov model continued to provide an extremely close fit to the data, whereas models which assumed a gradual increase in associative strength, such as those of Bush and Mosteller, or Spence, were unable to predict these sequential data.

Nevertheless, the Markov model, which assumes abrupt transitions between different states, is almost certainly applicable only to those situations, such as studies of eyelid or flexion conditioning, where the measure of conditioning is whether or not a CR occurs on a given trial. The rate or amplitude measure employed in studies of salivary conditioning, for example, does not behave in the way required by the Markov model. The data from Anrep's two dogs, shown in Fig. 2.1a, suggest a gradual increase in the amplitude of salivary CRs, rather than any abrupt transitions between unconditioned, intermediate and conditioned states.* It may well be more reasonable, then, to suppose that changes in the associative processes underlying conditioning occur relatively gradually. With rate or amplitude measures, it is possible to detect these gradual increments in associative strength; with discrete, probability measures of responding, it is not.

B. Extinction

Reinforcement of a CR by the presentation of a UCS increases the strength of the CR. Withdrawal of reinforcement, Pavlov discovered, led to the gradual disappearance of the CR. The operation, and later the result, was termed extinction. Fig. 2.2 illustrates the course of extinction of each of the three types of CR shown in Fig. 2.1. Although the three response systems appear to extinguish at more or less comparable rates, this is likely to be largely coincidental, since the rate of extinction of any one response system is affected by several variables, most notably by the distribution of trials (Pavlov, 1927, pp. 52–3). The results shown in Fig. 2.2 were obtained with relatively short intervals between successive extinction trials; when very long intervals (24 h) are used, extinction proceeds much more slowly and irregularly (Pavlov, 1927, pp. 52–3; Konorski, 1967, p. 37). The distinction in Pavlovian terminology is between acute (massed trials) and chronic (spaced trials) extinction. The effect of introducing a long interval after a series of massed extinction trials is also shown in Fig. 2.2: after the CR has been apparently completely extinguished, a long interval results in its reappearance. This is Pavlov's spontaneous recovery.

---

* It is important to note that Anrep's data are for individual subjects. A group acquisition curve may show a gradual increase in the average probability of a CR even though the data for individual subjects show the type of abrupt transition consistent with the Markov model's analysis (cf. Estes, 1959).

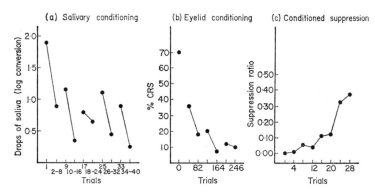

FIG. 2.2. Extinction of classical CRs. Trial 0 represents performance at the end of acquisition. Connected points in (a) and (b) represent trials within a daily session; disconnected pointed represent transitions between days. (a) Salivary CRs in dogs. (*After* Wagner *et al.*, 1964.) (b) Eyelid CRs in rabbits. (*After* Schneiderman *et al.*, 1962.) (c) Conditioned suppression in rats. (*After* Annau and Kamin, 1961.)

C. Generalization and Discrimination

"If a tone of 1000 d.v. [Hz] is established as a conditioned stimulus, many other tones spontaneously acquire similar properties, such properties diminishing proportionally to the intervals of these tones from the one of 1000 d.v. Similarly, if a tactile stimulation of a definite circumscribed area of skin is made into a conditioned stimulus, tactile stimulation of other skin areas will also elicit some conditioned reaction, the effect diminishing with increasing distance of these areas from the one for which the conditioned reflex was originally established (Pavlov, 1927, p. 113)."

The phenomenon was termed generalization. In spite of the above introductory statements, many of the actual instances of generalization given by Pavlov were between qualitatively different stimuli, such as tactile and thermal stimuli. More quantitative studies of generalization along simple physical dimensions, intended to study the exact "form" of the generalization gradient, were left to American investigators (e.g., Hovland, 1937a, b). Until very recently, however, with the studies of Siegel *et al.* (1968) and Moore (1972) of the generalization of the rabbit's nictitating membrane response along the dimension of auditory frequency, the data on the generalization of classical CRs have been extremely sparse.

Differentiation or discrimination, according to Pavlov, occurs when a CS is reinforced and a neighbouring stimulus is not reinforced. Although the neighbouring stimulus, which we may refer to as CS−, initially elicits a CR, continued nonreinforcement results in extinction of its CR, while continued reinforcement of CS+ maintains its CR. Discrimination, therefore, is regarded as the result of combined acquisition and extinction, a view that continues to be extremely influential.

Discriminations may be formed between stimuli varying along a physical dimension, such as different tones, or between various combinations of qualitatively different stimuli: CS+ might be the combination of a tone and a tactile stimulus, while CS− might be the tactile stimulus alone (Pavlov, 1927, p. 143). In due course the tactile stimulus fails to elicit a CR. (This paradigm has been extensively studied in the case of instrumental learning by Jenkins and Sainsbury, 1969, 1970; see p. 569.) What is interesting about this case is that it makes explicit something that must presumably be occurring in the simplest conditioning experiment. As was noted above, it is sometimes implied that all that is happening during the acquisition of a CR is that a CS is presented and reinforced. But the CS is presented against a complex set of background stimuli arising from the experimental situation, all of which must be assumed to be impinging on the subject. The CS might therefore more properly be described as a compound, one of whose components, the set of background stimuli, occurs alone in the absence of reinforcement. Letting A stand for the experimenter's CS, and X for background stimuli, conditioning may be described as involving a discrimination between AX+ and X−. We should therefore expect to observe during early conditioning trials some change in the subject's behaviour during the intertrial interval. This is indeed the case: Sheffield (1965, p. 314) reported that his dogs salivated during the intertrial interval, and that the frequency of such bursts of salivation declined during the course of training; and a routine observation in studies of conditioned suppression is that the rat's overall rate of lever pressing declines initially and only slowly recovers (e.g., Annau and Kamin, 1961).

We may have here one instance of what was described above as "general adaptation", and was suggested as contributing to the initially slow rate of acquisition of a CR. In studies of conditioned suppression, for example, conditioning is measured as a difference in rate of responding in the presence and in the absence of the CS; if suppression is initially conditioned both to the CS and to background stimuli, then early trials of conditioning will fail to reveal differential conditioning to the CS. Not until suppression to background stimuli extinguishes will more suppression be observed in the presence of the CS than in its absence.

D. External Inhibition and Disinhibition

Pavlov (1927, pp. 44–7) stressed the importance of a constant and well-controlled experimental environment. If a novel, extraneous stimulus was presented either shortly before or at the same time as a CS, the CR on that trial was disrupted. "The interpretation of this simple case does not present much difficulty. The appearance of any new stimulus immediately invokes the investigatory reflex and the animal fixes all its appropriate receptor organs upon the source of disturbance . . . The investigatory reflex is excited and the conditioned reflex is in consequence inhibited (Pavlov, 1927, p. 44)." Since the investigatory reflex, or, as it is now more usually called, the "orientation reaction" or the "orienting reflex" (Lynn, 1966; Sokolov, 1963), is only elicited by relatively novel stimuli, a repeatedly presented external inhibitor rapidly loses its effect (Pavlov, 1927, p. 46).

When an extraneous stimulus is presented during the course of extinction, the strength of the CR on that trial may be increased (Pavlov, 1927, pp. 61–2). This phenomenon, called disinhibition, cannot satisfactorily be explained in the same way as external inhibition, for if the external stimulus elicits an orienting response which competes with the CR, it could hardly increase the strength of a partially extinguished CR. Indeed, there is evidence that the disinhibitory effect of an extraneous stimulus depends little if at all upon the novelty of that stimulus (Brimer, 1970). Disinhibition also occurs during the course of habituation (p. 39): in this instance it has been rather convincingly argued by Thompson and Spencer (1966) and Groves and Thompson (1970) that the phenomenon must be ascribed to a sensitizing or energizing effect of relatively strong stimuli. Since there are data showing that disinhibition is a function of stimulus intensity (Brimer, 1970), and since stimulus intensity has a facilitative effect on the acquisition of CRs (p. 41), the explanation, although postulating some rather poorly understood processes, has some merit.

II. CATEGORIES OF CONDITIONING

The examples used to illustrate the basic phenomena of conditioning have been derived from a number of conditioning procedures differing in a variety of ways. In this section an attempt will be made to elaborate some of these differences; the intention, however, is not so much to provide an exhaustive list of conditioning procedures (a job done adequately elsewhere, e.g. by Hull, 1934, and Hilgard and Marquis, 1940), as to classify some of the important differences between different procedures.

A. Motivational Significance of UCS

Pavlov distinguished between alimentary and defensive reflexes: for the former the UCS was food, for the latter acid, and in both cases the

UCR was salivation. There is now a longer list of what we shall call appetitive and aversive UCSs, and we can also distinguish two further classes of UCS.

1. *Appetitive*

Food as a UCS has been used to condition not only salivation but also general activity in pigeons (Slivka and Bitterman, 1966) and rats (Zamble, 1967). When the CS is the illumination of a pigeon's response key, food as a UCS can also be used to condition key pecking (Brown and Jenkins, 1968; Williams and Williams, 1969; Jenkins and Moore, 1973).* Water has been used as a UCS to condition licking in rats (Boice and Denny, 1965; DeBold *et al.*, 1965; Patten and Deaux, 1966), and jaw movements in rabbits M. C. Smith *et al.*, 1966; Gormezano, 1972). Finally, access to a sexual partner as UCS has been used to condition courtship behaviour in quails (Farris, 1967) and pigeons (Rackham, 1971; cited by Moore, 1973).

2. *Aversive*

The most extensively used aversive UCS is electric shock. Applied to the forepaw of a restrained dog, it is used to condition leg flexion (Bekhterev, 1932; Brogden, 1939b; Soltysik and Jaworska, 1962; Wahlsten and Cole, 1972); to the cheek or across the eye of a rabbit, to condition eyelid or nictitating membrane closure (Brelsford and Theios, 1965; Smith, 1968); to the feet of the freely moving rat to condition suppression of appetitively reinforced lever pressing (Estes and Skinner, 1941; Annau and Kamin, 1961). It is also used for the conditioning of such autonomic responses as pupillary dilation (Gerall and Obrist, 1962), GSR and vasomotor responses (White and Schlosberg, 1952; Fromer, 1963), and changes in heart rate (Black, 1965, 1971). In these cases, however, there is some question whether the autonomic changes recorded by the experimenter are being directly conditioned, or whether they may not be an indirect consequence of changes in other systems (p. 104). A puff of air to the cornea is the most usual UCS in eyelid conditioning in human subjects (Ross and Hartman, 1965), although a loud noise can also serve as an effective UCS for this response (Hilgard, 1931), and an airpuff is also often used in eyelid and nictitating membrane conditioning in rabbits (Schneiderman *et al.*, 1962; Schneiderman and Gormezano, 1964).

As noted above, Pavlov used acid as a UCS for salivary conditioning (see also Ost and Lauer, 1965). Pavlov (1927, pp. 35–7) also reported the use of morphine as a UCS for conditioning salivation and nausea (see also

---

* Because pigeons are usually shaped to peck keys in instrumental experiments, Brown and Jenkins referred to their procedure for classically conditioning the key-pecking response as "auto-shaping".

Crisler, 1930), while Spragg (1940) reported data which can be construed as evidence of conditioned reduction in withdrawal symptoms in addicted chimpanzees when injections of morphine were used as the UCS: injections of saline solution, under conditions similar to those holding when subjects had received morphine injections, led to temporary relaxation and loss of withdrawal behaviour. Finally, a number of recent studies have demonstrated the conditioning of components of fighting or aggressive behaviour in various animals; they have included studies of Siamese fighting fish, using a mirror image as the UCS (Thompson and Sturm, 1965), and of rats (Lyon and Ozolins, 1970) and squirrel monkeys (Hutchinson *et al.*, 1971) using electric shock as the UCS.

What is the basis for the distinction between appetitive and aversive conditioning? At one level, it is clearly an intuitive distinction between pleasant (Thorndike's "satisfying") and unpleasant (Thorndike's "annoying") stimuli. Following Thorndike, therefore, one could provide an operational classification of stimuli into these two classes, appetitive UCSs being those which an animal works for or does nothing to avoid, aversive UCSs being those which an animal avoids or escapes from. This would be to rely on instrumental behaviour in order to provide a classification of classical conditioning. Konorski (1967) has provided an alternative, theoretically derived, basis for the distinction. He argues that appetitive and aversive conditioning depend upon different, mutually antagonistic, motivational systems, and that the difference between the two types of conditioning is reflected in the fact that interference occurs between them: if a stimulus has been turned into an appetitive CS it will be difficult to turn it into an aversive CS, and vice-versa (Konorski and Szwejkowska, 1956; see Fig. 2.4 below).

The phenomenon of "counter-conditioning" also indicates that appetitive and aversive UCSs may inhibit one another. Pavlov (1927, pp. 29–31) reported that if a relatively mild electric shock was used as a CS signalling food, the UCR to shock was gradually inhibited until the only response elicited by shock was salivation. Similarly, fear responses conditioned to a stimulus associated with shock may be most rapidly suppressed by feeding the animal in the presence of that stimulus (e.g., Klein, 1969). And just as a CS signalling shock will suppress instrumental, food-rewarded responding (conditioned suppression), so a CS signalling food may suppress instrumental avoidance responding (Bull, 1970; these studies are discussed in more detail on p. 83).

3. *Interoceptive*

This is a somewhat unsatisfactory category of UCSs, ranging from electrical stimulation of the motor cortex (Loucks, 1935; Doty, 1969) or cerebellum (Brogden and Gantt, 1937), which elicits various limb movements which can be conditioned, to such stimuli as temperature change eliciting vascular contraction or dilation, water intake eliciting diuresis, or a change in the CO_2 content of inhaled air eliciting changes in respiration (Bykov, 1957; Razran, 1961). One of the interesting questions concerning some of these experiments is how far the stimuli used as UCSs should be regarded as either appetitive or aversive (p. 94).

4. *Neutral*

The most comprehensive, as well as theoretically neutral, description of the operations of an experiment on classical conditioning is that the experimenter presents two stimuli, S_1 and S_2, in some specified temporal relationship. Restrictions that S_2 must be of intrinsic motivational signifi-

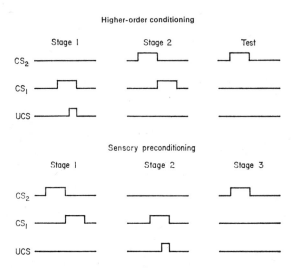

FIG. 2.3. Design of experiments on higher-order conditioning and on sensory preconditioning. In both types of experiment CS_1 is paired with a UCS, CS_2 is paired with CS_1, and evidence of conditioning is provided by the occurrence of CRs to CS_2. The difference between the two procedures lies in the order of Stages 1 and 2.

cance, or that it must be the UCS for some readily measurable UCR, may serve to exclude at least two types of experiments which, on any reasonable definition, constitute instances of conditioning.

The first case is that of higher-order conditioning (Pavlov, 1927, pp. 33-4). The paradigm for higher-order conditioning is illustrated in Fig. 2.3. The experimenter pairs CS_1 with a given UCS, and then pairs

B

CS_2 with CS_1, finally testing for the occurrence of a CR to CS_2. Pavlov stated that such conditioning, although real, was relatively slight in extent, and that third-order conditioning (pairing CS_3 with CS_2) was usually impossible, unless the original UCS was a strong electric shock. Highly reliable second-order conditioning has been obtained in several recent experiments (e.g., Davenport, 1966; Kamil, 1969; and Rizley and Rescorla, 1972) all of which have used conditioned suppression in rats.

The fragility of higher-order conditioning has been traditionally explained by the suggestion that nonreinforced presentations of CS_2 and CS_1 result in extinction of responding to CS_1, which therefore ceases to act as an adequate UCS. Shapiro et al. (1971), however, observed transitory higher-order conditioning of salivation in dogs, but found that even when no CRs occurred to the CS_2–CS_1 compound, presentation of CS_1 alone still elicited salivation. Responses to CS_1, therefore, had not been extinguished, and the transitory nature of higher-order conditioning must be due to some other factor. As Shapiro et al. note, one need not look far for the other factor: if CS_1 is paired with a UCS, and CS_2 is then paired with CS_1 but without being followed by the UCS, then CS_2 is a reliable signal that the UCS is not going to occur. Indeed the paradigm conforms quite closely to that used for turning CS_2 into a conditioned inhibitor (p. 34), and Herendeen and Anderson (1968) have shown that while a small number of CS_2–CS_1 pairings may establish CS_2 as a higher-order CS, extended training may indeed turn it into a conditioned inhibitor. Pavlov, it should be noted, did not present CS_2 and CS_1 simultaneously in his studies of higher-order conditioning; the presentation of CS_2 always preceded that of CS_1 by several seconds, and Pavlov regarded this as an essential precaution for preventing the development of conditioned inhibition. Whether this is true, and if so why, is not clear.

A second category of conditioning experiment with apparently neutral stimuli is sensory preconditioning (Brogden, 1939c), or sensory conditioning as it is called by Russian workers (Razran, 1961). The paradigm, also illustrated in Fig. 2.3, is similar to that of higher-order conditioning, but reverses the first two stages. That is to say, CS_2 and CS_1 are first presented together, and only after these pairings are completed is CS_1 paired with a UCS. The final test phase, as in higher-order conditioning, measures CRs to CS_2. Again as with higher-order conditioning, appropriate controls are necessary to ensure that the occurrence of CRs to CS_2 is due not simply to generalization from CS_1 to CS_2, but to the learned association between them established in the first stage of the experiment.

In Brogden's original experiment, dogs received 200 trials on which a buzzer and a light were simultaneously presented, and then received leg-flexion conditioning with one of these stimuli signalling shock to the paw.

Test trials to the other stimulus produced an average of 9.4 flexion CRs, whereas a control group that did not receive buzzer-light pairings made an average of 0·5 CRs to the untrained CS. The effect appeared reliable, but not very large, a conclusion endorsed by Seidel (1959) in a review of several other early studies. More recent experiments, however, have suggested that Brogden's original study chose less than optimal values of two parameters. Firstly, Brogden presented CS_1 and CS_2 simultaneously, whereas Hoffeld et al. (1958) and Wynne and Brogden (1962) found much better sensory preconditioning in cats with a 4 sec interval between the onsets of CS_2 and CS_1 than with simultaneous presentation. Secondly, Brogden gave 200 CS_2–CS_1 pairings; Hoffeld et al. (1960) in another study with cats, found that the greatest number of test CRs occurred in subjects given only 4 CS_2–CS_1 pairings; additional groups given anywhere between 8 and 800 pairings showed a substantially smaller amount of sensory preconditioning. Two experiments using conditioned suppression in rats have found highly reliable sensory preconditioning after 16 CS_2–CS_1 pairings (Prewitt, 1967; Tait et al. 1969); Prewitt found a slight decline in the magnitude of the effect when the number of pairings was increased to 64. The theoretical significance of these results is discussed below (p. 97).

B. Preparatory and Consummatory Responses

In both appetitive and aversive conditioning experiments, stimuli paired with food or shock can be shown to acquire a variety of CRs. Konorski (1967) has argued that we should distinguish between two main classes of CRs. First, there are diffuse, preparatory CRs, such as restlessness or excitement in appetitive conditioning, and conditioned suppression or changes in heart rate in aversive conditioning; these preparatory CRs should be distinguished from precise, consummatory CRs, such as salivation or licking in appetitive conditioning, and leg flexion or eye blinking in aversive conditioning. Preparatory conditioning is regarded as the direct conditioning of motivational states (corresponding to what Mowrer, 1960a, called fear and hope), the symmetrical effects of which are indexed, for example, by the elevation and suppression of instrumental avoidance responding by CSs previously paired with shock or food (Sidman et al., 1957; Grossen et al., 1969). Consummatory conditioning involves the learning of specific, adaptive CRs and can occur only against a background of prior preparatory CRs. Any given experiment will typically involve both types of conditioning: thus early in salivary or leg-flexion conditioning, the animal is restless in one case and struggles a great deal in the other (this is the phase of preparatory conditioning), while after extended training the animal is calm and the consummatory CR occurs at full strength (Konorski, 1967, pp. 276, 285).

Is this distinction a valid one? There are, indeed, two apparent differences between preparatory and consummatory CRs. Preparatory CRs appear to condition more rapidly (see Fig. 2.1), and they can also be conditioned with a very much longer interval between the onset of CS and the presentation of the UCS: conditioned suppression experiments routinely use a CS–UCS interval of up to three minutes, while eyelid conditioning may be impossible at intervals of more than a second or two. This latter point is discussed at greater length below (p. 65). For the present we shall consider the differences in rate of acquisition.

A comparison between a single study of conditioned suppression and one of the conditioned eyelid response (as provided in Fig. 2.1) can hardly establish that the two types of CR are acquired at different rates. There are far too many procedural differences between the two studies (e.g., species, intensity of UCS, spacing of trials) which may have been responsible for the difference in outcome. However, it is perfectly possible to record different CRs simultaneously in a single experiment, and when this has been done, the results have been relatively consistent: preparatory CRs begin to occur after a few conditioning trials, consummatory CRs only after more extended training. Apart from the studies of Konorski mentioned above, this conclusion has been supported by several experiments on aversive conditioning: Liddell et al. (1935) reported that changes in respiration occurred in sheep before any signs of a flexion CR; Lynch (1966) and Newton and Gantt (1966) were able to measure a significant heart rate CR in dogs after a single conditioning trial, but could find no evidence of a flexion CR until many more trials had been given; similar results have been obtained with cats by Bruner (1969). Yehle (1968) gave rabbits discriminative conditioning to different tones and found significant heart rate conditioning within the first day of training, as well as differential responding to CS+ and CS−, but a much more gradual appearance of nictitating membrane CRs, with rather little evidence of discrimination until after four days of conditioning.

Although this evidence seems quite clear, it is less clear that it establishes the existence of any profound difference between preparatory and consummatory conditioning. For one thing, there is evidence that different preparatory responses may condition at different rates. deToledo and Black (1966) and Parrish (1967), for example, found that suppression of lever pressing conditioned more rapidly in rats than did changes in heart rate. Secondly, there is a relatively simple, albeit less adequately substantiated, explanation of all of these observed differences. Early in conditioning, the main effect of an electric shock to the foreleg of a restrained animal is to elicit a great deal of struggling, yelping, and disorganized activity. If this is the UCR to shock on early trials, it is not surprising that the CR that first emerges will also be diffuse struggling and a general emotional response, nor is it surprising that these CRs will interfere with and prevent the occurrence of the more precise flexion CR. It is only later in conditioning that a precise flexion UCR appears without the general struggling, and it is only at this point that flexion CRs can be expected. The rule, therefore, may be very simple: the response that occurs as CR at any stage of conditioning will depend on the response that occurs as UCR at that stage. Whether this constitutes a sufficient explanation of the differences between the rates of conditioning of different response systems is not at all clear. At least, however, the idea merits experimental analysis.

C. CS–UCS Relationships

So far we have discussed only what may be called "excitatory" conditioning, where a positive correlation exists between CS and UCS. In the simplest case, the correlation is $+ 1 \cdot 0$: the UCS is presented if and only

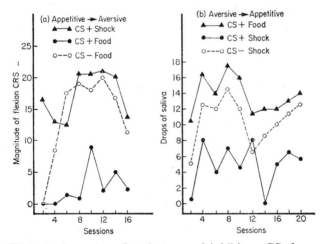

FIG. 2.4. The transformation of excitatory and inhibitory CSs for one class of reinforcer into a CS for a different class of reinforcer. A CS+ for food elicits flexion CRs only after a large number of pairings with shock, whereas a CS− for food may be turned into an aversive CS relatively rapidly. Similar results occur when a CS+ and CS− for shock are paired with food. (*After* Konorski and Szwejkowska, 1956.)

if the CS has been presented. But other relationships are possible. The most important of these is that defined by "inhibitory" conditioning, where a negative correlation exists between CS and UCS, so that in the extreme case (a correlation of − 1·0), the occurrence of the CS is a reliable signal that the UCS is not about to occur. This distinction between excitatory and inhibitory conditioning, it should be noted, is orthogonal to that previously made between appetitive and aversive conditioning. A CS may signal the presence or absence of an appetitive UCS such as food, or equally the presence or absence of an aversive UCS such as shock. Indeed, Konorski (1967) has shown that the mutually antagonistic relationship between appetitive and aversive conditioning is found only when an excitatory appetitive CS is turned into an excitatory aversive one, or vice-versa. As Fig. 2.4 shows, an inhibitory appetitive CS may be turned into an excitatory aversive one, and an inhibitory aversive CS into an excitatory appetitive one, much more rapidly (Konorski and Szwejkowska, 1956).

1. *Excitatory Conditioning*

The standard case of excitatory conditioning is where the UCS occurs if and only if the CS occurs. Less perfect correlations may, however, be programmed. Numerous experiments have reduced the probability of the CS being followed by the UCS, i.e. have programmed "partial reinforcement", and have found a decrease in the rate of conditioning and sometimes a lower level of conditioning at asymptote. Fitzgerald (1963) and Wagner *et al.* (1964), studying salivary conditioning in dogs, found that partial reinforcement led to lower levels of performance throughout training than did consistent reinforcement. However, Thomas and Wagner (1964) and Leonard and Theios (1967b), studying eyelid conditioning in rabbits, found a lower rate of acquisition but comparable levels of performance at asymptote. Conditioned suppression in rats also proceeds more slowly, but to the same asymptote under partial reinforcement as under consistent reinforcement (Brimer and Dockrill, 1966; Wagner *et al.*, 1967a). It may be that these apparent discrepancies simply reflect differences of measurement; conditioned suppression rapidly reaches a ceiling of complete suppression, and the probability of a CR in eyelid conditioning in rabbits also tends to 1·0. In salivary conditioning, however, the measure of CR strength is an amplitude measure, and as such is relatively unbounded. Asymptotic differences can be observed only in the absence of ceiling effects. Some results reported by Brogden (1939b) support this argument. Brogden studied salivary conditioning in dogs, varying the probability of reinforcement between 0·20 and 1·0. Although the amplitude of salivary CRs varied substantially with this variation in reinforcement, the probability that some salivation would occur on each trial was virtually unaffected.

A second way of reducing the correlation between CS and UCS in excitatory conditioning is to increase the probability of the UCS occurring in the absence of the CS. This manipulation has been far less extensively studied. An experiment on conditioned suppression in rats by Rescorla (1968b), however, suggests that increasing the probability of the UCS in the absence of the CS, and decreasing the probability of the UCS in the presence of the CS, may have similar effects on conditioning. Suppression ratios for several of the experimental groups of Rescorla's experiment are shown in Fig. 2.5. The argument presented so far has not suggested that there is anything surprising in these data. And yet they in fact indicate that subjects in Group 0·4–0·2, who received four times as many pairings of CS and UCS as those in Group 0·1–0, nevertheless showed *less* suppression

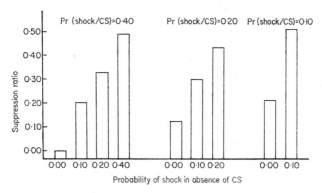

FIG. 2.5. Conditioned suppression in rats as a function of the probability of shock in the presence and absence of the CS. The probability of shock in each case is calculated over 2 min intervals (the length of the CS). Thus Group 0·4–0·2 received an average 0·4 shocks during each presentation of the CS, and 0·2 shocks during each 2 min interval in the absence of the CS. Suppression ratios are calculated over four unreinforced test trials. (*After* Rescorla, 1968b.)

to the CS: an increase in the probability of the UCS in the absence of the CS appears to outweigh the actual number of CS–UCS pairings.

Similar results have been reported in a study of conditioned key pecking in pigeons (Gamzu and Williams, 1971). When illumination of the key light signalled an increase in the probability of food being presented, pigeons rapidly learned to peck the key; but if the probability of food was the same when the key light was turned off as when it was turned on, pecking rapidly extinguished.

One possible explanation of Rescorla's results is to point out that conditioned suppression was measured as a ratio of responding during the CS to responding in an equivalent interval immediately preceding each CS. Thus a decline in the suppression ratio with any increase in the probability of shocks in the absence of the CS, may simply have reflected a decrease in rate of responding in the absence of the CS. Unfortunately, Rescorla does not present the absolute rates of responding which would permit a check of this possibility, but the suggestion is not particularly convincing since conditioning trials were given in an entirely separate apparatus from that used to test the effects of the CS on lever pressing, and subjects were also given two sessions of food-reinforced lever pressing between the end of conditioning and the start of testing. In Gamzu and Williams' experiment, moreover, absolute rates of CRs during illumination of the key light revealed no evidence of conditioning when the CS signalled no increase in the probability of food.

The most convincing analysis of these data, although stressing the importance of conditioning to background stimuli, differs in one important respect from this first possible explanation (Rescorla and Wagner, 1972). Simple excitatory conditioning may be regarded as discriminative conditioning between a compound CS+(AX), consisting of the CS (A) plus background stimuli (X), and CS−(X) consisting of the background stimuli alone. When the correlation between CS and UCS is +1·0, the X stimuli alone signal no UCS, and A should acquire all the excitatory strength available. But as the probability of the UCS in the absence of A increases, so X alone should acquire greater excitatory strength. Studies of compound conditioning have shown that when the CS consists of an explicit compound, AB, the presence of B may detract from the amount of conditioning accruing to A: the presence of one stimulus may, in Pavlov's terminology, "overshadow" another (p. 46). The magnitude of this overshadowing effect depends upon the reinforcement schedules associated with each of the component stimuli, such that the higher the correlation between B (presented alone) and the UCS, the more it will detract from conditioning to A. This, then, is presumably what was happening in Rescorla's and Gamzu and Williams' experiments: as the probability of shock or food in the presence of X alone increased, so the amount of conditioning to A on AX trials decreased.

2. Control Groups in Excitatory Conditioning

When a positive correlation between CS and UCS is arranged, we observe an increase in CRs to the CS during the course of training. But what is to say that these CRs are a consequence of the experimentally arranged contingency between CS and UCS? Two other factors have been suggested which may contribute to changes in behaviour to the CS. These two are sensitization and pseudo-conditioning (Gormezano, 1966; Razran, 1971).

The term "sensitization" refers to the possibility that a CS may itself reflexly elicit certain responses, which may be potentiated by exposure to the UCS. In some cases, moreover, the response reflexly elicited by the

CS may resemble the CR that is to be conditioned. Notable examples of conditioning situations where this may happen include studies of human eyelid conditioning, where a visual CS may elicit closure of the eyelid (cf. Hull, 1934), and studies of GSR conditioning, where *any* novel stimulus elicits the GSR. In this last case, the problem is particularly troublesome since presentation of the CS alone, on a test trial, will constitute a novel event sufficient to elicit a GSR, thus making it singularly difficult to discriminate between sensitization and genuine conditioning.

A good example of a sensitized increase in the responses reflexly elicited by the CS is provided in an experiment by Harris (1943a). Harris recorded the activity of rats in a stabilimeter cage using a buzzer as CS and shock as UCS. The buzzer elicited some responding when presented alone prior to conditioning, but elicited even more responding in a group that had just been exposed to a series of presentations of the UCS alone.

The term "pseudo-conditioning" refers to the possibility that the UCR to the UCS may come to be elicited by other stimuli in spite of the absence of any association between them (Grether, 1938). One reason for this might be that the UCR generalized to stimuli similar to the UCS. But it is also possible that repeated exposure to electric shock, for example, might produce some change in internal state which would cause any external stimulus to elicit escape or avoidance reactions. As such, pseudo-conditioning might well be a valuable and adaptive form of behaviour (Evans, 1966a, b; Wells, 1968, pp. 154–8).

Although pseudo-conditioning may be readily observed in invertebrates, such as polychaete worms (Evans, 1966a, b) or octopuses (Young, 1960), reports of significant pseudo-conditioning in intact vertebrates are relatively rare and seem to be confined to studies utilizing severe shocks (Razran, 1971, pp. 65–72). One example of pseudo-conditioning in rats, clearly attributable to generalization from the UCS to the test stimulus, is provided by Wickens and Wickens (1942). They placed rats in a box with two compartments. Exposure to shock in one compartment elicited running into the second compartment. The subjects were then presented with a light which had never been paired with shock, to see if running would still be elicited. The subjects were divided into four groups, trained with either an abrupt or gradual onset of the shock and tested with either an abrupt or gradual onset of the light. Fifteen of nineteen subjects in the abrupt–abrupt and gradual–gradual groups responded to the light; only three of eighteen subjects in the remaining two groups showed any pseudo-conditioned responses.

With these possibilities in mind, experimenters studying classical conditioning have routinely used a variety of possible control procedures intended to ensure that any change in the behaviour of an experimental group, exposed to paired presentations of CS and UCS, can be unambiguously attributed to the formation of an association between the two. Experiments on conditioned suppression, for example, usually include a small number of "pre-test" trials to the CS alone: initial presentation of the CS usually causes mild, unconditional suppression which, however, rapidly habituates. The standard control group used in most experiments is one which receives presentations of both CS and UCS, but on different trials, so that the two are never paired. Other control groups sometimes employed include one which receives the CS only and another which receives the UCS only. Schneiderman *et al.* (1962) and Schneiderman and Gormezano (1964), for example, employed all three control groups in studies of eyelid and nictitating membrane conditioning in rabbits: in no case did more than a negligible number of CRs occur in the absence of specific CS–UCS pairings.

Although such results are unambiguous in the sense that the difference between experimental and control groups must be attributed to the formation of some set of associations, they do not unambiguously point to the location of those associations, for they do not necessarily establish a true baseline of the probability of a CR occurring to the CS in the absence of any association between CS and UCS. Prokasy (1965) and Rescorla (1967a) have recently pointed out that the standard control group, which is exposed to both CS and UCS, but never to the two together, does in fact experience a specific contingency between them: whenever the CS occurs, the UCS will not occur for some minimal period of time. This control procedure, therefore, specifies a negative correlation between CS and UCS and may, as a consequence, provide the opportunity for inhibitory conditioning to occur.

How then can we obtain a true baseline where no association between CS and UCS will be formed, and from which we can detect deviations in the probability of a CR that can be ascribed to excitatory or inhibitory conditioning? The solution proposed by Prokasy and Rescorla is the use of a "truly random" control group. For such a group, CS and UCS are programmed according to independent, random schedules, but no constraints are imposed to exclude their occasional coincidence. In fact, the probability of the UCS is the same in the presence and in the absence of the CS. Since there is no correlation between CS and UCS, Prokasy and Rescorla argued, no association between the two will be established.

Several studies have confirmed this prediction. In three experiments on conditioned suppression in rats, Rescorla (1968b; 1969b) found no evidence of suppression to a CS whose presentation had been uncorrelated with shock. The results for three of the groups in one of these experiments were shown in Fig. 2.5: Groups 0·4–0·4, 0·2–0·2, and 0·1–0·1 were all exposed to equal probabilities of shock in the presence and absence of the CS, and none showed any reliable suppression to the CS. Gamzu and Williams (1971) observed no conditioned key pecking in their pigeons when illumination of the key light was completely uncorrelated with the delivery of food. Rescorla (1966) and Weisman and Litner (1969) trained dogs and rats to avoid unsignalled shocks and then studied the effects of superimposing CSs that varied in their relationship to shock. In both experiments, three conditions were examined: an excitatory CS was positively correlated with shock; an inhibitory CS was negatively correlated

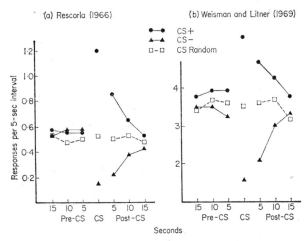

FIG. 2.6. The effects of superimposing various CSs upon a baseline of unsignalled avoidance responding. CS+ had been correlated with an increase in the probability of shock, CS − with a decrease in the probability of shock, and CS Random had been uncorrelated with shock. (a) Shuttle box avoidance in dogs. (*After* Rescorla, 1966.) (b) Wheel turning avoidance in rats. (*After* Weisman and Litner, 1969.)

with shock (since subjects were exposed to both CS and UCS, but never to the two together, this procedure is similar to that frequently used as a control condition); and a random CS was completely uncorrelated with shock. The results of the two experiments are shown in Fig. 2.6. In both studies, the excitatory CS elevated rate of responding; the inhibitory CS suppressed avoidance responding, and the random CS had no discernible effect on responding at all.

In spite of these results, however, it appears that there are some conditions under which random presentations of a CS and UCS may produce significant conditioning to that CS. Kremer and Kamin (1971), Kremer (1971), and Quinsey (1971) all observed significant levels of conditioning to a random CS in studies of conditioned suppression. Several factors may be responsible for the discrepancies between the outcomes of these different experiments. The data of Rescorla's studies of conditioned suppression, for example, were averaged over the four unreinforced test trials of each day; Kremer and Kamin, on the other hand, observed that a highly significant level of suppression on the first test trial might completely extinguish in fewer than four trials.

Kremer argued that one important factor determining whether a CR uncorrelated with shock would control a significant level of suppression, was the overall rate at which the CS and shock had been presented. He found significant suppression to a CS after subjects had been exposed to 20 random presentations of the CS and UCS in a two-hour session, but no evidence of suppression if the rate of presentation had been only four per session. One reason for the importance of this factor has been suggested by Benedict and Ayres (1972): a high rate of presentation of CS and UCS may permit more fortuitous pairings of the two at the outset of training, and hence generate a temporary, spurious contingency. They found that, in the absence of such an initial positive correlation between CS and shock, even a high rate of presentation resulted in no conditioning to the CS.

Another possibility is that when subjects are exposed to a sufficiently dense schedule of random CSs and UCSs, they may react to the overall contingency between sessions. In a typical study of conditioned suppression, for example, rats are initially trained to press a lever for food, are then given several sessions during which a CS and shock occur at unpredictable intervals, are retrained on the lever-pressing response, and are finally tested for suppression to the CS. Although there may be no correlation *within* any given session between the CS and shock, there is a perfect correlation *between* sessions: shock is delivered only in those sessions which also contain presentations of the CS. It is not entirely unreasonable to expect that under some conditions this contingency might come to affect the subjects' behaviour: Dweck and Wagner (1970), indeed, have shown that a deliberate attempt to establish the contingency (by alternating sessions containing CS and UCS with sessions containing neither) resulted in significant conditioning to an otherwise random CS whose occurrence within each session bore no temporal relation to the UCS.*

It does, therefore, seem reasonable to conclude that if a CS predicts no change in the probability of a UCS, no association between the two will

---

* It is worth noting that in studies of conditioned enhancement of avoidance responding, where no evidence of conditioning to a random CS has been detected, subjects are exposed to shock before the outset of training with the random CS and UCS: the introduction of the CS, therefore, is not correlated with the introduction of the UCS.

be established. Some degree of contingency between CS and UCS is necessary for the development of excitatory conditioning. As Kremer and Kamin (1971) have argued, however, this does not mean that a positive contingency between CS and UCS is sufficient to establish conditioning. Although the contingency between a CS and UCS remains the same whether their presentations are separated by one second or one hour, only the former interval will typically result in the development of significant conditioning (p. 61). Some minimal degree of temporal contiguity between CS and UCS is also necessary for the establishment of any association between the two.

We shall discuss below why temporal contiguity appears to be necessary for conditioning. One possible reason for the importance of a contingency between CS and UCS has already been discussed. If the probability of the UCS is as great in the absence of the CS as in its presence, then situational stimuli alone are equally valid predictors of the occurrence of the UCS, and may come to overshadow the CS (Rescorla and Wagner, 1972). Whether this is a sufficient account is, perhaps, questionable. As we shall see, there is evidence that a CS which predicts no change in the probability of a UCS does not remain neutral: animals exposed to random presentations of a CS and UCS appear to learn that the two are uncorrelated, and such learning may severely interfere with the establishment of any association between the two during subsequent conditioning trials (p. 40).

3. *Inhibitory Conditioning*

The problem in measuring excitatory conditioning is to find an appropriate baseline, deviations from which can be unambiguously attributed to the association between CS and UCS. The additional problem in measuring inhibitory conditioning is to find an effect to measure: not only may any control procedure provide the opportunity for unsuspected associative effects, it may also produce a baseline of essentially zero CRs, and in these circumstances it would be impossible to observe any *decrease* in CR probability attributable to inhibitory conditioning. Inevitably, therefore, the detection of inhibitory conditioning relies on less direct techniques. Several have been suggested, not all of which are satisfactory.

Pavlov argued that the decline in CRs observed during extinction cannot be simply attributed to a loss of excitatory conditioning, but must be due to the accumulation of inhibition. Two phenomena were used to support this argument—spontaneous recovery and disinhibition. In each case, the argument ran, the reappearance of the CR implied that the initial excitatory conditioning had not been lost; it had been merely overlaid by a new process, inhibition, which dissipated in time to produce spontaneous recovery, or could be disrupted temporarily by an extraneous stimulus to produce disinhibition. While both these arguments imply that the disappearance of CRs during extinction cannot be taken as evidence for the complete loss of the original association between CS and UCS (i.e., the subject cannot be assumed to be returning to an initial, naive state), neither argument is entirely convincing. Spontaneous recovery may be attributed partly to the reinstatement of some components of the stimulus complex to which the CR was conditioned: for example, although the subject has been exposed to a series of successive, closely spaced, nonreinforced trials, replacement in the apparatus after a long interval is a situation previously (on successive acquisition days) associated with reinforcement (Skinner, 1950; Estes, 1955; see Chapter 8). Disinhibition, as we have seen, may be partly attributed to a dynamism effect and furthermore occurs when virtually any procedure, not all of which can be regarded as inhibitory, is used to reduce the probability of responding (Brimer, 1970).

A second finding thought by Pavlov (1927, pp. 302–3) to indicate that a CS negatively correlated with a UCS acquired inhibitory properties, was the observation that after extended nonreinforcement of a CS—, subsequent reinforcement turned it into an excitatory CS only with the greatest difficulty. This observation has been confirmed by Konorski and Szwejkowska (1952b) and Szwejkowska (1959) in studies of salivary conditioning, by Hammond (1968) and Rescorla (1969b) in studies of conditioned suppression, and by Marchant et al. (1972) in a study of nictitating membrane conditioning in rabbits.

The problem with this procedure is to distinguish two possible causes of retardation of excitatory conditioning (Rescorla, 1969a; Hearst, 1972). A CS might be slow to acquire excitatory strength either because nonreinforcement had turned it into an inhibitory stimulus, and excitation and inhibition are incompatible processes, or because the subject has learned that it signals nothing of motivational significance (no reinforcement) and now does not attend to it. Nonreinforced exposure to a CS prior to any conditioning experience does, of course, retard subsequent conditioning to that CS (the phenomenon of latent inhibition). Since it can be unambiguously shown that such nonreinforced pre-exposure does not in fact result in inhibitory conditioning, but must be attributed to some attentional effect (p. 38), it becomes logically necessary to have additional criteria for the ascription of specifically inhibitory properties to a nonreinforced stimulus.

The clearest demonstration of inhibitory conditioning is the procedure developed by Pavlov (1927, Chapter 5) which was described by him as "conditioned inhibition", and by Rescorla (1969a) as the "summation" procedure. In this, the inhibitory effect of a CS— is assessed by measuring the extent to which the CR to an established CS+ is suppressed when the CS+ is presented in compound with the CS—. In one of Pavlov's experiments, for example, a tone CS was reinforced when presented alone and not reinforced when presented in compound with the ticking of a metronome. A second CS +, a rotating object, had also been trained, and when presented alone elicited 18 drops of saliva. When the rotating object was presented in conjunction with the metronome (the conditioned inhibitor), no CR occurred (Pavlov, 1927, p. 76). Since controls indicated that the effect could not be attributed to external inhibition, which is, anyway, an effect due to the presentation of a novel stimulus, the results provide relatively clear evidence that a CS predicting the omission of a UCS becomes inhibitory. Konorski (1948), and Szwejkowska and Konorski (1959) have provided additional examples in studies of appetitive conditioning; and Hammond (1967), Bull and Overmier (1968), Reberg and Black (1969), Rescorla (1969b), and Marchant et al. (1972) have provided examples of this procedure from studies of aversive conditioning. A similar test procedure has produced similar evidence of inhibition in studies of instrumental discrimination learning (e.g., Cornell and Strub, 1965; Brown and Jenkins, 1967; Lyons, 1969b; see Chapter 9).

Instead of superimposing a CS— on a CS+ in order to measure the inhibitory effects of the former, it is possible to rely on the fact that in certain situations a high rate of responding may be maintained in the absence of an explicit CS+. Dogs or rats, trained in a shuttle box to avoid unsignalled shocks, continue to respond at a steady rate in the absence of any specific signal for shock. With such a preparation, the inhibitory properties of a CS— can be directly measured by its ability to suppress this steady rate of responding. This was the technique used by Rescorla (1966) and Weisman and Litner (1969), whose results were shown in Fig. 2.6; similar results have been obtained by Rescorla and LoLordo (1965) and LoLordo (1967).

Finally, Wagner et al. (1967b), Thomas (1971), and Thomas and Basbaum (1972), using electrical stimulation of the motor cortex to elicit limb movements, or of the hypothalamus to elicit rage reactions, have measured conditioned inhibition by showing that the threshold level of current required to elicit a UCR is significantly raised when a CS— is applied at the same instant as the UCS.

The demonstration that a particular stimulus significantly decreases the probability of a particular CR may be taken as sufficient grounds for characterizing that stimulus as an inhibitory CS. It remains to consider what training conditions are required to establish a CS as a conditioned inhibitor. We have so far simply suggested that it should be negatively correlated with a UCS, but there are several different procedures for establishing such a contingency.

In Pavlov's experiments, a conditioned inhibitor was a stimulus which, when presented in conjunction with an otherwise reinforced CS+, signalled nonreinforcement. A similar procedure has been used in several recent studies (e.g., Rescorla and LoLordo, 1965, Wagner, 1971; Marchant *et al.*, 1972). If stimulus A is reinforced when presented alone, and the compound of A plus B is not reinforced, B becomes a reliable inhibitor. Wagner, studying eyelid conditioning in rabbits and conditioned suppression in rats, found that the greater the excitatory strength of A before nonreinforcement of the AB compound, then the greater the amount of inhibition conditioned to B in a fixed number of nonreinforced trials. The implication is that the effectiveness of inhibitory conditioning is a function of the prevailing expectation of reinforcement.

This implication has been confirmed by Rescorla (1969b), using a second procedure for establishing inhibitory conditioning. In an experiment on conditioned suppression in rats, he programmed a zero probability of shock in the presence of a CS−, and exposed different groups to different probabilities of shock in the absence of any explicit stimulus. When the inhibitory effect of this CS− was assessed by presenting it in conjunction with an excitatory CS+, Rescorla found that the greater the probability of shock in the absence of CS− during training, the greater the inhibitory effect of CS− during testing. A similar result has been reported by Hammond and Daniel (1970).

A third procedure for establishing conditioned inhibition is differential conditioning. Exposure to randomly alternating trials, with stimulus A reinforced and stimulus B not reinforced, is usually sufficient to turn B into a reliable inhibitor (Konorski and Szwejkowska, 1952b; Szewjkowska and Konorski, 1959; Rescorla and LoLordo, 1965; LoLordo, 1967). There are, however, some important exceptions to this generalization which suggest that differential conditioning may not be the most effective procedure for establishing inhibitory conditioning. If a CS− has been reinforced during an earlier stage of training, then extended discriminative conditioning, even though sufficient to extinguish all CRs to CS−, may not be sufficient to turn it into an effective conditioned inhibitor. The reinstatement of reinforcement, for example, results in more rapid reconditioning to such a stimulus than to a novel CS (Konorski and Szwejkowska, 1950; 1952a). Moreover, if a CS− has itself never been reinforced, but has merely elicited generalized CRs by virtue of its similarity to CS+, reconditioning may be faster than to a novel CS: the rate of reconditioning is, in fact, a positive function of the similarity of the CS− to CS+ (Szwejkowska, 1959). Both of these conclusions have been confirmed in more recent studies employing summation tests to measure inhibitory conditioning. Reberg (1970) found that differential conditioning would not establish CS− as a conditioned inhibitor if it had previously been reinforced; and Thomas and Basbaum (1972) found that only if CS− was sufficiently dissimilar to CS+ that it elicited no generalized CRs during the course of differential conditioning, did it reliably decrease the probability of responses elicited by another stimulus.

At first sight, these results appear to contradict the conclusion, derived from the data of Wagner (1971) and Rescorla (1969b), that inhibitory conditioning is a function of the prevailing expectation of reinforcement. If CS− has been previously reinforced, or is very similar to CS+, this can only increase the expectation of reinforcement in its presence and, it seems to follow, should therefore increase the inhibitory effects of non-reinforcement. The point is, however, that if CS− is to be an effective conditioned inhibitor, its inhibitory strength must be greater than its excitatory strength. Although, for example, greater generalized excitation from CS+ may result in stronger inhibitory conditioning to CS−, this would do no more than cancel out the greater generalized excitation. It would not be sufficient to increase the *net* inhibitory value of CS−. It appears, indeed, from the data of Konorski and Szwejkowska (1952b) and Thomas and Basbaum (1972), that differential conditioning may establish a CS− as a conditioned inhibitor when it is sufficiently distinct from CS+ to elicit no generalized excitation at all. The suggestion is, therefore, that it is not the generalization of excitation from CS+ that ensures that non-reinforcement of CS− will establish it as an effective conditioned inhibitor. As Wagner and Rescorla (1972) have argued, it must be the excitation conditioned to situational cues that provides the additional expectation of reinforcement against which nonreinforcement of CS− results in sufficient inhibitory conditioning to outweigh any generalized excitation from CS+.

The conclusion suggested by studies of discriminative conditioning is that if a stimulus has a past history of reinforcement or generalized excitation, then nonreinforcement in the context of discriminative conditioning may not establish that stimulus as a conditioned inhibitor. Further evidence supporting this generalization has been summarized by Rescorla (1969a) who has reviewed a number of studies which suggest that the extinction of a former CS+ is never sufficient to turn it into an effective inhibitory stimulus. In simple extinction, there appears to be a second factor operating against the establishment of a significant level of conditioned inhibition. Not only does the CS have a prior history of reinforcement, but an extinction schedule provides at best a rather weak negative correlation between a CS and reinforcement. In all other procedures, nonreinforcement of CS− is given in the context of continued reinforcement in the absence of CS−, or else CS− is presented in conjunction with a known CS+ and the compound is not reinforced. In an extinction schedule, no further reinforcement is given and CS− is presented alone. The importance of this factor has been reasonably well documented: Reberg (1970) found that concurrent reinforcement of CS+ was necessary in order to establish CS− as an effective inhibitor after differential conditioning; while, in a subsequent study, Reberg (1972) found that nonreinforcement of a former CS+ in the absence of concurrent reinforcement of another stimulus (i.e., simple extinction) resulted in a stimulus that showed evidence of excitation rather than inhibition in a summation test.

In conclusion, therefore, inhibitory conditioning may be most readily measured by presenting a CS− in a situation which reliably elicits a CR, and recording a decrease in the probability of the CR. The operations for establishing inhibitory conditioning may be said to depend upon arranging a negative correlation between a CS and UCS. More precisely, two general procedures have been shown to be adequate: the experimenter may schedule reinforcement in the absence of a particular stimulus and no reinforcement in its presence; or he may present the stimulus in conjunction with an already established excitatory CS and not reinforce the compound. Simple nonreinforcement of a particular stimulus presented on its own has never been shown to produce significant inhibitory conditioning. The implication is that inhibitory conditioning is a consequence of a decrease in the probability of reinforcement from the level expected by the subject (Konorski, 1948; Wagner and Rescorla, 1972).

4. *Latent Inhibition*

If a stimulus becomes a reliable conditioned inhibitor only if it signals the omission of an expected reinforcer, then the presentation of a CS, without reinforcement, but before the subject has been exposed to any reinforcement in the experimental situation, would not be expected to turn that stimulus into a conditioned inhibitor. Nevertheless, Lubow and Moore (1959) used the term "latent inhibition" to describe their finding that 10 nonreinforced presentations of a CS prior to any conditioning significantly retarded the acquisition of a leg flexion CR to that CS in sheep and goats. Fig. 2.7 shows the results of an experiment by Lubow *et al.* (1968) in which movement of the pinna in rabbits was conditioned to a tone, shock to the ear serving as the UCS: 40 nonreinforced presentations of the tone preceded conditioning for the pre-exposed group, and this treatment both retarded acquisition of the CR and facilitated its extinction. The basic result has since been repeatedly confirmed for a variety of other conditioning procedures: for conditioned suppression by

Carlton and Vogel (1967), Anderson *et al.* (1969), Lubow and Siebert (1969), and Kremer (1971); for eyelid conditioning in rabbits by Siegel (1969a, b); and for tail flexion in rats by Chacto and Lubow (1967). The magnitude of the effect varies with the number of nonreinforced presentations of the CS (Lubow, 1965; May *et al.*, 1967; Siegel, 1969a); and the effect generalizes to similar stimuli (Carlton and Vogel, 1967; Siegel, 1969b). There is some suggestion that the magnitude of the effect is greatest during the early stages of conditioning (Chacto and Lubow, 1967; Lubow *et al.*, 1968, see Fig. 2.7; James, 1971b). This last result may be due partly to a ceiling effect: as control subjects start to approach asymptote, so the difference between them and pre-exposed subjects may start

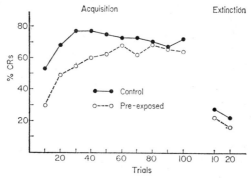

FIG. 2.7. Latent inhibition: the effect of nonreinforced pre-exposure to the CS on the conditioning and extinction of pinna responses in rabbits. (*After* Lubow *et al.*, 1968.)

to disappear. Nevertheless, it is noteworthy that several studies of one-trial conditioned suppression have observed extraordinarily large effects of pre-exposure. Lubow and Siebert (1969), for example, gave a single conditioning trial on which a tone was paired with shock, and then presented the tone to thirsty rats drinking at a water tube. Control subjects, never pre-exposed to the tone, showed market disruption of drinking, averaging 228·8 sec to complete 10 licks at the tube while the tone was sounded: subjects that had received 70 nonreinforced pre-exposures to the tone before the one conditioning trial averaged 1·65 sec to complete 10 licks. Similar results were reported by Carlton and Vogel (1967) in a similar study of one-trial suppression in rats.

That nonreinforced pre-exposure to a stimulus significantly interferes with subsequent conditioning to that stimulus is, therefore, well established. There has been less agreement about the explanation of the phenomenon. As was noted above, the finding that nonreinforcement of a CS retards subsequent excitatory conditioning may mean either that the CS has acquired inhibitory properties which specifically interfere with excitatory conditioning, or that subjects have learned to ignore a stimulus which predicts no reinforcement. Although Lubow and Moore (1959) suggested that nonreinforced pre-exposure might turn a stimulus into an effective inhibitor, the fact that conditioned inhibition appears to depend on the omission of an expected reinforcer makes this suggestion somewhat implausible.

Conditioned inhibition cannot be regarded simply as a case of subjects failing to attend to stimuli signalling nonreinforcement, since the presentation of a CS−, negatively correlated with reinforcement, in conjunction with a known CS+, significantly reduces the magnitude of the CR elicited by that CS+. Conversely, it can be shown that latent inhibition is not due to the establishment of inhibitory or competing responses, which specifically interfere with the development of excitatory conditioning, because nonreinforced pre-exposure to a CS not only fails to turn that stimulus into a conditioned inhibitor as revealed by a summation test, but actually retards inhibitory conditioning.

Pavlov (1927) had noted that the disruption of a CR by the presentation of an extraneous stimulus in conjunction with the CS (termed external inhibition) depended on the novelty of the external inhibitor. Familiar stimuli (i.e., stimuli to which the subject has been repeatedly exposed in the absence of reinforcement) eventually "become lost upon the organism and lose all their inhibitory properties (p. 46)". More formally, Reiss and Wagner (1972) in a study of eyelid conditioning in rabbits, found that increasing the amount of nonreinforced pre-exposure to a stimulus decreased rather than increased its ability to disrupt CRs when it was presented in conjunction with a CS+. In a study of conditioned suppression, Rescorla (1971) gave rats nonreinforced pre-exposure to a tone, followed by differential conditioning in which reinforced trials to a light were interspersed with nonreinforced trials to a light–tone compound. Although both pre-exposed and control subjects showed similar acquisition of conditioned suppression to the light, pre-exposed animals continued to suppress in the presence of the light–tone compound for a much longer time than controls. Rescorla concluded that pre-exposure significantly interfered with inhibitory conditioning to the tone. Hearst (1972) has similarly reported that nonreinforced pre-exposure may significantly interfere with instrumental discrimination learning, whether the pre-exposed stimulus serves as S+ or S−.

These results argue strongly against interpreting latent inhibition as a consequence of the conditioning of inhibitory or interfering responses during nonreinforced pre-exposure. If such exposure produces a stimulus less able to disrupt established CRs, and less able to enter into new associations (whether excitatory or inhibitory), this suggests that, in some sense, animals are less likely to attend to a pre-exposed stimulus than to a novel stimulus. One interpretation of such a notion is suggested by Carlton and Vogel's (1967) description of latent inhibition as habituation. The term "habituation" refers to the observed decline in a response elicited by a given stimulus, when that stimulus is repeatedly presented without reinforcement (Harris, 1943b; Thompson and Spencer, 1966; Groves and Thompson, 1970; Hinde, 1970). Operationally, the procedures of experiments on habituation and latent inhibition are identical; the only difference is that in the former case the experimenter records a change in the subject's behaviour during exposure to the stimulus, while in the latter case he records an effect on the subject's subsequent behaviour. It is, therefore, tempting to suggest that it is the habituation of such responses as the orienting reaction that underlies the phenomenon of latent inhibition. Sokolov (1963) has presented some evidence consistent with this idea. Recording the GSR as an indicant of the orienting reaction, he observed its habituation over the course of a series of nonreinforced presentations of a tone; subsequent conditioning to the tone was significantly retarded and the emergence of CRs depended on the complete recovery of the orienting reaction.

Stronger evidence of the relationship between habituation and latent inhibition could be provided by more studies which simultaneously measured both phenomena, and also by studies of latent inhibition which varied those parameters, such as intertrial interval, known to affect habituation. One such parameter has been studied more than once. A general feature of most habituated responses is that they show spontaneous recovery. It is thus a matter of considerable interest that the effects of pre-exposure on subsequent conditioning do not seem to depend upon the length of time elapsing between the two: pre-exposure appears to retard conditioning equally, whether conditioning begins immediately or not until a week after pre-exposure (Siegel, 1970; Crowell and Anderson, 1972). This suggests that latent inhibition may be caused partly by some process more permanent than that typically measured in studies of habituation, although this point should be accepted only with some caution, since it is clear that recovery from habituation proceeds at widely different rates in different response systems (Hinde, 1970, pp. 298–301).

There is, therefore, reason to believe that latent inhibition may represent more than the simple habituation of attention to a nonreinforced stimulus. There is some evidence that an additional process may be involved, specifically, that animals may actively learn to ignore such a stimulus. The critical observation is that the presentation of a stimulus in the absence of all reinforcement is neither the only, nor the most effective, procedure for interfering with subsequent conditioning. Several studies have shown that the addition of random presentations of the UCS to the pre-exposure stage may very severely retard subsequent conditioning. Gamzu and Williams (1971) found that pigeons exposed to random illuminations of a key light and presentations of food before classical conditioning of the key pecking response, were still responding at a lower rate than control birds after 35 sessions of conditioning. Kremer (1971) studied the acquisition of conditioned suppression in rats that had been exposed either to the CS alone or to random presentations of the CS and shock. The results for these two groups, together with those of a control group, are shown in Fig. 2.8; it is clear that random presentations of CS

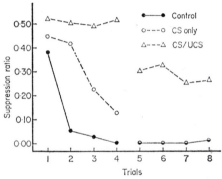

FIG. 2.8. Latent inhibition: the effect of nonreinforced pre-exposure to the CS, and of random presentations of CS and UCS, on the acquisition of conditioned suppression in rats. (*After* Kremer, 1971.)

and UCS had the more deleterious effect on conditioning. This result has been confirmed by Mackintosh (1973), who was also able to show that this effect was specific to the reinforcer used as UCS in conditioning: conditioned suppression was retarded by exposure to random presentations of the CS and shock significantly more than by random presentations of the CS and water; conversely, conditioned licking was more retarded by random presentations of the CS and water than by presentations of the CS and shock.

These results can hardly be due to any simple habituation of attention. They suggest, rather, that animals may specifically learn that a particular CS and UCS are uncorrelated (that the CS predicts no change in the probability of the UCS), and that this learning interferes with the establishment of an association between the two during subsequent conditioning. This idea has a number of implications which we shall discuss in due course.

III. PARAMETERS OF CONDITIONING

In this section we discuss the major variables determining the strength of conditioning.

A. CS Variables

1. *Intensity or Salience of the CS*

According to the Pavlovian "law of strength" (Razran, 1957), the relative intensities of the CS and UCS affect the strength of conditioning in the following way: for any given UCS, strength of conditioning first increases and eventually decreases as CS intensity is increased, and the optimal CS intensity increases directly with UCS intensity. A simpler empirical generalization was formulated by Hull (1949); strength of conditioning was said to increase monotonically with CS intensity. Hull explained this generalization by postulating a factor of "stimulus intensity dynamism" (V) which varied directly with CS intensity and was multiplied with habit strength to yield excitatory potential.

Although some early studies cast some doubt on the validity even of Hull's weaker empirical generalization (Gray, 1965b), several more recent experiments reviewed by Gray establish that the phenomenon is a real one. Barnes (1956), for example, found better leg-flexion conditioning in dogs with an 80-db tone than a 60-db tone. Kamin and Schaub (1963) found substantial differences in the rate of acquisition of conditioned suppression between three groups of rats trained with different intensities of white noise as the CS; in a second experiment, they found even larger differences between their two extreme intensities when they used trace CSs. Kamin and Brimer (1963) confirmed this result, but found no evidence of a decline in strength of conditioning with a very intense CS, nor any sign of the interaction between CS and UCS intensities predicted by the law of strength. Kamin (1965) has suggested that the decline in conditioning with an extremely strong CS, reported in some Russian studies reviewed by Razran (1957), may have been a consequence of their experimental procedure: since training was conducted with a CS of medium intensity, performance on test trials with a strong CS may have suffered from generalization decrement.

That conditioning does improve with the intensity of the CS, however, does not necessarily require the type of explanation proposed by Hull. Intensity might be only one of several related factors that affected conditioning. Pavlov (1927, pp. 56–7, 141–2), for example, observed that stimuli from different modalities differed in their effectiveness as CSs for salivary conditioning in dogs. Similar observations have been made in other Russian studies of conditioning (e.g., Razran, 1971, p. 145), and in entirely different preparations for the study of conditioning. When rats are poisoned after drinking water with a distinctive flavour, they subsequently show an aversion to that particular taste; certain flavours, however, may be much more effective CSs for conditioning this aversion than others: Kalat and Rozin (1970), for example, found that rats developed aversions to a solution of casein hydrolysate more readily than to a salt solution, and more readily to the salt than to sucrose. Finally, studies of instrumental discrimination learning, in both rats (Kendler, 1971) and pigeons (Baron, 1965), have frequently observed more rapid learning for some classes of stimuli than for others.

Theories of both classical conditioning and instrumental discrimination learning must accommodate these observations. Rescorla and Wagner (1972), for example, introduce a parameter, representing the "salience" of a stimulus, which determines rate of conditioning to that stimulus. Attentional theories of discrimination learning assume that different stimuli may command different initial probabilities of attention (Zeaman and House, 1963; Lovejoy, 1968; Sutherland and Mackintosh, 1971). Although there has been some tendency to assume that differences in salience might reflect some ordering of an "attentional hierarchy" for different modalities (Baron, 1965), Pavlov was surely right in supposing that similar principles govern differences in the effectiveness both of stimuli from different modalities, and of stimuli of different intensity from the same modality.

If, however, tones are usually more effective CSs for dogs than are lights, this cannot be meaningfully attributed to differences in the physical intensities of the particular tones and lights used. There is no single physical scale for measuring the relative intensities of auditory and visual stimuli. Moreover, the ordering of different modalities is quite obviously as much dependent on the subject as on the stimuli in question: pigeons, unlike dogs, might condition more readily to visual than to auditory stimuli. Differences between modalities, in fact, are surely attributable to differences in the sensitivity of the subject's sensory apparatus, and this, of course, suggests that the effects of intensity within a modality should also be attributed to differences in discriminability.

Perkins (1953) and Logan (1954) pointed out that more intense stimuli are more readily discriminated from the background. More specifically, they argued that the intensity of a stimulus may affect conditioning because the discriminability of the CS determines the amount of generalization from the background stimuli alone (which are never reinforced) to the combination of CS and background (which is reinforced). Kendler (1971) has proposed a similar analysis, in terms of generalization between

S+ and S—, to account for the effects of salience on the rate of instrumental discrimination learning.

Two lines of evidence establish beyond any doubt that much of the effect of stimulus intensity on simple conditioning should be attributed to differences in discriminability, rather than to the sort of dynamism effect proposed by Hull (1949). In the first place, such effects are reliably obtained only when the experimental procedure requires a discrimination between stimulus and background. As we have argued, any classical conditioning experiment requires such a discrimination. But in instrumental learning experiments, reinforcement may be programmed independently of any stimulus (or the stimulus may remain on continuously). Two experiments (Gray, 1965a; Perkins, 1953) have shown that stimulus intensity affects instrumental learning only if discrimination of the stimulus from the background is required. Even more conclusively, several classical conditioning studies have employed a *decrease* in stimulation (e.g., background noise or light) as the CS and have found that conditioning depends upon the change in stimulation, rather than on the absolute intensity of the CS (Kamin, 1965; Logan and Wagner, 1962). Kamin, for example, had a constant 80-db noise on during intertrial intervals in a study of conditioned suppression, with the CS for different groups being a decrease in noise intensity to 0, 45, 50, 60, or 70 db. The greater the decrease, and therefore the less intense the CS, the faster conditioning occurred.

At first sight, these results suggest that the effects of stimulus intensity can be explained entirely in terms of discriminability. But life is rarely so simple. Kamin (1965) has gone on to show that when the magnitude of the change between CS and background is equated, there remains a small effect which may be ascribable to intensity *per se*. He compared the results of the five groups described above, for whom the CS consisted of a decrease from an 80 db noise, with the results of an additional five groups, for whom the CS consisted of an increase from a background of either 0, 45, 50, 60, or 70 db to a CS of 80 db noise. The results are shown in Fig. 2.9. Although there was no difference between increase or decrease conditions when the magnitude of the change was large, with smaller changes there was significantly better conditioning in subjects for whom the CS was an increase to a higher intensity rather than a decrease to a lower intensity.

One problem with interpreting this finding is that the presentation of a continuous loud noise during an intertrial interval may cause sensory adaptation, and hence reduce the sensitivity of the auditory system to small changes in intensity. If this happened in Kamin's experiment, then animals exposed to the 80 db noise during the intertrial interval would have experienced more sensory adaptation than those exposed to less intense noises, and the consequent difference in sensitivity might have been sufficient to produce the small differences in rate of conditioning that were observed. A similar argument may be advanced to account for the results of an experiment reported by Gormezano (1972). Gormezano found that eyelid conditioning in rabbits proceeded more rapidly when the CS consisted of the onset of an intermittent tone rather than its offset. Although the use of an intermittent stimulus may have reduced sensory adaptation, in the absence of independent psycho-physical data on the discriminability of increases and decreases in stimulation it would be hazardous to attribute the results too confidently to a dynamism effect.

One should not, perhaps, dismiss the possibility of such an effect out

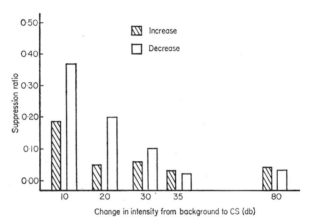

Change in intensity from background to CS (db)

FIG. 2.9. The effect of the absolute intensity of the CS, and of the change in intensity from background to CS, on conditioned suppression in rats. The greater the change in intensity from background to CS, the more rapidly conditioning occurs; but for small changes in intensity, an increase in intensity results in better conditioning than does a decrease. The data represent the mean suppression ratios over Days 2–6 of conditioning. (*After* Kamin, 1965.)

of hand. The mere labelling of the phenomenon, however, does little to explain it. Perhaps very intense stimuli have effects on "arousal" as suggested by Groves and Thompson (1970) when considering disinhibition of an habituated response (p. 16). The fact, noted by Brimer (1970), that the effect of stimulus intensity on disinhibition is much smaller than its effect on conditioning can, of course, be explained by the observation that the latter effect is largely due to the additional factor of discriminability.

However this may be, it is apparent that the main effect of intensity is to increase the discriminability of the CS from the background. In many ways, the simplest interpretation of this finding is that proposed by Logan (1954) and Perkins (1953) in terms of generalization of excitation and inhibition between reinforced and nonreinforced stimuli. Recent evidence, however, suggests that additional factors may be involved. Grice and Hunter (1964) compared two intensities of CS in an eyelid conditioning study with human subjects. They found very little difference in rate of conditioning between two separate groups, one conditioned with the strong CS, the other with the weak CS. However, a third group, receiving alternate trials with the strong and weak CS, showed markedly better conditioning with the stronger stimulus. This finding has been replicated in a number of human studies (reviewed by Grice, 1968), and in one experiment on eyelid conditioning in rabbits (Frey, 1969). Gormezano (1972), however, found no evidence of such an effect. Grice and Frey have interpreted their results in terms of signal detection theory (Green and Swets, 1966). They suggest that when subjects are exposed to a constant CS they adopt a higher or lower decision criterion appropriate to the intensity of the CS, and their performance, therefore, is relatively unaffected by the intensity of the CS; but when subjects are exposed to both high and low CSs they adopt an intermediate criterion, which enhances the probability of detecting the strong CS and decreases the probability of detecting the weak CS. Grice's analysis depends upon McGill's (1963) analysis of latency mechanisms and may be more appropriate to discrete, short-latency CRs such as the eyeblink, than to preparatory CRs such as conditioned suppression. Indeed, the major effect of intensity in a within-subject design is on the latency of the CR (Leonard and Monteau, 1971). It remains to be seen, however, whether diffuse CRs such as conditioned suppression fail to show greater within-subject than between-subject effects of intensity.

2. *Compound Conditioning*

Although the strength of conditioning to a particular CS may be affected by its own salience or intensity, it is also apparent that it is dependent on the salience, intensity and other characteristics of other stimuli with which it is presented in compound. As a first approximation, there appears to be some competition for the acquisition of associative strength, such that the stronger the conditioning to one component of a compound CS, the less conditioning will occur to other components. A good illustration of this general principle is given in Fig. 2.10, which shows the results of an experiment on GSR conditioning in cats (Wickens *et al.*, 1963). In this experiment, a constant component, CS_2, always preceded the UCS (a 50-msec shock) by 550 msec, and the variable component, CS_1, preceded the UCS by either 700, 1050, or 1250 msec. Both CS_1 and CS_2 always overlapped and terminated with the shock. Conditioning to the CS_1–CS_2 compound and to each component was assessed on a series of test trials.

Conditioning to the compound remained the same under all conditions, but although the relationship between CS_2 and UCS remained constant, the amount of conditioning accruing to CS_2 varied widely, being inversely related to the amount of conditioning accruing to CS_1.

It was Pavlov (1927, pp. 141–4) who first demonstrated that the strength of conditioning to one stimulus critically depended upon whether it was presented alone or in a compound. When a compound CS, consisting of either tactile and thermal stimuli or auditory and visual stimuli, was conditioned and subjects were tested with each component in isolation, only one of the components was found to elicit a CR. In these two instances, the animals failed to respond to the thermal and visual components presented in isolation, although Pavlov argued: "it is obvious . . . that the

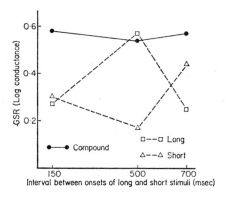

FIG. 2.10. Compound conditioning in cats. The magnitude of the GSR conditioned to a compound GS remains constant regardless of the interval between the onset of CS_1 and CS_2; but there is an inverse relationship between the strength of conditioning to the two components. (*After* Wickens *et al.*, 1963.)

ineffective components . . . could easily be made to acquire powerful conditioned properties by independent reinforcement outside the combination (p. 142)." Similar results were obtained even when both stimuli were in the same modality, but only when one of the stimuli was less intense than the other. Pavlov concluded on the basis of a number of experiments that the stronger of two component stimuli would "overshadow" the weaker to a greater or lesser extent dependent on the relative intensities of the two stimuli (Pavlov, 1927, pp. 142–3, 269–70).

The occurrence of overshadowing, and its dependence on the relative intensities of the component stimuli, has been confirmed in recent studies of conditioned suppression (Kamin, 1969; Mackintosh, 1971a) and instrumental discrimination learning (Miles and Jenkins, 1973). Kamin's study, for example, employed five different groups of rats. For Group L, the CS was light; for Group N, an 80 db noise; for Group n, a 50 db noise; and Groups LN and Ln were conditioned to compound stimuli consisting of the light and either the 80 db or the 50 db noise. Suppression to L was essentially complete in all groups trained with L either alone or in compound with a noise. In other words, L was not apparently overshadowed by either N or n. Suppression to N was virtually complete in Group N, but significantly less (suppression ratio if 0.25) in Group LN: the light, therefore, partially overshadowed the more intense noise. Group n also showed virtually complete suppression to the less intense noise, but Group Ln showed no suppression to n and, indeed, subsequently acquired suppression no more rapidly than a naive control group. The light, therefore, completely overshadowed the less intense noise.

Razran (1965) cites a number of other Russian studies which support this general position. A further claim made by Razran on the basis of studies by Palladin, Platonov and Zeliony, is that if sufficient compound training is given, then *neither* component when tested in isolation will elicit a CR; with overtraining, "configural" conditioning to the compound occurs and the isolated components are ineffective. It is difficult to interpret these results unambiguously for the normal procedure in these experiments was to provide repeated, nonreinforced test trials with the components, interspersed with continued, reinforced training with the compound, and this would have encouraged the formation of a discrimination. There have, however, been occasional studies to which this objection does not apply (Platonov, cited by Razran, 1965; see also Thomas *et al.*, 1968, in a study with successively presented components in instrumental learning by pigeons; and Booth and Hammond, 1971, in a study of conditioned suppression in rats). This suggests that under some, as yet unspecified, conditions configural learning may occur (see Baker, 1968, for a discussion of this issue).

Overshadowing of one cue by another is affected not only by their relative intensities, but also by what Wagner (1969b) has called their relative validities—a more valid cue, i.e. one which better predicts the occurrence of reinforcement, will overshadow a less valid one. A series of experiments by Wagner *et al.* (1968) established this conclusion in conditioned suppression in rats and eyelid conditioning in rabbits, as well as in instrumental discrimination learning in rats. Their basic experimental design is shown in Table 2.1. For both groups in the experiment, Correlated and Uncorrelated, the light was followed by reinforcement on 50 per cent of trials. The difference between the two conditions was that for the Correlated group one tone, T_1, always predicted reinforcement while the other, T_2, predicted nonreinforcement, and for the Uncorrelated group,

T_1 and T_2 did not predict the occurrence of reinforcement any more reliably than L. In the absence of any more valid predictors of reinforcement (i.e., in the Uncorrelated group), L acquired substantial control over responding; in the presence of a more valid predictor (i.e., in the Correlated group), L was completely overshadowed.

Wagner and Dweck (see Wagner, 1969b) varied the validity of the overshadowing cue by varying the probability of unsignalled reinforcement occurring in its absence. Rats in a conditioned suppression experiment received a number of shocks signalled by a compound stimulus $(A + B)$, and an equal number of shocks either signalled by A alone, or occurring in the absence of any CS. Stimulus A overshadowed B to a significantly greater extent when all additional shocks were signalled by A (i.e., when

TABLE 2.1

Design of experiment by Wagner *et al.* **(1968)**

Groups	Stimuli	
Correlated	T_1L+	T_2L-
Uncorrelated	$T_1L\pm$	$T_2L\pm$

T_1, T_2 designate tones; L, light.

A was a good predictor of shock), than when the additional shocks were unsignalled.

A third variable which affects overshadowing of one cue by another is the extent to which subjects receive prior training on the overshadowing cue in isolation. This phenomenon has been studied extensively by Kamin (1968, 1969) in experiments on conditioned suppression. He has found that sufficient training on A alone before AB compound training, may result in virtually no conditioning occurring to B. He has referred to the effect as "blocking" of one element of a compound by prior training on the other element. The complete design and results of Kamin's basic experiments are shown in Table 2.2, from which it can be seen that the test procedure used (measuring suppression to the added element on a nonreinforced test trial) failed to find any evidence of conditioning to B; whether B was light or noise, the amount of suppression was the same as that shown by animals never given pairings of B with shock. In subsequent experiments, Kamin reported that a relearning measure might reveal some conditioning to B on compound trials, but the effect was small, and the conditioning appeared to be confined to the *first* compound trial: additional compound trials added no further associative strength to B.

Two further results reported by Kamin should be mentioned, since they suggest that blocking of B by prior training with A occurs only if reinforcement is unchanged on compound trials. In one experiment, after receiving 16 reinforced trials with A, rats were exposed to eight *non*-reinforced trials to the AB compound. Clear evidence was obtained that some inhibitory conditioning occurred to B. As noted above (p. 34), this result is quite reliable: indeed Wagner (1971) has shown that the more reinforced training given with A alone, the greater the inhibitory conditioning to B on nonreinforced compound trials. In a second experiment, Kamin showed that if shock intensity was increased (from 1 to 4 ma) on

TABLE 2.2

Design and results of experiments by Kamin (1968, 1969)

Groups	Stage 1 16 reinforced trials	Stage 2 8 reinforced trials	Test 4 nonreinforced trials	Suppression ratios on first test trial
NL	—	NL	L	·05
N --> NL	N	NL	L	·45
N only	N	N	L	·44
NL	—	NL	N	·25
L --> NL	L	NL	N	·50
L only	L	L	N	·49

L designates light; N, noise.

compound trials, then significant excitatory conditioning occurred to the added element. Provided that the addition of B to the compound predicts some change in reinforcement, then significant conditioning to B can occur.

What is the interpretation of this array of data on overshadowing and blocking? Hull (1943, pp. 219–20) attempted to explain Pavlov's overshadowing data in terms of "generalization decrement." In Kamin's overshadowing experiment, for example, Group LN, trained with the compound, showed less suppression on test trials with N alone than did Group N, trained and tested with N alone. But for Group LN, unlike Group N, the transition from training to testing was marked by a change in the CS. Studies of generalization show that after conditioning to one CS, the probability of a CR being elicited by other stimuli is an orderly function of the magnitude of the difference between training and test stimuli. This decline in performance is presumably a consequence of removing some of the stimulus elements that had been conditioned to the response. In the present instance, a stronger claim is being made: the removal of stimulus A on test trials is said not only to remove that portion of total CR strength that was due to A, it must also be supposed to obscure the CRs conditioned to B. Why this should be so is not clear, and Hull's appeal to "afferent neural interaction" does little more than assert that it is true.

In fact there is abundant evidence that no appeal to generalization decrement is sufficient. As Kamin (1969) noted for the case where the least intense noise was completely overshadowed by the light (Group Ln), there is corroborating evidence that the noise acquired no associative strength. During initial training, Group Ln showed no faster acquisition of conditioned suppression than Group L; if the addition of the noise to the CS compound failed to affect rate of initial learning, the most natural interpretation is to say that this is because no conditioning occurred to the noise. Furthermore, in experiments that have varied the relative validity of A and B, as well as experiments on blocking of B by prior training on A, comparisons of the amount learned about B are made between different subjects, all of whom have been trained with the AB compound and tested with B alone, and for all of whom, therefore, generalization decrement from training to testing must be constant.

Kamin (1969), considering his basic blocking result, suggested that no conditioning occurs to B on compound trials because the UCS is already fully predicted by the presentation of A: only surprising reinforcers are effective. Something of this idea has been formally expressed by Rescorla and Wagner (1972) and Wagner and Rescorla (1972). They suggest that a given UCS will support only a given level of conditioning: as the associative strength of a compound CS approaches this asymptote, so there will be a decrease in the magnitude of increments accruing to all elements forming part of the compound. In the limiting case, once asymptote has been reached there is no further conditioning to any element, even to one newly added, whose own associative strength is therefore zero. Revusky (1971) has expressed a rather similar idea in terms of associative interference. If there is competition between stimuli for association with a given reinforcer, then to the extent that one stimulus has become a signal for that reinforcer, its presence on subsequent conditioning trials with the same reinforcer will prevent other stimuli from becoming effective signals.

As Kamin (1969) has argued, the strongest evidence to support this type of analysis is precisely the failure of blocking when reinforcement is changed. If shock intensity is increased on compound trials then some conditioning will occur to the added element, since a stronger shock can support a higher asymptote of conditioning. If the addition of a new element to the compound is correlated with the omission of reinforcement, then inhibition can be conditioned to the added element, since the asymptote limiting further inhibitory conditioning is an expectation of no reinforcement rather than of continued reinforcement. Not the least virtue of this analysis is that it provides symmetrical accounts of the conditions required to produce excitatory and inhibitory conditioning (Wagner and Rescorla, 1972). Inhibitory conditioning, we saw, depends upon the omission of an expected reinforcer; increments to inhibition, in other words, are an increasing function of the prevailing expectation of reinforcement against which nonreinforcement occurs. In just the same way, increments to excitation are regarded as an increasing function of the extent to which reinforcement is *not* expected so that, in the limiting case, the presentation of a completely expected reinforcer results in no excitatory conditioning at all.

A second approach is to assimilate the data on overshadowing and blocking to a number of phenomena of instrumental discrimination learning which have been thought to point to the operation of a mechanism of selective attention (Sutherland and Mackintosh, 1971). The idea here is that there is a limit not on associative strength or the effectiveness of reinforcers as is implied by Kamin, Rescorla and Wagner, and Revusky, but on the number of stimuli that can simultaneously be attended to and hence enter into new associations. Although the idea that animals have a limited capacity to attend to incoming information has proved interesting and useful (as will be evident in Chapters 9 and 10), one may question whether this capacity is so limited as to prevent simultaneous processing of the small number of long-lasting and relatively salient stimuli used, for example, in Kamin's experiments. This idea also has to add further assumptions in order to deal with the successful learning of the added component when conditions of reinforcement are changed. The critical importance of this maintenance of the original reinforcer suggests that in some sense it must be the redundancy of the added component that prevents its acquiring any associative strength: its introduction into the experimental situation signals no change in reinforcement. This formulation suggests the assimilation of blocking to latent inhibition and learned irrelevance (p. 36); in both cases, it might be suggested, animals learn that a particular stimulus is not correlated with any change in reinforcement, and hence come to ignore that stimulus (Mackintosh and Turner, 1971; Mackintosh, 1973). However, these issues, and others that arise in the interpretation of experiments on instrumental discrimination learning, are discussed at greater length in Chapter 10 (p. 583). For the moment, we shall simply accept as a valid empirical generalization that there is some competition between stimuli for the acquisition of associative strength.

c

B. CS–UCS Relevance

It is one thing to accept that the strength of conditioning to a given CS is affected not only by its own salience or intensity, but also by the salience, validity and past history of other stimuli signalling reinforcements; it seems, however, quite another thing to be forced to accept that the extent to which a stimulus may become associated with a given reinforcer depends upon the nature of that reinforcer. One of the more important recent discoveries in the study of animal learning has been that if animals are exposed to a compound stimulus signalling reinforcement, the element of this

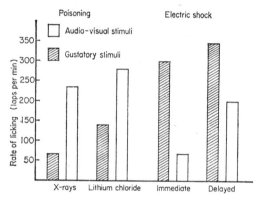

FIG. 2.11. Conditioning depends on the nature of the CS and of the UCS. An external CS is readily associated with an aversive UCS, such as shock, but not with poisoning; whereas a gustatory CS is readily associated with poisoning, but not with shock. Conditioning is measured as a decrease in the amount of water consumed during a test period, when the water is given the flavour present during conditioning, or when its ingestion is accompanied by the visual and auditory stimuli present during conditioning. (*After* Garcia and Koelling, 1966.)

compound associated with the reinforcer may be critically affected by the type of reinforcer used.

The classic demonstration of such an effect was provided by Garcia and Koelling (1966). They trained rats to drink water which was given two sets of characteristics: (1) it had a particular taste (either salt or saccharin); (2) its ingestion produced a particular set of external stimuli (a light and a clicker were operated whenever the rat made contact with the drinking tube). Although animals were permitted to continue drinking ordinary water, the ingestion of this particular water was established as a signal for aversive reinforcement. One group received an electric shock contingent on drinking; another group was exposed to X-irradiation after drinking, or was injected with lithium chloride, treatments which resulted in severe sickness. After these different treatments, animals were tested with the elements of the compound in isolation: they were given the opportunity to drink either the distinctly flavoured water unaccompanied by the light and click, or to drink ordinary water in the presence of these external stimuli. The results, presented in Fig. 2.11, show that animals associated the flavour of the water with sickness but not with shock, and the auditory and visual stimuli with shock but not with sickness.

Since each set of stimuli was an entirely effective CS for one type of reinforcer, Garcia and Koelling's results cannot be attributed to differences in their salience or discriminability. Their basic observation has been confirmed for other classes of stimuli; Garcia *et al.* (1968), for example, found that rats would associate the taste of food with sickness but its visual characteristics with shock; and Rozin (1969) confirmed that rats would associate the taste of a liquid with sickness much more readily than the location in which they had drunk the liquid. Although Revusky and Garcia (1970) have characterized the learning occurring in these situations as instrumental, arguing that the response of drinking or eating is being punished in the presence of a particular discriminative stimulus, the fact that rats may develop aversions to a flavour which they have been exposed to without ingestion (Domjan and Wilson, 1972) suggests that a characterization in terms of classical conditioning may be more appropriate.

Evidence that rats can associate the taste of a particular substance with later sickness has been available for some time (e.g., Rzoska, 1953; Barnett, 1963, pp. 47–51). The additional point made in Garcia's experiments is that flavours will be associated with only certain types of aversive consequence. Furthermore, although flavours can with difficulty be established as signals for shock (Garcia *et al.* 1970) or visual stimuli as signals for sickness (Rozin, 1969), it is noteworthy that whereas an association between flavour and sickness can be established when the interval between ingestion and poisoning is several hours (see below), conditioning between flavour and shock, or between visual stimuli and sickness, is dependent on a close temporal contiguity between the two events.

Most discussions of these studies have suggested that the rat's ability selectively to associate flavours with sickness over very long intervals of time represents a specialized adaptive system which may not obey the conventional laws of learning derived from typical laboratory studies of learning (e.g., Garcia and Ervin, 1968; Rozin and Kalat, 1971). Rozin and Kalat, for example, suggest a point very similar to one made by Garcia *et al.* (1970), when they write:

Taste does not become a *signal* for poison in the sense that tone or light becomes a signal for shock. In taste-aversion learning, the animal's perception of the taste itself or of its affective value may change. . . . The taste itself may become aversive or unacceptable, as if it were unpalatable. .... By contrast, stimuli associated with shock do not themselves become aversive; they evoke little avoidance outside the training situations (Rozin and Kalat, 1971, p. 478).

One rationale for this argument is that rats poisoned after exposure to a particular flavour will refuse to ingest a similarly tasting substance wherever they find it. If they are shocked in the presence of an external stimulus, on the other hand, they are likely to show fear of that stimulus only if exposed to it in the original training situation (or one very similar). But this observation seems to be no more than a consequence of the fact that external stimuli, including situational stimuli, are much less readily associated with poisoning than are flavours, while all external stimuli, including situational stimuli, are readily associated with a reinforcer such as shock.

In fact, the list of parallels between taste-aversion learning and other conditioning paradigms is long and impressive. Although one of the most striking features of associations between flavours and poisoning is that they may be established over intervals of several hours, it remains true that the strength of conditioning, as in other paradigms, is an inverse function of the interval between CS and UCS (Garcia, *et al.*, 1966; Smith and Roll, 1967; Revusky, 1968; Nachman, 1970; see Fig. 2.14 below). Conditioning between flavour and sickness, as between other pairs of events, is a positive function both of the intensity of the CS (Kalat and Rozin, 1970) and of the intensity of the UCS (Revusky, 1968). Numerous investigators have observed an effect analogous to latent inhibition in taste-aversion learning: a novel flavour is much more readily associated with sickness than is a familiar flavour (Revusky and Bedarf, 1967; Wittlin and Brookshire, 1968; Nachman, 1970). Finally, Revusky (1971) has observed both overshadowing and blocking in taste-aversion learning: exposure to a second flavour before poisoning may interfere with the conditioning of an aversion to a first flavour, and the magnitude of this effect is increased if the second flavour has already been established as a signal for poisoning.

These results suggest that it may be premature to regard taste-aversion learning as a unique form of conditioning with its own, unique laws. This conclusion is not intended to detract from the importance of Garcia's findings. At the very least, there remain two respects in which his data violate traditional principles. First, associations may be formed between a flavour and sickness when the two are separated by minutes or hours rather than seconds. Secondly, rats associate tastes rather than auditory or visual stimuli with subsequent sickness, although the same auditory or visual stimuli, and not tastes, are associated with electric shocks. The first of these points is discussed in more detail below (p. 66). the second raises equally important questions. If the nature of the reinforcer determines which of several stimuli become associated with it, it is clear that differences in the effectiveness of different CSs cannot be attributed solely to differences in their salience or discriminability.

Although the clearest examples of the specificity of stimulus to reinforcer have come from studies of taste-aversion learning in rats, the principle is undoubtedly of greater generality. Many birds learn to reject unpalatable food substances on the basis of their visual characteristics (Capretta, 1961; Brower, 1969); moreover, both chicks (Moore and Capretta, 1968) and quail (Wilcoxon *et al.*, 1971) form aversions to the colour rather than to the flavour of water if its ingestion is followed by aversive consequences. More critically, Shettleworth (1972a) found that chicks would associate visual rather than auditory stimuli accompanying drinking with the unpalatable taste of the water, but would associate the same auditory rather than visual stimuli with electric shock in a conditioned suppression paradigm.* Finally, Foree and LoLordo (1973) have shown that if pigeons are trained to press a treadle in the presence of a compound visual and auditory stimulus, the relative importance of the two stimuli depends upon the nature of the reinforcement for responding. When tested with the component stimuli in isolation, pigeons that had been responding for food rewards continued to respond in the presence of the light but showed little control by the tone while pigeons trained to avoid shock continued to respond in the presence of the tone, but showed little control by the light.

Although the results of these experiments may undermine simple theories of learning, there is a sense in which they are not surprising. Few biologists would be reluctant to admit the possibility that animals may be especially adapted to learning those particular contingencies that are of significance in their normal habitat. Animals might generally be expected to associate the taste of food with subsequent sickness, and the selective pressures producing such a tendency would be particularly strong for an omnivorous, opportunist feeder like the rat. The phenomenon of Batesian mimicry should certainly lead one to expect that birds will associate the visual characteristics of their prey with its palatability. And as Foree and

---

* To add to the complexity of these results, Shettleworth found that although shocks were associated with auditory rather than visual stimuli in conditioned suppression, chicks punished by shock for drinking would associate the shock with the visual rather than the auditory stimuli. Part of this difference was correlated with the location of shock: the auditory stimuli were completely ineffective when shock for drinking was delivered through the water, but did become associated with shock for drinking when it was delivered through the grid floor. As Shettleworth argues, however, it seems probable that the relative effectiveness of visual and auditory stimuli was partly determined by the nature of the activity the birds were engaged in.

LoLordo (1973) pointed out, if the feeding behaviour of birds is usually guided by visual stimuli, there is reason to believe that auditory stimuli, such as alarm calls, may be more important determinants of escape or flight.

To point to the adaptive value of these constraints on learning, however, is not the same as explaining the processes underlying them. Must we suppose that animals come into the world with differences in the associability of certain classes of stimuli with certain classes of reinforcer already wired into their systems? There is no *a priori* reason to reject this possibility but other possibilities may be worth exploring. According to the argument from adaptation, it makes sense for rats to associate flavours with poisons because poisoning is usually a consequence of having eaten a particular type of food. But if the environment is constrained in this way, then it is possible that the reason why rats are ready to associate flavours with sickness is not because they come into the world with a particular set of predispositions, but because they are exposed to this natural correlation between gustatory stimuli and internal states, and learn about it.

It is true that the rat's tendency to associate flavours with sickness and external stimuli with shocks is inconsistent with any theory that explains differences in the rate of conditioning to different CSs solely in terms of differences in the fixed salience of each CS. But we have already seen reason to reject such a theory. Nonreinforced pre-exposure to a given stimulus reduces the associability of that stimulus with any subsequent change in reinforcement. The salience or discriminability of a stimulus, therefore, is only one determinant of the rate of conditioning to that stimulus. The extent and nature of the subject's past experience with the stimulus exerts an equally important effect on conditioning. Moreover, although nonreinforced pre-exposure to a stimulus reduces the associability of that stimulus with any subsequent reinforcer, we noted earlier that exposure to random presentations of a stimulus and a given reinforcer might result in an ever greater decrease in the associability of that stimulus was that particular reinforcer (Mackintosh, 1973). Rats exposed to random presentations of tone and shock, for example, found it very difficult to learn later that the tone was a signal for shock, although quite ready to associate it with the delivery of water. During their lifetime, rats may also learn that changes in visual or auditory stimulation are uncorrelated with changes in their internal state, whereas changes in gustatory stimulation do predict such changes. Thus their tendency, as adults, to associate flavours rather than visual or auditory stimuli with poisoning may be a consequence of their ability to learn about and generalize from the correlations to which they have been exposed. Whether such an analysis is sufficient to account for all dispositions to associate only certain stimuli with certain reinforcers may well be doubted; it is, however, worth further study.

C. CS–UCS Interval

1. *Varieties of Temporal Relationship*

In standard excitatory conditioning, the temporal relationship between CS and UCS may vary widely. Fig. 2.12 illustrates four types of temporal relationship that have been distinguished by different names. In simultaneous conditioning, CS and UCS both begin and end simultaneously; in delay conditioning, the CS begins before the UCS and usually overlaps with it, ending at the same time (although in some experiments the CS ends at the moment of onset of the UCS); in trace conditioning, the CS both begins and ends before the onset of the UCS and in backward conditioning, the CS usually begins at the moment of offset of the UCS,

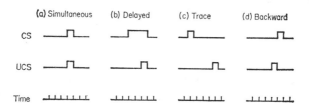

Fɪɢ. 2.12. Schematic diagram of the temporal relationship between CS and UCS in (a) simultaneous; (b) delay; (c) trace; and (d) backward conditioning.

although it may equally begin some seconds after the termination of the UCS. In a related procedure the CS may begin some seconds before the end of a relatively long UCS and both stimuli may end simultaneously. In delay and trace conditioning, the interval between the onset of CS and UCS is referred to as the inter-stimulus interval (ISI). In simultaneous and backward conditioning, it should be noted, CRs can be measured on test trials only when the CS is presented alone. This may also apply to delay conditioning if the ISI is very short, and such test trials must be used if the ISI is shorter than minimal latency of the CR.

Pavlov used the term simultaneous conditioning as a synonym for delay conditioning where the ISI was less than a few seconds. In the more exact definition given here, the procedure has only rarely been used, and when used has not resulted in significant conditioning (e.g., Asratyan, 1965, p. 159; Bitterman, 1964; Smith *et al.*, 1969). According to Pavlov, the stimulus that becomes the effective CS in ordinary conditioning is the state of afferent activity occurring immediately before the onset of the UCS. This supposition, of course, implies that neither simultaneous nor backward conditioning will be effective procedures. As we shall see, there is little evidence that contradicts this implication.

Pavlov's analysis also implies that delay conditioning should be much more effective than trace conditioning, since with a trace procedure the afferent state of affairs immediately preceding the onset of the UCS will be similar to that characterizing the intertrial interval. The only difference would be that just before the UCS there may be some trace, after-effect, or memory of the occurrence of the CS some seconds earlier. Pavlov reported that "long-trace reflexes", where the ISI was usually a minute or more, were hard to form, relatively unstable when formed and furthermore showed wide generalization (Pavlov, 1927, p. 113). All of these conclusions have been confirmed by Ellison (1964); using a 1 sec trace CS, he found little difference between trace and delay procedures in salivary conditioning when the ISI was 8 sec, but that delay conditioning was significantly superior at a 16 sec ISI. Ellison suggested that the reason why a long-trace procedure results in such extensive generalization might be because the effective CS in such a procedure consists of stimuli associated with various patterns of "mediating" behaviour elicited by the trace CS and serving to span the interval between offset of the CS and onset of the UCS. Schneiderman (1966) compared delay and trace conditioning of the rabbit's nictitating membrane response at ISIs ranging from 0·25 to 2·00 sec, and confirmed that a trace procedure was less effective than a delay procedure particularly at the longer intervals. Kamin (1965) also compared trace and delay conditioning procedures, using conditioned suppression in rats, and found delay conditioning to be superior to trace conditioning only when the ISI was 3 min. With a 1 min ISI, a 1·5 sec trace CS produced as effective conditioning as a delay procedure. With a 3 min ISI not only did a 1·5 sec trace CS result in virtually no conditioning but trace CSs that were either 175 or 179·5 sec long (i.e., leaving a gap between offset of CS and onset of UCS of only 5·0 and 0·5 sec) resulted in much less efficient conditioning than the delay procedure.

2. Backward Conditioning

Pavlov (1927, p. 27) at one time argued that backward conditioning did not occur (although, as noted below, making a somewhat different claim elsewhere). From time to time, however, studies apparently demonstrating positive results have been reported (e.g., Switzer, 1930, in eyelid conditioning; Wolfle, 1930, with finger withdrawal; Champion and Jones, 1961, with the GSR). None of these studies has stood up well to analysis and replication: Trapold et al. (1964) and Smith et al. (1969) found no evidence of significant backward conditioning of finger withdrawal or eyelid CRs in humans or rabbits; Grether (1938), Harlow (1939), Fitzwater and Reisman (1952), and Prokasy et al. (1962) have shown that backward groups do not respond more than a pseudo-conditioning control group. With GSR experiments, an additional problem arises; a backward conditioning procedure necessitates the use of test trials with the CS alone to measure the CR, but the CS presented alone without being preceded by the UCS is a novel event, and the occurrence of a GSR on the first few test trials is most simply thought of as a component of the orientation reaction. In support of this interpretation, all studies of backward conditioning of the GSR (Champion and Jones, 1961; Prokasy et al., 1962; Trapold et al., 1964; Zeiner and Grings, 1968) agree that "conditioning" is maximal on the first one or two test trials, and thereafter tends to decline.*

Pavlov (1928, pp. 381–2) and Konorski (1948, pp. 19, 135–6) argued that a backward CS might actually become inhibitory with sufficient training. Several recent studies of aversive conditioning in animals have confirmed this suggestion. Kamin (1963), James (1971a) and Siegel and Domjan (1971), using rats in conditioned suppression, Moscovitch and LoLordo (1968), using dogs and CSs imposed on a baseline of avoidance responding, and Siegel and Domjan (1971), using rabbits and conditioned eyelid responses, have all found that a CS presented just before or just after the offset of shock inhibits conditioned fear. Kamin and James measured this as an increase in the baseline of appetitively reinforced responding; Moscovitch and LoLordo as a decrease in the baseline of aversively reinforced responding; while Siegel and Domjan measured inhibitory conditioning by the retardation of subsequent excitatory conditioning. None of this is surprising. A backward conditioning procedure satisfies the conditions outlined above for establishing inhibitory conditioning; the presentation of the CS signals a period (the intertrial interval) during which the UCS will not occur. Moscovitch and LoLordo were able to show that this contingency was indeed necessary for the development of inhibition to a backward CS.

A slightly different backward procedure may also result in inhibitory conditioning. Segundo et al. (1961) exposed cats to variable durations of pulsed shock, presenting a tone CS 2 to 5 sec before the termination of shock. After a number of such trials the tone elicited what may be regarded as conditioned relaxation, a pattern of responding which included inhibition of the cortical EEG elicited by the shock. Zbrozyna (1958) has similarly

---

* An additional complication (deriving from the use of students as subjects) has been noted by Zeiner and Grings (1968): subjects that reported expecting the CS to signify something shows a markedly higher level of GSR responding than did less suspicious subjects.

shown that a stimulus presented shortly before the removal of food would eventually come to inhibit eating.

Although these results clearly imply that the most important association formed to a CS presented at the termination of a UCS is a forward, inhibitory one (the CS predicting either the termination or absence of the UCS), the occurrence of such inhibitory conditioning may in fact counteract or obscure some initial excitatory conditioning. Pavlov (1928, p. 38), indeed, on the basis of a study of Vonogradov, did eventually suggest that backward conditioning might initially produce some CRs, although they would be "of small magnitude and short-lived",* and with continued training the CS would become inhibitory. Razran (1956) cites a number of studies which suggest an initial increase in CR probability followed by a decline, but the possibility that this inhibitory conditioning obscures some genuine excitatory associations is not supported by the evidence. Moscovitch and LoLordo (1968) eliminated inhibitory conditioning to a backward CS by ensuring that it did not predict a period of time during which no UCSs would occur, but failed to find any evidence of excitatory conditioning to the CS.

Moscovitch and LoLordo also included a group for whom the CS followed the UCS by 15 sec. According to Razran (1956), longer UCS–CS intervals are more likely to result in significant backward conditioning; but in Moscovitch and LoLordo's study, the delayed CS was just as inhibitory as one presented at a shorter interval after the UCS. The studies cited by Razran to support his argument do not appear to have controlled either for pseudo-conditioning effects or for the possibility that the results are due to a forward trace connection being formed between the CS presented on one trial and the UCS presented on the next trial.† The remaining Russian studies thought by Razran to prove that backward conditioning is a genuine phenomenon, showed only that backward UCS–CS pairings might lead to relatively slow extinction of the forward CRs initially established to that CS, or, if interspersed with forward CS–UCS pairings in initial training, might produce more rapid acquisition of a CR than simple nonreinforced presentations of the CS. Neither of these procedures, of course, demonstrates the *establishment* of CRs during backward conditioning, and the extinction data are readily interpreted as another instance of the fact that occasional, randomly presented reinforcements will markedly retard the course of extinction (p. 408). All in all, there is little reason to accept the reality of backward conditioning.

* This is Razran's translation of a phrase that appears in Gantt's translation as "insignificant and evanescent" (Razran, 1956).

† Asratyan (1965, pp. 150–58) and Razran (1971, pp. 93–4) cite some more recent Russian studies purporting to show backward conditioning, but there is still no evidence that they have satisfactorily controlled for both of these possibilities.

3. *The Interstimulus Interval in Delay Conditioning*

As mentioned earlier, Pavlov (1927, p. 104) assumed that the effective CS (i.e., the set of stimuli that actually entered into association with the UCS) consisted of the pattern of afferent activity immediately preceding the onset of the UCS. He also assumed, correctly enough, that the actual pattern of activity instigated by a stimulus would change as the stimulus was prolonged: through sensory adaptation, it will tend to decrease. From this it is possible to derive two predictions. First, when a relatively long delay CS is used, the CR will tend to occur most strongly toward the end of the CS (i.e., shortly before the UCS). If the early and later parts of the CS are discriminably different, it is the latter pattern of activity that is associated with the UCS. The second prediction is that conditioning will be more efficient with shorter rather than longer CSs, since the longer the CS the weaker the pattern of activity it will produce at the moment of onset of the UCS and (via the known effects of CS intensity) the less rapidly conditioning will occur. Pavlov derived only the first of these predictions. Hull (1943, pp. 167–9), following Kappauf and Schlosberg (1937), derived the second. Both have been amply confirmed.

(*a*). *Inhibition of delay.* The delay in the occurrence of maximal CRs until towards the end of a long CS was described by Pavlov as "inhibition of delay", and he provides numerous examples of the phenomenon (1927, pp. 88–103). Ellison (1964) and Williams (1965) have also reported inhibition of delay in salivary conditioning; Parrish (1967) obtained inhibition of delay of heart rate CRs in rats; Bitterman (1964) reported significant effects in the conditioned activity of goldfish with an aversive UCS; Zielinski (1966) observed a rather gradual appearance of the effect in conditioned suppression; and Smith (1968) observed that the maximum amplitude of the nictitating membrane CR in rabbits coincided relatively closely with the moment of onset of the UCS.*

Although the fact that CRs occur just before the onset of the UCS is consistent with the Pavlovian analysis, an alternative explanation has been

---

* Latency of initiation of CRs was somewhat less affected by ISI in Smith's study, and a further finding, which has been amply documented in other eyelid studies (e.g., Schneiderman and Gormezano, 1964; Schneiderman, 1966; Smith *et al.*, 1969), was that at all ISIs CR latency decreased as training progressed: in other words, the *initiation* of the CR tended to occur closer and closer to the onset of the CS rather than the UCS. This decrease in latency presumably reflects an increase in strength of the CR during acquisition: provided that it occurred more rapidly than the shift in peak CR amplitude towards the onset of the UCS, it would not be incompatible with the Pavlovian analysis. When the ISI is shifted from a shorter to a longer value, CR latency tends to increase appropriately (e.g., Prokasy and Papsdorf, 1965; Leonard and Theios, 1967a). Since these shifts have been effected after substantial training, latencies have presumably become asymptotically short before the shift.

suggested. Boneau (1958), Ebel and Prokasy (1963), and Prokasy (1965), have argued from their data on eyelid conditioning in human subjects that a CR occurring at this point will be better reinforced than one occurring at the onset of the CS: a CR coinciding with the delivery of the UCS, unlike one occurring at the onset of the CS, will attenuate the aversive effects of the UCS by ensuring that the puff of air will fall on a closed eye. This "response-shaping" analysis of classical conditioning will be discussed at greater length in the next chapter (p. 98). Since, however, the rabbit's nictitating membrane response occurs at the onset of the UCS whether the UCS is a puff of air or an electric shock (Smith, 1968), and since it is hard to see how the response succeeds in attenuating the shock, response-shaping may play little or no role in the timing of CRs. It might also be possible to assess the importance of such an instrumental contingency, by comparing the effects of different ISIs on subjects exposed to pairings of CS and UCS, but unable to respond because they were immobilized by curare.

That the onset of a sufficiently long delay CS may, with sufficient training, actually become inhibitory, was inferred by Pavlov from the observation that an extraneous stimulus, presented shortly before the CS, would disinhibit the CR—i.e., cause an increase in the rate of salivation during the early part of the CS. Rescorla (1967b) was able to show that if a long delay CS paired with shock was superimposed upon ongoing avoidance responding in dogs, the rate of responding *declined* at the onset of the CS, and rose to a peak as the CS continued.

Pavlov (1927, p. 92) reported that inhibition of delay developed much more slowly to an intermittent CS than to a continuous CS. This is consistent with the idea that it is adaptation to a prolonged stimulus that enhances the discriminability of early and late parts of the stimulus. Further support for this aspect of the analysis comes from studies which have deliberately manipulated the difference between the beginning and end of a long CS, e.g. by comparing conditioning to a 2 min tone with conditioning to a CS consisting of a 1 min light immediately followed by a 1 min tone. Kamin (1965) and Brahlek (1968) have demonstrated that a constant CS produces much greater conditioned suppression at its onset than does a varied CS, and Williams (1965) has shown that the same holds for salivary conditioning.

(b). *The optimal ISI.* The Pavlovian analysis, as developed by Hull, implies that conditioning will occur most readily with a very short ISI, with the lower limit being presumably set by neural conduction times, or at any rate by the time required for maximal activity to occur in whatever centres are responsible for the elaboration of associations. Numerous studies have been undertaken to determine the optimal ISI. Kimble concluded that their "most typical finding has been that the optimal interval is in the quarter-second range on either side of 0·5 seconds" (Kimble, 1961, p. 158). The conclusion is gratifyingly close to Hull's estimate, although since the estimate is based on electrophysiological studies of the eel (Adrian, 1928), while the empirical data are derived from studies of finger withdrawal and eyelid conditioning in humans, one might question the basis of the coincidence.

As we shall see below, the idea that there is a single, optimal ISI, valid for all conditioning studies, is a gross over-simplification. Before considering the problems raised by the apparent variations in the optimal ISI in different response systems let us examine one such system, the rabbit's nictitating membrane response, for which there are now extensive data

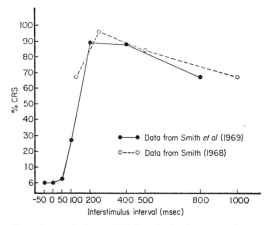

FIG. 2.13. Conditioning of the rabbit's nictitating membrane response as a function of the interval between onset of CS and UCS. (*After* Smith, 1968; *and* *Smith et al.*, 1969.)

(Gormezano, 1972). Initial studies (Schneiderman and Gormezano, 1964; Schneiderman, 1966) varied ISI between 250 and 4000 msec, and found that conditioning, measured as the percentage of CRs, was best at the shortest interval—250 msec was significantly better than the next shortest interval, 500 msec. Two subsequent studies (Smith, 1968; Smith *et al.*, 1969) showed that ISIs shorter than 250 msec led to a decline in conditioning: in the former experiment 125 msec was inferior to 250 msec; in the latter, 100 msec was inferior to 200 msec. The overall results of these two studies are shown in Fig. 2.13. Under the conditions of this entire series of studies (the CS a tone, the UCS either a corneal airpuff or a paraorbital shock, CRs measured either during training or on test trials to the CS alone), the optimal ISI is in the region of 200 to 250 msec.

The Pavlovian analysis implies that ISIs shorter than this produce less effective conditioning because the pattern of afferent neural activity initiated by the CS takes some finite period of time to attain its maximum amplitude. From this it follows that if the CS were not an external signal, but direct stimulation of afferent pathways, conditioning at very short ISIs might be improved. Patterson (1970) has confirmed this prediction by using stimulation of the inferior colliculus as the CS in a study of nictitating membrane conditioning; he obtained relatively good conditioning at an ISI of 50 msec, a value that produced no conditioning at all in the Smith *et al.* study.

At very short ISIs, CRs can be measured only on test trials given with the CS alone: the minimum latency of the nictitating membrane response is of the order of 70–80 msec, while the average latency is over 100 msec. This has a further important implication. Smith *et al.* (1969) observed significant conditioning with a 100 msec ISI and Patterson (1970) with a 50 msec ISI, although in each case the median CR latency observed on test trials was well over 100 msec. A response-shaping interpretation of the optimal ISI implies that conditioning will not occur at very short ISIs because the CR cannot occur before the onset of the UCS and therefore cannot be reinforced (in the case of an aversive UCS) by attenuating its effects. In these two experiments, highly significant conditioning occurred when most CRs were initiated after the onset of the UCS and therefore on this interpretation would not have been reinforced at all.

This is not to deny that the latency of a CR may have some effect on the optimal ISI, nor that differences in latencies of different CRs may have some bearing on variations in the optimal ISI. It is possible to provide a quite different interpretation of these facts. If a given CR never occurs with a latency shorter than, say, 2 sec, then it can be recorded only if the CS is presented for more than 2 sec. If conditioning is given with an ISI of 500 msec, then this CR must be measured on test trials on which a 2- or 3-sec CS is presented without the UCS. But such a test trial represents a relatively drastic departure from the conditions of training, both in the duration of the CS and in the absence of the UCS, and it may well be that, by generalization decrement, CRs will be markedly disrupted. Hence it will be found that the optimal ISI for this CR, unlike the rabbit's nictitating membrane response with its latency of 100–200 msec, is substantially greater than 500 msec. From this analysis, indeed, it would be expected to be of the order of 2 to 3 sec.

Studies of other response systems have usually reported optimal ISIs substantially greater than that required for eyelid or nictitating membrane conditioning. Other discrete skeletal responses require relatively short ISIs; McAdam *et al.* (1965) found that conditioning of a leg flexion in cats was most effective with an ISI of 500 msec—this being better than either 250 or 1000 msec. Other response systems, however, have been found to condition better at much longer ISIs. The normal ISI used in Pavlov's studies of salivary conditioning was between 10 and 30 sec (Konorski, 1948, pp. 7–8). Ost and Lauer (1965) compared ISIs of 2, 5, 10 and 15 sec in a study of salivary conditioning to acid as the UCS and reported that "the highest response rates appear in the five- and ten-second groups (p. 203)". Boice and Denny (1965), studying conditioned licking in rats, found that an ISI of 1 sec or less produced less conditioning than one of 2 or 4 sec and that there was a dropping off in performance at an ISI of 6 sec. Gormezano (1972), studying conditioning in rabbits with water as the UCS, observed comparable levels of conditioning at ISIs ranging from 250 msec to 4 sec. Church and Black (1958) found equally good heart rate conditioning in dogs, whether the ISI was 5 or 20 sec; Black and Black (1967) studied ISIs ranging from 500 msec to 10 sec in a study of heart rate conditioning in rats and found that conditioning was optimal at the 5-sec interval. Bitterman (1964) found that conditioned fear in goldfish was little affected by variations in ISI between 800 msec and 10 sec. Dyal and Goodman (1966), using a procedure similar to conditioned suppression in rats, found equally good conditioning with ISIs ranging from 500 msec to 1 min while Kamin (1965) reported that a 3 min ISI produced only slightly worse conditioned suppression than a 1 min ISI.

This list of studies suggests that discrete, skeletal CRs such as eyelid and flexion responses are the only ones that are optimally conditioned at ISIs substantially shorter than a second. This conclusion is greatly strengthened by some further studies reported by Schneiderman (1972). Meredith and Schneiderman (1967) and Vandercar and Schneiderman (1967) simultaneously recorded nictitating membrane and heart rate CRs from rabbits and confirmed that the former response system was better conditioned at ISIs of less than a second, while the latter was better conditioned at intervals of up to 6·75 sec—an interval at which essentially no nictitating membrane CRs could be detected. The direct comparison of separate systems undertaken in these experiments provides much more compelling evidence of the reality of these differences than can any number of comparisons of the outcomes of different experiments.

A second problem, noted by Gormezano (1972), making interpretation of all these studies difficult, is that different response systems are recorded in different ways. Conditioning of eyelid and flexion responses is measured by the percentage of trials on which a response of greater than criterion amplitude occurs; conditioning of heart rate, salivation, licking and suppression is assessed by a rate measure, which records the total amount of activity elicited by the CS. Within limits, the number of licks or drops of saliva that occur to a CS can hardly fail to increase as the duration of the

CS increases. With rate measures such as these, therefore, it is hardly surprising that good conditioning occurs at relatively long ISIs. This point is not a trivial one: when a comparable measure is taken of the nictitating membrane response (the duration and magnitude of membrane extension averaged over a 2 sec presentation of the CS alone on a test trial), it is found that conditioning even of this response system improves as the ISI is increased from 125 msec to 1 sec (Smith, 1968). In other words, the optimal ISI for the nictitating membrane response depends upon the response measure used, being 250 msec for a percentage CR measure and 1 sec for an overall amplitude measure.

Schneiderman (1972) has argued, however, that at even longer ISIs (e.g., 6·75 sec) it is impossible to detect nictitating membrane CRs whatever the measure used, and it certainly seems unlikely that one would observe conditioning in such a system with an ISI of several minutes (as is routinely used in studies of conditioned suppression). This, then, does create a problem for the Pavlovian analysis. We have argued that one reason why different responses might appear to condition at different ISIs is because differences in latencies of different CRs may create recording problems. The Pavlovian analysis implies that conditioning should decline as ISIs increase, because a longer CS creates a less intense trace at the moment of onset of the UCS. While the analysis receives nice confirmation from the finding that an intermittent CS permits better conditioning of the nictitating membrane response at a relatively long ISI (Gormezano, 1972), it has no ready explanation for why the ISI function for discrete skeletal CRs should decline so much more sharply at longer intervals than that for conditioned suppression or salivation. According to the response-shaping view (Prokasy, 1965), the reason why the eyelid response cannot be conditioned at longer intervals is because it is reinforced only if it immediately precedes the UCS and at longer intervals the temporal discrimination required to ensure such accurate timing of the response is impossible. Schneiderman (1972) has suggested that one reason why CRs, such as salivation, heart rate and suppression can be conditioned at longer ISIs may be because they are responses of relatively longer duration than discrete flexion or eyelid responses and can still overlap with the UCS even when they are elicited by a CS several seconds before the onset of the UCS. The implication is that some overlap between the CR and the UCS is a necessary condition for reinforcement, although the nature of the reinforcement remains obscure.

4. *Conditioning with Very Long Intervals between CS and UCS*

The problems posed by the results just reviewed are very slight in comparison with those raised by the studies of Garcia, Revusky, and others, which have shown that the conditioning of an aversion to a particular flavour is possible with an ISI of several hours. A number of earlier observations had foreshadowed this conclusion. Pavlov (1927, p. 35) reported that several features of the UCR to an injection of morphine, including nausea, salivation and vomiting, were readily conditioned to the stimuli accompanying the administration of the injection although these physiological reactions to morphine do not normally occur for several minutes after an injection. The association between the stimuli accompanying the injection and the subsequent nausea, therefore, was established in spite of the relatively long delay separating the two events. Further documentation of this point was provided by Spragg (1940) in his experiments on morphine addiction in chimpanzees. As noted earlier (p. 18), Spragg was able to show that the reduction of withdrawal symptoms following an injection was a conditioned reaction to the stimuli accompanying the delivery of the injection. Animals given an injection of saline solution under identical conditions relaxed immediately and remained relaxed for up to half an hour. This suggests that the physiological consequences of a normal injection of morphine were also delayed by a similar period.

Early studies of sickness induced by irradiation also suggested that conditioning occurred to stimuli accompanying the exposure to radiation, even though the effects of radiation could not be detected until minutes or even hours later (e.g., Garcia et al., 1955). But the first unequivocal evidence that an association between a flavour and sickness could be established in spite of an interval of several hours between the two was provided by the studies of Garcia et al. (1966), Smith and Roll (1967) and Revusky (1968). Although exposure to radiation or injection of lithium chloride or apomorphine does not induce sickness for at least several minutes, in each of these studies the delay between CS (flavour) and UCS (poisoning) was conservatively estimated as the time between the termination of exposure to the flavoured substance and the administration of the sickness-inducing treatment. Although all three studies found that conditioning was more effective at shorter ISIs, the range of intervals that would support conditioning was strikingly longer than had previously been thought. The main results of the studies by Garcia et al. and Smith and Roll are shown in Fig. 2.14.

These findings have frequently been confirmed: there can be no reasonable doubt that rats are able to associate the flavour of a novel substance with subsequent sickness over very long intervals of time (Revusky and Garcia, 1970; Rozin and Kalat, 1971). The problems posed by these findings differ from those discussed in the preceding section. There the question was why some CRs could be conditioned at ISIs which were too long to support the conditioning of others; here the most acute problem seems to be one of memory: how can the rat associate its present sickness with an event that occurred several hours earlier? And if rats are able to remember a single exposure to a given flavour over such intervals of time, why should the conditioning of suppression, for example, to a brief tone be impossible when the interval between tone and shock is a few minutes?

The adaptive significance of the rat's ability to associate flavours and sickness over very long intervals is obvious enough. If selective association between flavours and sickness is a natural adaptation to the fact that the two are naturally correlated, it remains true that ingestion of a particular

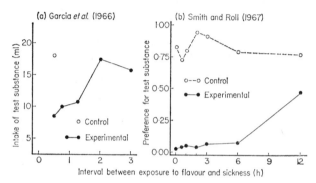

FIG. 2.14. Taste-aversion conditioning in rats as a function of the interval between administration of CS and UCS. Experimental subjects were made sick after ingesting the test substance (saccharin-flavoured water), while control subjects were simply exposed to the test substance. In Garcia et al.'s experiment, the UCS was an injection of apomorphine, and conditioning was measured by the amount of the test substance drunk in a 10 min period. In Smith and Roll's experiment, the UCS consisted of X-irradiation, and conditioning was measured by providing subjects a choice between saccharin-flavoured and plain water. (*After* Garcia et al., 1966; *and* Smith and Roll, 1967.)

substance is not usually followed by instantaneous gastric disorder and the ability to form the relevant association must be geared to this natural delay.

As we have noted before, however, to point to the presumed adaptive significance of a particular associative capacity is not the same as explaining the processes underlying the capacity. One of the more common reactions to these data has been to suggest that the formation of associations between flavours and sickness over very long intervals must be a consequence of some particular properties of gustatory stimuli. Perhaps flavours, unlike typical exteroceptive stimuli, produce unusually persistent stimulus traces, so that animals can still taste a particular substance several hours later when poisoning takes effect. Or perhaps the sickness induced by poisoning causes vomiting or regurgitation, and hence re-exposes subjects to the taste of a previously consumed meal. The problem with the latter explanation is that rats do not vomit when they are sick (Garcia and Ervin, 1968); against the former possibility a number of objections may be made. In the first place, aversions can be formed over moderately long delays (at least half an hour) to characteristics of foods and liquids other than their flavour. Nachman (1970) found that rats could associate the temperature of their drinking water with later sickness, and quail form aversions to the visual characteristics of food when sickness is induced up to half an hour later (Wilcoxon *et al.*, 1971). Moreover, Rozin (1969) showed that if rats were exposed on alternate days to different concentrations of casein or saccharin solution, and then poisoned after drinking one of the two concentrations, they formed an aversion to that particular concentration and not to the other. It is difficult to see how the stimulus traces of different concentrations of a particular solution could still be discriminably different half an hour after ingestion. Finally, and perhaps most conclusively, several studies have shown that exposure to a second flavour in the interval between ingestion of a first substance and subsequent poisoning may not abolish the aversion formed to the first. Indeed, if the second flavour is familiar, it will not itself become associated with poisoning (because of latent inhibition) and will then interfere hardly at all with the formation of an aversion to the first (Revusky and Bedarf, 1967; Wittlin and Brookshire, 1968; Revusky, 1971).

Although, therefore, the ability to associate flavours with sickness over very long delays does not seem to be a consequence of the persistence or reinstatement of gustatory stimulus traces at the moment of sickness, it may not be necessary to assume that taste-aversion learning obeys entirely different laws from other classes of conditioning. It might be more profitable to examine these other situations in an attempt to understand why conditioning is not more often obtained with intervals between CS and UCS longer than a few seconds or minutes. Revusky (1971), advocating such a strategy, has argued that the reason why UCSs such as food or shock are readily associated only with CSs preceding their delivery by a short interval, is because intervals between CS and UCS of an hour or more would permit other intervening events to become established as signals for the UCS. When a rat is poisoned after exposure to a novel flavour, there are few if any events intervening between ingestion and sickness which will be associated with the sickness. This is because, by the principle of relevance, only flavours are readily established as signals for gastric disorders. If the rat consumes no novel substance after exposure to the first flavour, there will be nothing (except eventual forgetting) to prevent that flavour being associated with sickness. On the other hand, if a tone and shock are separated by more than a few minutes, the interval between the two will contain any number of events, including changes in the subject's behaviour as well as changes in external stimulation, all of which are as readily associable with shock as the tone. Indeed, since these events occur closer in time to the shock, they will, all else being equal, be more easily associated with the shock and, by the principle of overshadowing, will tend to prevent conditioning to the tone. Thus the possibility of conditioning with long delays between CS and UCS in taste-aversion learning is a secondary consequence of the fact that only flavours are readily associated with sickness, and subjects are not therefore normally exposed to as many potentially overshadowing or interfering events as they would be if a comparable period of delay was interposed between a tone and a shock.

In support of this argument, Revusky (1971) has shown that exposure to a novel flavour in the interval between ingestion of one substance and poisoning does indeed interfere with the formation of an aversion to the first substance. A similar result has been reported by Kalat and Rozin (1971). Since, however, they found that exposure to three novel flavours in a 30 min interval between ingestion of a test substance and poisoning did not completely prevent the formation of an aversion to that substance, Kalat and Rozin argued that Revusky's analysis must be inadequate. Their argument does not seem entirely convincing. If Revusky's account is correct, it must be conceded that we have no idea how many potentially interfering events may occur within a 30 min interval between the presentation of a tone and a shock, nor how many would be necessary to prevent conditioning to the tone. One may, however, hazard the guess that it is more than two or three.

D. UCS Variables

1. *UCS Intensity*

Pavlov (1927, pp. 31–2) claimed that the strength of conditioning was directly related to the intensity of the UCS: indeed this statement forms part of the Pavlovian law of strength. Numerous confirming instances can be provided. Wagner *et al.* (1964) found more rapid salivary conditioning and greater resistance to extinction when the UCS consisted of six food pellets rather than one. Ost and Lauer (1965) found better salivary conditioning with a stronger solution of acid. Annau and Kamin (1961) and Kamin and Brimer (1963), varying the intensity of shock used as UCS in conditioned suppression, found that intensities below 0·85 ma resulted in relatively slow conditioning to a relatively low asymptote, and although acquisition was not affected by increases in intensity above this level (presumably due to a ceiling effect), rate of extinction was inversely related to UCS intensities up to 2·91 ma. Smith (1968) found that conditioning of the nictitating membrane response of rabbits was directly related to the intensity of shock used as UCS, although rather surprisingly no differences appeared in extinction; numerous studies of eyelid conditioning in human subjects have reported better conditioning with stronger airpuffs (see Spence and Platt, 1966).

Pavlov seemed to imply that a weaker UCS led to worse conditioning because it failed to elicit an adequate UCR. This is, almost certainly, an insufficient explanation. Pavlov's evidence (1927, p. 32) was that if a dog was not hungry, food would elicit little unconditional salivation, and no conditioning would occur. While this is true (e.g., Finch, 1938a), recent evidence suggests that increasing satiation may have a drastic effect on salivary CRs, while still leaving the UCR relatively unaffected (Soltysik, 1971). Bruner (1969) used the weakest shock that would reliably elicit a flexion UCR in cats and found no evidence of conditioning with this UCS at all; other instances of a UCS which reliably elicits a UCR but may fail to support conditioning are provided in Chapter 3 (p. 92).

Although part of the effect of UCS intensity in aversive conditioning may be that it contributes to a higher overall level of arousal or drive (Spence, 1956), it is certain that differences in the reinforcing effects of stronger and weaker UCSs must also be invoked (Spence et al., 1958 a, b). This is not surprising, although the details of how the greater reinforcing effectiveness of stronger UCSs is actually translated into better conditioning has aroused its share of controversy (Burstein, 1965; Spence and Platt, 1966). Burstein argued that a more intense UCS increases the proportion of subjects that condition at all, while Spence and Platt argued that a more intense UCS increases the strength of conditioning in all subjects. It is known that in situations other than eyelid conditioning with human subjects, a stronger UCS will lead to stronger conditioning—even when all subjects condition to the weaker UCS; stronger shocks increase resistance to extinction of conditioned suppression after all subjects have reached asymptotic levels of suppression in acquisition (Annau and Kamin, 1961). In the case of eyelid conditioning, the question may be analysed in terms of a specific model, such as the Markov model proposed by Theios and Brelsford (1966b). Use of this model will reveal the parameters that need to be changed to represent changes in UCS intensity. Prokasy and Harsanyi (1968), employing a variant of a model proposed by Norman (1964), found that changes in UCS intensity were best represented by changes both in the number of trials before subjects started conditioning and in the asymptotic level of conditioning.

2. UCS Probability: Partial Reinforcement

If the UCS is presented following only a proportion of presentations of the CS, a partial reinforcement schedule is defined. Partial reinforcement has been studied extensively in instrumental learning and the fact that a reduction in the probability of reinforcement in acquisition increases subsequent resistance to extinction [the partial reinforcement effect (PRE)] is one of the most studied as well as most robust phenomena of animal learning (Chapter 8). There are conditions under which the PRE may fail to occur in instrumental learning (e.g., substantial initial training with consistent reinforcement; a very small reward with a very long intertrial interval) but they are relatively few. Studies of partial reinforcement in classical conditioning have been rarer and in general have failed to demonstrate a PRE with anything like the regularity obtaining in instrumental studies. Indeed, Kimble (1961, pp. 103–4) suggested that this is one variable that differentiates classical from instrumental conditioning.

There is some evidence that partial reinforcement may not only have different effects on extinction in the two procedures, but also different effects on acquisition. In instrumental learning experiments, partial reinforcement may slightly interfere with initial acquisition of the instrumental response, but has sometimes been found to lead eventually to superior performance (e.g., Weinstock, 1958; see p. 160). In classical conditioning experiments, on the other hand, partial reinforcement may have a prolonged deleterious effect on acquisition, as noted above (p. 24), in salivary conditioning, with an amplitude measure of CR strength which does not impose ceiling effects, there is good evidence that partial reinforcement produces less effective conditioning than consistent reinforcement.

The reliability of this difference between classical and instrumental experiments should not be exaggerated. It should not be thought that partial reinforcement of instrumental running in an alley inevitably produces faster running speeds: *this* effect, unlike the effect of partial reinforcement on extinction of instrumental running, is notably elusive, and the conditions governing its appearance remain obscure (p. 160). Equally, as we shall see, partial reinforcement does not always have such deleterious effects on the acquisition of classically conditioned responses.

Partial reinforcement has interfered with the acquisition of salivary CRs in studies by Brogden (1939b), Fitzgerald (1963), Wagner et al. (1964) and Sadler (1968). In aversive conditioning it has led to inferior conditioning of general activity in goldfish (Berger et al., 1965) and of the startle response in rats (Wagner et al., 1967a). In all these cases an amplitude or rate measure of conditioning has been used but this does not seem to be sufficient to produce this result. Fitzgerald et al. (1966b), for example, found no terminal difference in heart rate conditioning between consistently and partially reinforced dogs, and Slivka and Bitterman (1966) found marginally superior conditioning of general activity in pigeons to a food UCS with a 50 per cent reinforcement schedule.

In eyelid conditioning, as noted before, terminal acquisition performance is relatively unaffected by differences in reinforcement schedule (Thomas and Wagner, 1964; Leonard and Theios, 1967b; Vardaris and Fitzgerald, 1969), although similar studies with human subjects (reviewed by Ross and Hartman, 1965) show greatly inferior performance with partial reinforcement. Brogden (1939b) found little effect of probability of reinforcement in leg flexion conditioning in dogs. In conditioned suppression experiments, partial reinforcement has been found to have little effect on either rate of acquisition or terminal performance (Brimer and Dockrill, 1966; Willis, 1969). Whether these results are due to a ceiling effect is not known; it would be interesting to study the acquisition of conditioned suppression with a weak UCS that does not produce complete suppression within half a dozen trials.

If these acquisition data seem confusing enough, it is even harder to bring any order into the extinction data. In part the problem is one of measurement. How does one define resistance to extinction? If partial reinforcement results in a lower level of performance in acquisition, then it is hardly surprising if it also results in a lower level of performance in extinction. It is tempting, in such cases, to define resistance to extinction as the *rate* at which performance declines in extinction; but although consistently reinforced animals may now appear to extinguish faster, this may be because a higher level of performance in acquisition simply provides room for a greater decline in performance during extinction.

Differences in resistance to extinction, therefore, can only be unequivocally demonstrated either when there are no gross differences in terminal acquisition performance or when, given such differences, the group that starts extinction at a higher level not only declines more rapidly, but actually shows inferior performance at a later stage of extinction.

With this restriction, we can proceed to examine the results of different classes of experiment. Studies of salivary conditioning yield ambiguous results. Terminal acquisition performance is profoundly depressed by partial reinforcement; in one study, no cross-over appeared in extinction (Wagner et al., 1964); in a second study, there was some evidence of a cross-over (Fitzgerald, 1963). No eyelid conditioning experiment employing animals as subjects has produced a significant PRE. Thomas and Wagner (1964) found no difference in terminal acquisition or in extinction between consistently and partially reinforced rabbits. Vardaris and Fitzgerald (1969) found that partial reinforcement of the eyelid response in dogs tended to depress performance in both acquisition and extinction (although neither effect was significant). Studies of the conditioning of startle responses or general activity are more ambiguous. Wagner *et al.* (1967a) found no PRE with the startle response in rats; Berger *et al.* (1965) found an occasional slight suggestion of a PRE with general activity to an aversive UCS in goldfish, but only under the most favourable of conditions. Slivka and Bitterman (1966) obtained a relatively clear PRE in pigeons when the UCS was food, and Fitzgerald *et al.* (1966b) obtained a highly significant PRE in conditioning heart rate in dogs to an aversive UCS. Finally, studies of conditioned suppression in rats have produced somewhat conflicting data, but the results of Brimer and Dockrill (1966) and Hilton (1969) leave little doubt that a substantial PRE may be obtained. Wagner *et al.* (1967a) observed a PRE only when conditioning was conducted while rats continued to lever press for food, and no effect when the lever was removed during conditioning trials; Hilton, however, was unable to replicate this difference. Scheuer (1969) found no PRE in her study of conditioned suppression but this is not too surprising since partial reinforcement was preceded by extensive training with consistent reinforcement. Such a procedure reduces the PRE both in instrumental experiments (Sutherland *et al.*, 1965) and in experiments on conditioned suppression (Hilton, 1969).

Although it is clear that partial reinforcement *may* increase resistance to extinction of a classically conditioned response, the generality of the effect leaves much to be desired and even when an effect does occur, it is often relatively small. Furthermore, in nearly all studies which have obtained a PRE, partially reinforced subjects received the same number of reinforcements in acquisition as consistently reinforced subjects (and therefore, on a 50 per cent schedule, twice as many actual trials). The only clear exception to this is the study by Brimer and Dockrill (1966). In instrumental experiments, partially and consistently reinforced groups are routinely equated for the number of trials received in acquisition rather than for the number of reinforcements; in spite of receiving fewer reinforcements, partially reinforced subjects show much greater resistance to extinction.

The general implication of this discussion has been that there are differences in the effects of partial reinforcement in classical and instrumental conditioning. Let us first consider the significance of apparent differences in acquisition. In a typical classical conditioning experiment, the behaviour of the subject immediately before the delivery of reinforcement is measured. In a typical instrumental experiment, we measure a relatively long chain of responses preceding the delivery of reinforcement. This chain, of course, can and has been broken down into smaller components. A usual observation (Goodrich, 1959; Haggard, 1959; Wagner, 1961) is that behaviour closer to the delivery of reinforcement is less likely to be facilitated by partial reinforcement and that running speed in the terminal section of an alleyway is consistently depressed by a 50 per cent reinforcement schedule. In the light of this it becomes distinctly less surprising that classically conditioned responses recorded just before the delivery of reinforcement are also depressed by partial reinforcement. What is needed is a study of the effects of partial reinforcement in classical conditioning when the CS consists of a series of discriminably different stimuli delivered in the same sequence on every trial. It is certainly possible that partial reinforcement would enhance conditioning to earlier stimuli in the chain (i.e., that it would diminish inhibition of delay) and such a finding would go a long way to resolving the apparent discrepancies between the classical and instrumental results.

A similar argument cannot be applied to extinction effects: there is no evidence that the PRE in instrumental experiments varies with the distance from the goal of the response being measured. Nevertheless there remains a second, perhaps more fundamental, reason why the evidence presently available is insufficient to prove a difference between classical and instrumental experiments. Partial reinforcement effects, both in acquisition and extinction, do depend to some extent on the values of various parameters: although, therefore, comparisons across different experiments may suggest that classical and instrumental experiments are affected in different ways by schedules of reinforcement, this may still not necessarily be the case. The conclusion must remain uncertain until an experiment is specifically designed to compare the two procedures under conditions made otherwise as comparable as possible. An ideal preparation would be one in which animals are trained on an instrumental response while a simultaneous record is made of salivary or other classical responses (a number of such procedures are mentioned below, p. 223).

IV. SUMMARY

Classical conditioning may be defined operationally as a training procedure in which the experimenter arranges contingencies between stimuli without regard to the subject's behaviour. In spite of the simplicity of these operations, however, the processes involved in any experiment on classical conditioning may be numerous and complex. Although this chapter has deliberately avoided discussion of supposedly basic theoretical issues, it has involved some relatively complicated arguments and has covered too much detail at too rapid a pace for it to be easy to condense, or useful to summarize every observation. It will be more useful to stress a small number of more general points, several of which will recur in later chapters.

The reinforcers used in classical conditioning, as in instrumental learning, may be characterized as appetitive or aversive. These appear to be antagonistic in the sense that a stimulus established as a signal for one type of reinforcer may suppress behaviour maintained by the other, and that the stimulus may be transformed from a signal for one into a signal for the other only with great difficulty.

Stimuli may also be characterized as excitatory or inhibitory, depending on whether they are correlated with an increase or decrease in the prevailing probability of reinforcement. Excitatory conditioning can be measured as an increase in the probability, rate or amplitude of CRs from some putative zero baseline; inhibitory conditioning, being measured as a decrease in responding, requires a baseline above zero to be detected. Whether excitatory and inhibitory conditioning are to be regarded as fundamentally different processes, as Pavlov, at least, certainly assumed, has not been seriously considered. It will be sufficient for most purposes to assume that animals may learn either positive or negative contingencies, without further speculation on the possible differences underlying such learning.

Just as the UCS signalled by a CS may be neither appetitive nor aversive but neutral, so the correlation between a CS and UCS may be neither positive nor negative but zero. Exposure to such a contingency may not only provide the most useful control procedure against which to assess associative conditioning, but also appears to have a marked effect on the associability of that CS and UCS if they are subsequently paired. There are, indeed, a number of factors which may affect the ease of establishing a given CS as a signal for a given UCS. At the simplest level, this will be a function of the discriminability or salience of the CS. More interestingly, the phenomenon of overshadowing implies that the extent to which one CS will be associated with a UCS may depend upon the presence of other, competing, stimuli; the results of experiments on taste-aversion learning suggest that it may depend upon the nature of the CS and UCS; and the results of experiments on latent inhibition and blocking suggest that it may depend upon the nature and extent of the subject's past experience with that CS. If prior exposure to a given CS in the absence of reinforcement (and, more particularly, prior exposure to uncorrelated presentations of a CS and UCS) significantly retard conditioning to that CS, this suggests that, in addition to learning about the excitatory or inhibitory relations between a CS and UCS, animals may also learn about the significance of particular stimuli as potential predictors of reinforcement.

Other determinants of conditioning include the intensity and probability of the reinforcer and particularly the degree of temporal contiguity between

CS and UCS. The most interesting problem raised by studies of the interval between CS and UCS in conditioning is that some discrete skeletal responses, such as leg flexion or eyelid CRs, appear to be conditionable only at intervals of less than a few seconds, while other response systems, such as salivation and conditioned suppression, may be conditioned at intervals up to a few minutes; in yet other cases, such as taste-aversion learning, associations may be formed between a CS and UCS separated by several hours. The reasons for these discrepancies are not very well understood.

CHAPTER 3

Classical Conditioning: Theoretical Analysis

I. INTRODUCTION

For no very good reason, the theoretical analysis of classical conditioning never succeeded in creating the sound and fury that surrounded other theoretical issues in animal learning. Perhaps, as Black (1971) suggests, this was because the basic operations of the classical conditioning experiment appeared so simple that it was thought an equally simple matter to understand the processes involved. As Black also suggests, the optimism implied by such an attitude is unjustified: we shall see that theoretical understanding is far from perfect. Indeed it could hardly be otherwise, for the experimental analysis needed to decide most of the critical theoretical issues has been at best sporadic. Contrary to received opinion, an absence of theoretical concern and controversy has not in fact produced the research needed to answer interesting questions. We still do not know, for example, whether Pavlov's analysis of classical conditioning is correct, or whether, and how critically, instrumental reinforcement contingencies affect the outcome of classical experiments. The reason for this ignorance is that the necessary experimental analysis has not been undertaken.

In this chapter we shall consider four main issues: the nature of the associations formed in classical conditioning; the conditions required for the formation of such associations; the nature of reinforcement in classical conditioning; and finally, the relationship between classical conditioning and instrumental learning. These questions are not, it should be stressed, independent of each other; as will become clear, their separation is artificial and largely a matter of convenience for purposes of exposition.

II. THE ELEMENTS OF ASSOCIATION

In a typical classical conditioning experiment, a CS is presented shortly before a UCS which reliably elicits a UCR. In due course a response, usually bearing notable similarities to the UCR, begins to occur to the CS in anticipation of the UCS. Pavlov's account of what is happening here was that an initially indifferent stimulus, the CS, comes to elicit the response belonging to another stimulus, the UCS: one stimulus becomes a substitute for another. This "stimulus-substitution" theory provides an account of why the subject makes one particular response to the CS rather than another; it does not, however, say anything about the nature of the associations formed that produce the change in behaviour to the CS. Pavlov took little interest in this question, but seems to have assumed that the association was between the centres of neural activity produced by presentation of CS and UCS (e.g., Pavlov, 1927, pp. 36–8). In this he has been followed by other Russian and East European workers (e.g., Asratyan, 1965, pp. 149–51; Konorski, 1948, p. 87; 1967, pp. 265–7). The closest equivalent in Western terminology is the assumption that stimulus–stimulus (S–S) associations are formed (Tolman, 1934; Schlosberg, 1937; Birch and Bitterman, 1949; Bindra, 1972).

The crudest alternative to this analysis was Watson's reflex-arc theory of association (Watson, 1914). The clearest alternative was the S–R account provided by Guthrie (1935). According to Guthrie, conditioning results when a response, the UCR, occurs in close temporal proximity to a stimulus, the CS (in more exact statements of this analysis, Guthrie suggested that the effective stimulus might be produced by proprioceptive feedback from responses to the CS); this contiguity of stimulus and response is sufficient to cause an association between the two such that, in the absence of subsequent interference, that stimulus will in future elicit that response. Hull (1943) was also relatively clear that the association formed was between CS and UCR, although he assumed that the temporal coincidence of the two was not sufficient to guarantee the formation of an association in the absence of the reduction of some drive state. The question whether S–S or S–R associations are formed in the course of a classical conditioning experiment has been the subject of a certain amount of experimental analysis. The question has usually been formulated as though the alternatives posed were mutually exclusive. This is not, of course, necessarily so and we shall see that evidence can be brought to bear against a literal interpretation of either theory.

A. Conditioning with the UCR Blocked

A straightforward interpretation of S–R association theory is rather simply discredited. If one of a variety of methods is used to block the occurrence of responding during CS–UCS pairings, this will not prevent the appearance of full-scale CRs when the blocking agent is removed.

Crisler (1930) and Finch (1938b) injected dogs with atropine (a drug which blocks salivation), paired a CS with either acid or morphine as UCS and, when the effects of atropine had worn off, observed substantial salivary CRs to the CS. Light and Gantt (1936) and Beck and Doty (1957) paired a CS with shock to a limb while the ventral roots innervating the limb had been crushed and observed significant flexion CRs upon subsequent testing. Solomon and Turner (1962) have reviewed a number of earlier studies in which subjects were conditioned under curare, and themselves definitively established that administration of curare (a drug which blocks skeletal musculature) does not prevent the subsequent appearance of appropriate skeletal CRs when subjects are tested after the drug has ceased to have an effect. Finally, Moore (1973) has shown that pigeons exposed to a key light paired with food, but with access to the key prevented by a wire screen, will peck vigorously at the key as soon as the screen is removed.

How much do these studies establish? Certainly, it must be accepted that the overt occurrence of a response is not necessary for conditioning. There remain, however, two arguments which S–R theorists have frequently fallen back on. First, it can be said, the elimination of overt responding does not necessarily imply the elimination of central nervous activity normally linked with the occurrence of the overt response. Perhaps this central activity is the "response" which enters into association with the CS. The question here is whether such an argument does not abolish the distinction between S–S and S–R analyses. If it is not the UCR elicited by a UCS that becomes associated with the CS but some antecedent neural event, how does this position differ from one which postulates the formation of an association between neural correlates of the CS and UCS? The distinction does not seem to be one which will yield rapidly to experimental analysis.

A second argument which could be advanced is to point out that the elimination of some overtly recordable responses does not necessarily imply the elimination of others. Changes in heart rate, for example, are conditionable while skeletal responses are completely blocked; Black (1965) successfully conditioned changes in heart rate in dogs paralysed by d-tubocurarine; Yehle et al. (1967) conditioned changes in heart rate and blood pressure in rabbits paralysed by flaxedil. Similar responses, it could be argued, must have been conditioned in Solomon and Turner's dogs, while Light and Gantt specifically noted that their animals exhibited general restlessness during the course of conditioning.

The force of this argument is weakened by evidence which suggests that changes in heart rate are not themselves directly conditioned, and could not be the mediating cause of changes in skeletal activity. On the contrary, changes in heart rate appear to be the consequence of changes in either overt skeletal activity or (in paralysed subjects) in the central neural correlates of that activity (p. 103). Furthermore, the argument depends upon establishing appropriate links between the responses that are presumed to occur during the course of conditioning (e.g., the general restlessness observed by Light and Gantt) and the CR that actually appears during testing. Such links are by no means always self-evident: for example, there is good reason to suppose that the conditioning of general restlessness interferes with, rather than mediates, the appearance of a flexion CR (p. 23).

One recent version of the sort of theory envisaged in this argument is that proposed by Rescorla and Solomon (1967). They have suggested that the pairing of a CS and UCS results in the conditioning of a central motivational state, which mediates the appearance of a variety of overt CRs. Whether such a mediational analysis is reasonable, and if so whether a central motivational state is usefully regarded as a response, will be discussed in the following section.

B. Conditioning of Central Motivational States

When a stimulus, paired with shock according to a classical conditioning procedure, is superimposed upon ongoing, food-reinforced, instrumental responding, such responding is typically suppressed. This conditioned suppression paradigm has been referred to repeatedly in the preceding chapter as an instance of classical conditioning. Yet several questions may be asked about this procedure.

If conditioning trials are given while the subject is engaged in instrumental responding, it is clearly possible that, although no instrumental contingency is programmed, the instrumental response may by chance occur just before a shock is delivered and hence be punished. Although there is some evidence that such adventitious response–shock pairings may have some effect on the progress of conditioned suppression (e.g., Gottwald, 1967), the fact that complete suppression is routinely observed after conditioning trials have been given with the instrumental manipulandum removed from the apparatus, proves that instrumental contingencies do not play a critical role in conditioned suppression (examples of this general method of studying conditioned suppression—sometimes described as conditioning off the baseline—include: Kamin *et al.*, 1963; Rescorla, 1968b; Hilton, 1969; Hoffman and Barrett, 1971).

Even if it is accepted that conditioned suppression does not necessarily depend upon such instrumental contingencies, there remains the question why a CS paired with shock should have the effect of suppressing appetitive instrumental responding. A strict proponent of an S–R theory of conditioning would tend to argue that some overt component of the UCR to shock, conditioned to the CS, must actively interfere with the emission of the instrumental response.

What is the response, elicited by shock, that is conditioned to the CS in studies of conditioned suppression? Brief, inescapable shocks, delivered to the feet of freely moving rats in an enclosed chamber, elicit flinching followed by prancing and jumping responses (Kimble, 1955; Hoffman et al., 1964; Reynierse et al., 1970). The response that becomes conditioned to a CS preceding such a shock, however (or in the absence of any discrete signal, to the background, apparatus cues), is usually described as crouching or freezing (Blanchard and Blanchard, 1969a; Reynierse et al., 1970).

It seems quite reasonable to suppose that a CR such as freezing or crouching will indeed interfere with most instrumental responses. These observations, therefore, are entirely consistent with an analysis of conditioned suppression in terms of overt interfering responses, and do not suggest the need for any appeal to more central conditioned states. Nevertheless, although the nature of the peripheral responses elicited by a CS signalling shock must certainly be taken into consideration, there is ample evidence that additional processes must be postulated in order to provide a complete account of conditioned suppression and related phenomena.

If the response conditioned to a CS signalling shock is one of immobility or freezing, then the presentation of such a CS should not only invariably suppress appetitive instrumental behaviour, it should be equally effective in suppressing aversively motivated instrumental behaviour. There is good evidence against both of these propositions. There are exceptions to the general rule that an aversive CS always suppresses appetitive responding. Blackman (1968), for example, trained rats to press a lever for food on an interval schedule of reinforcement in the presence of one stimulus and to space their lever presses at least 6 sec apart in order to obtain food in the presence of another stimulus (this constitutes a DRL schedule of reinforcement; see p. 165). The presentation of a CS signalling shock reliably suppressed responding on the interval schedule at all levels of shock intensity, but when the intensity was 1 ma or less resulted in a significant *increase* in rate of responding on the DRL schedule. The interpretation of changes in rate of responding on DRL schedules is undoubtedly fraught with difficulties (p. 382) and this increase in rate of lever pressing was probably mediated by the suppression of "collateral" behaviour which had been responsible for the spacing of lever presses. Nevertheless, if the only effect of the CS had been to cause subjects to freeze, it could hardly have resulted in an increase in the rate of lever pressing.

As was noted in the preceding chapter, a CS paired with shock typically enhances rather than suppresses instrumental avoidance responding. There are certainly exceptions to this generalization, but the range of conditions under which it has been confirmed is by now substantial. This includes studies of rats either turning a wheel to avoid shock (Weisman and Litner, 1969), or running in a shuttle box (Grossen and Bolles, 1968; Kamano, 1970; Scobie, 1972); studies of dogs in shuttle boxes (Rescorla and LoLordo, 1965; Rescorla, 1966), or pressing panels (LoLordo, 1967); and studies of monkeys pressing a lever (Sidman *et al.*, 1957). The exceptions seem to fall under two headings. Scobie (1972) found that the use of a very strong shock as UCS (in particular, the use of a shock more intense than that being used in avoidance training) might result in suppression rather than enhancement of avoidance responding (see also Bryant, 1972). Secondly, it is possible that if rats are required to press a lever to avoid shock, they will be more likely to show suppression than if they are trained to turn a wheel or run in a shuttle box. Although Blackman (1970) found that an aversive CS might sometimes enhance rate of lever pressing as an avoidance response, Roberts and Hurwitz (1970) and Hurwitz and Roberts (1971) observed rather consistent suppression. This may be related to the difficulties often experienced in establishing lever pressing as an avoidance response in the first place (p. 340).

In spite of these exceptions, however, it is clear that a CS signalling shock will usually elevate rather than suppress the rate of avoidance responding. If the effects of an aversive CS depend on the nature of the motivation for instrumental responding, they can hardly be mediated solely by the pattern of peripheral responses conditioned to that CS. They are more plausibly interpreted in terms of some central, possibly motivational, mediator. The ideal test of this proposition would involve training two groups of subjects to make the same instrumental response, one to obtain food, the other to avoid shock, and then exposing both groups to a CS signalling shock. The CS should suppress the food-reinforced response, but enhance the avoidance response. Herrnstein and Sidman (1958) have reported just this result in an experiment where monkeys were trained to press a lever either to obtain a reward of orange juice or to avoid shock. More recently, Scobie (1972) trained rats either to press a lever for food or to run in a shuttle box to avoid shock; when the two groups received classical pairings of a CS and shock, identical training parameters resulted in suppression of the appetitive response and enhancement of the avoidance response.

Although rate of avoidance responding may be enhanced rather than suppressed by the presentation of a CS signalling shock, there is good evidence that it may be suppressed by the presentation of a CS signalling

D

food (Grossen *et al.*, 1969; Bull, 1970; Davis and Kreuter, 1972).* To test whether this difference might be a consequence of differences in the peripheral CRs conditioned to aversive and appetitive CSs, Overmier *et al.* (1971a) trained dogs to avoid shock in a shuttle box, and then super-imposed a tone upon ongoing avoidance responding. For one group, the tone had been established as a signal to press a panel to obtain food; for a second group, it had been established as a signal to press the same panel to avoid shock. Although all animals had learned to make the same response to the tone, when it had been a signal to obtain food it suppressed avoidance responding in the shuttle box but when it had been a signal to avoid shock it enhanced the rate of responding.

An aversive CS undoubtedly elicits characteristic patterns of peripheral CRs; in many situations and subjects, these CRs may include freezing; but it is clear that, unless the UCS is very intense, the effects of superimposing such as CS upon instrumental responding cannot be interpreted simply as a consequence of an interaction between the instrumental response and these peripheral CRs. Conditioned suppression of food-reinforced re-sponding and conditioned enhancement of avoidance responding, there-fore, must be mediated by other consequences of classical aversive con-ditioning. According to theorists such as Mowrer (1960a), Konorski (1967), Rescorla and Solomon (1967), and Estes (1969b), these consequences are to be regarded as conditioned motivational states. Shock as a UCS elicits a characteristic emotional reaction, which will be conditioned to a CS signalling shock, and it is this reaction which indirectly affects instru-mental responding by either augmenting or inhibiting the motivation underlying the instrumental response.

In the present context, according to the version of two-factor theory propounded by Rescorla and Solomon (1967) for example, the associations formed during classical conditioning are still between stimuli and responses, even though the response in question may be central and observed only indirectly through its effects on overt behaviour. A theory which implies that peripheral CRs may be only imperfect indicators of some central conditioned state clearly represents a rather substantial departure from the type of S–R theory envisaged by Watson or Guthrie. It is obvious that the distinction between S–S and S–R theories of association has once more been eroded. While this is true (and all to the good) it is still perhaps possible to find some points at issue and to bring some evidence to bear on them.

* In an ideally symmetrical universe, given that avoidance responding is enhanced by an aversive CS and suppressed by an appetitive CS, it would follow that appetitive instru-mental responding, which is suppressed by an aversive CS, would be enhanced by an appetitive CS. In general, however, it is not (p. 225).

A study by LoLordo (1967) provides some evidence that conditioned enhancement of avoidance responding may depend more upon the conditioning of an emotional reaction than upon the formation of any direct association between a CS and aversive UCS. LoLordo found that rate of avoidance responding in dogs was enhanced not only by the presentation of a CS signalling shock, but also by a CS signalling an intense, and presumably aversive noise. If stimuli signalling either shock or noise have similar effects on avoidance responding, it is not unreasonable to argue that this is because they have both become associated with similar emotional responses.

A more recent study by Overmier *et al.* (1971b) provides evidence pointing rather strongly to the importance of associations with the specific UCS used in conditioning. Dogs were trained to avoid shock in the presence of one signal by pressing a panel to the left of their head; in the presence of another signal they were required to press a panel to the right of their head in order to avoid shock. For one group, the two stimuli consistently signalled two different shocks, one to one hind leg, the other to the other. For a second group, there was no correlation between signal and location of shock. The former group learned the discriminative avoidance responses significantly more rapidly than the latter. Since the two shocks differed not in intensity, but only in where they were applied, it is very difficult to see how they could be said to have elicited discriminably different emotional reactions. The fact, therefore, that a unique relationship between signal and shock reliably facilitated the establishment of different responses to each signal, implies that associations must have been formed between particular stimuli and particular shocks rather than, or in addition to, associations between stimuli and central motivational states.

C. Sensory Preconditioning

Perhaps the most convincing evidence that the pairing of two stimuli is sufficient to produce an association between the two, rather than between one of the stimuli and the response elicited by the other, is provided by studies of sensory preconditioning (p. 20). In such experiments, joint presentations of CS_2 and CS_1, followed by pairings of CS_1 with a UCS, result in the appearance of significant CRs to CS_2 when it is presented alone on subsequent test trials. The most natural interpretation of this finding is that subjects form associations between CS_2 and CS_1 during Stage 1 of the experiment. Stimulus–response theorists must search hard to uncover the responses that are being associated with stimuli here.

If they have failed, it has not been for want of trying. Fig. 3.1 illustrates the connections that have been suggested (e.g., by Osgood, 1953, pp.

461–2; and Kimble, 1961, p. 217). The basic assumptions are as follows: the stimuli used in sensory preconditioning experiments—although relatively neutral events such as lights, tones and buzzers—themselves elicit particular responses; the pairing of CS_2 and CS_1 in Stage 1 results in the formation of an association between CS_2 and R_1 (the response elicited by CS_1); the pairing of CS_1 with the UCS in Stage 2 results in the formation of associations between CS_1, R_1, S_1 (the stimuli produced by R_1) and the experimenter's UCS. Thus, in the final test session, CS_2 will elicit R_1 and

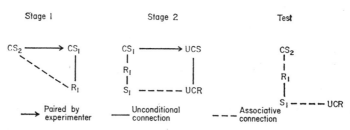

Fig. 3.1. A possible S–R analysis of sensory pre-conditioning. An association is formed initially between CS_2 and the response elicited by CS_1; when CS_1 is paired with a UCS, associations are established between the stimulus consequences of the response to CS_1 and the UCR, such that CS_2 will later elicit both the response to CS_1 and the UCR.

therefore the UCR to the experimenter's UCS. Although the experimenter does not usually trouble to record these responses,[*] they exist and serve to mediate the test responses taken as evidence of successful preconditioning.

According to Kimble "the mediation hypothesis appears to be worth everything it costs the S–R position in terms of theoretical concessions" (pp. 217–18) because it permits new predictions. For example, "if the first phase of such studies is, indeed, an ordinary classical conditioning procedure, the variables important for classical conditioning should be important for it" (*loc. cit.*). It is hard to see why this unsurprising prediction should be unique to S–R theory. The problem with the explanation is that of specifying the nature of the responses unconditionally elicited by CS_1 and CS_2 in such a way that the analysis does not become vacuous. A

---

[*] An exception to this general statement is a study by Adamec and Melzack (1970). They indeed recorded a tendency for subjects to orient toward the site of CS_1 during the presentation of CS_2 in Stage 1 of their experiment and in their case, since CS_1 was located immediately at the site of the reinforcer (the UCS was milk), it becomes reasonable to assume that the subsequent facilitation of conditioning to CS_2 (a tone) after CS_1–UCS pairings, was indeed mediated by a tendency for preconditioned subjects to orient towards the reinforcement site when CS_2 was presented. The special arrangements involved here, however, make it dangerous to assume that such specific mediation is a general rule.

study by Kendall and Thompson (1960) illustrates the problem for here the two CSs were a 2000 and a 250 Hz tone produced by the same loud-speaker. When CS_1 is, say, a tone and CS_2 a light, it makes sense to say that the orienting responses elicited by the tone, such as pricking up of the ears and turning towards the loudspeaker, might become conditioned to the light (or vice-versa). However, it is difficult to see why two different tones should not elicit the same orienting reaction; if this were so, the paired presentation of the two tones could hardly increase the probability with which one elicited the orienting response to the other. Since Kendall and Thompson observed a highly significant effect of paired presentations of CS_1 and CS_2 by comparison with a control group which received unpaired presentations of the stimuli, such pairings must have established an association between some much more specific, central effects of the two stimuli. The S–R theorist may, of course, insist that it was the cortical response to one tone which became conditioned to the presentation of the other, but it is not immediately obvious that such an assertion reflects anything more than a determination to use the word "response" as the only admissible name for the second term of an association. What is clear is that the position thus defended differs in no significant way from that of Konorski.

D. Changes in Significance of UCS

Another way of attempting to elucidate the nature of the associations formed in classical conditioning is to pair a CS and UCS until a reliable CR is elicited and then to alter the effectiveness or significance of the UCS, before finally testing the subject's responses to the CS. If the initial pairings of the CS and UCS establish an association between the two stimuli, then it would be expected that the response to the CS in the final test would be affected by any intervening operation that had altered the subject's response to the UCS. If, on the other hand, an association had been established between the CS and the response initially elicited by the UCS, the sub-sequent modification of the response to the UCS might be expected to leave this original association unaffected, so that the CS would still elicit the original CR.

A study by Harlow (1937) illustrates the nature of the argument. Conditioned fear reactions were initially established in monkeys by pairing a buzzer as CS with a paper "blow-out", which elicited a reliable fear response. Subjects were then repeatedly exposed to the blow-out, either alone or in conjunction with food, in an attempt to habituate their fear of this stimulus. In spite of the success of this procedure, presentation of the buzzer was still sufficient to elicit an undiminished fear reaction. If the initial pairing of buzzer and blow-out had simply established an association between the two stimuli, the habituation of fear to the latter should have been sufficient to eliminate, or at least attenuate, the subjects' fear of the former. The results suggest, therefore, that conditioning depended upon the formation of an association between the buzzer and the fear response elicited by the blow-out, and that this association was unaffected by the subsequent habituation of fear to the blow-out itself.

A recent study of conditioned suppression in rats failed to confirm Harlow's conclusions (Rescorla, 1973). Rescorla paired a CS with an intense burst of noise and found that the CS rapidly came to suppress lever pressing for food. If subjects were habituated to the noise, however, a subsequent test session revealed a significant reduction in suppression to the CS. The reason for this discrepancy is not, at present, clear.

One situation in which it should be relatively easy to alter the significance of a UCS is in experiments on higher-order conditioning (p. 19). Here, after initial pairings of CS_1 and UCS, CS_1 itself is used as a UCS to establish conditioning to CS_2. Testing with CS_2 alone may reveal significant evidence of CRs similar to those elicited by CS_1. It will, however, be a simple matter to alter the significance of CS_1 between initial pairings of CS_2 and CS_1 and subsequent testing with CS_2 alone. In a study described by Konorski (1948, p. 107) dogs were initially given salivary conditioning with CS_1 signalling food which was followed by paired presentations of CS_2 and CS_1. Before subjects were tested with CS_2 alone, however, CS_1 was turned into an aversive CS by being paired with shock to the paw. Although CS_1 now elicited a flexion CR, CS_2 continued to elicit a salivary CR.

In view of the difficulty of turning an appetitive into an aversive CS, subsequently discovered by Konorski and Szwejkowska (1956), and the frequent regressions from the new CR back to the old which they noted (Fig. 2.4, p. 23) it is unfortunate that Konorski's description of this experiment is extremely brief and cursory. However, Rizley and Rescorla (1972) have provided support for Konorski's conclusions in a study of conditioned suppression in rats. After initial pairings of CS_1 and shock, rats received paired presentations of CS_2 and CS_1. Before being tested with CS_2 alone, however, they received a series of nonreinforced presentations of CS_1; although these were sufficient to result in substantial extinction of suppression to CS_1, suppression to CS_2 was barely affected.

The implication of the results of Konorski and Rizley and Rescorla is that higher order conditioning may depend upon the establishment of S–R associations. It is as though Rizley and Rescorla's rats remembered that they had been afraid shortly after each presentation of CS_2 and continued to show suppression in its presence even when suppression to CS_1 was extinguished. Although these conclusions are consistent with those suggested by Harlow's data, Rescorla's own study of simple conditioning suggested that habituation of fear to a UCS did result in some loss of fear to a first order CS. On the basis of these results, Rescorla (1973) has argued that it is only higher-order conditioning that depends upon S–R associations: in first-order conditioning, subjects form associations between CS and UCS.

One objection to the design of all these experiments is that they seem based on the assumption that only one type of association could be formed in any one situation. As was suggested earlier, there is no *a priori* reason why animals should not associate a CS both with a UCS and with their reaction to that UCS. If it is argued that there may be a difference of emphasis, rather than a categorical distinction, between the associations formed in first-order and those formed in higher-order conditioning, Rescorla's conclusion might be more acceptable. Even so the data presently available are far too confused, and the amount of supporting evidence far too fragmentary, for it to be accepted with any great confidence. It is possible, for example, that the differences between the outcomes of some of these studies may have been due to the use of different UCSs. If Harlow (1937) obtained evidence of the maintenance of fear to a CS in spite of the habituation of fear to the UCS with which it had been paired, while Rescorla (1973) found that fear of a CS was partially extinguished by this procedure, this may have been because the UCS used by Harlow was apparently much more aversive than that used by Rescorla. It has frequently been suggested in other contexts that conditioned fears established by the use of very severe shocks may be partially irreversible (Solomon *et al.*, 1953; see Chapter 6). Rizley and Rescorla's (1972) observation of the maintenance of suppression to CS_2 in spite of the extinction of suppression to CS_1 was also, of course, made when a relatively strong shock was used as the UCS. Although Rescorla (1973) reported some evidence for a similar retention of suppression to a higher-order CS_2 after habituation of fear responses to the loud noise UCS, this result should perhaps be treated with some caution, since the level of suppression maintained to the higher-order CS by a control group was relatively slight.

There have, unfortunately, been no detailed reports of the retention of higher-order appetitive conditioning after extinction of the first-order association. Nevertheless, there is considerable evidence from apparently analogous instrumental experiments which suggests that such retention is not the rule. A first-order CS paired with food can be used not only to establish second-order classical conditioning to a second CS, but also to serve as a conditioned reinforcer for instrumental responses on whose occurrence it is made contingent (see Chapter 5). There is ample evidence that the effectiveness of a conditioned reinforcer can be diminished by extinguishing the original association between CS and food, and that even if extinction is undertaken without the opportunity for instrumental responding, subsequent testing may reveal an immediate decline in the strength of that response. Observations to this effect have been reported in conventional studies of conditioned reinforcement (e.g., Coate, 1956) and also in experiments on latent extinction (see Chapter 8) where, after a rat has received reinforcement in a particular goal box, subsequent non-reinforcement in that goal box, without the opportunity for instrumental responding, can be shown to have an immediate effect on the strength of earlier components of the instrumental chain (e.g., Seward and Levy, 1949; Gonzalez and Shepp, 1965). Such results suggest, albeit indirectly, that analogous experiments on higher-order appetitive conditioning might also reveal substantial extinction of CRs to CS_2 after extinction of responses to CS_1.

E. Conclusions

It is too early to provide any definitive assessment of how altering a subject's responses to a UCS may affect the responses of that subject to a CS previously paired with the UCS. It is apparent that in some circumstances, perhaps especially in cases of higher-order conditioning or of conditioning with a highly aversive UCS, modification of the subject's reactions to the UCS leaves the originally established CRs relatively unaffected, thus suggesting that associations had been formed between the CS and the response originally elicited by the UCS. Under other conditions, however, there is equally good evidence that habituation of responses to a UCS may result in an immediate decline in the strength of previously established CRs, thus suggesting that the initial associations had been formed between the CS and UCS.

It would doubtless be possible to rephrase these conclusions so that they conformed to the language of S–R theory. However, little would be gained by such a procedure, other than to obscure some possibly important distinctions. Stimulus–response theory has often been thought to provide a more rigorous analysis of, or at least a more rigorous language for the description of, experimental data on learning. At the very least the evidence reviewed in earlier sections suggests that conditioning can be described as the formation of S–R associations only at the cost of interpreting the term "response" to refer to central representations of stimuli, neural correlates of responses and unrecorded emotional states which may bear no more than the most indirect relation to the overt CRs recorded by the experimenter.

The available data are most simply described by saying that when the presentation of one stimulus is made contingent on the presentation of another, animals may detect this relationship and learn to associate the occurrence of the first with the occurrence of the second (and perhaps also with the occurrence of any responses or emotional reactions reliably elicited by the second). The clearest analysis of the associative learning underlying classical conditioning, then, is that which would be expected in view of the nature of the experimenter's operations. Exposure to a contingency between two stimuli results in the formation of an association between them.

There is a further merit to such an analysis. If it is assumed that exposure to a positive correlation between a CS and UCS ensures that subjects learn that the UCS follows the CS, then exposure to a negative correlation between CS and UCS may be assumed to ensure that they learn that the UCS does not follow the CS. The symmetrical operations of excitatory and inhibitory conditioning, therefore, can be regarded as setting the occasion for symmetrical learning processes. Stimulus–response theory has always had some problem in dealing with inhibitory learning. It has usually fallen back on an analysis in terms of interfering responses, assuming that the omission of reinforcement leads to a decline in the probability of a particular CR only because it elicits a competing response (Guthrie, 1935, Wendt, 1936). While this is not the place for an extended discussion of this issue (see Chapter 8), a good case can be made for the suggestion that interference theory has maintained its popularity, not because its proponents have consistently succeeded in pointing to observed instances of competing responses, nor because they have been able to show that any such competing response was causally responsible for the disappearance of a CR during inhibitory conditioning, but largely because the theory provides the only analysis which permits retention of basic S–R assumptions.

III. THE CONDITIONS OF ASSOCIATION

According to Pavlov, a conditional reflex is reinforced, i.e., the probability that the CS will elicit a CR is increased, whenever the CS is followed by a UCS that reliably elicits a UCR. The requirement that the UCS elicit a reliable UCR was intended to exclude certain cases: appetitive salivary conditioning, for example, is possible only if the animal is sufficiently hungry to salivate at the presentation of food. As we have already noted, although it is true that conditioning depends upon drive level (e.g., DeBold et al., 1965), it is not possible to attribute this entirely to the effect of drive on the magnitude or reliability of UCRs (p.71): a decrease in drive may have a much greater effect on CRs than on UCRs. The implication of this observation is that the occurrence of a reliable UCR is not sufficient to ensure reliable conditioning. Although this conclusion is supported by several of the studies which will be discussed in the following sections, it remains to be seen whether it is critically damaging to a Pavlovian interpretation of conditioning.

A. Failures of Conditioning despite a Reliable UCR

We saw that the strength of the UCS had a direct effect on the strength of conditioning. In the extreme case, a UCS may be too weak to support any detectable conditioning at all, but several studies have shown that such an ineffective UCS may still be strong enough to elicit a UCR. Konorski (1967, p. 284) states that when acid is used as the UCS for salivary conditioning, a weak concentration may elicit salivation but totally fail to establish conditioning. Bruner (1969) found that the weakest level of shock that would reliably elicit leg flexion in cats was insufficient to establish flexion CRs. Stimuli such as lights and buzzers do in fact elicit certain UCRs, but a number of studies have failed to find any evidence of conditioning when they are used as the UCS. Bruner (1965) found that a bright light would elicit nictitating membrane closure in rabbits, but would not support conditioning unless accompanied by an airpuff. Stolz (1965) found that although a 2 sec buzzer elicited reliable vasomotor responses, it would not support conditioning, although Fromer (1963) found significant vasomotor conditioning when a shock was used as the UCS. Gerall and Obrist (1962), using cats, and Gerall *et al.* (1957), using students, obtained significant pupillary conditioning when shock was used as the UCS, but none when the UCS was a decrease in the level of illumination, even though such a stimulus reliably elicits pupillary dilation.

Colavita (1965) prepared dogs for salivary conditioning with acid as the UCS by providing them with an external fistula. For one group, the acid injected into the mouth passed out through the external fistula and never entered the stomach; a second group had acid inserted directly into the stomach; a third had acid injected into the mouth and passing out of the fistula, but simultaneously had acid injected into the stomach; while for a final control group the fistula was connected up so that acid injected into the mouth passed via the fistula into the stomach. The results are shown in Fig. 3.2. They are relatively clear: only if the acid entered both mouth and stomach (i.e., only in the last pair of groups) did reliable salivary conditioning occur. Acid injected into the mouth, but allowed to pass out through an external fistula, elicited copious unconditional salivation but failed to sustain significant conditioning. Acid injected directly into the stomach produced neither UCRs nor CRs.

The implication of all these studies is quite consistent: stimuli without motivational significance to the subject may elicit UCRs, but will not necessarily support conditioning.* Salivary conditioning to acid, for example, will occur only if a relatively strong solution of acid actually enters the stomach. It is difficult to resist the interpretation that the *function* of salivary CRs is to dilute acid and thereby protect the stomach. From this

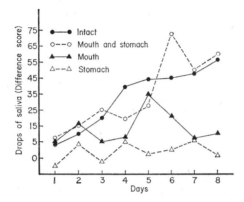

FIG. 3.2. Salivary conditioning in dogs, with the delivery of acid as the UCS. Conditioning proceeds normally provided that the acid is both injected into the mouth and reaches the stomach (Intact, and Mouth and stomach groups), but there is no reliable evidence of conditioning when acid is injected into the mouth without reaching the stomach, or when it is injected directly into the stomach. (*After* Colavita, 1965.)

it is but a short step to assuming that this function underlies the reinforcing mechanism for the formation of such CRs. Although there are reasons for attempting to resist this interpretation, it will be necessary, before entering into the argument, to examine two other areas of research: first, studies which have employed direct stimulation, usually of the nervous system, as the UCS and secondly, studies of sensory preconditioning.

B. Direct Stimulation of Effector System as the UCS

Literally interpreted, the idea that the sole requirement for conditioning is the reliable elicitation of a UCR, implies that conditioning would occur

---

* It is interesting to note a parallel between these results and those of some studies of conditioned reinforcement. In some cases stimuli which reliably elicit consummatory responses, such as pecking, chewing, or swallowing may not act as effective reinforcers unless they are accompanied by an event of motivational significance, such as the delivery of food to the stomach (Sterritt and Smith, 1965; Holman, 1969).

when a response was elicited by direct stimulation of the response system. Stimulation of peripheral effector systems has, however, been shown to be clearly ineffective. Pilocarpine, which acts directly on the salivary gland to elicit salivation, will not support salivary conditioning (Kleitman, 1927; Finch, 1938c); nor will direct stimulation of the motor pathway support flexion conditioning (Hilgard and Allen, 1938). These results are entirely consistent with those of Pavlov; the failure of conditioning need not be attributed to the motivational insignificance of the UCS, but to the fact that it does not act on the central nervous system.

If this explanation is correct, then conditioning should be possible if a response is elicited by direct stimulation of the central nervous system, e.g. of the motor cortex. The classic study of this question is that of Loucks (1935). In this experiment three dogs received about 600 pairings of a 1 sec buzzer with electrical stimulation of the sigmoid gyrus eliciting leg flexion. None of the subjects showed any signs of conditioning.

These negative results have been cited frequently (e.g., by Spence, 1951) as evidence against a contiguity interpretation of conditioning and as implying that a motivationally significant UCS is necessary for conditioning to occur. Even this argument could be upheld only by ignoring other studies which had been successful: Brogden and Gantt (1937), for example, used stimulation of the cerebellum as a UCS, being careful to employ an intensity of stimulation well below that which appeared to elicit any overt signs of pain, and obtained significant flexion conditioning within five to fifteen trials. More recent studies, most notably those by Giurgea and Doty, have also been successful (Doty and Giurgea, 1961; Doty, 1969).

Why then was Loucks unable to obtain conditioning? Doty (1969) suggests that Loucks' use of a very short intertrial interval and possibly rather restless subjects may have produced too much variability in the form of the UCR to generate consistent CRs. A more ingenious explanation has been advanced by Wagner et al. (1967). They obtained significant conditioning using Loucks' general procedure, but noted that Loucks' dogs were suspended in a hammock while theirs were standing on a platform. They suggested that under their conditions the UCS might cause animals to overbalance and that the CRs involved "a postural adjustment which served to minimize the abruptness of forcefulness of the UCR (p. 196)". Thomas (1971) has provided some evidence consistent with this analysis.

One would, however, like to see a direct comparison of the effects of cortical stimulation as UCS in freely standing and suspended subjects before accepting this as a definite explanation of Loucks' failure. The argument does not appear to be entirely convincing: if, for example, stimulation elicits flexion of the right foreleg, then the most successful postural adjustment would presumably be the extension of the other legs rather than flexion of the right foreleg. Furthermore, the analysis will not serve to account for numerous successful studies. Doty and Giurgea (1961) successfully conditioned flexion of the forearm, elicited by cortical stimulation, in a seated and restrained monkey; they also observed successful leg flexion conditioning in dogs when the CR was completed (i.e., the leg was extended again) before the onset of the UCS.

The conclusion from these studies, then, is that a UCS without obvious motivational significance may support reliable conditioning. The contradiction with the studies cited in the preceding section is obvious. To resolve the contradiction it will be necessary to undertake some direct comparisons: if direct stimulation of the motor cortex of a dog produces reliable flexion conditioning, will a threshold shock to the leg which elicits the same flexion response also support conditioning? And if not, what other differences can be detected between the two types of UCS? Do the animals prefer the successful UCS to be signalled, but show no preference for whether the unsuccessful UCS is signalled or not?

C. Sensory Preconditioning

In sensory preconditioning experiments, animals appear to form associations between a CS and a "UCS" (i.e., between CS_2 and CS_1) when the latter is a stimulus of such little motivational significance as a light or a tone. The apparent fragility of early demonstrations of the phenomenon enabled Spence (1951) to argue that equivocal results did not necessarily pose insuperable problems for theory; as we have seen, the equivocation was due not to the insubstantial nature of the phenomenon, but to the use of less than optimal procedures for its demonstration (p. 21). Sensory preconditioning experiments may not produce very large effects but they are undoubtedly reliable and their obvious interpretation is that animals can learn to associate motivationally neutral events.

The implication has, of course, been resisted. It is always possible to insist that the CS_1 of a sensory preconditioning study must have some motivational properties (Spence, 1951) and in support of this one can even point to the fact that rats sometimes learn to press a lever when the sole consequence of this response is that a brief light is turned on (Kish, 1966). This particular question may not be worth the effort of resolving, and concentration on it seems to have obscured a rather more interesting implication suggested by these experiments. Studies of preconditioning demonstrate that when learning is measured by a *subsequently* trained response, the simple pairing of a tone and a light appears to be sufficient to produce an association between the two. However, from the studies reviewed in Section A, it would appear that when learning is measured by *concurrent* changes in some response system, no learning is demonstrable: pairing a CS with a light or a buzzer produces no conditioning of the pupillary or eyelid response elicited by the light or buzzer.

What are we to make of this apparent discrepancy? The first solution is to deny its reality. One could argue that the amount of learning demonstrable in a sensory preconditioning experiment is small, and that it may indeed be reduced by extensive training (Hoffeld *et al.*, 1960). Perhaps significant conditioning of flexion, eyelid, or pupillary responses occurred on early trials in Bruner's (1965, 1969) and Gerall and Obrist's (1962) experiments, but rapidly disappeared. Inspection of their published data does not in fact support this contention, but this may not be crucial. The data of Colavita shown in Fig. 3.2 do indeed suggest that some conditioning of salivation occurred on early trials when acid was injected into the mouth and prevented from reaching the stomach.

A second way of dismissing the apparent discrepancy is to point out that it rests upon comparisons between experiments differing in subjects, stimuli, interstimulus interval and, doubtless, innumerable other details of experimental procedure. What is needed is a set of experiments specifically designed to compare the incidence of sensory preconditioning between a light and a buzzer with the occurrence of, say, eyelid conditioning when the same two stimuli are used as CS and UCS. Without such a direct comparison we cannot be absolutely confident that the failure of eyelid conditioning in Bruner's experiment (1965) was due to a failure of response elaboration rather than a failure of association.

It is still possible, however, that the discrepancy is real. In that case we must suppose that animals in both types of experiment learn that one stimulus follows another, but that the formation of such an association is not sufficient to produce a CR to the first stimulus. This would imply that a motivationally significant UCS is not required for the formation of an association between CS and UCS, but only for the elaboration of a CR to the CS.

D. Conclusions

At least some of the results reviewed in the preceding sections are consistent with a non-Pavlovian analysis of classical conditioning. One could argue that they show that CRs will occur only when motivationally significant UCSs are used and that this implies that CRs are reinforced only if they are followed by motivationally significant consequences. While this can be accepted, it seems worth stressing that the conclusion can as easily be resisted. It is equally possible to provide a Pavlovian analysis of most of the data under review.

Let us assume that the pairing of two neutral stimuli is sufficient to establish some degree of association between them. Even so, however, we know that animals rapidly habituate to the repetitive presentation of neutral stimuli and that, in all probability, this habituation underlies the fact that continued pairing of CS_1 and CS_2 in a sensory preconditioning experiment yields no further increments in conditioning. We know further that the establishment of a stable CR such as the flexion or eyelid response often requires a substantial amount of training and that the number of trials before the first CR appears is inversely related to the strength of the UCS (p. 71). With a very weak UCS, therefore, it is entirely probable that habituation will set in before measurable conditioning has occurred. If this analysis is correct, one might expect that although a weak UCS would be insufficient to *establish* conditioning, it might add to the strength of CRs occasionally reinforced with a strong UCS. In confirmation of this, Bruner (1965) found that if rabbits received an airpuff as the UCS on 50 per cent of trials, the presentation of a light as the UCS on the remaining trials produced a significantly higher overall level of conditioning than was attained when the CS was presented without any UCS on the remaining trials.

Two further points are worth making. One of the more valid generalizations about classical conditioning is that CRs, even if they resemble the UCR very closely, are usually weaker and of lesser amplitude (e.g., Hilgard, 1936; Kimble, 1961, p. 53). If this is so, then it is hardly surprising that a threshold UCS, only just strong enough to elicit a UCR, should never result in the appearance of a detectable CR.* Finally, if conditioning is essentially a matter of association of CS and UCS, it does not necessarily follow that all the responses elicited by the UCS will automatically come to be elicited by the CS. It may be the case that certain response systems are simply not conditionable in this sense. Young (1965) has argued that the absence of pupillary conditioning when a change in illumination is used as the UCS is a consequence of the fact that pupillary responses cannot be directly conditioned. According to this argument, the use of shock as the UCS results in the conditioning of a general emotional reaction which happens to include, as one component, an increase in pupillary dilation.

If any or all of these arguments are accepted, then the results reviewed in the preceding sections must still be regarded as inconclusive. More experimental analysis will be needed to understand why weak or apparently neutral UCSs are often ineffective for conditioning.

---

* Perhaps a more sensitive measurement, such as the threshold technique used by Thomas (1971), or the summation procedure used by Reberg (1972), would in fact reveal some evidence of conditioning.

IV. THEORY OF REINFORCEMENT IN CLASSICAL CONDITIONING

A. Three Theoretical Analyses

The discussion of the preceding section can only, in the present state of knowledge, remain somewhat inconclusive. There is, indeed, some suggestion that a motivationally significant UCS is often necessary to produce reliable conditioning, but even if we were to accept this conclusion we should be far from understanding the role of the UCS in conditioning, the reason why a particular set of CRs appears, or the relationship between the CR and UCR. We shall now consider these more obscure theoretical questions.

It is possible to distinguish at least three different theoretical positions. The first is that of Pavlov, usually known as the "stimulus-substitution" theory. This theory states that a conditional reflex is reinforced by the presentation of the UCS or more specifically, that the pairing of CS and UCS leads to the formation of an association between them, such that the CS alone comes to elicit the set of responses normally elicited by the UCS. At the very least, then, the CR consists of components of the UCR and any function that it may serve, such as "preparing" the subject for the UCS, is purely incidental.* The CR, therefore, is not reinforced by its consequences. It is simply because the UCS elicits a UCR that its presentation increases the probability of the CR.

The Pavlovian theory may be contrasted with two variants of a position that seeks to interpret classical conditioning in terms of a general law of effect. According to both of these variants, CRs occur to the CS because they are followed by a rewarding state of affairs, in just the same way that instrumental responses are said to be strengthened because they are followed by rewarding events. Both versions of the theory can be found in the writings of Hull.

The first variant may be called the "superstitious reinforcement" theory (the term, of course, is Skinner's, 1948). According to Hull (1943), stimulus–response connections were strengthened whenever, in the presence of a particular stimulus, a response happened to be followed by a rewarding or drive reducing event. Classical conditioning satisfies this paradigm, since although the experimental operations make presentation of the UCS contingent upon a stimulus rather than upon a response, the

---

* Incidental, that is, as far as the mechanism of reinforcement is concerned. It is obviously not biologically incidental: the fact that animals are conditionable, i.e., that components of the UCR may occur in anticipation of the UCS, may well be of survival value.

fact remains that the response elicited by the UCS invariably occurs and is invariably followed by the reduction of a drive (e.g., the ingestion of food if the UCS is food, the termination of shock if the UCS is shock). The fact that there is no contingency between response and reinforcement is irrelevant: constant conjunction is sufficient to ensure that responses elicited by the UCS in the presence of the CS become strengthened and will soon occur in anticipation of the UCS. As with stimulus-substitution theory, the presumption is that the CR will normally consist of components of the UCR; however, in principle *any* response that happened to occur in the presence of the CS would be strengthened by the reduction in drive associated with the UCS, and would come to form part of the CR complex. It is not, of course, necessary to hold to a drive-reduction theory of rein-forcement in order to offer this analysis of classical conditioning. Skinner (1948) observed that if food were presented to pigeons at regular intervals, irrespective of their behaviour, they would come to emit regular, although idiosyncratic, patterns of behaviour just before the food was delivered. He argued that the absence of an experimental contingency between behaviour and reinforcement should not affect the workings of the law of effect: once a particular pattern of behaviour had occurred once or twice just before the delivery of food, it would become sufficiently strengthened to be emitted regularly and would thereby be further strengthened. The absence of the contingency made the behaviour "superstitious", but none the less strengthened by reinforcement. If a specifically arranged contingency between response and reinforcement is unnecessary for the strengthening of a response, then the fact that no such contingency is arranged in a classical conditioning procedure does not necessarily imply that a different principle of reinforcement must be invoked.

The third analysis, which we may call the "response-shaping" theory, argues that, whatever the experimenter's intentions, there is in fact a specific contingency between responses and reinforcement in a classical conditioning experiment. Hull (1929) noted the possible functional value of responses to a UCS which antedated that UCS: if the response to shock began to anticipate the delivery of shock, then, provided the response involved flight rather than freezing, the animal might avoid the shock altogether. In a classical conditioning experiment, of course, this is the one thing the subject cannot do but it is still possible to argue that the CRs that develop somehow serve to reduce the severity of an aversive UCS and that the form taken by the CR will depend upon this implicit contingency (hence the name "response shaping" for this theory). The general form of the analysis, then, is this: although the delivery of the UCS is programmed independently of the subject's behaviour, the *value* of the UCS may be affected by the occurrence of a CR. The aversiveness to the subject of an aversive UCS may be diminished by an appropriate CR; the attraction of an appetitive UCS may be increased by an appropriate CR. In a few instances the argument has some validity: in eyelid conditioning, an anticipatory lid closure may serve to protect the cornea from the puff of air; in salivary conditioning, anticipatory salivation may serve to make dry food more palatable. From this analysis, clearly, there is no reason to expect any particularly close similarity between CR and UCR: those CRs will develop which are most effective in increasing the value of the UCS.

The evaluation of these three theories can conveniently be divided into a discussion of two questions—the nature of the response in a classical conditioning experiment and the nature of the reinforcing event.

B. The Nature of the CR

If CRs are superstitiously reinforced according to the law of effect, there is relatively little constraint on their nature. If they are shaped by differential reinforcement so as to optimize the value of the UCS, then they might be recognized as somehow preparing the subject for the delivery of the UCS. The specification of what CR should emerge is at best intuitive and most often provided only after the fact. By contrast, stimulus-substitution theory makes explicit and strong predictions about the relationship between CR and UCR. Not surprisingly, then, it is the only theory confidently said to have been disproved. Tolman (1932), Hilgard (1936), Zener (1937), Culler (1938), Hilgard and Marquis (1940), Osgood (1953), Kimble (1961) and Black (1971) are only a few of the writers who have argued that stimulus-substitution theory is at best incomplete, at worst irrelevant, because CRs may differ from UCRs. Relatively few voices have been raised against this opinion: Hull (1934) noted that with the apparent exception of respiratory CRs, the resemblance between CR and UCR was very close; while Konorski (1967, pp. 268–70) has provided a relatively detailed analysis of some of the similarities.

1. *Specific Similarities and Dissimilarities between CRs and UCRs*

In many conditioning experiments, the recorded CR does indeed closely resemble the recorded UCR. Salivation in dogs to food or acid; leg flexion in dogs to shock to the paw; licking in rats to water; pecking in pigeons to food—all of these responses elicited by the UCS are routinely recorded as the CR to a stimulus preceding the UCS. The common argument that closer examination of the CR and UCR always reveals differences is very often untrue. Kellogg (1938) obtained tracings of leg flexion responses in a dog. As can be seen from Fig. 3.3a, the CR and UCR are indistinguishable. The CRs produced by morphine injections as the UCS include numerous components of the UCR to morphine, such as salivation, vomiting and shivering (Kleitman and Crisler, 1927). The form and force of a pigeon's pecking CRs are determined by whether the UCS is food or water: in the former case, the pigeon emits sharp, brief pecks with the beak opening just before impact, as in the consummatory response to food; in the latter case, responses are softer, and longer, with the beak almost closed, and are accompanied by licking and swallowing movements, as in the consummatory response to water (Jenkins and Moore, 1973). Culler *et al.* (1935)

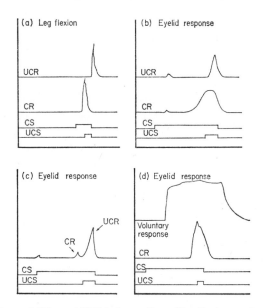

FIG. 3.3. Tracings of CRs and UCRs in studies of leg flexion and eyelid conditioning. (a) Leg flexion CR and UCR in dogs. (*After* Kellogg, 1938.) (b) Eyelid UCR and atypical CR in a single human subject. (*After* Hilgard and Campbell, 1936) (c) Eyelid CR and UCR in a human subject. (*After* Hilgard, 1936.) (d) CR and "voluntary" eyelid response in a human subject. (*After* Kimble, 1964.)

trained dogs on an avoidance contingency with a tone followed by shock to the paw unless a flexion response of sufficient amplitude occurred during the tone. On early trials, when the subject fails to avoid, an avoidance experiment simply involves a classical contingency. The authors noted:

> The animal begins, in five or ten trials, to show the first symptoms of conditioning; symptoms which are often indistinguishable from the activity elicited by the shock itself . . . Contrary to much recent opinion, we have been more impressed by the similarity than by the dissimilarity of the two. The CR is in most cases reduced in amplitude; but in some cases the two

are virtually indistinguishable. A dog may show precisely the same signs of disturbance when the tone begins as when he actually feels the shock; in fact, so realistic is the animal's distress that we have been deceived into thinking that the shock was accidentally being applied along with the tone (pp. 223-4).

By contrast, specific examples of CRs entirely different from UCRs are not always so impressive as has been argued, and, where impressive, may sometimes be interpreted without departing from stimulus-substitution theory.

Salivary CRs are sometimes said to be less viscous than, and to contain different constituents from, salivary UCRs. Since, however, the viscosity and constituents of salivation depend upon the rate of salivation and change during the course of prolonged salivation (Gormezano, 1966), these differences may only reflect the fact that the UCR is larger and lasts longer than the CR. Such quantitative differences, even if here having qualitative by-products, are hardly sufficient evidence against the essential similarity of CR and UCR.

Hilgard (1936), Hilgard and Marquis (1940) and Kimble (1961) all argued that the CR and UCR in human eyelid conditioning are recognizably different. Some of the differences, such as those in latency or amplitude, seem again to be only quantitative. Others are of doubtful validity: Hilgard (1936) and Hilgard and Marquis (1940, p. 38) reproduced tracings of CRs and UCRs from an experiment by Hilgard and Campbell (1936). These are shown in Fig. 3.3b. The difference in slope of the response is substantial. The CR depicted here, however, appears to be quite atypical and was specifically acknowledged as such by Hilgard (1936, p. 380), who observed that it represented the "occasional" response pattern of one subject late in conditioning. Figure 3.3c shows more typical CRs and UCRs recorded by Hilgard and Campbell (1936) and Fig. 3.3d shows CRs and "voluntary" lid closures recorded by Kimble (1964, p. 36). Both these CRs resemble the unconditional eyeblink rather closely, although in the former case the difference in amplitude is considerable. The voluntary response depicted in Fig. 3.3d raises one of the more intractable problems in studies of eyelid conditioning in human subjects. Since closure of the eyelid, timed appropriately, will prevent a puff of air from striking the cornea, the typical eyelid conditioning experiment, unless it uses electric shock as the UCS, obviously permits differential response-contingent reinforcement to occur. The assumption has been that responses subject to this contingency will be voluntary and that they may be distinguished from true CRs by differences in latency (Spence and Ross, 1959) or slope (Hartman and Ross, 1961). The problems with these criteria have been amply documented by Gormezano (1965) and Goodrich (1966). For present purposes, the problem must be that response-contingent reinforcement may occur in eyelid conditioning and may interact with a simple stimulus-substitution principle to determine the form of responses classified as CRs. The special nature of eyelid conditioning, however, should make one pause before generalizing from this conclusion, itself no more than speculative, to other instances of conditioning.

A final, commonly cited example of differences between CRs and UCRs comes from heart rate conditioning. Black (1971), following several earlier authorities, has pointed out that although the UCR to shock is uniformly an increase in heart rate, CRs are sometimes an increase and sometimes a decrease in rate. The problem with heart rate as a dependent variable in conditioning studies is that it is clearly affected by a variety of other responses. Zeaman and Smith (1965) have presented evidence rather strongly suggesting that, in human experiments, deceleration as the CR may depend upon changes in respiration during the CS: when subjects were trained to hold their breath throughout the duration of a 20-sec CS, the CR was a uniform acceleration in heart rate; with normal respiration the CR was a brief acceleration followed by deceleration.

In animal studies, there is good evidence that the direction of the change in heart rate during a CS signalling shock depends on whether the CS elicits an increase or a decrease in overall activity. An accelerative CR may occur in unrestrained rats when the CS elicits running and jumping (e.g., Black and Black, 1967); a decelerative CR occurs either in restrained animals (e.g., Fitzgerald et al., 1966a), or when the CS is imposed upon ongoing appetitive activity, as in a conditioned suppression experiment, and therefore produces an overall decrease in activity (de Toledo and Black, 1966; Parrish, 1967). A particularly clear example of this comes from an experiment by Borgealt et al. (1972). They found that a CS paired with shock resulted in an acceleration of heart rate, but that when the same CS was superimposed upon ongoing lever pressing, it both suppressed responding and resulted in a deceleration of heart rate. Similar results have been reported by Black and de Toledo (1972), and by Sutterer and Obrist (1972). Black and de Toledo studied conditioned suppression in rats and obtained uniformly high correlations between decreases in activity and heart rate deceleration during the CS; they also trained dogs to avoid shock either by pressing a pedal or by standing absolutely still. In the former case heart rate accelerated during the warning stimulus, in the latter case it decelerated. Sutterer and Obrist gave discriminative classical aversive conditioning to dogs with the same CS+ and CS− presented in two different experimental chambers. In one of the chambers the dogs had also been trained to press a panel for food and the CS+ therefore suppressed ongoing activity; in the other, CS+ elicited some activity. The results for one dog are shown in Fig. 3.4: when CS+ elicited activity, heart rate accelerated; when CS+ suppressed activity, heart rate decelerated. In general the correlations between changes in activity and changes in heart rate are striking. Taken together, these results suggest that changes in heart rate are not themselves directly conditioned but may be solely a consequence of conditioned changes in other response systems such as general activity (a possibility suggested earlier by Smith, 1954, and Shearn, 1961).

It must not be supposed that changes in heart rate are solely produced by proprioceptive feedback from overt skeletal responding. As noted earlier, it is well documented (e.g., Black, 1965) that accelerative heart

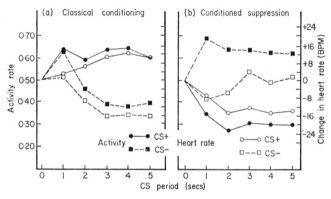

FIG. 3.4. The relationship between changes in heart rate and general activity during aversive conditioning. When CS+ suppresses activity, it results in a decrease in heart rate; when it increases activity, it results in an increase in activity. Similarly, CS− affects heart rate in the same direction as it affects activity. (*After* Sutterer and Obrist, 1972.)

rate CRs will occur in dogs that are completely curarized. There are two possibilities: first, that heart rate does change directly as a consequence of CS–UCS pairings but that these accelerative CRs are often obscured in unparalysed animals by the stronger influence of skeletal activity on heart rate; second, that heart rate is not directly conditioned but is influenced by central correlates of skeletal activity. Black and de Toledo, who favour the second alternative, suggest that the hippocampal correlates of voluntary activity reported by Vanderwolf (1971) may underlie the changes in heart rate. If this is so, then although changes in heart rate may be a perfectly useful measure of learning, they will only be indirect: like the pupil (p. 97), the heart itself may be unable to learn.

In conclusion, then, the evidence against stimulus-substitution theory is surprisingly sparse. There is often a close similarity between CR and UCR, and where there is not this may be because the response being studied is either unusually susceptible to shaping by differential reinforcement or is largely determined by other CRs which are more directly conditioned.*

There is no point in insisting that CRs will always be found to be the same as the responses elicited by the UCS. The preceding arguments were directed more towards showing that many commonly accepted examples of differences between CR and UCR do not stand up well to closer scrutiny. It is far too early to say whether this will be generally true in all cases. What is undoubtedly required is a more careful analysis of other response systems. For example, if rats are placed in a small chamber within which they are allwed to move around, the delivery of shock to the feet elicits flinching, followed by prancing or jumping. The response conditioned to a CS signalling such a shock, however, is usually described as freezing or crouching (p. 82). Whether freezing is one component to the flinching response, or whether CR and UCR are here quite different, is simply not clear at the present time.

2. *Diversity of CRs during a CS*

A second common argument against stimulus-substitution theory is that the experimenter, by recording only the salivary responses in conditioning to food, virtually ensures that the CR will resemble the UCR. The scope for detecting differences between two drops of saliva is not great, but it is implied that sufficiently detailed recording of an unrestrained animal's behaviour during the CS would reveal a whole range of responses which did not occur to the UCS and which could all be described as evidence that the animal expected, and was preparing for, the arrival of the UCS. The *locus classicus* for this argument is an article by Zener (1937), who made such a recording, and concluded:

> The lack of identity in effector terms of conditioned and unconditioned behaviour even in the training situation, but especially when the situation is varied, is considered to be incompatible with all stimulus substitute theories of conditioning. Assertions of identity of behaviour can only be maintained if observation is restricted to the secretory component (p. 402).

Zener's thesis was that less restricted observation revealed behaviour during the CS "anthropomorphically describable as looking for, expecting, the fall of food with a readiness to perform the eating behaviour which will occur when the food falls" (p. 393).

---

* Heart rate may not be the only response system more influenced by other responses than by direct conditioning; it is possible that the same applies to changes in respiration. Here, too, as pointed out by Tolman (1932), Hull (1934) and Hilgard (1936), the changes that occur to the CS are not necessarily the same as those that occur to the UCS (Upton, 1929; Wever, 1930).

An examination of Zener's protocols does not provide particularly overwhelming support for these claims. During the CS, animals were observed to turn their head sometimes towards the CS (a bell) and sometimes towards the food tray; they might oscillate between the two, but would generally increase their orientation towards the food tray as the CS continued. Similar orientations of the entire body, including approach towards the bell or towards the food tray, were also observed. Chewing and licking responses occurred, but not, according to Zener, as frequently as Pavlov implied. Sometimes the dogs showed restless behaviour, and sometimes they panted and yawned.

The only components of this pattern of behaviour that could reasonably be described as a preparation for eating are the orientation and approach towards the food tray. Orientation and approach towards the bell are hardly the most natural indicators of an expectation of the imminent arrival of food in the tray: they look more like a tendency to treat the CS as if it were the UCS. Chewing and, of course, salivation form part of the complex of responses elicited by food in the mouth. Finally, it is not obvious what interpretation Zener puts on panting and yawning—neither of which, on the face of it, are obvious forms of preparation for eating.

Zener argued that approach to either bell or food tray is inconsistent with stimulus-substitution theory, since neither formed part of the UCR. If true, this can have been only because the dog's nose was already in the food tray when the food was delivered: approach to the tray was surely part of the response elicited by delivery of food early in training. Indeed, Razran (1971, p. 20) cites several Russian studies said to show that the UCR of approaching the odour of food may develop at an earlier age than that of salivation. It remains to consider why the dogs should sometimes have approached the bell and sometimes the food tray. Zener seems to imply that approach to the tray is the decisive indicator of an expectancy and therefore the decisive evidence against stimulus-substitution theory. Ironically enough, it has recently been argued that although the form of the pigeon's pecking response conditioned to a stimulus signalling food may closely resemble the UCR to food, and although it appears to be strengthened by a Pavlovian principle of reinforcement, the fact that it is directed towards the key light that serves as the CS is decisive evidence against any Pavlovian explanation (Williams and Williams, 1969; Staddon and Simmelhag, 1971). Moore (1971), however, has persuasively argued that CRs directed towards either the CS or the site of an appetitive UCS are equally interpretable in Pavlovian terms. Pavlov himself certainly noted that dogs would approach and lick the CS (Pavlov, 1941, pp. 120, 150), and had no doubt that the occurrence of such CRs conformed to the principles of stimulus-substitution theory: a dog that licks the CS is treating the CS as though it were the UCS. Equally, if a CR is directed towards the site of the UCS this is surely an instance of behaviour, usually elicited by the UCS, beginning to be elicited by the CS. In any particular situation, whether CRs will be directed to the CS or to the site of the UCS, may well depend, as Moore argues, on procedural and experimental details. A localized CS is more likely to elicit directed CRs than a diffuse visual or auditory stimulus. Even a localized CS, however, must compete with the site of the UCS for control of directed responses. If the site of the UCS is available and visible during intertrial intervals, it may elicit CRs which will then extinguish. However, if the intertrial interval is spent in total darkness and a light is used as the CS, then the site of the UCS will be visible only during CS periods and will be just as reliable a signal of the delivery of food as the experimenter's CS. Under these conditions, one might expect the site of the UCS to control directed CRs more effectively than the nominal CS. This is exactly what has been observed in some unpublished studies of auto-shaping that I have undertaken: if the chamber is illuminated during intertrial intervals, pigeons direct their pecking responses at the key light (the nominal CS); if it is dark, they tend to peck at the magazine (cf. Wasserman, 1973). If there is no external CS at all, but food is presented at regular intervals, pigeons end up pecking the wall near the magazine just before the delivery of food (Staddon and Simmelhag, 1971).

Zener's remaining observations were that the recorded CRs included too much panting and yawning and not enough chewing and eating movements. This may well be related to his use of a relatively long, 15 sec, CS. Pavlov (1941, p. 120) seems to have implied that the motor components of the UCR to food such as chewing would become conditioned only with a short CS (cf. the discussion of the effect of interstimulus interval on different CRs, p. 65). The restlessness observed by Zener often occurred at the onset of the CS: the fact that similar patterns of behaviour were observed in extinction suggests that during conditioning they may have been evidence of inhibition of delay. The real problem posed here for stimulus-substitution theory is one of accounting for the apparent differences in the CRs, such as conditioned suppression or enhancement, that appear when relatively long CSs are used and the discrete, skeletal CRs, such as leg flexion or eyelid closure, that appear only when very short CSs are used. Konorski (1967) is the only writer who has shown an awareness of this problem in his distinction between preparatory and consummatory conditioning (p. 21).

3. *Conclusions*

In spite of almost universal rejection, stimulus-substitution theory is strikingly resilient: the variety of responses elicited by a CS closely matches the variety of responses elicited by the UCS. Apparent exceptions become less convincing on closer inspection and attempts to read into an animal's behaviour signs of expectation and preparation tell us more about the experimenter's preconceptions than the subject's behaviour. It is interesting that it was precisely because of the temptation to indulge in such interpretations that Pavlov deliberately ignored the skeletal components of CRs and concentrated on the salivary component (Pavlov, 1927, pp. 17–18).

From what has been said, however, it is clear that stimulus-substitution theory is not without its problems. The most reasonable conclusion is not that it should be rejected out of hand, but that the problems mentioned should be studied more thoroughly: some of those that have been earlier thought to be intractable have proved surprisingly amenable to analysis; it is possible that the remainder can also be dealt with.

It remains true that dogs do not routinely bite off, chew and swallow the lights used as CSs in experiments on salivary conditioning. The furthest they go is occasionally to lick them.* In general, a CS may not come to elicit all of the behaviour elicited by the UCS. This is probably because the nature of the behaviour elicited in anticipation of the UCS will depend upon the nature of the stimuli available: certain responses require certain supporting stimuli. Monkeys or rats, when exposed to electric shocks, will attack a live or stuffed conspecific (p. 18). Such aggressive behaviour has not been reported in standard experiments on conditioned suppression. Presumably, therefore, it depends upon the presence of suitable stimuli (cf. Azrin et al., 1967; Galef, 1970). The same principle may serve to explain why the consummatory behaviour elicited by a simple visual stimulus paired with food does not usually include all components of the UCR to food.

It does not seem particularly difficult to accept the idea that the nature of the CR conditioned to a CS might depend upon the extent to which that CS supports particular components of the UCR. There is excellent evidence that the nature of the UCR elicited by a UCS, and hence the nature of the CRs conditioned by that UCS, is greatly affected by the method of presenting the UCS, and the situation in which it occurs. Food delivered into a tray, for example, elicits approach responses from a dog as well as salivation; injected into the mouth, it elicits only salivation. The CRs conditioned to a stimulus signalling food are equally dependent on the method of presenting the food (Kierylowicz, et al., 1968). Similarly, grid shock delivered to the feet of a rat confined in an enclosed chamber

---

* But turkeys, porpoises and whales, according to Breland and Breland (1966, pp. 68–9), have been known to swallow coins, rubber balls and inner tubes when these stimuli had been associated with food reinforcement.

elicits flinching and jumping, and the response conditioned to a stimulus signalling such a shock consists of freezing and crouching (p. 82). But if the shock is administered from a discriminable and localized source (Blanchard and Blanchard, 1969b, 1970) or in an apparatus providing the possibility of escape to a place of safety (Wickens and Wickens, 1942; McAllister and McAllister, 1962a, b; Baum, 1966), both the responses elicited by shock, and the responses conditioned to any associated stimuli, consist of escape or flight rather than freezing. In some situations, both freezing and escape responses may be conditioned: Sheffield (1948) recorded both types of response when guinea-pigs were shocked in a rotating wheel.

It has often been claimed that the evidence thought to support stimulus-substitution theory is based on too narrow a concentration on limited components of the CR. It is possible, however, that the evidence thought to contradict stimulus-substitution theory is based on too narrow a concentration on limited components of the UCR. At the very least, our understanding of classical conditioning, and of its importance, might be greatly advanced by a more careful and extensive analysis of the range of responses elicited by appetitive and aversive reinforcers under various conditions.

C. The Nature of Reinforcement

Perhaps the most important, and certainly the least tractable, question that is raised by the analysis of classical conditioning concerns the process of reinforcement responsible for the appearance of the CR. Pavlov's theory of reinforcement, as we have noted, is extremely simple: the presentation of an effective UCS, by reliably eliciting a UCR, increases the probability that components of that UCR will be elicited by preceding stimuli. In contrast to this, several theorists have argued that the law of effect is sufficient to account for all changes in behaviour resulting from associative learning: both classical CRs and instrumental responses are assumed to be modified by their consequences. It is clear, therefore, that the question at issue here merges into the question of the relationship between classical and instrumental conditioning. If a single principle of reinforcement were sufficient to explain all learning, there might be little reason left for regarding the operational distinction between classical and instrumental experiments as of any fundamental significance.

1. *Relationship between CR and UCS*

According to stimulus-substitution theory, the UCS strictly determines the nature of the CR: only responses elicited by the UCS, or components or by-products of such responses, will become conditioned to the CS. Some similarity between CRs and UCRs would also be expected from the law of effect if only because, in the absence of other competing behaviour, the responses elicited by the UCS always occur in the experimental situation and are therefore available to be strengthened by subsequent reinforcement. But if other responses reliably occurred in close proximity to reinforcement, they too would be strengthened. In any situation, therefore, where the CS itself reliably elicited certain responses it is hard to see why, according to the law of effect, these responses should not themselves be strengthened by the reinforcement provided by the UCS.

Konorski and Miller (1937) were the first to see the implications of the law of effect in such a case. Suppose, they argued, the CS were a moderately weak electric shock which unconditionally elicited flexion of the leg and this CS was paired with the delivery of food. If the delivery of food reinforces all immediately preceding responses, it is clear that the flexion response should be strengthened. According to the experiments of Erofeeva (described by Pavlov, 1927, pp. 29–31) this does not happen:

> A strong nocuous stimulus—an electric current of great strength—was converted into an alimentary conditioned stimulus, so that its application to the skin did not evoke the slightest defense reaction. Instead, the animal exhibited a well-marked alimentary conditioned reflex, turning its head to where it usually received the food and smacking its lips, at the same time producing a profuse secretion of saliva (Pavlov, 1927, pp. 29–30).

Experiments by Brogden (1939a) and Davydova (1967) have confirmed this conclusion. Davydova, studying salivary conditioning in dogs with food as the UCS, used electrical stimulation of the motor cortex as CS. Although this CS initially elicited leg flexion, she found that the threshold for eliciting this response to the CS increased significantly during the course of conditioning. Thus even though the flexion response was reliably followed by food, its probability decreased rather than increased. Brogden employed a somewhat different procedure, and since his study has usually been cited as evidence for the law of effect (e.g., by Hilgard and Marquis, 1940, p. 86; Osgood, 1953, pp. 321–2), it is worth describing the procedure and outcome in detail. Dogs were trained to give a flexion response to a bell in order to avoid shock. Once the response was established the shock was disconnected and the subjects were then reinforced with food, provided that they gave a flexion response in the presence of the bell. In spite of the fact, therefore, that instrumental contingencies of reinforcement operated in both phases of the experiment, Brogden observed considerable disruption of performance when the reinforcement changed from shock avoidance to food presentation.

As food continued to appear in the food-box after each flexion response to the bell, there was a tendency for flexion to be of short duration or completely absent, replaced by the immediate turning of the head to the food-box in anticipation of food. . . . It was necessary in several cases to return to reenforcement with shock in order to reestablish a persistent flexion for the duration of the bell (Brogden, 1939a, pp. 47–8).

More recently, Konorski and Szwejkowska (1956) have shown that appetitive CSs may be turned into aversive CSs, and vice-versa, and that although such learning may be slow and protracted, the time course of the transfer is similar to that required for turning an excitatory CS into an inhibitory one. In the end, the response elicited by the CS is determined by the current UCS, although, according to the law of effect, the substitution of shock for food as UCS should only have strengthened the already established salivary CR.

It is difficult to resist the conclusion that the most important point about classical conditioning is entirely missed by any account which attempts to reduce conditioning to the operation of superstitious reinforcement of responses that happen to antedate the occurrence of reinforcement. In the absence of a specific contingency between some other response and reinforcement, the presentation of a reinforcing stimulus determines the class of responses that will be strengthened: indeed, as we shall see later, even the existence of such a contingency is not always sufficient to override this effect. This conclusion is not discredited by any apparently contrary evidence from the study that did most to establish the principle of superstitious reinforcement. Although Skinner (1948) claimed that the occasional presentation of food to hungry pigeons would ensure the development of whatever idiosyncratic pattern of behaviour each bird happened to be emitting when reinforcement occurred, subsequent analysis has suggested that this conclusion is somewhat misleading. When Staddon and Simmelhag (1971) repeated Skinner's experiment, they found that if they presented food independently of the pigeons' behaviour at fixed intervals of time, although different birds came to fill much of each interval with idiosyncratic patterns of behaviour, as the moment for food reinforcement approached, all birds showed an increasing tendency to peck at the wall near the food magazine. Pecking, of course, is the response elicited by food.

2. Reinforcement as Drive Reduction

Historically, the most influential application of the law of effect to classical conditioning was probably Hull's. According to Hull (1943), reinforcement in classical conditioning, as in instrumental learning, is dependent upon the reduction of some drive: effective UCSs are those whose onset (in the case of food or water) or offset (in the case of an electric shock or puff of air) involve the reduction of a drive. While this analysis may seem plausible enough in the case of appetitive conditioning, in aversive conditioning it makes the paradoxical assumption that it is the *termination* of the aversive UCS that is the reinforcing event: fear is said to be conditioned to a warning stimulus because a fear response is elicited by the aversive UCS and is then stamped in by the drive reduction resulting from the termination of the aversive UCS.

This analysis has been challenged by Mowrer (1947, 1950, 1960a). At least three pieces of evidence suggest that it cannot be correct. The first, most popular, and least convincing experimental attack has been to vary the duration of the aversive UCS on the assumption that if reinforcement depends upon offset of the UCS, the longer the UCS the greater the delay of reinforcement. Mowrer and Solomon (1954) compared conditioning to a 3 or 10 sec shock and found little or no difference between the two groups. Several studies of GSR, heart rate and eyelid conditioning with humans have confirmed this conclusion (Bitterman *et al.*, 1952; Wegner and Zeaman, 1958; Runquist and Spence, 1959). If a longer UCS leads to greater delay of reinforcement, it should decrease the rate of conditioning. On the other hand, as Runquist and Spence (1959) pointed out, a longer shock may generate a stronger drive, so that the reinforcement, although delayed, will be stronger. Since there does not seem to be any answer to this possibility, the uniformly negative results of these studies do not pose a crucial problem for drive-reduction theory.*

Although these studies, therefore, provide no conclusive evidence one way or the other, a series of studies initiated by Mowrer and Suter (Mowrer, 1950, pp. 278–93), Barlow (1952) and Mowrer and Aiken (1954) is more decisive. Mowrer and Aiken used a relatively long shock and a short CS, arranging the CS to coincide either with the onset or with the offset of the shock: they found, as did Barlow, that a CS paired with the onset of shock was more effective in disrupting appetitively reinforced lever pressing than one paired with its offset. Davitz (1955) confirmed this result, with precautions taken to ensure that drive reduction could not have occurred immediately after the onset of the shock by the rat's jumping

---

* A more recent study by Strouthes (1965) claimed that an increase in length of UCS did actually retard the conditioning of fear, but the measure employed to assess fear of the CS was so idiosyncratic that it is difficult to accept this conclusion. Rats were placed in an alley, the CS was turned on when the door of the start box was opened and turned off when the rat entered the goal box. Strouthes assumed that strong fear of the CS would interfere with running and indeed found that longer intervals between the onset of the CS and offset of the UCS during conditioning led to faster alley speeds during testing. It is difficult to see, however, why greater levels of fear should not have led to faster running, since the CS was turned off as soon as the subject had traversed the alley.

off the electric grid. If, as Hull's theory requires, it is the offset of shock that is the reinforcing event in aversive conditioning, then the closer the CS occurs to this point, the stronger the expected conditioning. These results, therefore, provide strong evidence against drive-reduction theory. The results of an experiment by Segundo et al. (1961) strengthen the argument. They found that a CS occurring just before the offset of a shock of variable length would inhibit fear responses. The implication is clear: both the onset and offset of shock are reinforcing events which will support conditioning to associated stimuli. They differ, however, in the nature of the conditioning they will support: a CS paired with the onset of shock elicits fear; a CS paired with the offset of shock elicits relief (the term used by Mowrer, 1960a) or relaxation (Denny, 1971). This in turn implies that no *single* change in drive level is the necessary condition for reinforcement: conditioning, as Mowrer argues, can occur to either an increase or to a decrease in drive.

3. *Response-contingent Reinforcement in Classical Conditioning*

One way of escaping this conclusion is to argue that reinforcement in aversive conditioning occurs because the CR succeeds in reducing the aversiveness of the UCS. Aversive conditioning is viewed as avoidance learning: closure of the eyelid before the onset of the airpuff, for example, is thought to be reinforced because it succeeds in preventing the airpuff striking the cornea; salivation before the injection of acid into the mouth is reinforced because it prevents undiluted acid entering the stomach. As was noted above, this analysis may also be applied to appetitive conditioning: salivation may also enhance the taste or digestion of dry food.

Relying on these few but plausible examples, a number of writers (e.g., Culler, 1938; Prokasy, 1965; Perkins, 1968) have argued that all classically conditioned responses may be reinforced in this way. But the fact that *some* responses *may* be reinforced by their consequences in classical conditioning experiments does not imply that all CRs are strengthened by response-contingent reinforcement. It is equally easy to provide instances of CRs to which no very plausible instrumental contingency can be ascribed. It is relatively difficult to see how flexing the leg succeeds in reducing the aversiveness of shock to the paw or how closing the eyelid or nictitating membrane reduces the aversiveness of paraorbital shock. It is even harder to see how an attack on an inanimate object reduces the aversiveness of shock, or how salivation and nausea reduce the aversiveness of an injection of morphine. Nor is it easier to see the general benefits resulting from appetitive CRs: while salivation may improve the taste of dry food, a rat seems to gain little by making licking movements (or a rabbit by making jaw movements) before the delivery of water, and there seems to be no advantage to the pigeon in pecking a key light which signals the delivery of food.

The fact that some CRs are adaptive is hardly surprising: most biological systems share this property. It does not imply that the mechanism of reinforcement in the individual organism involves selective strengthening of adaptive responses and elimination of maladaptive or indifferent responses. In fact, the plausibility arguments of this section can be supported by some more direct evidence to which we can now turn.

4. Omission Training of CRs

However the law of effect is applied to classical conditioning, whether the coincidental occurrence of a reward is said to strengthen immediately preceding responses superstitiously or whether CRs are said to be differentially reinforced to the extent to which they affect the value of the UCS, a clear prediction is that the probability of a CR will be affected by deliberately arranging a contingency between CR and UCS. A long series of studies, going back to those of Schlosberg (1934, 1936), has examined this question, originally in an attempt to discover whether classical or instrumental contingencies of reinforcement were more important. The design required to answer this question seemed simple enough. One group of subjects was exposed to a straightforward classical contingency, with a CS signalling a shock regardless of the subject's behaviour in the presence of the CS, while a second group was exposed to an omission schedule, where the CS signalled a shock only if no designated CR occurred in the presence of the CS. With an aversive reinforcer, the omission schedule is more naturally described as an avoidance schedule: the subject can avoid shock by responding appropriately during the CS. The term omission schedule, however, initially suggested by Sheffield (1965), has the virtue of greater theoretical neutrality and of applying more naturally to experiments using appetitive reinforcers. Here, a classical schedule might involve presenting a CS which signalled food regardless of the subject's behaviour during the CS, while on an omission schedule the CS would signal food only if no CR occurred on that trial.

The Pavlovian principle of reinforcement implies that the classical contingency will result in more efficient learning than the omission schedule. On the omission schedule, the CS may initially signal reinforcement quite consistently but as soon as CRs start to occur, the probability of the UCS declines and the subject is essentially on a partial reinforcement schedule. As we have seen (p. 72), classical conditioning is usually affected adversely by partial reinforcement.

If classical CRs are reinforced by their consequences, however, the situation should be entirely different. If the delivery of an appetitive reinforcer is contingent upon the absence of a particular response, then it is hard to see how that response could ever become established. Fortuitous occurrences of the response would lead to the omission of reward and the response should rapidly be suppressed. Thus if dogs were rewarded with food in the presence of a stimulus only if they refrained from salivating, they could learn not to salivate. On the other hand, when an aversive reinforcer is used, one might expect the omission of shock to be a rewarding event: hence an omission schedule of aversive reinforcement should maximize the rewarding effects of a CR.* If closure of the eyelid in classical conditioning is really reinforced because it attenuates the aversiveness of the UCS, performance can only benefit from the addition of a contingency which specifies that the UCS will be omitted on trials when the subject blinks.

The evidence appears at first sight confusing, for different studies have yielded different outcomes. In spite of these conflicts, however, there is good reason to conclude that classical CRs are not simply modified by their consequences in the manner required by the law of effect. In too many cases, omission schedules of appetitive reinforcement have resulted in the acquisition and maintenance of responses whose sole effect is to prevent the delivery of food or water; other studies employing aversive reinforcers have frequently found that the addition of an avoidance contingency may decrease rather than increase the probability of a CR.

The earliest studies were those of Schlosberg (1934, 1936). In two experiments, Schlosberg conditioned rats to an aversive UCS—in one study the CR was tail flexion, in the other, leg flexion. In neither experiment did learning occur very reliably, but in both cases the classical schedule with shock occurring on every trial led to a higher level of responding than the omission schedule. Schlosberg explicitly noted that the omission schedule led to a cyclical fluctuation in the probability of a CR: as soon as CRs started to occur, the shock was omitted and they extinguished; when CRs failed to occur shock was presented, leading to a renewed increase in the probability of responding. This is precisely what a Pavlovian analysis would lead one to expect.

Schlosberg's conclusions, however, have not been substantiated in the majority of other studies in which shock has been used as the aversive reinforcer. Although there are considerable discrepancies between the outcomes of different studies of leg flexion conditioning, several experiments in which a running response has been conditioned either classically

---

* This ignores the problem, sometimes thought to be insoluble, of how the omission of any event, whether aversive or not, could come to act as a reward. The problem is not really so intractable as it has been made out to be (see Chapter 6). In any event, since some responses are reliably established by avoidance schedules, the problem must, in one way or another, be considered.

E

or on an omission schedule have uniformly found the omission schedule to be more effective. The most widely cited study is that of Brogden *et al.* (1938), in which guinea-pigs ran in a rotating wheel. On the omission schedule, animals rapidly learned to run on 100 per cent of trials; on the classical schedule, animals never learned to respond on more than 50 per cent of trials. Brogden *et al.* concluded that running must have been reinforced by the omission of shock. Hunter (1935) had earlier shown that rats would learn to run in a circular maze more efficiently on an omission than on a classical schedule; a similar outcome was reported by Bolles *et al.* (1966) with rats required to run in a rotating wheel, and by Kamin (1956, 1957a) with rats running in a shuttle box.

More recently, Wahlsten and Cole (1972) compared the effects of

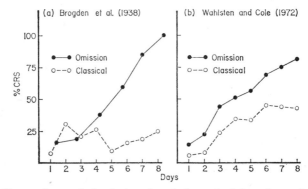

FIG. 3.5. Comparisons of classical and omission schedules of aversive reinforcement. (a) Wheel running in guinea-pigs. (*After* Brogden *et al.*, 1938.) (b) Leg flexion conditioning in dogs. (*After* Wahlsten and Cole, 1972.)

classical and omission schedules on leg flexion conditioning in dogs. They too found that the omission schedule produced a very much higher level of responding than the classical schedule. The differences between the two schedules in some of these experiments have been substantial. Fig. 3.5, illustrating the results of Brogden *et al.* and Wahlsten and Cole, shows that an omission schedule may lead to essentially perfect performance, while a classical schedule may maintain only a low probability of responding.

These results, then, clearly suggest that a response which is only imperfectly established when its occurrence has no explicit consequences, is learned very efficiently when a specific instrumental contingency is added. The fact that some responses are affected by the explicit instrumental contingency of an avoidance schedule, however, does not imply that all CRs are established by implicit instrumental contingencies. Schlosberg's results, therefore, stand as some evidence that flexion CRs in rats may be established, albeit imperfectly, without the benefit of any such implicit instrumental contingency. Moreover, the failure to establish a running response reliably when a classical schedule is employed is not entirely surprising: in many situations, the CR conditioned to a stimulus signalling shock is freezing rather than running. In a detailed replication of the experiment of Brogden et al., Sheffield (1948) analysed his subjects' performance on a trial-by-trial basis, with particular attention to the question of the UCR elicited by shock on each trial. A Pavlovian analysis, Sheffield argued, implies that the response whose probability will increase on trial n + 1 is that which occurs as the UCR to shock on trial n. The trial-by-trial analysis confirmed this prediction: when subjects ran in response to the UCS on trial n, the probability of a running CR on trial n + 1 increased by 0·14; when the UCS elicited freezing on trial n, the probability of running on trial n + 1 declined by 0·18.

The results of Wahlsten and Cole's (1972) study are more surprising.* Flexion conditioning has traditionally been treated as a standard classical procedure, but their finding that the flexion response is so readily affected by an explicit instrumental contingency clearly raises the possibility that even on an ostensibly classical schedule, responding is reinforced because it somehow reduces the aversiveness of the UCS. As Wahlsten and Cole note, however, the fact that a response *can* be modified by an explicit instrumental contingency does not prove that it *is* modified by an implicit instrumental contingency during classical conditioning. Several observations suggest that the explicit instrumental contingency results in a radical re-structuring of the animal's behaviour rather than merely exaggerating the effects of an already effective, implicit instrumental contingency. Kellogg (1938), in an earlier study of omission training of the dog's flexion response, noted that very early in conditioning (when the omission schedule is essentially the same as a classical schedule), the form of the flexion response was indistinguishable from the UCR elicited by shock (see Fig. 3.3). Once the avoidance contingency began to take hold, however, the response changed to a new form, involving a more rapid, but smoother, lifting of the leg. Wahlsten and Cole have amply documented this point. They recorded latencies of responding in both classical and

---

* One of the more surprising features of their results is the ineffectiveness of their classical schedule. Classical conditioning of flexion CRs in dogs routinely results in an asymptotic probability of responding of over 0·90 (e.g., Brogden, 1939b; Whatmore et al., 1946). In Wahlsten and Cole's experiment, the asymptote was 0·46. Moreover, although Whatmore et al. reported that when dogs were concurrently trained to flex one leg on an omission schedule and the other on a classical schedule the avoidance response interfered with the classical CR, their data suggest that when animals were trained on one schedule alone, the classical schedule produced equally rapid learning.

omission trained subjects, both before and after each subject attained an arbitrary criterion of learning. As can be seen from Fig. 3.6, the performance of the two groups before reaching criterion was relatively similar: as soon as they attained criterion, however, subjects on the omission schedule showed an abrupt decline in latency of responding, whereas subjects on the classical schedule continued to respond, as before, relatively slowly. The implication is that before reaching criterion, both omission and classical subjects are reacting to the situation in much the same way, but that as soon as the avoidance schedule takes hold, the behaviour of the omission group suddenly changes.

There is a further reason for suspecting that when leg flexion responses are trained on a classical schedule, they are not being affected by any

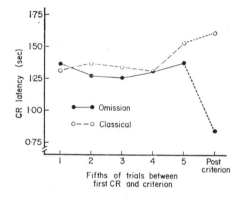

FIG. 3.6. Latency of leg flexion CRs in dogs trained on classical and omission schedules, both before reaching criterion and after criterion has been reached. (*After* Wahlsten and Cole, 1972.)

implicit instrumental contingency. If a classical schedule does permit the operation of such implicit instrumental contingencies, then animals that have received extended training on a classical schedule should show appropriate changes in behaviour when the UCS is modified by the experimenter on occasional test trials. In one of their experiments on flexion conditioning in sheep, however, Liddell *et al.* (1935) reported that a decrease in the intensity of shock on one trial reliably decreased the amplitude of the CR on the next trial. Soltysik and Jaworska (1962), in a study of leg flexion conditioning in dogs, interspersed two kinds of test trials among continued retraining trials. When the UCS was completely omitted on some test trials, the latency of the CR increased on the following trial by 15–25 per cent. When the UCS was increased in intensity, however, latencies decreased by 15 per cent. A reduction in the intensity of shock, or its complete omission, therefore, weakens rather than strengthens the CR; an increase in intensity of shock, which would appear to be a procedure analogous to instrumental punishment, strengthens rather than weakens the response.

Although, therefore, leg flexion responses can be affected by instrumental contingencies, the available data suggest that the imposition of such a contingency has detectable effects on the form and latency of the response and that, when trained on a classical schedule, there is little reason to believe that the response is being modified by its consequences. Studies of eyelid conditioning permit an ever stronger conclusion. Although we have frequently noted that closure of the eyelid must in fact protect the cornea from exposure to a puff of air, studies of eyelid conditioning in both humans (Logan, 1951; Kimble et al., 1955) and rabbits (Gormezano, 1965; Gormezano and Coleman, 1973) have uniformly found that the deliberate imposition of an omission or avoidance schedule produces a level of responding notably inferior to that maintained by a classical schedule of reinforcement. Such a finding seems to imply that the response does not benefit from an explicit instrumental contingency and a fortiori, therefore, that it is not being maintained by any implicit instrumental contingency when trained on a classical schedule.

There is one possible objection to this conclusion. Kimble et al. (1955) pointed out that consistent presentation of a UCS on a classical schedule may produce superior performance, not because the CR is reinforced by presentation of the UCS but because the subject's motivation is maintained at a higher level. If motivation is affected by the overall frequency of the UCS, then a proper comparison between classical and omission schedules must somehow arrange to equate this factor, while still specifying a contingency between CR and UCS only for the omission group. It would seem to be possible to satisfy this requirement by using a "yoked-control" procedure, where subjects are run in pairs. One member of each pair is on an omission schedule while the other, yoked, member receives a UCS, regardless of its own behaviour, only on those trials when the omission subject fails to respond and thus also receives a UCS.

The comparison of omission and yoked-control subjects has usually shown that the omission schedule produces slightly superior performance, both in studies of eyelid conditioning in human subjects (Moore and Gormezano, 1961), and in studies of nictitating membrane conditioning in rabbits (Gormezano, 1965). However, the difference may not be very large, and does not occur under all conditions. Fig. 3.7 (p. 122) illustrates this point from Gormezano's study of nictitating membrane conditioning: when the CS–UCS interval was short, subjects on the omission schedule attained a higher level of performance than those on the yoked schedule; when the interval was long, however, there was some suggestion of a difference in the opposite direction.

Even if it were generally true that yoked-control subjects conditioned less effectively than subjects trained on the omission schedule, this would not necessarily prove that eyelid or nictitating membrane responses were being reinforced by the avoidance of the UCS. As Church (1964) and Gormezano (1965) have pointed out, the yoked-control procedure contains an inherent source of bias which, even if the instrumental contingency were entirely ineffective, would still tend to favour omission subjects at the expense of yoked subjects. The simplest way of expressing the argument is as follows. We shall assume that the CR in question is affected solely by Pavlovian reinforcement, but that subjects differ in their conditionability (or sensitivity to Pavlovian reinforcement). If a subject trained on an omission schedule conditions rapidly, it will receive relatively few UCSs; a subject that conditions relatively slowly, on the other hand, will receive many more UCSs. Each will therefore tend to receive the number of reinforcements necessary to maintain an intermediate level of responding. For their yoked partners, however, there is nothing to ensure that the number of UCSs received will match their differences in conditionability. A rapid conditioner might receive a large number of reinforcements but since the effectiveness of reinforcement tends to decrease as response strength increases, their effect will be partly wasted. A slow conditioner, however, might receive too few reinforcements to establish reliable conditioning at all. In view of this inherent bias against yoked subjects, the fact that they sometimes condition as effectively as subjects on an omission schedule suggests that eyelid and nictitating membrane responses are relatively insensitive to instrumental contingencies.

Finally, several studies have examined the effects of imposing an omission schedule on an appetitive CR, and have observed the reliable acquisition and maintenance of responses when the only consequence of such responses is a net loss of food or water to the subject. Sheffield (1965) showed that if a dog was trained on an omission schedule, where a CS signalled food provided that the dog refrained from salivating in the presence of the CS, a reliable salivary CR was established and maintained for 800 trials. Patten and Rudy (1967) gave 275 trials of omission training to rats, with the CS signalling the delivery of water to a tube in the rat's mouth provided that the rat did not lick in the presence of the CS. The schedule resulted in the rapid acquisition of licking, which was maintained on over 50 per cent of trials throughout the experiment. Gormezano and Hiller (1972) have reported similar acquisition of jaw movement CRs when rabbits were trained on an omission schedule with water as the UCS.

Several experiments have studied omission training of the pigeon's key pecking response, with illumination of the key light signalling food, provided that the pigeon does not peck the illuminated key (Williams and Williams, 1969; Herrnstein and Loveland, 1972; Schwartz and Williams, 1972a, b; Moore, 1973). In all of these experiments, pecking was established and maintained at a relatively steady rate for several hundred trials. In a variant on this procedure, Jenkins (1973) trained pigeons on a discrimination problem, where the key might be illuminated with a green dot on S— trials, and with a green and a red dot on S+ trials. The pigeons learned the discrimination rapidly, withholding pecks on S— trials, and pecking only on S+ trials. These pecks were consistently directed at the specific feature distinguishing S+ from S— (in the present example, at the red dot); and even the introduction of a specific contingency which caused the subject to lose food on any trial on which it pecked at the red dot, was not sufficient to prevent the appearance and maintenance of such pecks. This contingency, of course, constituted a form of omission schedule, with the more lenient requirement that it was only the direction, not the occurrence, of key pecks that was specified by the schedule.

The finding that appetitive CRs may be established in spite of their adverse instrumental consequences makes it hard to believe that the reason why they are established on an ordinary classical schedule is because they have some implicit beneficial consequence. One further way to assess the sensitivity of an appetitive CR to an instrumental contingency would be to compare the effects of an omission schedule with that of a yoked-control schedule. Even if the omission schedule is sufficient to establish responding, one might reasonably expect that subjects whose responses were consistently followed by the omission of an appetitive reinforcer might learn to respond less frequently than subjects exposed to the same overall frequency of reinforcement but without this explicit contingency. Gormezano and Hiller (1972), however, comparing omission and yoked subjects in their study of jaw movement conditioning in rabbits with water as the reinforcer, found that omission subjects responded at a marginally *higher* rate than yoked subjects. Their results are shown in Fig. 3.7, together with the comparable results of Gormezano's study of nictitating membrane conditioning in rabbits. As was noted earlier, if subjects differ in their sensitivity to Pavlovian reinforcement, then a yoked-control design is inherently likely to produce a higher level of conditioning in omission than in yoked subjects. While this difference is in the same direction as that predicted by the law of effect when an aversive reinforcer is used, it is in the opposite direction to that which would be expected were appetitive CRs affected by an instrumental contingency. Gormezano and Hiller's data, therefore, suggest that jaw movement CRs in rabbits are established by Pavlovian reinforcement, and are almost completely insensitive to an instrumental contingency.

As we shall have occasion to argue in more detail later, it is hardly to be expected that all responses subject to Pavlovian reinforcement should be completely unaffected by instrumental contingencies. It is not entirely surprising, therefore, that Schwartz and Williams (1972a) should have found that when pigeons were reinforced on an omission schedule, they pecked at a slightly slower rate than when they were on a yoked schedule. It must be remembered, however, that when key pecks are subject to an omission schedule, it is necessarily the location, rather than the occurrence of pecks, that is subject to the contingency: pecks on the key may lead to

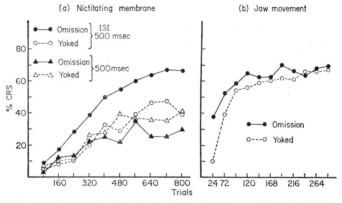

FIG. 3.7. Comparisons of classical and omission schedules of reinforcement in rabbits. (a) Nictitating membrane conditioning with short and long intervals between CS and UCS. (*After* Gormezano, 1965.) (b) Conditioning of jaw movement CRs. (*After* Gormezano and Hiller, 1972.)

the loss of reinforcement, but pecks on the wall next to the key are neither recorded nor punished. It is more surprising, therefore, that pigeons should direct any pecks at all at a key light in such circumstances than that they should peck on the key at a marginally lower rate. A similar argument may apply to the finding, reported by Schwartz and Williams (1972b), that when pigeons are trained on an omission schedule, they tend to emit pecks of shorter duration than those that typically occur on a classical schedule. As Moore (1973) has noted, the differential reinforcement of key pecks of different durations is surprisingly ineffective, and detailed observations reveal that changes in duration are usually a consequence of changes in the location of a peck: short pecks, for example may occur because the pigeon pecks at the edge of the key, touching it with only one tip of its beak. The shorter pecks observed by Schwartz and Williams in their omission trained birds may have been due partly to a similar redirection of pecking.

5. Conclusions

It is difficult to quarrel with the assessment reached by Sheffield (1965): the establishment and maintenance of any appetitive CR by an omission schedule shows that this response is relatively insensitive to instrumental contingencies. Thus it seems unlikely that the establishment of responses such as salivation, licking, or pecking on an ordinary classical schedule can reasonably be attributed to any instrumental contingency implicit in the situation. A similar conclusion is suggested by the observation that an omission schedule is relatively ineffective in establishing eyelid or nictitating membrane conditioning. However easy it is to see the beneficial consequences of blinking just before a puff of air strikes the cornea, or of salivating just before eating dry food, and however plausible it may therefore be to assume that these consequences are responsible for the establishment of such responses in a classical conditioning experiment, intuitive arguments such as these cannot reasonably be accorded the same weight as the empirical observations from studies of omission training. The evidence of these studies, moreover, reinforces the conclusions suggested by several earlier sets of studies. Instrumental interpretations of classical conditioning, although frequently attempted on grounds of economy and common sense, have not been particularly successful. Hull's idea that classical CRs are superstitiously reinforced, because they happen to be followed by rewarding events, is unable to account for the relationship between CR and UCS, and has problems in accounting for aversive conditioning at all. The response-shaping theory is both intuitively more plausible and clearly more successful in solving some of the problems encountered by the notion of superstitious reinforcement. It is also able to predict that different response systems may condition at different ISIs (p. 66), and that CRs may not be elaborated unless a motivationally significant UCS, is used (p. 96); even in these cases, however, the theory is by no means consistently supported by the evidence, and apart from these two sets of predictions, the theory seems to rely very largely on considerations of plausibility.

A Pavlovian account of classical conditioning is not free of problems. The relationship between CR and UCR, although closer than has often been thought, may not be quite as simple as that assumed by the principle of stimulus substitution. It is surprising how little serious experimental analysis of this question has been undertaken: there seems to have been an unfounded assumption that the observations reported by Zener (1937) had definitively answered all the important questions. It is certainly fair to argue, however, that stimulus-substitution theory has been rejected out of hand on quite inadequate evidence and that our present understanding of the determinants of the forms of both CRs and UCRs is distinctly limited.

It is also reasonable to conclude that there is considerable evidence to support a Pavlovian interpretation of the nature of reinforcement in classical conditioning. If responses such as closure of the nictitating membrane in anticipation of a puff of air, movements of the jaw or licking in anticipation of water, or pecking or salivation in anticipation of food, are relatively unaffected by their consequences when these are explicitly programmed, it is hard to see how they could be established because of their consequences when no such consequences are explicitly programmed. The only alternative is to assume that they are established, as Pavlov argued, because they are elicited by the UCS. This conclusion, however, should not be taken to imply that these or other CRs are completely insensitive to instrumental contingencies, let alone that the Pavlovian principle of reinforcement is universally effective. It is difficult to dismiss the possibility that the form of certain CRs may be modified by a response-shaping process and even that some CRs may not occur at all unless they produce rewarding consequences. This raises a possibility that will form the focus of discussion in the final section of this chapter. The possibility is that response systems may differ in their susceptibility to instrumental contingencies of reinforcement; the implication of this possibility is that no hard and fast distinction between classical and instrumental processes can be upheld.

V. THE RELATIONSHIP BETWEEN CLASSICAL CONDITIONING AND INSTRUMENTAL LEARNING

A. Statement of the Problem

As was noted in Chapter 1, early students of animal learning made little of the distinction between the operations of experiments on classical conditioning and those on instrumental learning, and it was those who were clearest about the existence of these operational differences who were also the first to suggest that further differences might be involved (Miller and Konorski, 1928; Konorski and Miller, 1937; Schlosberg, 1934, 1936, 1937; Skinner, 1935, 1937, 1938). However, as Rescorla and Solomon (1967) have pointed out in a review of the issues, the dominant trend in most American theorizing has been to assume that the operational distinction between stimulus-contingent and response-contingent reinforcement signifies no further difference of substance. Indeed this is the one point held in common by the classic theories Guthrie, Hull, and even Tolman (although in 1949 Tolman revised his views and argued that different associations were formed in classical and instrumental experiments). In principle, of course, such a "one-factor" theory is perfectly tenable: there is obviously no *a priori* reason why the experimenter's ability to arrange his experiments in different ways should reflect any underlying difference in the mechanisms of learning involved in the two situations. However, there are no *a priori* reasons for rejecting such a view and "two-factor" theories have been advanced by Schlosberg (1937), Skinner (1938), Konorski (1948), Mowrer (1947, 1950) and Rescorla and Solomon (1967). Most of these analyses have appeared to assume that it must be possible to draw a sharp distinction between the processes of classical and instrumental conditioning, corresponding to the apparently sharp distinction between the contingencies of the two types of experiment. An alternative view, suggested, for example, by Catania (1971a), is that although there are different processes, they are better viewed as differences of degree than of kind.

Rescorla and Solomon have discerned several points at issue between one- and two-factor theories. Essentially, however, these must reduce to the question whether the operational distinction between classical conditioning and instrumental learning implies that different learning processes are engaged in the two situations. One possibility is that different associations are formed. We have argued that the simplest analysis of associative learning in classical conditioning is to say that the subject learns about the contingency specified by the experimenter, i.e., learns to associate CS and UCS. An attractively simple and symmetrical analysis of instrumental learning would be to say that the subject again learns about the contingency specified by the experimenter, i.e. learns to associate a response with a reinforcer, often in the presence of a particular set of stimuli. This is not the place to justify this analysis of instrumental learning (see Chapter 5), but it does seem clear that whereas an association between CS and UCS would be logically sufficient to generate classical conditioning, the different contingencies of an instrumental experiment, where the reinforcer occurs in the presence of a stimulus only if a specified response occurs, would mean that a simple stimulus-reinforcer association might well be insufficient. The view that animals will learn to associate any set of events that occur together in their environment and the plausible assumption that an animal's own behaviour constitutes one class of event, provide the rationale for this associative analysis of instrumental learning. We shall see below that one of the factors which appears to determine whether a particular response is subject to instrumental contingencies is precisely the amount of feedback available to the subject from the occurrence of the response. This may reasonably be thought to influence the extent to which the response can enter into associations with other events.

We shall provisionally assume that different events may be associated with reinforcement in classical and instrumental experiments and that this constitutes one difference between the two procedures. The second, more important, but also more complex, issue dividing one- and two-factor theories concerns the nature of reinforcement. Pavlov and Thorndike postulated different principles of reinforcement: Pavlov's principle of stimulus substitution states that the occurrence of a reinforcing stimulus in a particular context will cause the responses elicited by that reinforcer to antedate its delivery; Thorndike's law of effect states that the occurrence of a reinforcing stimulus in a particular context will strengthen any response that happens to precede its delivery. Thus one common suggestion has been that classical CRs are modified by Pavlovian reinforcement, while instrumental responses are modified by the law of effect. The problem then reduces to one of identifying classical and instrumental responses. The easy solution would be to argue that classical CRs occur in operationally defined classical experiments, while instrumental responses occur in operationally defined instrumental experiments. But this suggestion, although attractively simple, is clearly inadequate. As we have already noted, the fact that the experimenter arranges no contingency between response and reinforcer in a classical experiment does not rule out the possibility that classical CRs are subject to implicit instrumental contingencies. Conversely, the fact that the experimenter does programme an explicit contingency between a response and reinforcement does not imply that any change in the probability of that response is caused by that instrumental contingency. A hungry pigeon may be trained to peck a key by arranging that every peck directed towards the key will be rewarded with food. But such a procedure will also specify a classical contingency between the sight of the key and the receipt of food. Since pecking may be established both in the absence of any instrumental contingency and in the face of an adverse instrumental contingency, one may reasonably wonder whether this implicit classical contingency is not the more important (Moore, 1973).

It follows, therefore, that two-factor theory must establish independent criteria for determining when an organism's behaviour will be modified by classical, and when by instrumental, contingencies. The question is usually put by asking which response systems are subject to classical reinforcement, and which are modified by the law of effect. One-factor theory, on the other hand, must deny that there are two principles of reinforcement and must either assume that all responses are modified by their consequences or that all are subject to Pavlovian reinforcement. At the very least, the results of experiments on omission training, reviewed in the preceding section, suggest that there are some response systems that are not very readily modified by their consequences. Thus a viable one-factor theory would seem to be forced to conclude that it is the Pavlovian principle of reinforcement that is of universal validity. Although this possibility has only a few adherents, it is, perhaps, not quite so easily dismissed as has usually been thought. As will become apparent later in the present chapter, the most widely studied instrumental responses are precisely those which resemble Pavlovian CRs. Moreover, the problems involved in elucidating the role of implicit classical contingencies in an instrumental experiment are no less complex than those involved in elucidating the possible contribution of implicit instrumental contingencies in studies of classical conditioning (see Chapter 5).

Nevertheless, there are good reasons for questioning whether a Pavlovian principle of reinforcement is sufficient to account for all changes in behaviour. The clearest evidence again comes from studies of omission training. The fact that some responses seem relatively unaffected by the contingencies of an omission schedule was used to argue that they must be established by Pavlovian reinforcement. By the same token one may argue that responses which *are* reliably affected by an omission schedule must, by definition, be affected by their consequences in the manner implied by the law of effect. Although eyelid CRs are not very effectively established by an omission schedule of aversive reinforcement, other responses, most notably the response of running in a rotating wheel or shuttle box, are learned very rapidly on such a schedule. It is true, as Sheffield (1948) argued, that one reason why running is not reliably established on a classical schedule, may be that it is not reliably elicited by shock: when shock elicits freezing, the probability of running on future trials decreases. But this argument, although plausible and important, does not suggest why running, by comparison with closure of the eyelid or flexion of the tail or leg, should be so rapidly and efficiently established on an omission or avoidance schedule. The finding that rats or guinea-pigs can learn to run in a rotating wheel on every trial to avoid shock is quite inconsistent with any Pavlovian analysis of reinforcement. The Pavlovian analysis predicts, as observed in studies of eyelid conditioning, that responding will never be established consistently but will go through a series of alternating cycles between acquisition and extinction.

The point was made most directly in Wahlsten and Cole's (1972) study of leg flexion conditioning in dogs. Although classical and omission schedules at first produced very similar performance, with CRs being initiated shortly before the onset of the UCS, as soon as animals received sufficient exposure to the contingencies of the omission schedule they

showed an abrupt increase in the probability of responding, with a concomitant decrease in the latency of responses on each trial. It is difficult
to resist the conclusion that this abrupt change in the behaviour of subjects
on the omission schedule was caused by the instrumental contingency.
Initially, subjects learned that the CS signalled shock to the paw and this
was sufficient to produce some tendency to flex the leg in anticipation of
the shock. Eventually, however, they learned that the CS signalled shock
only in the absence of a flexion response and this learning resulted in an
abrupt increase in the probability of responding.

Experiments on omission schedules, therefore, may provide a means of
disentangling some of the complexities surrounding the question of the
relationship between classical and instrumental conditioning. In the first
place, the omission schedule may define criteria for assessing the effectiveness of classical and instrumental contingencies: responses subject to
instrumental reinforcement, unlike those subject to Pavlovian reinforcement, should show appropriate changes on an omission schedule. Furthermore, although unfortunately there have been no studies systematically
designed to compare the sensitivity of different response systems to
classical and omission schedules, the studies reviewed in the preceding
section did suggest several conclusions: some responses, such as leg
flexion in dogs, may be affected by both classical and instrumental reinforcement; other responses, such as blinking, salivating or licking, may be
affected more by classical reinforcement; yet other responses, such as
running, may be affected more by the law of effect.

B. Response Systems Susceptible to Classical and Instrumental Reinforcement

If neither the Pavlovian nor the Thorndikian principle of reinforcement
is vacuous, there should be some distinction between the occasions when
they operate. Specifically, it has been assumed that some responses are
subject to one principle and others to the second. It has often been
further assumed that this distinction is a clear and rigid one. One particular basis for the distinction—that between responses mediated by
autonomic and those mediated by skeletal musculature—has unfortunately
pre-empted much of the discussion.

1. Autonomic and Skeletal Responses

Miller and Konorski (1928), in the first paper to argue that classical and
instrumental experiments differed both in the operations and in the
processes involved, suggested that instrumental learning might involve
only skeletal musculature, while classical conditioning might apply to both

autonomic and skeletal responses. This was subsequently elaborated by Skinner (1938) and Mowrer (1947) who stated that autonomic responses might be modifiable only by classical reinforcement and skeletal responses only by instrumental reinforcement.

The distinction between skeletal and autonomic responses tends to overlap with others that have been advanced as underlying the classical—instrumental dichotomy. Many autonomic responses are involuntary (Skinner, 1938) and diffuse or preparatory (Schlosberg, 1937), while many skeletal responses are voluntary and precise or adaptive. The distinction also tends to coincide with many of the responses actually studied in operationally defined classical and instrumental experiments. The large majority of instrumental experiments employ skeletal responses such as lever pressing, key pecking, running in shuttle boxes, traversing alleys and mazes. Many classical experiments record changes in such autonomic responses as salivation, vasoconstriction, GSR and heart rate, and even where changes in skeletal responding provide the data, as in studies of conditioned suppression, it is assumed that they reflect changes in some underlying motivational state which may be autonomically mediated.

Nevertheless, the coincidence is very far from perfect. In the first place, there is good evidence that many skeletal responses may be modified by Pavlovian reinforcement. Numerous Pavlovian experiments record changes in skeletal response systems: the classic examples are leg flexion and eyelid conditioning. Other, more recently established, instances include nictitating membrane closure, jaw movements and licking, while older examples include the knee jerk. We have seen that key pecking in pigeons is subject to Pavlovian reinforcement, as are fighting and aggressive behaviour and courtship displays. Furthermore, although we may reject Smith's (1954) argument that all autonomic conditioning is in fact mediated by skeletal responding, there is evidence that changes in some autonomic responses such as heart rate may be mediated by neural correlates of skeletal activity (p. 104).

The only possible argument against the conclusion suggested by these examples would be to say that all studies of classical conditioning using skeletal responses have been classical in name only: the operations have been those of classical conditioning, but response-contingent reinforcement was always responsible for modifying the responses studied. Even this will not do. Several responses, including nictitating membrane closure in rabbits, pecking in pigeons, licking in rats, and jaw movements in rabbits, have been among those CRs whose resistance to an omission schedule of reinforcement has provided the strongest evidence against any instrumental analysis of classical conditioning. At best, some skeletal CRs may be partially modified by the law of effect, but the same might be true of some autonomic CRs.

Not only, then, are some skeletal responses subject to Pavlovian reinforcement, it now seems possible that some autonomic responses may be subject to instrumental reinforcement. Since Kimble stated that "for autonomically mediated behaviour, the evidence points unequivocally to the conclusion that such responses can be modified by classical, but not instrumental training methods" (Kimble, 1961, p. 100), evidence for the instrumental learning of autonomic responses has been diligently sought, often with apparent success (Katkin and Murray, 1968; Miller, 1969; Black, 1971).

This should not be construed as implying that modification of autonomic responses by instrumental contingencies is either as rapid or as readily obtained as modification of some skeletal responses, nor that the effect is always a large one, nor even that when modification does occur it can be unequivocally interpreted as an instance of instrumental, autonomic learning. There have been numerous failures to obtain evidence of autonomic instrumental learning, many of them casually reported (e.g., Skinner, 1938, p. 112; Mednick, 1964, p. 52), but others more fully documented (e.g., Kimmel and Hill, 1960; Mandler et al., 1964; Stern, 1967). There have been cases where, although statistically significant differences between experimental and control groups have occurred, the basis for the difference has not been an increase in response probability in the experimental group but a decrease in response probability in the control group (e.g., Fowler and Kimmel, 1962; Greene, 1966). Finally, even when more substantial increases in response probability occur as a consequence of instrumental contingencies, two serious problems of interpretation remain.

The first is to ensure that changes in behaviour are indeed a consequence of the instrumental contingency, rather than of an implicit classical contingency. If a rat is trained to increase its heart rate in response to a signal in order to avoid shock, the classical contingency between signal and shock might be sufficient to produce heart rate acceleration as a CR. One answer to this objection has been to employ a yoked-control procedure in which a yoked subject receives shock whenever the master subject fails to respond. Both subjects are exposed to the same classical contingency, but only the master subject is exposed to the instrumental contingency. As was noted earlier, however, if subjects differ in their sensitivity to the classical contingency, the yoked-control design contains an inherent bias in favour of the master subject (Church, 1964). That this is no mere theoretical argument is suggested by some results reported by Kimmel and Sternthal (1967); when they carefully matched experimental and control subjects for responsivity of the GSR to shock, they found no significant differences between experimental and yoked subjects when experimental subjects were able to avoid shocks by changes in GSRs, although an earlier study

(Kimmel and Baxter, 1964), using the same procedures but with unmatched subjects, had obtained highly significant differences.

The second problem is that even if changes in behaviour are due to instrumental contingencies, autonomic responses may not have been directly affected. They may have changed only indirectly, as a consequence of some change in skeletal activity (Skinner, 1938, pp. 113–14). Heart rate or GSR changes might occur because the subject learned to be active or stay still, to breathe rapidly or slowly. Even if no overt changes in skeletal activity can be detected, it is possible that neural changes normally correlated with skeletal activity might mediate autonomic changes: since there is good evidence that classically conditioned changes in heart rate may be mediated in this way, it would not be surprising if the same were true of instrumental changes. Finally, in experiments with human subjects, we should have little hesitation in admitting the possibility that the subject will think of something that affects the autonomic system in the way required by the experimental contingency. If paid for salivating, I might learn to think of lemons.

Not all of these problems are entirely insoluble. For example, DiCara and Miller (1968a, 1969a), using curarized rats, and Black (1971), using curarized dogs, found it possible to establish either an increase or a decrease in heart rate as an instrumental response reinforced by the avoidance of shock. Shock as a UCS can hardly have simultaneously elicited both increases and decreases in heart rate; successful learning, therefore, must have been due to the instrumental contingencies.

The fact that avoidance learning was established under curare indicates that changes in heart rate cannot have been a direct consequence of overt skeletal activity. As Black (1971) has pointed out, however, it is harder to rule out the possibility that such changes may be mediated by central neural correlates of skeletal activity; and Black's own experiments indicated that such mediation was probably of some importance. A similar conclusion is suggested by the results of an experiment by Goesling and Brener (1972). Whether mediation is always necessary is, of course, a different matter.

One strategy adopted by DiCara and Miller in order to rule out such an interpretation has been to record simultaneously from a number of different autonomic systems, while providing instrumental reinforcement for changes in one only. The reasonable argument is that:

> It is difficult to account in terms of central impulses to skeletal responses for the variety and specificity of visceral learning that has been demonstrated in curarized rats: intestinal contractions independent of heart rate (Miller and Banuazizi, 1968), formation of urine by kidney independent of heart rate and blood pressure (Miller and DiCara, 1968), blood pressure

independent of heart rate (DiCara and Miller, 1968b), and differential vasomotor responses in the two ears (DiCara and Miller, 1968c) (DiCara and Miller, 1969b, p. 162).

The evidence presented is undoubtedly a testament to the experimenters' persistence and ingenuity. Nevertheless, the force of these arguments is merely to persuade, rather than to command assent, and recent evidence that some of these effects may be difficult to replicate (Miller and Dworkin, 1974) suggests that one should not be too readily persuaded. It is, perhaps, premature to reject out of hand the possibility that autonomic changes are always mediated by central correlates of skeletal activity (Black, 1971). Moreover, it would be a mistake to forget the qualifications made earlier: even if instrumental autonomic learning is possible, it is still relatively slight, requires careful shaping and sometimes fails to occur at all.

In spite of this perhaps excessively cautious conclusion, it cannot reasonably be maintained that the distinction between autonomic and skeletal responses forms the basis for the distinction between the domains of classical and instrumental principles of reinforcement. Any such argument is directly contradicted by numerous demonstrations of the classical conditioning of skeletal responses, and must take a relatively strong position against the possibility of the instrumental training of autonomic responses.

2. *Differences in Response-produced Feedback*

It has sometimes been suggested that one reason why autonomic responses are not readily susceptible to instrumental reinforcement is that they involve little or no feedback. Although the absence of proprioceptive feedback from autonomic responses should not be exaggerated, it may prove profitable to explore this dimension as a primary cause of differences in the effects of classical and instrumental reinforcement. It may be that any response generating little feedback can only with difficulty be trained by instrumental means.

The strongest argument in favour of this supposition is, as was noted above, a theoretical one about the nature of instrumental learning. The earliest and simplest version of a theory stressing the role of response-produced stimuli in instrumental learning was that proposed by Miller and Konorski (1928) and Konorski and Miller (1937). More recently, Mowrer (1960a) and Estes (1969a, b) have proposed theories also stressing this factor. In Konorski and Miller's experiments, a dog was fed in the presence of a signal whenever its leg was moved passively by the experimenter. In due course, the dog learned to flex its leg actively as soon as the signal was presented. Their analysis of this result was that the stimulus compound comprising signal plus feedback from flexion of the leg was the only reinforced state of affairs and the animal would therefore learn to produce it. A response that generated little or no feedback, therefore, would be learned only with difficulty.*

There is little direct evidence on the importance of response-produced feedback in instrumental learning. Brener and Hothersall (1966) reported that the learning of autonomic responses may be facilitated by the addition of a response-produced signal. Taub and Berman (1969) showed that instrumental avoidance learning in monkeys was quite markedly impaired by total de-afferentiation of the responding limb, together with the elimination of all re-afferent feedback. Less directly, Turner and Solomon (1962) reported that college students may never learn to avoid a shock delivered to their toe if the response required consisted of flexing either that or another toe, but they did learn rapidly to pull a lever to avoid the shock.

Further indirect evidence is provided by a comparison of several of the studies of classical and omission schedules reviewed in the preceding section. The one response that was established rapidly and consistently on an omission schedule but not on a classical schedule, was running in a rotating wheel, circular maze or shuttle box. The difference between the effects of instrumental contingencies on running and on tail flexion in rats certainly suggests that the amount of feedback generated by a response may determine the extent to which the occurrence of that response can be associated with appropriate consequences.

A final point about the role of response-produced feedback in delineating the domains of classical and instrumental reinforcement is that there could

---

* It should be noted that there is more than one potential source of feedback from the occurrence of a response. In addition to kinaesthetic feedback, based, in the case of skeletal responses, on muscle-spindle, tendon and joint receptors, there is also good evidence that information about responses is derived from the monitoring of efferent commands (Howard, 1966). Furthermore, different responses may produce different amounts of change in the external environment, or re-afferent information. Running, for example, may result in a substantial change, for it usually moves the organism from one place to another. This would clearly be less true, however, of running in a rotating wheel. The response of pressing a lever, although leaving the organism in much the same place, may be accompanied by some visual and auditory stimuli. Responses such as closing an eyelid or flexing a leg, however, must usually produce very little change, and autonomic responses essentially no change in the external environment. We shall see in Chapter 5 that the distinction between external, re-afferent changes and direct, kinaesthetic feedback from responding, may be of some theoretical importance. If a particular instrumental response always causes a particular change in the subject's external environment, then any instrumental contingency between that response and reinforcement can also be described as a classical contingency between reinforcement and the stimuli produced by the response. It may then be possible to treat the change in instrumental behaviour as a consequence of the classical conditioning of approach or escape responses to the external stimuli produced by the response.

hardly be any suggestion of a rigid distinction between responses producing no feedback and responses producing a great deal. The hypothesis implies that we are dealing with a dimension of susceptibility to the two principles, i.e., that responses differ in degree, not kind. Very few responses, therefore, should be totally unaffected by instrumental contingencies and even among responses normally regarded as instrumental, one might still expect to find differences in ease of training. In this respect, the hypothesis is similar to the final basis for the distinction which we shall discuss.

3. *Reflexiveness*

In order to record changes in behaviour to the CS in a classical conditioning experiment, the reinforcer (the UCS) must reliably elicit a response (the UCR), components of which then occur to the CS. If this stimulus-substitution view of conditioning is accepted, it follows that classically conditioned responses can only be responses reliably or reflexively elicited by the reinforcer. In an instrumental experiment, on the other hand, the relationship between response and reinforcer is, in principle, arbitrary. If this is so, then one basis for the distinction between responses that are classically conditionable and those that can be instrumentally trained may be their reflexiveness. Reflexive responses, i.e., those that are uniformly and reliably elicited by the reinforcing stimulus with a short latency and high probability, may be subject to classical conditioning. Nonreflexive responses, i.e., those that are not necessarily elicited by the reinforcing stimulus, may be subject to instrumental training. The suggestion has been elaborated by Turner and Solomon (1962) in the context of an analysis of avoidance learning in human subjects.

The concept of reflexiveness, like that of amount of feedback, defines a dimension along which responses may vary continuously, rather than a property possessed by some responses but not by others. As before, the implication is that there is a continuum of responses modifiable by classical and instrumental reinforcement, rather than a hard and fast distinction between two categories of response. A second point to be noted is that reflexiveness is a characteristic not of a response viewed in isolation, but of a response with respect to a particular stimulus. An important, new implication of this view is that the nature of the reinforcer used in any given experiment determines which responses will be classically conditioned and which can be trained instrumentally. A nice, although once again indirect, illustration of this point is provided by a comparison of the outcome of two experiments on the omission training of salivary responses. As mentioned above (p. 120), Sheffield (1965) attempted to train dogs not to salivate in the presence of a signal: the reinforcement for not salivating was food and learning failed to occur. Miller and Carmona (1967), on the other hand, reinforced dogs for a decrease in rate of salivation in the presence of a signal; the reinforcer was water and all subjects learned successfully. Unfortunately, as so often before in this chapter, we are forced to rely on comparisons between different experiments, with all their attendant dangers; but it is possible that the factor responsible for the successful learning in Miller and Carmona's experiment and its absence in Sheffield's, is that salivation is reflexively elicited by food. In Miller and Carmona's experiment, therefore, the instrumental contingency could operate unopposed, while in Sheffield's experiment, its effects were counteracted by a Pavlovian process of reinforcement which ensured that the signal continued to elicit salivation.

Other studies examining the effects of classical and omission schedules have provided evidence suggesting that Pavlovian reinforcement may largely affect relatively reflexive responses. Although differences in the sensitivity of different responses to omission schedules may be correlated with differences in the amount of feedback accompanying the response, there is reason to believe that differences in sensitivity to classical schedules may be correlated with differences in reflexiveness. The reason why running was rapidly established on an omission schedule in the study of Brogden et al. (1938) may be because running produces substantial feedback, but the reason it was so inefficiently established on a classical schedule is surely, as Sheffield (1948) argued, because it was not reliably elicited by shock. Since freezing was also elicited by shock on some trials, freezing responses were also conditioned on the classical schedule and interfered with running. Similarly, the reason why leg flexion CRs are not always well established on a classical schedule may be because shock sometimes elicits incompatible extensor responses or diffuse struggling (Girden, 1938; Wahlsten and Cole, 1972). Other responses such as flexion of the toe, which is reliably elicited by shock to the toe, or closure of the eyelid or nictitating membrane, which is reliably elicited by a puff of air, will be more reliably conditioned on classical schedules.

It should not, however, be thought that an organism's behaviour can be neatly partitioned into two categories, reflexive and nonreflexive, with the former being modified by classical reinforcement and the latter by instrumental reinforcement. For not only is reflexiveness a dimension along which responses vary, rather than a property which is present or absent, there is excellent evidence that much of the behaviour typically studied in instrumental experiments may be elicited by the reinforcers used in those experiments. The most striking instance of this is the pigeon's key peck, a response which has probably been more extensively studied in instrumental experiments than any other (Ferster and Skinner, 1957, alone recorded approximately a quarter of a million instances of this response). As we have seen, not only can key pecking be trained by classical operations, it is also relatively resistant to the instrumental contingencies of an omission schedule (p. 121). As Moore (1973) has documented, attempts to modify such characteristics of key pecking as its force or duration by instrumental contingencies are surprisingly unsuccessful. In general, the topography of key pecking is not readily, if at all, affected by instrumental contingencies. Nor is key pecking an isolated example. Just as pecking is a response unconditionally elicited by the sight of food in birds, so, we may reasonably assume, is approach a response unconditionally elicited by the sight of food in most animals. Running down an alleyway may be regarded as approaching sets of stimuli paired with the presentation of food (i.e., appetitive CSs), and it is not entirely surprising, therefore, that simple running in rats is relatively insensitive to the equivalent of an omission schedule. Logan (1960), Rashotte and Amsel (1968a), and Amsel and Rashotte (1969) required rats to run in an alley at a *slower* than specified speed in order to obtain food in the goal box and found that even after extended training, their subjects continued to miss about half of the potentially available rewards by running too fast. Williams and Williams (1969) also noted that in other situations rats run faster than is required by the instrumental contingency. Just as key pecking in pigeons will emerge when classically conditioned, so rats placed in a distinctive goal box and fed, will then run rapidly down an alley leading to the goal box (Gonzalez and Diamond, 1960; see p. 211). Lever pressing, the other response studied in instrumental experiments with rats, involves approach and manipulation, both of which are presumably components of the unconditional response to food. Once again, recent evidence suggests that lever pressing can be established by a classical contingency in much the same way that key pecking can be auto-shaped in pigeons (Peterson *et al.*, 1972).

Three of the most commonly studied appetitive instrumental responses, then, are all, to a greater or lesser extent, elicited by food. The clear implication is that if reflexiveness is the dimension along which responses differ in their susceptibility to classical conditioning, then there is no hard and fast distinction between classical and instrumental responses. This implication receives confirmation of another sort from the outcome of instrumental experiments in which the response used bears no relation to the reinforcer, i.e. is at the extreme nonreflexive end of the continuum. Key pecking and lever pressing are admirable responses for instrumental experiments using appetitive reinforcers, but they are learned only with considerable difficulty as escape or avoidance responses (e.g., Hoffman and Fleshler, 1959; Meyer *et al.*, 1960; see p. 340). Bolles (1970, 1971) has summarized this and other evidence which supports the conclusion that efficient avoidance learning tends to occur only when the response required forms part of the subject's "species-specific defense reactions", i.e. is one of the set of responses unconditionally or reflexively elicited by aversive stimulation.

In the case of appetitive reinforcement, the argument is strikingly documented by the observations of Breland and Breland (1961, 1966). In attempting to train a variety of animals to perform various tricks, they frequently found that although they might have some initial success, the animals' behaviour eventually became progressively less efficient in spite of the maintenance of the instrumental contingency. When a pig, for example, was trained to pick up a wooden token and drop it into a container (insert a coin into a piggy-bank), the animal would "repeatedly drop it, root it, drop it again, root it along the way, pick it up, toss it in the air, drop it, root it some more, and so on (Breland and Breland, 1961, p. 683)". A racoon trained to perform a similar task would spend his time rubbing the coins together instead of inserting them into the slot. Needless to say, such behaviour was indulged in only at the expense of a substantial increase in the delay of reinforcement. In other cases, learning never became efficient in the first place: a cow was trained to chase a man and engage in a bull fight with him. Although the animal learned the required movements, no increase in hunger drive would induce her to speed up her performance to the desired pace. Herbivores, unlike carnivores, feed, and therefore engage in food related activities, at a slow but steady pace (Breland and Breland, 1966, pp. 74–5).

Another example is provided by the work of Sevenster (1968, 1973), attempting to establish biting as an instrumental response in sticklebacks. The male fish may be trained rapidly to swim through a small ring in order to gain access to a female, but if the instrumental response required involves biting at a glass rod, clear evidence emerges of a conflict between the instrumental requirement and the behaviour elicited by the presentation of the reinforcer. In the presence of the female, the male normally initiates courtship behaviour in the form of a "zig-zag" dance. After exposure to the female on one trial, the male rapidly returns to the rod to initiate the next trial, but instead of biting it he directs the zig-zag dance towards it. As Shettleworth (1972b) has pointed out in a review of these and other studies, the idea that reinforcers may be used interchangeably to establish any instrumental response, although pleasingly simple, is certainly wrong.

Breland and Breland summarized the problems encountered in their studies as being a matter of "instinctive drift". Learned behaviour was said to drift towards instinctive behaviour. In the present context, the point could be rephrased in the following way. The reinforcer used in any instrumental experiment is also a potential classical UCS. A set of responses will therefore be reflexively elicited by the reinforcer and be classically conditioned to stimuli reliably accompanying its delivery (e.g., the tokens in the situation used with the pig and the racoon). If these classically conditioned responses are incompatible with the performance of the instrumental response specified by the experimenter, then instrumental learning will be inefficient. In this sense, therefore, the potency of the Pavlovian principle of reinforcement dictates that responses completely unrelated to the reinforcer, and, in particular, responses incompatible with behaviour elicited by the reinforcer, will be instrumentally trained only with the greatest difficulty. Not only are nonreflexive responses unavailable for classical conditioning, but if sufficiently unrelated to the reinforcer they may be equally unavailable for instrumental training.

C. Conclusions

The data reviewed in the preceding sections have been interpreted consistently as pointing towards certain conclusions. So that the reader may recognize the bias inevitably present in such an exercise, it will be as well to summarize these conclusions quite explicitly.

The reinforcing stimuli used in experiments on learning, whether the experiment is operationally classical or instrumental, typically have motivational significance to the subject and typically elicit certain identifiable patterns of behaviour. The Pavlovian principle of reinforcement implies that components of these behavioural patterns will come to be elicited by stimuli antedating the delivery of the reinforcer. There can be little reasonable doubt of the validity of this principle. However, response-contingent reinforcement is also an effective method of changing behaviour and its effect cannot in all instances be readily reduced to the operation of Pavlovian reinforcement. In this sense, both classical and instrumental principles of reinforcement must be accepted. But the difference between classical conditioning and instrumental learning is, in fact, hardly more than could have been anticipated from a description of the two kinds of experiment. A sufficient description of a "pure" case of classical conditioning is that the subject associates CS and UCS, and, as a consequence of this association, responses elicited by the UCS come to be elicited by the CS. A sufficient description of a "pure" case of instrumental learning is that the subject associates a particular response with the reinforcer, and as a consequence of this association the subject's behaviour may be modified so as to increase the probability of obtaining the reinforcer. In

practice, of course, both kinds of association and both principles of reinforcement may operate in a single situation.

Pavlovian reinforcement will tend to produce behaviour in anticipation
of the delivery of the reinforcer that is related to the behaviour elicited by
the reinforcer. Provided that the behaviour elicited by the reinforcer is the
same as the behaviour specified by the experimenter as necessary for the
delivery of the reinforcer, there may be little need to appeal to anything
more than simple Pavlovian processes to account for the change in the
animal's behaviour even if the experiment is operationally instrumental.
In such a case, it is possible that no response-reinforcer associations will
occur and that the animal's behaviour will not be modified because it
generates reinforcing consequences. However, if the response elicited by
the reinforcer is different from that specified by the experimenter's
instrumental contingencies, then instrumental learning is required and
may occur. Provided that the response specified by the experimenter is
one generating sufficient feedback to enable the subject to associate its
occurrence with the delivery of the reinforcer, and provided that it is not
incompatible with behaviour elicited by the reinforcer, the subject's
behaviour will be modified so as to satisfy the experimenter's requirements.

In any given experimental situation, changes in behaviour may well be
produced by a combination of stimulus-reinforcer and response-reinforcer
associations and of Pavlovian and Thorndikian reinforcement. There are
doubtless few pure cases of classical conditioning where response-produced
consequences have no effect on the animal's behaviour; equally, there are
perhaps no cases where the associations and modifications required for
instrumental learning occur without some Pavlovian conditioning also
occurring. Which of the two sets of processes will predominate in any
given case will vary in a continuous fashion as a function of the amount
of feedback generated by the response under study and the relationship
between the response and the reinforcer. If this conclusion shows greater
dependence upon two-factor theory than upon one-factor theory, it should
be recognized as a highly qualified version of two-factor theory. For, in
effect, the argument states that not only is any difference between classical
and instrumental conditioning a graded rather than an abrupt one, but
it is also largely a consequence of the differences in the operations involved
in the two kinds of experiment. Where no change in behaviour is required
by the experimenter and where the reinforcer elicits a specific set of
responses, then changes in behaviour are likely to be a consequence only
of Pavlovian processes; where the delivery of the reinforcer is contingent
on a change in behaviour, it is possible that the appropriate change will
depend upon the formation of an association between behaviour and
reinforcement.

VI. SUMMARY

The theoretical issues raised by the study of classical conditioning have included questions about the nature of the associations formed, the conditions necessary for their formation, the nature of the CR and the process of reinforcement responsible for its establishment. On the answers given to these questions will depend the answer given to the question concerning the relationship between classical and instrumental conditioning.

The most general associationist view would hold that animals may associate any set of events that happen to be correlated in time. Much of what has traditionally passed for the theoretical analysis of learning has been specifically concerned with denying such a view and to argue, in its stead, that only certain classes of events can be associated. Such a position can be maintained only at the cost of elaborate redefinitions. In the face of the phenomena of sensory pre-conditioning, of evidence of the occurrence of conditioning in the absence of overt CRs and of effects of conditioning on instrumental responding which cannot have been mediated by overt CRs, S–R theory must redefine the response to include neural correlates of sensory events and both neural correlates and motivational antecedents of overt responses. On the other hand, the results of studies which have examined the effects on conditioning of subsequent modification of the subject's reaction to the UCS suggest that associations may be formed not only between the CS and UCS, but also between the CS and the responses originally elicited by the UCS. The nature of the factors determining which associations will predominate is not yet clear.

Studies of sensory preconditioning also suggest that associations may be formed between relatively neutral events. There is, however, some evidence implying that CRs are elaborated only when the UCS is of some motivational significance to the subject. Although this conclusion would be consistent with the idea that CRs are reinforced by their affective consequences, alternative explanations are possible.

The nature of the CR in any study of conditioning is determined by the nature of the UCS. The strongest interpretation of this relationship is that provided by stimulus-substitution theory, which states that the CR consists of components of the UCR elicited by stimuli signalling the occurrence of the UCS. Although stimulus-substitution theory has been dismissed often, several investigators have noted striking similarities between the behaviour elicited by CS and UCS, and have reported that changes in the nature of the UCS, or the manner of its presentation, have entirely appropriate effects on the nature of the CR. An alternative position, somewhat weaker than that taken by stimulus-substitution theory, would be to say that the UCS determines a set of potential CRs, even though none might be identical to any UCR elicited by that UCS. Thus the delivery of an aversive reinforcer would ensure the conditioning of a set of fear reactions, which might include freezing or flight, although these responses might differ from the set of pain reactions elicited by the UCS itself.

Although the procedures employed in experiments on classical conditioning involve the delivery of reinforcers independently of the subject's behaviour, it has frequently been suggested that classical CRs are in fact reinforced by their consequences. One possibility is that the presentation of an appetitive UCS, or the removal of an aversive UCS, acts as a reward to stamp in any response that happens to occur during the CS. A second, more plausible, suggestion is that appropriate CRs are differentially reinforced because they increase the value of an appetitive UCS, or decrease the aversiveness of an aversive UCS. Although it is difficult to rule out the possibility of such implicit instrumental contingencies, the available evidence, particularly from studies of omission schedules, suggest that many CRs, including nictitating membrane responses to a puff of air, salivation or pecking for food and licking for water, are relatively insensitive even to explicitly programmed instrumental contingencies.

The implication is that classical CRs are reinforced, as Pavlov had supposed, simply by virtue of being elicited by the UCS. If this is accepted, then it follows that the law of effect cannot be regarded as a sufficient principle to account for all associative conditioning. Since there is, however, reason to believe that some responses can be modified by their consequences in the manner suggested by the law of effect, it seems reasonable to accept the contention of two-factor theory that the operational distinction between classical and instrumental experiments may correspond to some difference in the processes of learning and reinforcement involved in the two situations. Traditionally, two-factor theory has argued that there is a sharp discontinuity between the response systems subject to Pavlovian reinforcement and those subject to the law of effect. A preferable alternative would be to stress that the differences are of degree, not of kind. The Pavlovian principle of reinforcement will ensure that, regardless of the operationally defined contingencies of any particular experiment, the responses elicited by the reinforcer will tend to occur in anticipation of that reinforcer. If one or more of these responses is similar to the instrumental responses required by the experimenter, then the Pavlovian principle of reinforcement will be sufficient to produce changes in behaviour usually regarded as instrumental. Where this does not happen, the sole predictor of reinforcement will be some new change in the subject's behaviour. Instrumental learning may be thought of as involving the learning of this contingency between response and reinforcement.

The available evidence suggests that, where required, instrumental contingencies may be learned, provided that the necessary responses produce adequate feedback and that it is not incompatible with the responses unconditionally elicited by the reinforcer.

CHAPTER 4

Instrumental Learning: Basic Operations

I. INTRODUCTION AND DEFINITIONS

The argument of the preceding chapter suggested that the operational distinction between experiments in which reinforcers are presented contingent on a stimulus (classical conditioning) and those in which they are presented contingent on a response (instrumental learning) may not coincide with any abrupt difference in the mechanisms of learning involved. It is more plausible to suppose that associations can be formed between any set of contiguous events and that there is only a graded difference in the sensitivity of different response systems to instrumental contingencies. This does not necessarily mean that an experiment on classical salivary conditioning will always produce results similar to those obtained in a study of instrumental lever pressing. The differences in the contingencies involved may be sufficient to ensure differences in the way the measured responses react to different operations and variables. Indeed, this is often the case (e.g., Weisman, 1965). Bindra (1972) has shown in principle how differences in contingencies may generate different experimental outcomes in spite of the fundamental identity of the basic learning processes. This, then, must be part of the justification for treating instrumental and classical experiments separately.

This chapter outlines some of the simpler operations of instrumental experiments, provides a sampling of the more important and interesting parameters that affect instrumental performance and, finally, extends the analysis to simple instances of choice behaviour. First, some definitions.

In studies of simple instrumental learning, the experimenter records some change in the latency, speed, probability, or rate of occurrence of a specified response, when the delivery of a particular reinforcing event is made contingent upon the occurrence of that response. The responses most commonly studied are alley running or lever pressing in rats and key pecking in pigeons, although other mammals and birds (and occasionally fish and a few randomly chosen invertebrates) have been trained to pull levers, press panels, grasp objects, pull strings that open doors, draw in strings that have food attached to the end, swim through doors and dig through sand.

The experimenter may either arrange a series of discrete trials on which the subject may respond, or permit the subject to respond at its own rate in a free-operant procedure. In discrete-trial experiments, a rat might be placed in the start box of an alley at the beginning of each trial and removed at the end. If the subject is permitted to remain in the apparatus, the manipulandum may somehow be made inaccessible or a situation arranged where responding is improbable: the rat's lever may be mechanically withdrawn from the box, or the response key of a pigeon's chamber (or the entire chamber) may be darkened. Response strength is measured by a latency, speed or probability measure. In free-operant experiments, the subject is left in the apparatus with the manipulandum always available and is free to press the lever or peck the key continually. Response strength is usually measured in terms of rate of responding over time. Much nonsense has been written on the relative merits of latency, speed or rate measures, and hence on the supposed superiority of one procedure over the other. Skinner (1966) has argued that rate of responding is the best measure for an experimental analysis of instrumental behaviour, since other measures are arbitrary, vary noisily from trial to trial and tell us more about the apparatus or procedure than the organism. We shall not take so narrow a view. Different procedures and different measures do not always yield similar results from apparently similar operations: to take one example, simultaneous positive contrast occurs in free-operant experiments using a rate measure, but probably not in discrete-trial experiments using a latency measures (p. 384). We do not always understand the reasons for these differences, but it is equally true that the same procedures may produce different outcomes in different types of apparatus and that we are no less ignorant about the causes of these differences. Overtraining, for example, is more likely to reduce resistance to extinction in discrete-trial experiments using a runway, than in those using a lever box (p. 423). Such ignorance may be unfortunate, but the existence of such differences must tell us a great deal about the limits of the conditions under which certain effects operate and may even tell us something about the underlying processes involved. Furthermore, the analysis of instrumental behaviour would be the poorer, and its conclusions less valid, if its data came exclusively from a single, standardized experimental situation: we should never know whether a particular phenomenon was simply a consequence of some peculiarity of the experimental situation if we had no information about its generality across situations.

Both discrete-trial and free-operant procedures may provide the subject either with a single response to make or a single manipulandum to operate, or with two or more simultaneously available options (e.g., two alleys, levers, or keys to choose from). The simplest comparison is between the single alley and the T-maze. In the alley, learning is measured as an increase in speed of running to the goal box, this being the only response recorded by the experimenter. In the T-maze, the subject may turn either left or right at the choice point and learning is usually measured as an increase in probability of the correct response, i.e., that response which leads to the goal box containing reinforcement. The distinction between situations studying single responses and those studying choice behaviour does not, it should be stressed, coincide with the distinction between nondiscriminative and discriminative situations. Discrimination learning may be programmed with either a single response available or with two or more responses. In the former situation, as in differential conditioning, positive and negative stimuli are presented successively and, in the simplest case, responding is reinforced in the presence of the positive stimulus and extinguished in the presence of the negative stimulus. This procedure may be termed successive discrimination learning. In simultaneous discrimination learning, on the other hand, two or more stimuli are simultaneously presented to the subject, with responses to only one of the two stimuli being reinforced: simultaneous discriminations, therefore, involved a choice between two alternatives. Both successive and simultaneous discriminations may be programmed using discrete-trial or free-operant procedures: in the terminology of the free-operant tradition, the successive discrimination is referred to as a multiple schedule, the simultaneous as a concurrent schedule.

Finally, instrumental responses may be strengthened by either positive or negative reinforcement. Positive reinforcement is provided by the presentation of food, water, or some other appetitive UCS; in instrumental experiments, appetitive reinforcers are frequently referred to as rewards. Negative reinforcement is provided by the opportunity to escape from shock, cold, or some other aversive UCS. The distinction between positive and negative reinforcement may, of course, be procedural rather than fundamental: according to drive-reduction theory, for example, both types of reinforcer are based on a diminution of an aversive drive state. Two other instrumental contingencies have been studied: in punishment training, the delivery of an aversive reinforcer, rather than (as in escape training) its removal, is made contingent upon a response; in avoidance training, the delivery of an aversive reinforcer is made contingent upon a failure to perform a specified response.* The analysis of punishment and avoidance has traditionally involved a somewhat more complex theoretical

---

* Signalled avoidance training is, of course, equivalent to the procedure previously called omission training.

approach than that thought sufficient to account for simple positive and negative reinforcement. Although this difference in approach can now be seen to have been largely a historical accident, clarity of exposition will be served by respecting this convention and treating punishment and avoidance in a later chapter.

II. BASIC PHENOMENA

A. Preliminary Training and Shaping of Instrumental Responses

The presentation of a reinforcing event contingent upon the occurrence of a particular response normally increases the strength of that response. The typical instrumental experiment, however, usually incorporates some more or less elaborate pretraining phase before the instrumental contingency is imposed. In operant experiments, animals are usually given magazine training i.e., are trained to associate the noise made by the operation of the magazine with the delivery of food and to approach and eat from the magazine when it operates; in alleyway experiments, they are often placed in the goal box and allowed to eat there. Alleyway studies also sometimes permit some preliminary habituation to the apparatus, in the form of a period of unrewarded exploration of the alley, while operant studies frequently establish the final instrumental response via a set of gradual approximations—a procedure termed "response shaping". The principle of utilizing successive approximations to a terminal goal is a valuable and important one: discriminations between extremely close pairs of stimuli, such as neighbouring tones, can be established by initial training on stimuli further apart on the same continuum (p. 593). Relatively complex problems can be taught to subjects thought to be incapable of solving them by careful programming of a set of approximations to the final problem (e.g., Sidman and Stoddard, 1966); complex chains of responses can be shaped by initially training with the terminal response in the chain and successively adding new requirements to the beginning of the chain (Kelleher, 1966a). It is not at all certain, however, that shaping, in the sense intended, plays any important role in the production of a simple, single, response such as the key peck of pigeons. Since key pecking is rapidly classically conditioned when the illumination of the key precedes the delivery of food, it seems probable that the procedures involved in shaping the key peck (which include operating the magazine when the pigeon is standing close to the key, when it looks at the key and when it turns or moves its head towards the key) in fact achieve their effect simply because they ensure that visual stimulation from the key is paired with the delivery of food. In this case, response shaping may simply be a procedure for ensuring rapid classical conditioning to a localized CS (Moore, 1973).

B. Acquisition

The acquisition of instrumental responses, like that of classical CRs, has been studied by a variety of theorists interested in the possibility of elaborating formal models for the description of the acquisition process. In both cases, however, the course of acquisition appears to depend very much on the nature of the experimental situation and on the measure of

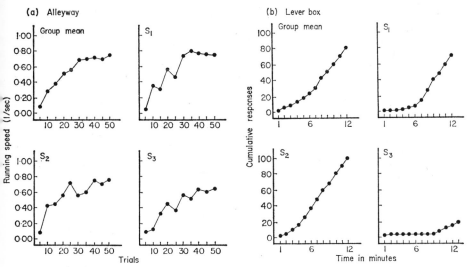

FIG. 4.1. Acquisition of instrumental responding in discrete-trial and free-operant situations. (a) Rats running in an alleyway. The first panel shows a group mean learning curve; the remaining panels show the individual curves of a few selected subjects. (*After* Mackintosh and Lord, 1973.) (b) Rats pressing a lever. The first panel shows a group curve, and the remaining panels the curves of some individual subjects. (*After* Estes, 1959.)

response strength. Figure 4.1 shows group and individual acquisition curves for two single-response situations. Figure 4.1a shows first a group learning curve from an alley experiment with rats (Mackintosh and Lord, 1973), followed by the learning curves of a few selected individual subjects. Response strength is measured as speed of running in the alley, averaged over blocks of trials; the group curve is a simple negatively accelerated function and, as can be seen, although there may be exceptions such as subject 3, this group curve is reasonably representative of the individual

F

functions. Figure 4.1b illustrates the acquisition of lever pressing in rats; the data are from Estes (1959, pp. 388–90). The first panel shows the mean learning curve for a number of subjects, some of whose individual curves are shown in subsequent panels. The measure in this case is simply the cumulative number of responses occuring in time; this cumulative record is a common measure in free-operant experiments, having the virtue, as Skinner has argued, that rate of responding is directly represented by the slope of the curve. The group learning curve here is positively accelerated. This is inevitable: since it shows the cumulative number of responses, it could hardly be any other shape; a negatively accelerated cumulative record would represent a progressive *decline* in rate of responding. Nevertheless the initially gradual slope illustrates clearly enough that increments to rate of responding are small at first and increase rapidly, thus showing some

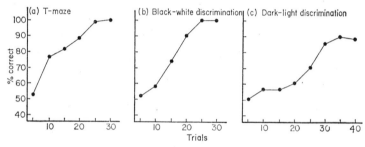

Fig. 4.2. Acquisition of instrumental discriminations. (a) Spatial discrimination learning in a T-maze. (*After* Clayton, 1969.) (b) and (c) Brightness discrimination learning in a discrimination box. (*After* Mackintosh, 1969.)

positive acceleration. This group curve is, again, reasonably characteristic of individual curves, although some animals, such as subject 3, show a relatively abrupt transition from a low to a high rate.

   Fig. 4.2 shows three group learning curves from choice situations. The first panel shows simple T-maze learning (Clayton, 1969); the second shows learning of a simultaneous black–white discrimination, and the third shows the learning of a more difficult discrimination between a dark and a light grey (both from Mackintosh, 1969). In all three cases the measure presented is the probability of a correct response. The most striking feature of the three curves is the gradual change in the shape of the learning curve: as the problem becomes harder, so the learning curve shows an increasing tendency towards an initial phase of positive acceleration.

   Theoretical analysis of these sorts of data has, just as in classical conditioning, concentrated more on the problem of devising general theories of the acquisition process, than on the more analytic questions of the relationship between response measure and learning function or the importance of differences in experimental procedure.

If we ask whether learning is gradual or abrupt, the initial answer must be that this depends upon the nature of the response measure. Speed of running or rate of lever pressing increase in a gradual fashion and the individual data illustrated in Fig. 4.1 show that the incremental nature of the group curve does not conceal a series of all-or-none changes on the part of individual subjects. To a first approximation, however, this is exactly what they do succeed in doing in the case of choice data. Clayton (1969) has shown that the group learning curve shown in the first panel of Fig. 4.2 is accurately represented by a Markov model postulating a one-step transition from unlearned to learned states: when the protocols of individual subjects are examined, it is possible to show that the probability of an error does not decrease gradually, but in fact remains constant until the animal abruptly enters a criterion run of correct responses. In visual discrimination experiments, such as those whose data are shown in Fig. 4.2, it has been known for a long time that the majority of subjects respond steadily at chance for a greater or lesser number of trials before shifting relatively abruptly to a nearly perfect level of performance (Lashley, 1929, p. 135; Spence, 1936; Sutherland and Mackintosh, 1971, pp. 87–8). As Sutherland and Mackintosh (p. 138) have noted, however, individual learning curves may show a more gradual increase in the probability of correct responses on more difficult visual discriminations.

Even when individual learning curves show abrupt changes from one level of performance to another, it is possible, as it was in the case of classical conditioning, to argue that this is a consequence of a crude measure of response strength. The molecular changes underlying the molar change from one probability of a correct response to another may occur more gradually; alternatively, abrupt changes in response probability may be due to the existence of a threshold, or to the existence of interfering responses which successfully compete with the correct response until they are reduced to a lower strength. This last suggestion seems particularly plausible in the case of choice experiments: one habit interfering with the learning of a simple T-maze is the rat's spontaneous tendency to alternate from one side to another on successive trials (Dember and Fowler, 1958); in visual discriminations, rats tend to develop position habits which interfere with selection of the positive stimulus (Sutherland and Mackintosh, 1971, pp. 88–96).

If it is assumed that learning is gradual and incremental, it can be further asked what is the true shape of the learning curve. Although some of the curves shown in Figs 4.1 and 4.2 are negatively accelerated, others are not. Spence (1956, pp. 76–83) has argued that instrumental learning, like classical conditioning, typically shows an initial phase of positive acceleration. It seems entirely possible that in both classical and instrumental experiments, differences in learning curves can be explained by similar differences in experimental procedures and situations. Just as we argued that the positive acceleration of some classical conditioning curves might be caused by the practice of pre-exposing subjects to the CS in the absence of reinforcement, thereby reducing the probability of associating that CS with subsequent reinforcement (p. 36), so it seems plausible to suppose that differences in choice learning curves as a function of discriminability of the alternatives, as illustrated in Fig. 4.2, may be a consequence of different initial probabilities of attending to the relevant stimuli (Zeaman and House, 1963). It should also be noted that instrumental experiments employ a variety of pretraining procedures which may include any combination of handling and taming the subject, exposing the subject to the apparatus in the absence of reinforcement, feeding the subject in the goal box of an alleyway, or magazine training in an operant chamber. Some of these procedures are operationally equivalent to pre-exposing the subject to a CS; others may inadvertently reinforce behaviour incompatible with the subsequently required instrumental response. Until the exact effects of these different procedures have been examined, the quest for a typical or true learning curve will be of questionable value.

III. PARAMETERS OF INSTRUMENTAL LEARNING: SINGLE RESPONSES

The major emphasis in studies of instrumental acquisition has been on factors of drive and reinforcement.

A. Drive

In general, and not very surprisingly, increases in drive level lead to increases in speed or rate of responding. The most popularly studied situation has been the alleyway: here, increases in hunger have led to greater speed of running whether pellets of food or solutions of sucrose have been used to provide reinforcement (Cotton, 1953; Hillman *et al.*, 1953; Cicala, 1961; Stabler, 1962); increases in thirst have led to faster running for water (Kintsch, 1962); increases in the intensity of shock present in the alley from which the subject can escape by running to the goal box also lead to faster running (Campbell and Kraeling, 1953). Discrete-trial studies of lever pressing show that greater hunger produces shorter latencies of responding (Ramond, 1954). Free-operant studies have also shown that rate of responding is directly affected by drive level, both in rats (Clark, 1958; Carlton, 1961) and pigeons (Ferster and Skinner, 1957, p. 365; Shull and Brownstein, 1968). Here, however, the effect depends upon the schedule of reinforcement used: the above studies employed interval schedules; Carlton found a much slighter effect when continuous reinforcement was used; Sidman and Stebbins (1954), and Ferster and Skinner (1957, pp. 71–7) found that on a fixed-ratio schedule drive affected only the length of the pause after each reinforcement, not the rate of responding once responding has started.*

The most common interpretation of these effects of drive level is to suppose that drive acts as a general energizer of behaviour (Hull, 1943; Hebb, 1955). The implication is that drive simply acts to potentiate whatever set of responses is currently most probable. Two points suggest that this analysis is probably incomplete (Estes, 1958). In the first place, many analytic studies have shown that the effect of increases in drive is not so much to increase the vigour with which a rat runs down an alley, as to increase the probability that it will run rather than engage in other, competing activities (Cotton, 1953). Secondly, there is ample evidence that a given level of drive does not simply affect current performance, but may continue to have an effect long after the subject's drive level has been changed: animals initially trained under a high drive will persist in running faster than those trained under a lower drive, even after both groups are shifted to the lower level of drive (e.g., Zaretsky, 1966; E. D. Capaldi, 1971a).† The implication is that drive is not simply an energizer of current performance, but may affect what an animal learns, for example by affecting the incidence of competing behaviour, and will thus continue to have an effect on performance even under a different level of drive.

B. Magnitude of Reinforcement

The effect of variations in reinforcement is as general and as unsurprising as that of variations in drive (Pubols, 1960). Speed of running in an alley is increased by an increase in the number of food pellets or weight of food used as reward (e.g., Crespi, 1942; Zeaman, 1949; Armus, 1959; Bower and Trapold, 1959; Logan, 1960, pp. 52–3; Wagner, 1961; Hill and

---

* Definitions and descriptions of the various schedules of reinforcement used in free-operant experiments are provided below (p. 164).

† A long series of studies has examined this question by investigating the effects of drive level during acquisition on performance in extinction. Although some studies have found that acquisition drive may affect performance in extinction (e.g., Theios, 1963a), other, probably more reliable, studies (e.g., Leach, 1971) have failed to replicate this result. At best, there is a slight effect of acquisition drive during early extinction trials (e.g., Barry, 1958). E. D. Capaldi (1971a) has directly shown that testing in extinction provides a much less sensitive measure of the persisting effects of an earlier drive level than does continued reinforced training.

Wallace, 1967; Gonzalez and Bitterman, 1969; Roberts, 1969; Wike and Chen, 1971; Daly, 1972), an increase in the concentration of sucrose or saccharin solution (Goodrich, 1960; Kraeling, 1961; Snyder, 1962; Rosen, 1966), an increase in the volume of water (Kintsch, 1962), or an increase in the amount by which shock is reduced when the goal box is entered (Campbell and Kraeling, 1953; Bower et al., 1959). Latency of lever pressing in a discrete-trial situation is also decreased by increases in the amount of food or concentration of sucrose used as reward (Likely, 1970); while free-operant studies with rats have found that rate of responding on interval schedules varies directly with both the number of pellets (Meltzer and Brahlek, 1968), concentration of sucrose (Guttman, 1953; Stebbins et al., 1959) and amount and flavour of a liquid reward (Hutt, 1954). Pigeons, however, show little effect of variation in the duration of access to food when responding on interval schedules (Keesey and Kling, 1961), unless they are rapidly alternated between the two values (Shettleworth and Nevin, 1965). Catania (1963b) and Neuringer (1967) have shown that although rate of responding may not vary with magnitude of reinforcement, pigeons learn to choose a stimulus correlated with a longer access to food over one correlated with a shorter access.*

This relatively tidy picture has been confused by attempts to provide more detailed analysis and interpretation. When rats run in an alley for large or small amounts of food in the goal box, it is not clear whether the important determinant of differences in speed of running is the number of pellets, the weight of the food, the amount of consummatory activity, the time spent in the goal box, or any combination of these factors. More recently, some doubts have been expressed whether magnitude of reinforcement affects rate of learning, or asymptotic speed of running, or both.

The first question has prompted studies specifically comparing different ways of varying reinforcement. Traupmann (1971) and Daly (1972) have suggested that a large amount of food given in the form of a large number of small pellets may produce faster acquisition but the same terminal speed of running as the same amount given in the form of a single large pellet. Studies using sucrose or saccharin solutions as reinforcers have attempted to vary the effectiveness of reinforcement by varying the concentration of

---

* These findings are of some generality. The effect of a particular variation in reinforcement has often been detected more readily by concurrently exposing a single subject to several values of reinforcement than by exposing several different subjects each to a single value (e.g., Lawson, 1957). Recent free-operant experiments have extended this generalization by showing that the more rapidly a subject is alternated between different values of reinforcement the greater the difference in rate of responding to the more and less favourable reinforcer (Shimp and Wheatley, 1971; Todorov, 1972). This presumably represents a contrast effect (Chapter 7).

the solution and thus without varying the visual characteristics or the time involved in the consumption of the reward. As noted above, increased concentrations of sucrose or saccharin reliably increase speed of running in an alley and rate of pressing a lever at all levels of training.

Several recent studies (e.g., Black, 1969; McCain et al., 1971; Campbell et al., 1972) have found that although large rewards may increase speed of running early in training, the differences between groups receiving different rewards may eventually disappear. The implication accepted by these investigators is that the size of reward may affect rate of learning without affecting the asymptote of response-strength. This is probably an unjustifiable assumption. Several of the studies cited at the beginning of this section found that differences between groups receiving different rewards were maintained even when training was continued for 100 trials or more. This was true whether reinforcement was manipulated by varying the number of pellets, the weight of a single pellet, duration of access to food, amount of water, or concentration of sucrose or saccharin. With free-operant procedures, differences are maintained over many sessions. The occasional disappearance of the difference with extended training in alley studies may well represent ceiling effects on speed of running rather than a common asymptote of response strength. When animals receive several trials in a single session (as in the studies of Black, 1969, and McCain et al., 1971), there is also the danger that large rewards may produce more rapid satiation and hence a lower level of drive.

According to the law of effect, reinforcers increase the strength of responses on which they are contingent. The simplest interpretation of these data, then, is to assume that the asymptotic habit strength of a given response is an increasing function of the magnitude of the reinforcer which follows it; it is possible to assume, in addition, that the increment to habit strength produced by a single reinforcement is also an increasing function of the magnitude of the reinforcing event. Hull (1943) originally assumed that reinforcement affected learning in this way and although he abandoned this assumption, it still retains some adherents (e.g., Capaldi, 1967a; Rescorla and Wagner, 1972). The evidence that caused Hull to change his analysis came from the studies by Crespi (1942) and Zeaman (1949), in which rats were initially trained with one magnitude of reinforcement and then shifted to another. Such shifts in reinforcer were found to produce relatively abrupt shifts in performance: subjects trained with a large reinforcer and then shifted to a small one would take only one or two trials to slow down to the speed attained by subjects trained throughout with the small reinforcer. Although there is evidence which suggests that it is sometimes possible to detect persisting effects of a previous reinforcer (e.g., Spear and Spitzner, 1968), the demonstration of such abrupt shifts in performance led Hull (1952) and Spence (1956) to abandon the original version of the law of effect and to separate learning of the required instrumental response (habit) from learning about the reinforcer contingent upon that response (incentive motivation). In plain terms, consistent with an expectancy theory, the animal is said to learn both what response to make and what reinforcer will follow the occurrence of that response. A given reinforcer is then said not to affect the strength of any learned connection between stimulus and response, but rather to motivate a given level of responding: shifts in reinforcer lead to appropriate shifts in performance as soon as the subject has learned that the change has occurred. The theoretical implications of these data on shifts in magnitude of reinforcement are discussed at greater length in the next chapter (p. 214).

C. Interaction of Drive and Reinforcement

Increases in both level of drive and magnitude of reinforcer lead to faster responding. It is possible to ask whether these two effects are independent and additive, or whether they interact in such a way that the effect of a given variation in reinforcement is itself dependent upon the current level of drive. The question, although apparently quite meaningful, does not seem to be of very great moment and is difficult, if not impossible, to answer. In spite of this, it has generated much experimental investigation since Hull (1952) argued that the effects interacted, whereas Spence (1956) suggested that they were simply additive.

In principle, the question should be answered easily, requiring no more than a large factorial experiment in which several levels of drive are combined with several magnitudes of reinforcement, with an analysis of variance being performed to test for a significant interaction of the two factors. Several such experiments have been performed in which no such interaction occurred (e.g., Reynolds and Pavlik, 1960; Pavlik and Reynolds, 1963). In other experiments, however, a significant interaction has been detected, especially when either or both factor is allowed to take on an extremely low value, i.e. when the subjects are satiated or run with no reinforcer (Seward et al., 1958; Seward et al., 1960). In these studies, the major portion of the interaction was indeed due to the inclusion of the zero drive or reinforcement groups, although other studies have reported a significant interaction in the absence of such groups (e.g., Kintsch, 1962; Stabler, 1962). In some cases the interaction, although falling short of significance, has been in the predicted direction (e.g., Weiss, 1960). The overall conclusion, therefore, which should surprise no one, is that variations in drive or reinforcement are often less effective at low values of the other factor: which is to say that, however hungry, a rat will run down an alleyway relatively slowly if there is no food in the goal box; or, however great the amount of food in the goal box, the rat will run relatively slowly if completely satiated.

In spite of this apparent partial victory for Hull, and complete victory for common sense, it is worth noting that no pattern of results must necessarily force rejection of either theoretical position. The reason is simple: the two theories are specifying a relation between drive, reinforcement and a theoretical term "reaction potential"; neither is required to specify a definite relation between speed of running and reaction potential. Only if this latter relation were linear would the observed data tend to discredit both positions. If Hull assumed a negatively accelerated function relating running speed to reaction potential, he would perfectly predict the general trend of the evidence—an interaction between drive and reinforcement at lower values, but little or no interaction at higher values. Equally, however, Spence could predict this outcome by assuming an initially positively accelerated function at low values followed by a negatively accelerated function at higher values of reaction potential. It follows that both theories could predict any other outcome by choosing other functions.

D. Delay of Reinforcement

The experimenter typically arranges that a reinforcer is presented as soon as the designated response has occurred: food is available in the goal box of an alleyway as soon as the rat enters the goal box: the magazine automatically delivers a pellet as soon as the lever is pressed; shock in the alleyway is turned off immediately the subject enters the goal box.

If reinforcement is delayed, performance is usually adversely affected. Logan (1960, p. 46), Renner (1963), and Sgro et al. (1967) have shown that rats will run down an alley more slowly if a food reward is delayed by values ranging from 3 to 30 sec. Interestingly enough, Logan and Spanier (1970) found that delay of a water reward had a much slighter effect on performance than did an equivalent delay of food. Skinner (1938, p. 73) reported little effect of short delays (up to 4 sec) between lever pressing and the delivery of food, but, as he pointed out, the scheduling of delays of reinforcement in a free-operant situation is a complex matter: if a second response occurs in the delay interval scheduled between a first response and its reinforcement, either the reinforcement must be cancelled and the first response is therefore not reinforced at all, or the second response is effectively reinforced at a delay shorter than the scheduled interval. In discrete-trial situations with a retractable lever, short delays of reinforcement (up to 10 sec) markedly increase latencies of responding (Perin,

1943; Harker, 1956). Finally, Fowler and Trapold (1962) showed that speed of running down an alleyway to escape from shock was inversely related to the delay between entrance into the goal box and the termination of shock: in this case delays of reinforcement as short as 1 or 2 sec had marked effects on performance.

Hull (1943) originally interpreted the effect of delay of reinforcement as an effect on habit strength: an increase in delay of reinforcement, like a reduction in size of reinforcement, was assumed to produce a weaker learned connection between the stimulus situation and the instrumental response. Subsequently, when the effect of magnitude of reinforcement was re-interpreted as an effect on performance rather than on learning, a comparable re-interpretation of the effect of delay was provided: longer delays of reinforcement were not assumed to produce less effective learning, but to reduce the motivation responsible for translating learning into performance.

It is not clear why Hull made this change. Although there is, as we shall see, evidence that delay of reinforcement does reduce incentive, it might also be thought of as interfering with associative learning. Both Spence (1956) and Mowrer (1960a) have maintained such an associative interpretation. According to Mowrer, the establishment of an association between a response and reinforcer in an instrumental experiment, just like the establishment of an association between a stimulus and reinforcer in classical conditioning, is made harder by the interposition of a time interval between the two. Spence has argued that the effects of delay will depend upon the amount of competing behaviour that occurs during the critical interval. Again this is reasonable enough: the difficulty of establishing an association between a response and a reinforcer over a long interval must be greatly increased if several other responses occur during the interval, for each of these conflicting responses will be more readily associated with the reinforcer.* Spence (1956, pp. 155–63) cites a number

---

* The argument has been put more colourfully by Revusky and Garcia (1970). Their thesis is that associations between events widely separated in time fail to get established only because intervening events become associated with the reference events. To illustrate the thesis they suggest a thought experiment:

While you are reading this paper, you find $100 on the floor. Presumably, this functions as a reward for you. The $100 was left by an insane billionaire experimenter because, two hours ago at lunch, you ate gooseberry pie for dessert instead of your usual apple pie. The experimenter wanted to increase the future probability that you would eat gooseberry pie. It is very unlikely that this experiment will be successful . . . Hundreds of events are bound to occur during the two hours between consumption of the gooseberry pie and receipt of the $100 . . . The odds are very great that you have associated one of these intervening events with the $100 and that this would have drowned out the association with the still earlier gooseberry pie. (Revusky and Garcia, 1970, p. 20.)

of studies which demonstrate that the opportunity to make competing responses during the delay interval has a marked effect on performance: when rats ran down an alleyway into a very narrow goal box which forced them to maintain their orientation towards the feeding cup, performance was hardly affected by a 10 sec delay of reinforcement.

One difference between Mowrer and Spence in their analysis of delay of reinforcement is the time scale envisaged, i.e., the length of delay over which associations are assumed to be formed. Spence (1947) had earlier argued that the maximum interval was of the order of seconds and that apparent exceptions were due to the operation of intervening conditioned reinforcement. If a rat is trained to run in an alley to a goal box where reinforcement is delivered only after a delay of a minute, the goal box stimuli will become associated with reinforcement and therefore provide immediate, conditioned reinforcement for the behaviour involved in running in the alley. The only rigorous way, it has been thought, of avoiding this sort of problem is to study choice learning as a function of delay of reinforcement. It should be possible to equate the stimulus consequences of any choice and once one has ensured that the correct response does not produce stimuli that predict the occurrence of delayed reinforcement, then there will be no immediate conditioned reinforcement for one response over the other and it should be possible to obtain a pure measure of the effect of delay of reinforcement.

A series of three experiments, imposing progressively greater control over the possibility of such conditioned reinforcement, obtained progressively less efficient learning at longer delays. The results are illustrated in Fig. 4.3. Wolfe (1934) trained rats in a T-maze, detaining the rats in ostensibly similar delay chambers following correct and incorrect choices. Perkins (1947) randomly interchanged the two delay chambers and the steeper gradient obtained in his study presumably reflects the fact that Wolfe's two delay chambers were not identical. Finally, Grice (1948a) trained rats on a simultaneous black–white discrimination, delaying them in the same delay chambers following both correct and incorrect choices. Since the positive stimulus was equally often on the left and on the right, no one position was consistently correct. Grice argued that his procedure was therefore more rigorous than that used in the earlier T-maze studies. The left and right turns that constitute correct and incorrect responses in a T-maze produce discriminably different proprioceptive stimulation which persists for some time; thus the stimulation produced by a correct turn, by becoming associated with reinforcement, might provide a source of immediate conditioned reinforcement for the correct response. As can be seen from Fig. 4.3, Grice obtained a very much steeper delay of reinforcement gradient than had been obtained in the T-maze studies. Indeed, three of five subjects trained at the 10 sec delay failed to solve the discrimination within 1000 trials. As a further test of the role of proprioceptive stimuli as potential sources of conditioned reinforcement, Grice trained one group of rats with a 5 sec delay of reinforcement, but required them to respond in discriminably different ways in the delay box following correct and incorrect choices. The addition of this response requirement resulted in a significant increase in speed of learning.

While one can accept that the presence, in the delay interval, of stimuli consistently associated with reinforcement may well serve to improve learning with delayed reinforcement, it does not necessarily follow that learning is quite impossible in the absence of such stimuli. Indeed,

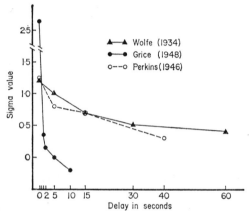

FIG. 4.3. The effect of delay of reinforcement on discrimination learning. Wolfe and Perkins trained rats on a spatial discrimination; Grice trained rats on a brightness discrimination Sigma values are derived from the proportion of correct choices over a given block of training trials. (*After* Grice, 1948a.)

Spence's subsequent analysis in terms of interference suggests that the opportunity for interference is a more important factor than the occurrence of conditioned reinforcement. Mowrer (1960a, p. 360) argued that Grice's brightness discrimination was too complex a problem to provide a sensitive measure of the limits of association across a delay, and Lawrence and Hommel (1961) have provided some strong support for this argument. Other studies have reported evidence of successful learning at intervals very much longer than those employed by Grice. A recent study by Lett (1973) showed that rats could learn a spatial discrimination with delays of reinforcement of several minutes. After each choice, subjects were removed from the apparatus, placed in a holding cage for the duration of the required delay and then replaced in the start box with the door closed. If their last choice had been correct, they were fed in the start box; if it had been an error, they received another trial. Since reliable learning occurred even with an 8 min delay of reinforcement, and animals were handled extensively during the delay between choice and reward, it is hard to believe that immediate conditioned reinforcement was provided by a particular pattern of proprioceptive stimulation maintained throughout the interval.

Revusky (1973) has shown that the colour of the goal box of an alley can serve as a discriminative stimulus signalling whether the following trial will be rewarded, even when the interval between trials is 5 min. Studies of conditioned suppression have reported rapid trace conditioning with intervals between offset of CS and onset of UCS of at least a minute (Kamin, 1965). More recent evidence from studies of conditioned aversions, of course, has shown that rats can associate flavours with sickness across intervals of several hours (p. 67). Studies of alternation learning, discussed in the next section, show that rats can associate the outcome of one trial with reinforcement of the next trial, even if trials are spaced up to half an hour apart. It is difficult to see why the associations between responses and reinforcers required in instrumental experiments should be impossible across intervals longer than a few seconds. The fact that learning is possible with long delays of reinforcement in some situations but not others has prompted Revusky (1971) to argue that the important theoretical problem is not so much to find sources of conditioned reinforcement which enable associations to be formed over intervals of a few seconds, but to find the sources of interference which prevent the formation of associations over even longer intervals.

The delay between response and reinforcement, it may be accepted, adversely affects the formation of an association between the two, the extent of this effect being a function of the opportunity for the formation of competing associations. As Spence (1956) accepted, however, it seems probable that Hull's interpretation of the effect of delay on incentive is also partially correct. Discrimination training between stimuli associated with shorter and longer delays of reinforcement produces contrast effects analogous to those produced by variations in magnitude of reinforcement (p. 393) and animals prefer alternatives associated with shorter delays of reinforcement, a preference which can be counteracted by varying the magnitude or probability of reinforcement associated with the longer delay (Logan, 1965; McDiarmid and Rilling, 1965; Chung and Herrnstein, 1967).

E. Schedules of Reinforcement in Discrete-Trial Experiments

1. Random Schedules

Reinforcement may not only be delayed, it may be omitted altogether on certain trials, thus defining a schedule of partial reinforcement. Studies using food as the reward for running down an alley have sometimes reported the rather surprising finding that rats run faster when reinforced on a random 50 per cent of trials than when reinforced on 100 per cent of trials (e.g., Weinstock, 1958; Goodrich, 1959; Haggard, 1959; Wagner, 1961). More accurately, speed in the initial part of the alleyway (latency of leaving the start box and running through the alley) is increased, although speed of running into the goal box itself is usually decreased by partial reinforcement. The explanation of this finding that has been most generally accepted is that provided by frustration theory (Spence, 1960; Amsel, 1967): the occurrence of reinforcement in the goal box is said to establish a classically conditioned anticipation of reinforcement; the occasional omission of reinforcement then frustrates this expectancy and frustration becomes conditioned to stimuli preceding the goal box. One of the properties of frustration is that it is assumed to provide a source of motivation and occasional nonreinforcement may therefore lead to a higher overall level of drive. Since, according to Hull–Spence theory, drive is a general source of energy, potentiating any learned response, partially reinforced animals may respond more vigorously. Since frustration also elicits competing responses, these may interfere with running as the animal approaches the goal, leading to slower running in later sections of the alley. (Frustration theory in general is discussed at greater length in Chapter 5.)

One problem with this explanation is that the phenomenon it seeks to explain is embarrassingly elusive: although there can be no reasonable doubt that partial reinforcement does sometimes increase speed of running, there have been numerous studies in which the effect has failed to occur (e.g., Bacon, 1962; Brown and Wagner, 1964; Amsel et al., 1966; Mikulka and Pavlik, 1966), none of which has succeeded in elucidating the conditions responsible for these differences in outcome. Furthermore, several studies (e.g., Gonzalez and Bitterman, 1969; Mackintosh, 1970a) have obtained the effect only under conditions which suggest that it may sometimes be an artefactual consequence of differences in drive level: consistent reinforcement may lead to more rapid satiation and hence to slower running on the later trials of each session.

A recent study by Rashotte et al. (1972) has emphasized the unsatisfactory state of our present understanding. After exposure to 160 trials of consistent reinforcement, different groups of rats were shifted to either a 75, 50 or 25 per cent schedule of reinforcement for a further 176 trials and were then returned to consistent reinforcement for a final 80 trials. Overall, there was some tendency for groups to increase their speed of running when shifted to the 75 per cent or 50 per cent schedule, but to slow down when shifted to the 25 per cent schedule. Unfortunately, in the 50 per cent group at least, there was no evidence of a return to baseline performance when the consistent schedule of reinforcement was reinstated. Moreover, there was no suggestion that the increase in speed of running in either the 75 per cent or 50 per cent groups was greater in earlier sections of the alley than in later sections. Finally, and most importantly, the averaged performance of each group served to conceal a bewildering variety of individual changes, with some subjects speeding up, others slowing down, and yet others showing no change in speed. The authors concluded that "no rat in this experiment performed as frustration theory would expect". (p. 203).

The variability of their data prompted Rashotte *et al.* to suggest that a schedule of partial reinforcement might inadvertently provide differential reinforcement for different speeds of running. If a rat happened to run fast on a few rewarded trials and slowly on a few unrewarded trials early in training, this might be sufficient to produce an overall increase in speed of running. They reported that careful examination of their data revealed some support for this suggestion in some individual cases. Wilton (1972) has suggested that it may be the uncertainty rather than the frustration produced by random reinforcement that is responsible for any increase in speed of running. Whatever the merits of these suggestions, it must be obvious that no theoretical analysis can be regarded as satisfactory until it has accounted for the variability of the effect from one experiment to another, or from one subject to another in the same experiment.

2. *Regular Schedules*

In typical experiments on partial reinforcement, the schedule of reinforced and nonreinforced trials is quasi-random. When the schedule is determined according to some fixed sequence, more systematic changes in behaviour may be observed. If rats are trained on a regular schedule, with reinforced (R) and nonreinforced (N) trials occurring in strict alternation, they initially tend to run faster after R trials than after N trials (Tyler *et al.*, 1953; Bloom, 1967). Since this result is also obtained with random schedules of reinforcement (Tyler *et al.*, 1953; McCain, 1966), it presumably reflects the local incremental and decremental effects of reinforcement and nonreinforcement. With continued training on the alternating schedule, however, performance may begin to reverse, until eventually subjects may come to run faster after N trials than after R trials, and thus run faster on rewarded trials than on unrewarded trials (e.g., Tyler *et al.*, 1953; Capaldi, 1958).

Such appropriate performance must depend on the ability to discriminate when reward is available. The question is how this discrimination is achieved. In some circumstances it may be possible for rats to smell from the alley whether the goal box is baited or not; while Ludvigson and

Sytsma (1967) have presented evidence suggesting that whenever a group of rats is run on the same schedule, subjects may use cues provided by the preceding animal in the group as a reliable indicator of the presence or absence of reward. Although these cues may be important in some cases, as has been shown by Amsel et al. (1969), there is good evidence that they are not necessary for the appearance of the discrimination. Their importance can be tested by comparing the performance of subjects trained on a regularly alternating schedule with that of subjects trained on a random schedule (e.g., Tyler et al., 1953; Capaldi, 1958; Flaherty and Davenport, 1972). They can be eliminated by appropriate arrangements for the delivery of reinforcement, and randomization of the order of running different subjects (e.g., Harris and Thomas, 1966; Bloom and Malone, 1968). And they are also eliminated in studies of discrete-trial lever pressing, where the subject remains in the apparatus between trials and reinforcement is delivered from a magazine (e.g., Gonzalez et al., 1966a; Heise et al., 1969).

The pattern of responding observed in studies such as these implies that subjects must be using the outcome of one trial as a cue to predict the outcome of the next (Capaldi, 1967a). On a regularly alternating schedule of reinforcement, of course, an N outcome on one trial is a perfect predictor of an R outcome on the next and an R outcome predicts that the next trial will be nonreinforced. There are several lines of evidence which add weight to the conclusion that the outcome of the preceding trial is the cue used to predict the outcome of the next. Perhaps the most straightforward support is provided by the finding that appropriate patterning is affected by the length of the intertrial interval. Several studies have shown that rats will learn to respond appropriately on single-alternation schedules much more readily when the interval between trials is short rather than long (Katz et al., 1966; Heise et al., 1969). Capaldi and Stanley (1963) found better patterning with a 15 sec interval between trials, rather than a 2 or 10 min interval. Although they found that patterning improved again when the intertrial interval was increased to 20 min, this result must be treated with some reserve, since the conditions under which this group was run seem to have been such as to permit the use of cues provided by the preceding subject in the group, and a subsequent study, run under otherwise similar conditions, did not report any evidence of patterning at a comparable intertrial interval (Capaldi and Minkoff, 1967).

Although the discrimination between R and N trials may be easier when the interval between trials is short, there is unequivocal evidence that with sufficient training significant patterning can occur at intervals ranging from 15 to 30 min (Bloom and Malone, 1968; Flaherty and Davenport,

1972; Pschirrer, 1972). Capaldi and Spivey (1964), indeed, reported evidence of significant patterning when the interval between trials was 24 h, but this is another result which should be treated with considerable caution since several subsequent studies have failed to obtain any sign of such an effect (Surridge and Amsel, 1966, 1968; Amsel *et al.*, 1969).

If rats use the outcome of one trial to predict the outcome of the next, this information must be remembered over the length of the intertrial interval and it is not surprising that some forgetting should occur as the length of this interval is increased. One potent cause of forgetting should be proactive interference: the memory for the outcome of the last trial might be expected to suffer interference from the opposite outcome of the preceding trial.* In accordance with this suggestion, Flaherty and Davenport (1972) noted that when they trained rats on a single-alternation schedule with a 15 min interval between trials, discriminative performance was better on the second trial of each day than on subsequent trials. On the second trial, there is no other preceding trial to interfere with remembering the outcome of the first trial of each day. Mackintosh (1971b) reported rather similar results in a study with goldfish.

Several studies have examined schedules of reinforcement more complex than single alternation of R and N trials. Except in rare instances of extensive training with highly massed trials (e.g., Heise *et al.*, 1969), the discriminative performance that emerges appears to depend exclusively on the cues provided by the outcome of the immediately preceding trial. For example, although a double-alternation schedule (RRNNRRNN) could, in principle, be responded to appropriately, by learning that two R trials predict an N trial and so on, the outcome of one trial cannot be predicted by the outcome of only the preceding trial, and rats generally show no evidence of reliable discrimination on such a schedule (Bloom and Capaldi, 1961). When the schedule consists of the regular sequence RNNRNN, rats learn to respond rapidly on R trials and slowly on the first N trial of each pair but continue to run rapidly on the second N trial of each pair (Capaldi and Senko, 1962). Although the first N trial is reliably predicted by an R trial, both the second N trial and the R trial are preceded by an N trial, and no single event, therefore, can serve as a cue to discriminate between the two. Several studies have shown that the difficulty of learning the appropriate discrimination on schedules such as these is due to the ambiguity of the cue provided by the preceding trial. Pschirrer (1972) and Likely (1970) trained rats on RRNRRN schedules and showed that the use of different rewards on the two R trials enabled subjects to learn to run slowly on the N trial and rapidly on both R trials. E. J. Capaldi (1971) has shown that

---

* The possibility of interference between competing memory traces is discussed at greater length in Chapter 8.

rats can learn to respond appropriately on a double-alternation schedule if differently coloured alleys are used on alternate trials in such a way as to establish a reliable relationship between the outcome of one trial and the outcome of the preceding trial in that alley.

The fact that patterning based on the outcome of the preceding trial occurs readily enough at intertrial intervals of at least several minutes adds force to the argument presented above that learning can occur with relatively long delays of reinforcement. The notion that animals can associate only perfectly contiguous events cannot be seriously maintained: they do, after all, have memories.

IV. FREE-OPERANT SCHEDULES OF REINFORCEMENT

Just as reinforcement may not be scheduled on all trials in a discrete-trial experiment, so it may not be scheduled for all responses in a free-operant experiment. Free-operant experiments, however, permit much greater latitude in the scheduling of reinforcement, and the development and analysis of such schedules has been one of the central concerns of operant conditioners, sometimes indeed assuming an almost mystical importance in their thinking. For some of these reasons, it will be as well to devote a separate section to the discussion of such schedules.

A. Descriptions of Schedules

A free-operant schedule of reinforcement specifies the occasions on which a subject's responses will be reinforced. We can distinguish between simple schedules, in which a single schedule is continuously in effect, and higher-order schedules, in which the subject is exposed to more than one simple schedule in a single session.

The two most common simple schedules are ratio and interval schedules, each of which can be either fixed or variable. In a fixed-ratio (FR) schedule, the subject must complete a fixed number of responses in order to obtain each reinforcement, with the delivery of reinforcement resetting the counter to zero: in an FR 10 schedule, for example, every tenth response is reinforced. In a fixed-interval (FI) schedule, reinforcement is delivered for the first response after a given interval of time has elapsed, with the delivery of reinforcement resetting the clock to zero: in an FI 30 sec schedule, for example, reinforcement is available 30 sec after the last reinforcement. On variable-ratio (VR) and variable-interval (VI) schedules, the number of responses required, or seconds elapsing, before reinforcement is delivered varies around the mean value specified by the schedule.

On a VR 10 schedule, for example, an average of 10 responses is required for each reinforcement, but the requirement may range, in an unpredictable sequence, from 1 to 20. On a VI 30 sec schedule, reinforcement is available on the average after 30 sec have elapsed since the last reinforcement but may be available as shortly as 5 sec later, or not until 60 sec have elapsed.

The remaining simple schedules are those in which reinforcement is explicitly contingent upon the pattern and spacing of responses emitted by the subject. On a DRH schedule (differential reinforcement of a high rate of responding), reinforcement is contingent upon the occurrence of a response within a specified period of time since the preceding response. On the more commonly studied DRL schedule (differential reinforcement of a low rate of responding), reinforcement is contingent upon the occurrence of a response separated from the preceding response by at least some specified interval: on a DRL 10 sec schedule, for example, only responses preceded by a 10 sec pause are reinforced; if the subject emits a second response within 10 sec of the first, the clock is reset and he must now wait 10 sec from the second response.

Simple schedules may be combined in various ways to produce higher-order schedules. The simplest is a mixed schedule: a mixed (FR 10 FR 100) schedule, for example, specifies that reinforcement is sometimes delivered after 10 responses, but that the animal must sometimes complete 100 responses before obtaining reinforcement. A multiple (FR 10, FR 100) schedule is the same as the mixed schedule, except that different external stimuli are correlated with each of the simple schedules involved. As noted above, a multiple schedule in which one of the component schedules is extinction constitutes the free-operant analogue of a successive discrimination in discrete-trial situations. In addition to mixed and multiple schedules, there are tandem and chained schedules. In a tandem schedule, reinforcement is delivered only upon completion of the second simple schedule: the subject must satisfy the requirements of the first schedule in order to enter into the second schedule. A chained schedule is to a tandem schedule as a multiple schedule is to a mixed schedule: in other words, a chained schedule is a tandem schedule in which different external stimuli signal which part of the schedule is in effect at the moment. Thus a chain (FR 10, FI 30 sec) schedule, the subject must respond 10 times in the presence of one stimulus, at which point the stimulus changes and the first response after 30 sec have elapsed is reinforced. Two further higher-order schedules are concurrent schedules, in which two or more responses are simultaneously available to the subject with different schedules in effect for each response, and conjunctive schedules in which reinforcement is delivered only when the requirements of two simple schedules have been satisfied. In a conjunctive schedule, the subject may work towards the satisfaction of both requirements at the same time: for example, in a tandem (FR 10, FI 30 sec) schedule, the subject must first satisfy the ratio requirement, at which point the interval schedule comes into effect and the first response after 30 sec is reinforced; in a conjunctive (FR 10, FI 30 sec) schedule, on the other hand, reinforcement is delivered as soon as the subject has completed 10 responses, provided only that 30 sec have elapsed since the last reinforcement and the 30 sec interval is timed from the moment of the last reinforcement, not from the moment when the ratio requirement is completed.

We can now turn to a more detailed examination of how various schedules affect instrumental responding.

B. Interval Schedules

1. *Fixed Interval*

The data from experiments on schedules are usually presented in the form of cumulative records, which plot the cumulative number of responses emitted by the subject against the passage of time. Fig. 4.4 shows a number of cumulative records of pigeons trained on FI schedules.

The first panel shows performance of a pigeon very shortly after being shifted from continuous reinforcement (sometimes abbreviated to CRF)

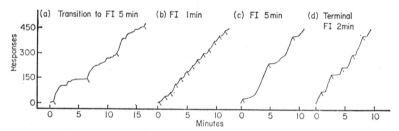

Fig. 4.4. Performance of pigeons on FI schedules of reinforcement. (a) Performance shortly after the transition from continuous reinforcement to FI 5 min. (b) and (c) Performance on FI 1 min and FI 5 min. (d) Performance on FI 2 min after extended training. (*After* Ferster and Skinner, 1957.)

to an FI 5 min schedule. The development of behaviour here shows some striking parallels with the behaviour occurring in an alleyway on a regularly alternating schedule of reinforcement. As can be seen in the first panel, reinforcement initially increases rate of responding, but nonreinforcement produces a gradual decline in rate of responding. After extended training, however, the pattern is reversed. The occurrence of reinforcement serves as a signal for nonreinforcement and little responding occurs, but as time passes the rate of responding increases. The second and third panels show relatively stable performance on FI 1 min and FI 5 min schedules. The characteristic terminal pattern of responding on any FI schedule is a pause following reinforcement, followed by a more or less gradual increase in rate of responding until the next reinforcement is obtained (Ferster and Skinner, 1957, pp. 135–9). The resulting curve is referred to as an FI scallop. The length of the post-reinforcement pause (as a comparison of the second and third panels shows) varies directly with the length of the FI (Schneider, 1969). If sufficient training on an FI schedule is given, the relatively smooth scallop of the second and third panels changes into the relatively abrupt transition shown in the final panel of Fig. 4.4: after reinforcement, no responding occurs until the subject starts responding at its terminal rate (Ferster and Skinner, 1957, p. 158; Schneider, 1969).

2. *Variable Interval*

A VI schedule produces a relatively steady rate of responding, with little of the scalloping of an FI schedule (Ferster and Skinner, 1957, Chapter 6). Different methods of programming VI schedules produce slight differences in local rates of responding. If the probability of reinforcement becoming available remains constant as the time since the last reinforcement increases, then rate of responding is quite steady; with the more commonly used arithmetic series of intervals,* probability of reinforcement increases, and there is a corresponding slight increase in rate of responding, with time since the last reinforcement (Catania and Reynolds, 1968). The overall rate of responding on a VI schedule varies inversely with the mean interval between reinforcements (Clark, 1958; Herrnstein, 1961; Catania and Reynolds, 1968).

C. Ratio Schedules

1. *Fixed Ratio*

It is not, in general, as easy to shift from CRF to a high FR schedule as it is to shift to a long FI schedule: instead, an FR 100 schedule must be approached by increasing the FR requirement relatively slowly (Ferster and Skinner, 1957, p. 42). The characteristic patterns of responding produced by different FR schedules are shown in Fig. 4.5. In general, there is a short pause following each reinforcement, and this is followed by an abrupt transition to a very high rate of responding; it is quite common

---

* An arithmetic VI 30 sec schedule is generated by having a number of intervals ranging from, say, 10 to 50 sec, each of which appears equally often in a random sequence.

for pigeons to peck at a rate of three or four responses per second. In fact, the transition is not always an abrupt one to a high terminal rate; there are numerous instances in which a more gradual increase in rate has been observed and careful recording shows that there is some decline in rate of

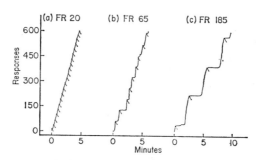

FIG. 4.5. Performance of pigeons on FR schedules of reinforcement. (*After* Ferster and Skinner, 1957.)

responding towards the end of each ratio (Davison, 1969b; Platt and Senkowski, 1970a). The length of the post-reinforcement pause, as can be seen from Fig. 4.5, varies directly with the size of the ratio requirement, with terminal rate of responding being more nearly invariant (Ferster and Skinner, 1957, pp. 49–57; Felton and Lyon, 1966).

2. *Variable Ratio*

Variable-ratio schedules of reinforcement produce a high overall rate of responding, with little or no tendency to pause following reinforcement. Occasional breaks in responding occur, especially on high VR schedules, and especially when a large number of responses has occurred without a reinforcement (Ferster and Skinner, 1957, p. 403).

D. Theoretical Analysis of Schedules of Reinforcement

The behaviour maintained by these simple schedules of reinforcement is reasonably orderly. Although individual subjects may differ widely in their overall rate of responding, the particular pattern of responding produced by each schedule is relatively consistent. This does not, however, mean that the processes responsible for these characteristic patterns of responding are either simple or well understood. On the contrary, they are complex and have proved extremely difficult to analyse. There is reason to suspect that this is because the very nature of a free-operant experiment permits the simultaneous operation of too many variables for it to be easy to disentangle their independent contributions.

In a typical free-operant experiment, the subject can be thought of as repeatedly emitting the same response in an unchanging environment. The data obtained from such an experiment are usually represented as rates of responding and differences in rate of responding at different points in time. Since animals may be assumed to be sensitive to the passage of time, and thus to be able to discriminate between different points in time since the last reinforcement or last response, it seems reasonable to assume that many schedules may come to exert characteristic effects on behaviour because they explicitly or implicitly specify differences in the probability of reinforcement at different points in time. The most obvious temporal distribution of responses is observed in FI schedules, where there is an explicit correlation between the probability of reinforcement and the passage of time since the last reinforcement. Other schedules, such as DRL, in which explicit temporal contingencies are arranged, may also produce more or less appropriate effects on behaviour, and analysis in terms of implicit temporal contingencies has also been applied to differences in overall rates of responding maintained by various schedules which do not, at first sight, appear to provide any temporal contingencies.

Different schedules may not only arrange different probabilities of reinforcement at different points in time, they may also specify different delays of reinforcement for different patterns or rates of responding. Furthermore, the delivery of a reinforcer may not only directly increase the probability of immediately preceding responses or patterns of responding, it may also provide delayed reinforcement for earlier responses and serve as a UCS to elicit further patterns of behaviour. The problem of unravelling these different effects is indeed formidable.

1. *Interpretation of Differences between Fixed and Variable Schedules*

On FI and FR schedules, animals eventually learn to stop responding for some interval of time following each reinforcement, the length of this post-reinforcement pause being proportional to the length of the interval between reinforcements on FI schedules. Neither VI nor VR schedules produce such marked post-reinforcement pauses.

Catania and Reynolds (1968) showed that local rate of responding on VI schedules is in fact affected by the momentary probability of reinforcement at different times. They found that an arithmetic VI schedule generated a pattern of responding showing a mild scallop: rate of responding tended to increase steadily as the time since the last reinforcement increased. When a number of very short intervals were added to the schedule, rate of responding increased immediately following reinforcement and when a constant probability schedule was used, rate of responding was extremely steady.

The implication is that, on any interval schedule, rate of responding varies, as time since the last reinforcement passes, in accordance with changes in the momentary probability reinforcement. On an FI schedule, the subject learns that the occurrence of reinforcement signals that no reinforcement will be available for x sec and therefore stops responding. As time passes, the subject finds it more difficult to discriminate between x sec and the time that has actually elapsed since the last reinforcement, and responding increases steadily.* The post-reinforcement pause, therefore, is due to a temporal discrimination, and the initiation of a steadily increasing rate of responding is due to a breakdown in the temporal discrimination. Several lines of evidence are consistent with this analysis.

Staddon and Innis (1969) and Kello (1972) trained pigeons on FI schedules and occasionally omitted reinforcements, or only turned on the magazine light without delivering food. Their results suggested that the stimuli accompanying the delivery of reinforcement come to serve as signals for a temporary low probability of reinforcement, and hence come to control a low rate of responding: they found that the greater the change from the set of stimuli normally accompanying reinforcement, the greater the increase in rate of responding at the beginning of the subsequent interval. Trapold et al. (1965) and Zamble (1969) have shown that if animals are exposed to a noncontingent FI schedule, where reinforcement is delivered at fixed intervals without any response being required of the subject, they immediately show a typical FI scallop when transferred to a standard, response-contingent FI schedule.

Dews (1962) and Morse (1966) have argued that FI scalloping is a consequence of differences in delay of reinforcement: responses early in each interval are reinforced only after a long delay, while responses towards the end of each interval are more immediately reinforced. Since response strength is a function of delay of reinforcement, responding should gradually increase throughout each interval. There is, of course, little difference between arguing that responses occur only rarely at the beginning of each interval because they are not reinforced, or because they are reinforced only after a long delay. However, the data reported by Trapold et al. and by Zamble suggest that a typical FI scallop may appear without

---

* As we have seen (p. 167), a sufficient amount of training on an FI schedule may eventually cause animals to respond at a constant high rate from about halfway through each interval. This may be, as Ferster and Skinner (1957, p. 134) have suggested, because once the scallop is developed the probability of reinforcement is (coincidentally) higher for higher rates of responding: low rates of responding, occurring early within each interval, are correlated with a low probability of reinforcement and may eventually extinguish. This possibility illustrates the complexity of schedule-produced behaviour. The experimenter has rather little control over the regularities that may come to control the subject's behaviour: these regularities may be a consequence of the behaviour itself.

subjects having been exposed for any length of time to any instrumental contingency. The FI scallop, after all, is in many ways analogous to inhibition of delay in classical conditioning (p. 61), with the delivery of one reinforcer serving as a trace CS for the next reinforcer. Morse (1966) has advanced the more curious argument that since the FI scallop can, in principle, be accounted for by differences in delay of reinforcement, it is unnecessary to appeal to any temporal discrimination at all. It is difficult to understand what could be meant here: different delays of reinforcement for responses emitted at different points in time could produce differences in rate of responding only if subjects were discriminating between those different points in time.

FR schedules also produce a post-reinforcement pause. This is, at first sight, more surprising: on FI schedules, responses immediately following reinforcement are completely wasted; on FR schedules, however, any pause following reinforcement inevitably postpones the delivery of the next reinforcement. Nevertheless, long FR schedules do, as a matter of fact, ensure that responses occurring shortly following reinforcement are never immediately reinforced. Just as on FI schedules, therefore, short intervals following reinforcement are correlated with a zero probability of reinforcement. An experiment by Killeen (1969) compared the length of the pause in pigeons trained on various FR schedules, and in yoked subjects, for whom reinforcement was made available whenever the FR bird obtained reinforcement. The yoked birds were in effect on an interval schedule, where the length of the interval, although not fixed, varied only within narrow bounds. Post-reinforcement pausing was indistinguishable in the two groups. The implication is that pausing on FR schedules and on FI schedules is indeed due to exactly the same temporal discrimination.

Neuringer and Schneider (1968) have provided relatively conclusive evidence for this analysis of the FR post-reinforcement pause. Pigeons were trained on an FR 15 schedule of reinforcement, but the overall interval between successive reinforcements was manipulated by turning off the key light, the sole source of illumination in the chamber, for short periods following each response. Since pigeons do not readily peck in the dark, the effect of this procedure was to slow down the rate of responding, and thus to increase the interval between reinforcements. It resulted in a marked increase in latency of responding, i.e. in the length of the post-reinforcement pause. However, the same procedure applied to an FI schedule had no effect on rate of responding—presumably because it had no effect on the interval between successive reinforcements.

The behaviour generated by FI and FR schedules, therefore, leaves little doubt that the delivery of reinforcement constitutes a stimulus event which may acquire discriminative control over responding and that animals

are capable of discriminating different times since the delivery of reinforcement. The implication that reinforcers may act as stimuli is one that
we have already come across in discussion of discrete-trial alternation
learning (p. 162); it also, as we shall see, has a bearing on a large number
of other areas.

2. Explicit Reinforcement of Different Rates of Responding

Fixed-interval schedules explicitly reinforce responses only after a fixed
time since the last reinforcement and the evidence suggests that the
control over responding established by such schedules depends upon the
formation of this temporal discrimination. Other schedules may explicitly
reinforce responses only after a fixed interval since the last response: a
DRL 10 sec schedule, for example, reinforces only those responses spaced
at least 10 sec apart. The question is whether the pattern of responding
maintained by such a schedule depends upon an analogous temporal discrimination. The overall rate of responding maintained by DRL schedules
is, indeed, relatively slow: in fact, a bimodal distribution of responses in
time tends to develop, with most responses clustering around the specified
interval and many of the remainder occurring in bursts at very short
intervals (Kramer and Rilling, 1970).

In addition to DRL and DRH schedules, other studies have examined
the possibility of explicitly reinforcing other distributions of responses in
time. Shimp (1967), having trained pigeons to respond on a VI schedule,
then reinforced only those responses preceded by pauses of a specified
duration. If reinforcement was contingent on a pause of between 0·9 and
1·2 sec between responses (usually described as an inter-response time or
IRT of 0·9 to 1·2 sec), such pauses increased reliably. Blough (1966)
developed a schedule which explicitly reinforced only those responses
preceded by pauses which recently had occurred less frequently than
would be expected by chance. Such a schedule, which in effect required
subjects to distribute their responses at random intervals, produced IRT
distributions which relatively closely matched an expected random
distribution.*

Reynolds (1966) demonstrated that the length of the pause preceding a
response may serve as a stimulus in a further way. Pigeons were required
to peck twice on a red key, the second peck turning the key blue for 30 sec;
reinforcement was available in the presence of the blue light (on a VI
schedule) provided that the interval between the two responses on the red

---

* The important exception, noted by Blough, was that although IRTs shorter than
0·8 sec were never reinforced, many responses continued to occur at these short IRTs.
Blough suggested that such responses may reflect particular forms or topographies of key
pecking, such as pecking with the beak open in such a way that the key is struck first by
the top part of the beak and then, a fraction of a second later, by the bottom of the beak.

key was at least 18 sec. Rate of responding to the blue stimulus varied directly with the length of the pause before the second response to red on that trial. Although, in fact, the majority of responses during the red stimulus occurred too rapidly to set up reinforcement during the blue stimulus, the subjects thus showed that they were well able to discriminate when reinforcement would be available.

All that has been said so far, however, is that the differential reinforcement of particular IRTs, specified, for example, by a DRL schedule, somehow modifies the animal's distribution of responses so as to increase the probability of reinforced responses. It may be that time since a response constitutes a simple stimulus dimension for rats and pigeons. However, when either rats (Wilson and Keller, 1953), or pigeons (Laties *et al.*, 1965) are trained on DRL schedules, experimenters have noted that they tend to develop stereotyped patterns of behaviour during the intervals preceding lever pressing or key pecking. One suggestion is that the cue for the emission of a reinforced response is not the passage of an appropriate amount of time since the previous response, but the occurrence of an appropriate amount of this "collateral" behaviour.*

In principle, a similar argument could be applied to the temporal pattern of responding maintained by FI schedules. It might be that the overall increase in the rate of key pecking or lever pressing during each interval is a consequence of the reinforcement of a particular pattern of behaviour, which includes different responses occurring between key pecks at the beginning of an interval, or different forms or topographies of key pecks at different points in time during each interval. The results of several experiments by Dews (1962, 1966) suggest that this is not an important determinant of the FI scallop. Dews was able to show that the introduction of brief black-outs at different points in an FI schedule, although sometimes totally suppressing key pecking during the black-out, had essentially no effect on the pattern of responding maintained when the key light was illuminated.

There is, however, some evidence that the behaviour maintained by DRL schedules may critically involve patterns of collateral responses. Laties *et al.* (1969) observed that some of their rats tended to gnaw at the grid floor or corner of a wall during intervals between lever presses; when a partition prevented access to the floor or corner, DRL performance was markedly disrupted. In other cases, rats started to gnaw on blocks of wood as soon as these were made available and this improved their DRL

---

* There is, it should be noted, good evidence that the amount of a specified pattern of behaviour (e.g., the number of lever presses or key pecks) can act as an effective cue for controlling subsequent behaviour (Rilling and McDiarmid, 1965; Pliskoff and Goldiamond, 1966), and that it is the number of responses, rather than the amount of time required to make them, that is often the effective stimulus (Rilling, 1967).

performance; as soon as the blocks were removed their performance declined. It is hard to resist the conclusion that in these instances the behaviour observed was not incidental, but served as a cue signalling the appropriate occasion for lever pressing.

It is difficult, however, to reach any conclusion of general validity. The fact that in some situations the occurrence of collateral behaviour is positively related to the accuracy of DRL timing does not imply that such behaviour is always necessary for the proper spacing of responding. Internal cues to the passage of time are presumably available to animals and they may, if other possibilities are excluded, gain effective control over behaviour. No single demonstration of the importance of collateral behaviour could rule out this possibility. On the other hand, the observation of such behaviour does suggest that a rather different interpretation of many instances of spaced responding could be advanced. As Blough (1966) noted, key pecks that occur at very short IRTs may simply be a consequence of a particular manner of pecking the key. It may be that the mechanism whereby differential reinforcement of specific IRTs gains its effect is not by reinforcing the response of pecking a key or pressing a lever in the presence of a specific, temporal stimulus, but by the reinforcement of a specific topography of key pecking or lever pressing (Reynolds and McLeod, 1970). Reynolds and McLeod suggested that a DRH schedule might teach pigeons to "hold their beaks open close to the key and 'peck' by rapidly nodding their heads so that the top and bottom halves of the beak alternately hit the key" (p. 97). A DRL schedule might selectively reinforce a different pattern of pecking—one which necessarily involved a longer interval between successive closures of the key. Even though, as we have seen, the requirements of the schedule can, and perhaps should, be stated as involving differential reinforcement of specific intervals between responses, it might be misleading to describe the change in behaviour as a change in the probability of responding at those specific intervals. It is possible that "the change has occurred in the topography of the behavior itself" (Reynolds and McLeod, 1970, *loc. cit.*). This possibility should be borne in mind when considering the differences in the patterns or rates of responding maintained by ratio and interval schedules.

3. *Differences between Interval and Ratio Schedules*

There appear to be two main differences between the effects of interval and ratio schedules. First, it is easier to establish responding on interval than on ratio schedules: an overall frequency of reinforcement which is sufficient, for example, to maintain responding on an FI schedule may not maintain responding on an FR schedule. The most striking illlustration of this point is provided by an experiment by Herrnstein and Morse (1958).

Pigeons trained on an FI 15 min schedule of reinforcement were averaging about 300 responses in each 15 min interval. When they were shifted to a conjunctive (FR x, FI 15 min) schedule, i.e. when they were *required* to emit at least x responses in order to obtain reinforcement after 15 min, rate of responding declined dramatically—even when the FR requirement was substantially less than the 300 emitted, on the average, in the absence of any requirement. These results are illustrated in Fig. 4.6: even requirements as small as FR 40 produced some decline in responding and at FR 120 or 240 responding nearly disappeared.

Herrnstein and Morse pointed out that although the average number of responses emitted on the FI schedule in the absence of any ratio requirement was about 300, there was considerable variation from one interval to

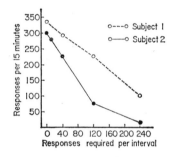

FIG. 4.6. The effect of adding a ratio requirement to performance maintained by an FI schedule. In the absence of any ratio requirement (Point 0 on the abscissa) subjects averaged about 300 responses in each 15 min interval. The addition of any ratio requirement, however, led to a sharp decline in the rate of responding. (*After* Herrnstein and Morse, 1958.)

another: sometimes subjects responded only 10 or 20 times in a 15 min interval, sometimes as many as 500 times. Now any interval schedule ensures an inverse relationship between the overall rate of responding and the probability that any one response will be reinforced: when only 10 responses occur in an interval, the probability of reinforcement per response is 0·1; when 100 responses occur, the probability of reinforcement per response declines to 0·01, If rate of responding is inversely proportional to probability of reinforcement, then interval schedules have a built-in correcting device: when rate of responding declines, probability of reinforcement increases and rate will increase; when rate increases, probability of reinforcement decreases and rate will decline again. In accordance with this, it has been noted frequently that on any long fixed-interval schedule, the number of responses per interval fluctuates in a regular manner, intervals with a small number of responses being followed by intervals with many responses, and vice-versa (e.g., Ferster and Skinner, 1957, pp. 158–9; Cumming and Schoenfeld, 1958).

A ratio schedule, of course, arranges no such inverse relationship between rate of responding and the probability of a response being reinforced. This latter probability is fixed by the value of the ratio. Furthermore, as rate of responding declines, so does the overall rate of reinforcement. If we were to suppose that rate of reinforcement directly determines rate of responding then a ratio schedule would have build-in positive feedback, leading either to an increase in rate of responding to the upper limit of the subject's capacity, or to the complete cessation of responding.

Two different principles have been suggested here and it is important that they be kept distinct: the first possibility is that responding may be affected by the *probability* of reinforcement for each response; the second is that responding may be affected by the overall *rate* of reinforcement, or probability of reinforcement per unit of time. The data generated by interval schedules provide evidence for the importance of probability of reinforcement per response as a determinant of responding; whether rate of reinforcement is an important factor is a question we shall return to below. To anticipate that discussion, it can be pointed out that a decline in rate of reinforcement may affect responding by increasing the delay of reinforcement for each response. In the meantime, we need to consider the second difference in the patterns of responding generated by ratio and interval schedules.

Provided that they maintain responding at all, ratio schedules generate very high rates of responding, apparently substantially higher than those maintained by interval schedules. Thomas and Switalski (1966), following a procedure first utilized by Ferster and Skinner (1957, pp. 399–405), trained one group of pigeons on a VR schedule and permitted pigeons in a second group to obtain reinforcement whenever a matched bird in the VR group was reinforced. This second, yoked group was in effect being trained on a VI schedule, with an overall frequency and distribution of reinforcement precisely matching that in effect for the VR group. Thomas and Switalski found no overlap between the rates of responding maintained by the two schedules: the slowest bird in the VR group responded more rapidly than the fastest bird in the yoked, VI group.

Why should ratio and interval schedules maintain different rates of responding? The most widely accepted explanation, originally proposed by Skinner (1938, pp. 275–84), and Ferster and Skinner (1957, pp. 133–4), is that the two classes of schedule effectively assign different probabilities of reinforcement to different rates of responding, or, what amounts to the same thing, differentially reinforce responses preceded by different IRTs. "The VR schedule produces a higher rate than the VI because the VR differentially reinforces relatively short IRTs while the VI differentially reinforces relatively long IRTs" (Reynolds, 1968, p. 65).

In spite of its authority, this statement is only half true. Although it may well happen that on a ratio schedule the majority of reinforcements occur following responses preceded by short IRTs, this is only because the majority of all responses on a ratio schedule are preceded by short IRTs. The *probability* that a response will be reinforced on a ratio schedule is fixed at the reciprocal of the ratio, and is quite independent of the length of the pause preceding that response. From what has been said earlier about the importance of contingency rather than frequency of reinforcement (p. 25), it does not seem likely that ratio schedules can be thought of as providing differential reinforcement for high rates of responding in this sense at all.

It is true, on the other hand, that an interval schedule does specify a relationship between the probability that a response will be reinforced and the length of the pause preceding that response. On an FI 30 sec schedule, for example, responses preceded by a pause of 30 sec are always reinforced. In general, the longer the subject waits before responding on an FI or VI schedule, the greater the probability that the next response will be reinforced (Revusky, 1962). What is true in logic is also true in fact: Williams (1968) and Dews (1969) have noted that interval schedules do indeed differentially reinforce responses preceded by relatively long IRTs.

It is possible, therefore, that this differential reinforcement of slow rates of responding implicit in any interval schedule is responsible for the general difference in the rates of responding maintained by interval and ratio schedules. Ferster and Skinner (1957, pp. 416–22) attempted to test the importance of this factor by arranging a schedule which eliminated the differential reinforcement of slow rates of responding specified by an ordinary interval schedule. They added a short ratio requirement onto the end of a long interval schedule, turning an FI 45 min schedule into a tandem (FI 45 min, FR 10) schedule, and observed a dramatic increase in the subject's overall rate of responding. However, Dews (1969), repeating the experiment with a much shorter interval schedule (FI 2 min), was unable to find any consistent increase in overall rate of responding when an FR 2 or FR 10 requirement was added onto the end of each interval.

Although it is true, therefore, that interval schedules implicitly reinforce responses preceded by relatively long pauses, while ratio schedules do not, it has yet to be definitively shown that this difference is responsible for the large difference in overall rate of responding typically shown on these schedules. Even if this were true, of course, it might be better to say that interval schedules generate a relatively slow rate of responding, not because they differentially reinforce long IRTs but because they reinforce patterns or topographies of behaviour that necessarily involve relatively long intervals between successive responses (Millenson, 1966). A second possibility which has often been suggested is that although the *probability* of reinforcement per response on a ratio schedule is independent of the rate of responding, the *rate* of reinforcement is directly related to the rate of responding (Herrnstein, 1970). The faster a subject responds, the greater the number of reinforcements obtained in any given interval of time.

The notion that rate of reinforcement might be an important determinant of rate of responding is one rather naturally suggested by free-operant experiments where the former can be varied at will by the experimenter and the latter by the subject, both within rather wide limits. It has even been suggested that rate of reinforcement may affect rate of responding on interval schedules and that the reason why a VI 30 sec schedule maintains a higher rate of responding than a VI 3 min schedule is that the former delivers reinforcement at a higher rate (e.g., Catania and Reynolds, 1968). Since, however, variations in rate of reinforcement on interval schedules are always correlated with variations in the probability of reinforcement per response, an appeal to rate may be superfluous. It is only on ratio schedules that rate of reinforcement may vary in spite of a fixed probability of reinforcement for each response.

It is true that a ratio schedule arranges a perfect correlation between rate of responding and rate of reinforcement. But rate of reinforcement is not necessarily the controlling variable, for it is equally true that the more rapidly a subject responds on a ratio schedule, the shorter the delay of reinforcement for each response. Although ratio schedules do not specify different *probabilities* of reinforcement for different rates of responding, they do specify different *delays* of reinforcement. Since subjects will choose between two responses on the basis of different delays of reinforcement (p. 186), it is reasonable to suppose that when two rates or patterns of responding are correlated with different delays of reinforcement, the pattern reinforced after the shorter delay will increase in frequency (Dews, 1962; Catania, 1971b).

Delay of reinforcement and rate of reinforcement are not the same variable. Killeen (1968), for example, studying pigeons on a concurrent chained schedule (see p. 242), found that subjects showed a preference for a key that predicted some very short delays before the first reinforcement, in spite of the fact that the overall rate of reinforcement predicted was no higher on this key than on the other. It should be possible to hold overall rate of reinforcement constant on a ratio schedule while permitting delay of reinforcement to vary inversely (in the usual way) with rate of responding. If a variable intertrial interval were introduced following each reinforcement and its length were adjusted so as to ensure a constant time between the start of one trial and the start of the next (i.e., its length varied inversely with the time required by the subject to complete the ratio), then one could see whether rate of responding increased in the absence of an increase in overall rate of reinforcement.

An appeal to overall rate of reinforcement as a determinant of responding is, as Herrnstein (1970) recognizes, likely to involve an appeal to some underlying variable such as "response strength" which is responsible for generating a particular rate of responding. Such a step does not, as yet, seem forced by the data. It is possible to account for patterns of responding on different schedules by a more molecular analysis in terms of the probability of reinforcement per response at different points in time, and in terms of delay of reinforcement.

4. *Maintenance of Response Strength*

There remains one further question to consider. Ratio schedules may maintain high rates of responding because they arrange an inverse correlation between rate of responding and delay of reinforcement. Other schedules, however, do not appear to provide any differential reinforcement for high rates of responding, but although subjects respond more slowly on such schedules than they do on a ratio schedule, they still respond very much more rapidly than is necessary to maximize reinforcement. On an FI 1 min schedule, for example, only one response a minute is required by the schedule; yet pigeons may respond on average as many as 50 times per minute.

The paradox is even more marked on DRL schedules: here rapid responding is not simply a waste of effort, it actually postpones the delivery of reinforcement. Yet Staddon (1965) gave pigeons a total of 255 $2\frac{1}{2}$-hour sessions of training on various DRL schedules ranging from DRL 5 sec to DRL 30 sec, and reported that the best level of performance of which his subjects were capable was to emit approximately 50 per cent of their responses at IRTs equal to or longer than the reinforced IRT. Even this level was attained by only two of the three pigeons in the experiment, and only at DRL values up to 20 sec. The third subject could manage no better than a median IRT of 10 sec on a DRL 20 sec schedule. On DRL 30 sec, performance completely broke down in all subjects. Kramer and Rilling (1970) provide many other examples of similarly inefficient performance. Reynolds (1966), in a study already described, found that pigeons were better able to discriminate the length of the pause between their successive key pecks than actually to space their responses sufficiently far apart to obtain reinforcement. The implication is that the inefficiency of DRL responding cannot simply be attributed to a failure of temporal discrimination.

G

This should come as no surprise. A DRL schedule bears a more than superficial resemblance to an omission schedule: in each case responding in the presence of stimuli associated with reinforcement postpones or prevents the occurrence of reinforcement. Just as pigeons are strikingly inefficient at inhibiting key pecking on an omission schedule (p. 121), so one might expect them to be inefficient at inhibiting key pecking on DRL schedules. Presumably the same reason is operating: pecking, as a component of the UCR to food, becomes classically conditioned to stimuli associated with the delivery of food. The implication is that if a different response, or perhaps a different reinforcer, were involved, DRL performance would improve. Several recent studies have confirmed this expectation. Hemmes (1970) found that pigeons performed much better on a DRL schedule when the response required was pressing a treadle rather than pecking a key. Kramer and Rodriguez (1971) trained rats on a DRL schedule for water reinforcement: performance was less efficient when subjects were required to respond by licking at the tube from which the water was delivered (a response which is readily classically conditioned), than when the response was touching a bar. Finally, Schwartz and Williams (1971) found that the performance of pigeons on a DRL schedule was very substantially improved by the provision of a second key to peck on. Subjects that were unable to inhibit pecking on the main key were able to shift the direction of their pecks to another key during the DRL interval and thereby increased the proportion of reinforcements received.

Although the pigeon's key peck is the prime example of an instrumental response whose occurrence is at least partly determined by a process of classical conditioning, we have argued that most successful instrumental responses, including lever pressing in rats, are easily established on instrumental contingencies, precisely because they are not arbitrary but are more or less closely associated with the reinforcer (p. 135). One reason, then, why both lever pressing and key pecking occur more frequently on interval and DRL schedules than the schedule requires, may simply be that they are generally successful instrumental responses and their probability of occurrence is determined partly by anticipatory conditioning. Other, more arbitrary, responses might be found which would be more "efficient' on interval schedules (i.e., would occur less frequently and therefore be less wasted), but the suggestion is that they would be harder to establish in the first place and would be *less* efficient on, for example, ratio schedules.

We have also touched on a second reason why interval schedules generate more responses than are necessary for maximizing reinforcement. Both interval and DRL schedules arrange an inverse relationship between the rate at which a response is emitted, and the probability that any one response will be reinforced. On either a DRL or FI 20 sec schedule, for example, if subjects space their responses at least 20 sec apart, then the probability of each response being reinforced is 1·0. The implication is that differential reinforcement of long IRTs may never be completely successful in establishing the ideally slow rate of responding, not because of the difficulty of the *discrimination* between reinforced and nonreinforced IRTs but because the probability of reinforcement per response increases as responding slows down to the specified rate and this increase in pro-bability of reinforcement leads to an increase in the probability of responding.

E. Conclusions

The behaviour generated by the major simple schedules of reinforcement is orderly and reproducible. But neither of these characteristics carries any implication about the simplicity of the causal factors responsible for generating such behaviour. The argument of the preceding sections may well have struck the reader as tedious and complex. It has certainly shown a tendency to appeal to new principles whenever the data seemed to require them. We have talked of overall probability of reinforcement per response; of delay of reinforcement; of discrimination between short and long intervals between responses and of time since reinforcement; of differ-ential reinforcement for responding at these different points in time; and of anticipatory, classically conditioned responding. Even so, it is relatively certain that the principles here invoked are insufficient to encompass the data on schedules already available (Jenkins, 1970). This state of affairs may well seem thoroughly unsatisfactory to the theorist; to the untheoretic-ally inclined, it may seem further inducement to ignore the problems of explanation and to return to the laboratory in order to generate further instances of orderly patterns of behaviour produced by newer and more exotic schedules.

It is possible that the picture will change; it is just conceivable that one or two new principles will be discovered that prove capable of explaining all behavioural results of different schedules. There is reason, however, to be sceptical. The fact is, surely, that however orderly the behaviour eventually produced by a particular schedule, it is the product of a complex interaction between the subject's past and present behaviour and the particular pattern of reinforcement received—what the subject does in-fluences what reinforcement is received, and the receipt of reinforcement can exert a variety of different effects on what will be done in the future. In this sense, schedules are very far from being simple. They are frighten-ingly complex and not necessarily at all well suited to an elucidation of the important processes underlying instrumental behaviour. Too many variables are permitted to vary in too many ways for it to be easy to achieve any analytic understanding of the operation of any one of them. This conclusion needs constant reiteration in the face of the argument that an understanding of schedules is basic to the understanding of all instrumental behaviour, and that intermittent schedules of reinforcement, and the patterns of behaviour they generate, are somehow the fundamental operations and phenomena of a proper science of behaviour. In contrast to the claim of Reynolds (1968, p. 60) that "schedules are the mainsprings of behavioral control and thus the study of schedules is central to the study of behavior", it is worth citing a more conservative estimate of the importance of schedules provided by Jenkins (1970):

> We have the option of whether or not to attempt an exhaustive analysis of schedules. The status of reinforcement schedules in experimental psychology is not coordinate with the status of reproduction in experimental biology. Reproduction is a given and, in the development of biological science, there has been no alternative but to analyze its mechanisms in detail. Schedules of reinforcement, on the other hand, are an invention and it is possible to choose whether or not to analyze in detail the effects they produce. There are interesting analogies between reinforcement schedules as arranged by psychologists and the circumstances of behaviour at large. The analogies are, however, probably not as close as popular treatments of reinforcement schedules may suggest. Neither men nor animals are found in nature responding repeatedly in an unchanging environment for occasional reinforcement. In any case, experimental arrangements that resemble natural occurrences are not necessarily the ones best suited to advance the development of a science. An important consideration in choosing phenomena for intensive analysis is simplicity of determination. Neither free operant nor discrete trial schedules are at all attractive in that respect. There is no need to allow the complexities of any given experimental arrangement to force upon us an extensive program of analysis (p. 107).

V. PARAMETERS OF INSTRUMENTAL LEARNING: CHOICE BEHAVIOUR

Although data from experiments on choice behaviour have been mentioned where relevant to the discussion of other topics, the primary concern so far in this chapter has been with experiments which permit the subject to make a single response and which provide some measure of the strength of this response, e.g., latency, speed, probability, or rate as a function of various experimental manipulations. In choice experiments, two or more responses are simultaneously available to the subject (two keys or levers, or two arms of a T-maze) and the measurement taken is often enough simply the percentage of choices of one alternative over the other. Free-operant choice experiments (concurrent schedules) can record the amount of time spent responding on one key or lever rather than the other; this is measured by a pair of clocks, one for each key, with one clock starting as soon as a response occurs to one key, and stopping as soon as a response occurs to the other key. Other measures of responding may, of course, be provided—speed of running in a T-maze, latency of responding in a discrete-trial lever-pressing situation, or rate of responding on each key in a concurrent schedule.*

In the remainder of this chapter we shall discuss the effects of certain basic motivational and reinforcement variables on choice behaviour. One of the main questions of interest here is how animals distribute their responses between two alternatives when each signals a different proba-bility, amount, or delay of reinforcement. The evidence suggests that models of choice behaviour must incorporate some "maximizing" assump-tion, for in most cases the more favourable of two alternatives tends to capture virtually all choices.

A. Drive

Just as an increase in drive increases speed or rate of responding in a single-response situation so, but much less reliably, it may increase the probability of selecting the correct alternative in a choice situation. Several studies of simple T-maze learning (Bronstein and Spear, 1972) or brightness discrimination learning (Eisman et al., 1956; Spence et al., 1959) have found significant increases in speed of learning as drive level is increased, but the effect is not very large, often falls short of significance (Wike et al., 1963) and frequently does not appear at all (Teel, 1952; Hillman et al., 1953; Lachman, 1961). Two variables that may influence the outcome of such experiments are the type of training procedure and the difficulty of the discrimination. The use of a forcing procedure that pro-vides experience with both alternatives may reduce or abolish the effect of drive (Teel, 1952; Spence et al., 1959; Lachman, 1961); higher drive may not facilitate learning if the discrimination is at all difficult. The importance of this variable may be overestimated by its elevation to the status of a law (the Yerkes–Dobson Law), but there is some reason to believe that very

---

* As it happens, the measure typically reported in experiments on concurrent schedules is the proportion of responses to each key: on the face of it, this would seem to be a second-order measure, being derived from the rate of responding on each key and the time spent responding to each key. In fact, the evidence suggests that in many if not most concurrent schedules, rate of responding on each key is relatively constant and differences in the proportion of responses to each key are almost entirely due to differences in the amount of time spent responding to each key (Catania, 1966; Baum and Rachlin, 1969; Stubbs and Pliskoff, 1969).

high levels of drive may actually interfere with, rather than facilitate, choice learning and that the level of drive at which this change occurs is lower for more difficult problems (Yerkes and Dodson, 1908). Unfortunately, although the law has been recently confirmed by Broadhurst (1957), there has been no systematic study of the role of problem difficulty with appetitive drives: Yerkes and Dodson used escape from varying intensities of shock; Broadhurst used a water maze and increased drive by varying the time that rats were confined underwater before the start of each trial. However, it is noteworthy that several studies, in which no discernible effect of drive on choice learning has been found, have employed relatively difficult visual discriminations (Miles, 1959) or multiple-unit mazes (Hillman et al., 1953). Ginsburg (1957) has provided very nice evidence that high drive facilitates learning of a reasonably easy discrimination in pigeons, but retards the learning of a complex conditional discrimination. The discussion of this interaction will be deferred until the next section, where we shall see that there is some evidence of a similar interaction between magnitude of reinforcement and difficulty of discrimination.

B. Magnitude of Reinforcement

1. Effect of Magnitude of Reinforcement on Choice Learning

Pubols (1960), reviewing studies of choice learning in single-unit or multiple T-mazes, or of simultaneous brightness discriminations, found only one experiment in which a larger reward led to faster learning (Reynolds, 1950) and five studies in which no effect was observed. More recent studies have altered that picture. There is now no question but that larger reinforcers usually produce more rapid learning of a position habit in a T-maze or discrimination box (Clayton, 1964, 1969; Clayton and Koplin, 1964; Ison, 1964; Theios and Blosser, 1965; Eimas, 1967; Waller, 1968; Singer, 1969), or of a simple black-white discrimination (Eimas, 1967; Hooper, 1967; Waller, 1968; Mackintosh, 1969; Singer, 1969).

The discrepancy between Pubols' conclusion, that magnitude of reinforcement has little or no effect on choice learning, and the conclusion, based on more recent evidence, that it usually does have a significant effect, may be due to several factors. As in the case of the effect of drive on choice behaviour, it should first be stressed that even in studies reporting positive results, the effect has often been small, and that it has required large numbers of subjects to demonstrate a significant difference. It is also clear that the effect depends upon certain procedural variables: Hooper (1967), for example, reported an effect when a noncorrection procedure was used, but not when subjects were permitted to retrace to the correct alternative after an error. Several studies in which no effect was observed have in fact used a correction procedure (e.g., Furchtgott and Rubin, 1953; Maher and Wickens, 1954). Reynolds (1950) found faster learning for a larger reinforcer when training rats on a two-choice position problem by a non-correction procedure, but in an earlier experiment (Reynolds, 1949) no effect when rats were trained on a black-white discrimination with a correction procedure. A second factor that may be important is the difficulty or complexity of the task. Mackintosh (1969) noted some tendency for performance on a relatively easy black-white discrimination to be more affected by magnitude of reinforcement than was performance on a more difficult brightness discrimination. It is relevant to note that several early failures (e.g., Furchtgott and Rubin, 1953; Maher and Wickens, 1954) used multiple-unit T-mazes rather than a single T-maze, while Lawson (1957), who found an effect of magnitude of reinforcement only in subjects concurrently exposed to both values of reinforcement, used a relatively difficult absolute brightness discrimination.

If these two factors are indeed involved in determining the outcome of this set of studies, it may be possible to reduce their effects to a common principle. A retrace correction procedure clearly minimizes the consequences of an error and there is evidence to suggest that, although sometimes benefiting performance, it also tends to increase confusion between the correct and incorrect alternatives (presumably by providing delayed reinforcement for errors), sometimes retarding learning (Hull and Spence, 1938; Kalish, 1946; Towart and Boe, 1965). If so, its effect may be the same as that of an increase in the similarity of the two alternatives. Now this effect is, of course, exactly the same as the interaction between problem difficulty and drive level noted in the preceding section. It may be that variations in either drive or reinforcement have one effect on the learning of easy choice problems, but another effect when the problem is more difficult. One fairly simple explanation is to point to the importance of competing responses or irrelevant stimuli. If high levels of drive and large magnitudes of reinforcement increase response strength, they should facilitate learning in a choice situation—provided that it is the *correct* response that benefits most from their effects. If, however, there are strong competing response tendencies, as there might be in a complex, multiple-unit maze, or salient irrelevant stimuli, as there might be in a difficult discrimination problem, then a high drive or large reinforcement may initially increase inappropriate response tendencies and only later, when these responses have been extinguished, come to facilitate appropriate choice behaviour. Under these conditions, both these variables would have conflicting effects on overall speed of learning and it is not surprising that the net result should so often be that it is impossible to detect any difference between different groups.

2. Choice between Different Magnitudes

If one goal box of a T-maze contains four pellets of food and the other three, rats will eventually learn to select the more favourable alternative (Hill and Spear, 1963b). However, learning progresses more rapidly if the alternatives are four and one pellets and this problem in turn is learned more slowly than the choice between four and zero pellets. Similar evidence that choice between two arms of a T-maze is affected by the magnitude of reinforcement on each alternative has been provided by Davenport (1962) and Clayton (1964), while Logan (1965) has shown that the preference for the more favourable alternative can be offset by increasing the delay of reinforcement on that side. It is important to note, however, that even where both alternatives provide some reinforcement, most animals end up consistently choosing the more favourable side of the T-maze. The implications of this are discussed below (p. 192).

Catania (1963b), studying pigeons on a concurrent schedule, showed that responses were distributed between the two alternatives in proportion to the relative magnitudes of reinforcement available on the two. This result was confirmed by Brownstein (1971), but Fantino et al. (1972) found that preference for the stimulus signalling longer access to reinforcement was not always very strong.

C. Delay of Reinforcement

1. Choice Learning as a Function of Delay of Reinforcement

We have already reviewed a number of studies demonstrating that both spatial and simultaneous visual discriminations are learned more slowly when reinforcement for a correct response is delayed—even by as little as 5 or 10 sec (p. 157). An interesting exception to this generalization is provided by a recent study by Carlson and Wielkiewicz (1972) which showed that rats trained on a conditional discrimination to press one lever in the presence of a tone and a second lever in the presence of a series of clicks, actually learned the discrimination more rapidly if reinforcement for correct responses on one lever was always delayed by 5 sec. It seems that the additional cue provided by the difference in the delay of reinforcement for responses to the two levers more than offset the deleterious effects of the delay.

2. Choice between Different Delays of Reinforcement

In both discrete-trial and free-operant situations, animals show a preference for the alternative correlated with a shorter delay of reinforcement (Anderson, 1932; Logan, 1965; Chung and Herrnstein, 1967; Killeen, 1968, 1970; McEwen, 1972). In both cases the preference is maintained even though the probability of reinforcement on the two alternatives is the same. In T-mazes, delay of reinforcement may also be manipulated by varying the length of the two arms, both of which lead to a reinforced goal box, rather than by imposing different delays after the subject has traversed one or other of two equal arms. Under these conditions animals learn to choose the shorter of the two paths to the goal (Yoshioka, 1929; Grice, 1942; Munn, 1950, pp. 385-8).*

In all of these studies, preference has been found to be affected more by the ratio of short to long delay, than by the absolute difference between the two delays. Thus rats learn to discriminate more rapidly between delays of 10 and 30 sec, than between delays of 30 and 60 sec, even though in absolute terms the latter difference is larger (Anderson, 1932). Similarly, the preference for the shorter of two paths to a goal depend more on the ratio of short to long paths than on their absolute difference (Grice, 1942).

These results are hardly surprising. Differences in delay of reinforcement may be assumed to influence choice behaviour by affecting the association between the consequences of each choice and subsequent reinforcement, and thereby affecting the incentive value of each alternative. The difference between these two incentives, therefore, will depend on the discriminability of the two delays. As Weber's law states, discrimination between different values of any physical continuum depends more on the ratio of the two values than on the absolute difference between them. An important implication of this for the analysis of concurrent interval schedules is that preference between different VI schedules should be determined not by differences between the arithmetic means of the two schedules, but by differences between some appropriate transformation such as their harmonic means (the means of the reciprocals of the sets of intervals). The arithmetic means of a VI schedule assumes that the difference between 10 and 60 sec is no greater than the difference between 60 and 110 sec. The harmonic mean assumes that the former difference is

---

* Varying the lengths of the two arms of a T-maze obviously causes variations not only in delay of reinforcement but also in the amount of effort required to obtain reinforcement and the number of discriminably different stimuli intervening between start box and goal box. There is evidence that rats will learn to select an alternative requiring a less effortful response (e.g., Thompson, 1944) and that pigeons prefer an alternative that does not impose particular requirements on the pattern of responding necessary to obtain reinforcement (Fantino, 1968). Although, however, a theory of maze learning such as that proposed by Deutsch (1960; see p. 199) would certainly expect that the apparent distance to the goal box would have an effect on the preference between two alternatives over and above any effect due to differences in delay, there is little evidence on the importance of this factor in maze learning, even if recent studies of concurrent schedules suggest that factors other than the temporal delay of reinforcement may significantly affect choice (e.g., Duncan and Fantino, 1972). Normally, of course, delay, effort and apparent distance will tend to co-vary and until more evidence is available on their relative importance in different situations, it will be better to treat them together.

much greater than the latter. The available evidence supports this latter assumption (e.g., Killeen, 1968; Davison, 1969a; McEwen, 1972).

Hull (1943, pp. 152-7), in applying his goal gradient principle to choice between different delays of reinforcement, assumed that the ratio of the short to long delay was not a perfect predictor of preference. Largely on the basis of data reported by Anderson (1932), he argued that the preference for the shorter delay would be greater when both delays were of intermediate length, than when both were short or both were long. Recent studies of concurrent schedules have tended to confirm at least part of this suggestion: McEwen (1972), for example, found that the preference for a 40 sec delay over an 80 sec delay was greater than the preference for a 5 sec delay over a 10 sec delay.

A final case where animals can be thought of as choosing between alternatives that lead to different delays of reinforcement, is in a multiple-unit T-maze. In such an apparatus animals are always allowed to retrace to correct their errors: when an animal enters a blind alley it is permitted to retrace to the choice point, and follow the true path. This means that every run through the maze, including those in which several errors are made, terminates with reinforcement. Entrances into blinds, however, increase the delay of reinforcement. If we assume that each such error adds a constant x sec to the time required to traverse the maze, then each error is reinforced after $t + x$ sec, and each correct choice is reinforced after t sec, where t is the time required to traverse the remainder of the maze. The difficulty of eliminating a particular error will, of course, depend upon the ratio of delay of reinforcement following an error to that following a correct response, i.e., on the ratio $t + x : t$. Since this ratio decreases as t increases, it follows that the difficulty of eliminating an error will increase as the distance from the goal increases. Errors will, therefore, be eliminated in a backward order, starting from the choice point nearest to the goal (Spence, 1932; Hull, 1952, pp. 275–305).

In general this is true (see Munn, 1950, pp. 376–85, for a review of many of the earlier studies). However, as both Spence and Hull recognized, additional factors may well operate to obscure the operation of this simple goal gradient and additional principles must be invoked to accommodate these divergent results. The most important such principle is that a response established at one choice point may generalize to others: if the last choice point is learned first, the response learned there will tend to generalize backward to earlier choice points; of course, responses learned at all choice points will tend to generalize to all others. Whether this generalization will facilitate or interfere with learning of the maze will depend upon whether the same or different responses are required at different choice points. The computation of these effects must, in the end, be based on numerous, more or less arbitrary, assumptions for we are dealing with differences in the strength of the simple association that would have been established at each choice point in the absence of generalization from other choice points, as well as differences in the amount of generalization between different choice points. Nevertheless, one would expect that choice points in the middle of the maze would receive more generalization, both forward and backward, than those earlier or later; since, however, responses are learned from the last choice point backward, the total strength of generalized responses would be greatest shortly after the middle (see Hull, 1952, pp. 159–67 for these derivations). The results of experiments by Hull (1947) and Sprow (1947) bear out these predictions. In both studies rats were trained in a four-unit maze, with four doors at each

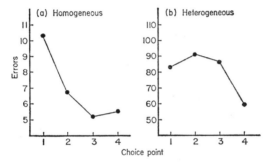

FIG. 4.7. A comparison of the number of errors made at each choice point of a four-unit maze, when the same position is correct at each choice point (homogeneous), and when a different position is correct at each choice point (heterogeneous). (*After* Sprow, 1947; and Hull, 1947.)

choice point. Fig. 4.7 shows the number of errors made at each choice point in each of the two experiments. In Sprow's experiment, the same response was required at each choice point; in Hull's, a different response was required at each. In Sprow's experiment, therefore, generalization facilitated learning, especially of the second and third choice points; in Hull's generalization interfered with learning, particularly at these two choice points. Fig. 4.7 shows that, as expected, in the former case the second and third choice points produced a small number of errors, while in the latter case they were learned extremely slowly.

D. Probability of Reinforcement

1. *Choice Learning as a Function of Probability of Reinforcement*

Rate of learning in a T-maze is usually retarded if choice of the correct alternative is reinforced on less than 100 per cent of trials (e.g., Clayton and Koplin, 1964; Hill *et al.*, 1962), although the effect may be small and not significant (e.g., Lehr, 1970). Similar effects have been observed for the learning of a simultaneous visual discrimination (e.g., Sutherland, 1966), although again several studies have reported no significant effect (Grosslight *et al.*, 1954; McFarland, 1966a). Even when learning is slowed down, however, animals still learn to select the correct alternative on essentially 100 per cent of trials in spite of inconsistent reinforcement (in addition to the studies cited above, see also Grosslight and Radlow, 1956; Kendler and Lachman, 1958).

2. *Choice between Different Probabilities of Reinforcement*

A point of rather greater interest than whether choice behaviour is affected by inconsistent reinforcement of the correct alternative, is how animals distribute their responses between two alternatives, when both are reinforced, but with different probabilities or on different schedules. The usual arrangement in discrete-trial experiments, known as studies of probability learning, is that one alternative is reinforced on a randomly chosen proportion of trials and the other alternative is reinforced on remaining trials. On each trial, therefore, reinforcement is available and in the studies of major interest some type of free- or forced-correction procedure is used such that if the subject's first choice on a given trial is of the alternative not reinforced on that trial, he is then permitted or forced to respond to the reinforced alternative. Only in this way can the experimenter guarantee exposure to the reinforcement schedule in effect on each alternative; if a noncorrection procedure were employed, an intended 50 : 50 reinforcement schedule, where each alternative is reinforced on a randomly chosen 50 per cent of trials, might be experienced as something entirely different by the subject. In the extreme case, a subject that chose one alternative consistently would be on a schedule indistinguishable from a 50 : 0 (or 50 : 100) schedule.

The free-operant analogue of experiments on probability learning exposes the subject to a concurrent schedule in which each stimulus is correlated with a different schedule. The most frequently studied case is a concurrent (VI, VI) schedule, in which one stimulus is correlated with a schedule of higher value than the other. As we shall see, the analogy with the discrete-trial case is imperfect and care must be taken in the interpretation of the data.

The interest shown in discrete-trial probability learning experiments arises from the light they throw on models of choice behaviour. The data from a group of rats learning a simple T-maze can be expressed as a trial-by-trial probability of the correct choice being made, a probability which increases as training continues. It is tempting to express this change in behaviour as due to a change in some underlying theoretical probability. Stimulus-sampling theory (Estes, 1959; Atkinson and Estes, 1963) can represent this behavioural change in a very direct manner: changes in the probability of a correct response are due to a change in the proportion of stimulus elements conditioned to the correct response. Now if the probability of turning right in a T-maze is directly dependent upon the proportion of elements conditioned to a right turn, what will happen when both right and left turns are reinforced, but with different probabilities? The occurrence of reinforcement is assumed to result in the conditioning of all elements sampled on that trial to the response made on that trial. If both responses are sometimes reinforced, then some stimuli must be conditioned to both, and the probability of choosing the more favourable alternative, i.e. the alternative reinforced with the higher probability (or majority stimulus) can never approach unity. Indeed, with certain simplifying assumptions, it will be predicted that animals will eventually come to distribute their responses between the two alternatives in such a way as to match the probability of reinforcement on each. On a 50:50 problem they should choose each alternative equally often; on a 70:30 problem they should choose the majority stimulus on no more than 70 per cent of trials.

It has, in fact, been argued that animals do show matching behaviour, that in discrete-trial experiments they come to choose the majority stimulus with a probability no greater than its probability of reinforcement (e.g., Estes, 1959) and that in free-operant concurrent schedules they distribute their responses between the two stimuli in proportion to the number of reinforcements received on each (e.g., Herrnstein, 1970). There are good reasons for objecting to both these assertions: in the discrete-trial case the statement is untrue; in the free-operant case, although behaviour may in fact be distributed in the way implied, such a distribution is only misleadingly described as matching. A much stronger case can be made for the argument that in both types of experiment, animals tend to distribute their choices in such a way as to maximize the momentary probability of reinforcement.

In discrete-trial probability learning experiments, maximizing the probability of reinforcement involves consistent choice of the majority stimulus (except in the case of a 50:50 schedule where any distribution of initial choices will produce the same probability of reinforcement). In simple spatial discriminations in T-mazes, rats tend to maximize: asymptotic choice of the majority side on, for example, a 70:30 schedule has ranged from 99·4 per cent (Roberts, 1966) to as low as 79·7 per cent in a two-lever box (Uhl, 1963), with the majority of studies (e.g., Bitterman *et al.*, 1958; R. N. Johnson, 1970) reporting asymptotes of over 90 per cent. The data have been reviewed by Sutherland and Mackintosh (1971, pp. 405–9) who argued that results thought to demonstrate matching in spatial probability learning by rats cannot in general be taken at face value.

In simultaneous visual problems, asymptotic choice of the majority stimulus tends to be somewhat lower. Although both rats (Bitterman *et al.*, 1958; Roberts, 1966) and pigeons (Mackintosh *et al.*, 1971) have been reported to choose the majority stimulus of a 70 : 30 problem on just over 90 per cent of trials, asymptotic performance is more frequently somewhat below 90 per cent (e.g., Mackintosh, 1970b; Bitterman, 1971). Again, however, there is little or no evidence of matching behaviour in rats; although pigeons have been reported to match in some studies (e.g., Bullock and Bitterman, 1962), they more frequently exceed matching levels (Graf *et al.*, 1964; Shimp, 1966; Mackintosh *et al.*, 1971). The only extensively studied animals that have shown relatively consistent matching have been fish, such as goldfish or African mouthbreeders (Bitterman *et al.*, 1958; Behrend and Bitterman, 1961; 1966; Mackintosh *et al.*, 1971).

In general, then, studies of probability learning have found that the distribution of an animal's choices between two reinforced alternatives does not match their relative probabilities of reinforcement, but shows a clear tendency towards consistent selection of the more favourable alternative. Furthermore, there is reason to believe that failure to maximize completely is often not so much due to a failure to select the correct response, as to a failure to attend to the relevant stimuli (Shimp, 1966; Mackintosh, 1970b). On this hypothesis, at least some choices of the minority stimulus may represent not a momentary preference for that alternative, but a choice determined by irrelevant stimuli: in visual discriminations, rats and pigeons sometimes respond systematically to one position or another, or to the most recently reinforced position (Mackintosh, 1970b; Mackintosh *et al.*, 1971). To the extent that this is true, it follows that the mechanism of response selection as such may not operate in a probabilistic, matching manner, but is rather one which tends to resolve decisions between two reinforced alternatives in favour of the more favourable alternative. As we have noted, exactly the same conclusion applies to cases where animals choose between two different magnitudes of reinforcement: here too, maximizing is the rule.

In studies of concurrent schedules, behaviour that has been described as matching, is, as Shimp (1966) has argued, more accurately thought of as maximizing. Fig. 4.8 shows the results of a study on concurrent (VI, VI) schedules reported by Herrnstein (1961). The subjects of this experiment may be said to be matching in the sense that they distributed their responses between the two stimuli in accordance with the distribution of reinforcements received on the two stimuli. Similar results have been reported in numerous other experiments (e.g., Catania, 1963a, 1966; Herrnstein, 1970). The interpretation of these results, however, is complicated by two factors. First, a standard procedure in studies of concurrent schedules is to arrange the programme in such a way that the subject's first response on a given key after shifting from the other will not be reinforced. This is achieved by specifying a change-over delay (COD) which ensures that responses on one key will not be reinforced for several seconds after the change. Without such a procedure, pigeons show a distressing tendency simply to alternate between the two keys, thus dividing their responses equally between them. The longer the COD, not surprisingly, the more rarely the subject alternates. Even though such studies

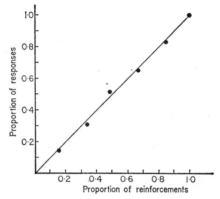

FIG. 4.8. Average performance of a group of pigeons trained on a Concurrent (VI, VI) schedule. The proportion of responses to one key closely approximates the proportion of reinforcements received on that key. (*After* Herrnstein, 1970.)

have found that matching is obtained over a relatively wide range of values (Shull and Pliskoff, 1967), this is by no means always true (Fantino *et al.*, 1972a). In this sense, therefore, matching is a consequence of the value of the COD employed by the experimenter.

A second point was emphasized by the results obtained by Shull and Pliskoff when long CODs were used. As CODs increased up to 20 sec, so subjects became less and less likely to alternate responses. This did not, however, destroy the matching relationship, for as the proportion of responses to one alternative increases, so, necessarily, does the proportion of reinforcements actually received on that alternative. In the extreme case, a subject might respond exclusively to one alternative, collect all reinforcements on that alternative, and still be said to be matching. But this would be matching only in a trivial sense. Nor is this mere idle speculation. Animals respond to both alternatives of a concurrent schedule, only if the component schedules are interval schedules. On a concurrent (FR, FR) schedule, pigeons confine all their responses to the alternative correlated with the smaller ratio (Herrnstein, 1958).

These findings suggest the possibility that subjects on concurrent schedules tend to respond at any moment to that alternative offering the higher probability of reinforcement. On a concurrent (FR, FR) schedule, this must *always* be the alternative correlated with the smaller ratio. The crucial point about concurrent (VI, VI) schedules is that both VI schedules continue to run while the subject is responding to only one of the alternatives. Since, on any interval schedule, the probability of reinforcement increases with time since the last reinforcement, it must necessarily be the case that the probability of reinforcement becoming available on the other alternative will eventually surpass the probability of reinforcement for continued responding to the same alternative. Hence, a shift to the other alternative is in accordance with a maximizing principle. If pigeons are trained on concurrent FI or VI schedules which do not run simultaneously, so that this condition no longer holds, they tend to select consistently the alternative correlated with the shorter average delay before reinforcement (Killeen, 1970; Fantino and Duncan, 1972; McEwen, 1972). When two VI schedules do run simultaneously, however, the distribution of responses shown in Fig. 4.8, although conventionally called matching, is in fact the distribution which maximizes the momentary probability of reinforcement (Shimp, 1966). That pigeons do in fact distribute their responses so as to maximize reinforcement was further shown by Shimp (1966), when he varied the momentary probability of reinforcement on two keys, and observed that the pigeons tracked this probability (see also Fantino and Duncan, 1972). A similar tracking of reinforcement has been observed in a discrete-trial study by Williams (1972).

The conclusion suggested by these observations recalls one reached in a discussion of formal models of instrumental learning at the outset of this chapter (p. 149). Simple choice behaviour in a T-maze may appear at first to be well represented by a Markov model which postulates an all-or-none transition from unlearned to learned states. While it is true that the probability of a correct choice in a T-maze may shift abruptly from 0·5 to 1·0, it does not follow that this is due to an equally abrupt change in the set of associations underlying choice behaviour. Other instrumental experiments, which provide a less restricted measure of responding, agree in showing a relatively gradual increase in response strength with reinforced practice. These two observations can be reconciled by postulating some sort of threshold model of choice behaviour. A subject may shift more or less abruptly from equal choice of two alternatives to consistent choice of the more favourable of the two, as soon as some minimal difference between their associations with reinforcement has been established. This agrees nicely with the implication of studies of probability learning, that consistent choice of the more favourable alternative is maintained provided that there is some minimal difference between the probabilities of reinforcement that have been associated with the two alternatives.

VI. SUMMARY

Instrumental experiments arrange contingencies between responses and reinforcers. Among the more important determinants of instrumental learning, therefore, will presumably be variations in the parameters of the reinforcer, and in the degree of contingency and temporal contiguity between response and reinforcer.

Instrumental learning is usually studied by measuring changes in the probability, latency, or rate of a single response, or measuring an increase in choice of one alternative over another. In the former case, both when subjects are required to respond on a series of discrete trials, and when they are free to respond at their own rate in a free-operant situation, performance improves with an increase in the subject's level of drive, varying directly with increases in the size or quality of reward, and inversely with an increase in the delay of reward. Although the main features of these results can be interpreted in terms of the law of effect (by assuming, for example, that larger, more immediate rewards strengthen S–R connections more effectively than smaller, delayed rewards) more detailed analyses tend to suggest alternative explanations.

In free-operant experiments, the major focus of study has been on the rate and pattern of responding maintained by various schedules of intermittent reinforcement. The delivery of reinforcement may be made to depend either on the passage of time, as in interval schedules, or on the completion of a certain number of responses, as in ratio schedules. In both cases the requirements may be fixed, or variable. More complicated schedules may make reinforcement available only if the subjects maintain a particular rate of responding or interval between successive responses. On fixed-ratio and fixed-interval schedules, there is evidence that the delivery of a reward not only reinforces preceding responses, but also serves as a stimulus signalling a temporary change in the probability of reinforcement. (A similar conclusion is suggested by the results of certain discrete-trial experiments, where the subject may learn that a reward on one trial signals that the next trial will not be reinforced.) Although it is possible to reinforce particular rates of responding, it remains uncertain whether such schedules achieve their effects by establishing appropriate temporal discriminations, or whether the subject learns particular patterns or topographies of responding: if reinforced for a slow rate of responding, for example, a pigeon may learn to peck the key only after a certain number of pecks on the wall beside the key. Ratio schedules maintain very high rates of responding, presumably because they arrange a correlation between rapid responding and short delays of reinforcement.

Studies of choice behaviour have shown that the rate at which a subject learns to choose one alternative over another (or, in some cases, the preference for one over the other in a steady-state experiment) is a function of the subject's level of drive, and the magnitude and delay of reinforcement for correct responses. Some of these effects are less readily observed in choice situations than when a single response is being measured, perhaps because many choice situations contain irrelevant cues, and a large reward may strengthen responses to irrelevant cues and thus interfere with learning.

When animals are required to choose between two alternatives, both signalling reinforcement but of different magnitudes or with different probabilities or delays, they tend to maximize their payoff by selecting the more favourable of the two relatively consistently.

CHAPTER 5

Theoretical Analysis of Appetitive Instrumental Learning: Incentive, Conditioned Reinforcement and Frustration

I. LAW OF EFFECT: EMPIRICAL AND THEORETICAL

The discussion of the last chapter on the acquisition of instrumental responses was intentionally kept at a relatively empirical level, eschewing detailed theoretical analysis. The operations of an instrumental experiment involve the presentation of certain reinforcing stimuli contingent upon the occurrence of certain responses, often in the presence of certain specified stimuli. An empirical description of such experiments, therefore, rather naturally uses the language of the law of effect: it is demonstrably the case that the probability of a response being made is usually increased if the presentation of an appetitive reinforcer is made contingent upon the occurrence of that response and that the scheduling of the reinforcer, the delay between response and reinforcer, and the magnitude of the reinforcer all have effects on this change in probability. In this loose, and more or less trivial sense, the descriptive account of the preceding discussion could be said to accept the empirical law of effect.

It is time to examine such a position more closely. In several cases where more detailed analysis was undertaken in the preceding chapter, the account provided tended to depart rather drastically from anything that would have been recognized by Thorndike as approximating to the law of effect. The fact that changes in instrumental responding can often be described by saying that response-contingent reinforcement increases the probability of that response does not force acceptance of the law of effect as a theoretical explanation of this change in probability. The reasons why behaviour can be modified by its consequences might be altogether different from anything Thorndike had in mind.

Thorndike (1911) and Hull (1943) assumed that reinforcers increased the probability of a response upon which they were contingent because they strengthened the learned connection between that response and the stimulus situation in which it occurred. Skinner (1938), although rejecting S-R associationism in favour of the idea that discriminative stimuli "set the occasion" for instrumental responses rather than becoming themselves associated with such responses, equally ascribed to the view that the role of reinforcement in an instrumental experiment was to strengthen antecedent responses. We can, then, distinguish two propositions: first, that changes in behaviour are due to the formation of an association between the reinforced response and the stimulus situation in which it occurs and secondly, that the function of reinforcement is to strengthen antecedent responses. Evidence may be brought to bear on each of these assumptions.

II. THE ELEMENTS OF ASSOCIATION IN INSTRUMENTAL LEARNING

As Estes (1969a) has argued, S-R association theory has gained its many adherents largely from considerations of parsimony: "Since changes in performance are most simply describable in S-R terms, it has seemed to learning theorists of many generations that it should be simplest also to represent learning in terms of changes in S-R tendencies or dispositions" (p. 163). But, continues Estes, since the theory in its original simple form cannot seriously be defended, and since any adequate S-R theory must have recourse to inferred, unobserved and possibly unobservable responses, it might be better "to face up squarely to the question of whether greater overall simplicity might not be achieved within a theory which conceives learning to be basically a matter of learning relations between stimulus events rather than the strengthening and weakening of the connections of responses to stimuli" (op. cit. p. 165).

A. Association between Stimuli is Sufficient to Produce Appropriate Instrumental Behaviour

1. Learning without Instrumental Contingencies

The operational definition of an instrumental experiment is that a contingency is arranged between the subject's behaviour and a reinforcing event. This contingency may be restricted to hold only under certain conditions, e.g., only when the key light is turned on in a pigeon box, or when the rat is placed in a black alley rather than a white alley. This restriction implies some contingency between certain stimulus events and the reinforcer. It may be the case that such a stimulus-reinforcer contingency is sufficient to generate behaviour by the subject that closely approximates the experimenter's response requirement; if the subject naturally produces appropriate responses as a consequence of the stimulus-reinforcer contingency, then the formation of an association between stimulus and reinforcer is all that needs to be postulated in order to account for the observed change in behaviour.

This argument amounts to saying that not only may classical conditioning be part of what occurs in an operationally instrumental experiment, it may be all that is necessary to produce the appropriate change in behaviour. The evidence for this argument is that the absence of the response-reinforcer contingency may in fact make no difference to the experimental outcome. Pigeons come to peck illuminated keys, irrespective of whether reinforcement is contingent upon pecking or simply upon the illumination of the key (Brown and Jenkins, 1968; p. 17). Rats will approach a goal box in which they have been fed irrespective of whether they received reinforcement only after running to the goal box or when they were simply placed directly in the goal box (Gonzalez and Diamond, 1960). Nor need an analysis that appeals to classical conditioning in this way be confined to such simple situations: a classical chaining process can be envisaged, which would enable the analysis to encompass more complicated sequences of instrumental responses. The establishment of the correct sequence of responses in a multiple-unit T-maze could be said to depend upon the classical conditioning of approach responses to the stimuli of the goal box, and so on back through the maze. Stimuli produced by the true path, then, would in turn elicit approach responses such that once the subject had approached the correct path beyond the first choice point, it would be exposed to a more strongly conditioned set of true path stimuli, which lay closer to the goal, and therefore approach them. Just such a theory of maze learning has been proposed by Deutsch (1960), and Bindra (1969, 1972) has also argued that much instrumental learning may be interpreted in this way.

The accounts of Deutsch and Bindra of maze learning may be regarded as modern versions of the theory of learning advanced by Tolman (1932). One of the most direct sources of support for such a theory is that provided by evidence of response equivalence. Rats that have been rewarded for running through a maze show essentially perfect transfer when extensive cerebellar lesions prevent their running normally (Lashley and Ball, 1929), or when the maze is filled with water and they must swim through to the goal box (MacFarlane, 1930). Equally, if they have been exposed to the true path without the opportunity to respond, by being pulled through in a trolley, they may run through correctly when given the opportunity later (Gleitman, 1955; McNamara *et al.*, 1956; Dodwell and Bessant, 1960).

2. *Observational Learning (Imitation)*

Thorndike (1911) accepted that the strengthening of S–R bonds by reinforcement was a sufficient explanation of the development of instrumental learning, largely because he was unable to find any evidence of behaviour that would force the postulation of different learning processes. Only when an animal made a response and that response was followed by reinforcement, could Thorndike detect any increase in the probability of that response. A variety of experiments "seemed to show that the chicks, cats and dogs had only slight and sporadic, if any, ability to form associations except such as contained some actual motor impulse. . . . They could not, for instance, learn to do a thing from having been put through it by me" (p. 222). The opportunity to watch a trained conspecific operate the latch of a problem box in no way facilitated subsequent learning. Thorndike also failed to find any evidence of imitative learning in cebus monkeys, concluding rather cautiously: "Nothing in my experience with these animals, then, favours the hypothesis that they have any general ability to learn to do things from seeing others do them. The question is still an open one, however, and a much more extensive study of it should be made" (*loc. cit.*).

He was taken at his word: numerous later investigators provided a variety of primates with the opportunity to learn how to escape from a problem box by watching a trained subject operate the mechanism. These studies have been reviewed by Spence (1937b) and Hall (1963), but, as Hall argued, many laboratory situations have been singularly ill designed for the study of imitation. Only rarely, for example, has serious thought been given to the problem of suiting the demonstrator to the learner. Observations on the structure of the natural social groups of many primates suggest that imitation of one sibling by another, or of a mother by her offspring, would be much more likely to occur than imitation of one strange adult by another. In fact, the semi-naturalistic observations of changes in the feeding habits of Japanese monkeys still provide some of the most striking examples of imitative learning (Imanishi, 1957; Miyadi, 1964). Animals learned to accept new foods, such as caramels wrapped in paper, by observing others eat them. One young female developed a new method of getting the sand off sweet potatoes left on the beach by washing them in the sea and also a way of separating grains of wheat from sand by floating the wheat in a stream. In both cases these new habits eventually spread to nearly all members of the group: but they were first taken up by siblings and close associates of the originator and only gradually spread to other members.

In principle, laboratory studies of imitation should permit a more precise specification of what is being learned by the imitator. In practice this has not always been true, if only because many experiments have lacked necessary controls, and many experimenters have been confused as to what it was they were trying to show. More recent discussions (e.g., Thorpe, 1963, pp. 132–6; Hall, 1963; Aronfreed, 1969) may help to resolve some of these confusions.

On the one hand, there is evidence that in many animals processes of "social facilitation" may have important effects on behaviour. Katz (1937), for example, reported that satiated chicks will start eating as soon as they see other, hungry chicks eating; while hungry, restless chicks will calm down, even though not fed, if put with a crowd of satiated chicks. Other examples of this include observations that the presence of another rat will reduce fear (Davitz and Mason, 1955), facilitate exploration (Hughes, 1969) and increase responding in extinction (Treichler et al., 1971). It is also possible that the mere presence of a conspecific may somehow facilitate learning (e.g., Gardner and Engel, 1971), or, as Spence (1937b) suggested, that the actions of the demonstrator may attract attention to important features of the experimental problem.

On the other hand, it is reasonable to suppose that there may be several components to imitative learning. An observer might learn, from watching a demonstrator, that certain stimuli or certain responses are associated with certain outcomes; over and above this, the observer might, when given the opportunity, *imitate* the demonstrator's behaviour, regardless of the observed outcome of the demonstrator's behaviour. The latter possibility may be of greater interest to child psychologists; the former, perhaps more appropriately called observational learning, is more relevant to the point at issue here.

It is not easy to distinguish empirically whether a subject has learned by observation what response to make in a particular situation, or what is the significance of certain stimuli. For example, later investigators, such as Warden and Jackson (1935), found considerably more evidence than had Thorndike that monkeys would learn the solution to puzzle box problems by observing a successful demonstrator. A recent study by John et al. (1968) suggested that cats may learn either to press a lever in the presence of a signal to obtain milk, or to perform a signalled avoidance response, in part by watching a successful demonstrator perform these responses. Even if one accepts these results at their face value (and the controls employed by John et al. leave something to be desired), it is impossible to say whether observation of the demonstrator has established the significance of certain responses, or the significance of particular discriminative stimuli.

The clearest demonstrations of observational learning, in fact, have been those explicitly designed to provide evidence that animals may learn the significance of certain stimuli. Kohn and Dennis (1972) found that the learning of a simultaneous visual discrimination by rats was facilitated by prior opportunity to watch a demonstrator solve the problem. Although exposure to the stimuli produced some general facilitation, it is clear that the observers acquired specific information from their experience since observation of a demonstrator trained on the *reverse* of the test problem significantly retarded subsequent discrimination learning.

Recent experiments by Myers (1970), employing rhesus monkeys trained on multiple schedules, and by Groesbeck and Duerfeldt (1971), with rats trained on a visual discrimination, have also provided relatively good evidence of observational learning; but the most elegant demonstrations remain those of Darby and Riopelle (1959) and Riopelle (1960). In each of these experiments, two rhesus monkeys sat facing each other across the stimulus display board of a WGTA*; on each of a long series of discrimination problems one or other monkey acted as demonstrator, choosing one stimulus object and being rewarded for correct responses, while the other observed the demonstrator and was allowed to choose after having observed the demonstrator's choice on the first trial. Over the course of 1000 problems, including 500 as observer and 500 as demonstrator, observer monkeys learned to choose the correct stimulus about 75 per cent of the time on their first trial: a single observation trial was sufficient to produce highly significant learning. Perhaps even more interestingly, the observer's performance was considerably more accurate when the demonstrator's first-trial choice was incorrect. The observer, therefore, was clearly not *imitating* the demonstrator, nor (as Spence had suggested for the case of imitation in the puzzle box) simply tending to manipulate objects recently manipulated by a conspecific. The observer must be assumed to have learned the relationship between stimuli and reinforcement, and such learning must be assumed to be sufficient to mediate appropriate instrumental behaviour.

3. Classical Mediation

Darby and Riopelle (1959) pointed out that the observational learning of their rhesus monkeys was formally rather similar to the learning that

---

* Or Wisconsin General Test Apparatus (Harlow, 1959). When used for monkeys, the subject is usually confined behind a wire screen, in front of which is placed a tray containing one or more stimulus objects, each of which covers a food well. The subject responds by displacing one of the objects; the food well covered by the correct object contains a reward.

occurs in delayed response experiments, where the experimenter baits one
of two stimuli in view of the subject and then permits the subject a choice
between the two. In general, learning by observation may well be interpretable in terms of the same processes as those involved in studies of
classical mediation. Bower and Grusec (1964), for example, initially
trained rats to press a lever for water on a VI schedule; then, with the
lever absent, they gave discriminative classical conditioning in which a tone
was established as a CS+ and a clicker as CS− for water reinforcement;
finally, with the lever present again, they gave instrumental discrimination

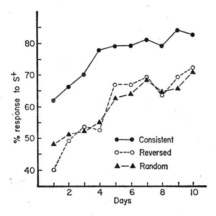

FIG. 5.1. Transfer from discriminative classical conditioning to instrumental discrimination learning. The proportion of responses made in the presence of S+
during the course of training on a free-operant discrimination, when S+ and S−
retain the same significance from classical to instrumental stages, when their
significance is reversed, and when the stimuli had been uncorrelated with the
delivery of reinforcement during classical conditioning. (*After* Mellgren and Ost,
1969.)

training with the tone and click serving as discriminative stimuli. Just as in
Kohn and Dennis' (1973) experiment, subjects for whom the stimuli
retained the same significance between training and testing learned the
instrumental discrimination significantly faster than those for whom the
significance of the stimuli was reversed. Similar results have been reported by Trapold *et al.* (1968) and by Mellgren and Ost (1969). In this
last experiment, discriminative conditioning was either consistent with
the subsequent instrumental discrimination problem (i.e., CS+ became
S+), or reversed (i.e., CS− became S+); for a third group, CSs were
presented randomly, without relation to reinforcement. The results for
these three groups are shown in Fig. 5.1: at the outset of instrumental
training the three groups were clearly separated, with the consistent group

performing above, and the reversed group below, chance. Not surprisingly, in view of the deleterious effects of random presentations of a stimulus and reinforcer (p. 40), the reversed group rapidly came to perform as accurately as the random group.

Trapold and Winokur (1967) found that establishing a tone as a classical CS for food significantly facilitated lever pressing in rats in a discrete-trial situation, when the tone signalled the onset of each trial. Bull and Overmier (1968) found that establishing a tone as a classical CS for shock facilitated performance in a discrete-trial avoidance situation when the tone was added to the usual warning signal. These two results are reminiscent of those reported by John *et al.* (1968), where performance on signalled avoidance or appetitive instrumental responding was facilitated by observation of a conspecific being exposed to the appropriate contingencies.

These parallels favour the argument that many if not most studies of imitation are in fact studies of observational learning: the observer does not learn to imitate the demonstrator, but learns, from watching the demonstrator's performance, the relationship between certain stimuli and reinforcement. Perhaps the major role of the demonstrator is to attract the observer's attention towards the relevant stimuli: it is worth noting that studies of classical mediation have usually used diffuse, auditory stimuli which require less in the way of specific orientation of the subject's receptors for their adequate perception than do the localized stimulus objects often used in experiments on observational learning.

B. Learning about the Consequences of Responses

The fact that *some* transfer occurs between exposure to stimulus-reinforcement contingencies and subsequent instrumental performance, does not necessarily imply that instrumental learning can always be reduced to a matter of the formation of associations between stimulus events. In neither classical mediation nor observational learning studies, after all, do animals typically show instantaneously perfect instrumental performance; in many such experiments, indeed, subjects receive independent training on the instrumental response and are simply required to show that classical conditioning leads to some control by the CS over the subject's instrumental behaviour. An instrumental experiment arranges a contingency between some behaviour and reinforcement, often in the presence of a specified stimulus: unless the stimulus-reinforcement aspect of this contingency is sufficient to generate the appropriate responses *classically*, then subjects must learn to modify their behaviour so as to conform to the experimenter's requirements. The presentation of a tone, in the presence of which a dog will be reinforced for turning its head to the right, does not tell the dog that a right turn rather than a left (or any other potential pattern of responding) is the required response. It would be unreasonable to argue that associations involving responses do not enter into instrumental learning.

All of these arguments seem plausible and, since they are consonant with the dominant S–R tradition of modern learning theories, their force has only rarely been questioned. Since the early formulations of Tolman (1932), very few theorists have supposed that associations between sensory events could ever be sufficient to account for the appearance of all instrumental behaviour. Nevertheless, these arguments merit careful scrutiny.

The fact that classical pairings of a stimulus and reinforcer do not necessarily ensure immediately appropriate performance of discriminative instrumental responses, is neither surprising nor crucial. The addition of the instrumental contingency, as Bindra (1972) has pointed out, will almost certainly change the contingencies between stimuli and reinforcement to which the subject is exposed. A classical contingency might mean that a tone was a reliable signal for food; the addition of an instrumental contingency, requiring the subject to press a lever in the presence of the tone in order to obtain food, would ensure that the tone alone was no longer a reliable signal of reinforcement. The subject is reinforced only when stimulated both by the tone and by the sight of, and contact with, the lever. Appropriate instrumental responding might be established because stimuli produced by approach to, and contact with, the lever, themselves become signals for reinforcement. As we have pointed out before (p. 132), there is reason to believe that among the responses most susceptible to instrumental contingencies are precisely those whose performance is necessarily accompanied by changes in external stimulation. A rat can press a lever only when it is in a particular area of the apparatus and making contact with the lever; even more clearly, the response of running through an alley or maze is accompanied by a change in intra- and extra-maze stimuli.

It is one thing to point out that instrumental contingencies typically involve more complex stimulus–reinforcer contingencies than those operating when a CS is paired with a UCS in a simple, Pavlovian experiment. It is another thing to show how these more complex associations would be sufficient to generate appropriate instrumental performance. As Moore (1973), however, has argued, the responses required in many instrumental experiments often involve little more than approaching stimuli associated with appetitive reinforcers, and avoiding stimuli associated with aversive reinforcers. Such approach and avoidance responses form part of the unconditional reactions elicited by these reinforcing events and it follows, therefore, that a Pavlovian mechanism of reinforcement would be quite sufficient to generate appropriate instrumental behaviour in these cases. Moreover, a response such as key pecking in pigeons, which clearly involves more than simple approach to the key, turns out on closer analysis to be hardly distinguishable from the consummatory responses elicited by appetitive reinforcers (Jenkins and Moore, 1973).

It is probably not easy in all cases to ascertain whether this sort of analysis is sufficient. There do, however, seem to be numerous situations where behaviour is altered by instrumental contingencies in a way which suggests that more specific learning about the contingency between responses and reinforcers must have occurred. Rats, for example, learn to run in a running wheel, both to avoid shock (Bolles et al., 1966) and to obtain food (Bernheim and Williams, 1967) or rewarding brain stimulation (Williams, 1965). While a running response is no doubt involved in normal approach and avoidance behaviour, the fact is that the performance of the response in this situation is not accompanied by approach towards stimuli associated with appetitive reinforcers, or escape from stimuli associated with aversive reinforcers. Nor can the learning of the true path through a maze always be reduced to the conditioning of approach responses to stimuli associated with appetitive reinforcement. Although rats may often learn to solve a maze problem by approaching in turn a series of stimuli always closer to the goal, they can also learn to make a left or right turn at a choice point in the absence of any difference between the external stimuli contingent upon different turns (e.g., Scharlock, 1955; see Chapter 10). Moreover, dogs can learn to raise one foot both in order to obtain food (Konorski, 1948, pp. 211–46) and to avoid shock to that foot (Kellogg, 1938; Wahlsten and Cole, 1972); monkeys have been trained to press a lever to avoid shock, even when their arm is totally de-afferented and they are unable to see it (Taub and Berman, 1968). Finally, of course, there is some evidence that autonomic responses can be modified by instrumental contingencies (p. 130).

It is implausible to analyse any of these examples of instrumental behaviour in terms of the classical conditioning of appetitive or aversive properties to particular environmental stimuli and the consequent elicitation of approach towards, or escape from, such stimuli. It seems more reasonable to assume that some changes in behaviour may occur as a consequence of the establishment of associations directly involving particular responses. Stimulus-response theory, of course, has never denied this; it has, however, adopted a particular view of the nature of these associations, namely, that the appropriate response is associated with the stimuli present at the time of reinforcement and that these associations are stamped in by reinforcement. An alternative view, which will be argued here, is that in instrumental experiments responses may be associated with subsequent reinforcement just as stimuli are associated with reinforcement in classical conditioning. In general, it will be argued, animals may learn about the consequences of their behaviour, i.e. the relationship between responses and other events, in much the same way that they learn about the relationship between any events in their environment. Although, as we have just conceded, it is often not clear whether a particular case of instrumental learning requires the postulation of such associations, or whether it may depend only on associations between stimuli and reinforcers, it will perhaps be more convenient to talk as though instrumental contingencies between responses and reinforcers did establish associations between the two, just as it was most convenient to talk as though the contingencies of a Pavlovian experiment always established associations between stimuli and reinforcers.

Three classes of experiment suggest that instrumental learning is not a matter of associating a particular response with a particular antecedent stimulus and strengthening this association by reinforcement. Studies of latent learning suggest that, when running through a maze, rats learn that a particular turn at a choice point is followed by a particular set of stimuli; studies of irrelevant incentive learning suggest that learning about reinforcement can be conceptualized in a similar manner, i.e. that rats learn that a particular reinforcer follows a particular response at a choice point; and studies of contrast effects imply that animals learn to anticipate the reinforcer contingent on a response rather than to perform a response which is then stamped in by reinforcement.

1. *Latent Learning*

Studies of latent learning were originally designed to test the law of effect, i.e. to see whether reinforcement was necessary for the strengthening of S–R associations rather than to elucidate the nature of the associations themselves (Tolman, 1932, pp. 343–4). Some latent learning experiments, however, have provided evidence on this second question. In such experiments, rats are given preliminary exposure to a maze without food or water present and may demonstrate that they have learned something during this ostensibly nonreinforced exposure by performing relatively efficiently as soon as some reinforcer is subsequently introduced. The prototypes of such experiments were the studies of Blodgett (1929) and Tolman and Honzik (1930) in which rats, given a number of trials in multiple-unit mazes with no food in the goal box, showed an abrupt decline in errors when food was introduced, performing within one or two trials as accurately as subjects that had received food in the goal box from the outset of training. The results are shown in Fig. 5.2.

A subsequent study by Muenzinger and Conrad (1954) suggested the important conclusion that part of what is happening in such an experiment is that nonreinforced exposure to a maze may cause habituation of fear or startle responses to what would otherwise be a novel situation and that it is this loss of fear that is responsible for the unusually rapid rate of error reduction when reinforcement is eventually introduced. They found that one or two days' exposure to the *mirror-image* of the test maze was just as effective in facilitating subsequent learning as was an equivalent amount of exposure to the test maze itself. Although it is true that rats can learn the reversal of a multiple-unit maze very rapidly (Dabrowska, 1963a, b),

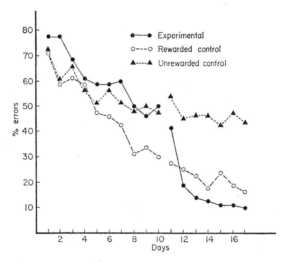

Fig. 5.2. Latent learning in a multiple-unit T-maze. The number of errors made during each daily run through a 14-unit maze, when rats received reward or no reward in the goal box, and the effect of introducing reward on Day 11 to previously unrewarded experimental subjects. (*After* Tolman and Honzik, 1930.)

and although continued exposure to the mirror-image maze did eventually interfere with subsequent learning, these results do suggest that facilitation of subsequent learning after one or two days exposure may have been due to some general habituation to the apparatus. Unfortunately, no other studies of this general design have controlled for this possibility so that their results must be treated with considerable caution.

Even if we ignore this complication, however, we are still left with several possible interpretations of Blodgett's results. Tolman wished to infer that rats learned the sequence of stimuli provided by different paths of the maze, or, in his terminology, formed a "cognitive map" of the maze; but it was equally possible for Hull (1952) to suggest that the reason why pre-exposed rats immediately make the correct set of turns to get to the goal box after they have found food there is because pre-exposure has resulted in considerable strengthening of the correct sequence of responses, even though this learning is not fully translated into performance until food is introduced. Although Hull's later analysis involved a departure from his earlier, more straightforward, adherence to the law of effect, it did preserve the S–R associationist view. This S–R analysis depends upon arguing that ostensibly nonreinforced trials still provide some differential reinforcement for true-path responses and that these responses are therefore strengthened. That such differential reinforcement is possible, it would be fruitless to deny: either removal from the goal box, or the aversiveness of entering and having to turn round in narrow blind alleys, might be sufficient to strengthen true-path responses. Moreover, some progressive elimination of errors is readily detectable in the pre-exposure phase of most latent learning experiments (McCorquodale and Meehl, 1951; Fig. 5.2 shows that the effect occurred in the experiment of Tolman and Honzik, 1930).

There are two studies employing multiple-unit mazes, whose results are not readily explained by this analysis. Kimball *et al.* (1953) found that their rats tended to increase rather than decrease the number of entrances into blind alleys during nonreinforced exposure to a six-unit maze and that subjects making more cul-entries during preliminary exposure showed better performance on subsequent test trials. In this case, therefore, there is little reason to believe that good test performance depended upon prior strengthening of the true-path response. An early study by Herb (1940) found that pre-exposed rats would perform well on test trials, even when food was made available in a blind rather than in the goal box. If the normal effect of pre-exposure is to strengthen the true-path response (and Herb observed a significant decline in blind entries during pre-exposure), it should have interfered with performance in this case.

A more decisive test of this analysis of latent learning can, however, be provided by simplifying the experimental situation. In a multiple-unit maze, a single, true path leads to the only goal box. In a T-maze, on the other hand, two alternative paths lead to two different goal boxes. If rats are exposed to a single-unit T-maze, with equal experience of both arms and both goal boxes, it is hard to see how this could lead to the establishment of a preference for one turn rather than the other at the choice point and it is always possible to control for any fortuitous preferences by counterbalancing. The demonstration of latent learning in a T-maze, therefore (i.e., the demonstration that subjects can learn which turn leads to which goal box), could not be interpreted in terms of prior strengthening of the appropriate response. In two separate experiments, Seward (1949) allowed rats to explore a T-maze containing two distinctive goal boxes which were not visible from the choice point. He then placed subjects in one of the goal boxes and fed them there. When they were returned to the start box, 54 out of 64 rats chose the arm of the T-maze leading to the goal box in which they had been fed. Tolman and Gleitman (1949) demonstrated the converse of this learning: after rats had been given an equal number of rewarded trials to the distinctive goal boxes of a T-maze, they were placed in one of the goal boxes and shocked. When placed in the start box, 22 of 25 rats chose the arm leading to the other goal box. As we shall see, these studies raise questions about the role response-contingent reinforcement in instrumental learning; for the moment, however, the important implication is that the subjects must be assumed to have learned which goal box followed which turn at the choice point. The associations formed in a maze do not simply, if at all, consist in the formation of connections between one response rather than another and the stimuli present at a choice point; rather, rats associate one response at a choice point, or one set of choice point stimuli, with one goal box, and another response or set of stimuli with the other.*

The final simplification of the latent learning experiment involves the use of a straight alley leading to a single goal box. Subjects may be given a number of exploratory trials in the alley and goal box, and then placed in the goal box and fed. Latent learning would be demonstrated if such subjects ran more rapidly when replaced in the start box of the alley, than did control groups not fed in the goal box, or groups fed in some different box.

If the goal box were visible from the start box, then such a result could be interpreted as a simple instance of classical conditioning: pairing goal box stimuli with food is sufficient to strengthen a response (approaching the CS) similar to that elicited by the UCS (approaching food). When the goal box is invisible from the start box, an interpretation in terms of classical conditioning must appeal to sensory preconditioning (p. 20). During pretraining, an association might be formed between the goal box and successive stimuli in the alley. When the goal box is associated with reinforcement, stimuli from the alley also acquire the capacity to elicit approach behaviour. Alternatively, of course, we could assume that pre-

* This conclusion is not seriously invalidated by the fact that appropriate choice behaviour is not invariably shown in experiments modelled after those of Seward and of Tolman and Gleitman (e.g., Denny and Davis, 1951; Minturn, 1954). Successful learning must obviously depend upon the distinctiveness of the two goal boxes (Seward, 1949); it would also presumably depend upon the amount of reinforcement given on placement trials and on whether *differential* reinforcement was given (e.g., whether shock trials in one goal box were alternated with non-shock trials in the other). Seward et al. (1952) have shown that successful performance also depends upon the time elapsing between goal box placement and test trial.

training established associations between the goal box and a particular pattern of responding in the alley.

In principle, it might also be possible to explain such performance in accordance with S–R theory. Spence (1956. p. 147) developed the following analysis. He assumed that rewarded placements in the goal box would result in the classical conditioning of incentive motivation (K) to the goal box stimuli; now since conditioned incentive motivation is assumed to facilitate ongoing behaviour, any generalization between goal box and start box would ensure that incentive motivation was elicited when the animal was placed in the start box, and this generalized motivation would ensure rapid running. Stimulus generalization, instead of a learned association, is thus invoked to account for the transfer between goal box and start box. Simple as this analysis seems, it leads, as Bitterman (1957) pointed out, to some strikingly implausible predictions. Not only would one expect facilitation of running to depend on the physical similarity of start and goal boxes; one must also predict that an animal initially run from, say, a black start box to a white goal box should show greater facilitation if rewarded placements were given in a black box similar to the start box, than if they were given in the goal box normally found at the end of the alley. Gonzalez and Diamond (1960) have disproved both of these predictions: facilitation of running was found to occur only when feeding took place in the goal box previously found at the end of the alley; furthermore, facilitation was actually inversely related to the similarity between start and goal boxes. Both of these results, which have been frequently confirmed (e.g., Furedy and Champion, 1963; Heathcote and Champion, 1963; Gonzalez and Shepp, 1965), clearly indicate that facilitation cannot be due to the generalization of some motivational state.

In conclusion, the results of experiments in which rats are exposed to a maze before receiving some such reinforcement as food in the goal box, irrespective of the light they may throw on the role of reinforcement in learning, suggest very strongly that animals are capable of learning that their behaviour has certain consequences. Responses must be assumed to be associated with their consequences, rather than being stamped in by certain classes of consequence.

2. Irrelevant Incentive Learning

If animals can learn that a right turn in a T-maze is followed by such and such a goal box, it seems plausible to suppose that they could also learn that the response was followed by food in that goal box. It should not be difficult to discriminate between such an analysis and that implied by S–R effect theory, according to which food in the goal box directly strengthens the antecedent turning response. Experiments on irrelevant

H

incentive learning were designed as a way of answering this question. The idea was to see if rats, satiated for food but exposed to a T-maze, one of whose goal boxes contained food, would demonstrate that they had learned the location of food by choosing appropriately as soon as they were made hungry.

There is little doubt that, under these simple conditions, reliable learning occurs: positive results have been reported, for example, by Spence *et al.* (1950), Seward *et al.* (1950), and Kendler and Levine (1953) (see Thistlethwaite, 1951, and MacCorquodale and Meehl, 1954, for reviews). With variations in this procedure, however, it is possible to find conditions under which such learning cannot be demonstrated. In several studies, animals have been motivated explicitly by inducing drive conditions irrelevant to the reinforcement in one goal box. Although rats made thirsty, or forced to swim through a maze, may still learn the location of food even though they are not hungry (e.g., Bendig, 1952; Norton and Kenshalo, 1954), there is some evidence that a strong irrelevant thirst drive may interfere with appropriate choice of the food arm on a subsequent test (Johnson, 1952). When, in addition to being made thirsty during initial training, animals are rewarded by presentation of water in either goal box, then learning about the location of food becomes somewhat less likely (e.g., Grice, 1948b), although it certainly can occur (e.g., Walker *et al.*, 1950; Rollin, 1958), especially when the food is located some distance before the goal box containing water (McAllister, 1952; Thistlethwaite, 1952). The one condition under which no learning has been demonstrated is when thirsty rats are run in a T-maze with water in one goal box and food in the other (being forced to the food side on 50 per cent of trials), and are then tested hungry but not thirsty (e.g., Spence and Lippitt, 1946). This final result may, perhaps, be explained by some plausible assumptions. When one choice has been consistently reinforced during training and the other has consistently led to the frustrating experience of finding food instead of water, it is not entirely surprising that rats should show little preference for the food arm on subsequent tests.

Very few studies have explored the possibility of demonstrating irrelevant incentive learning for incentives other than food and water, or under drive states other than hunger and thirst. Given the interactions between these two drives (Verplanck and Hayes, 1953), this is unfortunate. One interesting exception, however, is a study by Krieckhaus and Wolf (1968). Thirsty rats were trained to press a lever, receiving sodium solution as reinforcement although they were not at the time sodium deficient. When they were subsequently satiated for water, but made sodium deficient, they pressed the lever 170 times in extinction. A control group, reinforced with water instead of the sodium solution in the first phase of the experiment, made only 54 responses in extinction. Additional control groups, both in this study and in a subsequent replication by Khavari and Eisman (1971), ruled out several alternative explanations, leaving it reasonable to conclude that the rats had learned that lever pressing led to the presentation of a particular reinforcer, even when they had no need for that reinforcer. It is also interesting to note that in a preliminary T-maze study, Krieckhaus and Wolfe found no significant evidence of appropriate performance on a test trial when employing the type of design which had yielded negative results when used by Spence and Lippitt (1946).

Plausible explanations of the factors affecting irrelevant incentive learning are not hard to find. Tolman (1948) suggested that a strong drive would narrow the range of learning, i.e. decrease the probability of learning about stimuli unrelated to that drive (a principle for which there is a moderate amount of evidence from other sources; see Sutherland and Mackintosh, 1971, pp. 123–6). A second possible reason for failures of irrelevant incentive learning is that the animal is required to demonstrate its learning under conditions differing from those in force at the time the learning occurred. A hungry rat is required to choose the food arm of a T-maze when the opportunity to learn the location of food was provided under conditions of satiation. A change in drive may disrupt appropriate performance: if a second drive has also been changed (i.e., if training was given under conditions of thirst), then one might expect even greater disruption.

Both Hull (1952, pp. 148–50) and Spence (1951, pp. 278–80) took some pains to explain successful performance in irrelevant incentive experiments in terms of S–R theory. Although such explanations may be possible, they are notably complex: they involve, for example, the supposition that implicit consummatory responses are elicited by the sight of a reinforcer such as food; that these responses become conditioned to the stimuli of the arm leading to food; that stimuli associated with these consummatory responses become conditioned to the response of turning into that arm; and that, finally, an increase in hunger selectively increases the vigour of consummatory responses for food, thereby also increasing the salience of stimuli associated with that consummatory response and hence increasing the probability of approaching the food arm. The explanation is not particularly successful in explaining the conditions affecting performance in these experiments and has been subjected to a number of detailed criticisms by Deutsch (1960, pp. 84–100).

3. Contrast Effects

More decisive evidence that reinforcers are events that are associated with responses, rather than agents responsible for the formation of associations between responses and antecedent stimuli, comes from experiments on contrast effects. A literal reading of the law of effect implies that reinforcers invariably strengthen responses upon which they are contingent, regardless of the animal's previous or concurrent experience with other reinforcing events. Although it may be assumed that increments to response strength are a decreasing function of current response strength, or that different reinforcers differ in their effectiveness, such qualifications do not seriously affect this general position. In studies of contrast effects, however, it can be shown that the effectiveness of a given reinforcer varies with the subject's exposure to other reinforcers, with the consequence that a single

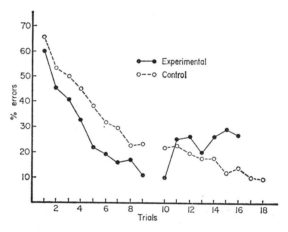

FIG. 5.3. The effect of a change in reward on performance in a multiple-unit maze. The substitution of sunflower seeds for the preferred bran mash on Day 10 led to an immediate increase in errors in the Experimental group, to a level above that attained by control subjects reinforced throughout with the sunflower seeds. (*After* Elliott, 1928.)

pellet of food presented contingent upon running down an alley may decrease response strength rather than increase it if the subject has previously received ten pellets in the goal box.

Although Crespi (1942) is often credited with the first demonstration of contrast effects in animal experiments, and although his studies did indeed provide the first parametric investigation with rats, earlier experiments by Elliott (1928) and Tinklepaugh (1928) (reviewed by Tolman, 1932, pp. 71–6), and by Nissen and Elder (1935) and Cowles and Nissen (1937) had shown what was to be expected. Elliott trained hungry rats for one trial per day for 10 days in a 14-unit maze with either sunflower seeds or bran mash as the reward. As can be seen from Fig. 5.3, the bran mash was a more effective reinforcer than the sunflower seeds, leading to faster running and fewer errors over Days 1–10. From Day 10 on, the bran mash group received sunflower seeds as the reward in the goal box. This substitution of a less favourable for a preferred reinforcer produced an immediate disruption of performance. Tolman, commenting on Elliott's observation that these rats did not spend all their time in the goal box eating, but started looking around as though searching, argued that such searching behaviour should "be taken as empirical evidence for, and definition of, an immanent expectation of the previously obtained bran mash" (p. 74). Tinklepaugh's monkeys were trained on a delayed-response problem, in which the experimenter, in view of the monkeys, baited one of two food wells; after a delay period, during which a screen was lowered between monkey and food wells, the subject was required to choose which well to respond to. Either banana or lettuce served as an adequate reinforcer for this problem, but if the female monkey had seen a well baited with the preferred banana and then found the less preferred lettuce on the test trial, clear evidence of a disappointed expectation could be seen.

> She extends her hand to seize the food. But her hand drops to the floor without touching it. She looks at the lettuce but (unless very hungry) does not touch it. She looks around the cup and behind the board. She stands up and looks under and around her. She picks the cup up and examines it thoroughly inside and out. She had on occasion turned toward observers present in the room and shrieked at them in apparent anger. (Tinklepaugh, 1928, p. 224).

It would be churlish to deny that such behaviour indicates the disappointment of an expectation that the food well would contain the reward with which it had been baited. Nissen and Elder (1935) and Cowles and Nissen (1937) confirmed Tinklepaugh's results with chimpanzees in a delayed-response task, and accepted this type of explanation.

Crespi's demonstration was that a decrease in the magnitude of reward (in more recent experiments this has usually been achieved by changing the number of pellets) produces rapid disruption of runway performance in rats, such that animals, shifted from a large to a small number of pellets, temporarily perform below the level of a control group trained throughout with a small reward. This finding is now known as a successive negative contrast effect (Dunham, 1968). In simultaneous contrast experiments, animals are concurrently exposed to two different conditions of reinforcement, with different stimuli signalling the different reinforcers. For example, rats might receive 10 pellets of food for running down a black alley, but only 1 pellet on interspersed trials in a white alley. Under these conditions, speed of responding in the alley correlated with the single pellet is significantly slower than that of a control group receiving a single pellet in both alleys (Black, 1968).

The data and theoretical analysis of contrast experiments are the subject of more detailed analysis in Chapter 7. For the moment it is sufficient to point to the problems they raise for any version of the law of effect. If reinforcers simply serve to strengthen the connection between antecedent responses and the stimulus situation in which they occur, it is hard to see how the effectiveness of a small, or less preferred, reinforcer should be dependent upon the subject's previous or concurrent experience with other reinforcers. It is difficult to resist the implication that experience with a large, preferred reinforcer establishes an expectation that appropriate behaviour in this situation will produce this reinforcer, and that the occurrence of a smaller, less preferred reinforcer is disappointing, frustrating or inhibitory.* The implication, then, is that the role of reinforcement in instrumental learning is not to strengthen antecedent responses; reinforcers do not increase the strength of an association between stimulus and response; they are themselves associated with those responses.

Although he did not accept this last conclusion, Hull modified his initial adherence to the law of effect in the light of Crespi's data. In his later theorizing, Hull (1952) assumed that although reinforcers served to strengthen S–R connections, they also affected performance (reaction potential) by determining the value of K, incentive motivation, which directly determined the vigour of all ongoing response tendencies. This solution has all the appearance of an unhappy compromise: it is not clear why reinforcing stimuli should be assigned both of these properties, and Spence (1956), who had always accepted the analysis of reinforcement in terms of incentive motivation (p. 134), rapidly abandoned any reliance on the law of effect (p. 151). The general trend of recent theorizing has strongly supported Spence: Mowrer (1960a), Miller (1963), Logan and Wagner (1965), Sheffield (1966), Estes (1969a, b) have all argued that reinforcing events do not serve to strengthen particular connections, but to channel behaviour into one path rather than another, to energize ongoing behaviour, or to lead to the completion of response sequences that are accompanied by increases in incentive level.

C. Parallels Between Stimulus-Reinforcer and Response-Reinforcer Associations

The parallel between the response-reinforcer contingency of instrumental experiments and the stimulus-reinforcer contingency of Pavlovian experi-

---

* To describe this as an inhibitory operation is to remind the reader that an earlier discussion of inhibitory conditioning came to the conclusion that significant inhibition depends upon the occurrence of nonreinforcement in the context of an expectation of reinforcement (p. 34).

ments, suggests the possibility that parallel types of associations underlie learning in the two situations. Explicit statements of such a suggestion have been made by Revusky (1971) and Seligman *et al.* (1971). In instrumental learning, Seligman *et al.* argued, subjects "learn about the outcome of their acts: they can learn that some act produces reinforcers (acquisition), that withholding some act produces reinforcers (differential reinforcement of other behaviour, DRO), or that some act no longer produces reinforcers (extinction)' (p. 368).

Additional parallels between classical and instrumental experiments are not hard to find. Studies of delay of reward in instrumental learning are similar to classical studies of trace conditioning (p. 57). In experiments on latent learning, the arrangement of a contingency between a response and a neutral stimulus results in a change in the probability of that response when the stimulus is subsequently associated with reinforcement; the paradigm is similar to that used in experiments on sensory preconditioning (p. 20). In experiments on conditioned reinforcement (see below, p. 233), a stimulus previously associated with reinforcement is used to reinforce a new instrumental response; the paradigm is similar to that used for the study of higher-order conditioning (p. 19). Two further examples are worth analysing in some detail.

1. *Learning that Responses Predict No Change in Reinforcement*

Seligman *et al.* (1971) were particularly interested in studies which have deliberately arranged that there should be no contingency between responses and reinforcers. In these experiments, animals are exposed to usually unsignalled, always inescapable, shocks, i.e. reinforcers which occur without the subject being able to control them in any way. Their argument was that just as "traditional operant experiments . . . can be interpreted to mean that S learns that his response controls the reinforcer, . . . (so) we think that S can learn that it cannot control the reinforcer" (p. 350).

The idea that animals can learn that reinforcers occur independently of their responses is used to explain the phenomenon of "learned helplessness". Seligman and Maier (1967) exposed dogs to a series of intense, unsignalled and inescapable shocks before training them on a signalled avoidance task in a shuttle box. During avoidance training a signal came on at the beginning of each trial; if the animal failed to jump over the barrier, shock came on 10 sec later; both signal and shock continued for a further 50 sec, unless the animal responded. Not only did those dogs initially exposed to the uncontrollable shock fail to learn to avoid shock, they failed to learn reliably to escape and thus continued to receive shock for 50 sec on nearly all trials. Although not all subsequent studies have necessarily found such a striking impairment, substantial interference has been observed in a wide variety of conditions and animals including, for example, rats (Mullin and Mogenson, 1963; Weiss *et al.*, 1968) and gold-fish (Padilla *et al.*, 1970) (see Maier *et al.*, 1969, and Seligman *et al.*, 1971, for reviews).

One plausible interpretation of this interference is that although shock is, as far as the experimenter is concerned, unrelated to any behaviour on the part of the subject, certain responses might be strengthened during exposure to uncontrollable shock and might subsequently interfere with the acquisition of avoidance responding. One might appeal either to super-stitious reinforcement of behaviour that happened to precede termination of shock, or to classical conditioning of responses such as freezing that are elicited by shock. Although it is difficult to rule out these possibilities, Seligman and Maier (1967) found that if, instead of exposing subjects to uncontrollable shocks, they explicitly trained them to escape shock by pressing a panel, this caused little or no interference with subsequent avoidance learning in a two-way shuttle box. Moreover, Maier (1970) explicitly trained dogs to remain immobile (i.e., to freeze) in order to escape shock and found only transient interference with subsequent avoidance learning in the shuttle box. Weiss *et al.* (1968), however, although finding that rats learned to avoid shock in a shuttle box more slowly after exposure to inescapable shock, also observed a significant correlation between the amount of freezing elicited by inescapable shock and the extent to which exposure to such a shock interfered with subsequent avoidance learning. It is, furthermore, always conceivable that a response consistently rein-forced during escape training will extinguish faster, and hence provide less interference during subsequent avoidance learning, than a response that was randomly reinforced during exposure to inescapable shock.

Nevertheless, these explanations do not, on balance, seem entirely con-vincing. Seligman *et al.* (1971) concluded that animals can learn that changes in their behaviour are uncorrelated with changes in reinforcement and argued that this learning interferes with the subsequent establishment of an association between a particular response and reinforcement. There is a close parallel between this analysis and the account provided earlier of the effects of latent inhibition and random presentations of CS and UCS in classical conditioning (p. 39). Just as a stimulus which is uncorrelated with changes in reinforcement is only with great difficulty established as a CS for that reinforcer during subsequent conditioning, so, when changes in behaviour are uncorrelated with changes in reinforcement, any subse-quent association between behaviour and reinforcement is hard to establish.

One possible consequence of an extended exposure to different correla-tions between particular stimuli and particular reinforcers is that some reinforcers may be much more readily associated with some stimuli than with others. Rats, for example, learn to associate the taste, rather than the visual characteristics of their food, with subsequent poisoning (p. 52). It is tempting to speculate that a similar process might be responsible for some examples of inefficient instrumental learning. Several instances have been noted previously where the reinforcement of a particular response might not succeed in establishing that response (p. 137). One possible cause of several of these failures, in particular many of those recorded by Breland and Breland (1961, 1966), may have been that the response being trained was incompatible with the behaviour elicited by the reinforcer. This may not, however, always be the most convincing analysis: it is possible that other failures of instrumental learning are a consequence of an extensive past history during which changes in the response in question were never correlated with changes in that reinforcer. For example, Thorndike (1911, pp. 47–8), Bolles and Seelbach (1964), Konorski (1967, pp. 466–7) and Shettleworth (1973) have shown that responses such as yawning, scratching and grooming in animals such as dogs, cats, rats and gerbils, may be only imperfectly established by instrumental contingencies of reinforcement. The problem is certainly not caused by a low operant level of responding preventing adequate exposure to the instrumental contingency; nor does it seem entirely plausible to suppose that it is due to any serious incompatibility with the responses elicited by the reinforcer. It is possible, therefore, that animals have had extensive opportunity to learn that maintenance activities such as these are normally uncorrelated with changes in the probability of appetitive reinforcers, such as food, or aversive reinforcers, such as intensely loud noises, and therefore find it difficult to associate the two, even where there is a contingency between them. A similar analysis might apply to the difficulty of changing autonomic responses by instrumental contingencies.

2. Overshadowing and Blocking

In classical conditioning, the association formed between a given stimulus and reinforcer is not only a function of the contiguity and contingency between the reinforcer and that stimulus, it also depends on the contiguity and contingency between the reinforcer and all other stimuli presented at the same time (p. 46). The association between a stimulus and reinforcer may be overshadowed or blocked if other more valid or more salient stimuli are presented simultaneously. If, therefore, response–reinforcer associations underlie instrumental behaviour in the same way that stimulus–reinforcer associations underlie classically conditioned behaviour, it should be equally possible to demonstrate the overshadowing and blocking of response–reinforcer associations.

Konorski (1948, pp. 218–19) has reported some observations which suggest that a stimulus–reinforcer association may overshadow a response–reinforcer association and hence prevent the appearance of instrumental learning. A dog could be trained to flex its leg if the foot was lifted passively by the experimenter and the subject was then fed. After a sufficient number of such trials, the dog started to flex its leg spontaneously. If, however, dogs were exposed to this contingency only in the presence of a particular stimulus, that stimulus was turned into a reliable appetitive CS, eliciting a profuse flow of saliva, but there was little evidence of the formation of an association between leg flexion and food. In order to

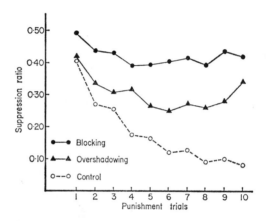

FIG. 5.4. Punishment of instrumental responding, when the occurrence of each punished response is accompanied either by a stimulus previously established as a signal for shock (blocking group), or by a novel stimulus (overshadowing group), or by no external stimulus. The effectiveness of the punishment procedure is measured by calculating a suppression ratio, comparing the number of responses in the 5 min interval following each shock, with the number of responses in a 5 min baseline period. (*After* St. Claire-Smith, 1970.)

obtain a reliable flexion response in the presence of the signal, it was necessary to intersperse reinforced trials, on which the passive flexion was accompanied by the signal, with nonreinforced trials on which the signal was presented alone. This observation seems analogous to the finding, reported by Egger and Miller (1962), that one external stimulus may overshadow another unless it is deliberately turned into a less reliable predictor of reinforcement (p. 249).

St. Claire-Smith (1970) has undertaken a more systematic study of overshadowing and blocking of instrumental learning. After training rats to press a lever for food, he gave a series of classical conditioning trials in which a very brief light or noise was paired with shock. Finally, with the lever back in the chamber, animals received a shock, contingent upon lever pressing, every five minutes. In one group of subjects (blocking group) the stimulus previously paired with shock was presented for 190 msec coincident with each punished response; for a second, over-shadowing group, the stimulus not paired with shock was presented coincident with each punished response; for the third, control group, the punished response was not accompanied by any external stimulus. The results are shown in Fig. 5.4; it can be seen that the rapidity with which response–contingent shock suppressed responding depended on whether the response was accompanied by another stimulus and whether that stimulus was already a reliable predictor of shock. The implication is that the learning underlying suppression of responding by punishment is an association between that response and the punishing stimulus, and that this association, like those between external stimuli and reinforcers, can be overshadowed or blocked by the concurrent presentation of other, more salient or valid signals of reinforcement.

D. Conclusions

The results of most experiments on instrumental learning can readily be analysed in S–R terms, i.e. by supposing that the association between the required response and the stimulus situation is strengthened by reinforcement. However, they are equally consistent with the supposition that animals can learn to associate any events that occur together, including stimuli with other stimuli, stimuli with reinforcers, responses with stimuli and responses with reinforcers. When additional evidence is brought to bear on the question, the attempt to maintain the purity of S–R theory, i.e., to insist that all instrumental learning is based on the formation of stimulus–response connections, becomes increasingly more difficult. Exposure to contingencies between stimuli and reinforcers is sometimes sufficient to generate appropriate instrumental behaviour and often facilitates appropriate instrumental performance; experiments on latent learning suggest that animals learn about the consequences of their behaviour; the effects of variations in such parameters of reinforcement as delay, magnitude and quality, suggest that animals learn to anticipate these characteristics of the reinforcers they receive; finally, there are sufficient parallels between the operations and outcomes of classical and instrumental experiments to suggest a parallel set of explanations—one involving the establishment of associations between stimuli and reinforcers, and the other the establishment of associations between responses and reinforcers. In view of all the evidence, the statement that an animal learned a particular response in an instrumental experiment because that response was followed by reinforcement should be construed as shorthand for the statement that an association was formed between the response and reinforcement and that this association underlies the increase in the probability of that response.

III. INCENTIVE THEORY

The classic objection to this analysis of the nature of instrumental associations is that it entails an unnecessarily complex analysis of instrumental performance. Stimulus–response theory, as Estes (1969a) among others has pointed out, has the virtue of simplicity in this sense at least: if we say that a response is stamped in by its reinforcing consequences, then we have already said why the probability of that response being made in that situation has increased. Other theories, by implication, have failed to solve this problem: to say, for example, that a subject has learned that a response is followed by a particular reinforcer is not to say that the probability of that response will increase.

Two points should be noted here. In the first place, the type of theory proposed by Deutsch (1960) and Bindra (1972) does not encounter this problem. If animals learn that a particular sequence of stimuli is associated with a particular reinforcer, then a Pavlovian mechanism of reinforcement will ensure that the behaviour elicited by that reinforcer is also elicited by stimuli associated with it. Just as animals approach food and escape from shock, so they will approach stimuli associated with food and escape from stimuli associated with shock. Such a theory does not need to appeal to a supplementary process to account for the evocation of responses.

The second point is that no serious S–R theorist now supposes that changes in response probability can be regarded as perfectly faithful correlates of changes in underlying S–R associations. No modern S–R theory fails to distinguish between learning and performance, or lacks constructs designed to translate one into the other. As we have already noted, Hull (1952), in his later theorizing, assumed that the activation of habits (learning) into reaction potential (performance) depended on a number of motivational constructs. The earlier distinction between learning and performance (Hull, 1943, pp. 238–40) had been suggested by the data of Pavlov (1927, p. 127) and Perin (1942), showing that the strength of both classically and instrumentally conditioned responses depended upon the organism's current level of drive. A given number of training trials did not inevitably lead to a given level of performance, for response strength could, in the extreme case, be reduced to zero by satiating the subject. In the later version (Hull, 1952, pp. 140–48), the results of studies of latent learning and shifts in incentive, where increases or decreases in reward produced abrupt changes in performance, led to the postulation of an incentive construct (K) acting in exactly the same way as the original drive construct. Both incentive motivation and drive level were assumed to act multiplicatively on habit strength to produce performance. With this modification, the theory could now explain why a very low drive level or a very small reward might similarly obscure the extent to which a subject had learned the true path through a maze.

Both Hull (1952) and Spence (1951, 1956) assumed that incentive motivation was based on classically conditioned consummatory responses (the fractional anticipatory goal response, or r_g).* In this general assumption they have been followed by many others (e.g., Rescorla and Solomon, 1967), who have proposed that appetitive instrumental behaviour is somehow mediated by classically conditioned motivational states. This theory has generated a substantial amount of research, but although some have professed to find support for the theory in some of the results of this research, there is serious reason to question this conclusion. It is true, for example, that classical CRs may be recorded in the course of an instrumental experiment: dogs have been shown to salivate while engaged in instrumental responding for food (e.g., Kintsch and Witte, 1962; Williams, 1965; Shapiro et al., 1966) and licking responses have been recorded in rats instrumentally responding for water (Deaux and Patten, 1964; Miller and DeBold, 1965). As most of these investigators recognized, however, these observations may establish only what was already obvious. An instrumental experiment provides the occasion for the classical conditioning of CRs to stimuli signalling reinforcement; but this does not prove that these CRs are an index of a central state which motivates instrumental responding. The evidence for this proposition is distinctly unsatisfactory.

A. Motivational Theories of Incentive

1. Incentive as a Source of General Drive

Hull and Spence assumed that incentive motivation, like drive, was a source of general motivation that served to activate all response tendencies

---

* The r_g mechanism, it should be stressed, was not simply a motivational construct in Hull's and Spence's theorizing. Indeed, its most important role was as a "pure stimulus act": it provided stimuli which could become associated with responses and which thus served as mediating associative links to account for such phenomena as irrelevant incentive learning (p. 213). The following discussion is not directed to this aspect of the theory, but only to the motivational role ascribed to r_gs. Their associative role is here ascribed to expectancies. It should also be noted that the statement that r_gs are classically conditioned did not implicate any Pavlovian process of reinforcement. Both Hull and Spence assumed that classical CRs were reinforced because they were followed by rewarding consequences.

indifferently. Although it is true that stimuli paired with food tend to elicit "general activity" in both rats and pigeons (Sheffield and Campbell, 1954; Slivka and Bitterman, 1966; Zamble, 1967, 1969), the relationship between general activity and general drive is surely no more than verbal. There is little reason to doubt that the behaviour measured in such experiments is directed towards obtaining food and if this is so, then its occurrence provides no evidence for the proposition that stimuli associated with food are capable of energizing any pattern of behaviour.

If a CS for food is a source of general drive, then it should activate that pattern of behaviour in the organism's repertoire which is momentarily dominant, irrespective of the relationship of that behaviour to food. There is good evidence that it does nothing of the sort. An early study, often cited as evidence of general drive, is that by Brown et al. (1951), showing that the presentation of a CS previously paired with shock, coincident with the elicitation of a startle response in rats, would increase the magnitude of the startle response. The close connection between startle and fear responses should make one pause before accepting these results as evidence for any very general drive theory; they do, however, suggest a simple test of the idea that a CS for food generates a state of general drive. If it does, then such a CS should be equally effective in augmenting the startle response. Not very surprisingly, tests of this idea have produced uniformly negative results (e.g., Trapold, 1962; Armus and Sniadowski-Dolinsky, 1966): a CS for food, so far from augmenting the startle response, tends to decrease it.

The presentation of an appetitive CS also tends to inhibit rather than facilitate ongoing avoidance responding (Bull, 1970; Grossen et al., 1969). Indeed such results formed but one part of the argument, reviewed earlier, that it is important to distinguish between appetitive and aversive reinforcers and motivational states, and that these systems tend to inhibit rather than facilitate one another (p. !18). What is at issue, here, is the whole Hullian concept of a general drive state: it should be apparent that the trend of recent thinking has been steadily away from such a position (e.g., Bolles, 1967; Hinde, 1970; see also p. 349). If this is accepted, then we can abandon any idea of incentive motivation as a source of general energy for all ongoing patterns of behaviour.

2. Incentive as a Source of Appetitive Drive

The rejection of Hull and Spence's account does not necessarily entail the rejection of all motivational theories of incentive. It remains possible that incentive should still be conceived as a motivational construct, although now a more specific one. A CS paired with food might still exert its influence over instrumental behaviour by increasing motivation, although now its effect would be confined to an increase in appetitive motivation, and it would activate only food-related responses. Such an account has, in effect, been proposed by Konorski (1967) and by Rescorla and Solomon (1967).

What sort of evidence is relevant to the assessment of such a theory? It is not enough to show that classical CRs may be recorded in an instrumental experiment, for this does not prove that such CRs are correlated with a conditioned central state which motivates instrumental responding. Indeed, as we shall see shortly, a closer examination of the actual correlation between classical CRs and instrumental responses suggests that the occurrence of the former could not possibly provide a measure of the strength of motivation for the latter. Nor do studies of transfer between differential classical conditioning and instrumental discrimination learning necessarily show that instances of positive transfer must be attributed to the *motivational* effects of the classical CSs. It is clear that the results reported by Bower and Grusec (1964), Trapold *et al.* (1968) and Mellgren and Ost (1969), described earlier (p. 203), are at least as easily explained in terms of associative transfer as in terms of motivational transfer (Trapold and Overmier, 1972).

One possible criterion for ascribing motivational properties to a CS paired with a particular reinforcer is the observation that its presentation to a subject engaged in instrumental responding maintained by that reinforcer may result in an increase in rate of responding. A CS signalling shock, for example, typically increases the rate of free-operant avoidance responding (Sidman *et al.*, 1957; see p. 83). A motivational analysis of positive incentive suggests that a similar result should be found in an appetitive experiment.

As it happens, the appetitive analogue of the experiment by Sidman *et al.* was first undertaken more than 20 years earlier, having been reported by Konorski and Miller in 1936 (see Konorski, 1967, pp. 371–2). Konorski and Miller trained a dog to press a panel for food and then presented stimuli that had been established as excitatory and inhibitory CSs for food. The results were dramatic: in the presence of CS−, the animal usually increased its rate of responding; in the presence of CS+, the animal stopped pressing the panel and started to salivate profusely.

This result was recently rediscovered by Azrin and Hake (1969): they too found that the presentation of a relatively brief (10 sec) appetitive CS would typically suppress appetitive instrumental responding. Several other recent studies have confirmed these findings, but have shown that reliable suppression may depend upon the use of a CS of less than 30 to 40 sec (Meltzer and Brahlek, 1970; Miczek and Grossman, 1971). It is tempting to explain these results by suggesting that a relatively short CS may elicit responses, such as approaching the site of reinforcement, which interfere with the performance of the instrumental response. This argument, however, is rendered somewhat less plausible by the observation that a short CS signalling rewarding brain stimulation will also suppress appetitive instrumental responding (Azrin and Hake, 1969; Hake and Powell, 1970; Van Dyne, 1971).

Meltzer and Brahlek (1970) found some evidence that the presentation of a long, 2 min CS might elevate rate of instrumental responding, but this finding was not confirmed by Miczek and Grossman (1971) and may be questioned for other reasons. Although their rats did indeed respond more rapidly in the presence of the 2 min CS than in its absence, rate of responding in the CS still fell well short of the baseline rate of responding measured before the start of conditioning. Estes (1943) had earlier presented evidence suggesting that a CS signalling food might elevate rate of lever pressing in rats, but testing was conducted while lever pressing was being extinguished, and this fact suggests a number of alternative explanations. One might well argue that if animals respond more rapidly when a stimulus previously signalling food is presented in the middle of an extinction session, this is not because their motivation for responding has increased, but because their expectation of food has changed. Since Bolles and Grossen (1970) were able to replicate Estes' results only when the appetitive CS was presented in extinction, the evidence for a motivational enhancement is less than overwhelming.*

The only unequivocal evidence that a CS signalling food may elevate the rate of appetitively reinforced responding is a study by LoLordo (1971). LoLordo trained pigeons to peck a white key on a VI schedule of food reinforcement; when a change in the stimulus projected onto the key was established as a CS+ for food, the pigeons significantly increased the rate at which they pecked the key in its presence. LoLordo et al. (1974) have pointed out that there is no need to interpret this finding as an instance of a motivational enhancement of instrumental responding, since key pecking is precisely the CR elicited in pigeons when a change in the illumination of the key is established as a CS for food (p. 17). When LoLordo et al. repeated LoLordo's original experiment, establishing a change in the key light as a signal for food, but with pigeons trained to press a treadle rather than to peck a key for food, the effect of the signal was to decrease the rate of instrumental treadle pressing, and to increase the probability of classically conditioned key pecking. Moreover, when Farthing (1971) performed a similar experiment, but used an auditory

---

* It is also possible, as Baum and Gleitman (1967) have suggested, that any enhancement of responding during extinction may be due to disinhibition rather than to an increase in motivation.

stimulus as the CS signalling food, he found no evidence that the presentation of the auditory stimulus would enhance rate of key pecking.

These studies, therefore, provide no support for the view that the presentation of a stimulus established as an appetitive CS affects instrumental responding by increasing appetitive motivation. Any effects of such a stimulus may be attributed either to the overt CRs which it elicits, or to the associations with reinforcement which have been established. This conclusion is supported by the results of a study reported recently by Trapold (1970), which suggest that the effects of a CS paired with a particular reinforcer are specific to that reinforcer. Rats were required to press one of two levers in the presence of a tone to obtain a pellet of food, and the other, in the presence of a clicker, to obtain sucrose solution. Animals that had received tone-food and clicker-sucrose pairings on previous classical conditioning trials learned this instrumental discrimination significantly faster than animals for whom the tone had been paired with sucrose and the clicker with food. In other words, the tone and clicker as CSs did not exert their control over instrumental responding by arousing any simple appetitive motivational state, but must be assumed to have produced more specific expectations of the reinforcer with which they had been paired.

If these results are accepted, then any motivational analysis of incentive must be regarded as at best incomplete, and at worst as missing much of the point (Trapold and Overmier, 1972). Stimuli paired with a reinforcing event become signals for that event rather than generators of a motivational state. This conclusion is strongly supported by the more detailed analytic evidence to be discussed in the following section.

B. Incentive as Expectation of Reinforcement

Any motivational analysis of incentive implies that the vigour of a dominant instrumental response associated with a particular reinforcer will co-vary with the strength of the classically conditioned central state elicited by a CS that has been paired with that reinforcer. Although this state need not be identified with such overt CRs as salivation or licking, it should presumably correlate with such indices of classical conditioning. From which it follows that the occurrence of overt instrumental and overt classical responses should be correlated. In general, this correlation has not been particularly impressive. When animals have been trained in a free-operant situation, and the experimenter has recorded both instrumental and classical responses, the momentary probability of one has not varied very much with the momentary probability of the other (e.g., Kintsch and Witte, 1962; Miller and DeBold, 1965). Even when the correlation has been high, as it was in a study by Shapiro *et al.* (1966), the implications for a motivational analysis are less than encouraging. These investigators trained dogs to press a panel for food on a DRL 2 min schedule (i.e., responses were reinforced only if spaced at least 2 min apart). Classical conditioning sessions in which a tone was paired with food were given separately. When the tone was presented during DRL sessions in the middle of the 2 min interval following a reinforced response, it sometimes resulted in an inappropriate (because too early) instrumental response. The interesting finding was that there was a nearly perfect correlation between panel pressing and salivation: only if the CS elicited salivation did it also produce an instrumental response. It is important, however, to remember that such instrumental responses were inappropriate: a DRL schedule specifies reinforcement for withholding responding until an appropriate interval has elapsed. The subjects in this experiment had learned to withhold responses rather effectively; the presentation of the CS, therefore, increased the probability of a non-dominant response. It is hard to see why an increase in appetitive motivation should have produced such an effect. There is, in fact, little or no evidence to suggest that DRL performance is greatly affected by drive level (Kramer and Rilling, 1970) and it is also worth remembering that Blackman (1968) has shown that the presentation of an *aversive* CS may also enhance rate of responding on a DRL schedule (p. 82). A more natural interpretation of the results Shapiro *et al.* obtained is to suggest that the presentation of the CS led subjects to expect that the time for the delivery of food, and therefore the time for responding, had arrived.

A closer analysis of the actual relationship between classical and instrumental responses, in other experiments where both have been recorded, reinforces these conclusions. If a classically conditioned motivational state instigates instrumental responding, then one might reasonably expect to observe peripheral CRs occurring before, rather than after, the initiation of instrumental responding. Under certain conditions, exactly the opposite outcome is observed. One general finding of studies on delayed conditioning is that stimuli sufficiently distant in time from the presentation of the UCS, especially if they are discriminable from the stimuli present immediately before the UCS, do not themselves elicit strong CRs. Indeed, they may actually become inhibitory—hence the Pavlovian term inhibition of delay (p. 61). Consider, then, two instrumental situations in which reinforcers are presented only occasionally: fixed interval and fixed ratio schedules. On each schedule reinforcers are delivered quite regularly; on the FI schedule the delivery of reinforcement depends on the passage of time since the last reinforcer; on the FR schedule the delivery of reinforcement depends upon the time required to complete the ratio requirement. Stimuli associated with a short passage of time since reinforcement, therefore, are relatively distant from the next reinforcer and would be expected to elicit no CR. On the FR schedule early responses in the ratio also reliably predict that reinforcement is not yet due and should also inhibit CRs. On the FI schedule, on the other hand, the number of responses emitted in each interval tends to vary rather widely (p. 175), so that both few and many responses become associated with reinforcement and should elicit CRs. These expectations are confirmed. Williams (1965), for example, recorded salivary responses in dogs instrumentally pressing a panel for food on either FR or FI schedules. The two schedules were matched for the average frequency of reinforcement. On neither schedule

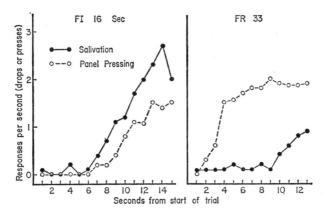

FIG 5.5 Concurrent measurement of salivation and panel pressing in a dog trained on FR and FI schedules of instrumental reinforcement. (*After* Williams, 1965.)

did animals salivate immediately after reinforcement, but salivation reliably accompanied the subsequent initiation of instrumental responding only on the FI schedule; on the FR schedule the dogs did not start salivating until they had completed a substantial fraction of the ratio requirement. The results for one dog trained under each condition are shown in Fig. 5.5.

The problem, then, is this: if salivation is an index of a central motivational state, the initiation of instrumental responses should be accompanied by, or even preceded by, an increase in salivation. Since salivation does not begin until well after instrumental responding has begun, what provides the motivation for the initiation of such responding on FR schedules? At the very least, incentive motivation theory should predict that initial responding on the FR schedule would be slower than initial responding on the FI schedule (where responding is accompanied by salivation). But ratio schedules typically result in a much higher rate of responding than interval schedules (p. 176), and Williams' data provide no exception to this rule.

Other studies have reported similar results. Ellison and Konorski (1964) trained dogs to press a panel in the presence of one signal; a ratio of nine presses produced a second signal which was followed 8 sec later by the delivery of food. They observed a virtually complete dissociation between instrumental panel pressing and classical salivary responding: when the first signal came on the animal pressed but did not salivate; when the second signal came on, panel pressing stopped and the animal started to salivate profusely. Similar results have been observed in alleyway experiments. If an animal is required to run in an alley for reinforcement in the goal box, then stimuli associated with being placed in the start box and with the initiation of running stand in the same relation to reinforcement as do stimuli associated with the post-reinforcement pause and initiation of responding on FR schedules. Deaux and Patten (1964) recorded licking responses from rats trained to run in an alley for water and found that these CRs were not elicited until the subject had approached or even entered the goal box.

These data make it difficult to see how the initiation of an instrumental chain of responses can be motivated by an incentive system whose operation is in any way directly mirrored by overt CRs. Further analysis suggests even more acute problems, for there is at least one factor whose effects on the vigour of instrumental responses are quite inconsistent with its effects on classical CRs. Inhibition of delay in classical conditioning is a consequence of a discrimination between the onset of a long CS and the stimulus situation at the moment of reinforcement: when discriminability is increased by presenting different CSs in succession (CS_1—CS_2—UCS), the amplitude of CRs at the onset of CS_1 is decreased. In instrumental experiments, however, the speed of responding at the initiation of a long response chain may actually be increased by an increase in the discriminability of stimuli present at the beginning of the chain from those present at the moment of reinforcement. Saltz et al. (1963) trained four groups of rats to run in an alley to a goal box for food; for two of the groups alley and goal box were the same colour (both black, or both white); for the remaining two groups, alley and goal box were different colours (one black and the other white). The two groups running to a goal box of a different colour from the alley showed significantly faster starting speeds. Donahoe et al. (1968) trained rats under discrete-trial conditions to press a lever for food on an FR schedule; they found that latency of responding on each trial was longer when a constant tone came on at the beginning of the trial and remained on until food was obtained, than under a second condition in which the tone changed as the subject progressed through the ratio requirement. Finally, the data of several latent learning experiments provide additional support for this conclusion. When rats are exposed, in the absence of reinforcement, to an alley and goal box and are then fed in the goal box, subsequent speed of running on test trials is much higher if the goal box and alley differ in colour than if they are the same colour (Gonzalez and Diamond, 1960). Once again, speed of initiation of instrumental responding is inversely correlated with the probable strength of any classically conditioned response system.

C. Conclusions

The presentation of a stimulus established as a classical CS for appetitive reinforcement does not appear to increase the vigour or rate of appetitive instrumental responding. Moreover, the initiation of a chain of instrumental responses cannot be attributed to any antecedent change in a classically conditioned motivational state. In the light of this evidence it seems necessary to reject the idea that instrumental behaviour is motivated in any simple way by a classically conditioned system of incentive motivation. The suggestion is hardly revolutionary: it has become increasingly apparent to a number of writers, such as Bolles (1972) and Trapold and Overmier (1972), that the theory of incentive motivation proposed by Hull and Spence was fundamentally unsatisfactory.

This conclusion was anticipated by Konorski (1967) who, on the basis of the observations reported by Konorski and Miller (p. 225) and Ellison and Konorski (1964), distinguished between long-term appetitive CRs and short-term consummatory CRs (see p. 21), arguing that instrumental behaviour was motivated not by the occurrence of consummatory CRs such as salivation, nor by any central motivational state correlated with such overt CRs, but rather by the long-term conditioning of an appetitive central drive state. Since appetitive and consummatory CRs are assumed to be incompatible, the occurrence of salivation may indicate a decrease rather than an increase in the motivation for instrumental responding.

Although employing radically different terminologies, the theoretical views of Mowrer (1960a) and Estes (1969a) contain some relatively similar ideas. No attempt will be made in the following discussion to adhere to either version of this general analysis. Consider the case of a hungry animal rewarded with food for the performance of some appropriate chain of instrumental responses. The occurrence of salivation in this situation may be assumed to reflect a learned association between food and a particular set of stimuli. On intuitive grounds it is hard to see why an animal should be expected to engage in more vigorous instrumental activity in the presence of stimuli which, it has learned, signal the imminent delivery of food. It is precisely in the absence of food that a hungry animal needs to be active: animals that were active in the presence, and inactive in the absence, of food would be unlikely to survive long enough to serve as subjects for study by psychologists. More precisely, it would seem that activity is appropriate when it is instrumental in leading an animal out of a stimulus situation in which it has learned food does not occur, into a situation in which food has previously been presented. Instrumental responding, therefore, should not be initiated because it is preceded by stimuli signalling food and eliciting salivation, but because performance of the instrumental chain is accompanied or followed by the occurrence of such stimuli and hence by an increase in salivation or anticipation of food.

The available evidence seems entirely consistent with this analysis. Williams' (1965) dogs, for example, trained on FI or FR schedules, neither salivated nor responded at the beginning of each trial, because the stimuli accompanying the start of a trial signalled reinforcement only after a long delay and instrumental responding would not cause any change in those temporal stimuli. After a short pause responding was initiated on the FR schedule, because feedback from responding had been established as a signal for reinforcement and the initiation of responding, therefore, resulted in increased anticipation of reinforcement. On the FI schedule, on the other hand, responding would be initiated only when it was accompanied by temporal stimuli signalling increasing proximity to reinforcement. Once initiated, rate of responding would tend to accelerate on both schedules, because responding was continually accompanied by events more closely associated with reinforcement. Other experiments, which have deliberately manipulated the similarity between stimuli present at the start of an instrumental chain and those present at the moment of reinforcement, have clearly shown that it is when an increase in the level of incentive accompanies the performance of a chain of instrumental responses that those responses are most rapidly initiated.

The motivation for instrumental responding, therefore, is not correlated with the *current* level of a classically conditioned state of incentive, but with the discrepancy between present and anticipated states [or the "relative valence" (Gonzalez and Diamond, 1960) of current and future, or proximal and distal, stimulation]. Incentives do not motivate instrumental behaviour unselectively. Indeed, it is not clear that anything is gained by introducing the concept of motivation at all. The initiation and maintenance of instrumental responses is a consequence of a past history in which such responding has been accompanied by an increase in proximity to reinforcement. The "motivation" for instrumental responding, therefore, can be reduced to the establishment of associations between responding and reinforcement or stimuli signalling reinforcement. Such stimuli do not instigate responding: as a consequence of past learning they act as goals for responding.

IV. CONDITIONED REINFORCEMENT

The argument of the preceding section advocated discarding the idea of incentive as motivation and replacing it with the idea of incentive as the anticipation of a goal. Stimuli associated with reinforcers do not motivate instrumental responses; they may become established as goals for instrumental responses. Their effect on instrumental behaviour, therefore, may be similar to that of unconditioned reinforcers; they may, in other words, serve as conditioned reinforcers. It should be pointed out, however, that this does not imply any commitment to a particular theory of instrumental reinforcement, let alone to any theoretical version of the law of effect. It says only that stimuli associated with primary reinforcers may acquire some of their properties.

The concept of conditioned incentive motivation was never, within Hullian theory, very satisfactorily reconciled with that of conditioned or secondary reinforcement. Since the operations for establishing stimuli as elicitors of incentive motivation or as secondary reinforcers were identical, being in both cases simply a matter of pairing them with primary reinforcement, the different roles assigned to these constructs within the theory threatened to destroy any internal consistency it may have possessed. The preceding argument, therefore, in conjunction with the conclusions about conditioned reinforcement that will be suggested in the present section, may serve to restore some theoretical consistency.

It may help to begin with two notes on terminology. First, Hull (1943) talked of secondary reinforcers. The practice of Skinner (1938), Kelleher and Gollub (1962) and Hendry (1969a) of talking of conditioned reinforcers is followed here: this has the advantage of suggesting the origin of their reinforcing properties. Secondly, although the term "conditioned reinforcer" is usually reserved for stimuli associated with appetitive reinforcers, and although the discussion of aversive reinforcement is largely postponed to the following chapter, it will be convenient to include some discussion of conditioned aversive reinforcers in this section.

A. Procedures for Establishing Conditioned Reinforcers

If a neutral stimulus is associated with food, it may subsequently satisfy the operational definition of an instrumental reinforcer by increasing the probability of responses upon which it is contingent. Before accepting, however, that such a demonstration is sufficient to ascribe conditioned reinforcing properties to the stimulus, a number of points must be established. First, the stimulus must exert its effects because of its association with primary reinforcement. Secondly, the increase in instrumental responding must be a consequence of the contingency between responding and conditioned reinforcer. Finally, it must be shown that instrumental responding is not being maintained by any previous or current association with primary reinforcement.

Several procedures have been used to study conditioned reinforcement, not all of which are sufficient to establish all of these points.

1. Maintenance of Responding in Extinction

Bugelski (1938) trained rats to press a lever for food, with each presentation of food accompanied by a click. When this click was presented contingent upon lever pressing in extinction, it significantly increased the number of responses emitted. Bugelski concluded, therefore, that the click had served as a conditioned reinforcer. Although this procedure was subsequently used in a number of studies, it suffers from one crucial defect. It is obvious that subjects extinguished in the absence of the click in Bugelski's experiment experienced a greater change between acquisition and extinction than those extinguished with the click. Any difference in resistance to extinction observed in such an experiment, therefore, may be attributed to generalization decrement. That this is not mere idle speculation is shown by the results of an experiment by Melching (1954). Melching trained several groups of rats to press a lever for food, with each response in acquisition being reinforced. Different groups received a buzzer coincident with the delivery of food on 100 per cent, 50 per cent or 0 per cent of acquisition trials and were then extinguished with or without the buzzer contingent on each lever press. The design of the experiment, together with the number of responses made by each group in extinction, is shown in Table 5.1. Groups 100–100 and 100–0 constitute, in effect, a replication of Bugelski's experiment, and produced comparable results; but the fact that animals trained without the buzzer extinguished more slowly with no buzzer (Group 0–0) than when the buzzer was added (Group 0–100) suggests that it was the change from acquisition to extinction that affected extinction performance in Groups 100–100 and 100–0. Finally, animals exposed to the buzzer on 50 per cent of acquisition trials extinguished equally slowly whether the buzzer occurred on all trials (50–100) or on no trials (50–0) in extinction. There is no evidence that the presence of the buzzer per se affected performance in extinction.

Other studies, in which a previously established response is extinguished with or without the presentation of stimuli that have accompanied reinforcement in acquisition, have also shown that performance may be significantly affected by generalization decrement. Bitterman *et al.* (1953) and Elam *et al.* (1954), indeed, found that the effects attributable to generalization decrement might be larger than those attributable to conditioned reinforcement. Even though this conclusion has not always been substantiated (e.g., Friedes, 1957; Paige and McNamara, 1963), it remains clear that this procedure for measuring conditioned reinforcement is entirely unsatisfactory. We shall not, in general, have occasion to refer to studies in which it has been employed.

TABLE 5.1

Design and results of experiment by Melching (1954)

Groups % of trials with buzzer		Number of responses in extinction
Acquisition	Extinction	
100	100	755
100	0	574
0	0	637
0	100	402
50	100	719
50	0	727

2. *The Establishment of New Responses*

A more valid procedure, which still involves testing in the complete absence of primary reinforcement, is to make the putative conditioned reinforcer contingent upon the performance of some new response in the organism's repertoire. The first study suggesting that stimuli associated with reinforcement might serve to establish some new response was provided by Grindley (1929), who observed that hungry chicks would learn to run down an alley when the only consequence of this response was the sight of unavailable food in the goal box. Subsequently, Skinner (1938, pp. 82–3) showed that rats, previously given magazine training in a lever box, would learn to press a lever if their response was followed by the sound of the magazine operating, even though no food was delivered.

One possible objection to these studies is that changes in rate or vigour of responding may be open to alternative interpretations. Grindley, for example, had no control group to show that the contingency between responding and the sight of food was responsible for the increase in speed of running. It is possible, for example, that a stimulus paired with reinforcement might increase rate of responding not because it reinforces that response, but because it has been established as a discriminative stimulus for instrumental responding or because it increases the subject's level of arousal or general activity. This might even be expected to happen because the nonreinforced presentation of a stimulus previously correlated with reinforcement now elicited frustration reactions. Whatever the explanation, Wyckoff *et al.* (1958) found that a stimulus paired with water in the absence of any lever in the box, would maintain subsequent lever pressing equally effectively, whether it was presented contingent upon a lever press, or contingent upon *not* pressing the lever.

One solution to this problem is to employ a yoked-control procedure, presenting a conditioned reinforcer contingent upon lever pressing in an experimental group and at the same frequency (but not contingent upon responding) in a control group. Employing such a procedure, Crowder *et al.* (1959) did indeed find a significant difference in rate of lever pressing between experimental and yoked groups. We have, however, argued that the yoked-control procedure may not be free of error (p. 120).

A second, more satisfactory, solution is to design experiments in such a way that the measured change in instrumental responding cannot be attributed to any change in level of arousal or general activity. One possibility is to measure choice behaviour—presenting a conditioned reinforcer contingent on one choice rather than another. Another possibility is to employ different schedules of reinforcement in a free-operant situation. Illustrations of each of these solutions can be provided. Saltzman (1949), for example, showed that after rats had been fed in a distinctive goal box, they would learn to make a particular turn at the choice point of a T-maze in order to get to the goal box, although no food was present in the box during maze learning. Kelleher (1961) trained pigeons to peck a key on a VI schedule and then presented the empty magazine, with its accompanying noise and light, contingent upon responding in extinction. However, he programmed a variety of schedules of magazine presentation during extinction, each of which produced characteristic patterns of responding. He was able to show, for example, that an FI schedule of magazine presentation resulted in an appropriate FI scallop, that a DRL schedule resulted in a slow rate of responding and that rate of responding subsequently increased when the schedule was changed to FR.

Kelleher was able to record the effects of a conditioned reinforcer in the absence of primary reinforcement over a relatively long period of time, but this was presumably because he was not so much studying the acquisition of an entirely new response as changes in the *pattern* of an old response under the effect of different schedules of conditioned reinforcement. That response, having been acquired on a VI schedule, would be relatively resistant to extinction. The problem of extinction is often a serious one: Saltzman's rats, for example, performed above chance in the

T-maze for less than 10 trials: by Trial 10 they were choosing the previously reinforced, but now empty, goal box on only 50 per cent of trials. This is hardly surprising: the association between a stimulus and reinforcement, upon which it must be assumed the conditioned reinforcing properties of that stimulus depend, is presumably subject to extinction when reinforcement is no longer presented. The very procedure used to provide an uncontaminated measure of conditioned reinforcement guarantees that the effect will be evanescent.

3. *Token Rewards and Chain Schedules*

One solution to this problem is to maintain the association between the conditioned and unconditioned reinforcers, but to take other steps to control or minimize the direct effect of unconditioned reinforcement on the instrumental behaviour under study. Wolfe (1936) and Cowles (1937), for example, trained chimpanzees to exchange poker chips as tokens for food and then used the tokens as conditioned reinforcers in a variety of discrimination problems. The association between tokens and food was maintained by permitting the chimpanzees to exchange them for food either after a certain number had been earned or at the end of a session; but it is reasonable to suppose that the substantial delay between instrumental responding and eventual presentation of food prevented any direct reinforcement. More recent experiments have used tokens as conditioned reinforcers for lever pressing on various schedules of reinforcement in both chimpanzees (Kelleher, 1957, 1958b) and rats (Malagodi, 1967a, b). Malagodi (1967b), for example, trained rats to press a lever 20 times for a marble (FR 20); the token was initially exchangeable for food immediately upon delivery, but later the token-exchange schedule was increased to the point where the subject was on average earning eight tokens before being allowed to exchange them. Although occasionally quite long post-reinforcement pauses developed, responding was in general maintained at a high rate.

Instead of providing the subject with a manipulable token as a conditioned reinforcer and requiring the subject to deposit the token in a special receptacle in order to obtain food (as in these traditional token-reinforcement experiments), it would be equally possible, without changing the essential nature of the experiment, to provide a change in the colour projected onto a pigeon's pecking key as the conditioned reinforcer and then simply to require some further pattern of key pecking in the presence of the conditioned reinforcer in order to obtain food. In other words, a pigeon might be required to peck 10 times (FR 10) at a blue key (S_1), with the tenth response turning the key red (S_2); in the presence of S_2, the first response after 30 sec (FI 30 sec) would produce food, after

238 THE PSYCHOLOGY OF ANIMAL LEARNING

which the key would turn blue again and the entire cycle would be repeated. The schedule would be called chain (FR 10, FI 30 sec) (see p. 165): responding in the presence of S_1 produces a stimulus, S_2, in the presence of which another schedule controls the delivery of food. In a chained schedule, therefore, S_2 is not only a discriminative stimulus signalling that a particular schedule of unconditioned reinforcement is in effect, it may also serve as a conditioned reinforcer to maintain responding in S_1. A chained schedule, therefore, is analogous to a token-reinforcement experiment, with the possibility of adding further components to the chain providing a parallel to the operation of increasing the number of tokens to be collected before exchange. In view of their greater experimental simplicity, it is not surprising that chained schedules have provided an increasingly popular method for studying conditioned reinforcement (Kelleher and Gollub, 1962; Kelleher, 1966a; Hendry, 1969a).

Two varieties of chained schedules have been studied. In the first variety, just described, different schedules may be programmed in each component, with a different stimulus presented for the duration of each component. In the second variety, the same schedule may be programmed in each component and, instead of marking each component with a unique stimulus, a very brief stimulus change may be programmed to mark the termination of each component. Thus if a pigeon were required to complete five FI 1 min components in order to obtain reinforcement, the key might be illuminated with a white light throughout each component, but the first response after one minute in each component would result in the illumination of the key with a red light for 0·5 sec, and the fifth such presentation of the red light would be followed by food. Because such schedules may involve very large numbers of components, it would be cumbersome to describe them as chain (FI 1, FI 1, FI 1, FI 1, FI 1) etc; instead the above schedule is described as a second-order schedule, FR 5 (FI 1), where the term in the bracket refers to the component schedule and the term outside the bracket specifies the schedule according to which component performance is reinforced (Kelleher, 1966a). Thus a VR 2 (FR 10) schedule would require a subject to complete ratios of 10 responses and deliver reinforcement on average every second time such a ratio was completed.

Studies of token rewards and chain schedules have provided clear evidence of behaviour being maintained by the presentation of initially neutral stimuli. It remains to consider whether the controls employed have been sufficient to prove that the effects of these stimuli should be attributed to conditioned reinforcement.

First, how far is responding for token rewards or on chain schedules dependent on the contingency between response and conditioned reinforcer? The answer here is unambiguous. Wolfe (1936) and Cowles (1937), for example, showed that their subjects would solve discrimination problems when tokens were contingent on correct responses. Such a result seems as secure as that of Saltzman (1949) against any interpretation in terms of noncontingent facilitation of performance. In the case of chain schedules, the question may be answered, as it was by Kelleher (1961), by varying the nature of the component schedules and observing appropriate changes in the pattern of responding. A study by Zimmerman (1969) provides a good example of this. Thirsty rats were trained to press a lever on an FI schedule, with the delivery of water signalled by a 3 sec tone and light. While water was still delivered on the FI schedule, the presentation of the tone–light compound was now programmed according to VI, VR or DRO schedules. Provided that the compound also continued to accompany each delivery of water, its presentation controlled patterns of responding appropriate to its schedule: subjects responded rapidly, for example, when the tone and light were presented on the ratio schedule and stopped responding on the DRO schedule.

The results of Zimmerman's experiment also go some way to answering a second question that may arise in studies of chain schedules. Since primary reinforcement is still presented, it is possible that responding is maintained by the eventual delivery of a primary reinforcer rather than by the presentation of the conditioned reinforcer. Zimmerman noted that the presentation of the tone–light compound changed his rats' pattern of lever pressing from one appropriate to the FI schedule of water delivery to one appropriate to the schedule of conditioned reinforcement. A second control procedure is possible. When a subject's responses in the presence of S_1 produce a second stimulus, S_2, in the presence of which responding is reinforced, it is possible to evaluate the effect that presentation of S_2 has on responding in S_1 by comparing performance on the chain schedule (in which different segments are signalled by different stimuli) with performance on the comparable tandem schedule (in which the stimulus situation remains constant in spite of changes in the basic schedule). In a tandem (FR 20, DRL 6 sec) schedule, for example, the subject must first make 20 responses and then the first response preceded by a 6 sec pause will be reinforced. Such a schedule generates a relatively slow and uneven rate of responding, quite different from that generated by the comparable chain schedule. Other comparisons of chain and tandem schedules (see Kelleher and Gollub, 1962) have confirmed that the contingency of S_2 presentation has a marked effect on the pattern of responding in S_1, over and above any effect of the eventual presentation of the unconditioned reinforcer. A particularly clear example of this is provided by the comparison of chain (FI, FI) schedules with tandem (FI, FI). The tandem schedule produces a slow, relatively uniform rate of responding; the chain schedule produces discrete scallops in each component. Fig. 5.6 shows some data obtained by Kelleher (1966b) in a comparison of tandem with chain schedules: the schedule was FR 15 (FI 4), with or without the presentation of a brief stimulus at the completion of each FI component.

Given that the change in stimulus at the completion of each component of a chain schedule can be shown to maintain appropriate responding within that component, it remains to ask whether this function depends upon the association of such a stimulus with primary reinforcement. There are good reasons for wondering how far this is always true. Consider a subject trained on the second-order schedule FR 15 (FI 4) in Kelleher's (1966b) experiment. In the chained schedule, the first response

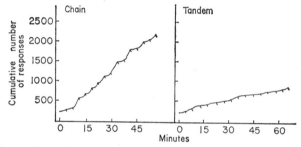

FIG. 5.6. Comparison of chain and tandem FI schedules. On the chain schedule the completion of each FI component is accompanied by a brief signal; on the tandem schedule the subject must complete the same number of FI components in order to obtain reinforcement, without any signal marking the end of each component. Each panel shows one complete cycle of the schedule (i.e. 15 FI components), terminating with reinforcement. (*After* Kelleher, 1966b).

after each 4 min interval produces a brief stimulus, and the first response after the fifteenth 4 min interval produces food. Compare this with the equivalent tandem schedule: if the subject fails to respond after 4 min, the final delivery of reinforcement will be delayed, but no response except the final, reinforced response produces any immediately detectable change in the subject's environment. It is hardly surprising that the rate of responding generated by such a tandem schedule is as slow and erratic as that shown in Fig. 5.6, whereas the performance on the comparable chain schedule shows sharp scallops within each component. Intuitively, however, one might not wish to attribute this difference to the *reinforcing* properties of the change in stimulus contingent upon the completion of each component in the chain schedule, but rather to the information such changes provide the subject about the structure of the schedule. The presentation of food at the completion of an FI schedule, after all, serves not only as a reinforcer of preceding responses, but also as a signal for a period of nonreinforcement (p. 170). It is not surprising, therefore, to find that in some operant studies of second-order schedules, responding in each component has been maintained by the presentation of a brief stimulus that has never been associated with reinforcement, and which could not therefore be acting as a conditioned reinforcer (e.g., Ferster and Skinner, 1957, pp. 67–71; Stubbs and Cohen, 1972).

Stubbs and Cohen, indeed, found that when pigeons were trained on a second-order FI schedule, the rate of responding and degree of scalloping within each FI component was entirely unaffected by whether the brief stimulus marking the end of each component was associated with reinforcement or not. Their conclusions, however, have not always been confirmed: Kelleher (1966b), de Lorge (1967), and Byrd and Marr (1969) all found that performance on a second-order FI schedule was better maintained if the stimulus presented at the end of each component was associated with primary reinforcement. The study by Zimmerman (1969), described earlier, although not strictly employing a second-order schedule, obtained clear evidence that the maintenance of responding by the tone–light compound depended upon the association between that compound and reinforcement. Finally, both Wolfe (1936) and, more extensively, Cowles (1937), in their studies of token rewards, found that distinctive tokens which were not exchangeable for food were significantly less effective reinforcers than the food tokens. It is reasonable to conclude, therefore, that although the tokens in these situations, and the stimuli presented at the end of each component of a second-order schedule, may have other effects on behaviour, their effectiveness as conditioned reinforcers is dependent on the maintenance of their association with primary reinforcement.

B. Variables Determining the Effectiveness of Conditioned Reinforcers

A long series of experiments has studied the effects of a number of different parameters on the reinforcing properties of conditioned reinforcers. Few of these results provide any reason to question the supposition that the strength of a conditioned reinforcer depends upon its degree of association with primary reinforcement. The effectiveness of a conditioned reinforcer increases appropriately with the number of pairings with primary reinforcement: this has been shown both when different groups are tested in extinction with a stimulus paired for different numbers of trials with reinforcement (Hall, 1951) and when a subject is shifted from a tandem to a chain schedule and an appropriate pattern of responding in S_1 gradually emerges (Ferster and Skinner, 1957, p. 689). Extinction of the association with primary reinforcement results in an appropriate decline in the effectiveness of the conditioned reinforcer (Coate, 1956; de Lorge, 1967; Zimmerman, 1969).

The initial effectiveness of a conditioned reinforcer increases with an increase in the probability of primary reinforcement previously associated with it (Mason, 1957; D'Amato et al. 1958). There is, however, good evidence that when conditioned reinforcement is tested in extinction, the persistence of the reinforcing effect of a stimulus previously associated with reinforcement is a decreasing function of the initial probability of primary reinforcement (Saltzman, 1949; D'Amato et al., 1958; Klein, 1959). This is, presumably, a case of partial reinforcement increasing resistance to extinction. The effect is not always very large, however, and some of the more striking claims, advanced by Zimmerman (1959), for the persistence of conditioned reinforcers established on an intermittent schedule of primary reinforcement have not stood up well to examination. Subsequent investigation has suggested that the effects ascribed to the presentation of the conditioned reinforcer did not in fact depend on the association of that stimulus with primary reinforcement at all (Wike et al., 1962).

Other variables that have been studied include delay and magnitude of primary reinforcement. A longer delay between presentation of conditioned and unconditioned reinforcers decreases the effectiveness of the conditioned reinforcer (Jenkins, 1950; Bersh, 1951; Stubbs, 1969); an increase in the magnitude of the unconditioned reinforcer (Butter and Thomas, 1958) increases the effectiveness of the conditioned reinforcer—although this result sometimes depends upon the use of a within-subject design, in which animals are exposed to different values of the unconditioned reinforcer, associated with different conditioned reinforcers (e.g., Wolfe, 1936; D'Amato, 1955).

D'Amato (1955) used a choice procedure in which rats received five pellets in one goal box, one pellet in another, and were then given the opportunity to choose between the two goal boxes situated at the end of opposite arms of a T-maze. It is also possible to measure choice between two conditioned reinforcers when these have been established by the use of a chain schedule. The schedule employed is known as a concurrent chain schedule, and is illustrated in Fig. 5.7. Two keys are available in the subject's chamber. At the beginning of a trial both keys are illuminated; responding on either key produces, usually in accordance with identical VI schedules on each key, a change in the stimulus on the key responded to. As soon as the stimulus has changed on that key, the other key is turned off, so that the subject must now continue to respond on the chosen key until reinforcement is obtained. Choice between the two keys during the initial link is taken as a measure of preference between the two schedules of reinforcement programmed on the two keys in the second link, or equally as a measure of the conditioned reinforcing properties of the two stimuli associated with those schedules.

Using this procedure, Herrnstein (1964a) and Autor (1969) showed that relative rate of responding to the two keys in the first link was directly related to the relative rate or probability of reinforcement in the second link, whether this was varied by using different values of VI, DRO, or VR schedules in the second link. Killeen (1970) found that choice between two keys depended upon the delay before reinforcement was available on each key in the second link and Neuringer (1967) showed that it

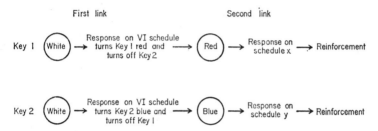

FIG. 5.7. Schematic diagram of the sequence of events on a concurrent chain schedule.

depended on the magnitude of reinforcement associated with each key in the second link.

It is, then, reasonably clear that stimuli associated with reinforcement are themselves able to affect the instrumental behaviour upon which their presentation is contingent and that their effectiveness is a lawful function of several variables usually observed to affect learning. There remain, however, two theoretical questions that have generated considerable dispute: first, what are the conditions necessary for establishing a stimulus as a conditioned reinforcer? Second, how is it that conditioned reinforcers affect the probability of responses on which they are contingent?

C. Conditions Necessary for Establishing Conditioned Reinforcers

1. *Conditioned Reinforcers as Classical CSs*

The normal procedure employed in establishing a conditioned reinforcer is to pair a neutral stimulus with a reinforcing stimulus. The operation involved would appear to be that of straightforward classical conditioning, and this parallel seems sufficient to justify the use of the term conditioned reinforcer. The operation should result in the formation of an association

I

between stimulus and reinforcement which can be indexed either by measuring the occurrence of CRs to that stimulus or by measuring its effectiveness as a conditioned reinforcer. One or two studies have provided these independent measures. Ellison and Konorski (1964) showed that a stimulus associated with food served both to elicit salivation in hungry dogs and to reinforce instrumental responses upon which its presentation was contingent. In one of his studies of token rewards in chimpanzees, Kelleher (1958b) wrote:

> Informal observations indicated that the Ss were very inactive at the start of each session. They became extremely active when they had numerous poker chips, and continually manipulated several poker chips with one hand. Often, they held several poker chips in their mouths and rattled them against their teeth by vigorous head movements. All this activity was accompanied by high rates of responding as well as the screaming and barking which usually occurred during daily feedings in the home cages (p. 288).

Interesting as these observations are, they should perhaps be treated with some caution. Cowles (1937, pp. 66–7) noted that chimpanzees would place in their mouth and bite not only the tokens which were exchangeable for food, but also the tokens which were not.

In the absence of further direct evidence of this sort, the suggestion that a stimulus becomes a conditioned reinforcer by virtue of becoming a CS can be indirectly tested only by attempting to correlate those procedures thought to affect classical conditioning with those affecting the strength of conditioned reinforcement. Several such lines of approach could be pursued, but few have been followed through to the point where they provide decisive evidence one way or the other. Jenkins (1950) employing a trace procedure, and Bersh (1951) a delay procedure, paired a neutral stimulus with the presentation of food, varying the interval between the onset of the two events. The results of these two experiments are shown in Fig. 5.8. In both studies an increase in interstimulus interval (ISI) produced a decline in the effectiveness of the conditioned reinforcer. Kimble (1961, pp. 183–4) commenting on these data, observed that although "the overall resemblance to the functions obtained in classical conditioning is quite apparent . . . substantial amounts of secondary reinforcement were obtained at intervals considerably longer than one would think possible from comparable studies of the interstimulus interval in regular classical conditioning". This qualification need not detain us too long since Kimble's judgement is based largely on studies of eyelid conditioning and ignores the numerous studies which have reported conditioning with other response systems at intervals of several minutes (p. 64).

A related approach is to examine the reinforcing effectiveness of various components of a compound stimulus which differ in their temporal relationship to reinforcement. Studies of serial conditioning employing relatively long delays imply that those elements of a compound occurring closer in time to the delivery of reinforcement might be more effective conditioned reinforcers than those occurring at an earlier point (p. 62). Egger and Miller (1962), in an experiment discussed in more detail below, found that if two stimuli preceded the delivery of food, then the first rather than the second might become the more effective conditioned reinforcer. The contradiction here is more apparent than real, not least because Egger and Miller's results appear to depend upon the amount of training given: in addition, the interstimulus intervals in Egger and Miller's experiment

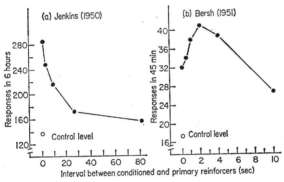

FIG. 5.8. The number of responses supported by a conditioned reinforcer during a test session, as a function of the interval between conditioned and primary reinforcers during training. (*After* Jenkins, 1950; and Bersh, 1951.)

were 2 or 3 sec, while the intervals between S_1 and S_2 in, for example, Williams' (1965) study of serial conditioning was 15 sec: serial conditioning with a very short ISI may produce stronger conditioning to S_1 than to S_2 (Wickens *et al.*, 1963; p. 45).

A final variable, whose study might throw light on the relationship between classical conditioning and conditioned reinforcement, is the schedule of reinforcement used to establish a stimulus as a conditioned reinforcer. A partial schedule of reinforcement often results in substantially less conditioning than a consistent schedule of reinforcement (p. 72). The data of Mason (1957) and D'Amato *et al.* (1958) from discrete-trial studies, and of Herrnstein (1964a) and Autor (1969) from free-operant studies, confirm that partial reinforcement also reduces the initial effectiveness of a conditioned reinforcer. As was noted above, however, there is good evidence that after primary reinforcement is withdrawn, a stimulus previously paired only intermittently with primary reinforcement may maintain its effectiveness as a conditioned reinforcer for a longer time than one associated consistently with primary reinforcement (e.g., Saltzman, 1949; Klein, 1959). The effect of partial reinforcement on the resistance to extinction of classical CRs, however, is considerably more ambiguous (p. 73). If further investigation establishes a real discrepancy here, this would present a serious problem for analysis.

2. Conditioned Reinforcers as Discriminative Stimuli

Skinner (1938, p. 245) assumed that a stimulus would become a conditioned reinforcer either if it were a classical CS, or if it were an instrumental discriminative stimulus which "set the occasion" for instrumental responding. Keller and Schoenfeld (1950, p. 236), following a study by Schoenfeld et al. (1950), suggested that the establishment of a stimulus as a discriminative stimulus was the necessary and sufficient condition for turning it into a conditioned reinforcer. Schoenfeld et al. trained rats to press a lever for food with, for one group, a light situated 12 in above the food magazine being illuminated after each lever press at the moment when the rat started eating. When the rats were extinguished, the presentation of the light contingent upon lever pressing had no effect on responding. This unsurprising result was said to contradict the view that a stimulus "which occurs repeatedly and consistently in close conjunction with a reinforcing state of affairs . . . will itself acquire the power of acting as a reinforcing agent" (Hull, 1952, p. 6, Corollary ii). More importantly, it also came to form the basis of the view that a stimulus would become a conditioned reinforcer only if it were a discriminative stimulus in the presence of which an animal was required to perform some response in order to obtain reinforcement. Since in the experiment of Schoenfeld et al., it was argued, the rat had already started to eat when the light was turned on, the light was not a discriminative stimulus for approaching the food magazine and this is why it was an ineffective reinforcer. The more plausible explanation—that a rat busy eating will ignore lights situated 12 in above its head—was itself ignored. It is worth remembering that strict simultaneity of CS and UCS results in no classical conditioning (p. 57).

Since most appetitive reinforcers require some minimal behaviour on the part of an animal if it is to benefit from them, for the animal must usually approach a food magazine or a water dipper, it is not necessarily easy to show that a stimulus in whose presence reinforcement is delivered without any response requirement can still become a conditioned reinforcer. However, intracranial stimulation has been used as the reinforcing event for establishing a conditioned appetitive reinforcer by Stein (1958) and Knott and Clayton (1966). Carlton and Marks (1958) have presented evidence suggesting that a tone paired with heat may serve as a conditioned reinforcer for cold rats. Stimuli paired with electric shock have been shown to become conditioned aversive reinforcers, capable of suppressing responses on which they are contingent (e.g., Mowrer and Aiken, 1954; Seligman, 1966): it is hard to see how any sense can be given to the notion that such a stimulus is a discriminative stimulus for shock. In general, there does not seem to be any evidence specifically implying that conditioned reinforcers must be discriminative stimuli in whose presence an animal must make some set of responses to obtain reinforcement, rather than simply classical CSs in whose presence reinforcement is delivered.

One line of evidence thought to be relevant to the discriminative stimulus hypothesis of conditioned reinforcement concerns the effect of discrimination training on establishing a stimulus as a conditioned reinforcer. Several experiments have suggested that it may not be sufficient to pair a particular stimulus with reinforcement; it may also be necessary to expose subjects to the omission of reinforcement in the absence of the stimulus. Ferster (1951) trained rats to press a lever for food in the continuous presence of a light and buzzer, but found that subsequent extinction was entirely unaffected by the presence or absence of these stimuli. In discrete-trial experiments, in which conditioned reinforcement is tested by seeing whether rats will learn to choose the arm of a T-maze leading to a goal box previously paired with food, neither Webb and Nolan (1953) nor McGuigan and Crockett (1958) found any evidence of conditioned reinforcement unless rats had both been fed in one goal box and placed in a different box without being fed. Although Saltzman (1949) observed reliable evidence of conditioned reinforcement in the absence of discrimination training, he too found that such training significantly increased the conditioned reinforcing effectiveness of a goal box. Finally, Notterman (1951) trained rats in a runway with a buzzer sounding on reinforced trials; he found that increasing the proportion of trials on which the buzzer was not sounded, and reinforcement was omitted, significantly increased the effectiveness of the buzzer as a conditioned reinforcer.

Although there are some exceptions to this generalization (e.g., Reynolds et al., 1963a, b), it is reasonable to conclude that discrimination training is often important for the establishment of a conditioned reinforcer. But what does this imply? Certainly not that conditioned reinforcement depends upon the existence of a response requirement for obtaining reinforcement in the presence of a discriminative stimulus: *that* relationship remains unaffected by scheduling trials without the discriminative stimulus and without reinforcement. The truth is that stimuli continually present in any experimental situation often exert no detectable influence on behaviour. Whether we are interested in the effects of stimulus intensity on performance (p. 43), or in the slopes of generalization gradients (p. 501), discrimination training is often necessary for the detection of control by a specific stimulus. As we argued, classical conditioning procedures automatically incorporate discrimination training, since the interval, spent in the apparatus, exposes the subject to the experimental situation without CS and without UCS. Discrimination training between the presence and absence of a specified stimulus is often necessary if that stimulus is to enter into association with reinforcement. The fact that it may also be necessary in order to establish that stimulus as a conditioned reinforcer does not imply anything special about conditioned reinforcement; it suggests only that conditioned reinforcement depends upon an association between the initially neutral stimulus and the primary reinforcer.

3. *Information Hypothesis*

Egger and Miller (1962, 1963) suggested that a stimulus must be informative if it is to become a conditioned reinforcer: in other words, it must be a reliable and nonredundant signal for primary reinforcement. They found that a brief, 1·5 sec stimulus preceding food would not become a conditioned reinforcer if food were also signalled by a 2·0 sec stimulus whose onset preceded the 1·5 sec signal by 0·5 sec. Their argument was that the 1·5 sec stimulus was redundant, since reinforcement was already predicted by the longer stimulus. Using the same length of stimuli, but shock instead of food as the primary reinforcer, Seligman (1966) confirmed that a 2·0 sec S_1 was a more effective conditioned aversive reinforcer than a 1·5 sec S_2. In another attempt to test Egger and Miller's hypothesis with aversive reinforcement, however, Ayres (1966) found no evidence that the earlier of two stimuli was associated more effectively with shock. Moreover, Thomas *et al.* (1968), in an experiment with pigeons and an appetitive reinforcer, found that the relative effectiveness of S_1 and S_2 critically depended upon the number of initial pairings with reinforcement. Only after an intermediate number of pairings was S_1 the more effective conditioned reinforcer; after a smaller number of trials S_2 was more effective; after a larger number of trials neither component was a strong conditioned reinforcer and only the S_1–S_2 compound maintained responding. Although the procedure used by Thomas *et al.* was not comparable to that of Egger and Miller, since S_1 terminated with the onset of S_2 instead of continuing through until reinforcement (a method of presentation that is analogous to turning S_1 into a *trace* CS), their results suggest some caution in interpreting Egger and Miller's data.

We have mentioned that when longer stimuli are used there will be less reason to expect that S_1 would be the more effective stimulus. However, even where S_1 was 15 sec and S_2 11 sec, Borgealt et al. (1972), studying conditioned suppression in rats, found that S_1 was somewhat more effective than S_2. However, when S_1 terminated with the onset of S_2 (when S_1 was a trace CS), a typical serial conditioning effect (i.e., stronger conditioning to S_2) was observed. These results emphasize, yet again, the parallel between conditioned reinforcement and classical conditioning.

It is not at all certain that any of these results should be attributed to a concept of "informativeness". As several investigators (e.g., Thomas et al., 1968, and Borgealt et al., 1972) have pointed out, the results may be attributable to generalization decrement. If S_1 precedes and overlaps S_2, then S_2, unlike S_1, never appears on its own during training: test trials with S_2 alone, therefore, represent a greater departure from conditions of training than do test trials with S_1 alone. The fact that considerable compounding occurs (animals sometimes respond only to the S_1–S_2 compound during testing) is also derivable from principles of generalization decrement.

Egger and Miller (1962, 1963), however, established a second, more valid, principle: they found that under conditions where S_1 became a stronger conditioned reinforcer than S_2, the effectiveness of S_2 as a conditioned reinforcer could be increased by alternating, during initial training, reinforced presentations of S_1–S_2 with nonreinforced presentations of S_1 alone. In their words, this turned S_1 into an unreliable signal for reinforcement, so that S_2, instead of being redundant, became the only reliable predictor. This finding is reminiscent of the sort of data repeatedly referred to in Chapter 2: the effectiveness of a given CS depends not only on its own correlation with reinforcement, i.e. its own validity, but also on its validity relative to that of other stimuli impinging on the subject at the same time. Experiments by Wagner (1969b) and Wagner et al. (1968) have demonstrated the generality of this conclusion, both in a variety of classical conditioning situations and in instrumental discrimination training.

4. Conclusions

Egger and Miller's experiments are important not because they tell us anything peculiar to conditioned reinforcement, but precisely because they imply that the conditions necessary for establishing conditioned reinforcers are the same as those required to establish any association between a stimulus and reinforcement. Just as we saw that the role of discrimination training in establishing conditioned reinforcers finds parallels in the effects of such training in instrumental generalization gradients and in classical conditioning, so the effects of redundancy and relative validity can be documented in all of these situations. Other, concomitantly presented, stimuli will compete with a particular stimulus for association with a given reinforcer and, if more salient or more valid, will reduce the associative strength of that stimulus. This association with reinforcement may be measured by recording CRs elicited by the stimulus as a CS, or by requiring instrumental responses in the presence of that stimulus as a discriminative stimulus, or by presenting the stimulus contingent upon instrumental responses as a conditioned reinforcer. In each case the same general result will hold. Conditioned reinforcers, therefore, acquire their effectiveness by being associated with primary reinforcers: the necessary and sufficient conditions are the same as those required to establish any association between stimuli and reinforcers.

D. The Nature of Conditioned Reinforcement

With appropriate experimental preparations, there is unequivocal evidence that conditioned reinforcers may increase the probability of responses on which they are contingent, that their effectiveness can be attributed to this contingency, and that it depends upon the establishment of an association between conditioned and unconditioned reinforcers. In this operational sense, stimuli associated with primary reinforcers may themselves become effective reinforcers of instrumental responding.

It is still possible to ask how such an effect operates. Two answers can be distinguished. One is to assert that a stimulus associated with a primary reinforcer acquires certain of the properties of that reinforcer. In just the same way that a stimulus associated with food acquires the capacity to elicit salivation, so such a stimulus acquires the capacity to reinforce instrumental responses. This does not, of course, imply that either class of reinforcer affects instrumental responding in ways envisaged by the law of effect. Both, presumably, achieve their effects either by eliciting approach responses in a Pavlovian manner, or by becoming associated with responses and hence motivating that pattern of responding which is accompanied by an increase in incentive. Conditioned reinforcers, like primary reinforcers, act as goals for instrumental responding.

A second position argues that conditioned reinforcers should not be thought of as themselves providing goals for behaviour: they reinforce instrumental responses only to the extent that they provide information about primary reinforcement. This suggestion should not be confused with Egger and Miller's (1962) hypothesis considered in the preceding section. Egger and Miller were proposing an account of the conditions necessary to establish a stimulus as an effective conditioned reinforcer. The view to be considered here is that conditioned reinforcers establish and maintain instrumental responses because they provide information or reduce uncertainty. The impetus for this view has come from studies of observing responses, in which the presentation of a stimulus, signalling whether reinforcement is available or not, is made contingent on a so-called observing response.

1. *Observing Responses*

In a typical experiment on observing responses, reinforcement is made available according to some random sequence determined by the experimenter. The subject's behaviour cannot affect the delivery of reinforcement, but some specified response—the observing response—results in the

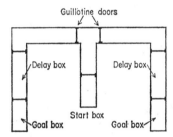

FIG. 5.9. Floor plan of a maze for studying observing responses. On each trial the subject is detained in the delay chamber on the side chosen on that trial. On one side the colour of the delay chamber, invisible from the choice point, is correlated with the availability of reinforcement; on the other side, there is no correlation between the two. (*After* Prokasy, 1956.)

presentation of stimuli whose value is correlated with the probability of reinforcement occurring. Observing responses, therefore, reduce uncertainty about whether, when or where reinforcement is to be delivered. Since animals learn to make observing responses, it is to be presumed that the stimulus change contingent on the observing response is an effective conditioned reinforcer and since the stimuli provide information about the occurrence of primary reinforcement, it is possible that it is precisely this information that is the basis of their effectiveness (Berlyne, 1960; Hendry, 1969b).

A few examples will illustrate the nature of these experiments. Prokasy (1956) trained rats in the maze shown in Fig. 5.9; on each trial the rat would be detained in a delay section for 30 sec before being permitted to run through into the goal box. Food was available on a random 50 per cent of trials in each goal box, and the only difference between the consequences of a left or right turn at the choice point was that the delay chamber on one side was black on reinforced trials and white on non-reinforced trials, whereas on the other side the colour of the delay chamber bore no relationship to whether or not reinforcement was programmed for that trial. Ten out of twelve rats learned to go to that side of the maze where the colour of the delay chamber was correlated with the occurrence of reinforcement.

The procedures employed in typical operant studies of observing responses are shown in Fig. 5.10. In Wyckoff's (1952) experiment, a pigeon's pecks at a white key were sometimes reinforced at the end of a 1 min period and sometimes not. The subject was thus being trained on a mixed (FI 1, Extinction) schedule. However, if the pigeon stood on a treadle, the colour of the key changed to red if reinforcement was

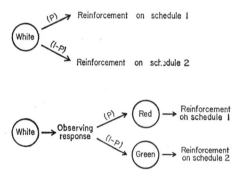

Fig. 5.10. Design of operant experiments on observing responses.

going to be delivered on that trial and to green if no reinforcement was due. Following this observing response, therefore, the subject was being exposed to a multiple (FI 1, Extinction) schedule. After each reinforcement, the key was illuminated with white light and the sequence began again. The observing response produced stimuli signalling whether reinforcement was available or not, but it did not affect the probability of obtaining reinforcement. A similar procedure has been employed by Kelleher (1958a) using chimpanzees, with various schedules in effect, and by Kendall (1965) and Hendry (1969b) with pigeons, when long or short FR schedules were in effect. In all cases subjects learned to make the observing response, thus providing stimuli signalling the schedule in effect on that trial (or, if one prefers, changing a mixed schedule into a multiple schedule).

Bower *et al.* (1966) used a choice procedure: reinforcement was scheduled for responses on either key of a pigeon chamber on either FI 10 sec or FI 40 sec. The schedule in effect on a given trial was determined by the experimenter and was not affected by the subject's behaviour. However, a response to one key turned that key red on FI 10 sec trials and green on FI 40 trials, while a response to the other key turned it yellow regardless of the schedule in effect on that trial. Subjects showed a strong preference for the "informative" key. Similar results have been obtained by Hendry (1969b) using FR 10 and FR 90 schedules in place of the short and long FI schedules.

Observing response experiments can also utilize aversive reinforcement: in a study by Lockard (1963), for example, a rat in a two-compartment box received random electric shocks in either compartment, but in one compartment a small light signalled the impending shock, whereas in the other the light and the shock were entirely uncorrelated. The animals rapidly developed a preference for the side on which they received the warning stimulus, even though this did not enable them to avoid the shock.

2. *Theoretical Analyses of Observing Responses*

To the extent to which the occurrence of an observing response does not affect the experimenter's predetermined schedule of reinforcement, it is reasonable enough to suggest that such responses may be reinforced because they produce stimuli which provide information about the delivery of reinforcement. According to this analysis, therefore, observing responses are reinforced because they provide information, or reduce uncertainty.

A second possible analysis, suggested for example by Perkins (1968), is that whatever the experimenter's intention, the occurrence of an observing response does change the effective schedule of reinforcement acting on the subject. In Lockard's experiment, for example, the shock may have been delivered in both compartments, but the rat that received warning of the shock may have been able to minimize its aversiveness by a variety of postural adjustments. It is true that Perkins *et al.* (1966) repeated Lockard's experiment, except for delivering shock via permanently attached ear clips, and still obtained similar results, but it is still conceivable that some preparatory response could reduce the pain of such a shock. A related argument can be applied to the results of some appetitive experiments. In the situation studied by Wyckoff (1952) it is possible to argue that making observing responses saves the subject from subsequent wasted responding: the animal that knows that reinforcement will not be available on this trial, or will not be available until some time has elapsed, will not waste its efforts in making a long series of unreinforced responses. The importance of this factor has been seriously questioned by Bower *et al.* (1966) and by Steiner (1967). Steiner, for example, showed that baboons would learn to make an observing response, signalling which of two noncontingent FI schedules would be in effect on that trial, where food was delivered after a specified interval without any response requirement. Nevertheless, as in the aversive case, it is always possible to insist that, when a signal is provided, some unspecified preparatory response can be made which increases the effectiveness of reinforcement: salivation, for example, may improve the taste, or aid the digestion, of dry food.

A third possible analysis of observing responses has been suggested by Wyckoff (1959) and Bower *et al.* (1966). This analysis, instead of appealing to information value to account for the effectiveness of conditioned reinforcers, appeals to conditioned reinforcement to account for the strengthening of observing responses. Animals learn observing responses, it can be argued, because the conditioned reinforcing value of the stimuli contingent upon making the observing response is greater than the value of the stimuli contingent upon not making the observing response. The argument can be presented by applying it to the situation studied by Prokasy (1956). In that experiment the observing response (choice of the arm leading to the correlated delay chamber) was followed on 50 per cent of trials by a stimulus which always signalled reinforcement, and on the remaining trials by a stimulus which signalled no reinforcement. The other response produced, on every trial, stimuli which signalled 50 per cent reinforcement. If one were to assume that the effectiveness of a conditioned reinforcer signalling consistent reinforcement was more than twice as great as the effectiveness of one signalling only 50 per cent reinforcement, then the observing response, even though reinforced on only half the trials, would still receive a greater amount of conditioned reinforcement and should therefore increase in probability. Analogous assumptions can be made to explain the results of those experiments where mixed long and short FI or FR schedules are in effect and the observing responses produce stimuli correlated with the schedule operating on that trial. If the value of a stimulus signalling a short delay (or ratio requirement) on every trial is more than twice as great as that of a stimulus signalling equally probable short and long delays (or ratios), then the fact that the observing response is followed only by its effective conditioned reinforcer on 50 per cent of trials will still be sufficient to ensure that it gains strength.

The associationist analysis states, in effect, that observing responses are reinforced only when they signal the preferred event, but that this reinforcement is sufficiently strong to outweigh the fact that it does not occur on all trials. The analysis has also been applied to the case where an observing response produces warning of impending shock. Here, as Mowrer (1960b, pp. 189–90) and Seligman (1968) have suggested, the observing response is not reinforced because it sometimes provides the warning stimulus, but because it more often produces a *safety stimulus*, i.e. the absence of the warning signal. In the uncorrelated compartment of

Lockard's experiment, shocks occur both in the presence and in the absence of the light; in the correlated compartment, however, if the light is off then no shock will be delivered, and the animal is therefore safe.

Although there is nothing logically incoherent about this general analysis, it seems to rely on some apparently arbitrary assumptions. Specifically, we must assume that animals scale the value of stimuli associated with reinforcement in such a way that the average value of a stimulus associated equally often with more and less preferred events will be less than half that of a stimulus consistently associated with the preferred event. Although this assumption is sufficient to explain why observing responses will be learned, it would appear to stand in need of some independent validation.

As it happens, there is good evidence which supports this assumption and simultaneously argues strongly against both the preparatory response hypothesis and the view that animals seek to reduce uncertainty. In a number of experiments, animals have been required to choose between two alternatives, when choice of one alternative is randomly followed by a more favourable, or by a less favourable reinforcer in the absence of any correlated signal, while choice of the other alternative results in a constant, intermediate reinforcer. For example, Leventhal et al. (1959) trained rats in a T-maze, with choice of one side leading to either two or zero units of food (on a random 50 : 50 basis), while choice of the other side was always reinforced with a single unit. Pubols (1962a) performed a similar experiment, varying delay of reward between, for example, zero and 30 sec on one side and holding it constant at 15 sec on the other. In both experiments rats rapidly learned to choose the alternative with the varied reinforcement.

The implication is that the value of the most favourable reinforcer (two units or zero delay) is more than enough to offset the disincentive of the least favourable condition (no food or long delay), with the consequence that the combination of most and least favourable conditions is better than the constant, mean reinforcer. Now this is entirely consistent with the assumptions required to explain the reinforcement of observing responses: in that case, the argument was that a stimulus correlated with the more valuable reinforcer would become a sufficiently powerful conditioned reinforcer to offset the disincentive of the stimulus correlated with the less valuable reinforcer. In both cases the assumption is that, over at least a considerable range, the incentive value of reinforcement is a positively accelerated function of the magnitude, immediacy, or other measure of the reinforcer.

Several studies employing concurrent chain schedules have confirmed the generality of the results of the T-maze studies of Leventhal et al. and

Pubols. Herrnstein (1964b) found that pigeons preferred a stimulus correlated with both short and long FI schedules to one constantly associated
with an FI schedule whose value was the mean of the long and short
schedules; and Killeen (1968) and Davison (1969a) have confirmed this
result. Fantino (1967), meanwhile, found that pigeons preferred a stimulus
correlated with short and long FR schedules to a stimulus always associated with the mean FR schedule.

The fact that animals often prefer unsignalled variations in reinforcement to a constant intermediate value of reinforcement provides little
encouragement for the view that observing responses are reinforced by
reduction in uncertainty: if uncertainty were undesirable, animals would
presumably choose the constant reinforcement. Similarly, if observing
responses were reinforced because they permitted the elaboration of appropriate preparatory responses, it is hard to see why animals should choose
a condition which makes such preparation impossible.

Further support for the associationist analysis may be provided by
examining the range of conditions over which animals will both learn
observing responses and show a preference for varied rather than constant
reinforcement. Leventhal *et al.* found that when the reinforcement in their
constant condition was large (1·0 gm of food), rats showed little preference
for the varied alternative (correlated with 2·0 gm or no food); the preference developed only when the absolute amount of reinforcement used was
relatively small. Similarly, Pubols found that the rate at which rats developed a preference for the varied-delay alternative was directly related to
the length of delay on the constant alternative: when the constant delay
was 20 sec (and the variable delay 0 or 40 sec) animals learned to select
the varied side in 6·2 days of training; when the constant delay was 2 sec
(and the varied delay 0 or 4 sec), they required 11·0 days to learn. Thus,
when the constant, intermediate reinforcer is reasonably large or immediate, the difference in value between it and the most favourable value on
the variable alternative may not be great enough to outweigh the difference
between the intermediate and least favourable reinforcer. In the case of
observing responses, there is some evidence that they may be unstable or
hard to establish if the less preferred schedule is still relatively favourable.
Hendry (1969b) was unable to obtain reliable observing behaviour in pigeons
when they were trained on a mixed (FR 10, FR 40) schedule, but found
an immediate increase when the schedule was shifted to mixed (FR 20,
FR 80). Some direct comparisons of the effects of different schedules in the
two situations might establish a more reliable correlation between the two.

There have been several other attempts to bring evidence to bear on
these different theoretical analyses of observing responses. Hendry (1969b)
and Wilton and Clements (1971a) have argued that if observing responses

are reinforced because they reduce uncertainty, they should extinguish when the uncertainty is removed from the situation. If observing responses have been established by the presentation of stimuli signalling whether FR 10 or FR 90 is in effect, then they should disappear when the FR 90 component is dropped, and only the more favourable schedule is retained. Both experimenters reported exactly this result and concluded that it argued strongly against any associationist analysis: if the observing response had been reinforced by the presentation of stimuli signalling the more favourable schedule, the removal of the less favourable schedule and its associated stimulus should only have strengthened the response. The argument does not seem convincing: the associationist analysis must be interpreted to be saying that observing responses are reinforced provided that they sometimes *change* the stimulus situation from one correlated with a mixed (i.e., intermediate) schedule of reinforcement, to one correlated with a more favourable schedule of reinforcement. Dropping the less favourable component from the mixed schedule not only abolishes the information provided by the observing response, it also ensures that the stimulus correlated with the mixed schedule now becomes an equally reliable signal for the more favourable schedule. With or without an observing response, therefore, subjects not only have all the information they might want, they are also exposed to a stimulus signalling the favourable schedule. Both analyses are equally able to predict that under such conditions the observing response will extinguish.

It is also possible to provide some test of the relative merits of different analyses, by considering the different predictions they make about the precise events that reinforce observing responses. If observing responses are reinforced because they reduce uncertainty, then the rats in Prokasy's experiment, for example, should have been reinforced for entering the correlated delay chamber not only on trials when the delay chamber signalled that food was available, but also on trials when the other colour of the delay chamber signalled that no food was going to be given. Both the preparatory response and the conditioned reinforcement analyses imply that the observing response would have been reinforced only on trials when the positive delay chamber was entered: the other delay chamber is not a reinforcer at all. In the aversive case all three analyses made different predictions. According to the informational analysis, Lockard's rats would have been reinforced for entering the correlated compartment both when the signal was on and when it was off, for in either case they received information about whether shock was imminent. The preparatory response hypothesis implies that the observing response was reinforced only by presentation of the warning stimulus, in the presence of which the rats could learn to brace themselves for shock. The associationist analysis states that the observing response was reinforced only when it was followed by the absence of the warning stimulus.

These contrasting predictions have been examined in a few recent studies. Dinsmoor *et al.* (1969) trained pigeons on a random alternation of a VI, food-reinforced, schedule and the same schedule with shock added on an FR schedule. An observing response turned the key red when shock was in effect and green when no shock was programmed; in the absence of an observing response, no signal was given to indicate which schedule was in effect. They found that the latency of observing responses was significantly increased on trials following a shock trial. Furthermore, if the two schedules remained in effect as before but the observing response produced no signal on shock-free trials (i.e., the key remained white) while still producing a change to red on shock trials, the birds would either not learn to make the observing response or would not maintain it if once it had been learned. These data suggest that the occurrence of the signal correlated with shock was at best a weak reinforcer for the observing response and may, in fact, have provided no reinforcement at all. In this sense they are inconsistent with analyses that appeal to the importance of either preparatory responses or uncertainty reduction; they suggest, therefore, that observing responses are reinforced because they produce stimuli which provide stronger conditioned reinforcement than that correlated with the stimuli contingent upon not responding.

Dinsmoor *et al.*'s conclusion may have depended upon their use of shock in one component, for it is possible that the aversiveness of a stimulus correlated with shock may be sufficient to outweigh its value as an informative signal. It is, therefore, important to know whether similar results would be obtained where the alternative schedules specified short or long delays of appetitive reinforcement, or simply the presence or absence of food. A study by Kendall and Gibson (1965) provides some evidence on this point. Pigeons were trained with a mixed (FR 50, FI 2 min) schedule of food reinforcement in effect on one key, while observing responses on a second key turned the food key blue on FR trials and red on FI trials. The two birds learned to make consistent observing responses. These were maintained when the two schedules remained in effect, but the observing response only produced the FR stimulus (leaving the key unchanged on FI trials); observing behaviour extinguished to about 10 per cent of its initial rate when the only consequence of such responses was the production of the red stimulus on FI trials. Since the FR schedule permitted a much higher rate of reinforcement, it is clear that the FI schedule was less favourable and that observing behaviour was not maintained when its only consequence was the appearance of a stimulus correlated with the less favourable alternative.

Although Schaub (1969) has presented evidence which he interprets to mean that the presentation of a stimulus signalling extinction may reinforce an observing response, his procedure was very unusual and any inference from his data far too indirect to enable one to accept his conclusions without serious question. There is, on the other hand, excellent evidence that pigeons will learn to make a response that *terminates* a stimulus signalling extinction, even though such a response has no effect on the delivery of primary reinforcement (Rilling *et al.*, 1969; Terrace 1971; p. 264). A recent study by Dinsmoor *et al.* (1972) shows unequivocally that if pigeons are trained on a mixed (VI, Extinction) schedule, observing responses are maintained only if they produce a stimulus correlated with the VI component. Whenever the only effect of the observing response was to produce a stimulus correlated with extinction, all subjects rapidly stopped responding on the observing key.

3. *Conclusions*

None of the studies under review has provided convincing evidence that observing responses are reinforced because they provide information or reduce uncertainty. It is possible that future research will provide such evidence. It is also true that one should hesitate to reject this theory out of hand, simply on the basis of experiments on pigeons and rats: it is possible that primates would provide much more favourable evidence, as a recent study by Lieberman (1972) suggests. Nevertheless, even if such evidence were provided, it would not affect the conclusion that conditioned reinforcers gain their effectiveness largely because they are associated with primary reinforcement, rather than because they provide information about its availability. Observing responses seem to be reinforced because the conditioned reinforcing value of the stimuli contingent upon such responding is greater than that contingent upon not responding. There is, moreover, unequivocal evidence that stimuli providing information about the omission of appetitive reinforcement or the occurrence of aversive reinforcement, become themselves aversive. It is, therefore, only natural to infer that stimuli that are associated with appetitive reinforcement become themselves attractive and therefore act as goals for instrumental behaviour.

V. INHIBITION IN APPETITIVE INSTRUMENTAL LEARNING: FRUSTRATION THEORY

A. Nonreinforcement and Frustration

We have argued that appetitive reinforcers exert their effects on instrumental responses not by strengthening a set of connections between such responses and the situation in which they occur, nor yet by simply providing a source of motivation which facilitates ongoing behaviour. Their most important effect is to act as goals or to establish other stimuli as goals for instrumental responses: animals learn the relationship between their behaviour and its consequences, and responses that have previously been associated with an increase in proximity to reinforcement or with stimuli associated with such an increase, will be repeated on future occasions. A symmetrical account can be provided for the effects of nonreinforcement, as, for example, in extinction. We could say that just as responses associated with increments in appetitive reinforcement are facilitated, so responses associated with decrements in appetitive reinforcement are inhibited.

This is not the place for a detailed evaluation of phenomena and theories of extinction. We shall see (in Chapter 8) that much of the decline in response probability occurring in extinction need not be attributed to any such inhibitory learning at all, but may be explained in terms of generalization decrement. The purpose of the present discussion is to round out the argument of this chapter by considering evidence bearing on the idea that nonreinforcement, in a situation previously associated with reinforcement, may generate a state of inhibition or frustration and that such inhibition or frustration may lead to a decline in the probability of an appetitively reinforced instrumental response.

This view, associated with frustration theory (Amsel, 1958, 1962), is to be contrasted with the accounts of the decline in responding observed in extinction that are associated with early S–R theory. The three main accounts were Guthrie's interference, or competing response, theory (Guthrie, 1935). Hull's theory of reactive inhibition (Hull, 1943) and Skinner's suggestion that nonreinforcement might simply lead to a loss of excitation (Skinner, 1938). Two main features distinguish inhibition or frustration theory from these earlier theories. The first is that inhibition or frustration is assumed not to be an automatic consequence of nonreinforcement, but only of nonreinforcement in a situation previously associated with reinforcement. "The important implication of frustration theory in the present context is that nonreward is frustrating and aversive *in proportion to the degree of anticipation of reward*" (Wagner, 1969c, p. 161). The reasons for the appearance of competing responses in extinction were never entirely clear unless they were assumed to be elicited by a frustrated expectation of reinforcement and at this point, of course, we have a version of frustration theory. Reactive inhibition, according to Hull (1943), so far from being generated only by nonreinforcement in the context of reinforcement, was an automatic consequence of all responses, whether reinforced or not.

To the extent to which nonreinforcement in the context of anticipated reinforcement can be shown to have particular consequences not attributable to the mere absence of reinforcement, it is clear why a discussion of frustration theory is relevant to some of the preceding arguments of this chapter. Such a demonstration would provide further evidence for the view that response-contingent reinforcement does not simply stamp in a response, but becomes an anticipated consequence of that response. The argument has already been advanced in the discussion of negative contrast effects (p. 214).

The second defining feature of frustration theory, as the quotation from Wagner suggests, is that the omission of an expected appetitive reinforcer is assumed to be an aversive event, with important similarities to the presentation of an aversive reinforcer such as shock.* This assumption provides an important connecting link between this and the following chapter. If the omission of an appetitive reinforcer is aversive, then such an event, if made contingent on a response, would define a procedure for punishing that response. In general, the argument suggests the following parallels in instrumental experiments: the omission of an appetitive reinforcer and the presentation of an aversive reinforcer, are punishing events which would be expected to lead to a decrement in instrumental responding; conversely, the presentation of an appetitive reinforcer and the omission of an aversive reinforcer are reinforcing events which would be expected to increase the probability of prior instrumental responses.

B. The Aversive Effects of Omission of Appetitive Reinforcers

The present section reviews a variety of studies which have sought evidence on these two defining features of frustration theory: that non-reinforcement in the context of reinforcement has effects over and above any attributable to the mere absence of reinforcement, and that these effects are aversive, i.e., similar to those produced by such paradigmatic aversive reinforcers as electric shock.

1. *Response Suppression*

One of the several defining characteristics of aversive reinforcers is that their presentation suppresses appetitively reinforced instrumental responses. This is true whether the aversive reinforcer occurs independently of the subject's behaviour, as in experiments on conditioned suppression, or whether its presentation is contingent on a particular response, as in experiments on punishment. In the latter case, however, the suppressive

* Skinner (1950) later assumed that nonreinforcement, in addition to leading to a loss of excitation, might also have disruptive, emotional effects.

effects are normally even greater (p. 275). A popular field for study, there-fore, has been to see whether the omission of an appetitive reinforcer, or a stimulus correlated with such omission, will, if made contingent on a response, reduce the probability of that response. But the fact that animals will not repeat responses that are followed by the omission of food establishes very little: since they will not often repeat responses fol-lowed by neutral stimuli, the effect of the omission is not necessarily aversive, nor necessarily dependent upon a frustrated anticipation of food. There is a further problem: in most studies to date, the effects of response-contingent omission of food have been studied while the animal continues to receive reinforcement for another response, or in the presence of another stimulus. Thus, a decline in the response followed by omission of food is necessarily accompanied by an increase in the opportunity to make the food-reinforced response. As Leitenberg (1965) and Wagner (1969c) have recognized, therefore, an appeal to the *punishing* effects of omission of food is redundant: "it seems adequate and more parsimonious to explain these findings by just saying: the pattern of behavior followed by most positive reinforcement is most strengthened" (Leitenberg, *op. cit.*, p. 431). Although it should, in principle, be possible to design an experiment that avoids these problems, no one so far seems to have done so (Leitenberg, 1965).

However, since appetitively reinforced instrumental responding can be suppressed by a stimulus signalling noncontingent shock, it might also be possible to detect some suppression by the presentation, independent of responding, of a stimulus signalling a period of nonreinforcement. If, for example, while an animal was responding on a VI schedule of appetitive reinforcement, a stimulus signalling a *later* period of nonreinforcement were to suppress responding during its presence, this would presumably result in a loss rather than a gain of reinforcement for the subject and provide strong evidence of the aversive properties of the omission of appetitive reinforcement. Several experiments have investigated this pos-sibility. Their results, however, have not been entirely consistent. Ferster (1958) using chimpanzees and Leitenberg (1966) using pigeons both observed an acceleration of instrumental responding during a stimulus predicting omission of reinforcement. Holmes (1972), however, using a shorter CS, and giving much more extended training than Leitenberg, did eventually observe virtually complete suppression of responding in pigeons. Of two studies with rats, one obtained no more than marginal evidence of a decline in responding (Kaufman, 1969), while the other observed significant, although not complete, suppression (Leitenberg *et al.*, 1968). Finally, Zbrozyna (1958), studying dogs, has found suppression of the consummatory response of eating in the presence of a stimulus signalling the withdrawal of food. Although, therefore, the suppression produced by a stimulus signalling a period of extinction is very much less than that typically produced by a stimulus signalling shock, there is undeniable evidence that such a stimulus may be sufficiently aversive to produce mild suppression of instrumental responding.

2. Elicited Responses

We have seen that shock often elicits what may loosely be called aggressive behaviour in various animals (p. 18). There is considerable evidence that omission of appetitive reinforcement may do the same.

Azrin *et al.* (1966) trained pigeons for a series of sessions during which periods of acquisition and extinction of food-reinforced key pecking were alternated. A "target" pigeon was confined in a compartment at one end of the chamber. Although little aggression occurred when the birds were first put into the apparatus and before the initial period of food reinforcement, the first transition from food reinforcement to extinction produced a burst of attacks on the unfortunate target bird. Attacks decreased in frequency during the course of each extinction session and increased with the number of reinforcements in the immediately preceding acquisition session. Such a pattern of aggressive behaviour suggests that it was elicited by the omission of expected reinforcement. Thompson and Bloom (1966) reported that hooded rats (but not albinos) would attack another rat at the outset of an extinction period and Hutchinson *et al.* (1968) showed that squirrel monkeys will bite a rubber hose under similar conditions. On some intermittent schedules of reinforcement (FI and FR), there are particular moments during the schedule when the probability of reinforcement is low. Immediately after reinforcement, animals stop responding and it can hardly be doubted that this is because the occurrence of reinforcement provides a signal for a period of time during which no reinforcement is available (p. 170). Aggressive behaviour has been observed in both pigeons (Gentry, 1968; Flory, 1969) and squirrel monkeys (Hutchinson *et al.*, 1968) immediately after reinforcement on such schedules.

A stimulus correlated with shock, unlike a stimulus correlated with food (p. 224), will also potentiate the startle response elicited in rats by a very loud auditory stimulus (p. 310). Wagner (1963a) found that a stimulus correlated with omission of food would also potentiate the startle response. Appropriate control conditions showed that it was only a stimulus signalling omission of anticipated food that had such an effect.

3. Escape from Frustration

Animals will learn to escape from shock (p. 152), or from stimuli previously paired with shock (p. 311). Similar behaviour occurs when escape from frustration is the reinforcing event.

The classic demonstration of escape from nonreinforcement is a study by Adelman and Maatsch (1956). One group of rats was trained to run an alley for food in the goal box, and then extinguished. During extinction they were permitted to escape from the goal box by jumping up onto a ledge. A control group, which received equivalent exposure to the goal box, but never with any food present, was placed in the goal box during extinction trials, and also permitted to jump up onto the ledge. The results were very clear: animals that had previously found food in the goal box were much more likely to escape from it than were control animals never previously reinforced in the goal box.

Escape from stimuli associated with the omission of reinforcement has been shown for rats trained on partial reinforcement in an alley by Wagner (1963a) and Daly (1969), for pigeons during successive discrimination training by Rilling *et al.* (1969) and Terrace (1971), and for pigeons and rats trained on FR schedules of reinforcement by Azrin (1961) and Thompson (1964).

Wagner and Daly trained rats to run in an alley, with food in the goal box on only 50 per cent of trials. A compound auditory and visual signal was presented on nonreinforced trials, while a control group received similarly nonreinforced exposure to the stimulus in a situation not previously associated with food. The subjects were then tested by being placed in a shuttle box; when the stimulus was turned on the animal could terminate it by jumping from one compartment to the other. In both experiments, subjects for whom the stimulus had signalled omission of anticipated food jumped significantly faster than control subjects, and Daly observed a significant increase in speed of jumping over a series of test trials.

Rilling *et al.* (1969) and Terrace (1971) trained pigeons on multiple schedule discriminations (e.g., VI, Extinction). In both studies, pigeons learned to peck a second key when the sole consequence of such a response was to turn off the stimulus displayed on the food key for a brief, fixed, period. Such responses were confined to S− periods (the stimulus correlated with extinction) and tended to occur at the outset of these periods. Since the only effect of these responses was to remove the stimulus correlated with extinction, without in any way changing the schedule of reinforcement or hastening the appearance of S+, and since such responses to the second key did not occur when they had no programmed consequence at all, it is reasonable to conclude that they were reinforced by escape from S−. Terrace trained a control group of pigeons with a fading procedure, which ensured the development of the discrimination without the occurrence of any errors to S−. Terrace (1972) has argued that such a procedure may prevent S− becoming aversive, and in the present experiment the effect of this procedure was to abolish escape responses on the second key .This observation was not, however, confirmed by Rilling *et al.* (1973).

4. *Effects of Drugs*

Dollard and Miller (1950, pp. 185–7, 369–78) suggested that such drugs as alcohol and the barbiturates may exert their effects on behaviour by reducing reactions to aversive events. Administration of these drugs interferes with avoidance of, escape from, or suppression of behaviour by, electric shock (Miller, 1961; Barry and Miller, 1965); experiments with rats and cats have shown that both alcohol and sodium amytal selectively disrupt the avoidance component of an approach–avoidance conflict (Conger, 1951; Bailey and Miller, 1952; Miller, 1961; Grossman and Miller, 1961; Barry *et al.* 1962a).

If the omission of anticipated appetitive reinforcement is aversive, then these drugs should also interfere with the course of simple extinction and discrimination learning. Barry *et al.* (1962b) showed that administration of either drug resulted in much slower extinction of food-reinforced alley running; this result has been confirmed for sodium amytal by Gray (1969) and Ison and Pennes (1969). The effects on discrimination learning are somewhat less clear. Wagner (1966) reported that after rats had learned a successive discrimination, performance was disrupted by an injection of alcohol, but this might simply be a matter of disruption due to any change in state; Miller (1961) reported rather little disruption of a discrimination when rats were injected with sodium amytal, although Ison and Rosen (1967), training rats from the outset under control and drugged conditions, found that sodium amytal significantly retarded the learning of a successive discrimination, specifically by causing subjects to maintain responding to $S-$.

5. *After-effects of Aversive Stimulation*

Although the presentation of shock, or of stimuli signalling shock, invariably suppresses ongoing appetitive behaviour, whether contingent on that behaviour or not, there is some evidence that the *after-effects* of such aversive stimulation may actually invigorate appetitive behaviour. Rats that have just been exposed to a relatively mild electric shock, will show an increase in their consummatory behaviour when given the opportunity to eat or drink (Amsel and Maltzman, 1950; Amsel and Cole, 1953). The effect, it should perhaps be stressed, is not necessarily a very large one, nor entirely reliable: initial exposure to shock may depress rather than enhance consummatory behaviour (Moyer, 1965) and shock depresses rather than enhances the consumption of unpalatable substances (Strongman, 1965).

But although the result does not necessarily provide evidence, as has often been thought, for the Hullian notion of general drive, it is striking enough to be taken seriously and a possible interpretation and further discussion will be given later (p. 352).

Whatever the interpretation of this after-effect of aversive stimulation, there is evidence that the omission of expected appetitive reinforcement has exactly the same effect on subsequent appetitive behaviour. The classic experimental situation for the demonstration of this invigorating effect of omission of food is the double alleyway of Amsel and Roussel (1952). Rats are trained to run down the first alley into a goal box in which food is available; this goal box also serves as the start box for a second alley and

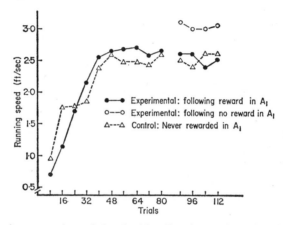

Fig. 5.11. A demonstration of the double-alley frustration effect. Over the last four blocks of trials, speed of running in the second alley is increased by the omission of reward in the first alley (A_1) for experimental subjects, and on these unrewarded trials experimental subjects run faster than control subjects never rewarded in A_1. (*After* Wagner, 1959.)

after a detention interval the rat is permitted to run the second alley and receive further reinforcement in the second goal box. If subjects have consistently received reinforcement in the first goal box, then the omission of this reinforcement should be a frustrating event and might be expected to facilitate subsequent appetitive behaviour (in this case, running down the second alley). This is precisely what Amsel and Roussel (1952) observed: when rats sometimes found food in the first goal box and sometimes did not, their speed of running the second alley was greater after nonreinforcement than after reinforcement.

Wagner (1959) ran the control condition which establishes that this "frustration effect" is a consequence of the omission of an expected appetitive reinforcer. He initially trained one group of rats with food always available in the first goal box and a second group with food never available (both groups always found food in the second goal box). After 76 trials he omitted the food in the first goal box for the first group on a random 50 per cent of trials. Running speeds in the second alley for both groups throughout the experiment are shown in Fig. 5.11: the omission of expected food, rather than the mere absence of food, is the critical event in producing the frustration effect.

Although the frustration effect is also a somewhat problematic phenomenon, being not entirely reliable and open to a number of different interpretations, these problems will not be discussed further here. Frustration effects are closely related to transient contrast effects usually observed in discrimination learning and a more detailed analysis of the two sets of results is postponed to Chapter 7. For the moment, it is sufficient to point to this final parallel between the effects of aversive reinforcement and omission of appetitive reinforcement.

6. *Conclusions*

When food is omitted from a situation in which an animal has received it in the past, or in the presence of stimuli similar to those which have previously accompanied its delivery, the omission has been shown to produce a wide range of effects strikingly parallel to those associated with the presentation of an aversive reinforcer such as shock. Perhaps the most important feature of such results, at least for present purposes, is that all these effects can be observed only when food is omitted in a context in which it has previously been presented. This implies that the presentation of an appetitive reinforcer is not simply an operation performed by an experimenter to strengthen some other set of associations; it must itself be an event that enters into association with antecedent stimuli or responses. These results, then, provide the final link in the development of the argument advanced in this chapter: the law of effect must be replaced as an explanation of instrumental reinforcement in favour of a theory which stresses the facilitation or enhancement of responses which have been associated with an increase in incentive. They also suggest a set of plausible and symmetrical analyses for the case of aversive reinforcement: if aversive reinforcers have effects on behaviour which are the opposite of the effects of appetitive reinforcers, then the reason why response-contingent presentation of an aversive reinforcer in experiments on punishment suppresses that response may be because animals learn the relationship between their behaviour and this aversive event, and increases in aversive reinforcement are suppressive. Conversely, the reason why response-contingent omission of an aversive reinforcer increases the probability of that response in experiments on avoidance learning may be because animals learn to expect the event and the omission of an expected aversive event facilitates behaviour.

VI. SUMMARY

It has been recognized frequently that the absence of any contingency between a subject's behaviour and the delivery of reinforcement in experiments in classical conditioning, does not necessarily imply that CRs are established without regard to their consequences. It has been less widely recognized that the presence of such a contingency in instrumental experiments does not imply that the responses established in such experiments are always reinforced by their consequences. Instrumental experiments necessarily contain implicit classical contingencies and many supposedly instrumental responses may be Pavlovian in origin. It is probable that pigeons peck keys in large part because pecking is a classical CR elicited by stimuli associated with food.

In many instrumental experiments, the performance of the required response is accompanied by changes in various external stimuli. Although it may appear that the contingency between response and reinforcer is responsible for the change in the subject's behaviour, the effective contingency may be that between these external stimuli and reinforcement. Stimuli associated with appetitive reinforcers may come to elicit approach responses: a rat traversing the correct path through a maze may be approaching in turn a set of stimuli, each of which is more closely associated with reinforcement than the preceding stimulus.

If instrumental responses are defined as those directly affected by their consequences, then it remains to be asked how those consequences come to exert control over a subject's behaviour. The answer provided by Thorndike and Hull was that rewarding consequences strengthen the connection between preceding responses and the situation in which they occur. There is a considerable body of evidence inconsistent with the analysis provided by this law of effect. Studies of latent learning, irrelevant incentive learning, and contrast effects all suggest that animals form associations between their responses and the events contingent on those responses. Thus instrumental learning, just like classical conditioning, may typically involve the establishment of associations between two events when a correlation is arranged between them. Evidence of learned helplessness and of overshadowing and blocking of the association between response and reinforcer in instrumental experiments provides further support for this parallel analysis of classical and instrumental conditioning.

The law of effect has been thought to provide a more parsimonious account of instrumental learning than any provided by an expectancy theory, on the grounds that it requires no supplementary principle to account for the translation of learning into performance. A rejection of the law of effect, therefore, has usually been accompanied by the postulation of some variety of incentive theory. That proposed by Hull and Spence suggested that a classically conditioned state of incentive provided a source of motivation which activated ongoing responses. A long series of studies has provided little support even for modified versions of such a theory. Instrumental responding is not typically enhanced by the noncontingent presentation of a stimulus associated with an appetitive reinforcer, nor are chains of instrumental responses initiated when such stimuli are presented. Rather it must be supposed that the motivation for instrumental responding (if motivation be required) is precisely the expectation of an increase in incentive. Those responses are initiated whose occurrence has previously been associated with increasing proximity to reinforcement.

Stimuli associated with reinforcement thus act as goals for responding rather than as instigators of responding. An association with reinforcement may give an initially neutral stimulus the capacity to reinforce responses on which it is contingent. To show that a stimulus has been established as an effective conditioned reinforcer for a given response, it is necessary to show that any increase in the probability of that response is due to the contingency between response and conditioned reinforcer, and that the effect depends upon the association between the conditioned reinforcer and primary reinforcement. By no means all studies of conditioned reinforcement have satisfied these requirements; nevertheless there remain a sufficient number of valid studies in which new responses or new patterns of responding have been established solely because they have been followed by the presentation of a stimulus previously, or concurrently, associated with primary reinforcement. The operations for establishing a stimulus as a conditioned reinforcer are those of classical conditioning, and there is no evidence to contradict the assumption that the processes involved are Pavlovian. A stimulus paired with food elicits some of the responses elicited by food: it also acquires another of the properties of food —that of reinforcing the instrumental behaviour of a hungry animal.

Experiments on observing responses, where an animal is given the opportunity to produce stimuli signalling whether or when reinforcement will be available, have been thought to show that conditioned reinforcement may depend upon the reduction of uncertainty. Although it is possible that animals seek out information, there is no existing evidence from studies of observing responses, which either requires this assumption, or contradicts the assumption that stimuli are effective conditioned reinforcers simply to the extent that they are associated with primary reinforcement.

The omission of an appetitive reinforcer has effects on behaviour that are frequently the opposite of those produced by its presentation. Such an event may suppress ongoing instrumental responding, or elicit escape or aggressive responses rather than approach. In certain important respects it appears to be aversive.

CHAPTER 6

Instrumental Learning: Avoidance and Punishment

I. INTRODUCTION

The study of aversively motivated instrumental behaviour has had a curiously different history from that of appetitively motivated behaviour. As we have seen in the preceding chapters, instrumental learning for food or water has been most commonly treated in the general terms of S–R reinforcement theory: a particular response is learned in a particular situation because it is followed by a reinforcing event. This theoretical analysis has been only gradually, and doubtless not completely, abandoned in favour of various versions of incentive theory, most of which have appealed to classically conditioned motivational states or classically conditioned anticipation of reinforcement to account for the initiation, maintenance and direction of appropriate instrumental behaviour, and for its loss with nonreinforcement. The analogous account of avoidance learning, the two-factor theory advocated by Mowrer (1947) and Miller (1948), has a much longer history; while the study of punishment has, for long periods of time, eschewed the use of the law of effect even at a descriptive level. Ironically enough, just as the analysis of appetitively motivated behaviour has begun to discard the law of effect, the analysis of aversively motivated behaviour has shown signs of moving in the reverse direction, with several recent writers advocating a return to Thorndike's earlier views. One of the aims of this chapter will be to account for these diverse trends, and to attempt some general integration of aversive and appetitive instrumental learning.

At the outset, it is worth reiterating the systematic parallels between the operations of experiments on punishment and avoidance and those on appetitive acquisition. In a study of punishment, the presentation of an aversive reinforcer (such as an electric shock) is contingent on the occurrence of a particular response, just as the presentation of food is contingent on a particular response in a typical appetitive instrumental

experiment. The difference in outcome—that in the former case the probability of the response usually declines, while in the latter case it usually
increases—must presumably be attributed to the difference between appetitive and aversive reinforcers. It should be noted, of course, that this
difference obviously necessitates some additional differences in experimental procedure: experiments on punishment, much like experiments
on inhibition, must first establish a high baseline probability for the
reference response (usually by prior appetitive reinforcement) in order to
study the decremental effects of the punishment procedure.

A punishment contingency usually suppresses responding; an avoidance
contingency usually increases the probability of responding. It is tempting,
therefore, to counter the parallel between the operations of punishment
and appetitive reinforcement with an underlying parallel between the processes involved in avoidance and appetitive reinforcement. Nor is this
parallel hard to find. We have already noted that the omission of an appetitive reinforcer has effects similar to the presentation of an aversive reinforcer; it would not be surprising if the omission of an expected aversive
reinforcer, which is the contingency specified in an avoidance schedule,
had effects similar to the presentation of an appetitive reinforcer.

II. BACKGROUND TO THE STUDY OF PUNISHMENT

A. The Fate of the Law of Effect

The first theory of punishment was proposed by Thorndike (1913).
Punishers were events symmetrical to rewards. Just as the positive law of
effect stated that responses followed by a satisfying state of affairs increased
in probability, so the negative law of effect stated that responses followed
by an annoying state of affairs decreased in probability. Furthermore, just
as the increase in probability in the former case was a matter of the formation of new bonds or connections (or the strengthening of old bonds or
connections) between the response and the situation in which it occurred,
so the decrease in probability in the latter case was a matter of the loss
or weakening of such bonds or connections. These two sets of propositions
constitute the empirical and theoretical laws of effect: the empirical law
states that response-contingent events change the probability of a response;
the theoretical law purports to explain these changes in probability in terms
of changes in the strength of S–R bonds. It is important to keep this
distinction in mind.

We have seen how the theoretical version of the positive law of effect
has been eroded in the past two decades. The negative law of effect was
abandoned much earlier, and not only in its theoretical form. The process
was started by Thorndike (1931, 1932a), continued by Skinner (1938) and

Estes (1944), and consecrated by Skinner (1953). The message attributed to these writers, partly as a consequence of a highly selective reading of what they had said, was that punishment does not work. It is at this point that the distinction between empirical and theoretical versions of the law must be carefully maintained. Refutation of the empirical version of the law would involve showing either that the presentation of a supposedly punishing stimulus contingent upon a response does not in fact reduce the probability of that response, or that such reduction as does occur is not dependent on the contingency between response and punisher. Not all the arguments involved were in fact directed to either of these questions; they were not arguments against the empirical version of the law, but against its theoretical version. Each of these writers argued that on those occasions that a punisher did decrease the probability of a response, it did so not by weakening any S–R connection, but rather by causing a temporary suppression of the response. We shall examine this argument below. The first point, however, is to examine the evidence bearing on the empirical law.

The supposedly negative evidence was never particularly impressive. Thorndike (1931, p. 45; 1932a, p. 288) reported that the probability that a human subject would repeat a particular verbal response was reduced little, if at all, by the experimenter's announcing that it was wrong. In experiments with animals, he reported that a 30 sec confinement following an incorrect choice did little to reduce the probability of that choice (Thorndike, 1932b). This is less than overwhelming. Nor, it should be stressed, was Thorndike unaware that other experimenters had found that more intense punishing stimuli (electric shocks) could have a much more profound effect: he cited the study of Warden and Aylesworth (1927) which showed that punishment of errors significantly improved discrimination performance in rats (Thorndike, 1931, p. 46). Faced with this evidence, he retreated to the weaker position of rejecting the theoretical version of the law.

Estes (1944) confirmed that both strong and mild electric shocks reduced the probability of a response on which they were contingent, but in two studies (Experiments I and J) found that the contingency between response and shock was not always very important: the probability of a response was equally affected whether or not shock was contingent on that response. In one of these studies, indeed, he found that the effects of punishment would dissipate if the animal were simply confined in the apparatus without the opportunity to respond. "The conclusion", he argued, "seems inescapable that at least a great part of the initial effect of punishment is due simply to correlation of the negatively reinforcing stimulus with the experimental situation" (pp. 25–6).

For these results, the conclusion does indeed seem reasonable. But, in spite of the claims made on his behalf by some modern writers (e.g., Rachlin and Herrnstein, 1969, p. 83), Estes did not assert that the contingency between response and punisher was always irrelevant and showed in one study (Experiment L) that although punishment of one response initially caused equal suppression of another, subsequent recovery sessions revealed clearly that the punished response was more enduringly suppressed. Although, therefore, Estes argued that a contingency between response and shock was not necessary for suppression to be produced, he cannot be construed as advocating complete rejection of the empirical law of effect. Like Thorndike and Skinner, however, Estes did argue that response suppression was not due to any weakening of S–R connections.

B. The Status of the Empirical Law of Effect

The conclusion foreshadowed in Estes' 1944 monograph—that although a response could be suppressed in the absence of any contingency between that response and a punishing stimulus, response-contingent punishment is often even more effective—has been amply confirmed since that time. Indeed, it is by now fair to say that Estes' data underestimated the importance of the contingency.

  The empirical study of punishment has made very substantial progress in the past 10 to 15 years and where earlier books on animal learning could touch on the topic lightly before passing on, there is now a substantial body of evidence on the effects of response-contingent shock. Reviews of much of this evidence have been provided by Church (1963, 1969) and by Azrin and Holz (1966).

  Experiments with rats and pigeons in various types of apparatus have amply shown that response-contingent shock suppresses that response and that the magnitude of the suppression is directly related to the intensity of the shock (e.g., Azrin, 1960; Karsh, 1962; Appel, 1963; Boe and Church, 1967; Camp et al., 1967; Church et al., 1967; Cohen, 1968), and to its duration (Church et al., 1967). While such results do not necessarily imply that the contingency between a response and shock critically determines the extent to which the response is suppressed, the finding that the degree of suppression is an inverse function of the delay between response and shock does support this notion (Azrin, 1956; Kamin, 1959; Camp et al., 1967; Baron et al., 1969).

  Several recent studies, specifically designed to assess the effect of the contingency between response and shock, have confirmed its importance. The exceptions to this generalization do not seem to be very significant: Hunt and Brady (1955), for example, used shocks of such intensity that responding was completely suppressed regardless of the contingency between response and shock. Hoffman and Fleshler (1965) used such strange procedures that their results are difficult to interpret. In the very large majority of studies, however, contingent shock has produced substantially more suppression than noncontingent shock (Azrin, 1956; Boe and Church, 1967; Camp *et al.*, 1967; Schuster and Rachlin, 1968; Church *et al.*, 1970). Two of these studies are worth describing in some detail. Schuster and Rachlin trained pigeons on a concurrent chain schedule (p. 242). Pecking in the first link on either key produced stimuli correlated with the second link. Identical VI schedules of food reinforcement were in effect on each key in the second link, but shocks were delivered independently of responding on one key, while response-contingent shock was programmed on the other key. Response-contingent shock resulted in much greater suppression of responding, but the pigeons chose between the two keys largely on the basis of the overall probability of shock on each. If the response-contingent schedule programmed fewer shocks, they would choose that key in the first link, even though responding in the second link was more suppressed.

Church *et al.* (1970) assessed the effects of the contingency between response and shock by comparing conditioned suppression with discriminated punishment. Rats were trained to press a lever and to pull a chain for food on VI schedules; then, with only one manipulandum present, they were exposed to a stimulus which signalled either response-independent shock (conditioned suppression), or a signal in whose presence a response was followed by shock (discriminated punishment). In spite of the greater frequency of shocks in the conditioned suppression procedure, the discriminated punishment procedure resulted, on subsequent test trials, in greater suppression of that response. However, if animals were tested with the other manipulandum present, the two procedures resulted in essentially equal amounts of suppression. The results of two replications are shown in Fig. 6.1. It can be seen that response-contingent shock resulted in greater suppression of that response than did noncontingent shock, but that when subjects were tested with the other manipulandum, the contingency had no effect. Church *et al.* concluded, as had Estes (1944), that shock may suppress all ongoing appetitive responding, whether it is contingent on responding or not, but that the addition of a contingency between a response and shock may result in additional suppression which is specific to the punished response.

The evidence briefly reviewed here admits of little doubt. It is no longer possible to argue that punishment is an ineffectual procedure for suppressing responses, or that response-suppression is an artefactual consequence of noncontingent, emotionally derived disruption. As an empirical,

K

statement, the negative law of effect is as true as the positive law. This is
not to deny that there are numerous exceptions to both. Just as it is
relatively difficult to increase the probability of such responses as yawning
or scratching in dogs and cats (p. 219), or to decrease the probability of
responses such as salivating in dogs or pecking in pigeons (p. 120), by the
contingent presentation of food, so it is possible to find cases where
response-contingent shock does not suppress responding. Morse *et al.*
(1967) found that squirrel monkeys, exposed to regular, noncontingent
presentations of shock, tended to pull and bite on a restraining leash; if
this response was then punished on an FI 30 sec schedule, they developed

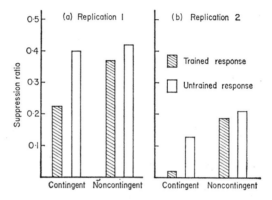

FIG. 6.1. Suppression produced by a stimulus signalling either response-contingent
or noncontingent shock, when the stimulus is presented to rats performing either
the punished response or a different response. The previously punished response
is more suppressed by the signal for punishment than by the signal for noncontingent shock; but a different response is equally suppressed by both stimuli.
(*After* Church *et al.*, 1970.)

a typical FI scallop, showing an *increase* in the rate of responding as the
moment approached when the response would be punished. McKearney
(1969) found that squirrel monkeys, after a long history of being trained
to press a response key to avoid shock, continued to press the key when
the only consequence of doing so was the delivery of shock on an FI 10
min schedule (they also showed a typical FI scallop while doing so).
Walters and Glazer (1971) found that gerbils would stop digging if
punished for that response, but that a similar punishment, contingent on
their standing up on their hind legs and posturing in a characteristically
alert manner, merely increased the probability of such posturing. These
exceptions to the negative law of effect, like the exceptions to the positive
law, are of the utmost importance and are discussed in some detail in later

sections of this chapter. As empirical generalizations, however, both positive and negative versions of the law remain generally true and there are no grounds for arguing that the positive law is better substantiated than the negative.

III. THEORIES OF PUNISHMENT

A. Symmetrical Analyses of Reward and Punishment

The fact that response-contingent shock reduces the probability of that response, in apparently much the same way that response-contingent food increases the probability of a response, may well have theoretical implications. But it should be clear that acceptance of any *theoretical* negative law of effect is not necessarily one of those implications. The preceding chapter has spelled out a long list of arguments against acceptance of the theoretical version of the positive law: the presentation of an appetitive reinforcer contingent on a response, rather than strengthening any S–R connections, may establish associations between response and reinforcer, and the consequent anticipation of the reinforcer may be sufficient to initiate and maintain an appropriate pattern of behaviour. Perhaps a similar argument will be seen to hold for the case of punishment. For the most tempting implication of the symmetrical effects of reward and punishment is that the theorist should look for symmetrical explanations of their effects. Thorndike was evidently impressed by the asymmetry inherent in his rejection of the negative version of the theory, and his retention of the positive version:

> Annoyers do not act on learning in general by weakening whatever connection they follow. If they do anything to learning they do it indirectly by informing the learner that such and such a response in such and such a situation brings distress . . . Satisfiers *seem* to act more directly (Thorndike, 1931, p. 46; italics in original).

The solution, as Estes (1969b) has argued, is to reject this second assumption:

> I believe now that I was right back in the early 1940s, both in assuming that the effects of reward and punishment should be essentially symmetrical, and also in concluding that interpretation of punishment requires a separate process rather than a simple weakening of associative strength. I think that where I went wrong, in the illustrious company of Thorndike, Skinner, and Hull, among others, was in assuming the effects of reward to involve a simple, direct strengthening of associative connections, and thus in looking for the wrong kind of symmetry (p. 65).

There are, then, two possible theories of punishment which would succeed in preserving a symmetry between the effects of rewards and punishers. The first, and apparently simpler, analysis is provided by the negative law of effect. Just as rewards produce learning by strengthening S–R connections, so punishers produce "unlearning" by weakening connections: the presentation of an aversive reinforcer contingent on a response results in a weakening of the association between that response and antecedent stimuli.

A second, and apparently more complex, theory appeals to what has been called "negative incentive" (Logan, 1969), or to the anticipation of aversive reinforcement (Mowrer, 1956; Estes, 1969b). Just as rewarding an instrumental response may result in the formation of an association between that response and the appetitive reinforcer, with a consequent facilitation of responses associated with an increase in proximity to appetitive reinforcement, so the punishment of an instrumental response may result in the formation of an association between that response and the aversive reinforcer, with a consequent suppression of responses associated with an increase in proximity to aversive reinforcement. The earliest version of such an analysis was that proposed by Estes (1944). Estes noted the similarity between the effects of response-contingent shock in studies of punishment and the effects of response-independent shock in studies of conditioned suppression (Estes and Skinner, 1941). He argued that appetitive instrumental responding was suppressed by the emotional state of fear or anxiety that was in the one case conditioned to apparatus cues (and specifically to the CS) and in the other case was conditioned both to apparatus cues and, presumably, to stimuli more directly linked with the occurrence of the punished response. Nevertheless, as Estes (1969b) has noted, the later form of the theory differs from the earlier, not only in now forming part of a general theory of instrumental reinforcement but also in two other respects: in the later version, Estes (1969b, pp. 71–2) is, properly, more explicit about the association between response and aversive reinforcer during punishment, thus stressing the difference between noncontingent and response-contingent shock; he also faces up to some of the problems involved in the punishment of escape and avoidance responses (see below).

B. Competing Response Theories

In its theoretical version, the negative law of effect talks of the weakening of S–R connections, or of *unlearning*; but it is not, upon reflection, entirely clear what unlearning might look like. Nor is it entirely obvious why appetitive reinforcers alone should be able to support learning, while aversive reinforcers do not. It seems, on the face of it, more plausible to suppose that both appetitive and aversive reinforcers provide the necessary conditions for learning. However, it is demonstrably the case that the probability of a punished response declines rather than increases, and since S–R theory permits only responses to be learned, the dilemma can be resolved only by supposing that what is learned in an experiment on punishment is some *other* response, whose increase in strength competes with and displaces the punished response. The disappearance of the punished response, then, becomes a second-order consequence of the primary effect of punishment, which is to provide a source of reinforcement for some other pattern of behaviour. This can be recognized as the same line of reasoning as that which led to explanations of extinction in terms of competing responses.

Three competing response theories of punishment can be distinguished. One distinction between them lies in the way in which the delivery of the punishing stimulus is assumed to reinforce the competing response. The simplest version is that proposed by Guthrie (1935):

> Sitting on tacks does not discourage learning. It encourages one in learning to do something else than sit. It is not the feeling caused by punishment, but the specific action caused by punishment that determines what will be learned. To train a dog to jump through a hoop, the effectiveness of punishment depends on where it is applied, front or rear. It is what the punishment makes the dog *do* that counts (p. 158).

The essential argument here is that shock is a UCS which elicits a variety of skeletal responses, including flinching, jumping, running, and fighting. A process of Pavlovian conditioning, therefore, may ensure that such responses will be conditioned to stimuli present in the experimental situation, or, in particular, to stimuli (such as the sight of the response lever) closely accompanying the performance of the punished response. If these conditioned responses are incompatible with the punished response, their establishment by Pavlovian reinforcement will guarantee the disappearance of the punished response.

In effect, this analysis suggests that the existence of an instrumental contingency in an experiment on punishment does not necessarily imply that any observed change in behaviour is due specifically to instrumental learning. The point should be a familiar one (p. 205). Just as the delivery of an appetitive reinforcer in an operationally instrumental experiment may change the subject's behaviour by eliciting approach responses which happen to satisfy the experimenter's requirements, so the delivery of an aversive reinforcer may change a subject's behaviour by eliciting particular responses that prevent the occurrence of the response punished by the experimenter. In this sense, therefore, this version of a competing response theory of punishment appeals to processes similar to those thought to account for at least some cases of rewarded learning.

Two other theories of punishment have appealed to competing responses. In their case, no analogy will be found with viable theories of appetitive reinforcement. Dollard and Miller (1950, p. 75) pointed out that when one response is followed by punishment, any new responses that occurred after the onset of the aversive reinforcer might be superstitiously reinforced because they happened to be followed by its termination. An instrumental rather than a Pavlovian process of reinforcement is invoked, and the theory is usually described as the escape hypothesis. The most obvious objection to this hypothesis is that the use of very brief shocks would presumably preclude the possibility of any response being elicited by the shock before the shock was terminated. Yet brief shocks may be entirely effective punishers.

The third, and most complicated, version of a competing response theory is the avoidance hypothesis advanced by Mowrer (1947), Dinsmoor (1954) and Solomon (1964). The argument usually involves the application of two-factor theory of avoidance to the case of punishment (or "passive avoidance" as Mowrer, 1960a, calls punishment). Two-factor theory of avoidance learning is outlined in greater detail later in this chapter. As applied to the case of punishment, the theory supposes that a motivational state of fear or anxiety becomes conditioned, via Pavlovian reinforcement, to stimuli antedating the punished response. If the subject, instead of continuing with the punished response, emits some incompatible response, the set of stimuli to which the subject is exposed will be changed from one signalling a high probability of shock, and therefore aversive or fear-eliciting, to one signalling a low probability of shock, and therefore relaxing or non-aversive. These incompatible responses will therefore be reinforced by a decrease in fear; thus they will increase in probability and successfully displace the punished response. The most obvious objection to this hypothesis would seem to be that it appeals to an extraordinarily complicated chain of events to account for what is, on the face of it, a rather simple phenomenon.

C. Punishing Stimuli as External Inhibitors or Discriminative Stimuli

When an animal has been trained, in the absence of shock, to respond for food, the introduction of a punishment contingency changes the set of stimuli affecting the subject from that which prevailed when responding was first established. On the assumption that shock is a salient stimulus, even if it had no other properties one would expect to observe some disruption of responding. Abrupt changes in stimulation often disrupt previously established responses. Pavlov (1927) referred to the effect as external inhibition (p. 15); other theorists have talked of generalization decrement.

There is one procedure sometimes used for the study of punishment which particularly invites analysis in terms of the stimulus properties of the punisher. Skinner (1938, p. 154) and Estes (1944, Experiments A and B), having trained rats to press a lever for food, studied the effects of mild punishment, produced by a "slap" from the lever or a weak shock, on responding in extinction. In these circumstances, it is tempting to argue that the punishing stimulus disrupted lever pressing not only because it was a novel stimulus, but also because it may have acted as a stimulus signalling the onset of extinction. As may be remembered, Melching (1954) found that the introduction of a buzzer contingent on responding in extinction produced a marked decrease in resistance to extinction (p. 234).

There is no question but that electric shocks, like appetitive reinforcers, can act as stimuli both in classical conditioning (p. 110) and in instrumental learning. The clearest demonstration of such an effect in an instrumental experiment is provided by a study by Holz and Azrin (1961). Having initially trained two pigeons to respond for food on a VI schedule, Holz and Azrin punished each response with a moderately intense shock while continuing to present food on the VI schedule. Once the subjects had stabilized at a somewhat suppressed rate of responding under these conditions, periods of VI food plus consistent shock were alternated with periods during which neither food nor shock was delivered. Efficient discrimination learning was established: the two birds maintained a steady rate of responding in the presence of shock, but did not respond at all during periods free of shock. Response-contingent shock acted as a discriminative stimulus for VI food.

In a subsequent experiment, Holz and Azrin were also able to show that a mild response-contingent shock, which did not initially produce much suppression, could be turned into a stimulus signalling extinction. If the shock was made contingent on responding during extinction periods, which alternated with periods of VI food reinforcement and no punishment, responding eventually ceased when followed by shock. The effect of shock on responding, therefore, depended on whether it signalled the availability or omission of food, and, Holz and Azrin argued, the effects of shock in Estes' experiment may have been a consequence of the fact that shock was there established as a discriminative stimulus signalling the onset of extinction.

Although it is not clear that anyone has seriously proposed that the effects of administering a supposedly punishing stimulus may be entirely understood by reference to its signalling properties, the possibility that shock may serve as a signal for appetitive reinforcement or its absence, or even as a signal for further shock or its absence, is one that must be taken seriously and may well account for a variety of apparent anomalies in the study of punishment.

IV. EMPIRICAL EVALUATION OF THEORIES OF PUNISHMENT

There is, perhaps, something depressing about this apparent proliferation of theories of punishment. Such a reaction may be intelligible, but it should not really come as any great surprise that such a variety of suggestions has been proposed. Many of these suggestions are plausible enough, in the sense that they seem to appeal to reasonable processes securely enough established from other areas of research. It is possible that many of the results of experiments on punishment can be fully understood only by a simultaneous appeal to several of these proposals. What should be stressed is that few of them should be thought of as *general* theories of punishment, applicable to all experimental situations. This is perhaps most clear in the case of the final hypothesis considered above, which appealed to external inhibition or the correlation between shock and the availability of appetitive reinforcement. This account cannot even explain the central fact about punishment, namely, that response-contingent shock suppresses behaviour more effectively than noncontingent shock.

The following discussion will be organized around these theories, rather than around particular empirical areas. In the case of each theory, it can be asked whether the theory is useful as an explanation of some features of the data on punishment and whether it can reasonably be regarded as a general theory of punishment.

A. Punishers as Stimuli

The disruption of responding produced by a punishing stimulus cannot simply be reduced to the disruptive effects of any salient stimulus, if only because response-contingent shock is more effective than noncontingent shock. Two further points reinforce this conclusion. First, the effects of an external inhibitor normally wear off with repeated presentation; the suppressive effects of a sufficiently severe punishing stimulus, however, increase with repeated presentation, even if the response continues to be reinforced (e.g., Camp *et al.*, 1967; Church *et al.*, 1967). Secondly, although mild electric shocks may act as discriminative stimuli in exactly the same way as any other external stimulus, as the intensity of the shock increases so suppressive effects start overriding signalling effects. Holz and Azrin (1962) compared the effects of response-contingent shock and a brief, response-contingent change in key colour as discriminative stimuli. When the shock was mild, both stimuli served equally effectively as signals for reinforcement and nonreinforcement, maintaining equally high rates of responding in the former case and equally low rates of responding in the latter case. When the intensity of the shock was increased, however, rate of responding decreased whether the shock signalled food or the absence of food. An intense shock suppressed responding even when it was a signal for appetitive reinforcement.

A general theory of punishment, therefore, must allow that effective punishers have effects on behaviour over and above those attributable to other stimuli. But this is not to deny that shocks are stimuli, or that the results of some experiments are explicable in these terms.

1. *Punishment of the Correct Response in Discrimination Learning*

Muenzinger (1934) found that rats shocked for correct choices in a simultaneous discrimination, as well as rats shocked for errors, learned the discrimination faster than unshocked controls. A recent series of studies, reviewed by Fowler and Wischner (1969) and Fowler (1971) has delimited the conditions under which this effect occurs. Fowler et al. (1967), for example, found that shock contingent on a correct choice facilitated discrimination learning only when the discrimination was relatively difficult, or when a retrace correction procedure was employed which allowed the subject to retrace from the incorrect side to the correct side immediately after an error. Both of these conditions can be thought of as increasing the confusion between correct and incorrect alternatives (p. 185); these results, therefore, suggest that animals shocked for correct responses are receiving additional discriminative stimuli that help to differentiate relatively indiscriminable alternatives.

Although shocks may be stimuli, in this situation also it can be shown that they are aversive stimuli. Muenzinger et al. (1938) and Fowler and Wischner (1965) have shown that shocks for errors facilitate learning more than do shocks for correct choices; Fowler et al. (1968) have shown that as shock intensity increases, so shock for correct choices becomes less likely to facilitate learning. Interestingly enough, this last result was abolished in rats injected with sodium amytal.

2. *Adaptation to Shock*

Miller (1960) found that rats trained to run down an alley for food could be induced to continue running even when severe shocks were given in the goal box, provided they had been exposed to a series of shocks of gradually increasing intensity. Animals receiving the intense shock at the outset, however, showed complete suppression of running. Miller referred to this as training rats to resist stress. Karsh (1963) confirmed this finding, but also showed that rats initially punished with a strong shock and then shifted to a less intense shock, showed *greater* suppression after the shift than animals punished throughout with the weaker shock.

Church (1969) has described a set of studies by Raymond confirming and extending Karsh's results. In one of these experiments, rats were initially exposed to noncontingent shock of an intermediate intensity and were later punished for lever pressing with either a more or a less intense shock. If the punishing shock was more intense, previously exposed rats showed less suppression than controls; if it was less intense, they showed more suppression. Church suggests that these results must be due to generalization from one punishing stimulus to another. If rats have been exposed to one shock and have learned to respond in the presence of that shock at a given rate, they may be expected, when exposed to a second shock, to continue responding to this similar stimulus in a manner similar to that established during exposure to the first shock. If the first shock was sufficiently intense to produce considerable suppression, this will generalize to a second weaker shock and produce greater suppression than in a control group. If the second shock is even stronger than the first, previously exposed subjects will still tend to respond at a rate maintained by the first and will now show less suppression than a control group.

3. *Punishment of Avoidance and Escape Responses*

Church's argument, outlined above, amounted to saying that if an animal has been reinforced for responding in a particular way in the presence of shock, subsequent presentation of shock, even if as a punishment for a particular response, will tend to reinstate or maintain a similar pattern of responding. The argument can be applied not only to the case where responding was initially maintained by appetitive reinforcement and suppressed by punishment, but also to the case where responding was initially maintained by escape from, or avoidance of, aversive reinforcers. If an animal has learned to press a lever or run down an alley in order to avoid or escape from shock, then, from what has been said so far, it is reasonable to suggest that part of the stimulus complex maintaining this behaviour is precisely the presentation of shock. The results of an experiment by Coulson *et al.* (1970) provide nice support for this idea. They trained rats to press a lever in order to avoid unsignalled shocks on a free-operant avoidance procedure. The rats learned to avoid well but not perfectly and continued to receive a small number of shocks in each session. Each rat was then extinguished, either by omitting all shocks, or by discontinuing the avoidance contingency but continuing to present, without reference to the rat's behaviour, the same pattern of shocks that had been received in the last session of avoidance training. The results for two rats are shown in Fig. 6.2: simple extinction resulted in an extremely rapid loss of the avoidance response; but responding was maintained, and could be reinstated, by presenting noncontingent shock.

Similar results have been reported for squirrel monkeys by Kelleher *et al.* (1963), Stretch *et al.* (1968) and McKearney (1969). In these studies, fixed-interval, noncontingent, shocks were presented during extinction of avoidance; eventually the monkeys developed an FI scallop, increasing

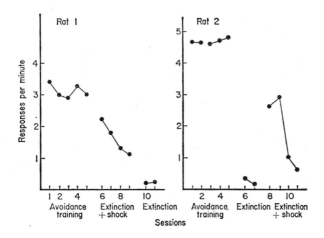

FIG. 6.2. The maintenance of free-operant avoidance responding during extinction by the occasional presentation of noncontingent shock. (*After* Coulson *et al.*, 1970.)

their rate of responding as the time for shock approached and responding at only a low rate following each shock. Once this pattern of behaviour had been established, the presentation of shock could be made contingent on responding on an FI schedule and the unfortunate beasts continued to respond, thus ensuring the delivery of regular, extremely severe shocks.

A variety of studies, employing both free-operant and discrete-trial procedures, have confirmed that the punishment of escape or avoidance responses may not only fail to suppress such responses, but may even, at least temporarily, enhance them (Fowler, 1971). In some discrete-trial studies, indeed, animals previously trained to escape or avoid shock by running down an alley into a safe goal box may continue to respond rapidly in extinction even when punishment is delivered on every trial (Gwinn, 1949; Brown, 1969). The effect, variously called self-punitive behaviour or the vicious-circle effect, seems to depend on a number of procedural details, but this is hardly surprising. Free-operant experiments have succeeded in maintaining avoidance responding reliably only on FI or VI schedules of shock delivery: if every response is punished (Black and Morse, 1961; Baron et al., 1969), or even if responding is punished on an FR schedule (Kelleher and Morse, 1969), responding may be facilitated briefly but is soon completely suppressed. In discrete-trial studies, although dogs show a transient increase in speed of responding when avoidance responses in a shuttle box are punished (Solomon et al. 1953), avoidance responding in rats in the shuttle box is immediately suppressed by response-contingent shock (N. F. Smith et al., 1966; Bolles et al., 1971; see Fig. 6.10, p. 339) and the degree of suppression is increased by a decrease in the delay of punishment (Kamin, 1959; Misanin et al., 1966). It is only when rats are trained to escape or avoid shock in an alleyway that punishment of the response in extinction can significantly facilitate responding.

If a rat has been trained to escape or avoid shock in an alley by running into a safe goal box, it is not entirely surprising that the delivery of shock *in the alley* on extinction trials might continue to elicit rapid running. Running in the presence of shock in the start box and alley has previously been reinforced by escape into the safe goal box; if punishment is delivered in extinction, then the subject will still be exposed to stimuli (shock in a section of the alley) relatively similar to those that established the behaviour during initial training, and running will still be reinforced by escape into the safe goal box. The procedure used to establish "self-punitive" behaviour in these experiments can thus be seen to be carefully designed so as to maintain the eliciting stimuli and reinforcing conditions of acquisition. As one would expect from this account, both escape and avoidance responding are facilitated to a greater extent if shock is delivered in a section of the alley near the start box than if it is delivered near the goal box (Martin and Melvin, 1964; Campbell et al., 1966a). Indeed, if shock is delivered actually in the goal box, avoidance responding is rapidly suppressed (Seligman and Campbell, 1965; Kintz and Bruning, 1967).

It is probable that additional principles are required to provide a complete explanation of these results and these will be further discussed in later sections. However, it seems plausible to suppose that one cause of the persistence of punished escape and avoidance behaviour is that the presentation of shocks reinstates some of the stimuli present when the behaviour was originally established.

B. Competing Response Theories

It can be accepted that the events used as punishers are stimuli and that part of their effects on behaviour may be explained by reference to this fact. But it is obvious enough that electric shocks are stimuli of a particular nature, which elicit certain patterns of behaviour, and whose termination or omission can be used to establish escape or avoidance responding. Perhaps these facts need to be considered before we can arrive at a complete account of punishment.

1. *Punishment of Responses Elicited by Shock*

The presentation of an appetitive reinforcer contingent upon the occurrence of one response, or upon the omission of another, is not always sufficient to ensure a reliable increase or decrease in the probability of that response (p. 137). The most reasonable account of many of these failures of the positive law of effect seemed to be to point to the possibilities of Pavlovian processes of reinforcement overriding the instrumental contingency. If subjects are required to make a response to obtain a particular reinforcer that happens to be incompatible with the responses elicited by that reinforcer, or if they are required to suppress a response that is in fact elicited by the reinforcer, then Pavlovian conditioning will interfere with instrumental learning. It is reasonable to suppose that similar effects will occur with such an effective UCS as electric shock. Punishment, therefore, may be ineffective if the punished response is one elicited by the punishing stimulus and avoidance learning may be ineffective if the required avoidance response is incompatible with the response elicited by shock (this latter case is discussed below, p. 340).

We have already noted that gerbils, punished for adopting a particular alert posture, show an increase in such posturing (Walters and Glazer, 1971) and that squirrel monkeys, punished for struggling on a leash on an FI schedule, show an increase in struggling as the time for the next punishing shock approaches (Morse *et al.*, 1967). Other studies have reported similar effects. Fowler and Miller (1963) trained rats to run an alley for food; with the exception of an unshocked control group, however, all subjects were shocked in the goal box as soon as they picked up the pellet of food. One group received shock to the forepaws, while a second group received shock to the hindpaws: the former shock was said to elicit a backward flinching response, the latter a forward lurch. Within each group, different subjects received shocks of different intensities. Over the range of intensities used, increasing intensities of shock to the forepaw resulted in increased suppression of running, while increasing intensities of shock to the hindpaws resulted in increased facilitation of running. Finally,

Melvin and Anson (1969) trained Siamese fighting fish to swim through an aperture for the opportunity to perform an aggressive display. They found that mild punishment of this swimming response resulted in a transient increase in rate of responding, accompanied by an increase in the vigour of the display.

Although all of these results are consistent with a simple Pavlovian analysis of punishment, it should be noted that they were all obtained either by using a relatively mild punisher or by scheduling only infrequent punishment. Where data are available, it appears that these and similar responses can be quite effectively suppressed by consistent, intense punishment. The punishment used by Walters and Glazer (1971), for example, was only a tone that had been previously paired with shock. Morse et al. (1967), although using a very intense shock (7 ma), delivered the shock on an FI schedule: subsequent studies (e.g., Azrin, 1970) have reported that when squirrel monkeys are punished for biting on a rubber hose with immediate, consistent shock, responding is rapidly suppressed. The range of shock intensities used by Fowler and Miller (1963) was relatively small, approximating to current values of 0·05 to 0·20 ma. Finally, Melvin and Anson (1969) showed that increasing intensities of shock reliably suppressed responding in their Siamese fighting fish, and Fantino et al. (1972b), using what was apparently a considerably more intense shock, found that punishment of the aggressive display in these fish resulted in rapid suppression of the display, with no more than the most transient initial increase in the vigour of the display.

At first sight, these results suggest that the suppressive effects of intense punishment cannot be understood in Pavlovian terms. If the reason why weak punishment facilitates rather than suppresses a particular response is because the punishing stimulus elicits that response, it hardly seems reasonable to explain the successful suppression of that response when more intense punishment is used by appealing to the same general analysis. Further consideration, however, may suggest that such an explanation is not so easily dismissed. Weak and intense shocks may elicit entirely different responses. A weak shock may elicit attack and biting; a strong shock may elicit struggling or freezing. As Fowler and Miller explicitly noted, shocks more intense than any they used would presumably have elicited freezing, and successful suppression of running, therefore, would be entirely compatible with an analysis in terms of elicited competing responses.

2. Punishment of Responses that Escape Shock

It does not sound as though it would be easy to arrange a punishment contingency in such a way that the punished response will also be instrumental in escaping from punishment. However, there is reason to believe that this may be exactly what happens in studies of "self-punitive" behaviour when rats, previously trained to run an alley to escape or avoid shock, are punished during extinction in one section of the alley. One of the most striking features of studies that have found that punishment facilitates running under these conditions is that they have all involved the delivery of shock to a particular section of the floor of the alley; in none has shock been turned on for a fixed period of time when the rat gets to a particular point in the alley. The implications of this difference are substantial: once a shock of fixed duration has been turned on, the subject's behaviour will have little effect on the shock received; but if shock is delivered over a 1-foot section of the alley, an increase in speed of running will ensure a more rapid escape from shock. As noted by Fowler (1971), the response that is punished is compatible with, if not identical to, the response that is instrumental in escaping from shock once it has been encountered. Although the importance of this contingency has not been directly compared, it is notable that studies that have eliminated the contingency by employing shocks of fixed duration have generally found that punishment suppresses rather than facilitates running (e.g., Campbell et al., 1966a; Misanin et al., 1966).

3. *Identification of Competing Responses*

It is one thing to say that punishment may fail to suppress a response if that response is similar to, or compatible with, behaviour elicited by shock or reinforced by escape from shock. It is another matter to argue that all cases in which punishment does suppress a response are to be accounted for by pointing to the incompatibility of the punished response and the behaviour elicited or maintained by shock. It does not follow from the cases considered above that successful suppression of one response is always a consequence of the establishment of a competing response.

It has often, and reasonably, been held against competing response theories that they fail to specify or record the responses to which they appeal. Competing responses are not usually observed; their occurrence is inferred from the observed decline in the punished response. There is much to be said for such a strategy, for when actual candidates for the role of competing response are observed and recorded, they have usually behaved in a manner that leaves much to be desired.

Dunham (1971) has reported some observations on the behaviour of gerbils permitted free access to food, water and paper (for shredding). When one of these activities was punished, another would increase in probability: thus if the animal were punished for eating, the proportion of time in each session spent eating would decline to zero and there would be a corresponding increase in the proportion of time spent shredding paper. While paper shredding can hardly be regarded as an activity elicited by shock, nor reinforced by escape from shock, it might, as Dunham pointed out, have been reinforced because it *avoided* shock. While the animal shredded paper, it could not eat and if it did not eat, it did not receive shock. Thus far, the argument has proceeded smoothly enough: another response has been recorded while a first response is punished; this second response becomes more probable and it is possible to specify a potential source of reinforcement for this increase in strength. But none of this proves that the decline in the frequency of the punished response is a consequence of the increase in the putative competing response. It is equally possible that the increase in the latter is a consequence of the independently produced suppression of the former.

Two facts supported this second interpretation of Dunham's gerbil data. First, a closer examination of the time course of the changes in probability of punished and competing responses revealed that in all cases the suppression of the punished response substantially preceded the increase in probability of the competing response. Secondly, in a subsequent study with rats, Dunham (1972) showed that punishment of one response (drinking) increased the probability of another (running in a wheel), but that this increase in running was simply a consequence of the unavailability of water. Running increased equally when the water tube was simply removed, and at no point did the proportion of time spent running exceed that measured in an initial baseline condition when no water was available.

In at least some cases, therefore, where a record of behaviour other than the punished response has been maintained, and such behaviour has been observed to increase, there is little reason to believe that this increase is a cause of the suppression of the punished response. It looks more like a consequence of the increase in time available for the competing response, once the punished response has been suppressed. The cause of this suppression must therefore be sought elsewhere. It is always possible, of course, to argue that other, unrecorded responses are elicited by shock and conditioned to stimuli antedating shock, in the way suggested by a Pavlovian analysis of punishment. Since an animal no longer engaged in the performance of a particular instrumental response is likely to fill the available time with some other activity, and since it is always possible to argue that inactivity or freezing is the response conditioned by shock, the theory may be difficult to discredit. Useful information might be gained by further analyses of the effects of punishing responses elicited by shock, for example by punishing leg flexion in dogs with a shock to the paw.

Nevertheless, the actual observation of potential competing responses has suggested that the suppression of one response is not always a simple consequence of its displacement by another. Moreover, it is clear that appetitive instrumental learning cannot always be reduced to the operation of implicit Pavlovian contingencies. It seems, therefore, reasonable to conclude that direct, instrumental learning is responsible for at least some of the effects of response-contingent shock.

C. The Negative Law of Effect

Several writers have implied that any appeal to a direct mechanism of response suppression heralds a return to Thorndike's discarded negative law of effect (e.g., Church, 1963; Boe and Church, 1967; Rachlin and Herrnstein, 1969). This is a serious confusion. As we have already seen, although they all pointed to instances of its inadequacy, Thorndike (1931, 1932a), Estes (1944), and perhaps Skinner (1938) were not so much concerned to reject the empirical version of this law, as to reject the *explanation* of the effects of punishment originally proposed by Thorndike (1913). As Estes put it:

> It is well known that a response can be weakened in strength as a result of an annoying after-effect regardless of the mechanism involved. But it remains to be determined whether the weakened response is actually eliminated from the organism's repertoire by punishment, or merely suppressed and capable of being released at its full original strength after punishment is discontinued (Estes, 1944, p. 1).

According to the negative law of effect, punishment involves the loss of S–R connections; it seems to follow, therefore, that a punished response, having been unlearned, must be lost never to return. If, however, it is assumed that animals learn to anticipate the aversive consequences of their responses, then such learning, like the anticipation of appetitive consequences, should be subject to extinction; once punishment is discontinued, therefore, the punished response should show some recovery, even if it is never again reinforced.

Both Skinner (1938, pp. 151–60), and Estes (1944, Experiments A to F) reported results which they regarded as contradicting this implication of the theoretical version of the negative law of effect. If a period of punishment was given at the outset of extinction, sufficient to suppress responding to a low rate, there was always some recovery of responding in later extinction sessions when punishment was terminated. "The rate of responding at the end of a period of punishment, regardless of severity or duration, is not a reliable index of the true latent strength of the response. Discontinuation of punishment is always followed by some recovery in strength" (Estes, 1944, p. 34).

If the punishment was sufficiently mild, and delivered for a sufficiently brief period of time, then both Skinner and Estes observed what has been called "compensatory recovery": after the termination of punishment, there was a sufficient increase in rate of responding over later extinction sessions to ensure that the total number of responses emitted in extinction equalled that of an unpunished control group. In Skinner's words: "all responses in the reserve eventually emerge without further positive reinforcement" (1938, p. 155). On the basis of their inability to obtain compensatory recovery even with mild shock, Boe and Church (1967) recommended a return to the negative law of effect. But this is quite unwarranted. Full compensatory recovery represents the end point of a continuum of possible results, all of which would provide *prima facie* evidence against the law of effect: any increase in rate of responding during extinction following the termination of punishment, whether or not sufficient to produce complete compensatory recovery, suggests that punishment may not have resulted in the permanent loss of S–R connections.

Recovery from punishment seems to imply that suppression of responding is due to the anticipation of shock, and that this anticipation is subject to extinction when shock is discontinued. It might, however, be possible to save the law of effect as the basic explanation of the effects of punishment by appealing to supplementary principles to explain recovery. As we saw earlier, Holz and Azrin (1961) and Azrin and Holz (1966) have argued that the procedure employed by Skinner and Estes in their experiments would have ensured that the punishing stimulus became a signal for the omission of food reinforcement. The recovery of responding when punishment was terminated, therefore, could have been simply a consequence of the removal of a discriminative stimulus for extinction and the reinstatement of some of the conditions of original training. To support this account, Azrin and Holz performed a similar experiment, but substituted a brief change in the colour of the pigeon's response key for the mild shock used in Estes' experiment. They observed a rapid loss of responding in the initial period of extinction when each response produced this change in the colour of the key, but an essentially complete recovery of the response when this condition was terminated, even though extinction remained in effect.*

* The argument here is similar to that suggested by Guthrie (1935) and Skinner (1950) to account for spontaneous recovery following extinction. Pavlov (1927) thought that the re-appearance of an extinguished CR indicated that the original excitatory tendency underlying the CR could not have been lost as a consequence of the extinction procedure, but must have merely been overlaid by a new, inhibitory, tendency. Guthrie and Skinner pointed out that spontaneous recovery might be due to the reinstatement of stimuli associated with the start of a session, whose association with reinforcement was never fully extinguished in a series of massed extinction trials. Although this may well be an important contributory factor, it is questionable whether it is sufficient to account for all the data on spontaneous recovery (p. 470).

A second factor which might produce recovery after the termination of punishment is the post-shock facilitation of appetitive behaviour discussed before: rats that have just been subject to noncontingent shock may show an increase in eating or drinking (p. 265). A similar, transient facilitation of instrumental behaviour has also been observed. Rachlin (1966), for example, trained pigeons on a multiple (VI, VI) schedule of food reinforcement, where responses in one of the two components were consistently punished. The shock used, however, was sufficiently mild to produce only a transient suppression of responding and, after a few sessions,

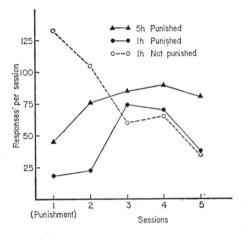

FIG. 6.3. Recovery from the effects of punishment scheduled during the first session of extinction. In the absence of further rewards, the two punished groups (differing only in the number of sessions initially received in acquisition) show a steady increase in rate of responding after punishment has been terminated. (*After* Estes, 1944.)

the birds were responding equally rapidly in both components. When punishment was terminated, rate of responding in that component showed a transient increase to a level significantly above the pre-shock baseline. A similar process operating in Estes' extinction experiments might be sufficient to account for the increase in rate of responding when shock was discontinued.

Both these explanations imply that the termination of shock should lead to an immediate increase in rate of responding, which might be sufficient to outweigh the loss of basic response strength demanded by the law of effect. After this transient increase, however, punished animals should revert to a lower rate of responding than unpunished controls. If recovery from punishment is due to extinction of the response–shock association, recovery should be slow even though, as Estes (1969b) points out, the fact that extinction of the response–food association is occurring simultaneously with extinction of the response–shock association makes exact prediction of the change in rate of responding a complex matter. In any event, a sufficiently detailed analysis should be able to distinguish between a recovery function due to the gradual extinction of a learned association and one due to a transient rebound or contrast effect, or to the removal of a stimulus signalling extinction. Although there has been little attempt to undertake the requisite analysis, the results of several of the experiments in Estes' (1944) monograph show that recovery may be quite slow. In one study (Experiment F), animals were given either one or five hours of acquisition training followed by five sessions of extinction. Two experimental groups were punished for responding during the first session of extinction, but not thereafter. A control group was never punished. As can be seen from Fig. 6.3, both punished groups responded at a higher rate in later sessions of extinction than in their first post-punishment session. Recovery spread out over this length of time can be interpreted only as extinction of the suppressive effects of punishment. Any such process is entirely incompatible with the process of unlearning envisaged by Thorndike in the original negative law of effect.

D. Anticipation of Aversive Reinforcement

As a UCS in studies of classical conditioning, shock can not only be used to condition such skeletal CRs as leg flexion, fighting, running, and freezing, it also has effects on behaviour that are not readily explained by changes in such CRs. The suppression of appetitively reinforced instrumental responding by a stimulus signalling noncontingent shock does not seem to be attributable solely to the conditioning of skeletal CRs to that CS (p. 82). Thus, stimuli signalling aversive reinforcers must be assumed to elicit some central motivational state of fear or anxiety, or to establish an anticipation or expectation of the aversive reinforcer, and either or both of these processes must be responsible for the suppression of instrumental responding typically observed in studies of conditioned suppression.

If a stimulus signalling noncontingent shock suppresses appetitive instrumental responding by arousing a state of fear, or by leading to the anticipation of shock, then it is only to be expected that the initiation and execution of a punished response should also become associated with an increase in fear or anticipation of shock. And just as increases in proximity to appetitive reinforcers facilitate responding, so increases in proximity to aversive reinforcers suppress responding. The effects of punishers can be explained in exactly the same way as the effects of rewards.

1. Punishment and Conditioned Suppression

In many ways, the clearest evidence for such a theory of punishment is precisely the phenomenon of conditioned suppression. It seems odd to appeal to entirely different processes to account for such similar phenomena. There is, of course, a difference between the two: punishment usually results in greater suppression of responding than noncontingent shock. But this difference can be readily understood by consideration of the difference in the contingency between stimulus and shock in conditioned suppression and between response and shock in punishment.

In one study of conditioned suppression, Rescorla (1968b), scheduled different probabilities of shock at random intervals both during a CS and in its absence. He observed that the greater the probability of shock during the CS, and the greater the difference between the probability of shock during the CS and the probability of shock in its absence, the greater was the suppression conditioned to the CS (p. 25). Consider, then, what would happen if two groups of rats were exposed to occasional presentations of a long, auditory, CS, in whose presence shocks of moderate intensity were delivered at an average rate of one per minute. For one group the only predictor of shock was this tone; for a second group, however, each shock was preceded by a brief light. In view of Rescorla's data, it is reasonable to expect that both groups would show some suppression to the tone, but that the second group would show complete suppression to the light. The difference between the conditions for these two groups seems analogous to the difference between conditioned suppression and discriminated punishment. In both procedures, a stimulus, analogous to the tone in the above example, predicts an increase in the probability of shock. But in discriminated punishment, unlike conditioned suppression, the initiation of the punished response stands in exactly the same relation to shock as does the light in the hypothetical experiment. The initiation of responding, in the presence of the discriminative stimulus, reliably predicts a very high probability of shock and therefore becomes an extremely effective suppressor of subsequent responding. More complete suppression is a consequence of a greater probability of shock.

2. Punishment of Aversively Motivated Responses

Stimuli signalling noncontingent shock may suppress appetitive instrumental responses either because they elicit a motivational state of fear or anxiety which suppresses the appetitive motivation maintaining responding, or because they lead to an anticipation of aversive reinforcement and such anticipation suppresses appetitive responding. The distinction between motivational and associative accounts may not in fact be a very precise one; it is certainly not one that admits of any simple, but decisive, experimental test. As was argued earlier (p. 85), the finding reported by Overmier *et al.* (1971b), that discriminative aversive conditioning proceeds more rapidly when discriminably different shocks are correlated with different stimuli, suggests that stimuli signalling shocks do not simply elicit an undifferentiated motivational state. Conversely, however, the finding, reported by LoLordo (1967), that a stimulus signalling a loud noise may have effects on avoidance responding that at least resemble those produced by a stimulus signalling shock, suggest that both stimuli may be eliciting just such a motivational state.

The point is one in which it will be necessary to return at several points in the present chapter. It is, however, clearly raised by experiments which have examined the effects of punishing aversively motivated instrumental responses. A stimulus signalling noncontingent shock, when superimposed upon free-operant avoidance responding is more likely to accelerate than to suppress such responding (p. 83). If the effect of response-contingent shock were just to establish that response as a reliable elicitor of fear, then one might expect that punishment of avoidance responding would simply increase the motivation for such responding and therefore inevitably facilitate responding. Although such facilitation can occur under certain conditions, it is far from being the general rule, and a theory of punishment unable to predict the suppression of escape or avoidance responding by response-contingent shock would be seriously deficient.

It is possible that some cases of paradoxical facilitation of escape or avoidance responding by punishment may be due to precisely such an increase in fear. Although many of these results may be largely accounted for by previously presented arguments, there is one which suggests that the stimulus maintaining performance in both acquisition and punished extinction is not simply the aversive event itself, but the motivational state elicited by the aversive event. Melvin and Martin (1966) trained rats to run in an alley to escape either a mild shock or a loud noise. In extinction, some animals received shock in the alley, some received noise and some neither. Punishment with shock facilitated running not only in animals previously reinforced by escape from shock, but also in animals reinforced by escape from noise; similarly, the performance of animals previously escaping from shock was facilitated by punishment with noise. Such transfer from one aversive event to another is similar to that observed by LoLordo (1967) and provides some evidence of mediation by fear.

The important fact remains that an escape or avoidance response whose execution is followed by immediate and consistent shock is almost invariably suppressed very rapidly. Reliable and persistent facilitation by response-contingent shock appears to occur only if the shock is scheduled infrequently, or if the shock is presented in such a way that continuation of the once initiated response ensures rapid escape from the shock. Since well-established avoidance responding tends to be facilitated rather than suppressed by an increase in fear, the effects of punishment cannot simply be reduced to the conditioning of fear to stimuli preceding or accompanying the punished response.

The problem here is similar to that encountered by motivational theories of positive incentive. Appetitive reinforcers do not affect instrumental behaviour simply by ensuring the conditioning of incentive motivation to the initiation of the appropriate instrumental response, if only because responding is initiated in the absence of all indices of such a motivational state (p. 231). The condition responsible for facilitating or reinforcing instrumental behaviour appears to be that such behaviour should be accompanied by an *increase* in conditioned anticipation of appetitive reinforcement. A similar argument is required to predict the effects of punishment: the condition for suppressing or punishing instrumental behaviour is that such responding be accompanied by an increase in conditioned anticipation of aversive reinforcement. The presentation of shock contingent upon the completion of an avoidance response will ensure that the series of component movements involved in the execution of the complete response become more strongly associated with shock, the later in the chain they appear. This argument, therefore, is clearly able to predict that avoidance responding will usually be suppressed by a punishment procedure. Although some anticipation of shock might be conditioned to background stimuli and to the initiation of responding, later components of the response will be more strongly conditioned to a more immediate shock, so that the execution of the full response will be accompanied by a steady increase in anticipation of shock. Such an analysis, therefore, predicts that punishment will normally be effective in suppressing aversively motivated behaviour, although it is easy to see that some association between shock and stimuli antedating the avoidance response would, at least temporarily, be able to facilitate the response.

E. Conclusions

Numerous suggestions have been advanced to explain the ways in which the delivery of a shock contingent on a response may affect the future probability of that response. Not all of these suggestions should be regarded as theories of punishment. For example, aversive reinforcers, like appetitive reinforcers, are stimuli and just as the delivery of food can serve as a stimulus to signal a change in the probability of reinforcement on free-operant FI schedules, or dicrete-trial alternation schedules, so the delivery of shock can serve as a stimulus to signal a change in the probability of either appetitive or aversive reinforcement.

Although response-contingent shock undoubtedly has further effects on behaviour, not all of these effects need be due to instrumental learning as such. As was noted in Chapter 5, many experiments employing appetitive reinforcers, although operationally described as instrumental, may in fact produce changes in behaviour that are not due to instrumental learning. A Pavlovian process of reinforcement may ensure the conditioning of responses to stimuli associated with the delivery of food that are sufficient to satisfy the experimenter's instrumental requirements. Similarly, punishment may often suppress a particular instrumental response because a Pavlovian process of reinforcement ensures the conditioning of responses incompatible with the punished response to stimuli signalling the delivery of shock. The importance of such a process is attested by several studies which have found that mild punishment may actually facilitate rather than suppress responding, when the response in question is one apparently elicited by the punishing stimulus.

Other theorists have argued that punishment is always due to the establishment of competing responses established by instrumental contingencies of escape or avoidance. There does not seem to be much evidence for the general importance of such contingencies. Moreover, although it is reasonable to argue that punishment may sometimes fail to suppress a response because the punishing stimulus tends to elicit the punished response, it is another matter to argue that *all* cases of successful suppression are due to the establishment of competing responses, whether these are maintained by Pavlovian or instrumental contingencies. Detailed analyses of apparent competing responses suggest that they could not play the role required of them.

Since appetitive reinforcers may directly maintain instrumental learning, it is plausible to suppose that aversive reinforcers may have similar instrumental effects. Although the positive and negative laws of effect suggest one interpretation of the symmetrical changes in behaviour produced by rewards and punishers, the reasons for rejecting the positive law of effect as an explanation of rewarded learning, discussed in Chapter 5, may apply equally to the case of the negative law of effect as an explanation of punishment. This would mean that the appropriate symmetrical analysis is that instrumental learning involves the establishment of associations between responses and their consequences: if those consequences are appetitive, responding is facilitated; if they are aversive, responding is suppressed. Although an analogy with the concept of incentive motivation produced by appetitive reinforcers can be found in the concept of a motivational state of fear produced by aversive reinforcers, it is not clear whether an appeal to motivational constructs provides any more enlightenment in the one case than it did in the other. It may be sufficient to say that responses accompanied by an increase in anticipation of aversive reinforcement are thereby suppressed.

V. THE STUDY OF AVOIDANCE LEARNING

A. Background

If response-contingent shock or punishment usually suppresses responding, the omission of shock contingent on the occurrence of an avoidance response usually increases the probability of that response. The omission of an aversive reinforcer seems to have effects on behaviour similar to those of the presentation of an appetitive reinforcer. The parallels here are suggestive, and it is a striking commentary on the compartmentalized nature of recent learning theory that the theoretical analysis of these phenomena should have progressed in such divergent ways.

The theoretical analysis of avoidance learning, as Herrnstein (1969) has recently complained, has long been dominated by two-factor theory. The longevity and success of such a theory, by comparison with the similar analysis of appetitively reinforced behaviour, may be partly deserved: there is, after all, some evidence that avoidance behaviour can be motivated by a classically conditioned motivational state in a way in which appetitive behaviour cannot be (p. 83). But the early appearance of a two-factor analysis of avoidance learning is more plausibly, if less charitably, attributed to an even earlier failure to distinguish between the procedures of classical aversive conditioning and those of instrumental avoidance learning. Although the distinction between the classical aversive procedure, in which the shock occurs regardless of the subject's behaviour, and the avoidance or omission procedure, in which the shock is omitted whenever a designated response occurs, was clarified and became the object of study during the 1930s (p. 124), the initial confusion had more enduring consequences. It ensured that avoidance learning was studied with procedures identical to those used in typical studies of classical aversive conditioning, with the sole addition of the response-reinforcement contingency, and it also ensured that avoidance learning would be interpreted in theoretical terms derived from the study of classical conditioning. Thus the typical avoidance experiment confronted an animal with a series of trials, each of which began with the presentation of a warning signal which, in the absence of a designated response, was followed by shock. The warning signal was regarded as a CS and the shock as a UCS; the interpretation of the outcome of this procedure stressed the idea that either some overt response, or some central motivational state, initially elicited by the UCS, came to be elicited by the CS, and thus either formed the avoidance response itself, or provided the motivation for the instrumental learning of some other pattern of behaviour. There are, however, other possible procedures for studying avoidance, and more than one other interpretation of the outcome of such experiments.

B. Procedures for Studying Avoidance

Avoidance learning has been studied in various types of apparatus: lever boxes, running wheels, shuttle boxes, jump-out boxes and alleyways. The avoidance response has involved pressing a lever; running in a wheel; running from one compartment to the other of a shuttle box, or jumping a hurdle separating the two compartments; jumping up onto a ledge; or running down an alleyway. In shuttle boxes, the animal may be required to run from one compartment to the other and then later run back to the first (neither compartment, in other words, is permanently safe). In one-way shuttling, on the other hand, the animal is removed from the apparatus following an avoidance response and placed in the same, dangerous, compartment at the beginning of the next trial.

The main procedural distinction in the study of avoidance learning, however, has been between discrete-trial, signalled avoidance and free-operant (often unsignalled) avoidance. In a typical discrete-trial experiment (Solomon and Wynne, 1953), a dog is placed in a shuttle box whose two compartments are separated by a hurdle. A warning stimulus is presented; after 10 sec, shock is turned on and remains on until the dog jumps into the other compartment. The response turns off the warning stimulus and the shock. If the response occurs before the shock is due (i.e., within 10 sec of the onset of the warning stimulus), the warning stimulus is turned off and no shock is delivered on that trial. In Solomon and Wynne's experiment, animals received 10 trials a day, with approximately 3 min between successive trials. Under these conditions, most dogs learned to avoid shock consistently within 10 to 20 trials. It should be noted that the response of jumping over the hurdle has several consequences in this experiment. It not only avoids shock but also produces escape from shock even when the animal fails to avoid the shock, and it terminates the external warning stimulus on all trials.

In typical free-operant procedures, the only programmed effect of the avoidance response is to avoid a scheduled shock. One typical procedure (Sidman, 1953a) programmes brief, inescapable, shocks on two independent interval schedules: in the absence of responding, shocks occur in accordance with one schedule; the occurrence of a response sets a second schedule into effect. In the standard Sidman avoidance procedure, the shock–shock schedule programmes shocks at regular intervals (e.g., every 10 sec); the response–shock schedule specifies that no shock can occur within a particular interval (again, say, 10 sec) following a response. Continued responding, therefore, can ensure the complete avoidance of all shocks, since each response postpones shock for 10 sec. In one variant on Sidman's procedure, Herrnstein and Hineline (1966) programmed shocks on variable rather than fixed intervals: in the absence of responding, shocks occurred on average six times per minute; when a response occurred, the frequency of shock was reduced to one per minute. After the animal had responded once, however, further responding did not postpone further shocks: a single response reduced the probability of shock, but sooner or later a shock would be delivered, which would have the effect of reinstating the higher density shock–shock schedule. A second variant on the original free-operant procedure was described by Sidman (1962a) as a fixed-cycle avoidance procedure: in this procedure, shocks were scheduled regularly at 15 sec intervals in the absence of responding; the occurrence of a response within any given interval avoided the shock scheduled for delivery at the end of that interval; although further responses within that interval had no programmed consequence, responding during the next interval would avoid the next shock, and so on.

All these free-operant procedures produce moderately efficient avoidance behaviour, although learning is distinctly slower than in some discrete-trial experiments and in some cases a certain proportion of animals may entirely fail to learn (Weissman, 1962). We shall see, indeed, that one of the most striking characteristics of the study of avoidance learning has been that different conditions produce extraordinarily large differences in rate of learning and efficiency of asymptotic performance.

VI. THEORETICAL ANALYSES OF AVOIDANCE LEARNING

A. Avoidance Responding as Anticipatory Responding to Shock

The simplest view of avoidance learning is that the aversive stimulus elicits or establishes certain responses and that these responses become strengthened to the point where they antedate the shock. At this point the experimenter calls them avoidance responses. According to one version of this account (attributable to Hull, 1929; 1943, pp. 73–4), when the shock is turned on, the animal engages in certain patterns of behaviour, one of which succeeds in turning the shock off. This escape contingency reinforces this particular response and it will eventually become strong enough to occur before the shock is delivered. The avoidance response is simply a reduced latency *escape* response. A Pavlovian analysis can be provided without any such appeal to reinforcement by drive reduction. The signalled avoidance situation is, as we have seen, modelled after a classical conditioning experiment: given this similarity, it is easy to suggest that the UCR to shock will in due course come to be elicited by stimuli preceding the shock in a straightforward Pavlovian manner, and that the avoidance response simply consists of these classically conditioned components of the UCR (Sheffield, 1948).

1. *Hull's Analysis*

Two sets of studies have clearly shown that avoidance responses are not always anticipatory escape responses. In the first place, reliable avoidance learning may occur when very brief, inescapable, shocks are used (e.g., D'Amato *et al.*, 1964; Bolles *et al.*, 1966). Secondly, several studies have explicitly required animals to make one response to escape from shock, once it has come on, and another to avoid the shock. Mowrer and Lamoreaux (1946) found that rats could learn to run to avoid shock, although only jumping and not running succeeded in turning shock off once it had started; Bolles (1969) has confirmed this result for a variety of different responses.

Bolles (1969) also showed, however, that although some avoidance responses, such as running, may be learned equally rapidly regardless of the nature of the escape response, other responses, such as turning, are learned only slowly if they are different from the escape response. In a later analysis, Bolles (1971) showed that this had been true even in Mowrer and Lamoreaux's data: in that experiment, animals learned to run to avoid shock whether running or jumping produced escape from shock, but jumping was learned rapidly only if jumping had also been the escape response. Other studies have suggested that these discrepancies may be quite general. Carlson and Black (1960), for example, found that if dogs were not permitted to escape from shock, they learned a hurdle-jumping avoidance response in a two-way shuttle box relatively slowly; but Theios and Brelsford (1966a) found that one-way avoidance learning in rats was only marginally affected by the prevention of escape responses.

Although some of these results suggest that an escape contingency may facilitate avoidance learning, they cannot be taken to imply that avoidance responses are established, as Hull's analysis requires, simply because they succeed in terminating a shock. In no case did animals totally fail to learn an avoidance response just because it was not also an escape response. The most reasonable interpretation of these data, therefore, is that an escape contingency may serve to increase the occurrence of responses that are initially of low probability. Responses that are already quite probable because they form part of the set of responses that tends to be elicited by shock will be learned readily as avoidance responses in the absence of an escape contingency. Responses that are unlikely to occur in the situation cannot be learned as avoidance responses until they occur, and in their case the escape contingency increases the chance that the avoidance contingency can take hold. The escape contingency, however, should not be thought of as a substitute for the avoidance contingency. Its role may be to increase the chances of an otherwise improbable response making contact with the avoidance contingency.

2. Pavlovian Analysis of Avoidance

A Pavlovian analysis of avoidance learning contains two cardinal propositions. First, successful avoidance responses are more or less directly related to the responses elicited by the aversive reinforcer. Secondly, the establishment of an avoidance response should not be directly dependent on the avoidance contingency as such. Both of these propositions, of course, will be familiar from previous analyses of appetitive instrumental learning and punishment.

One line of evidence consistent with the first of these propositions is provided by the observed variation in the rate at which different avoidance responses are learned. Just as rewards may be ineffective in establishing a response incompatible with the behaviour elicited by an appetitive reinforcer (p. 137), or punishment may be ineffective in suppressing a response elicited by an aversive reinforcer (p. 276), so an avoidance contingency may fail to establish a response incompatible with those elicited by an aversive reinforcer. As Bolles (1970, 1971) has stressed, not all responses in an animal's repertoire are equally amenable to avoidance training. The experimenter who selects an arbitrary response for avoidance training risks wasting his own time and causing his subject much pain. Rats, for example, may not avoid shock very efficiently if required to press a lever and pigeons can be trained to peck a key to avoid shock only with the greatest difficulty (p. 340).

Although cases of unsuccessful avoidance learning may often be attributed to the incompatibility between the response required by the experimenter and the responses elicited by shock, it does not follow that in all cases of successful avoidance the response in question is simply elicited by shock and conditioned to stimuli associated with shock. A literal interpretation of a Pavlovian analysis of avoidance seems to imply that the avoidance contingency is irrelevant to the appearance and maintenance of the avoidance response. It is the presentation of the aversive reinforcer, not its omission, that reinforces the response elicited by that reinforcer.

Much of the evidence bearing on this question was discussed in Chapter 3 (p. 115). It is apparent that in some cases this strict interpretation may be true. Some responses that are elicited by an aversive reinforcer, such as a twitch of the toe elicited by shock to the toe, may be quite impervious to an avoidance contingency (Turner and Solomon, 1962). In other cases, explicit comparisons of classical and avoidance schedules have revealed no additional learning attributable to the avoidance contingency. Gormezano (1965) showed that this held for the nictitating membrane response of rabbits; Schlosberg (1934, 1936) found no evidence that flexion responses in rats were reinforced by the omission of shock; and Woodard and Bitterman (1973) found that goldfish would swim from one compartment of a shuttle box to another just as readily when shocks were unavoidable as when a swimming response resulted in the omission of shock.

Nevertheless, it is clear that this conclusion does not hold for all response systems. When guinea-pigs or rats are studied in a running wheel (Brogden *et al.*, 1938; Bolles *et al.*, 1966), or rats are studied in a shuttle box (Kamin, 1956; 1957a), it can be shown that responding is established very much more reliably on an avoidance schedule than on a classical schedule. The avoidance contingency clearly has an effect on the animal's behaviour. Although this conclusion appears inconsistent with a strict Pavlovian analysis, such an appearance may be deceptive. It is still possible to question whether the avoidance contingency affects behaviour by reinforcing any instrumental response or rather by establishing one set of stimuli as a signal for shock and another set as a signal for the omission of shock. An animal given avoidance training in an alley or one-way shuttle box, for example, may learn that the start box and alley, or one compartment of the shuttle box, signal shock, while the goal box of the alley, or the other compartment of the shuttle box, signal safety. It would surely be consistent with a Pavlovian analysis to suggest that stimuli signalling shock may elicit running and stimuli signalling safety may elicit approach. This would in, effect, be similar to the argument, advanced in Chapter 5, that in some cases of appetitive instrumental learning, the subject's behaviour is not directly reinforced by the instrumental contingency, but is modified because certain stimuli become associated with food and elicit approach responses (p. 199). The main reason for the effectiveness of the avoidance contingency in some situations, therefore, may be that in the absence of the contingency no stimuli can be established as signals for the omission of shock and no escape response will be elicited. If a rat is shocked in an alleyway whether or not it runs into the goal box, the goal box will no longer be a place of safety.

In the end, therefore, the distinction between Pavlovian and instrumental analyses of avoidance learning may reduce to the question whether the avoidance contingency establishes certain external stimuli as signals for shock (which then elicit responses similar to those elicited by shock) and other external stimuli as signals for safety (which then elicit approach), or whether responses not necessarily elicited by shock may themselves be established as signals for the omission of shock and therefore be facilitated. The distinction may often be no easier to draw in the case of avoidance learning than it was in the case of appetitive learning. It might seem obvious enough that a response such as lever pressing is not elicited by shock, but the experimenter's and subject's definitions of a response may not always coincide. What the experimenter records as the depression of a lever sufficient to close a microswitch may be more appropriately described as a response of attacking and biting, elicited by stimuli associated with shock (Azrin *et al.*, 1964; Pear *et al.*, 1972). Although Black and Young (1972) found it possible to establish licking at a water tube as an avoidance response in rats, they noted that the avoidance contingency appeared to gain control over the subjects' behaviour by establishing the vicinity of the tube as a place of safety: once the subjects had approached the tube, licking was elicited as a response to thirst.

A Pavlovian analysis, however, is not very comfortably applied to unsignalled, free-operant experiments, where lever pressing or shuttling are reinforced by the omission of shock, even though they produce little change in external stimulation, and it is very difficult to see how some avoidance responses, such as head turning in dogs (e.g., Black, 1958; LoLordo, 1967) can be regarded as responses elicited by stimuli associated with shock. Moreover, even if shock does elicit wheel running in guinea-pigs or rats, or leg flexion in dogs, it is quite certain that these responses are affected by avoidance contingencies (Brogden *et al.*, 1938; Bolles *et al.*, 1966; Wahlsten and Cole, 1972). Since neither running in a wheel, nor flexing a leg produces any great change in external stimulation, neither can be said to involve approach towards stimuli established as safety signals.

Although, therefore, it may not always be possible to decide whether a particular avoidance response is established by an implicit Pavlovian contingency, the fact that some avoidance responses seem to be instrumentally reinforced means that a complete analysis of avoidance must include an account of the nature of this instrumental reinforcement. Two such theories will be considered.

B. Two-Factor Theory

There is nothing intuitively surprising about the observation that animals are capable of learning to avoid strikingly unpleasant events such as severe electric shocks. What must surprise the uninitiated observer is how reluctant psychologists have been to accept that animals may avoid shock efficiently because they have learned that the occurrence of a particular pattern of behaviour prevents the occurrence of the electric shock. The reluctance was explained by Mowrer (1947) as follows:

> It had previously been taken for granted by various writers that it is in some manner rewarding to an experimental subject to avoid a noxious unconditioned stimulus. It is easily seen that it is rewarding to escape from such a noxious stimulus. But how can a shock which is *not experienced*, i.e., which is avoided, be said to provide either a source of motivation or of satisfaction? (pp. 107–108).

The point was echoed by Schoenfeld when he asked "how the *non-occurrence* of an unconditioned stimulus can act as a reinforcement" (Schoenfeld, 1950, p. 83).

Two-factor theory, Mowrer claimed, was developed to resolve the paradox by showing how reinforcement could occur on trials when an animal avoided shock, without supposing that the "non-occurrence" of the shock itself was the reinforcing event. According to two-factor theory, the avoidance response was reinforced not because it avoided shock, but because it produced escape from fear.

The theory is most readily understood by applying it to the classical case of discrete-trial, signalled avoidance. In this situation, the warning stimulus is followed initially by shock on almost all trials. By a process of classical conditioning, the warning stimulus comes to elicit patterns of behaviour originally elicited only by the shock, including, most importantly, a central motivational state of fear or anxiety. Since fear is a drive state, it energizes all patterns of ongoing behaviour and any response that is followed by a reduction in fear will be thereby reinforced. Since, in the classical experiment, the avoidance response turns off the warning stimulus, it automatically produces a reduction in fear, and it is this that reinforces the response.

The development of two-factor theory was due to Mowrer (1947) and Miller (1948), although other writers made important contributions. Schoenfeld (1950), for example, also argued that avoidance responses were not reinforced by the omission of shock as such, but because they terminated a state of affairs that, by a process of classical conditioning, had become aversive. Schoenfeld talked about conditioned aversive stimuli, explicitly disavowing the assumption, made by Mowrer, that stimuli became aversive because they elicited a state of fear. Although it must be apparent that when a stimulus becomes aversive, this is because the subject rather than the stimulus has changed, there is something to be said for Schoenfeld's point. As we shall see, it is not clear whether stimuli become aversive because they elicit a motivational state of fear, or whether their correlation with an aversive reinforcer establishes an expectation or anticipation of that reinforcer. Rather than commit oneself to one or other alternative, it may be advisable to suspend judgement and simply talk of aversive stimuli.

Schoenfeld's analysis stressed two further points. First, he assumed that the stimuli which become aversive in an avoidance experiment include not only the warning stimulus explicitly manipulated by the experimenter, but also the proprioceptive stimuli associated with all patterns of behaviour other than the avoidance response. Secondly, Schoenfeld argued that avoidance responding was reinforced not only because it terminated this set of aversive stimuli, but also because it was accompanied by, or produced, a set of stimuli (including proprioceptive stimuli) that was explicitly unpaired with shock. The suggestion was that such "safety" stimuli might provide positive reinforcement for the avoidance response, in addition to the negative reinforcement produced by escape from aversive stimuli.

Many of the arguments advanced against a two-factor analysis of avoidance learning have been predicated on the assumption that the sole source of reinforcement to which such an analysis could appeal was the termination of a specific, external warning stimulus. As Schoenfeld's elaborations of the theory have suggested, such restrictions are misplaced. Stimuli signalling the omission of an expected shock will become conditioned inhibitors. The occurrence of such stimuli contingent on an avoidance response, therefore, should serve as effective reinforcement for that response. As several writers have subsequently pointed out, the termination of an external warning stimulus need not be the only way to reduce fear (Soltysik, 1963; Konorski, 1967; Weisman and Litner, 1969, 1971; Denny, 1971): avoidance responses will be reinforced provided their occurrence results in a change from a situation highly correlated with shock, and therefore eliciting a high level of fear, to a situation less highly correlated with shock, and therefore eliciting a lower level of fear. Properly understood, therefore, two-factor theory can be seen to involve the following propositions:—

(a) Pavlovian principles govern the conditioning of fear to stimuli correlated with aversive events such as electric shock. The strength of fear conditioned to a set of stimuli depends upon their correlation with shock. If shock occurs in the presence of A and not in its absence, then A will elicit fear and $\bar{A}$ will not. If shock occurs in the presence of A, but not in the presence of A + B, then B will become a conditioned inhibitor of fear.

(b) A response will be reinforced by a reduction in fear, i.e., if it changes the stimulus situations in the above examples from A to $\bar{A}$, or from A to A + B.

(c) The magnitude of reinforcement for a response depends upon the

L

degree of fear reduction. If A is correlated with a high probability of shock, and Ā or A + B with no shock, then reinforcement will be large; if Ā or A + B were simply correlated with a lower probability of shock, a response that changed A to Ā or to A + B would receive some (but less) reinforcement. A second way of varying the effectiveness of reinforcement will be to vary the discriminability of A and Ā or of A and A + B, for this will increase the generalization of fear from one stimulus situation to the other and thereby decrease the amount by which a change from one to the other will reduce fear.

C. Expectancy Theory

As the quotation from Mowrer's (1947) article implies, there have always been some writers who have resolutely insisted on upholding the common-sense view that what reinforces an avoidance response is the fact that it avoids an aversive event. Hilgard and Marquis (1940), following Brogden et al. (1938), were among the earlier exponents of this straightforward analysis:

> Learning in this [avoidance] situation appears to be based in a real sense on the avoidance of shock. It differs clearly from other types of instrumental training in which the conditioned response is followed by a definite stimulus —food or the cessation of shock. In instrumental avoidance training the new response is strengthened in the absence of any such stimulus; indeed, it is strengthened because of the absence of such a stimulus (pp. 58–9).

Mowrer and Schoenfeld's rhetorical question—how could the non-occurrence of a stimulus have any effect, let alone a reinforcing effect, on an animal—was considered, and answered, by Hilgard and Marquis, in the immediately following sentence:

> Absence of stimulation can obviously have an influence on behaviour only if there exists some sort of preparation for or expectation of the stimulation (p. 59).

The rhetorical question, it must be obvious, raises a false dilemma. We have already seen that the omission of a reinforcing stimulus may have profound effects on an animal: in appropriate circumstances it may produce inhibitory conditioning (p. 34) and elicit characteristic patterns of behaviour conveniently labelled frustration reactions (p. 260). Just as an avoidance response could logically be reinforced only by the omission of an *expected* shock, so inhibitory conditioning and frustration were dependent on the previous or concurrent establishment of an expectation of the reinforcing event. It is, clearly, the necessity of invoking concepts such as the expectation of reinforcement that has militated against the acceptance of this analysis of avoidance learning. Since the facts of inhibitory conditioning also require the postulation of just such a process, it is time to consider more seriously what are the general virtues and implications of this analysis of avoidance learning.

Instead of the development of a detailed expectancy theory of avoidance, what has in fact been offered is an attenuated version of such a theory by Sidman (1962b) and Herrnstein (1969). Herrnstein argues:

> Effective avoidance procedures include a common feature, so obvious as to be taken for granted, but possibly the sole necessary condition for avoidance. . . . The frequency of shock is reduced by the occurrence of the avoidance response, which is to say, an avoidance response avoids shock (p. 57).

Just as the rate of delivery of appetitive reinforcers was regarded as a primary determinant of reinforcement in appetitive instrumental experiments (p. 178), so here, the rate of delivery of aversive reinforcers is thought to be the primary determinant of the reinforcement of avoidance. But in both cases, the simplicity and elegance of the analysis depend upon a refusal to consider how these changes in rate of reinforcing events make contact with the animal. The statement that a reduction in shock frequency is reinforcing needs elaboration and expansion: a given frequency of shock is not itself reinforcing: it might become an effective reinforcer of a particular response when made contingent on that response, but presumably only if the subject had learned that a higher frequency of shock occurred in the absence of that response, i.e. had learned to expect shock when it failed to respond in that situation.

The suggestion is that animals associate a particular situation with the occurrence of shock and a particular response with the omission of the expected shock. By virtue of its association with shock, a particular situation or warning stimulus may become aversive; by virtue of its association with the omission of that shock or with a lower probability of shock, a particular response, or set of stimuli produced by a response, may become significantly less aversive. The transition from an aversive to a less aversive situation is reinforcing.

The analyses of avoidance learning proposed by Mowrer (1947) and Herrnstein (1969) are undoubtedly quite different. But when two-factor theory is modified to incorporate appropriate assumptions about conditioned inhibition of fear, and when Herrnstein's analysis is modified to incorporate necessary assumptions about expectation of shock, the two analyses that emerge resemble each other at more points than they differ: both assume that avoidance responses are reinforced because they change the situation from one that is highly aversive to one that is less aversive. The two analyses are not identical, but the differences between them are not what they may once have seemed. They reduce to questions which we have come across before. Are associations formed between a CS and UCS, or between the CS and central motivational state elicited by that UCS? Is it necessary to appeal to conditioned motivational states, such as incentive motivation for appetitive reinforcers or fear of aversive reinforcers, in order to account for the activation and reinforcement of instrumental behaviour? These questions may be difficult to answer, but to be able to ask the right questions must be regarded as some form of progress.

VII. ISSUES IN THE STUDY OF AVOIDANCE LEARNING

The remainder of this chapter will concentrate on the following issues: what motivates avoidance responding—is there evidence justifying the postulation of a state of fear or anxiety; what reinforces avoidance responding; what accounts for the extinction of avoidance responses; finally, what accounts for the extreme differences in the efficiency of avoidance learning under different conditions?

A. The Motivation for Avoidance Responding: Fear as a Learned Drive

In its original form, two-factor theory of avoidance learning argued that the pairing of a stimulus with shock, which was an inevitable part of any avoidance experiment, would initially ensure the classical conditioning of fear to that stimulus. Then, since fear was a source of aversive motivation, any response that succeeded in terminating these aversive stimuli would be reinforced. An important series of experiments by Miller, Brown and others provided the supporting evidence for these assumptions.

1. *Conditioned Fear as a Source of Motivation*

Brown *et al.* (1951) showed that the presentation of a CS, previously associated with shock, would increase the magnitude of the startle response elicited in rats by a loud noise (actually a pistol shot). There is, perhaps, some question whether this should be regarded as a *motivational* effect of the fear supposedly conditioned to the CS, or whether it is not simpler to assume that a set of skeletal responses, which include jumping, become conditioned to the CS and summate with the startle response elicited by the pistol shot. Kurtz and Siegel (1966) reported that the presentation of a CS paired with shock to the back of a rat, as opposed to its feet, completely failed to increase the magnitude of a startle response. They suggested that while shock to the feet elicited a jumping response, shock to the back elicited crouching, and any effect on the startle response could be attributed entirely to the conditioning of such skeletal responses. There is certainly some force to this argument, but there remain reasons for suspecting that motivational effects may also be operating: stimuli associated with other aversive events have been shown to augment the startle response (Wagner, 1963a) in ways which are hard to attribute to the conditioning of particular skeletal responses.

Subsequent experiments have shown that stimuli associated with shock have effects on ongoing avoidance responding which are rather easily thought of as motivational. Dogs and rats trained in a shuttle box on a Sidman avoidance schedule, will increase their rate of responding when a CS previously paired with shock is presented (p. 83). In addition to this relatively straightforward finding, other experiments have shown, beyond reasonable doubt, that avoidance responding in this situation can be affected by classically conditioned processes. The presentation of a stimulus specifically associated with the omission of an expected shock depresses the rate of avoidance responding (Rescorla and LoLordo, 1965; LoLordo, 1967; Bull and Overmier, 1968; Grossen and Bolles, 1968; Weisman and Litner, 1969); the presentation of a long delay CS, such as would be expected to result in some Pavlovian inhibition of delay, initially depresses rate of avoidance and increases rate of responding only after it has been on for some time (Rescorla, 1967b); the presentation of a short delay CS elevates rate of responding, which rapidly returns to normal after the CS has terminated, but the presentation of a trace CS elevates rate of responding even more after it has terminated (Kamano, 1970).

2. Reinforcement by Escape from Fear

There is, then, good evidence that avoidance behaviour can be influenced by some central state which appears to be conditionable in accordance with Pavlovian laws. The major assumption of two-factor theory has been that this is a motivational state, conveniently labelled "fear". If this assumption is temporarily granted, it is possible to proceed to the next step in the argument. This involves showing that animals will learn to escape from stimuli associated with shock; in terms of the theory, this is tantamount to showing that fear reduction is an effective reinforcer of instrumental responses. Experiments by Miller (1948), Brown and Jacobs (1949) and McAllister and McAllister (1962a, b) have sought evidence of such learning. Miller initially trained rats to escape from shock delivered in one compartment of a box by running into the other side. He then placed them in the initially dangerous compartment, with the door leading to the safe compartment locked. No further shocks were delivered, but the rats could still escape into the safe compartment if they learned to turn a wheel which opened the door. Thirteen of twenty-five rats succeeded in learning this new response, even though no further shocks were presented. The experiments by Brown and Jacobs and by McAllister and McAllister differed from Miller's in two respects. During initial training, no escape was allowed: the rats were simply locked up in one compartment and shocked. Secondly, the escape response was simpler than the wheel-turning response required by Miller: animals had only to jump over a hurdle to escape from the previously dangerous compartment. Under these conditions, rats learned fairly reliably to escape from a stimulus previously (but no longer) associated with shock. McAllister and McAllister employed a discrete warning stimulus during initial fear conditioning: they were able to show that fear had been conditioned both to this signal and to the more general cues of the apparatus.* According to two-factor theory, the processes operating in these experiments are very similar to those operating in a standard signalled avoidance experiment: the warning stimulus comes to elicit fear and the avoidance response is reinforced because it terminates the warning stimulus, or otherwise results in escape from the set of stimuli associated with shock. No additional processes need to be invoked.

3. *Probability of Responding in the Presence of Stimuli Signalling Shock*

Although free-operant experiments are usually scheduled so that responses at any time are equally effective in postponing shock, it is possible, while preserving this contingency, to present a warning signal immediately preceding each scheduled shock. When this is done, avoidance responses tend to be confined to the signal (Sidman, 1955b; Sidman and Boren, 1957; Keehn, 1959). This is, of course, exactly what would be predicted by two-factor theory; if shocks never occur in the absence of the signal, fear will tend to extinguish to background cues and be strongly conditioned to the signal. Hence appropriate motivation and reinforcement for responding are available only when the signal is turned on. Herrnstein (1969) and Keehn (1959) have made much of the fact that even if responding in the

---

* The reliable learning reported by Brown and Jacobs and by McAllister and McAllister becomes more striking when it is recalled that Seligman and Maier (1967) have shown that the prolonged exposure to inescapable shock may interfere with subsequent escape and avoidance learning. It is notable that Allison *et al.* (1967) found no reliable evidence of learning when they replicated Miller's experiment, but used inescapable shock to establish fear of one compartment. It is probable, as Seligman *et al.* (1971) have argued, that the interfering effects of inescapable shock depend on the complexity of the responses to be learned: exposure to inescapable shock may interfere with the learning of a wheel-turning response, but not with the simpler response of jumping over a hurdle.

absence of the signal serves to postpone the presentation of the signal and its associated shock, rats still tend to wait until the signal is turned on before responding. They have argued that if the warning signal becomes aversive, as two-factor theory requires, then rats should learn to avoid it. Since it is not avoided, it cannot be aversive, but must serve a discriminative function instead.

The argument seems unconvincing. According to two-factor theory, animals learn to make avoidance responses not because such responses postpone aversive events, but because they change the situation from one that is aversive to one that is less aversive. If shock never occurs in the absence of the signal, then the aversiveness of the background cues will probably be too low to maintain responding and the contingency in effect for responding in the absence of the signal will be irrelevant.

It is important to note that the warning signal in these experiments cannot, as Keehn and Herrnstein argued, have served as a discriminative stimulus which "sets the occasion for the avoidance response" (Herrnstein, 1969, p. 61). A discriminative stimulus for avoidance, according to Herrnstein, would be one in whose presence "the avoidance response reduces the amount of aversive stimulation; in its absence, the contingency is absent" (*loc. cit.*). If the aversive stimulation in question is simply the occurrence of shocks, then no such restriction on the avoidance contingency was imposed in these experiments. Responses both in the presence and absence of the warning signal had exactly the same effect on the delivery of shock. We shall see below that Herrnstein's analysis of avoidance learning is quite unable to account for the observed temporal distributions of responding in unsignalled free-operant experiments. In the present case, it seems unable to account for the distribution of responses when signals are presented.

4. Conclusions

In the absence of any restriction on the avoidance contingency, explicit or implicit, avoidance responses may still be concentrated in the presence of a stimulus correlated with shock. This suggests that it is not enough to say that avoidance responses are reinforced by a reduction in the frequency of shock: reinforcement depends upon a change in the situation, from one associated with a high probability of shock, to one associated with a lower probability. Such a conclusion is entirely consistent with the data reviewed in the earlier sections. The further implication of those data is that these associations are established in accordance with the principles of Pavlovian conditioning.

Two-factor theory is identified with the additional assumption that the effects of Pavlovian procedures on avoidance responding are mediated by the conditioning of a motivational state of fear or anxiety. The question is whether there is anything in the data which requires this assumption. Two sets of results have been thought to support a motivational interpretation. First, when animals have been trained in a free-operant avoidance situation, the noncontingent presentation of a stimulus associated with shock may elevate the rate of avoidance responding. Since it was precisely the failure to find such an effect in comparable studies with appetitive reinforcement that provided one reason for rejecting a theory of conditioned appetitive motivation (p. 227), it might seem reasonable to accept these results as evidence for a conditioned aversive drive. Before accepting this conclusion, however, it may be worth considering the nature of the contingencies programmed by free-operant avoidance schedules. Since such schedules arrange a positive correlation between the rate of responding and the proportion of shocks avoided, a well-trained subject will have learned that an increase in rate of responding is followed by a decrease in the probability of shock. If this has happened, it may not be necessary to appeal to an increase in motivation to account for an increase in rate of responding when an animal is exposed to a stimulus associated with an increase in the probability of shock.

The second major result thought to require the postulation of an intervening, motivational, variable is that animals will learn to escape from a situation associated with shock. This finding has traditionally been interpreted in terms of a drive-reduction theory of reinforcement: a situation associated with shock elicits fear and escape from that situation is reinforced by a reduction in fear. But in at least some of these experiments, as Bolles (1971) has pointed out, the escape response may simply be elicited by the aversive situation. Just as shock may elicit escape responses, so may stimuli associated with shock. Even where an appeal to instrumental reinforcement is required, it is only a prior commitment to a drive-reduction theory of reinforcement that necessitates the postulation of an aversive drive. If instrumental responses are reinforced when their occurrence is associated with an increase in proximity to appetitive reinforcers, it is reasonable to suggest they may also be reinforced if their occurrence is associated with a decrease in proximity to aversive reinforcers. If talk of conditioned incentive motivation is unnecessary in the former case, talk of conditioned fear may be unnecessary in the latter.

B. Reinforcement of Avoidance Responding

The preceding discussion has already touched on the question of the reinforcement of avoidance responses. It is time to examine this problem in detail.

1. *Reinforcement as Termination of the Warning Stimulus in Signalled Avoidance*

The traditional interpretation of two-factor theory has been to say that in the typical, discrete-trial, signalled avoidance experiment, the avoidance response is reinforced not because it avoids shock, but because it terminates the warning stimulus (e.g., Kimble, 1961, p. 269). This account has had the virtue of providing the impetus for some valuable experimental work. It also has the virtue of being partially true: there is good evidence that termination of a warning stimulus *is* reinforcing. But, as will be clear, termination of a warning stimulus is not a necessary condition for the reinforcement of an avoidance response; nor is there anything in two-factor theory to say that it must be.

Mowrer and Lamoreaux (1942) found that rats, trained in a shuttle box

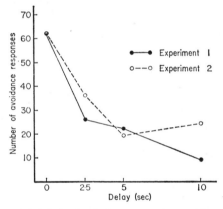

Fig. 6.4. A gradient of delay of reinforcement in avoidance learning. In two separate experiments the number of avoidance responses made by rats in a shuttle box depends upon the delay between the avoidance response and the termination of the warning signal. (*After* Kamin, 1957c).

with a buzzer as warning stimulus, learned the avoidance response much more readily when the buzzer was terminated by a response than when the buzzer was of a fixed, 5 sec duration. Kamin (1957b, c), in more extensive studies, varied the delay between response and termination of the warning stimulus from 0 to 10 sec. Although the shock avoidance contingency was identical for all groups, the effect of delaying the termination of the signal was substantial. The results are shown in Fig. 6.4: they give all the appearance of a typical delay-of-reinforcement gradient and it is hard to resist the conclusion that termination of the warning stimulus is at least an important source of reinforcement when rats are trained in a two-way shuttle box.

Two further experiments by Kamin (1956; 1957a) were designed as more critical tests of the proposition that, in this situation, avoidance responses were reinforced by termination of the warning signal rather than by avoidance of shock as such. The design of these experiments was deceptively simple. One group was trained in the shuttle box in the standard way, with responses both terminating the warning signal and avoiding shock. In a second group, responses only terminated the signal: the shock occurred 5 or 10 sec after the onset of the signal regardless of whether a response had occurred. In a third group, responses only avoided the shock: the signal always stayed on for 5 or 10 sec. Finally, for a fourth

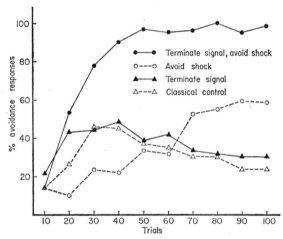

FIG. 6.5. The effect of termination of the warning stimulus, and of the avoidance of shock, on the acquisition of avoidance responding by rats in a shuttle box. (*After* Kamin, 1957a).

group, responses neither terminated the warning signal nor avoided the shock: in this condition, therefore, subjects were simply exposed to a classical contingency. The results of the second of these studies is shown in Fig. 6.5.

If the sole source of reinforcement in this situation were the termination of the warning stimulus, then one would expect that the group whose responses simply terminated the warning signal, without avoiding the shock, would learn as effectively as the control group. If, on the other hand, shock avoidance is the reinforcing event, then the group which only avoided shock should perform as well as the control group. In practice, neither of these outcomes was observed: when performance was assessed by counting the total number of responses made in 100 trials, both groups performed at an intermediate level much less effectively than the standard group but somewhat more effectively than animals whose responses had no effect on shock or warning signal. It is clear that both consequences of responding had some effect on response probability.

Kamin (1957a) interpreted his results as follows. The reason why the avoidance contingency affected response probability was not because the omission of shock was reinforcing, but because the occurrence of shock was punishing: if the shuttling response did not avoid shock, then it was punished and thereby suppressed. Hence animals whose responding only terminated the warning stimulus did not learn as well as the control group simply because they were being punished for responding. Nor, argued Kamin, should the learning displayed by animals, whose responding avoided shock but did not terminate the warning signal, be attributed to the avoidance contingency: the reinforcement in this case was the *delayed* termination of the warning signal.

There are two objections to this analysis. First, in order to explain one feature of the data, it appeals to a process of punishment without offering any analysis of how punishment works. Secondly, in order to explain another feature of the data, it ascribes reinforcing properties to the delayed termination of the warning stimulus when this event is not in fact contingent on any response. If one does not insist *a priori* that fear reduction must be correlated with termination of the warning signal, a more plausible account can be constructed. During early trials, before the first avoidance response, all subjects are shocked in the presence of the warning stimulus; during the intertrial interval, in the absence of the warning stimulus, they are never shocked. Although the apparatus cues may be partially associated with shock, therefore, the warning signal will presumably come to be more aversive than the apparatus cues alone. As soon as shuttling responses begin, however, each group's specific contingencies will begin to take effect. For the control group, shuttling terminates the warning signal and will therefore be reinforced by the removal of an aversive stimulus; since no shock is ever received in the absence of the warning signal, the avoidance response will always produce a situation less closely associated with shock than the situation prevailing in the absence of a response. For animals whose responses terminate the signal but do not avoid shock, however, things will be different. Responses will initially be reinforced by the removal of a stimulus correlated with shock; but since shocks now start occurring in the absence of the signal, the apparatus cues alone will eventually become as aversive as the warning signal and responding will no longer lead to any reduction in aversive stimulation. As is clearly shown in Fig. 6.5, the avoidance response will at first increase, but later extinguish.

Animals whose responses do not terminate the warning signal but do avoid the shock may conceivably, as Kamin suggests, obtain some super-stitious reinforcement from the fact that responding happens to be fol-lowed shortly by the termination of the signal. Since, however, other responses, such as stopping running, are likely to intervene between these two events, it is hard to believe that such reinforcement will be very effective. It seems much more reasonable to point out that for animals in this group, shocks occur in the presence of the signal only if no response has occurred; the combination of warning signal and a shuttling response, including any change in external stimuli dependent on the occurrence of shuttling, thus comprises a situation in which no shock is received. This should be sufficient to ensure that the response, and any stimuli accompanying it, will eventually be established as reliable conditioned inhibitors. The initiation of a shuttling response, therefore, is associated with a set of events reliably signalling the omission of an expected aversive reinforcer. By extension of the analyses previously developed to account for the effects of punishment, rewards and the omission of expected rewards, this condition will be sufficient to reinforce the shuttling response. In terms of two-factor theory, events signalling the omission of shock will become conditioned inhibitors of fear and responses that produce such events will be reinforced by a reduction in fear. Either way, however, this reinforcement will depend upon the gradual development of conditioned inhibition to the response and response-produced stimuli; unless these events are unusually salient (i.e. unless the response requires considerable effort, or produces a substantial change in the animal's environment) the development of inhibition will be slow, reinforcement initially will be minimal and learning of the response will be gradual. Fig. 6.5 confirms this expectation: animals whose responses avoided shock without ter-minating the warning stimulus learned slowly at first, but eventually surpassed the level of performance attained by animals whose responses only terminated the warning signal. A comparison of the overall number of avoidance responses misses the important fact that the course of avoidance learning in these two groups is quite different.

It is important to note that the outcome of this sort of comparison depends very much on the situation employed and the response studied. Bolles *et al.* (1966) found that when rats were trained in a running wheel as opposed to a shuttle box, termination of the warning signal was quite insufficient to establish responding, while animals whose running avoided shock but did not terminate the warning signal learned relatively rapidly. This could be explained in terms of the analysis offered here if one sup-posed that the prolonged running response that occurs in a running wheel constitutes a more salient event than does the brief response required in a shuttle box.

The important lesson of this analysis, however, is that there is no reason to equate reinforcement (even if this is still thought of as a reduction of fear) with the termination of a warning signal. Properly understood, as we saw earlier, two-factor theory is much more general than this. Avoidance responses are reinforced because their occurrence changes the situation from one associated with a high probability of shock to one associated with the omission of shock. The classic, signalled avoidance experiment can be seen as only one of several possible methods of arranging such a contingency. Since shock is scheduled in the presence of the warning signal, but not in its absence, termination of the signal will be reinforcing. But it is equally possible to arrange other changes in external stimulation contingent on a response. Bower *et al.* (1965), D'Amato *et al.* (1968b), and Bolles and Grossen (1969) have all shown that an avoidance response which terminates a warning signal may be learned more rapidly than one that does not, but that this difference may largely be eliminated if the avoidance response, instead of terminating the warning signal, produces a "safety signal" which stays on for the remaining duration of the warning signal. An avoidance response which produces a safety signal may be learned slightly more slowly than one which terminates a warning signal, but this, as we argued in the case of Kamin's experiment, is to be expected from the fact that it must take time before the safety signal becomes an effective conditioned inhibitor.

There is clear evidence that avoidance responding will be reinforced by a stimulus explicitly established as a Pavlovian conditioned inhibitor for shock. After training rats to turn a wheel on a free-operant avoidance schedule, Weisman and Litner (1969) gave differential Pavlovian conditioning, with a 5 sec tone signalling the omission of shock. Noncontingent presentation of the tone reliably suppressed the rate of ongoing avoidance responding (see Fig. 2.6, p. 29). When, however, the presentation of the tone was made contingent upon either an increase or a decrease in the baseline rate of responding, its effect depended on the contingency. The results are shown in Fig. 6.6: when the tone was contingent upon a high rate of responding (a DRH schedule), it increased response rate; on the DRL schedule, it decreased rate of responding. In both cases, therefore, it acted as a highly effective reinforcer and the intervening extinction sessions, during which the tone was not presented, resulted in an appropriate return to the baseline rate of responding.

Finally if avoidance responses are reinforced when they produce stimuli signalling a decrease in the probability of shock, then the magnitude of reinforcement for avoidance may be an increasing function of the discriminability of the change in stimuli contingent on the response, and a decreasing function of the anticipated probability of shock contingent on the response.

In conformity with the first of these predictions, Knapp (1965), training rats to jump out of one box into another to avoid shock, found that the greater the difference between the dangerous and safe boxes, the faster the animals learned the avoidance response. Similarly, Bower *et al.* (1965) showed that the effectiveness of a safety signal produced by the avoidance response in a signalled, discrete-trial, situation was dependent upon the discriminability of the signal. There does not appear to be any direct test of the second proposition in studies of discrete-trial signalled avoidance,

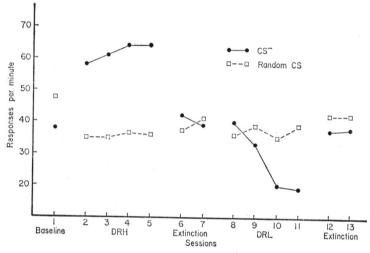

Fig. 6.6. The effect of presenting a stimulus (CS−), previously established as a signal for the omission of shock, contingent upon different patterns of free-operant avoidance responding. On a DRH schedule the stimulus increased rate of responding; on a DRL schedule it decreased rate of responding. (*After* Weisman and Litner, 1969.)

although it is entirely consistent with the data of many free-operant avoidance experiments (p. 323). Weisman and Litner (1971) have suggested that one reliable result of discrete-trial studies, which is consistent with this analysis, is the finding that avoidance learning is facilitated by the use of a long intertrial interval (Brush, 1962; Denny and Weisman, 1964). It is clear that the longer the intertrial interval in an avoidance experiment, the longer the period of time free of shock signalled by the occurrence of the avoidance response; from this analysis, therefore, it follows that an increase in intertrial interval may increase the reinforcement for avoidance.

2. *Avoidance Learning with a Trace Warning Signal*

If two-factor theory were really constrained to assume that the termination of an external warning signal was a necessary condition for the reinforcement of an avoidance response, then it would seem bound to predict that the use of a brief trace stimulus as the warning signal, which was terminated automatically several seconds before the scheduled delivery of shock, would make avoidance impossible. Although this strong prediction is not fully confirmed, and although two-factor theory does not in fact make any such strong prediction, it is a matter of some interest that discrete-trial avoidance with a trace warning stimulus may be strikingly inefficient.

Warner (1932), using rats in a one-way shuttle box, and Kamin (1954), using dogs in a two-way shuttle box, found very little learning of avoidance when a 1 sec or 2 sec trace stimulus preceded shock by 20 or 40 sec. Although Kamin observed relatively rapid learning when the interval between the 2 sec signal and shock was 5 or 10 sec, he attributed learning in these conditions to the fact that avoidance responses now often occurred *during* the 2 sec trace signal: under the conditions of his experiment, such responses did in fact terminate the signal.

Mowrer and Lamoreaux (1951) also studied avoidance learning in a shuttle box with a trace warning signal: they were not so much interested in seeing whether avoidance learning was possible with a trace signal, as in comparing the incidence of intertrial responses under conditions of trace and delay warning signals. Their argument was that, under trace conditions, shock is delivered in the presence of stimuli very similar to those present during the intertrial interval, whereas with a delay signal, the discrimination between the stimuli of the intertrial interval and those present when shock is delivered is relatively easy. Hence, with the trace signal, there should be greater generalization of fear or anticipation of shock in the intertrial interval, and a higher incidence of intertrial shuttling responses. This prediction was strongly confirmed, and has since been substantiated by several other experimenters (e.g. Kamin, 1954; Bolles *et al.*, 1966). Indeed, so high is the rate of intertrial responding under some trace conditions, that the incidence of actual avoidance responses (i.e. responses occurring between the trace signal and the scheduled shock) may be entirely attributable to this spontaneous rate of intertrial responding (Kamin, 1954).

The argument of the preceding section, however, should make it obvious that revised two-factor theory can point to potentially effective reinforcing events in trace-signalled avoidance learning: the trace of the warning signal provides a stimulus in the presence of which shocks are delivered; the occurrence of the avoidance response signals the omission of shock and may therefore act as a conditioned inhibitor of fear.

3. *Unsignalled Avoidance Learning*

If it is hard to find a set of external stimuli whose termination reinforces avoidance responding when a trace warning signal is used, it must be much more difficult to find any such source of reinforcement in unsignalled, free-operant avoidance learning. Yet rats, placed in a lever box, do learn to press the lever, albeit often rather slowly, if such a response postpones or cancels the delivery of unsignalled shock. If two-factor theory is to have anything useful to say about avoidance learning in this situation, it must allow that the occurrence of particular responses and response-produced stimuli can become associated with shock or the omission of shock.

In his initial investigations of unsignalled avoidance, Sidman (1953a, b) accepted the general analysis proposed by Schoenfeld (1950): the set of stimuli produced by all patterns of behaviour other than the required avoidance response became aversive, while stimuli dependent on the occurrence of the designated avoidance response were not associated with shock and therefore became reinforcing. Behaviour other than the avoidance response was associated with shock and therefore suppressed; while the avoidance response itself was associated with the omission of shock and therefore reinforced. This analysis was subsequently elaborated yet further by Anger (1963), who argued that the temporal regularities inherent in typical free-operant procedures would be expected to have important effects on conditioned fear or the anticipation of shock and also on the pattern of avoidance responding. In Sidman's original procedure, according to Anger, where the occurrence of a response always postpones the next shock by a fixed amount of time (the response–shock, or R–S, interval), the occurrence of that response predicts the delayed delivery of shock, in just the same way that a trace CS might predict the delayed delivery of any other reinforcer. As can be seen from the first panel of Fig. 6.7, therefore, the anticipation of shock will increase with the passage of time since the last response. Since, however, the occurrence of the response, and stimuli dependent on its occurrence, always signal a zero probability of shock, it follows (as can be seen from the second and third panels of Fig. 6.7) that the longer the subject waits before responding, the greater the anticipation of shock immediately before the response and therefore the greater the magnitude of reinforcement produced by the response. Hence, Anger concludes, one may expect the eventual development of a regularly spaced pattern of responding on typical Sidman avoidance schedules.

This analysis also, of course, accounts successfully for the major feature of avoidance learning under this schedule: that efficiency of avoidance responding increases as the R–S interval is made longer than the shock–shock (S–S) interval (Sidman, 1953b). The shorter the S–S interval, the greater the aversiveness of the situation in general; the longer the R–S interval, as Weisman and Litner's (1971) results imply, the greater the inhibition of fear produced by the occurrence of the response.

More recently, however, Sidman (1962b) and Herrnstein (1969) have argued that an appeal to two-factor theory is redundant and that avoidance learning in this situation is more economically accounted for by the suggestion that reinforcement derives from an overall reduction in the frequency of shock. It is clear that the shorter the S–S interval and the longer

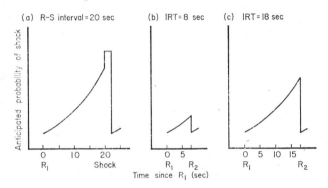

FIG. 6.7. Inferred changes in the anticipated probability of shock with the passage of time since a response, when subjects are trained on a free-operant avoidance schedule with an R–S interval of 20 sec. In (a) no response occurs within 20 sec of the first response; in (b) and (c) the interval between first and second responses (IRT) is 8 and 18 sec.

the R–S interval, the greater the reduction in frequency of shock produced by the response, so that such an explanation may seem equally satisfactory.

Herrnstein (1969) has argued vigorously that it is a great deal more than this. A two-factor analysis of unsignalled avoidance learning, he has contended, is redundant, and has "passed over the line into irrefutable doctrine" (p. 67). The basis for the latter assertion is obscure. The former seems to be justified by the following curious argument (earlier advanced by Herrnstein and Hineline, 1966). In some unsignalled free-operant experiments, "the experimenter does but one thing. He presents the shock at a specified rate, and the subject, by responding, also does but one thing, which is to alter the rate" (Herrnstein and Hineline, 1966, p. 430). From this it appears that appeal to any processes intervening between the experimenter's operations and the subject's behaviour is otiose. The one makes direct contact with the other, or, if not direct, then apparently in some manner that is no part of the psychologist's task to specify.

Free-operant avoidance schedules do not produce random distributions of responses in time. In the standard procedure, where each response postpones shock by a time equal to the R–S interval, there is good evidence that the temporal contingencies inherent in the schedule have some effect on performance. As training continues, the proportion of shocks successfully avoided increases, but the overall rate of responding decreases (Sidman, 1966). If lever pressing is the required avoidance response, some of this decline in rate of responding may be attributed directly to the decline in the number of shocks received, since there is good evidence that some of the responses recorded in such a situation are attacks or bites on the

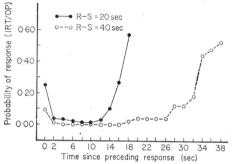

Fig. 6.8. Evidence of a temporal discrimination in a rat given extensive training on free-operant avoidance schedules. When the R-S interval was 20 sec, the probability of responding started to increase after 12 sec since the last response. When the R-S interval was increased to 40 sec, the probability of responding did not increase sharply until after 30 sec since the last response. (*After* Sidman, 1966.)

lever directly elicited by shock (Pear *et al.*, 1972). However, Pear *et al.* showed that rate of lever pressing declined even when such bites were recorded separately, and Riess and Farrar (1972) have reported a similar decline in rate of responding when rats are trained on a free-operant schedule in a two-way shuttle box. This overall decline in rate of responding, or increase in the interval between responses, seems readily attributed to the effect of the temporal contingencies pointed to by Anger (1963). The further finding that rate of responding decreases as the R–S interval increases (Sidman, 1953b) is also consistent with this analysis. When subjects are shifted from one R–S interval to another, rate of responding changes accordingly: a striking instance of this is shown in Fig. 6.8. The distribution of intervals between responses, or IRTs, may change relatively rapidly when a well-trained rat is shifted from one interval to another.

Since the probability of shock changes with the passage of time since the last response in a manner determined by the R–S interval, these observed temporal distributions of responding suggest that the subject's behaviour is affected by changes in the momentary probability of shock in the manner required by Anger's analysis. An experiment by Rescorla (1968a) provides additional support. After establishing a baseline rate of free-operant avoidance responding in dogs, Rescorla gave them a series of Pavlovian conditioning trials in which a 5 sec trace CS signalled shock 30 sec later. When this trace CS was superimposed upon ongoing avoidance responding, rate of responding was depressed during the CS and gradually accelerated after the termination of the CS. A second experiment sought to show directly the relevance of this finding to the temporal distribution of responding on free-operant avoidance schedules. Dogs were trained on such a schedule with a 30 sec R–S interval, but with the addition of a 5 sec stimulus produced by the occurrence of each response. This stimulus, of course, stood in exactly the same relation to shock as did the 5 sec Pavlovian trace CS employed in the first study; both stimuli signalled a 30 sec interval free of shock. When, at a later point in the experiment, the response-produced stimulus was no longer presented immediately after each response, but was simply presented by the experimenter while the animal was engaged in ongoing avoidance responding, it had exactly the same effect as the trace CS. In other words, if the experimenter arranges that an external stimulus is produced by each avoidance response on a free-operant schedule, then it can be shown that the stimulus is an effective conditioned inhibitor and that the probability of responding may increase with the passage of time since the presentation of the stimulus. It is hardly too fanciful to suggest that the occurrence of a response, or of stimuli automatically produced by a response, should play the same role in ordinary studies of free-operant avoidance.

The extent to which responses are always appropriately distributed on free-operant schedules, however, should not be exaggerated. Although latency of responding has been shown to be an increasing function of the interval between onset of warning signal and shock in studies of discrete-trial signalled avoidance (e.g., Low and Low, 1962), animals do not typically wait until just before the delivery of shock before responding. Indeed, Wahlsten and Cole (1972) found that the latency of avoidance responses was significantly shorter than the latency of comparable classical CRs. Thus Anger's analysis may exaggerate the role of temporal discriminations in many free-operant experiments: it should not be surprising if even a minimal anticipation of shock proves sufficient to motivate and reinforce avoidance responding. Certainly, as Hineline and Herrnstein (1970) have shown, there is no necessary correlation between the accuracy of a subject's temporal discrimination and the proportion of shocks avoided.

On the standard free-operant schedule, the probability of shock is determined precisely by the passage of time since the last response. On fixed-cycle schedules, where a single response in each cycle avoids the shock scheduled for the end of that cycle, subsequent responses within the cycle have no scheduled consequence. Thus the interval between the last response and shock may vary over a relatively wide range. A fixed-cycle schedule of shock avoidance does, however, bear an obvious resemblance to a fixed-interval schedule of appetitive reinforcement, where the delivery of one reinforcer signals the availability of the next after a specified interval of time. One might, therefore, expect to observe a similar FI scallop on the fixed-cycle avoidance schedule. On the avoidance schedule, however, a response in any interval results in the omission of shock at the end of that interval, so that as soon as successful avoidance occurs, the number of occasions on which it will be possible to detect evidence of appropriate scalloping decreases. Unfortunately, separate data are not available for performance in cycles preceded by a shock, and performance in cycles preceded by no shock. There is, however, evidence that when data are averaged over all cycles within a session, procedures which decrease the number of shocks avoided, and therefore increase the number of cycles initiated by a shock, produce much more pronounced scallops (Sidman, 1962a). As Sidman pointed out:

> The increasing shock frequency holds the key to an understanding of the changes that take place in the distribution of the animal's lever-pressing responses within the cycle. Once the animal has pressed the lever within the avoidance interval, thereby eliminating the shock that would have marked the beginning of the next interval, it no longer has a reference point from which to locate itself within the cycle. . . . For a temporal discrimination to be evident within the cycle, therefore, the animal must receive a substantial number of shocks (p. 103).

It remains to be seen whether the analysis proposed by Sidman (1962b) and Herrnstein (1969) can account for the emergence of free-operant avoidance, and for any temporal distribution of responding. Since patterning occurs in the absence of changes in external stimulation, and since one of the major purposes of their analysis is to obviate the need for postulating unobserved, uncontrolled stimuli, they would seem to have set themselves a hard task. Different probabilities of responding at different points in time cannot easily be accounted for without postulating some form of temporal discrimination. It is not enough to say, for example, that spaced responses are differentially reinforced for even if this were true, such differential reinforcement could not increase the probability of spaced responses unless the subject were able to discriminate such responses from others.

It is not even clear that differential reinforcement in the sense intended is in fact responsible for any of the temporal distributions observed. On a fixed-cycle schedule, for example, a response at any point in the cycle avoids the shock scheduled at the end of that cycle. Indeed, although most responses occur towards the end of each cycle, the majority of such responses are frequently wasted, since a single response is sufficient to

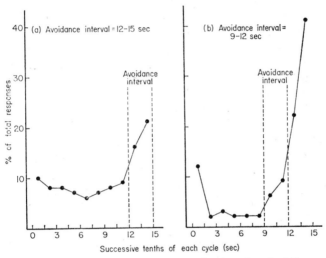

FIG. 6.9. The signalling function of shock on fixed-cycle schedules ensures that responses are concentrated toward the end of each cycle initiated by shock, whether (a) the avoidance contingency is confined to the end of the cycle, or (b) to an interval in the middle of each cycle. (*After* Sidman, 1962a.)

avoid shock, and further responses have no scheduled consequence. Differential reinforcement can, of course, be deliberately scheduled by arranging for the avoidance contingency to be in effect during only a limited interval within each cycle, so that responses occurring outside the correct interval do not succeed in avoiding shock (Sidman, 1962a). If this interval falls at the end of each cycle, responding tends to be concentrated at the appropriate point in time. That this is not a direct consequence of the programmed differential reinforcement, however, is forcibly suggested by the results shown in Fig. 6.9. When the 3 sec avoidance interval was moved forward within each cycle, the effect was not to produce an appropriate shift in the distribution of responses, but simply to increase the proportion of responses occurring in the *final* 3 sec of each cycle. As Sidman argued, limiting the avoidance contingency to a particular interval within each cycle did not succeed in differentially reinforcing responses at appropriate points in the cycle; rather, by decreasing the probability of a successful avoidance response, it increased the proportion of cycles initiated by a shock and therefore improved the temporal discrimination.

In the standard free-operant procedure, in which each response postpones shock for a fixed period, there is less difficulty in finding a potential source of differential reinforcement. As Sidman has pointed out: "The animal will gain more time free of shock with two well-spaced responses than by spacing two responses close together" (1966, p. 466). The suggestion is that rate of shock per response is lowest when responses are spaced just short of the R–S interval.

While this is true, it remains an open question whether this particular feature of the free-operant contingency is one that animals are sensitive to. One may also be pardoned for questioning whether Herrnstein's thesis that this account of avoidance learning is *simpler* than two-factor theory will stand up in the face of such a notion. Sidman's analysis, in effect, supposes that animals are not reacting to the short-term consequences of responding, for, in the short term, all responses (irrespective of the time since the preceding response) are followed by a shock-free period equal to the R–S interval. Rather, it must be assumed that animals are calculating variations in the overall rate of shock per session, and correlating these with variations in their own pattern of responding. There is, however, strong evidence that avoidance responses are maintained by their short-term consequences rather than by overall rate of shock per session. Bolles and Popp (1964) found that only one of fourteen rats learned to respond when a response did not avoid the next scheduled shock but did avoid later shocks. Baron et al. (1969) initially trained rats on a free-operant avoidance schedule and then presented occasional shocks contingent upon an avoidance response. These occasional response-contingent shocks were sufficient to suppress the overall rate of avoidance responding to the point where the overall rate of shocks was considerably higher than the rate that would have occurred had subjects maintained their original rate of avoidance responding. Finally, Hineline (1970) found that rats would learn to respond on a fixed-cycle schedule, in which the only consequence of a response was to retract the lever and postpone the delivery of shock to a later point in the cycle: avoidance learning occurred, in other words, even when overall frequency of shock was in no way affected by the occurrence of a response.

All of these results imply that an overall reduction in frequency of shock is neither a necessary nor a sufficient condition for the reinforcement of free-operant avoidance responding. In schedules which impose certain temporal regularities, it is the momentary probability of shock and the immediate consequences of a response that are the two important features responsible for the maintenance of responses and their distribution in time. Where no temporal regularity is specified by the avoidance schedule, no particular temporal distribution of responding would be expected, but this does not imply that the data generated by such a schedule will necessarily require an entirely different analysis. Herrnstein and Hineline (1966) introduced a schedule in which shocks occurred randomly with respect to time. In the absence of an avoidance response, shocks occurred with a given probability; the occurrence of an avoidance response reduced that probability to a lower value, until the next shock occurred, at which point the higher probability shock schedule was reinstated. Rats learned fairly readily to press a lever to reduce the probability of shock in this situation. Herrnstein (1969) argues that, according to two-factor theory,

> The avoidance response . . . must produce a "stimulus" that signals, on the average, an interval of shock-free time greater than the average shock-free interval correlated with the "stimulus" otherwise prevailing. These "stimuli" have no properties other than the very time intervals or shock rates they presumably signal. . . . The stimulus for two-factor theory here has but one attribute and that is its statistical relation to shock frequency. To detect these "stimuli", the animal must be reacting to shock frequencies in the first place (p. 59).

This is, perhaps, the basis for the suggestion that two-factor theory may have become "irrefutable". It is an odd argument, for it can only be construed as resting on the assertion that rats are unable to discriminate whether or not they have just pressed a lever, unless pressing the lever produces an explicitly programmed change in an external stimulus or a change in the schedule of reinforcement. A change in the probability of shock may be necessary to reinforce or punish lever pressing, but it is hard to believe that it is the basis for a discrimination between pressing the lever and failing to press the lever.

The only new feature of this random schedule of shock presentation is that no temporal discriminations will supervene to affect the distribution of responses. The occurrence of a lever press signals a lower, although still random, probability of shock than that signalled by the occurrence of a shock. There is good independent evidence that stimuli will differ in their aversiveness if they signal different probabilities of shock, even though shocks occur at random intervals in the presence of each stimulus. In an experiment on conditioned suppression described earlier (p. 25), Rescorla (1968b) found that a CS signalling shock with a probability of 0·4 per 2 min interval produced considerably greater suppression of appetitive lever pressing than one signalling shock with a probability of only 0·1 per 2 min interval.

4. Conclusions

Mowrer (1947) set the pattern for much of the theorizing that has been undertaken within the tradition of two-factor theory when he argued that avoidance responses were not reinforced by the omission of shock, but by a reduction in conditioned fear. It is as though these were two co-ordinate events which, by appropriate experimental design, could be put in opposition in order to discover which was the true reinforcer. This is seriously misleading. The statement that shock was omitted following the subject's avoidance response describes the operations performed by the experimenter; the statement that the avoidance response was followed by a reduction in conditioned fear is a possible inference about the consequences of those operations. If avoidance responding is reinforced by a reduction in fear, the external event responsible for such a reduction is a change in the stimulus situation from one associated with a high probability of shock to one associated with a low probability of shock. The usual way to ensure that stimuli will enter into such associations is to schedule shock in the presence of one set of stimuli and to omit shock in the presence of another. Thus it would be equally correct to say that avoidance responding is usually reinforced by the omission of a scheduled shock.

Two-factor theory, then, should be understood to be saying that the omission of scheduled shock affects a subject's behaviour because it establishes stimuli associated with this event as conditioned inhibitors of fear, while stimuli in whose presence shock occurs are conditioned excitors of fear. This is not so very different from saying that the subject learns to expect shock in the presence of one set of stimuli, and that a given response is followed by the omission of this expected shock. But we have seen that the analysis proposed by Herrnstein (1969), although perhaps masquerading as an expectancy theory, omits the crucial features of such a theory. The idea that avoidance responding is controlled by the presentation of discriminative stimuli in whose presence the avoidance contingency operates is simply insufficient to account for the observed occurrence of avoidance in the presence of stimuli preceding shock and for the temporal distribution of avoidance responses in unsignalled schedules. It is also clear that avoidance responses are not reinforced by a reduction in the overall frequency of shock, but by the omission of imminently expected shocks.

The dispute between a two-factor analysis of avoidance learning, and the view that avoidance responses are reinforced because they avoid shock, therefore, is not what it may once have seemed. The point at issue reduces to the more general one of whether instrumental behaviour is mediated by classically conditioned motivational states, or by the development of expectations about the consequences of responding.

C. Extinction of Avoidance Responding

Second only to the problem of how the *absence* of an event could reinforce avoidance responding has been the problem supposedly posed by the persistence of avoidance responding. The two problems share a number of features, not least the fact that both dissolve upon closer and more rational scrutiny. In the present case, we can distinguish between two questions: first, whether it is possible in principle to account for any degree of persistence of avoidance responding and secondly, whether in fact the observed persistence of such responding is amenable to theoretical analysis.

The problem of principle was originally much the same as the problem of accounting for the acquisition of the avoidance response in the first place. Any analysis that reduces avoidance responses to anticipatory escape or classically conditioned responses must predict that the emergence of the successful avoidance response initiates extinction of that response: as soon as subjects avoid shock, their responses are no longer reinforced, and must be expected to extinguish. Even two-factor theory, in its original guise, encounters what appear to be grave problems: if an avoidance response is motivated by the fear elicited by a warning signal and reinforced by a reduction in fear, then it will be motivated and reinforced only as long as the warning signal continues to elicit fear. But if the warning signal became aversive by virtue of being paired with shock, then as soon as successful avoidance occurs and this pairing is disrupted, fear of the warning signal should extinguish, leading to rapid extinction of the avoidance response itself.

1. *The Persistence of Avoidance Responding in the Absence of Shock*

Before considering whether two-factor theory is, in fact, bound by this simple argument, i.e. whether it necessarily predicts the immediate extinction of fear coincident with the emergence of the successful avoidance response, it will be more appropriate to examine the empirical evidence. Although it is commonly stated that avoidance responses are extremely resistant to extinction, the claim deserves more careful scrutiny than it commonly receives. It is usually accompanied by reference to the classic studies of Solomon *et al.* (1953). In one of these experiments, 13 dogs were trained on discrete-trial signalled avoidance in a shuttle box, with an extremely severe shock, adjusted so that it was just below the threshold for tetanization. After reaching a criterion of 10 consecutive avoidance responses, the animals received 200 extinction trials. Not only did no animal reach the arbitrary criterion of extinction (a single failure to respond within 2 min of the onset of the warning signal), but over the entire 200 extinction trials, only 11 responses occurred with latencies longer than 10 sec (the original interval between signal and shock), and the mean latency on the final day of extinction was *shorter* than the mean latency on the 10 criterion trials at the end of acquisition training.

Dramatic as these results are, it may be misleading to accept them as typical of the normally observed persistence of avoidance responses. A subsequent study by Brush (1957), although unfortunately presenting insufficient data to assess the overall course of extinction, reported that 3 of 25 dogs trained with a similar procedure, but with a less intense shock, reached the extremely stringent criterion of 10 consecutive failures to respond within 2 min in less than 200 trials. Other studies, using other procedures, have often observed relatively rapid extinction. Just about the only sense, indeed, in which it has been appropriate to talk of the persistence of avoidance responding has been in the context of an implicit comparison with the persistence of *escape* responses. Sheffield and Temmer (1950), employing rats in a one-way shuttle box, were the first to demonstrate greater resistance to extinction of avoidance than of escape responses, and their results have been confirmed by others (e.g., Santos, 1960). But in terms of the actual number of responses made in extinction, it is worth noting that Sheffield and Temmer's avoidance group required on average only 24·5 trials before reaching the experimenter's criterion of extinction (failure to respond within 10 sec on two consecutive trials), while in Santos' experiment the most persistent avoidance group was responding in less than 10 sec on no more than 25 per cent of trials after 40 extinction trials. In both of these experiments, avoidance training was in a one-way shuttle box, a situation in which learning occurs rapidly and efficiently. Using a similar type of apparatus, and defining an avoidance response as one which occurred within 2 sec of the onset of a warning stimulus, Page (1955) found that after training to a criterion of three consecutive avoidance responses, rats required fewer than 30 trials to reach a criterion of extinction of three consecutive failures to respond within 10 sec. Beecroft (1967), after training rats to a similar criterion to avoid shock by running down an alleyway into a safe goal box, found that extinction, defined as one failure to respond within 60 sec, occurred within 11 trials. Baum (1969), training rats to jump up onto a safe ledge, gave acquisition to the somewhat more rigorous criterion of 10 consecutive avoidance responses; the median number of trials required to reach the extinction criterion of one failure to respond within 5 min was 39.

If rats are trained to avoid shock in a two-way shuttle box, they show no greater persistence of responding in extinction. Kamin (1959) trained rats to a criterion of 11 consecutive avoidance responses, followed by 50 extinction trials. In one experiment, the median number of trials to a criterion of 10 consecutive failures to respond in less than 10 sec was 16·5; in a second, 46·5. Bolles *et al.* (1971), whose results are shown below in Fig. 6.10 (p. 339), found that a normal extinction procedure resulted in virtually complete loss of the avoidance response within 100 trials. Finally, the problem with lever pressing as an avoidance response has more often been its disappearance despite continued reinforcement, rather than its persistence in extinction (e.g., Coons *et al.*, 1960). Free-operant experiments, which have often employed lever pressing as the avoidance response, have sometimes been thought to show great persistence in extinction (e.g., Sidman, 1955a), but more recent evidence suggests that extinction may in fact be relatively rapid: Shnidman (1968) trained two rats with an S–S interval of 5 sec and an R–S interval of 20 sec; following at least 15 hours of training, at the end of which they were pressing the lever at the rate of 350 and 516 responses per hour, they were extinguished to a criterion of 15 min without a response; one subject made 64 responses, the other 183. An experiment by Coulson *et al.* (1970), whose results were shown above in Fig. 6.2 (p. 285), confirms that the omission of all shocks leads to relatively rapid extinction of free-operant avoidance in rats.

2. *Theoretical Analysis of Extinction of Avoidance by Removal of Shock*

It is possible to find experiments, other than that of Solomon *et al.* (1953), in which well-trained avoidance responses have proved extremely persistent. Seligman and Campbell (1965) observed little evidence of extinction within 150 trials after rats had been trained to avoid shock in an alley; Wilson (1973), training rats in a one-way shuttle box, observed very little decline in probability of responding over 60 extinction trials. And Wahlsten and Cole (1972) reported that some dogs, trained to avoid shock by flexing a leg, made runs of 1,000 consecutive avoidance responses without receiving a shock. Although Brush (1957) did not find very marked effects attributable to the intensity of shock used in acquisition, it seems probable that this may be one variable affecting the persistence of responding: there is good evidence from other procedures, such as conditioned suppression, that aversive conditioning with very intense shocks may significantly increase resistance to extinction (Annau and Kamin, 1961).

Nevertheless, avoidance responses do not generally persist indefinitely. The more mundane question, therefore, is whether there is anything in the data on the extinction of avoidance responses that poses particular problems for particular theories of avoidance. An experiment by Kamin *et al.* (1963) appears to show that a warning signal does in fact become less aversive as soon as rats learn successfully to avoid shock. Kamin *et al.*

trained different groups of rats to successively more stringent criteria of avoidance in a shuttle box (3, 9, or 27 consecutive avoidance responses), and then used the warning signal for shock as the CS superimposed on ongoing appetitively reinforced lever pressing. Conditioned suppression thus provided an independent measure of the aversiveness of the warning signal. Their results suggested that the warning signal was significantly less aversive for animals that had just completed 27 consecutive successful avoidance responses, than for those that had been trained to the less stringent criterion of only 9 consecutive avoidance responses. A subsequent study by Linden (1969) also suggested that extended avoidance training would result in some loss of conditioned suppression to the warning signal.

Before accepting the conclusion that these results raise insurmountable problems for theories of avoidance, it is worth pointing out that, in one sense at least, they are exactly what one might expect. Animals that are avoiding successfully receive no more pairings of warning signal and shock and this should lead to some extinction of the anticipation of shock conditioned to that signal. And so it does. The problem, it is implied, is that in spite of this decrease in the aversiveness of the warning stimulus, animals continued to perform the avoidance response. But this would be a problem only if the conditioned suppression measure revealed *no* evidence that the warning stimulus was still aversive; in the studies of both Kamin *et al.* and Linden, however, the most extensively trained subjects still showed a significant amount of suppression and there is no particular reason to suppose that this was insufficient to maintain avoidance responding.* An apparently more serious problem in Kamin *et al.*'s experiment is that there was a significant difference in the suppression ratios of the 9-trial and 27-trial groups, even though both groups were still consistently making the avoidance response. But we do not, in fact, have any independent measure of the strength of the avoidance response in these two groups: it is entirely possible that animals that have completed 27 consecutive trials without receiving shock would be responding with longer latencies, or would show less resistance to extinction, than animals that had last been shocked only nine trials earlier. Only if large differences in conditioned suppression were correlated with no differences in avoidance behaviour, would data such as these pose a serious theoretical problem.

---

* It is worth pointing out that conditioned suppression may not be a particularly sensitive measure of fear: intensities of shock that are sufficient to support escape or avoidance learning, e.g., intensities less than 0·5 ma, are insufficient to produce reliable suppression with the long CS–UCS interval typically used in studies of conditioned suppression (Cohen, 1968).

We have so far accepted, as though it were a necessary corollary of two-factor theory, the proposition that fear of a warning signal will rapidly extinguish as soon as the subject starts consistently avoiding shock. There is, however, reason to question this conclusion. For the situation correlated with shock when no response occurs is not the same as that correlated with omission of shock when the subject does respond. In any experiment where the avoidance response terminates an external warning stimulus, it follows that shocks occur only after a long stimulus, never after a short one. Since avoidance responses are usually permitted to continue terminating the warning stimulus in extinction, then provided the subject responds during extinction, the stimulus correlated with the omission of shock continues to be that correlated with the omission of shock during acquisition. If the subject discriminates between the long stimulus which was associated with shock and the short stimulus which was not, continued exposure to the omission of shock following the short stimulus may not be the most efficient procedure for extinguishing the former association. Even if the avoidance response never terminates the warning stimulus, it is possible, as Soltysik (1963) has suggested, that the occurrence of the avoidance response itself serves to protect the association between warning stimulus and shock from extinction. During acquisition, the warning stimulus alone was correlated with shock; the occurrence of an avoidance response in the presence of the warning stimulus signalled the omission of shock. If these two situations are discriminately different, then the omission of shock in extinction, provided the avoidance response occurs, may do little to extinguish the association between shock and the warning stimulus itself.

Although there is no evidence to show how well an association between shock and a long stimulus may survive the omission of shock contingent on the presentation of a shorter stimulus, Soltysik has provided some independent evidence to support his assumption that the occurrence of an avoidance response may protect the warning stimulus from extinction. In an earlier study (Soltysik, 1960), he had shown that 120 nonreinforced trials might result in relatively little extinction of responding to an appetitive CS if the CS was presented in conjunction with a reliable conditioned inhibitor on all extinction trials. Although this finding has not been directly confirmed in any study of aversive conditioning, Kamin (1969) found that extinction of conditioned suppression to an auditory CS was significantly retarded if the auditory stimulus was presented in conjunction with a novel visual stimulus on all extinction trials. Subsequently, Suiter and LoLordo (1971), also studying conditioned suppression, showed that nonreinforcement of a compound stimulus, consisting of one novel stimulus and one stimulus already established as a reliable conditioned inhibitor, resulted in relatively little inhibitory conditioning to the novel stimulus. The results obtained by Kamin and by Suiter and LoLordo will be recognized as the parallels, for inhibitory conditioning, of the phenomena of overshadowing and blocking which have been demonstrated repeatedly in excitatory conditioning (p. 46). Soltysik's argument, therefore, amounts to saying that in an avoidance experiment, the occurrence on extinction trials of an avoidance response previously established as a conditioned inhibitor or signal for the omission of shock, will tend to block any inhibitory conditioning to the warning stimulus.

There is thus both empirical and theoretical reason to believe that the typical procedure used to extinguish avoidance responding may be ill-designed to extinguish any association between the warning stimulus and shock. Either by terminating the warning stimulus before it has reached the length associated with shock, or by acting as a conditioned inhibitor, the occurrence of an avoidance response on extinction trials may tend to maintain the aversiveness of the warning stimulus. Although they do not serve to distinguish between these two possibilities, several studies have shown that extinction of avoidance may be reliably facilitated by preventing the occurrence of the avoidance response on early extinction trials. Different experiments have achieved this by locking the door of a shuttle box, removing the ledge from a jump-out box, or immobilizing subjects with curare; they have all, however, obtained similar results (e.g., Solomon et al., 1953; Page, 1955; Black, 1958; Baum, 1970). Black's experiment provides a good example. After training restrained dogs to turn their head to avoid shock, he injected one group of animals with curare and exposed them to 55 nonreinforced presentations of the warning signal. When subsequently given normal extinction training, these animals required only 45 trials to reach the criterion of extinction. Control animals, however, required an average of 287 trials to reach the criterion of extinction. It is clear that the 55 trials under curare were very much more effective than 55 ordinary extinction trials in leading to loss of the avoidance response.

It has been suggested that response prevention may facilitate the extinction of the originally established avoidance response by fortuitously permitting the establishment of a new competing response (Page, 1955). Although part of the effect observed in these experiments may be attributable to interference with the execution of the avoidance response, several studies have provided independent measures of the aversiveness of the warning stimulus following a number of trials on which the original avoidance response was prevented, and have confirmed that this procedure does ensure rapid extinction of the aversiveness of the warning stimulus (e.g., Shipley et al., 1971; Bersh and Paynter, 1972; Wilson, 1973). It remains to consider whether response prevention affects extinction by ensuring exposure to a warning stimulus of the duration correlated with shock during acquisition, or whether it is exposure to the warning stimulus in the absence of the avoidance response that is important. Katzev (1967) found that even if rats were free to respond in extinction, the introduction of a delay between responding and termination of the warning stimulus significantly facilitated extinction. The suggestion is that exposure to a long warning stimulus in extinction, even in the presence of an avoidance response, is sufficient to reduce the aversiveness of the warning stimulus. Shipley *et al.* (1971), on the other hand, employed a yoking procedure to equate exposure to the warning stimulus between one group of animals permitted to avoid and terminate the signal during an initial block of extinction trials and a second group prevented from responding during these trials. In spite of this similar exposure to the warning stimulus, response prevention significantly facilitated extinction of the avoidance response. It is indeed probable that both processes operate, and that their relative importance depends upon the experimental situation. The importance of response-contingent termination of a warning stimulus in the acquisition of avoidance responding varies from one situation to another (p. 318). It seems reasonable to suppose that where such a contingency is important in acquisition, the most important factor in extinction may be the duration of the stimulus correlated with the omission of shock. Where, however, the avoidance response is acquired without benefit from the termination of an external stimulus, the response itself must be established as a reliable conditioned inhibitor and it may be that it is the occurrence of the response on extinction trials that maintains the association between the warning stimulus and shock.

3. *Extinction of Avoidance by Response-Independent Shock*

It has often been noted that there is something odd about the normal procedure for programming extinction of avoidance responses. The experimenter disconnects his shock generator, but if well-trained animals continue to make the required avoidance response, there is no way that the experimenter's change of contingency could affect their behaviour. Is it any wonder, then, that avoidance responses are so resistant to extinction? The wonder is rather that they extinguish at all. If avoidance responses were simply reinforced by the omission of shock, the continued absence of shock in extinction would ensure their indefinite continuation. It must, of course, be supposed that it is the omission of an anticipated shock that reinforces avoidance responding and that it is the anticipation of shock that is extinguished under normal conditions.

In general, if avoidance responses are reinforced because they change the situation from one associated with a high probability of shock to one associated with a lower probability of shock, then responding might be expected to weaken because of the extinction of either of these associations. The typical procedure, in which shock is omitted regardless of the animal's behaviour, should lead to loss of the avoidance response via extinction of the association between the first situation and the high probability of shock. This has been the burden of the discussion up to this point. However, in order to eliminate avoidance responding by extinguishing the association between responding (or response-produced stimuli) and the omission of shock, it would be necessary to abolish the avoidance contingency not by disconnecting the shock generator, but by programming the same probability of shock when an avoidance response occurs as when it does not occur. Several investigators have, indeed, argued that this change in contingency is more appropriately regarded as extinction of the avoidance response (e.g., Davenport and Olson, 1968; Bolles et al., 1971). In one way, at any rate, it is more analogous to the operation of appetitive instrumental extinction than is the traditional procedure for extinction of avoidance. In appetitive experiments, as is shown in Table 6.1, the acquisition contingency involves the presentation of a reinforcing event (such as food) if a response occurs and no food in the absence of a response; the extinction contingency involves omission of food regardless of whether the subject responds or not. If the omission of an expected shock is the reinforcing event in avoidance, analogous to the presentation of food in the appetitive case, then the acquisition contingency for avoidance is directly parallel to the acquisition contingency for appetitive behaviour, with no shock substituted for food and shock substituted for no food. But in that case, the analogous extinction contingency for avoidance involves the presentation of shock regardless of whether a response occurs or not. Moreover, the appetitive analogue of the traditional procedure for extinction of avoidance, in which shock is omitted regardless of the subject's behaviour, is to present food regardless of the subject's behaviour.

Davenport and Olson (1968), training rats in a lever box, reported relatively rapid extinction of discrete-trial signalled avoidance when shock was presented at the end of the signal regardless of the rats' response. Bolles et al. (1971) specifically compared a number of different extinction procedures after rats had been trained to avoid signalled shock in a two-way shuttle box. Their results are shown in Fig. 6.10. It is clear that the traditional extinction procedure (complete omission of shock) produced a less rapid loss of the avoidance response than did the redefined extinction procedure, in which shocks were presented regardless of the subject's behaviour. As noted above (p. 286), the *reversal* of the avoidance contingency (presentation of shock contingent on the original avoidance response) resulted in an extremely rapid loss of responding.

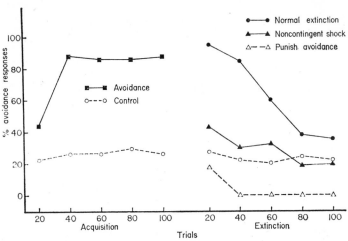

FIG. 6.10. The effect of different procedures for the extinction of avoidance responding. Punishment of avoidance responding resulted in the immediate suppression of responding. However, removing the contingency between the warning
signal and shock (normal extinction) was a less effective extinction procedure than
removing the contingency between responding and the omission of shock on each
trial (noncontingent shock). (*After* Bolles *et al.*, 1971.)

TABLE 6.1

**Contingencies involved in acquisition and in two procedures for
extinction of instrumental reward and avoidance learning**

	Appetitive Instrumental Learning	Instrumental Avoidance Learning
Acquisition	Response → Food No Response → No Food	Response → No Shock No Response → Shock
Extinction	Response → No Food No Response → No Food	Response → Shock No Response → Shock
Extinction II	Response → Food No Response → Food	Response → No Shock No Response → No Shock

M

4. *Conclusions*

The problem of the persistence of avoidance responding in extinction is, perhaps, less intractable than has been supposed. In the first place, avoidance responses often extinguish quite rapidly when shock is permanently omitted. Secondly, the occurrence of an avoidance response can be expected to interfere with the extinction of the association between warning signal and shock, and it is the integrity of this association that provides the motivation for further responding. Finally, avoidance behaviour can be very rapidly extinguished if the contingency between response and omission of shock is removed. None of these findings has yet been shown to require the postulation of novel, *ad hoc*, explanatory principles.

D. Variability of Avoidance Learning

Experimenters, in common with other human beings, are loth to admit failure. Only in the last decade has there been sufficient admission and analysis of the fact that several procedures for studying avoidance learning are astoundingly inefficient (Meyer *et al.*, 1960; Bolles, 1970, 1971). Rats learn to run from a start box to a safe compartment to avoid shock in less than five trials (e.g., Theios, 1963b), but may be trained for hundreds of trials on a signalled, lever-press avoidance task and end up avoiding the shock on fewer than 20 per cent of trials (Biederman *et al.*, 1964). Free-operant avoidance learning in rats is also notably less efficient when the avoidance response required is pressing a lever rather than running in a shuttle box (Riess, 1971). Pigeons can learn to avoid shock tolerably well in a shuttle box (Macphail, 1968), or if required to depress a treadle or lever (Foree and LoLordo, 1970; Smith and Keller, 1970), but require elaborate shaping if they are to learn to peck a key to avoid shock (Hineline and Rachlin, 1969). In short, any satisfactory analysis of avoidance learning must recognize that it is not sufficient for the experimenter to arrange suitable contingencies between some arbitrarily chosen set of stimuli, responses and shocks. The choice of one may critically restrict his freedom to choose others.

1. *Response-Produced Change in Stimulation*

We have argued that the reinforcement for an avoidance response depends not only on the different probabilities of shock associated with responding and not responding, but also on the discriminability of the stimulus change produced by responding. When this latter factor has been explicitly manipulated, for example by increasing the salience of a response-produced safety stimulus, it has been shown to have a pronounced effect on speed of learning (p. 319). It is also possible that one reason why some procedures are more effective than others for the establishment of reliable avoidance responding is that they differ in the extent to which the occurrence of an avoidance response changes the stimuli impinging on the subject. In a one-way shuttle box, alleyway or jump-out box, the avoidance response not only produces relatively salient proprioceptive stimuli, it also completely changes the external stimulus situation by leading the subject out of a situation in which shocks occur into a different situation (the safe side of the shuttle box, the goal box of an alleyway) in which shocks never occur. In any one of these classes of apparatus, rats typically learn to avoid shock in less than a dozen trials (e.g., Page, 1955; Knapp, 1965; Baum, 1966; Theios et al., 1966; Beecroft, 1967).

The avoidance response itself is the same in a two-way shuttle box as in a one-way box. It also produces a comparable change in external stimulation. In the one-way box, however, only one compartment is ever associated with shock, while in the two-way box shocks occur in both compartments. It is not surprising, therefore, that one-way avoidance is learned very much more rapidly than two-way avoidance: Theios et al. (1966), for example, reported that different groups of rats required between 5·4 and 10·3 trials to attain a criterion of 10 consecutive avoidance responses in a one-way shuttle box, and between 54·2 and 127·3 to attain the same criterion in a two-way shuttle box.

The response of pressing a lever involves relatively little intrinsic feedback and leaves the animal in much the same external stimulus situation as it was before the response occurred. By comparison with the response of running from one environment to another, lever pressing as an avoidance response, therefore, will receive relatively little reinforcement. An experiment by Masterson (1970) has shown that it is indeed this lack of change in stimulation that is at least partly responsible for the difficulty of establishing lever pressing as an avoidance response. A group of rats whose lever pressing not only avoided shock, but also gave them access to another chamber in which shock was never delivered, learned to press the lever relatively efficiently. In other words, if the response permitted escape from a set of external stimuli associated with shock to a set associated with no shock, learning was greatly improved.

2. Responses Elicited by Shock

The reinforcers used in instrumental experiments are also effective UCSs in classical conditioning: that is to say, they elicit responses which may come to be elicited by stimuli regularly antedating the reinforcer. Nowhere is this more important than in the case of electric shock. We have already seen that the effects of response-contingent shock in experiments on punishment cannot be fully understood without appeal to the classically conditioned responses which may come to be elicited by stimuli associated with the punished response. Exactly the same may be said about avoidance training. A discrete-trial signalled avoidance procedure has much in common with a classical conditioning procedure; indeed, until the avoidance response occurs, the two schedules are identical. It is probable, therefore, that early trials of avoidance training result in the conditioning of a set of classical CRs to the warning signal; if these CRs are incompatible with the experimenter's definition of an avoidance response, then it is reasonable to suppose that they will interfere with the establishment of efficient avoidance.

The major proponent of this type of argument in the context of avoidance learning has been Bolles (1970, 1971). Bolles has argued that, to a first approximation, only "species-specific defense reactions" will be readily established as avoidance responses. The aversive situation releases a set of innately determined, defensive reaction patterns; if one of these coincides with the experimenter's requirements for an avoidance response then this response will be selected—largely because the remaining set of defense reactions will be punished; if none of these coincides with the required avoidance response, then learning will never occur.

As we have seen, shock may elicit a variety of unconditional responses in rats, and an apparently even greater variety may become conditioned to stimuli acting as CSs for shock. The two most important observations are that if escape is possible, then rats tend to run out of an apparatus in which they are shocked but if escape is impossible, they tend to freeze or sometimes to attack a suitable target. It is relatively easy to see how this affects the observed incidence of avoidance responding in different situations. An avoidance response which involves running away from the aversive situation, as in a one-way shuttle box or alleyway, is learned very readily. One which involves remaining in the same place may be very difficult, since the freezing responses conditioned when escape is impossible may interfere with turning a wheel or pressing a lever. If the situation contains a target, such as a conspecific, likely to elicit aggressive attacks, then avoidance responding may be further disrupted (Logan and Boice, 1969). If the avoidance response actually involves approach towards a stimulus associated with shock, as when a pigeon is required to peck at a key whose illumination provides the warning signal for shock or when rats are required to press a lever situated directly beneath the light that serves as the warning signal, learning should be particularly difficult. As Moore (1973) has noted, Biederman et al. (1964) showed that the simple expedient of locating the warning light at the far end of the box from the lever dramatically improves the rate at which rats will learn to press the lever to avoid shock.

Two further studies serve to document the argument. Brener and Goesling (1970) trained rats either to remain immobile or to move around in order to avoid shock. Subjects required to remain immobile learned very much more successfully. Two additional groups, however, were yoked to the two avoidance groups, receiving shocks whenever an avoidance subject failed to meet the avoidance requirement. Both of these groups showed a progressive increase in immobility or freezing as training continued. Similarly, Grossen and Kelley (1972), having observed that exposure to noncontingent shock in an open field increased the proportion of time rats spent in close contact with the wall, went on to show that an avoidance response which required rats to jump onto a ledge was learned much more rapidly when the ledge was next to the wall of the apparatus than when it was in the centre.

As was noted earlier, even where the avoidance response is specified by the experimenter as pressing a lever or pecking a key, it is possible that the way in which the subject solves the problem set by the experimenter is by directing its instinctive attack or escape behaviour at the manipulandum. Rats may learn to depress a lever by attacking it (Moore, 1973), or even, perhaps, by freezing onto it (Bolles, 1971); pigeons may learn to make aggressive pecks at a key, or to depress a treadle by hitting it with a wing—a typical aggressive gesture (Smith and Keller, 1970). Thus even when the experimenter believes that he has selected a purely arbitrary response, the subject may still behave in ways dictated by the UCRs elicited by the reinforcer rather than by learning a new and arbitrary response–reinforcer contingency.

3. *Shock Intensity and Avoidance Learning*

Rate of avoidance learning is affected not only by the response requirement, but also, in a somewhat complex manner, by the intensity of the shock contingent on a failure to respond. It is clear, in fact, that the effect of shock intensity interacts with the response requirement. One-way avoidance learning is facilitated by increases in shock intensity (e.g., Moyer and Korn, 1966); while two-way shuttle box avoidance (Moyer and Korn, 1964; Levine, 1966) or lever-press avoidance (Biederman *et al.*, 1964; D'Amato and Fazzaro, 1966a) suffers from the use of too strong a shock. Theios *et al.* (1966) specifically compared the effects of shock intensity on one-way and two-way avoidance, and confirmed that stronger shocks facilitated one-way, but interfered with two-way avoidance. All of these studies have been with rats. Hineline and Rachlin (1969) have obtained some relatively similar results with pigeons: they found that the most efficient way to shape key pecking as an avoidance response was to start with very weak shocks, and gradually to increase their intensity.

Dogs, on the other hand, learn a two-way shuttle box avoidance response rapidly enough with a traumatically strong shock (Solomon and Wynne, 1953) and, over a fairly wide range, shock intensity has relatively little effect on avoidance learning in this situation for dogs (Brush, 1957).

The arguments advanced earlier in this section may help to clarify many of these relationships. An avoidance response compatible with classically conditioned flight or escape responses should only be facilitated by an increase in the intensity of the shock used. But when the avoidance response is inconsistent with the UCRs elicited by shock, then the stronger the UCRs the greater the interfering effect. Rats should find it only harder, therefore, to learn to avoid shock by pressing a lever, as stronger shocks elicit more insistent freezing. D'Amato et al. (1968a) provided good evidence for this analysis by showing that in the absence of any avoidance contingency, the incidence of lever pressing by rats was inversely related to the intensity of shock which they were experiencing in the situation. In an earlier study, D'Amato et al. (1967) had shown that once lever-press avoidance had been established with a moderately weak shock, subsequent increases in shock for failure to respond improved avoidance performance.

A related analysis may help to explain why rats find it harder to learn an avoidance response in a two-way shuttle box as shock intensity increases: shocks may reliably elicit escape only into a *safe* place. However, a second reason has been suggested by McAllister et al. (1971). They argued that since both sides of a two-way shuttle box are associated with shock, the use of stronger shocks may serve to raise the aversiveness of both sides; since the compartment in which the animal is positioned at the onset of a trial may be already asymptotically aversive, this increment in aversiveness will be felt mostly in the other compartment. Hence the move from one compartment to the other will be accompanied by only a slight reduction in aversiveness.

4. *Conclusions*

Two principles may be invoked to account for variations in rate of avoidance learning. Avoidance responses which result in little change in the set of stimuli impinging on the subject will receive only marginal reinforcement, and responses which are incompatible with the UCRs elicited by shock will be suppressed by the conditioning of the UCRs to the situation as a whole and in particular to stimuli immediately preceding shock. Both these principles find a fair degree of empirical support. The importance of this second factor confirms that Pavlovian contingencies may have profound effects on instrumental performance. The view that avoidance learning simply involves the formation of appropriate response–reinforcer associations, ignores the realities of animal behaviour. The reinforcers used by experimental psychologists release complex patterns of instinctive or consummatory behaviour (UCRs). This is particularly true of electric shocks, and the experimentalist ignores this truth at his peril.

VIII. SUMMARY

Few theorists have systematically attempted to provide a single, symmetrical account of appetitively and aversively reinforced instrumental learning. The analysis of punishment and avoidance has only rarely made contact with that of positively rewarded learning. This is curious, for it is easy to show that the presentation and omission of aversive reinforcers have effects on instrumental behaviour that are largely the mirror image of the effects of appetitive reinforcers. This symmetry suggests that instrumental learning may be regarded as the formation of associations between responses and the presentation or omission of reinforcement; that responses are strengthened when they are associated with the presentation of an appetitive reinforcer or omission of an aversive reinforcer and are suppressed when associated with the presentation of an aversive reinforcer or omission of an appetitive reinforcer. Furthermore, just as the changes in behaviour that occur in appetitive instrumental experiments may not always be due to instrumental learning, but may be a consequence of implicit Pavlovian contingencies and the establishment of responses elicited by the reinforcer, so the effects of schedules of punishment or avoidance may be partly a consequence of the Pavlovian reinforcement of responses elicited by shock.

Experiments on punishment have only recently established unequivocally that response-contingent shock will reliably suppress responding and that this suppression is partly a consequence of the contingency. The belated discovery that, in this empirical sense, the law of effect may apply as well for aversive reinforcement as for appetitive reinforcement has been thought to raise the possibility that the effects of punishment could be understood in terms of the theoretical law of effect. If appetitive reinforcers strengthen S–R associations, perhaps aversive reinforcers weaken those associations. In neither case, however, does the partial validity of the empirical law of effect provide any reason to accept the theoretical law. It is hard for a theory which states that punishment directly weakens S–R associations to account for all observed instances of recovery from punishment. Moreover, the parallels between the effects of punishment and of conditioned suppression suggest that in both cases animals associate antecedent events, whether responses or stimuli, with the delivery of shock and that an increase in the anticipation of aversive reinforcement suppresses ongoing behaviour.

Although the main effects of punishment may be attributed to the establishment of associations between responding and shock, it is clear that other effects can and do occur in typical studies of punishment. Like appetitive reinforcers, aversive reinforcers can be established as cues predicting the presentation or omission of further reinforcement: the punishment of responding in extinction, for example, may increase the rate of extinction by increasing the change from acquisition to extinction. Finally, since shock is an effective UCS in classical conditioning, it would not be surprising if in some cases the suppression of a punished response was brought about by the classical conditioning of competing responses. Certainly, where the response to be punished is itself elicited by shock, punishment may be relatively ineffective.

Most theoretical analyses of avoidance learning have attempted to explain avoidance without admitting that the omission of shock might be the event which reinforces successful responding. They have assumed that the absence of an event could not be regarded as a reinforcer. It is obvious that the omission of an event could have no effect, reinforcing or otherwise, unless the event itself was expected. As soon as this is accepted (and the phenomena of conditioned inhibition and frustration equally require such an assumption), the analysis of avoidance learning is considerably simplified.

Two-factor theory has suggested that avoidance responses are reinforced not by the omission of shock, but by escape from fear. This, however, is a false dichotomy: it is the correlation between a particular situation and the omission of shock which causes that situation to become less aversive and which may, therefore, be thought to reduce fear. Properly understood, two-factor theory does not differ from an expectancy theory in the set of experimental events assumed to be responsible for the reinforcement of avoidance responses. The only difference between the theories is that two-factor theory assumes that stimuli correlated with the presence and absence of shock control differences in a motivational state of fear, while an expectancy analysis eschews motivational constructs. The difference is the same as that between theories of appetitive instrumental learning which talk of incentive motivation and those which rely only on the anticipation of reinforcement.

The difference is not, in fact, important for an understanding of much of the data on avoidance learning. Avoidance responses are reinforced because their occurrence reduces the aversiveness of the experimental situation: this may be because they terminate stimuli associated with shock, because they produce stimuli associated with the omission of shock, or because they themselves are associated with the omission of shock. It is, however, important to note that an overall reduction in the frequency of shock is neither sufficient nor necessary to reinforce avoidance responding. The immediate consequences of a response are much more important than any overall change in the frequency of shock.

Avoidance responses are commonly thought to be exceptionally slow to extinguish. There are, indeed, recorded cases of their extreme persistence, but these seem to be the exception rather than the rule. Theoretically, the extinction of avoidance responding should depend either on the extinction of the association between the warning signal and shock, or on the extinction of the association between responding and the omission of shock. When the experimenter programmes extinction by omitting all shocks, only the former association can be expected to extinguish and there are, indeed, factors which may interfere with the required inhibitory learning. In order to extinguish the association between responding and the omission of shock, the experimenter must programme shock independently of the subject's behaviour. Such a procedure usually leads to the rapid loss of avoidance responding.

Finally, Pavlovian contingencies may be as important in studies of avoidance learning as in other operationally instrumental experiments. Avoidance learning may be easy if the required response is similar to that elicited by shock and difficult or impossible if the two responses are incompatible.

CHAPTER 7

Contrast Effects: Interactions between Conditions of Reinforcement

I. INTRODUCTION

The argument in much of the preceding two chapters has rested on assumptions about the reciprocal nature of appetitive and aversive reinforcers. Response facilitation or reinforcement was regarded as a consequence of an anticipated increase in appetitive, or decrease in aversive stimulation; conversely, response suppression or punishment was regarded as a consequence of an anticipated increase in aversive, or decrease in appetitive stimulation. This symmetry, in turn, relies on a set of presuppositions about mutual antagonism or inhibition between appetitive and aversive systems, a point of view most clearly enunciated by Konorski (1967).

A. Analysis of Transient Contrast as a Rebound Effect

The evidence for such a system of reciprocal inhibition between appetitive and aversive systems has already been documented (e.g., p. 18). The study of contrast effects is concerned with *transitions* between different reinforcing events, or stimuli correlated with different reinforcing events; in these cases, the evidence available since the time of Pavlov has suggested that quite different effects occur. The transition from a situation associated with aversive reinforcement or with the omission of appetitive reinforcement to one associated with appetitive reinforcement, is often accompanied by the facilitation of appetitive behaviour. An animal exposed to shock and then permitted to eat often shows unusually vigorous consummatory activity; an animal that has just found no food in a previously reinforced goal box will often run more rapidly down another alley to a second goal box (p. 266). Pavlov (1927, p. 188) attributed phenomena analogous to these, observed in classical conditioning, to a rebound effect. The terminology and theoretical analysis were borrowed from Sherrington

(1906). Sherrington observed that a reflex whose expression had been temporarily inhibited by the elicitation of an antagonistic reflex, would reappear with added vigour when the stimulus eliciting the antagonistic reflex was terminated. Apparently similar phenomena have been observed with more complex sequences of activities. Kennedy (1965), for example, has analysed the flying and settling behaviour of aphids in these terms: the elicitation of settling may inhibit flying, but after the removal of the eliciting stimulus, flying may be resumed with enhanced vigour.

Pavlov applied Sherrington's analysis to a number of effects which he described as instances of induction. The presentation of a CS− immediately before a trial with a CS+, Pavlov found, would sometimes result in an increase in the magnitude of the CR elicited by the CS+. The omission of an appetitive reinforcer, or the presentation of a stimulus signalling such omission, was assumed to inhibit appetitive centres; the removal of such stimuli would release the appetitive centre from the inhibitory influence and the activity of this centre would then rebound to a level temporarily above normal. Although this Pavlovian analysis has been viewed with cautious favour by a few recent investigators (e.g., Williams, 1965; Nevin and Shettleworth, 1966), the most generally accepted analysis of these phenomena, most explicitly elaborated by Amsel (1958, 1962, 1967, 1971), has been in terms of a Hullian theory of general drive. According to Amsel, aversive motivation produced by shock or the omission of food, like any other source of motivation such as hunger or thirst, feeds into a general motivational state which unselectively potentiates any pattern of ongoing behaviour. It is, of course, assumed that specific drive states have specific stimulus properties which may elicit, or become associated with, specific patterns of behaviour: such an assumption is clearly required by the demonstrable fact that depriving an animal of food is more likely to increase eating than sexual activity. But the energizing effects of all drives are entirely unselective and in this sense increases in fear or frustration may facilitate appetitive behaviour just as much as an increase in hunger. A variety of phenomena, therefore, including Pavlovian induction, the frustration effect and transient contrast effects, as well as the effect of shock on subsequent eating, are all attributed to the operation of such a general motivation system. Since such an idea runs counter to much of what has been assumed in earlier chapters, a digression to consider the merits of such a theory may be in order at this point.

B. Digression on the Theory of General Drive

The case against Hull's theory of general drive has been presented by several recent writers (Estes, 1958; McFarland, 1966b; Bolles, 1967;

Hinde, 1970). As we shall see, virtually the only counter-argument against this case relies upon just that set of data, such as the frustration effect, which can be interpreted as a rebound from inhibition.

1. *General Drive and General Activity*

Rats deprived of food or water have been said to show an increase in "general activity", and this general activity, by verbal sleight of hand, is said to be motivated by general drive. But the only defining characteristic of general activity has been that it is behaviour that activates the experimenter's recording device; there is no particular reason to suppose that the general activity recorded in pigeons during a CS signalling food (Slivka and Bitterman, 1966) is the same as the general activity recorded in goldfish during a CS signalling shock (Bitterman, 1964). In fact, changes in activity during deprivation depend on the nature of the apparatus, of the deprivation and of the subject.

Rats deprived of food and water increase the amount of time spent running in a wheel, but may show little or no change in activity in a stabilimeter cage (Weisner *et al.*, 1960; Treichler and Hall, 1962). Some records of activity in cages may actually show a slight decline during food deprivation (Strong, 1957), and it is common to observe a decline during water deprivation (Campbell and Cicala, 1962). If increases in activity during deprivation are recorded in stabilimeter cages, they are usually due to increases in responsiveness to changes in the environment; they are increases in reactivity, not in activity *per se* (Campbell and Sheffield, 1953; Teghtsoonian and Campbell, 1960). Finally, animals other than the rat show entirely different patterns of activity change during deprivation: Campbell *et al.* (1966b), for example, found that deprivation of food increased activity in chicks, had no effect in rabbits and produced a substantial decline in activity in hamsters. Hibernators, one should not be surprised to learn, do not show any increase in activity during deprivation (Cornish and Mrosovsky, 1965; Mrosovsky, 1971, pp. 61–4).

2. *Summation between Hunger and Thirst*

If hungry rats are trained to press a lever for food, their rate of responding during a subsequent extinction session will be a direct function of their current level of hunger (e.g., Perin, 1942; Koch and Daniel, 1945). Performance that is depressed by a decrease in hunger, however, can sometimes be facilitated by an increase in thirst; rats trained under food deprivation, but tested while satiated for food, may respond more rapidly if deprived of water (e.g., Webb, 1949; Brandauer, 1953). Such apparent summation between hunger and thirst drives has been attributed to their common effects on a state of general drive. The interpretation, however, is unwarranted. Verplanck and Hayes (1953) and McFarland (1964) have shown that hunger and thirst interact: animals deprived of water, for example, reduce their food intake and are therefore in a state of food deprivation.

Grice and Davis (1957) have shown that it is this thirst-induced increase in hunger that is responsible for summation effects. They trained hungry rats to press a panel for food and then tested them in extinction. Only one group was deprived of food during the interval between acquisition and extinction. The remaining three groups were all given free access to food; one of the three received unlimited water and was therefore

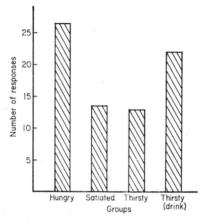

FIG. 7.1. The number of responses made during a session of extinction, when responding has been established under food deprivation, and subjects are tested either hungry, satiated, or thirsty. The fourth group, thirsty (drink), was deprived of water until 30 min before the test, and then permitted to drink, although now without food available. (*After* Grice and Davis, 1957.)

presumably satiated at the time of the test; one was deprived of water and was presumably thirsty; while the last group was deprived of water until 30 min before the test, at which point water was provided, but food removed.

The results of the experiment are shown in Fig. 7.1. The first point to note is that Grice and Davis were unable to replicate the findings of either Webb or Brandauer; satiated animals responded less in extinction than those still deprived of food, but there was no suggestion that this difference might be attenuated by making animals thirsty at the time of the test. The most important results, however, are those of the fourth group; these animals were deprived of water throughout most of the interval between acquisition and extinction, but, since they were permitted access to water before the test, they were presumably not thirsty during the test. The fact that they performed more vigorously than animals deprived of water throughout the interval, therefore, implies that thirst must inhibit rather than energize responses motivated by hunger. Thus the difference between the performance of this fourth group and that of the satiated group must mean that deprivation of water caused a reduction in food intake and therefore facilitated test performance by effectively increasing the subjects' level of hunger. Both of these implications are entirely consistent with the analysis suggested by Verplanck and Hayes and by McFarland.

3. *Summation of Appetitive and Aversive Drives*

McFarland (1966b) concluded that the only evidence favourable to the notion of a general drive came from demonstrations of the summation of appetitive and aversive drives. But most of the evidence supposed to show such summation comes from studies of the transition from one to the other, and this evidence is entirely consistent with a rebound analysis. The presentation of shock, or of stimuli associated with shock, while animals are *currently* engaged in appetitive activity, uniformly suppresses that activity. Studies of conditioned suppression (e.g., Annau and Kamin, 1961) and punishment (e.g., Boe and Church, 1967; Camp et al., 1967; Church, 1969) have observed suppression rather than facilitation of appetitive instrumental responding, down to the lowest intensity of shock used. Consummatory behaviour is also suppressed by stimuli associated with mild shock (e.g., Amsel and Cole, 1953).

There is equally little evidence of reliable facilitation of aversively motivated behaviour by increases in appetitive drives. Amsel (1950), and Dinsmoor (1958) found that hungry rats learned to escape from shock no faster than satiated rats; Frey and Sheldon (1970) found that food deprivation did not facilitate eyelid conditioning in rabbits; and in an extremely thorough series of experiments, Misanin and Campbell (1969) found no evidence that hunger or thirst had any effect on the rats' sensitivity or reactivity to shock, or on rate of avoidance learning. In two recent studies, indeed, food deprivation has been found to interfere with aversively motivated behaviour: Meyer et al. (1969) found that rats were slower to escape from shock when hungry and Leander (1973) found that food deprivation significantly suppressed the rate of free-operant avoidance responding.

The only evidence that deprivation of food may ever facilitate aversively motivated behaviour has come from two studies of escape from stimuli previously associated with shock (Amsel, 1950; Ley, 1965). In neither study, it should be noted, could any effect on the *learning* of the escape response be detected. Amsel's rats had already learned to escape from shock and this learning was unaffected by the level of appetitive drive: the only effect observed was that an increase in hunger retarded extinction of the escape response. Inspection of Ley's data suggests that deprivation had no effect on performance over the first 25 trials of training, but simply prevented extinction of the response over a second block of 25 trials. In view of the previously noted effects of deprivation on activity in a running wheel, it would be hazardous to attribute this observed increase in the persistence of a running response to any summation of drive states.

A recent study on the effects of hunger on the startle response in rats reinforces this sceptical attitude. Fechter and Ison (1972) found that the magnitude of the startle response elicited by a loud tone was reliably decreased by increases in hunger or thirst. "Reflex amplitude", they observed, "should provide the least complicated behavioral measure on which the drive mechanism can operate. That energization of the startle reflex by food deprivation cannot be demonstrated is, therefore, a particularly pronounced theoretical failure" (p. 122).

4. *Effects of Drugs*

Aversive stimuli affect appetitive behaviour in different ways depending on whether they are presented before, or during, the opportunity to engage in appetitive activity. The importance of this distinction is further suggested by evidence that certain drugs may affect behaviour in one situation but not in the other.

Alcohol and barbiturates greatly interfere with escape from, or avoidance of, aversive stimuli and with the suppression of appetitively reinforced behaviour produced either by shock or by nonreinforcement (p. 265). If the facilitation of appetitive behaviour that results from immediately prior exposure to shock or nonreinforcement were a consequence of the general motivational effects of these events, then such facilitation should also be abolished by these drugs. A long series of experiments has sought in vain for evidence to support this prediction: neither alcohol nor sodium amytal has shown any sign of attenuating or abolishing the frustration effect (Ludvigson, 1967; Ison *et al.*, 1967; Freedman and Rosen, 1969; Gray, 1969).

This series of failures has been recognized as posing a substantial problem for the traditional version of frustration theory. The solution suggested by the present analysis is to argue that the frustration effect is not produced by the process responsible for escape or punishment, but by a rebound from inhibition. Why such a process should not be affected by these drugs may well be puzzling, but to be able to point to a difference between those results affected by drugs and those not so affected, is at least a first step towards resolving the puzzle. The suggestion has the further advantage of being readily testable: it predicts that the facilitation of consummatory behaviour by prior exposure to shock, also a rebound effect, should also be unaffected by these drugs.

5. *Conclusions*

This has been a recognizably one-sided review of general drive theory. If it has failed to convince, it may as well be admitted that rejection or acceptance of the theory has little effect on the analysis of many contrast effects. Whether appetitive behaviour is temporarily facilitated by a previously elicited aversive state, because the residue of that state contributes to general drive or because its removal releases appetitive motivation from inhibition, leaves much of the application of frustration theory unaffected. But we shall come across exceptions to this statement. There is some evidence that transient effects may be symmetrical: in discriminative conditioning not only is responding to S+ temporarily facilitated by prior exposure to S−, but responding to S− is temporarily inhibited by prior exposure to S+. If the former effect is to be attributed to a temporary increase in general drive, we are left without an explanation of the latter. But both can be thought of as rebound effects—the first as an increase in excitation following release from inhibition, the second as an increase in inhibition following release from excitation. The temporary suppression in responding to S− in this case is an example of a negative contrast effect: as will be noted, any negative contrast effect (and there are other varieties, more reliably obtained) pose problems for the view that the omission of appetitive reinforcement is frustrating, and that frustration inevitably energizes ongoing behaviour. In what follows, it will be assumed that frustration is a synonym for a transient increase in excitation produced by a release from inhibition.

II. TRANSIENT CONTRAST EFFECTS

Contrast effects in learning were first observed by Pavlov (1927) who called them positive and negative induction. The term contrast effect (introduced by Skinner, 1938, p. 175) is perhaps more appropriate: the one feature common to all results subsumed under this name is that exposure to more than one condition of reinforcement exaggerates the difference between the performances maintained by each condition in isolation. Positive contrast occurs when exposure to nonreinforcement, or some less favourable condition of reinforcement, causes a subject to perform more vigorously for a reference reinforcer. Negative contrast occurs when exposure to a more favourable condition of reinforcement suppresses performance maintained by a reference reinforcer. Several varieties of

TABLE 7.1

Varieties of Contrast Effect

Classification		Definition	Examples
Transient Contrast	Positive	Performance to S_1 is enhanced by immediately preceding exposure to S_2 signalling a less favourable condition of reinforcement.	Pavlovian positive induction. Free-operant transient contrast. Double-alley frustration effect.
	Negative	Performance to S_1 is suppressed by immediately preceding exposure to S_2 signalling a more favourable condition of reinforcement.	Free-operant transient contrast.
Simultaneous Contrast	Positive	Performance to S_1 is enhanced if the reinforcement signalled by S_2 is less favourable than that signalled by S_1	Free-operant behavioural contrast: rate of responding to S_1 signalling a VI schedule increases if schedule signalled by S_2 changes from VI to extinction.
	Negative	Performance to S_1 is suppressed if the reinforcement signalled by S_2 is more favourable than that signalled by S_1.	Discrete-trial simultaneous contrast: performance for a small reward in S_1 is suppressed if subjects receive a larger reward in S_2.
Successive Contrast	Positive	Performance for a given reinforcer is enhanced by prior exposure to a less favourable reinforcer.	No unequivocal instances.
	Negative	Performance for a given reinforcer is suppressed by prior exposure to a more favourable reinforcer.	Crespi's "depression effect": rats shifted from a large to a small reward run less vigorously to the small reward than subjects that have never received the large reward.

contrast effects may be distinguished: the defining characteristics of the three main varieties are summarized in Table 7.1.

The first category to be considered is that of transient contrast effects. They are most easily illustrated by considering a subject trained on a discrimination between S_1, signalling a high probability of reinforcement, and S_2, signalling a low probability of reinforcement. During the course of discrimination training, performance to S_1 may be enhanced immediately following an S_2 trial, thus defining a transient positive contrast effect, and performance to S_2 may be suppressed immediately following an S_1 trial, thus defining a transient negative contrast effect. There are two senses in which these effects appear to be transient. First, they are usually present only while the subject is mastering the discrimination, eventually disappearing if sufficient training is given. Secondly, they depend upon a short interval elapsing between exposure to S_1 and exposure to S_2: performance to S_1 is enhanced only immediately following exposure to S_2.

A. Transient Contrast in Discriminative Classical Conditioning

Pavlov (1927, pp. 189–90) described an experiment by Foursikov in which a dog was given differential conditioning between a CS+, consisting of a metronome ticking at 76 beats per minute, and a CS− consisting of the metronome set at 186 beats. Trials with CS+ and CS− alternated at intervals of 10 to 20 min. When the presentation of CS+, on occasional test trials, was immediately preceded by CS−, the magnitude of the CR increased by 20–30 per cent. The effect occurred only with an interval between CS− and CS+ of less than 2 min and disappeared after prolonged discrimination training (although reappearing if a new CS·−, closer to CS+, was substituted for the old).

Pavlov did not, unfortunately, use this straightforward procedure to measure negative contrast. Although the probability of responding to a stimulus signalling nonreinforcement might be too low to detect evidence of further suppression, the procedure might still be satisfactory, provided that CS− simply signalled a lower value of reinforcement than that signalled by CS+. The procedure Pavlov did use to suggest that there was a process of negative induction symmetrical to positive induction, was less direct, and more open to objection. He measured the rate of excitatory conditioning to a former CS−, and reported that the ease of conditioning was markedly reduced if reinforced CS− trials were alternated with reinforced trials to the former CS+: excitatory conditioning to CS− is a reasonable enough measure of its inhibitory properties (p. 32) and the observed result is consistent with the idea that presentation of CS+ increases inhibition to CS−. Unfortunately, it does not necessarily imply any contrast-like process: since acquisition training involves alternate presentations of CS+ and CS−, the continued presentation of CS+ preserves some of the features of acquisition which are lost if CS+ is never presented. The results, therefore, are attributable to generalization decrement caused by the omission of CS+, rather than to a contrast effect caused by its presence. As the authors pointed out, the same interpretation can be applied to the comparable results of a study of instrumental discrimination learning in rats reported by Leonard et al. (1968).

A similar procedure can be used to study positive contrast in extinction: extinction of responding to nonreinforced presentations of CS+ may be compared when such trials are alternated with CS− trials and when CS− is omitted. A positive contrast effect has been reported for appetitive conditioning in rats (Senf and Miller, 1967), eyelid conditioning in rabbits (Frey and Ross, 1967) and instrumental discrimination learning in rats (Leonard et al., 1968; Ison and Krane, 1969). Once again, however, the result is, on the face of it, readily attributed to generalization decrement: continued presentation of CS− preserves some of the conditions of acquisition and should better maintain responding to CS+. Senf and Miller attempted to control for this possibility by giving subjects, at the end of discrimination training, a series of 16 trials to CS+ only. This may help to decrease any generalization decrement caused by the transfer from CS+ and CS− trials to CS+ trials only, but it is hard to see why it should eliminate it.*

Although, therefore, classical conditioning reveals reliable transient positive contrast, there is no satisfactory evidence of an equal and opposite negative effect. This should not be construed as a denial of the possibility of such an effect: it is simply that the appropriate experiments have not been done.

B. Transient Contrast in Instrumental Discrimination Learning

Transient contrast effects, both positive and negative, have been reported in experiments on instrumental discrimination learning, in studies using pigeons by Boneau and Axelrod (1962) and Nevin and Shettleworth (1966), and in studies using rats by Williams (1965) and Bernheim and Williams (1967). The results of one of Nevin and Shettleworth's experiments are shown in Fig. 7.2. They trained pigeons on a conventional multiple schedule, with alternating 3 min presentations of a green key

---

* Krane and Ison (1971) have shown more convincingly that rats will continue to run faster to S+ in extinction if trials with S− are interspersed with nonreinforced trials with S+, and that this effect cannot be attributed to generalization decrement. They have also shown, however, that it has nothing to do with a contrast effect as here defined.

light signalling a VI 2 min schedule of reinforcement, and a red key light signalling a VI 6 min schedule. Fig. 7.2 shows, for two subjects, the rate of responding during successive 30 sec periods in each stimulus: at the beginning of the red stimulus, rate of responding was temporarily lower than normal, thus defining a negative contrast effect; while at the outset of the green stimulus, responding was temporarily enhanced, thus defining a positive contrast effect. It was possible to detect negative contrast, since S— was correlated not with extinction but with a low probability of reinforcement sufficient to maintain some responding. Bernheim and Williams,

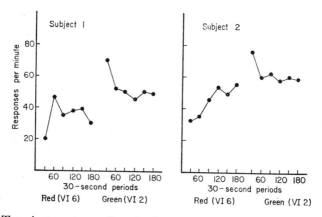

FIG. 7.2. Transient contrast effects in free-operant discrimination learning. Rate of responding is elevated at the onset of the stimulus signalling a more favourable schedule of reinforcement, and depressed at the onset of the stimulus signalling a less favourable schedule. (*After* Nevin and Shettleworth, 1966.)

indeed, found that negative contrast occurred more readily in their situation than did positive contrast.

A temporary elevation of instrumental responding following a period in which little responding occurred, or a temporary suppression of responding following a period of rapid responding, might be attributable to some fatiguelike process. Both Nevin and Shettleworth (1966) and Williams (1965) attempted to rule out such an explanation: the former experimenters observed transient negative contrast in a VI 5 schedule following exposure to a stimulus correlated with a higher probability of reinforcement, but which, owing to the use of a DRO schedule, controlled a lower rate of responding; Williams delivered free reinforcements in one component, locking the rats' running wheel to prevent responding, and still observed transient negative contrast in the next component. Neither of these procedures, however, rules out a second possible interpretation.

Following exposure to S+, subjects may be temporarily more satiated than after exposure to S— and differences in rate of subsequent responding may simply reflect these temporary changes in motivation. Williams (1965) used brain stimulation as the reinforcer in order to minimize this possibility, but one feature of the results of all these experiments provides the best evidence against any such interpretation. The nature of the immediately preceding stimulus did not have a permanent effect on the rate of responding at the beginning of the following stimulus: transient contrast effects did not appear until animals had received some exposure to the discriminative contingencies and the effects eventually disappeared with continued training. If changes in performance were due to momentary changes in motivation, they should presumably have appeared from the beginning of training and have been maintained indefinitely.

The transience of these contrast effects, therefore, provides the best reason for refusing to attribute them directly to changes in fatigue or motivation. They are also transient in a second sense: the effects of one condition of reinforcement on subsequent performance may be observed only immediately following exposure to that condition. Several lines of evidence support this conclusion. As can be seen from Fig. 7.2, it is the rate of responding at the beginning of each stimulus that is affected by exposure to the alternative stimulus. Boneau and Axelrod (1962) showed that rate of responding to S+ was higher when S+ followed S— than when S+ followed S+, and Wilton and Clements (1971b) have shown that rate of responding at the beginning of an S+ period is a direct function of the length of the preceding S— period. More direct evidence of this sense of transience has been provided by Mackintosh et al. (1972): they trained pigeons on a multiple (VI, extinction) schedule, with some subjects given a 10 sec interval between stimulus presentations and other subjects a 60 sec interval. Transient positive contrast, defined as elevation in rate of responding at the outset of S+ following an S— trial, was observed only in subjects trained with the short interval between stimuli.

Although the positive and negative contrast effects were equal and opposite in Nevin and Shettleworth's experiment with pigeons, in rats it has proved easier to obtain negative contrast than positive contrast. Williams (1965) found no reliable evidence of positive contrast at all; Bernheim and Williams (1967) found positive contrast only in rats trained on the more difficult of two discriminations; the more difficult discrimination resulted in a larger negative contrast effect than the easier discrimination, but reliable negative contrast still occurred in this second case. Although the reason for this difference between rats and pigeons is not entirely clear, we shall have reason to refer to it again.

C. Frustration Effects

We have already described the frustration effect: the speed with which rats run in an alley to a goal box containing food is increased if they have just been exposed to a goal box in which the expected food was not present (p. 266). With very few exceptions, the apparatus, procedures, and theoretical framework employed in studies of the frustration effect have tended to keep it isolated from the other contrast effects under consideration here. The frustration effect, however, is surely a transient positive contrast effect: exposure to nonreinforcement facilitates subsequent, reinforced responding. The effect is transient in both senses: repeated

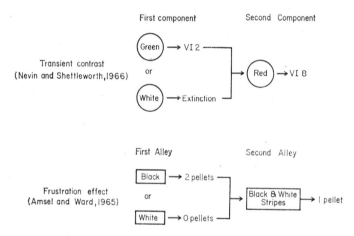

FIG. 7.3. Schematic outline of the design of Nevin and Shettleworth's study of transient contrast in free-operant discrimination learning by pigeons, and of Amsel and Ward's study of the double-alley frustration effect in rats. (*After* Nevin and Shettleworth, 1966; *and* Amsel and Ward, 1965.)

exposure to nonreinforcement in the first goal box produces an eventual decrease in speed in the second alley (McHose, 1963); and the effect disappears if too long an interval elapses between exposure to nonreinforcement and opportunity to run the second alley (Robinson and Clayton, 1963; MacKinnon and Amsel, 1964).

The parallel between transient positive contrast and the frustration effect can best be appreciated by comparing one of Nevin and Shettleworth's experiments (1966) with a study by Amsel and Ward (1965). The schematic designs of the two experiments are shown in Fig. 7.3. Nevin and Shettleworth trained pigeons on a "three-ply" multiple schedule (i.e. one involving three stimuli associated with three different schedules of reinforcement): periods in which the key was red and a VI 8 min schedule was in effect were alternated with periods of either a green key signalling a VI 2 min schedule, or a white key signalling extinction. Negative contrast was observed when rate of responding to the red key, signalling the VI 8 schedule, was temporarily suppressed following exposure to the green key, signalling the VI 2 schedule; positive contrast was shown when rate of responding to the red key was temporarily enhanced when the preceding trial had been an extinction trial signalled by white. Amsel and Ward employed a special double alleyway in which the first alley was either black or white: reinforcement was available in the first goal box only when the first alley was black. They observed that speed of running in the second alley was faster following nonreinforced trials in the white alley than reinforced trials in the black alley. In both experiments, sufficient training on the discrimination (between green and white keys for the pigeons, or black and white first alleys for the rats), abolished the contrast effects: performance in the constant component (the red key or the second alley) was eventually unaffected by the nature of the immediately preceding component.

The design of Amsel and Ward's experiment differs from that usually employed in studies of the frustration effect in the double alley. The standard procedure (Amsel and Roussel, 1952) employs a constant first alley, in which reinforced and nonreinforced trials are randomly alternated: speed of running in the second alley is faster following nonreinforcement in the first alley and, since there is no basis for discriminating when reinforcement will be omitted, the frustration effect persists indefinitely.*

In many experiments of this design, animals are at first consistently reinforced for responding in both alleys, and only later is reinforcement occasionally omitted in the first alley. While this procedure may not create problems in experiments employing a double alley, its use in a number of free-operant studies makes their data impossible to interpret. Staddon and Innis (1969), Staddon (1970), and Kello (1972) have argued that the occasional omission of reinforcement after exposure to consistent reinforcement may not only create a temporary state of frustration, it may also change the stimulus situation, and thus, by generalization decrement, tend to disrupt subsequent performance. In a double alley, such an effect would presumably serve to mask a frustration effect; in many free-operant situations, however, the delivery of one reinforcer may reliably signal that the

---

* The reason why the frustration effect disappeared in McHose's (1963) experiment is that animals were initially given consistent reinforcement in the first alley, and then consistently nonreinforced. Under these conditions, the continued omission of reinforcement was eventually no longer frustrating.

next reinforcement will not be available until later, and may therefore control a temporarily low rate of responding. This is particularly true of FI schedules. The observation reported by Scull *et al.* (1970), therefore, that the occasional omission of reinforcement may enhance rate of responding at the beginning of an immediately succeeding FI component, cannot be unambiguously interpreted as a frustration effect.

Even if this problem does not arise, there remain difficulties of interpretation in the standard double-alley procedure. As was first pointed out by Seward *et al.* (1957), any difference in speed of running in the second alley, following reinforcement or nonreinforcement in the first alley, may not be a consequence of facilitation of performance by nonreinforcement, but of a depression of performance by reinforcement: the consumption of food in the first goal box may result in a temporary state of satiation. Although this may not seem a serious danger when the reinforcement in the first alley is relatively small, there is direct evidence of a temporary, satiation-produced, decrement in speed of running in an alley when rats have just consumed a large number of pellets (e.g., Fox *et al.*, 1970). It follows that this design cannot be used to compare the effect of differences in the amount of initial reward on the frustration effect. If animals initially run to a large reward show a greater increase in speed following the omission of reinforcement than animals run to a small reward (Bower, 1962), this need not reflect greater frustration following omission of the large reward, but greater suppression following its consumption.

The procedure used by Amsel and Ward (1965) rules out this type of interpretation because the frustration effect in that experiment varied in magnitude during the course of the experiment. As was noted above in the case of transient contrast effects in free-operant discriminations, if the difference in speed of running in the second alley, following reinforcement or nonreinforcement in the first, were a consequence of the temporarily satiating effects of reinforcement, then they should have been in evidence from the first day of training and continued unabated throughout the experiment. In fact, reliable differences did not appear until animals had received about 20 to 30 training trials and they disappeared after the discrimination in the first alley had been learned. The procedure more commonly employed to control against this possible interpretation of the frustration effect is the between-subject design introduced by Wagner (1959). Wagner compared the performance of two different groups in the second alley. One group was never reinforced in the first alley, while the other was initially reinforced on all trials in the first alley and then reinforced on only 50 per cent of trials. Animals sometimes reinforced in the first alley ran faster in the second alley following nonreinforcement in the first, than did animals never reinforced in the first. Not only does this result show that it is only the omission of expected reinforcement that produces the frustration effect (p. 266), it also shows that the effect must be due to the facilitation of performance following unexpected nonreinforcement, rather than to the depression of performance following reinforcement.

Wagner's experimental design involves a comparison between different subjects following nonreinforcement in the first alley; the standard design compares the performance of the same group of subjects following nonreinforced and reinforced trials. Although Wagner's between-subject design appears to solve some problems, it has its own disadvantages. In the first place, perhaps because between-subject comparisons are less sensitive, it does not always produce significant results: a series of studies by McHose (1963), Hamm (1967), Ludvigson (1967) and Daly (1968), although reporting reliable within-subject differences, only rarely found evidence that experimental subjects, reinforced on some trials in the first alley, would run faster following nonreinforcement than control subjects that were never reinforced in the first alley.

A second problem is that if two groups of subjects receive different treatments in the first alley, it is possible that differences in their performance in the second alley are a permanent consequence of these different treatments, rather than a transient consequence of a momentary state of frustration or release from inhibition. In the within-subject comparison, on the other hand, differences in speed on different trials *must* be a consequence of the different events that occurred in the first alley on that trial. This argument has been elaborated by McHose (1970). It can be illustrated by consideration of a series of studies which sought evidence for a frustration effect produced by a decrease in the size of reward, rather than by the complete omission of reward, in the first goal box (Barrett *et al.*, 1965; McHose and Ludvigson, 1965; Daly, 1968). In these experiments, different groups were initially trained with different amounts of reward in the first goal box, and were then all shifted to a small reward. If this shift from a large to a small reward was frustrating, then subjects initially trained with large rewards should have run faster in the second alley than animals that always received a small reward in the first goal box. In general, they did not. In all of these experiments, however, animals given the larger rewards during initial training tended during this stage to run in the second alley significantly more slowly than animals given a small reward. While part of this difference may have been caused by a temporary decrease in motivation, it may also have been caused by a permanent negative contrast effect operating between the two alleys. Rats given a large reward in one alley and a small reward in another, run more slowly to the small reward than do control subjects given a small reward in both alleys (p. 387). Since the effect is independent of the distribution of trials, it cannot be ascribed to a temporary satiation effect. If this type of negative contrast operated in the double alleyway, it would have interfered with the appearance of a frustration effect in these experiments by permanently depressing speed of running in the second alley in animals initially given a large reward in the first.

Much the most satisfactory experimental design, therefore, for the demonstration of transient frustration effects in the double alleyway is that used by Amsel and Ward (1965), where a within-subject design is used, but the gradual appearance and eventual disappearance of differences in second-alley speeds rule out any interpretation in terms of temporary satiation following reinforcement. Unfortunately, few experiments have employed this design. In one example, Hug (1970b) showed that if rats were given regularly alternating reinforced and nonreinforced trials in an undifferentiated first alley, they would run down the second alley faster after nonreinforcement than after reinforcement. The frustration effect disappeared, however, as animals learned the alternating pattern of reinforcement in the first alley. In an experiment by Peckham and Amsel (1967), rats were trained with the first alley black or white, one colour signalling a large reward (eight pellets), the other signalling a small reward (two pellets). When reinforcement was occasionally omitted in the first goal box, Peckham and Amsel found some evidence that the omission of the large reward produced faster second alley speeds than did the omission of the small reward. The implication is that the omission of large rewards is more frustrating than the omission of small rewards. (The effect, however, was small and was not replicated in a study by McHose and Gavelek, 1969.) However, there was no good evidence that a frustration effect was produced by the delivery of the small reward itself. Although second-alley speeds were faster on trials when the small reward was delivered in the first goal box than on trials when the large reward was delivered, this difference was apparent from the first day of training and did not decline during the course of 256 acquisition trials. A frustration effect, as in Amsel and Ward's experiment, should have appeared only after subjects had come to expect the large reward and should have disappeared as the animals learned the discrimination in the first alley. The available evidence, therefore, suggests that the occurrence of a small reward in a situation previously associated with a larger reward may not be sufficient to produce a frustration effect.*

* Daly (1972) has shown that rats will learn to escape from a situation associated with a decrease in the size of reward, but, as has been pointed out already (p. 353), there is no reason to suppose that this sort of procedure is measuring the same process as that which produces transient contrast and frustration effects.

Two other results from conventional, within-subject designs are of interest and are not obviously open to alternative interpretations. Stimmel and Adams (1969) and Yelen (1969) have shown that the magnitude of the frustration effect increases with an increase in the number of consistently reinforced trials before omission of reinforcement. Bertsch and Leitenberg (1970) showed that speed of running in the second alley was increased not only by the omission of reinforcement in the first goal box, but also by mild punishment in that goal box. This establishes the continuity of the frustration effect with the after-effects of aversive stimulation in general.

D. Conclusions

There is good evidence that for a short time after the presentation of a stimulus correlated with a high probability of reinforcement, instrumental responding in the presence of a less favourable stimulus may be temporarily suppressed. This may be taken to define a transient negative contrast effect. The evidence for a transient positive contrast effect is more widespread: the amplitude of a classical CR is temporarily increased following exposure to a CS—; and rate or speed of instrumental responding to a stimulus signalling reinforcement is temporarily enhanced, both following the presentation of an S— and following omission of a previously presented reinforcer or the presentation of a mild shock. Not only are similar operations involved in producing Pavlovian positive induction, transient elevations in rate of responding to S+ in instrumental discriminations and the double-alley frustration effect, but all three results are transient or temporary in precisely the same way. Although more evidence is needed to establish the point, the most reasonable hypothesis, as Amsel (1971), for example, has noted, is that they may all be due to some common factor.

Symmetrical positive and negative contrast effects, as we have already suggested, point to a rebound interpretation; unfortunately, it is not yet clear whether symmetrical effects are the rule or the exception. The fact that positive contrast depends upon immediately preceding exposure to nonreinforcement, or to a stimulus correlated with a lower rate of reinforcement, also suggests a temporary rebound from inhibition.

When transient contrast is studied in discriminative situations, and defined as a difference in the rate of responding to one stimulus depending on the nature of the immediately preceding stimulus, there is good evidence that the effect disappears with continued discrimination training (Amsel and Ward, 1965; Nevin and Shettleworth, 1966). As will be noted in the following section, however, in free-operant discriminations at least, the disappearance of this difference is not so much due to an eventual decrease in rate of responding to S+ immediately after an S— trial, as to a gradual increase in rate of responding following another S+ trial. In other words, the transient contrast effect disappears because it becomes permanent.

One of the most important features of transient positive contrast is its dependence on the discriminability of S+ and S—. Pavlov (1927, pp. 193–4) reported that positive induction disappeared once a moderately easy discrimination had been learned, but reappeared when a stimulus closer to CS+ was substituted for the former CS—. Bernheim and Williams (1967) found no evidence of positive contrast when rats were trained on a discrimination between stimuli from different modalities, but found reliable effects in subjects trained on a discrimination between two auditory stimuli. The implication is that transient contrast may depend upon some degree of generalization between S+ and S—: where the two stimuli are sufficiently dissimilar to preclude any generalization, no contrast will occur. If this is true, then the conditions required to produce transient contrast are the exact reverse of those required for demonstrations of conditioned inhibition. Both summation tests and studies examining the retardation of subsequent conditioning have shown that an increase in the discriminability of CS+ and CS— is correlated with an increase in conditioned inhibition. It is questionable, indeed, whether simple discriminative conditioning is sufficient to establish a CS— as a conditioned inhibitor, if that CS— is sufficiently similar to CS+ for there to be substantial generalization between the two (p. 34). It is probable that this is because such a CS— elicits opposed excitatory and inhibitory tendencies, which would be likely to prevent the appearance of reliable inhibitory effects in a subsequent test. If this is so, it suggests that this elicitation of opposed tendencies is necessary for the appearance of contrast effects. There does, however, appear to be a limit to the increase in transient contrast produced by an increase in the similarity between S+ and S—. Mackintosh et al. (1972), for example, observed more reliable transient positive contrast in pigeons trained to discriminate between a vertical and horizontal line, than in pigeons trained to discriminate between a vertical line and one 15° from vertical. In the limiting case, where S+ and S— are identical, there is little evidence that the occasional omission of reinforcement is sufficient to produce an increase in rate of responding on the immediately following trial. It is noteworthy that in the two free-operant studies which have provided the least ambiguous evidence of a frustration effect (Dickinson, 1972; Hughes and Dachowski, 1973), rats were required to respond first on one lever and then on another. Here, the occasional omission of reinforcement for responding to the first lever may produce a transient increase in rate of responding on the second. Where, however, subjects have been required to response on a single lever, the occasional omission of reinforcement may not enhance rate of responding on the following trial (e.g., Platt and Senkowski, 1970b), or if such enhancement is observed (Scull *et al.*, 1970) it may be attributable to the removal of stimuli that controlled a low rate of responding.

It is possible that a transient contrast or frustration effect may occur in nondiscriminative situations. The observation that speed of running in an alley may be enhanced on early extinction trials if trials are very highly massed (e.g., Sheffield, 1949, 1950) is most easily interpreted as an analogue of the double-alley frustration effect (p. 421). But although Amsel (1971) has argued that transient contrast and frustration effects are unaffected by the discriminability of S + and S — (except in the sense that they disappear once a discrimination has been securely established, and will therefore persist longer with harder discriminations), the available evidence does not, on the whole, support his position. In particular, the finding that transient contrast or frustration may eventually decline as the difference between S + and S — is reduced to zero, points to the importance of discriminative factors. This conclusion is not, after all, very surprising: even if the omission of reinforcement produces a transient increase in frustration, it is also an inhibitory operation which must be assumed to decrease the overall response strength associated with a particular stimulus. If the subject is immediately required to respond again to the same situation, this increase in inhibition may well outweigh the temporary increase in frustration. But if the subject is required to respond to a discriminably different stimulus, the generalized inhibition may now be outweighed by the transient frustration and it is under these conditions that a transient contrast or frustration effect will be observed.

III. TRANSIENT AND PERMANENT CONTRAST EFFECTS

A. Definitions

The transient contrast effects so far discussed may be distinguished from more permanent interactions between different reinforcers or schedules of reinforcement. These interactions (hereinafter simply called contrast effects, with the adjective "transient" being added whenever reference is made to the phenomena discussed in the preceding section) may be studied under conditions relatively similar to those used to study transient contrast. In simultaneous contrast experiments, as is shown in Table 7.1 (p. 355), animals are trained to discriminate between two stimuli (S_1 and S_2), each of which signals a particular schedule, probability, magnitude, quality, or delay of reinforcement. Positive contrast is said to occur if animals respond more rapidly to S_1 when S_2 signals a less favourable schedule of reinforcement; negative contrast occurs if performance to S_1 is suppressed when S_2 signals a more favourable schedule of reinforcement. The effects are not transient: they may persist for a considerable number of sessions of discrimination training, and they do not necessarily depend upon the use of a very short intertrial interval, nor upon exposure to S_2 immediately preceding exposure to S_1.

The prototype of the successive contrast experiment is the depression effect discovered by Elliott (1928) and Crespi (1942): animals initially trained, in a constant stimulus situation, to respond for one reinforcer are later shifted to another. Positive contrast occurs when the second reinforcer is more favourable than the first and exposure to the first enhances performance after the shift: negative contrast occurs when performance for a less favourable reinforcer after the shift is depressed by prior exposure to a more favourable reinforcer. Although contrast effects do not usually persist indefinitely in this situation, they do not depend upon a purely transient after-effect of exposure to one condition of reinforcement: studies of negative contrast from shifts in size of reward are routinely performed with animals receiving only one trial a day.

B. Problems of Measurement

To demonstrate that the performance maintained by one condition of reinforcement is affected by exposure to another, it is necessary to compare the behaviour of experimental subjects with some control condition. Successive contrast experiments usually employ separate experimental and control groups: to assess whether a shift from a large to a small reward produces a negative contrast effect, the performance of subjects exposed to such a shift is compared with that of a control group trained throughout the experiment with the small reward. Simultaneous contrast experiments may also employ separate experimental and control groups: if experimental subjects receive a small reward in the presence of S_1 and a large reward in the presence of S_2, the presence of a negative contrast effect can be assessed by comparing their performance for the small reward in S_1 with that of a control group receiving the small reward in the presence of both stimuli.

Simultaneous contrast experiments have often dispensed with separate control groups and studied changes in the performance of individual subjects shifted from control to experimental conditions. In such experiments (typically free-operant studies of positive behavioural contrast), subjects are initially trained on a baseline condition (in which both S_1 and S_2 are correlated with the same schedule of reinforcement) and are then shifted to an experimental condition (in which the same schedule is maintained in the presence of S_1, but S_2 signals a different schedule). Contrast is then assessed by comparing performance in S_1 before and after the change in schedule on S_2. The advantage of such a procedure is the greater sensitivity of within-subject comparisons; the disadvantage is that one may not be justified in attributing any change in performance in S_1 to the change in schedule signalled by S_2, as opposed to the mere passage of time. The solution to this problem, it has usually been assumed, is to show that baseline performance has stabilized before the shift. But this is easier said than done, for it is not clear what range of variation constitutes stability, nor over how many sessions performance must remain within these limits. It is also necessary to note that the principle is one more often piously asserted than actually followed. In typical free-operant studies of contrast, pigeons are trained on a multiple schedule with S_1 and S_2 signalling identical VI schedules; "baseline" training is often continued for only 5 or 10 sessions before the schedule in S_2 is changed, and any ensuing change in performance to S_1 is attributed to the change in the schedule signalled by S_2. But there is excellent evidence that rats or pigeons trained on VI schedules continue to increase their rate of responding for at least 15 or 20 sessions (e.g., Smith and Hoy, 1954; Ellis, 1970; Yarczower, 1970; Mackintosh et al., 1972; Dickinson, 1972). It is true that some of the contrast effects observed in free-operant experiments are so large that it would be foolish to doubt their validity: pigeons shifted from a multiple (VI, VI) schedule to a multiple (VI, extinction) schedule may nearly double their rate of responding in the presence of S_1, and changes of this order of magnitude do not usually occur with continued training on VI alone. But this point is of little avail in the case of more analytic studies of contrast effects, which have programmed less drastic changes in the schedule of reinforcement, and observed correspondingly slighter changes in rate of responding. In the absence of more satisfactory control procedures; the conclusions often drawn from some of these studies can only be regarded as unproven.*

There is one more valid control procedure which has occasionally been used in within-subject experiments (e.g., Reynolds, 1961a; Bloomfield, 1967b). If subjects are returned to the baseline condition and an appropriate return to baseline rates of responding is observed, the attribution

---

* If the attitude taken here seems unduly churlish, it is worth remembering that for many years it was believed that a successive positive contrast effect produced by an increase in size of reward was as reliable a phenomenon as the negative contrast produced by a decrease in size of reward. The basis for this confident belief was the result of comparing the performance of shifted subjects with *extrapolations* of the supposedly stable performance of subjects trained before the shift with a large reward (Crespi, 1942; Zeaman, 1949). It is now clear that this confidence was entirely misplaced (p. 389).

of the change in performance in S_1 to the change in schedule in S_2 becomes considerably more secure.

C. Explanations

Simultaneous contrast experiments have employed both discrete-trial and free-operant procedures; the former have studied variations in probability, magnitude, quality and delay of reinforcement; the latter have rung changes on all the standard free-operant schedules of reinforcement. Successive contrast experiments have mostly employed alleyways and used discrete trials, but have employed the same variations in reinforcement as have discrete-trial studies of simultaneous contrast. The results of this array of experiments have been almost as variable as the procedures involved. Although there have been exceptions, however, the following generalizations, which will be more fully documented later, summarize much of the data. Free-operant studies of simultaneous contrast have routinely observed large and reliable positive contrast effects, but have been somewhat less successful in finding negative contrast effects. Although positive contrast has been produced by a variety of changes in schedule, it does not, for example, seem to be produced by a decrease in the size of reward signalled by S_2. Discrete-trial experiments, on the other hand, have only very rarely observed positive contrast effects, and then only for changes in probability of reinforcement; negative contrast, however, is reliably produced by a variety of manipulations in discrete-trial simultaneous experiments. Successive contrast effects seem to be somewhat more elusive than simultaneous effects: a decrease in size of reward reliably produces a negative contrast effect, but a decrease in quality, or an increase in delay, of reinforcement may fail to result in negative contrast. Positive contrast in successive experiments is as hard to find as in discrete-trial simultaneous experiments.

This diversity of outcome suggests that the single term "contrast effect" is in fact being used to describe a number of different situations which may have little in common. Any attempt to attribute all these results to the operation of a single principle seems a singularly inappropriate theoretical goal. For this reason, it does not seem worth paying too much attention to the most venerable explanation of contrast effects, which says that they reflect a relativistic assessment of the value of reinforcement: large, probable, rewards are said to look better to an animal used to small, improbable, rewards than to an animal used to even larger and more probable rewards (Collier and Marx, 1959; Bevan and Adamson, 1960). Such an account, readily derivable from adaptation level theory, implies that contrast effects should be both general and symmetrical: in the absence of ceiling or floor effects, all situations studied should reveal equal and opposite positive and negative contrast effects. Since they do not, a more profitable theoretical strategy may be to approach each situation independently.

IV. POSITIVE BEHAVIOURAL CONTRAST

In a typical study of positive behavioural contrast, pigeons are initially trained on a multiple schedule, in which S_1 and S_2 signal identical VI schedules; the schedule signalled by S_2 is then changed from VI to extinction, while S_1 continues to signal the original VI schedule; finally, in some experiments at least, the original VI schedule is reinstated in the

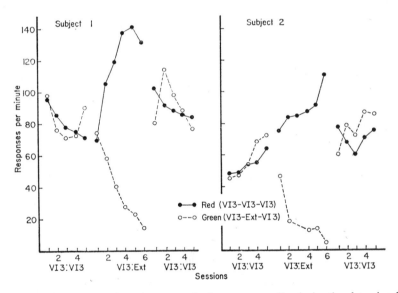

FIG. 7.4. Positive behavioural contrast in free-operant discrimination learning by pigeons. Rate of responding to S_1, signalling a VI 3 min schedule of reinforcement, increases when the schedule signalled by S_2 is changed from VI 3 min to extinction, and decreases when the schedule in S_2 return to VI 3. (*After* Reynolds, 1961a.)

presence of S_2. The results for two birds trained by Reynolds (1961a) on such a sequence are shown in Fig. 7.4. It is worth noting, in the case of the second bird, that had it not been for the decline in S_1 responding when the VI schedule was reinstated in S_2, one might have reasonably argued that the increase in rate of responding to S_1, when extinction was instituted in S_2, was no more than a continuation of the steady increase in rate of

N

responding shown during the VI baseline period. Thus these data, coming from the classic study of behavioural contrast, illustrate the dangers of the traditional, within-subject, control procedure. Relatively few other experiments have studied the effects of re-establishing the original baseline conditions and in some such cases the results have been somewhat disconcerting: Bloomfield (1967b) found no decline in rate of responding to S_1 when the VI schedule was reinstated in S_2 following 20 sessions of multiple (VI, extinction); however, when the birds were returned to multiple (VI, extinction) after 40 sessions of multiple (VI, VI), they showed a further increase in rate of responding to S_1.

A. Differences between Positive Behavioural Contrast and Transient Positive Contrast

The procedure for studying positive behavioural contrast may be exactly the same as that employed in studies of transient contrast: in Reynolds' experiment shown in Fig. 7.4, for example, S_1 and S_2 were regularly alternated, with no intertrial interval. In this case, therefore, the increase in the rate of responding to S_1 may have been a consequence of the immediately preceding presentation of a stimulus signalling nonreinforcement. Without further evidence, therefore, there would be no grounds for assuming that behavioural contrast differs from the transient contrast effects discussed earlier. At the very least, it is clear that a transient positive contrast effect must contribute to the overall increase in rate of responding to S_1.

Transient positive contrast disappears during the course of discrimination training and is measured as a short-term increase in rate of responding to $S+$ immediately following exposure to $S-$. If positive behavioural contrast is not to be reduced to a transient effect, it must be differentiated in either or both of these respects. While it is possible both that behavioural contrast may not be entirely permanent and that it may depend upon some degree of proximity of $S+$ and $S-$ trials, there is clear evidence that it cannot simply be reduced to a transient contrast effect.

Terrace (1966b), for example, has claimed that positive behavioural contrast may decline with extended discrimination training. A study by Bloomfield (1966), however, suggests that Terrace's conclusion may have depended upon his use of a "correction" contingency, according to which presentations of S_2, correlated with extinction, were not terminated until the subject had refrained from responding for 30 sec. When Bloomfield used this procedure, he observed a very rapid increase in rate of responding to S_1 when S_2 signalled extinction, followed by a decline, which still fell short of a return to baseline, within 10 sessions. When S_1 and S_2 were of fixed duration, however, he observed a steady increase in rate of responding to about the same eventual level as that attained after the initial decline in the correction procedure. Two other studies (Rilling et al., 1969; Hearst, 1971b), neither of which used a correction procedure, observed a positive contrast effect maintained in some subjects with little change over 60 sessions of training. The large, temporary increase observed by Bloomfield and Terrace with the correction procedure may well have been due partly to a transient contrast effect: transient effects depend upon the length of exposure to S— on the preceding trial, and the correction procedure ensures that S— presentations will be very long indeed during early sessions of discrimination training.

Whether or not positive behavioural contrast is a permanent consequence of free-operant discrimination training, there is additional evidence that it is at least more permanent than transient contrast. Nevin and Shettleworth (1966) recorded rate of responding throughout S_1 periods during multiple-schedule training, and observed that responding to S_1 was maintained above control levels long after the transient elevation of responding at the beginning of each S_1 period had disappeared. Mackintosh et al. (1972) have confirmed this observation.

Transient positive contrast is a matter of an increase in rate of responding to S_1 immediately following exposure to S_2. Terrace (1966a), training pigeons on a multiple schedule in which S_1 and S_2 were presented in a quasi-random sequence rather than in strict alternation, obtained evidence of a transient contrast effect by showing that, early in training, rate of responding to S_1 was higher after an S_2 trial than after another S_1 trial. Although this difference was no longer evident after 20 sessions, its disappearance was due not to a decline in responding after S_2 trials, but to an increase in rate of responding after S_1 trials. In other words, rate of responding to S_1 was elevated above the baseline level whether or not S_1 had been preceded immediately by S_2. The transient contrast effect disappeared because the increase in rate of responding to S_1 became permanent. Again, these observations have been confirmed by Mackintosh et al. (1972).

Positive behavioural contrast, therefore, is not simply a matter of a transient elevation in rate of responding following exposure to a stimulus correlated with nonreinforcement. It is, however, possible that the more permanent positive contrast effect may depend upon the prior occurrence of the transient effect. There is some evidence that positive behavioural contrast may be affected by the spacing of trials. The typical free-operant study of positive contrast employs an interval between stimulus presentations of 10 sec or less. By giving only a single, one-hour "trial" in each daily session, Bloomfield (1967b) effectively trained pigeons on a multiple schedule with a 23 h interval between S_1 and S_2 trials. Although observing some evidence of positive contrast, the effect was, as Bloomfield noted, very much smaller than the positive contrast obtained with massed trials.

The evidence on this issue is confusing. There are several results which suggest that positive contrast depends upon the proximity of stimuli signalling reinforcement and nonreinforcement. Wilton and Clements (1972) were unable to obtain positive contrast when pigeons were trained, as in Bloomfield's study, at a rate of one trial per day. Wilton and Clements (1971b) found little or no evidence of contrast in S_1 when pigeons were trained on a regular sequence of trials with three separate stimuli. Trials with S_1, signalling a VI schedule, always preceded trials with S_2, signalling extinction, while S_2 always preceded S_3, which signalled a second VI schedule. Reliable contrast occurred in S_3 (the stimulus following non-reinforcement), but there was hardly any contrast in S_1 (the stimulus presented only at the beginning of each day's session). Mackintosh *et al.* (1972) employed a between-subject design to assess positive contrast: a discrimination group received trials with S_1, signalling a VI schedule, randomly alternated with S_2, signalling extinction, while a control group received trials with S_1 alone. This comparison revealed both transient and permanent contrast effects when trials were spaced 10 sec apart, but no sign of either when trials were spaced 60 sec apart.

The results of these studies suggest that positive behavioural contrast might be the consequence of conditioning the high rate of responding to S_1 which is generated by a transient contrast effect. If S_1 and S_2, signalling different probabilities of reinforcement, are alternated in sufficiently rapid succession to produce a transient elevation in rate of responding to S_1 following exposure to S_2, this transient elevation may become conditioned to S_1 and thus be maintained throughout S_1 periods even after the transient, unconditional, effect has disappeared. Although the evidence for this suggestion is sketchy, we shall see that there are several further parallels between the conditions affecting transient and permanent positive contrast.

The argument, however, ignores another body of evidence, which suggests that positive contrast may be simply a consequence of reducing the proportion of time within each session occupied by a stimulus signalling reinforcement. On this account, contrast would be independent of the spacing of trials within a single session, although it might still not be expected to occur under the conditions of Bloomfield's (1967b) study, where a single "trial" occupied an entire session. Ellis (1970) reported significant positive contrast under conditions somewhat similar to those employed by Wilton and Clements (1971b) and where no transient contrast effects could have occurred: pigeons received daily sessions in which all trials with S_1, signalling reinforcement, always preceded all trials with S_2, signalling extinction. This was sufficient to increase rate of responding in S_1 to a higher level than that maintained by a control group which received no S_2 trials. Several experiments with pigeons have shown that rate of responding to S_1, signalling a VI schedule, increases when S_1 periods are simply alternated with periods of time-out (Reynolds, 1961a; Taus and Hearst, 1970; Mackintosh et al., 1972). No transient contrast effect occurs under these conditions: rate of responding to S_1 is not elevated at the beginning of S_1 trials (Mackintosh et al., 1972; Vieth and Rilling, 1972) and yet the overall rate of responding to S_1 is an increasing function of the length of this time-out or ITI (Taus and Hearst, 1970).

Perhaps the most attractive interpretation of these data is to point out that an increase in the interval between trials decreases the proportion of time in each session during which S_1 is present and signals reinforcement. This account suggests a parallel with the findings of Gamzu and Williams (1971) on autoshaping (p. 25): in that experiment, pigeons pecked at an illuminated key only when reinforcement was contingent on illumination of the key; if reinforcements occurred both when the key was dark and when it was illuminated, no pecking occurred at all. Gamzu and Schwartz (1973) have confirmed and extended these findings. They alternated the colour of a pigeon's key light from red to green and back. Without imposing any contingency between pecking and reinforcement at any stage of the experiment, they showed that when both colours signalled equal probabilities of noncontingent reinforcement, relatively little pecking occurred to either colour; but as soon as the probability of reinforcement in one stimulus was decreased, the rate of pecking in the other increased. The implication is that classically conditioned pecking directed toward the response key depends upon a change in the illumination of the key being differentially correlated with reinforcement. Perhaps the increase in rate of pecking, observed in contrast experiments with pigeons, is a consequence of an increase in the validity of S_1 as a signal of reinforcement whenever S_1 is alternated with periods of reduced reinforcement (Rachlin, 1973). Once again, the evidence for this suggestion is extremely sketchy, but it might help to explain why positive behavioural contrast is a phenomenon more reliably found when pigeons are trained to peck keys than when other preparations are used (see below).

B. Conditions Determining Positive Contrast

The standard procedure for studying positive contrast has been to train pigeons on a multiple (VI, extinction) schedule. The unspoken assumption has been that the phenomenon is of greater generality, and in one respect the assumption is valid. If pigeons are trained to peck a key on identical VI schedules in the presence of S_1 and S_2, the rate of responding to S_1 will increase when a variety of changes are made in the schedule signalled by S_2. But when other procedural details are varied, the generality of positive contrast becomes notably less impressive.

1. Subjects and Responses

A pair of studies by Smith and Hoy (1954), and Herrick et al. (1959), are frequently cited as demonstrations of positive contrast effects in rats. Neither can be seriously accepted. Smith and Hoy compared the rates of responding maintained in S_1 by a discrimination group, trained on a multiple (VI, 1·5 min, extinction) schedule, and a control group, for whom both S_1 and S_2 signalled a VI 3 min reinforcement schedule. Although such a procedure equates overall frequency of reinforcement, it does not equate frequency of reinforcement during S_1: the fact that the discrimination subjects responded more rapidly than the controls, therefore, is hardly sufficient evidence of a positive contrast effect. Herrick et al. observed a threefold increase in rate of responding to S_1 during the course of discrimination training, but had provided no more than two sessions of preliminary training before beginning discrimination training. Several more recent experiments have reported evidence of some positive contrast in rats (e.g., Beninger, 1972; Dickinson, 1972; Mackintosh et al., 1972), but the effect has often not been large and may fail to occur in all subjects: Pear and Wilkie (1971) found reliable positive contrast in only four of eight rats shifted from multiple (VI, VI) to multiple (VI, extinction).

It is clear that contrast effects can occur in rats, but the available evidence, although far from satisfactory, suggests that they may be less reliable than those occurring under supposedly comparable conditions in pigeons. But it may well be misleading to suppose that the conditions have been comparable. Both Hemmes (1973) and Westbrook (1973) have reported that when pigeons are required to press a lever rather than peck a key, the development of a discrimination is not accompanied by any increase in rate of responding to S_1. As noted above, these results suggest that positive contrast may depend partly upon the use of a response, such as key pecking in pigeons, which is strongly affected by classical conditioning contingencies. It is possible that an increase in rate of key pecking is a consequence of an increase in the validity of one particular stimulus projected onto the key when that stimulus is alternated with other conditions signalling no reinforcement. The suggestion could readily be tested by seeing whether pigeons would show as large a contrast effect when the stimuli signalling reinforcement and nonreinforcement were diffuse tones as when they were localized on the response key.

2. *Generalization between S_1 and S_2*

Hearst (1969a) trained several groups of pigeons to peck at a key with a vertical line. After several sessions of training with this S_1 alone, each group was trained on a discrimination between the vertical line as S_1, and either a blank key, a horizontal line, or a line 60° or 30° from the vertical as S_2. The average rate of responding to S_1 before the start of discrimination training, and on the day each group reached a criterion of discrimination learning, is shown in Fig. 7.5. The increase in rate of responding to S_1 was a direct function of the degree of similarity or generalization between S_1 and S_2.

Terrace (1963, 1966a) has reported that pigeons may be trained to discriminate between S_1 and S_2 without making any responses to S_2 at all.

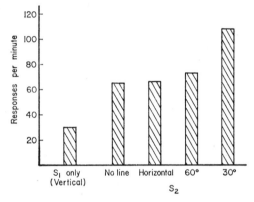

FIG. 7.5. The effect of the similarity of S_1 and S_2 on the magnitude of positive behavioural contrast. The terminal rate of responding to a vertical line is substantially greater when S_2 is a line 30 degrees from vertical, than when S_2 is a horizontal line, or a blank key. (*After* Hearst, 1969a.)

If, immediately after pecking is established to S_1, S_2 is gradually faded in, being presented at first at low intensities and for a short time, then birds may progress to a standard multiple (VI, extinction) procedure between S_1 and S_2 without making any responses to S_2. Terrace found that birds learning this discrimination did not show any increase in rate of responding to S_1 beyond the level of a control group trained with S_1 only. An errorless discrimination, in other words, did not result in a positive contrast effect. Since the procedures employed in establishing the discrimination are such as to prevent any generalized responses occurring in the presence of S_2, it seems natural to suggest that an errorless discrimination is one in which there is minimal generalization between S_1 and S_2. If this is accepted, Terrace's results fall at one end of the continuum implied by

Fig. 7.3: the amount of contrast in S_1 produced by nonreinforcement in S_2, is an increasing function of the generalization between S_1 and S_2.*

This conclusion is the same as that reached for transient contrast effects: Bernheim and Williams (1967) showed that transient contrast was greater when S_1 and S_2 were stimuli from the same modality than when one stimulus was visual and the other auditory. There was, however, evidence that a further reduction in the difference between S_1 and S_2 might decrease rather than increase transient contrast. Mackintosh *et al.* (1972) found less evidence of transient contrast in pigeons when S_1 and S_2 were lines 15° apart than when the difference in orientation was 90°. Moreover, some fragmentary data from studies of the frustration effect in lever boxes suggested that, in the limiting case where S_1 and S_2 are identical, transient contrast may be impossible to detect.

There are some data from studies of behavioural contrast that are consistent with these implications. Mackintosh *et al.* (1972) reported significantly less behavioural contrast in pigeons trained on the 15° line–tilt discrimination than in those trained on the 90° discrimination. It should be noted that the angular separation of the stimuli of the more difficult discrimination here was half that of the most difficult discrimination studied by Hearst (1969a). Pierrel and Blue (1967) undertook a more extensive analysis in an experiment with rats. They trained their subjects on a multiple (VI, extinction) schedule, where S_1 was a 100 db noise and S_2 was either 60 db, 90 db, or 100 db: the final group was thus trained on a mixed, rather than on a multiple, schedule. Rate of responding in S_1, or in equivalent intervals for animals on the mixed schedule, was a decreasing function of the intensity of S_2, with subjects trained on the 100–60 db discrimination responding about three times as rapidly to S_1 as animals trained on the mixed schedule.

Unfortunately, no single study has plotted contrast effects over the whole range of generalization between S_1 and S_2. Although the results of several different studies suggest that contrast may initially increase with increasing generalization between S_1 and S_2 and eventually decline as generalization becomes too great, this conclusion would be very much more secure if it could be established within the context of a single experiment.

3. *Changes in Probability or Magnitude of Reinforcement in S_2*

Positive contrast in S_1 is produced not only by correlating S_2 with an extinction schedule: a reduction to some nonzero probability of reinforcement in S_2, e.g. from a VI 1 to a VI 5 schedule, reliably increases rate of

---

* This analysis of Terrace's data may prove unnecessary: Halliday and Boakes (1974) have observed significant contrast during errorless discrimination learning.

responding to S_1 (Weisman, 1969). The total amount of reinforcement received in S_2 may also be reduced by decreasing the magnitude of each individual reinforcement. The available evidence suggests that this operation, however, does not produce a positive contrast effect. Shettleworth and Nevin (1965) varied size of reward for pigeons trained on multiple (VI, VI) schedules by varying the amount of time that the magazine remained operated. When duration of reinforcement was held constant in S_1, but reduced in S_2, the formation of a discrimination was evident from a decrease in rate of responding to S_2, but no positive contrast occurred in S_1. Although Kramer and Rilling (1969) found an increase in rate of responding to S_1 when access to reinforcement in S_2 was decreased from 4·0 to 0·5 sec, it is more than probable that the pigeons simply failed to obtain any food on at least some occasions when the magazine was operated for only 0·5 sec; this study, therefore, may well have inadvertently varied probability of reinforcement. In a study with rats, in which either probability or magnitude of reinforcement was varied in S_2, Mackintosh *et al.* (1972) observed positive contrast in S_1 only when S_2 was correlated with a lower probability of reinforcement than S_1, not when it signalled equally probable, but smaller, rewards.

If this discrepancy between probability and magnitude of reinforcement holds up under further investigation, it suggests a further parallel between positive behavioural contrast and one kind of transient contrast: there is no evidence that a decrease in the size of reward delivered in the first goal box of a double alleyway produces a frustration effect in the second alley (p. 364).

4. Changes in Probability of Reinforcement or Responding in S_2

The change from a multiple (VI, VI) to a multiple (VI, extinction) schedule arranges a decrease in the probability of reinforcement in S_2, and this decrease in reinforcement results in a decrease in responding to S_2. This observation has prompted the thought that the normal procedure for producing contrast confounds two potentially independent changes and that it is the task of an experimental analysis to unravel the two. A disproportionate amount of the research on free-operant positive contrast, accordingly, has been undertaken with a view to deciding whether increases in rate of responding in S_1 are caused by a decrease in rate of responding, or by a decrease in rate of reinforcement, in S_2. The large number of experiments performed to settle this issue has pointed unerringly to a single conclusion—that the question is the wrong one to ask. Positive contrast in S_1 has occurred in the absence of reduction in reinforcement and also in the absence of reductions in responding in S_2; it has also failed to occur when one or other has been reduced. Neither condition is sufficient, and neither is necessary, for the appearance of positive contrast.

Positive contrast may fail to occur when S_2 signals no reinforcement, but the discrimination between S_1 and S_2 is learned without errors; it may also fail to occur when S_2 signals a smaller reward rather than a reduced probability of reward. These two results suggest that a reduction in reinforcement in S_2 is not always a sufficient condition for a contrast effect in S_1. That this is not a necessary condition is suggested by the fact that contrast may occur when S_2 signals the same probability of reinforcement as S_1, but responses in S_2 are punished (Brethower and Reynolds, 1962; Terrace, 1968); a small contrast effect in S_1 has also been observed when the probability of reinforcement in S_1 and S_2 is equated, but the schedule in effect in S_2 is a DRL schedule (Terrace, 1968; Weisman, 1969).*

Even if a change in reinforcement in S_2 is neither sufficient nor necessary to produce contrast in S_1, there is good reason to believe that it is a more important variable than a change in the rate of responding in S_2. The use of DRL or DRO schedules has been a popular strategy for those intent on manipulating the probability of reinforcement and the probability of responding independently. Although responding on a VI schedule to S_1 may increase when S_2 signals a DRL or DRO schedule, the effect is small at best and does not occur in all subjects (Terrace, 1968), or in all experiments (Reynolds, 1961a; Nevin and Shettleworth, 1966; Nevin, 1968). In the only study which has systematically compared the relative importance of rate of reinforcement and rate of responding on DRL schedules in S_2 as determinants of contrast in S_1, Bloomfield (1967a) found that rate of responding in S_1 (signalling a VI schedule) increased as rate of reinforcement decreased in S_2, but decreased as rate of responding decreased. The studies by Reynolds and Nevin, moreover, make it clear that no contrast occurs in S_1 when the probability of reinforcement on the DRL schedule in S_2 is higher than the probability of reinforcement in S_1, no matter how slowly subjects respond in S_2.

Both Reynolds (1961b) and Bloomfield (1967a) have also assessed the relative importance of rate of responding and rate of reinforcement in S_2 by arranging various FR schedules. Again, positive contrast occurs in S_1 when the rate of reinforcement on the FR schedule in S_2 declines, even though the rate of responding in S_2 may be substantially higher than in S_1. Hemmes and Eckerman (1972) observed some increase in rate of responding to S_1, signalling a VI schedule, when the schedule signalled

---

* It is in cases like this that the problems with the typical within-subject design become more apparent. The increases in rate of responding to S_1 when pigeons were shifted from a multiple (VI, VI) to a multiple (VI, DRL) schedule were at best small, the baseline rates of responding in Terrace's experiment were extremely unstable, and neither Terrace nor Weisman returned their subjects to the baseline (VI, VI) schedule to see if responding to S_1 declined again. The evidence for positive contrast when responding is reduced by a DRL schedule, therefore, is less than overwhelming.

by S_2 was changed from VI to DRH, thus ensuring a very high rate of responding in S_2. At the very least, these results suggest that changes in rate of responding in S_2 are neither necessary nor sufficient to produce contrast in S_1.

One procedure, which has been claimed to produce results contradicting this conclusion, was developed by Brownstein and Hughes (1970). They trained pigeons on multiple (VI, VI) schedules and then added a second stimulus during S_2 periods which signalled the availability of reinforcement. The addition of this signal reduced rate of responding during S_2, for subjects now responded only when the added signal was presented. Since rate of responding during S_1 increased, the authors argued that contrast in one component was caused by a decrease in responding in the other. It would seem more reasonable to argue, however, that the effect of introducing the food signal into S_2 periods was to change S_2 from a stimulus signalling a given probability of reinforcement into one signalling extinction. The probability of reinforcement during S_2 alone declined to zero and since S_2 alone occupied most of the experimental time when S_1 was not presented, it is not surprising that contrast occurred in S_1.

Three additional sets of results show that a decline in rate of responding in S_2 is neither a sufficient nor a necessary condition for contrast in S_1. Contrast occurs in brain-lesioned rats and pigeons even when, as a consequence of the lesion, they fail to suppress responding during S_2 (Macphail, 1971; Dickinson, 1972). Conversely, the administration of chlorpromazine may not affect the suppression of responding to S_2, but does abolish any increase in rate of responding to S_1 (Bloomfield, 1972). Finally, if responding declines in S_2 because reinforcement is delivered without a response requirement, no contrast occurs in S_1 (Halliday and Boakes, 1972, 1974; Weisman and Ramsden, 1973).

5. Conclusions

It must be obvious that behavioural contrast cannot simply be attributed to a decrease in either reinforcement or responding in one component of a multiple schedule. There is, perhaps, an important lesson to be learned from this brief survey of what has been an intensively studied issue. The fact that free-operant schedules are so powerful that they enable the experimenter to manipulate at will such aspects of the experimental situation as rate of responding or rate of reinforcement, either together or independently, in no way guarantees that the variables chosen for manipulation will be the important ones for reaching any analytic understanding of the processes responsible for the behaviour under investigation. Just because it is possible to manipulate rate of reinforcement or rate of responding in a contrast experiment, it does not follow that either is the crucial factor involved in contrast.* It seems more reasonable to argue, as have Bloomfield (1969) and Premack (1969), that some rather less precisely defined variable such as "a change for the worse" in one component is responsible for an increase in rate in the other. A decrease in the probability of reinforcement, the delivery of shock, or a requirement for paced responding, might all produce contrast effects because they all constitute a change for the worse; the mystery may be taken out of such a concept by defining as "worse" any condition which, in a choice situation, is less preferred than the schedule in effect in S_1. Experiments employing concurrent schedules have shown that pigeons prefer a higher probability VI schedule to one delivering reinforcement at a lower probability (Herrnstein, 1964a; Autor, 1969), a schedule free of shock to one imposing shock (Schuster and Rachlin, 1968) and a VI schedule to one requiring paced responding (Fantino, 1968).

Even this analysis, however, is clearly incomplete. A schedule of reinforcement in S_2 less preferred than that scheduled in S_1 may be a necessary condition for a positive contrast effect in S_2, but it is certainly not sufficient. Halliday and Boakes (1972) reported, rather surprisingly, that pigeons prefer an ordinary VI schedule to one that delivers free, response-independent, reinforcement at the same rate; Neuringer (1967) showed that pigeons prefer reinforcement durations of 3, 4, 6, and 10 sec to a constant 2 sec duration; although the experiment has not been performed, it is hard to believe that pigeons would fail to show a preference for $S+$ signalling VI reinforcement over S_2 signalling extinction, just because they had learned to discriminate between S_1 and S_2 without errors. Yet if S_2 signals free reinforcement, or a smaller reinforcement, or has been discriminated from S_1 without errors, no increase may be observed in rate of responding to S_1.

The implication is that positive behavioural contrast will be understood only when we can specify the processes underlying it. In this sense, the search for easily defined, readily manipulated, independent variables,

---

* A related point should be made about the dependent variable chosen for study in contrast experiments. As soon as contrast is rigidly defined as an increase in rate of responding in one component of a multiple schedule and rate of responding is measured by the number of times the subject pecks the key or presses the lever, there is the risk that contrast will be reported when the experimenter observes an increase in the number of times a pigeon pecks the key during a stimulus signalling a DRL schedule of reinforcement (e.g., Reynolds and Limpo, 1968). But this is almost certainly a meaningless exercise: it is absurd to suppose that an increase in rate of key pecking or lever pressing when an animal is trained on a DRL schedule signifies the same as an increase in rate of responding on a VI schedule. As Bloomfield (1969) has pointed out in this context: "Certainly, the rate of pecking may be manipulated freely by appropriate scheduling, but only at the cost of making pecks harder to interpret" (p. 223).

which can be shown to constitute necessary and sufficient conditions for positive contrast, is likely to be without success. There may be more to be said for pursuing the two suggestions advanced earlier and investigating the extent to which behavioural contrast is a consequence of conditioning a transient contrast effect, itself due to a release from inhibition, or how far it is due to an increase in the relative validity of S_1, causing an increase in classically conditioned pecks directed at S_1. These suggestions may be less readily tested than an analysis which boldly specifies the independent variables that determine the effect and will certainly be regarded by some as typically undesirable instances of the search for mythical inner causes, rather than real (because manipulable) external causes. By searching for parallels with other procedures and phenomena, however, it is possible that one will succeed in specifying the processes responsible for one particular effect rather better than by an exhaustive specification of the necessary and sufficient conditions for its occurrence.

Measured by these standards, our present understanding of positive behavioural contrast is extremely limited. Very few attempts have been made to examine the parallels between transient and permanent contrast effects, or even to assess the contribution made by a transient positive contrast effect to the overall increase in rate of responding typically observed in studies of positive behavioural contrast. There is some suggestion that both effects may be similarly influenced by the interval between trials and the discriminability of S_1 and S_2. But even here the evidence is severely limited: there is a great need for more systematic studies of these and other variables.

The difficulty in finding evidence of behavioural contrast when pigeons receive only a single trial each day may indicate a relation between behavioural and transient contrast. Equally, however, when a trial with S_1 occupies an entire session, the shift from a multiple (VI, VI) schedule to a multiple (VI, extinction) schedule does not decrease the proportion of time within each session occupied by a stimulus signalling reinforcement. The typical procedure for demonstrating positive behavioural contrast may be thought of as increasing the correlation between S_1 and reinforcement, and this may increase the capacity of S_1 to elicit and direct classical CRs. On this view, the occurrence of behavioural contrast depends on the use of a preparation where the response studied is one elicited by and directed towards a stimulus correlated with reinforcement. There is some evidence that other preparations may be less likely to produce behavioural contrast, but it is still too early to assess the validity of this argument.

C. Positive Contrast in Discrete-trial Experiments

The preceding discussion of the conditions under which positive contrast occurs was deliberately confined to a consideration of free-operant, multiple schedule, experiments in which contrast is measured as an increase in rate of responding to S_1. The generality of positive contrast effects is further questioned when one departs from this paradigm. With two exceptions, studies of discrete-trial discrimination learning have failed notably to find any evidence of positive contrast in S_1 when S_2 is correlated with a lower value of reinforcement.

The majority of discrete-trial studies of contrast have trained rats on

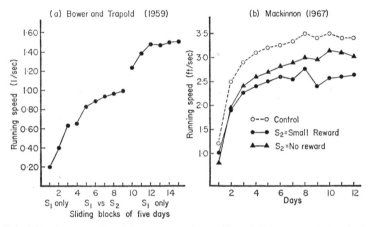

FIG. 7.6. The absence of positive contrast in studies of discrete-trial discrimination learning by rats.
(a) Speed of responding to S_1, signalling reward, is not greatly increased by the introduction of trials with S_2, signalling no reward; and the elimination of S_2 trials results in a sharp increase in speed of running to S_1. (*After* Bower and Trapold, 1959.)
(b) Speed of responding to S_1 is greater in a control group, receiving an equal reward in S_2, than in experimental groups, receiving either no reward or a smaller reward in S_2. (*After* MacKinnon, 1967.)

successive discriminations in black and white alleys, with one alley signalling one reinforcer and the other alley signalling another. These experiments have, more frequently than those employing free-operant procedures, compared the performance of an experimental group with that of a differently treated control group. When they have been designed to detect negative contrast effects they have been entirely successful, routinely observing, for example, that an experimental group exposed to a large reward in S_2 runs more slowly to a small reward in S_1 than a control group receiving the small reward in both S_1 and S_2 (see below). When designed to detect evidence of a positive contrast effect, however, they have been uniformly unsuccessful. The results of two typical experiments documenting this failure are shown in Fig. 7.6. Bower and Trapold (1959) performed a within-subject experiment: after preliminary training with S_1, signalling reinforcement, rats were given trials to S_1 (signalling reinforcement) alternating with trials to S_2 (signalling no reinforcement) and were finally shifted back to trials with S_1 alone; as can be seen, the final removal of S_2 was accompanied by an increase, rather than a decrease, in speed of running to S_1. MacKinnon (1967) trained rats with a large reward in S_1; for the control group S_2 was correlated with the same reward; for the experimental groups, S_2 signalled either a smaller reward or no reward at all. It can be seen that the experimental groups ran more slowly to S_1 than did the controls.

Several explanations have been offered for the discrepancy between these results and those typically reported in free-operant experiments. There may be merit in some, but it is extremely doubtful if any one is sufficient. Bower (1961) suggested that the general failure to find positive contrast effects in alley studies might be due to a ceiling effect: rate of responding in a free-operant situation is a relatively unbounded measure, but there is a limit to the speed at which rats can run in an alley. This cannot be a sufficient explanation: as can be seen from Fig. 7.6, it is not simply that nonreinforcement in S_2 fails to increase speed of running to S_1, it actually causes subjects to run more slowly than under control conditions when S_1 and S_2 both signal reinforcement.

Dunham (1968) pointed out that free-operant positive contrast occurred when probability rather than magnitude of reinforcement was varied; the majority of discrete-trial rat studies have varied magnitude rather than probability of reinforcement. Although this is true, it cannot provide a complete explanation. The results shown in Fig. 7.6 show no evidence of positive contrast when S_2 signals a lower probability of reinforcement—although there is a suggestion in MacKinnon's data that speed of running to S_1 was suppressed more when S_2 signalled a small reward than when it signalled no reward at all. Several other experiments have varied probability of reinforcement in S_2 without observing positive contrast in S_1 (e.g., Passey, 1957; Henderson, 1966; Ison and Krane, 1969).

Amsel (1971) has argued that positive contrast depends upon the use of a short interval between trials; in the majority of rat studies, this interval has been relatively long. This again is generally true, but there are exceptions that seem to rule it out as a complete explanation. Krane and Ison (1971) trained rats to discriminate between S_1 signalling reinforcement and S_2 signalling nonreinforcement; they varied the interval between successive trials from 30 sec to 10 min: they found no evidence that nonreinforcement in S_2 facilitated running in S_1.

A third possibility, which does not seem to have been very seriously entertained, is that the difference is a consequence of the use of rats instead of pigeons: there is good evidence to suggest that free-operant positive contrast appears more reliably with pigeons than with rats. The suggestion finds some support from the fact that the only two studies to have obtained positive contrast with a discrete-trial procedure (Jenkins, 1961; Terrace, 1963) did in fact use pigeons rather than rats. They also, it should be noted, used relatively short intertrial intervals and varied probability rather than magnitude of reinforcement. Perhaps all these factors contribute to the difference.

The results of an extensive series of experiments by Mackintosh (1974), however, casts some doubt on this conclusion. These experiments used a discrete-trial procedure, pigeons, and intervals between trials varying from 10 to 60 sec; no evidence of positive contrast was found. There was neither a permanent decrease in the latency of response to S_1 when S_2 signalled no reinforcement, nor a transient contrast effect: speed of responding to S_1 was not increased by the immediately preceding occurrence of a nonreinforced S_2 trial. Furthermore, there was evidence that the positive contrast effect reported by Jenkins (1961) was caused not by the presentation of nonreinforced trials to S_2, but by differences in the spacing of trials for experimental and control subjects.

A final possibility, not entirely unrelated to Bower's argument about ceiling effects, is that the latency measures used in discrete-trial experiments are simply not sensitive to contrast effects. In support of this argument, Mackintosh found that when pigeons were trained on a standard, free-operant multiple (VI, extinction) schedule, they showed large and reliable increases in rate of responding to S_1, but no decrease in the latency of the first response on S_1 trials. Positive contrast could be detected only with the rate measure, not with the latency measure. Whatever the explanation of this finding, it implies that the discrepancy between the results of free-operant and discrete-trial experiments may simply be a consequence of the use of different response measures. It is, in fact, possible to suggest several reasons why latency measures might not reveal evidence of a positive contrast effect. Latencies may, in part, reflect decision times and when S_1 and S_2 signal different schedules of reinforcement, subjects must presumably take some finite amount of time to decide which stimulus is being presented on a given trial (Mellgren et al., 1972). It is tempting to see a parallel here with the observation that human reaction times generally increase with an increase in the number of alternatives (e.g., Hick, 1952). A second possibility is that the increase in excitation produced by release from inhibition or by frustration may momentarily disorganize, rather than facilitate, the initiation of responding; but once responding has been initiated, it will occur more vigorously. This argument is related to an earlier observation about differences in response measures (p. 24): rate of pecking, like rate of salivation, is a relatively unbounded measure of responding; a discrete probability measure, such as whether the first peck is initiated or whether a flexion or eyeblink response occurs, is bounded and cannot vary over any substantial range at all. The argument is supported by the fact that although transient contrast effects have occurred in studies of salivary conditioning (p. 356), there is no evidence that they can be obtained in studies of eyelid conditioning (e.g., Gynther, 1957; Frey and Ross, 1967; Moore, 1972). Whatever the explanation, the results of discrete-trial experiments unequivocally show that positive behavioural contrast is not an automatic consequence of discrimination learning.

V. SIMULTANEOUS AND SUCCESSIVE NEGATIVE CONTRAST

A. Introduction

Two final classes of contrast effect remain to be discussed (see Table 7.1, p. 355). The standard examples of these effects have been studies of rats in alleyways employing relatively spaced trials. The sole difference between simultaneous and successive contrast studies is that in simultaneous experiments subjects are concurrently exposed, on randomly alternating trials, to one condition of reinforcement in one alley and a different condition in a second alley; in the successive case, subjects receive consistent experience in a single alley with one condition of reinforcement, before being shifted to consistent experience with a different condition, still in the same alley. But in the paradigm case where rats are exposed to different magnitudes of reinforcement, this difference in procedure seems to be of little or no consequence. In Figs. 7.7 and 7.8 are shown the results of two typical experiments: in Fig. 7.7 a study of simultaneous contrast and in Fig. 7.8 a study of successive contrast. In each study the performance of experimental subjects exposed to different rewards is compared with the performance of large- and small-reward control groups. Both the simultaneous and the successive experiments resulted in reliable negative contrast effects: animals exposed to a large reward, either concurrently or during pre-shift training, ran more slowly to the small reward than did a control group exposed only to the small reward. Neither experiment produced a positive contrast effect: animals exposed to a small reward did not run more rapidly to the large reward than control subjects exposed only to the large reward.

We have already discussed the absence of positive contrast effects in simultaneous studies: we need only note that at least two of the conditions of Bower's (1961) study of simultaneous contrast shown in Fig. 7.7—the

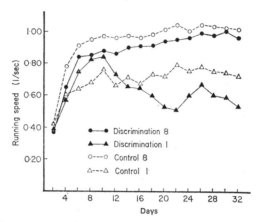

FIG. 7.7. Simultaneous negative contrast in rats. A discrimination group, receiving concurrent exposure to a large reward (8 pellets) in one alley and a small reward (1 pellet) in a second alley, responds more slowly for the small reward than does a control group receiving only the small reward. There is no evidence, however, of a comparable positive contrast effect. (*After* Bower, 1961.)

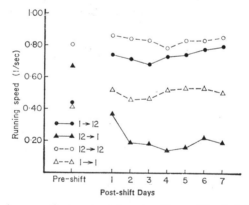

FIG. 7.8. Successive negative contrast in rats. Group 12→1, initially receiving a large reward (12 pellets) for running down an alley, responds more slowly for a small reward (1 pellet) than a control group (1→1) that receives the small reward during both pre-shift and post-shift trials. A group shifted from a small to a large reward (1→12), however, does not run faster than the control group (12→12) reinforced throughout with the large reward. (*After* Franchina and Brown, 1971.)

variation in magnitude rather than probability of reinforcement and the use of widely spaced trials—would have been sufficient to prevent positive contrast appearing. Positive contrast or elation was long thought to be as readily obtainable in successive experiments as negative contrast or depression, but this belief was based on the early results of Crespi (1942) and Zeaman (1949), both of whom assumed (probably unjustifiably) that stable, asymptotic, performance had been reached before subjects were shifted. Spence (1956, pp. 130–32) was the first to point out that the evidence for a successive positive contrast effect was decidedly suspect and he reported the results of three further experiments, in each of which a significant negative contrast effect appeared in the absence of any sign of positive contrast. In one of these experiments animals shifted from a small to a large reward appeared to run more slowly than a control group; this finding has been confirmed by Capaldi and Lynch (1967) and, as can be seen in Fig. 7.8, by Franchina and Brown (1971).

Just as the absence of positive contrast in simultaneous studies has been attributed to a ceiling effect, so there has been a persistent search for positive contrast in successive experiments under conditions assumed to obviate such problems. Several recent studies, however, have failed to find positive contrast under conditions where there was ample room for an increase in speed of running: Schrier (1967), Campbell et al. (1970) and Mellgren (1971a) shifted rats from a small to a large reward after 16 or fewer acquisition trials; although large-reward control subjects did not reach asymptote in less than 30 to 40 trials, animals shifted to the large reward consistently failed to run faster than the controls. DiLollo (1964) shifted rats from one to four pellets and found no evidence of positive contrast even though the four-pellet control group ran more slowly than animals receiving sixteen pellets. E. D. Capaldi (1971a) trained rats under both high and low levels of drive (thus ensuring, in the latter case, a speed of running well below any physical upper limit) and found no sign of positive contrast in either group.

Several studies which have purported to show successive positive contrast have not withstood closer scrutiny. Dunham and Kilps (1969) showed that the results of an earlier study by Collier and Marx (1959), apparently demonstrating symmetrical positive and negative contrast effects, were more plausibly regarded as a consequence of different levels of deprivation. Sgro and Weinstock (1963) reported a small positive contrast effect for rats shifted from a 15 sec delay of reinforcement to a zero delay, but had discarded 15 of the slowest runners in their experimental groups before the shift. Shanab et al. (1969) reported a positive contrast effect when they simultaneously changed size and delay of reward: experimental subjects were shifted from a small to a large reward while controls were maintained on a large reward, but coincident with this both groups were shifted from immediate to delayed reward. They found that after this shift in delay, animals initially exposed to the small reward ran faster than those exposed to the large reward. But since the delay was sufficiently long to produce an extinction-like, precipitous, decline in performance in both groups, this difference may perhaps be attributed to the known fact that large rewards decrease resistance to extinction (p. 427). In a subsequent experiment, Shanab and McCuiston (1970) found no positive contrast in subjects shifted from a small to a large reward when they had been run from the outset of training with delayed reinforcement.

This possibly one-sided survey should not conclude without mention of one or two recently reported demonstrations of successive positive contrast which are less easily dismissed (Marx, 1969; Mellgren, 1971b). It would, perhaps, be rash to assert too confidently that there is no such phenomenon. The available evidence, however, points strongly to the conclusion that there is no positive contrast effect equal and opposite to the negative contrast effect routinely observed when rats are shifted from a large to a small reward. Any analysis, such as that provided by adaptation level theory (Bevan and Adamson, 1967) which implies the existence of such symmetrical effects is either totally inappropriate, or at best has pinpointed a relatively unimportant process superimposed upon a much more important set of factors producing negative contrast only. What these factors are may be more readily determined after a brief review of some of the variables which determine the appearance and magnitude of contrast effects.

B. Variables Determining Negative Contrast Effects

1. *Studies Varying Magnitude of Reinforcement*

The most common method of varying reinforcement in studies of contrast has been to give different numbers of pellets of food, or different weights of food, to hungry rats. A shift to a smaller reward, defined in this way, produces reliable negative contrast in successive experiments, with the magnitude of the effect being a function of the size of the shift (e.g., Gonzalez *et al.*, 1962; DiLollo and Beez, 1966). In simultaneous contrast experiments, also, the speed of running to S_1 signalling a small reward is an inverse function of the size of reward signalled by S_2 (e.g., Ludvigson and Gay, 1966; Matsumoto, 1969).

Successive negative contrast is an increasing function of the number of pre-shift trials (Vogel *et al.*, 1966), drive level (E. D. Capaldi, 1971a) and abruptness of shift (Gonzalez *et al.*, 1962). However, it can be attenuated or abolished if subjects are given partially rather than consistently reinforced trials with the large reward before the shift (Mikulka *et al.*, 1967). The magnitude of the contrast effect is also affected by the interval between trials. Although contrast is reliably observed when trials are spaced 24 h apart, a much larger effect occurs when trials are relatively massed (Capaldi, 1972).

Successive contrast is often a transient effect in the sense that it may not persist for more than a small number of trials after the shift in reinforcement. This is hardly surprising: as experience with the new reinforcer continues, it would be difficult to see how behaviour could be indefinitely influenced by earlier experience with another reinforcer. Simultaneous contrast effects appear to persist for a long time; nor is there any evidence that they are affected by the intertrial interval. They do, however, depend critically upon continued exposure to the large reward in the same context: a control group that receives a large reward at the same time as the experimental group, but outside the experimental apparatus, shows no negative contrast effect (Maxwell *et al.*, 1969; Harris *et al.*, 1970). This finding, incidentally, provides good evidence that animals receiving a large reward in S_2 do not run more slowly to a small reward in S_1 simply because they are more satiated.

2. *Studies Varying Quality of Reinforcement*

Elliott's (1928) study of successive negative contrast (p. 214) varied type, rather than amount, of reward: rats trained with a more preferred reinforcer (bran mash) ran more slowly when shifted to the less preferred sunflower seeds than did a control group trained with sunflower seeds throughout. More recent studies have sought evidence of contrast effects by varying the concentration of sucrose or saccharin solutions used as reinforcers. Although higher concentrations produce faster running in acquisition, studies employing discrete trials in an alleyway have uniformly failed to find any evidence of a successive negative contrast effect when subjects are shifted from a high to a low concentration (Homzie and Ross, 1962; Rosen and Ison, 1965; Rosen, 1966). Successive negative contrast produced by a decrease in concentration has, however, been observed when rats are trained in a free-operant, lever-pressing, situation (Weinstein, 1970), or when they are simply permitted to drink the solutions from a bottle and the measure of responding is the rate of consummatory licking (Vogel *et al.*, 1968). The most obvious discrepancy between the two sets of studies is that negative contrast fails to occur when animals are trained on discrete, relatively widely spaced trials, but may occur when rate of responding for massed or continuously available reinforcement is measured. This is consistent with the observation that the magnitude of the contrast effect produced by a decrease in size of reward is also inversely related to the length of the intertrial interval (Capaldi, 1972).

The evidence from studies of simultaneous contrast is conflicting. Ison and Glass (1969) found no negative contrast effect when S_1 and S_2 signalled different concentrations of sucrose, although finding a reliable effect when the stimuli signalled different numbers of pellets. Other studies, however, have been more successful. Rossman and Homzie (1967) found a significant effect with different concentrations of xylose as the reinforcers. A recent study by Flaherty et al. (1973) provides convincing evidence that

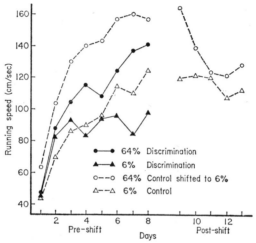

FIG. 7.9. Simultaneous and successive contrast for different concentrations of sucrose. During pre-shift training the discrimination group responds more slowly for the 6 per cent concentration than does the control group exposed only to 6 per cent sucrose, but does not respond more rapidly for the 64 per cent concentration than the control group exposed only to 64 per cent sucrose. When this latter group, however, is shifted to a 6 per cent concentration, there is no evidence of a successive negative contrast effect. (*After* Flaherty et al., 1973.)

variations in concentration of sucrose may produce simultaneous negative contrast without resulting in any successive contrast effect. Three groups of rats were trained in an alleyway: one group received 64 per cent sucrose solution on all trials; a second group received a 6 per cent solution on all trials; while a discrimination group received a 64 per cent solution in S_1 and a 6 per cent solution in S_2. Simultaneous contrast effects could be assessed by comparing the performance of the discrimination group with the performance of the two control groups, each trained with a single concentration. In order to test for the occurrence of a successive contrast effect, animals initially exposed only to the 64 per cent solution were shifted after 64 trials to the 6 per cent solution. The results are shown in Fig. 7.9: there was, as usual, no evidence of a simultaneous positive contrast effect, but clear evidence of simultaneous negative contrast. When the 64 per cent group was shifted to the 6 per cent concentration, however, they showed no signs of successive negative contrast.

3. Studies Varying Delay of Reinforcement

The implication suggested by studies varying sucrose concentration is that simultaneous negative contrast may occur when no successive contrast can be detected. This suggestion is further substantiated by studies varying delay of reinforcement. Beery (1968) found that speed of running to S_1 signalling a 10 sec delay of reinforcement was an inverse function of the delay of reinforcement concurrently experienced in S_2, thus defining a reliable simultaneous negative contrast effect. Similar results have been obtained by Chechile and Fowler (1973). Several studies, however, have failed to find evidence for a successive negative contrast effect when rats are shifted from immediate reinforcement to a 5 or 10 sec delay of reinforcement (Harker, 1956; Spence, 1956, pp. 155–63). More recent studies have reported more equivocal findings: Shanab and McCuiston (1970) found a slight and temporary negative contrast effect after shifting from immediate reinforcement to a 15 sec delay; Shanab (1971) found no significant effect after a shift to a 30 sec delay; while McHose and Tauber (1972) found a large and persistent effect when rats were shifted from a 10 sec delay to a 30 sec delay. Mackintosh and Lord (1973), however, in a study analogous to that of Flaherty et al. (1973) with sucrose concentration, explicitly compared the magnitude of simultaneous and successive negative contrast effects produced by a 15 sec delay of reward, and found significant simultaneous contrast but no evidence of a successive contrast effect.

4. Comparative Studies

All the studies on negative contrast discussed so far have employed rats, and all but two have employed alleys and discrete trials. How general are these negative contrast effects? Do they occur with other subjects and in other experimental situations? Free-operant multiple schedules would, in principle, provide a valuable procedure for studying simultaneous negative contrast with variations in reinforcement schedule; one might ask, for example, whether rate of responding to S_1 signalling a VI 5 min schedule would be slower if the schedule signalled by S_2 was VI 30 sec rather than VI 5 min. In practice, however, the question has not been definitively answered by the few experiments purporting to do so. Because of their penchant for a within-subject control procedure, most experimenters (e.g.,

Terrace, 1968; Weisman, 1969) have initially trained subjects on a multiple (VI 5, VI 5) schedule, and then shifted them to multiple (VI 1, VI 5). Quite apart from the usual problem of assessing the reliability of small changes in rate from unstable baselines, there is the added question whether simultaneous negative contrast ever occurs when subjects are initially exposed to the less favourable condition of reinforcement in the presence of both stimuli, and subsequently shifted to a more favourable condition in the presence of one stimulus. One line of evidence suggests that if the schedule signalled by S_2 changes from VI 1 to VI 5, while S_1 continues to signal VI 1, S_2 may become an inhibitory stimulus. Inhibitory gradients can be observed round S_2 after such training (Weisman, 1969; see p. 527) and this suggests that a negative contrast effect may be operating. The absence of direct evidence, however, is unfortunate, since a demonstration of simultaneous negative contrast using rate of responding in a free-operant situation would be of some value. We have already argued that latency measures may possibly be sensitive to the constraints imposed on a subject required to discriminate between stimuli signalling different reinforcers. If such an effect works against the demonstration of positive contrast in discrete-trial situations, it would work in favour of producing an artefactual negative contrast effect. It does not seem likely that discrete-trial negative contrast can be explained away entirely in this fashion, for this would not explain why the magnitude of the effect increases with an increase in discrepancy of reinforcement; but it is notable that a large part of the effect is attributable to an increase in the latency of initiating responding at the moment when the less favourable stimulus is first presented to the subject (Ludvigson and Gay, 1967).

Although there is no direct evidence that simultaneous negative contrast occurs in free-operant experiments, there is good evidence that it can be obtained in discrete-trial experiments with animals other than the rat. Brownlee and Bitterman (1968) trained pigeons in a discrete-trial discrimination and found that latency of response to a green key light signalling a single pellet was significantly slower when responses to a red key were reinforced with eight pellets rather than one pellet. More recently, Gonzalez and Powers (1973) and Cochrane et al. (1973) have reported that goldfish will also show a simultaneous negative contrast effect when size of reward is varied.

Reliable successive contrast effects, however, have not been reported for any animal other than the rat. Gonzalez (1971) reported that pigeons shifted from a large to a small reward showed a rapid decline in performance, but little or no sign of responding more slowly than a small-reward control group. Pert and Bitterman (1970) reported similar results for turtles. Several experiments have studied the effects of a shift from large to small reward in goldfish; the results have conflicted in details, but have agreed in showing no evidence of a significant negative contrast effect. Lowes and Bitterman (1967), and Gonzalez *et al.* (1972) found no decline in performance when fish were trained one trial a day and shifted from a reward of forty tubifex worms to one of four worms. Mackintosh (1971b) found only a marginal decline in performance after a shift from five to one pellets when subjects were trained with four trials per day. Raymond *et al.* (1972) found a fairly rapid decline in performance, but no evidence of negative contrast, while Cochrane *et al.* (1973), giving massed trials, also observed a significant decline in performance following a shift from large to small reward.

C. Theoretical Interpretations of Negative Contrast

The data reviewed above pose several problems of interpretation. First, there seems to be a considerable range of conditions which produce simultaneous but not successive negative contrast effects: variations in concentration of sucrose or delay of reinforcement, and variations in magnitude of reinforcement in several nonmammalian vertebrates, are all sufficient to produce one effect but not the other. This presents a serious problem for any attempt to provide a single explanation of both classes of effect. In spite of the similar outcome of studies varying size of reward in rats, it may be that simultaneous and successive contrast effects are produced by different factors. The process responsible for simultaneous negative contrast may not operate, or may operate only in attenuated form, in experiments on successive contrast. This would explain why successive negative contrast is a relatively elusive phenomenon but it would not, of course, explain why the effect does reliably occur under one particular set of conditions (when rats are shifted from a large to a small reward). This last observation poses an additional problem which would seem to require some account which is capable of differentiating between the consequences of different shifts in reward. Cumbersome as the final set of explanations promises to be, the pattern of results reviewed above does not seem to admit of any simple theoretical resolution.

1. *Inhibition and Simultaneous Contrast Effects*

In an earlier discussion (p. 216), negative contrast effects were attributed to an inhibitory process which was assumed to be triggered by the presentation of a relatively unfavourable reinforcer in a context previously or concurrently associated with a more favourable reinforcer. A rat given a large and immediate reward for running from a grey start box down a black alley, will expect the same large, immediate, reward whenever it is placed in the same start box. If running down a white alley is now associated with a small or delayed reward, the white alley will become an inhibitory stimulus signalling a less favourable condition of reinforcement.

Black (1968) has proposed an inhibitory analysis of negative contrast effects similar to that advanced here. One distinguishing feature of Black's analysis, however, is his assumption that in simultaneous contrast experiments the expectation of a large reward established in S_2 generalizes to S_1 because of the similarity of the two alleys or discriminative stimuli, and that it is this generalization which ensures that the presentation of a small reward in S_1 will be inhibitory. According to the analysis suggested here, on the other hand, inhibition occurs because the rat expects the more favourable reinforcer as soon as it is placed in the start box: generalization between the two alleys is not a necessary condition for the occurrence of contrast. There are several reasons for preferring this formulation to Black's. Ludvigson and Gay (1967) showed that a necessary condition for the occurrence of negative contrast was the use of a common start box on both S_1 and S_2 trials. Even more importantly, Harris *et al.* (1970) found evidence of a negative contrast effect when trials on which subjects received a small reward for running down an alley were interspersed with trials on which they were placed in the start box but then removed and fed a large reward in an entirely different situation. Rats simply given these large rewards, without being placed in the start box beforehand, did not show a negative contrast effect. These results strongly suggest that it is the establishment of an expectation of large reward in the start box, rather than in a similar alley, that is responsible for a negative contrast effect.

The difference between these two analyses could be further tested by studying the effects of the similarity of S_1 and S_2 on the magnitude of negative contrast effects. If, as Black suggests, negative contrast depends on generalization between S_2 and S_1, then it should tend to increase with increases in the similarity of the two stimuli. There may well be some limit to this, of course, since if S_1 and S_2 were indiscriminable, a large reward in S_2 would presumably facilitate rather than suppress performance for a small reward in S_1. The argument will be recognized as similar to that applied to transient and positive contrast effects: these effects, it was suggested, may first increase but will then decrease as S_1 and S_2 are made more similar (p. 378). In the case of transient and positive contrast there was at least some evidence to support the argument; in the case of simultaneous negative contrast there is none. Chechile and Fowler (1973) found relatively little effect of the discriminability of S_1 and S_2 on the magnitude of negative contrast: if anything, contrast appeared sooner and may have been more pronounced, with more discriminable stimuli. Moreover, the results of Harris *et al.*'s study suggest that negative contrast occurs when S_1 and S_2 are entirely different situations. A systematic comparison of the effects of the discriminability of S_1 and S_2 on the magnitude of negative and positive contrast and on inhibitory conditioning, might well reinforce the distinction between negative and positive contrast effects and strengthen the parallel between negative contrast and inhibitory conditioning. Conditioned inhibition to a CS— in simple discriminative conditioning is a decreasing function of the generalization between CS+ and CS— (p. 34). As was also noted in the discussion of conditioned inhibition, the process does not depend upon an expectation of the UCS generalizing from CS+ to CS—, but from such an expectation being conditioned to background apparatus cues which are presumably analogous to the start box cues of discrete-trial alleyway experiments.

One further conclusion drawn from studies of conditioned inhibition was that nonreinforcement of one CS, in the absence of continued reinforcement of another, may not be sufficient to turn it into a conditioned inhibitor. At the very least, concurrent reinforcement in the presence of another CS is a much more powerful way of generating inhibitory conditioning to a particular stimulus. The implication for contrast effects is straightforward. Since it is only in simultaneous contrast experiments that animals are concurrently exposed to two different reinforcers, the inhibitory analysis predicts that simultaneous negative contrast should be a more reliable and readily detectable phenomenon than successive negative contrast. This, of course, is exactly what the evidence suggests: successive contrast effects seem to occur considerable less reliably than simultaneous contrast effects.

These arguments suggest that the shift from a more favourable to a less favourable reinforcer in a successive paradigm may be less inhibitory than the concurrent exposure to more and less favourable reinforcers provided by the simultaneous paradigm. It is not suggested that the successive shift results in no inhibition. Daly (1972) showed that rats would learn to escape from a goal box associated with a decrease in size of reward in a successive experiment; Rosen *et al.* (1967) found that the successive contrast effect produced by a decrease in reward could be abolished by the administration of sodium amytal. Both of these results are consistent with an inhibitory analysis. Nevertheless, although there are independent reasons for supposing that a successive shift may be less inhibitory than concurrent exposure to different reinforcers, the inhibitory analysis seems unable to explain why simultaneous contrast effects occur with a wide variety of subjects and changes in reinforcement, while successive effects occur reliably only in rats, and only when they are exposed to shifts in the size of reward. It is one thing to say that successive effects should in general be smaller; it is another thing to explain the particular pattern of results observed in experiments on successive contrast. This suggests that further principles must be relied upon to account for some of the data on successive contrast.

It is impossible to leave this discussion of an inhibitory analysis of negative contrast without commenting on its relationship to the analysis in terms of frustration theory provided by Amsel (1967, 1971). The similarities between the two are, of course, considerable. There is, however, at least this difference: Amsel has attributed energizing, motivational, properties to frustration, so that the presentation of a reinforcer less favourable than that expected should not only interfere with ongoing behaviour by producing competing responses, it should also produce a state of aversive drive which will potentiate any dominant pattern of behaviour. The problem with this as an analysis of *negative* contrast, as Black (1968) has noted, is that it is hard to see why the interfering effects of frustration should continue to override the facilitative effects. It is usually assumed that frustration will become counter-conditioned to ongoing responding provided that, as is always the case in contrast experiments, such responding continues to be reinforced. It seems to follow, therefore, that only the energizing effects of frustration should be apparent and that responding to a stimulus signalling a small reward should eventually be facilitated, rather than suppressed, by concurrent exposure to a large reward. This argument seems to provide additional grounds for ascribing the energizing effects of frustration to a *release* from inhibition (p. 349). Positive contrast or frustration effects, therefore, are a transient consequence of an immediately preceding exposure to nonreinforcement. Since these conditions are not satisfied in typical studies of simultaneous and successive negative contrast, only inhibitory effects are observed in these experiments.

2. Successive Contrast Effects as Instances of Generalization Decrement

By pointing to factors known to affect the conditioning of inhibition, the inhibitory analysis outlined above successfully explains why simultaneous negative contrast is a more reliable phenomenon than successive contrast. But, as was noted, this does not resolve all problems, for there is one procedure—shifting rats from a large to a small reward—that produces a large and reliable successive negative contrast effect. Perhaps what is needed is an additional theoretical account to deal with this particular result.

Capaldi (1967a) has advanced such an account: he has proposed that successive negative contrast is simply an instance of generalization decrement. If rats receive a large reward for running down an alley over a series of trials, they will remember from one trial to the next the large reward received on preceding trials. Thus one aspect of the situation controlling responding may be the after-effects or memory of these preceding large rewards. When responding has been reinforced in the presence of one set of stimuli, a change in those stimuli often disrupts performance (generalization decrement). Similarly, if the subject remembers the reward received on preceding trials, and these memories become established as signals for further rewards, then a change in the reward received on one trial may disrupt performance on the next trial by a similar process of generalization decrement. An animal shifted from a large to a small reward, therefore, will gradually come to respond more slowly, as the expectation of a small reward is substituted for that of a large reward; but superimposed on this gradual decline in performance, there will be an abrupt, transient, decrement caused by the fact that the animal has never had the opportunity to associate the memory of preceding small rewards with the occurrence of further reinforcement.

Capaldi's analysis leads to several clear predictions. First, performance should be disrupted not only when a subject is shifted from a more favourable to a less favourable reinforcer, but when any discriminable change in reinforcer is made (with "discriminable" here meaning a change that produces discriminably different after-effects or memory traces). An experiment by Huang (1969) provides some support for this. Huang trained rats deprived of both food and water to run in an alley, with one group receiving food and the other water as the reward. Although the two reinforcers produced equivalent speeds of running by the end of the first stage of the experiment, animals shifted from food to water showed a significant negative contrast effect, running more slowly over the next 21 trials than animals that continued to run for water. It would be nice to see this experiment repeated with symmetrical shifts (e.g., from food to water, and from water to food) both disrupting performance. Such a demonstration would provide decisive evidence against an exclusively inhibitory analysis of successive contrast.

It follows from Capaldi's argument that a shift from a small to a large reward should also produce generalization decrement. Although such generalization decrement might not result in any overall decline in speed of running, since it would be counteracted by an increase in incentive, it should be detected as a failure for animals shifted from a small to a large reward to run as fast as large-reward controls for a number of trials. We have noted that exactly such an effect has been observed (p. 389).

A second prediction derivable from Capaldi's analysis is that successive contrast effects should be abolished, or at least attenuated, if shifted subjects have received some previous experience with the reward they are shifted to, or one similar to it. There is considerable evidence to support this (e.g., Spear and Spitzner, 1966, 1969). Capaldi and Lynch (1967) showed that rats trained with a small reward for 15 trials, followed by a large reward for 15 trials, showed no contrast effect at all when shifted back to the small reward for a final 15 trials. Capaldi and Ziff (1969) have argued that the reason why partial reinforcement given in pre-shift trials attenuates negative contrast is because animals that have been reinforced for running after nonreinforced trials will show less generalization decrement when shifted to a small reward, than will animals that have been reinforced for running only after reinforced trials.

Finally, if changes in reward disrupt performance on subsequent trials because the memory of the changed reward has never been established as a signal for reinforcement, it seems intuitively obvious that an increase in the length of the intertrial interval may decrease the extent to which performance is controlled by the memory of preceding rewards and therefore decrease the magnitude of any negative contrast effect. This expectation has been confirmed by Capaldi (1972). As will be shown in the following chapter in a discussion of the effects of the distribution of trials on rate of extinction (p. 422), it is possible to derive this prediction more formally from this sort of analysis. For the present it may be sufficient to rely on the intuitive derivation.

There remains, however, a critical problem for any analysis of successive contrast effects—that of explaining the wide range of conditions under which they fail to occur. If a shift from a large to a small reward disrupts performance because the memory of a small reward has never been established as a signal for further reinforcement, then why should a change in the concentration of sucrose used as the reinforcer, or a change in the delay of reinforcement, not have precisely the same effect? One possible, if arbitrary, answer is to assume that different concentrations of sucrose or different delays of reinforcement produce after-effects or establish memory traces that are either much more similar to each other, or much less likely to enter into control of instrumental responding, than are the traces of different sizes of reward.

The answer may appear worse than the problem it purports to solve. Assumptions as arbitrary as this cannot be seriously accepted. Fortunately, it is readily tested. A reinforcing event such as food or shock can be shown to act as a stimulus by arranging appropriate contingencies of reinforcement in its presence. An FI schedule establishes the delivery of reinforcement as a stimulus signalling a period of nonreinforcement and thereby ensures a low rate of responding after each reinforcement. In discrete-trial experiments, the possibility that the delivery of reinforcement on one trial can serve as a signal to control responding on the next trial can be tested by training a subject on a regularly alternating sequence of reinforced and nonreinforced trials. Reinforcement on one trial becomes a signal for nonreinforcement on the next, and rats show that they can remember the outcome of one trial, and use it as a stimulus to control responding on the next, by running rapidly on reinforced trials following nonreinforcement and slowly on nonreinforced trials following reinforcement (p. 162).

If different sizes of reward produce such discriminably different after-effects that the performance of a rat shifted from one size to another suffers disruption from generalization decrement, rats should learn an appropriate pattern of fast and slow running when trained on a regularly alternating sequence of large and small rewards. Several experiments have provided evidence of such learning (Capaldi and Cogan, 1963; Wolach et al., 1972; Likely, 1970). If, on the other hand, the reason why changes in concentration of sucrose, or in delay of reinforcement, fail to disrupt performance is because they do not produce any generalization decrement, such changes should also provide insufficient cues to support alternation learning. Cogan and Capaldi (1961) trained rats on a regularly alternating sequence of trials with a 20 sec delay of reinforcement and trials with immediate reinforcement, and found no evidence of patterning. The absence of patterning when immediate and delayed reinforcement are alternated depends upon the use of a relatively short delay of reinforcement. Burt and Wike (1963), who confirmed Cogan and Capaldi's findings, obtained significant patterning when reinforcement was delayed by 60 or 100 sec. The implication is that successive contrast will occur more readily if subjects are shifted to very long delays of reinforcement.

Likely (1970), training rats on regular sequences of trials reinforced with different concentrations of sucrose, also found little or no evidence of patterning. There was one exception to these negative results in Likely's experiments: with highly massed trials (a 10 sec interval), rats did learn to alternate for different sucrose concentrations; as we have seen, however, it is precisely when "trials" are highly massed, as in a free-operant situation, that a decrease in concentration of sucrose may produce a negative contrast effect.

Finally, there is evidence from comparative work that adds to this correlation between successive contrast and alternation learning. In at least one animal that has shown no evidence of successive negative contrast alternation learning is apparently relatively inefficient. Gonzalez et al. (1961) and Mackintosh (1971b) trained goldfish on a regularly alternating sequence of reinforced and nonreinforced trials for a total of 400 trials: in neither study was there any evidence of significant patterning. Zych and Wolach (1973) found no evidence of patterning after 196 trials in an alley. Although Gonzalez (1972) has obtained significant patterning after training goldfish for a very large number of trials with a large reward on R trials, the evidence still suggests that their behaviour is less readily controlled by traces of the outcomes of preceding trials than is the behaviour of rats (Mackintosh, 1971b).

3. Conclusions

Experiments on simultaneous and successive negative contrast differ in one, apparently minor, operational feature. In simultaneous experiments, subjects are exposed to a less favourable reinforcer while continuing to receive a more favourable reinforcer in the presence of another stimulus. In successive experiments, exposure to the more favourable reinforcer is terminated when subjects are shifted to the less favourable reinforcer. There is, however, reason to believe that this difference is of considerable importance. If negative contrast is caused by inhibitory conditioning, there is evidence from studies of classical conditioning that the procedures of the simultaneous experiment will result in stronger inhibitory conditioning than those of the successive experiment. Thus the observation that simultaneous negative contrast occurs more readily than successive negative contrast is not surprising.

Successive contrast effects are not, however, simply smaller in magnitude than simultaneous effects. Although there are some conditions under which they do not seem to occur at all, when rats are shifted from a large to a small food reward, successive contrast effects are routinely observed. This suggests that there may be an additional factor operating to produce a negative contrast effect in this one particular situation. One possibility is that this factor is a generalization decrement. Not only are there several results which are consistent with such an analysis of successive contrast, it also suggests why changes in size of reward in experiments with rats are more likely to produce contrast than are other preparations.

VI. SUMMARY

Although a reduction in reinforcement may be assumed to produce inhibitory learning, which is somehow opposed to the excitatory learning resulting from increases in reinforcement, there is good evidence that exposure to a less favourable reinforcer may sometimes enhance behaviour maintained by a more favourable reinforcer, and that exposure to a more favourable reinforcer may sometimes suppress behaviour maintained by a less favourable reinforcer. These interactions are usually called contrast effects.

One possible interpretation of contrast effects is that they reflect the subject's relativistic assessment of reinforcement: a small reward appears even smaller to a subject accustomed to a large reward; a large reward appears even larger to one accustomed to a small reward. There is nothing inherently implausible about this idea, and the idea that a small reward may be inhibitory to a subject expecting a large reward represents one version of such an account. The available evidence, however, suggests that contrast effects are neither as general nor as symmetrical as this analysis requires. A brief summary of the evidence suggests that most contrast effects are negative and may be a consequence of inhibitory conditioning. In addition to these permanent negative contrast effects, there are transient rebound effects produced by the after-effects of preceding reinforcers: the termination of exposure to reinforcement may produce a temporary increase in inhibition, while the termination of exposure to nonreinforcement may produce a temporary increase in excitation.

This analysis may prove much too superficial. The difficulty with making any general statements about contrast effects is that different experimenters have used different preparations to study a variety of different effects. Virtually no effort has been devoted to the systematic analysis of the differences in these procedures. Any attempt to develop a coherent account of contrast effects, therefore, may be premature.

Transient contrast effects may be defined as temporary increases in responding to S+ immediately after exposure to nonreinforcement, or temporary decreases in responding to S− immediately after exposure to reinforcement. They are best studied with a within-subject design, where the performance of a single subject may be shown to depend upon the immediately preceding condition of reinforcement. Pavlovian induction, transient operant contrast, and the double-alley frustration effect may all be instances of these effects.

The one reliable example of a positive contrast effect, other than that observed in studies of transient contrast, is the positive behavioural contrast observed in free-operant studies of discrimination learning in pigeons. Even here, unfortunately, few experimenters have troubled to distinguish between a transient elevation in rate of responding to S+ immediately after exposure to S− and a more permanent effect. There is some evidence which suggests that this behavioural contrast may represent a conditioned form of the transient contrast effect observed early in discrimination learning. In addition, however, it is possible that behavioural contrast may be a phenomenon relatively specific to the preparation typically used in its study. The effect seems less reliable in rats than in pigeons and does not occur when pigeons are trained to press a treadle rather than peck a key. The effect may, therefore, be due to an increase in the probability of food-elicited pecks being directed towards the response key in the presence of

o

a stimulus whose correlation with reinforcement has been improved by the programming of discrimination training.

In general, positive contrast is not observed in studies of discrete-trial discrimination learning, nor in studies of classical conditioning employing discrete, skeletal CRs such as the eyelid response. Even transient positive contrast has not been reliably observed in these situations. It is possible that release from inhibition may increase the amplitude or rate of a response such as salivation or key pecking, but may not decrease the latency with which these and other responses are initiated. Response systems that provide only latency measures, therefore, may be unsuitable for the demonstration of positive contrast effects.

Negative contrast effects, however, are reliably observed in studies of discrete-trial discrimination learning. It is unfortunate that they have not been studied in other situations, such as classical conditioning or free-operant discrimination learning, for the analysis presented here suggests that they should be entirely general. Simultaneous negative contrast is best regarded as an instance of inhibitory conditioning: the presentation of a reinforcer less favourable than that expected is sufficient to establish inhibitory conditioning to the stimulus signalling that reinforcer.

Inhibitory conditioning may also be partly responsible for the successive negative contrast effect sometimes observed when subjects are shifted from a more favourable to a less favourable reinforcer in a constant stimulus situation. Studies of inhibition in classical conditioning, however, suggest that a successive shift to a less favourable reinforcer results in less inhibitory conditioning than that produced by concurrent exposure to more and less favourable reinforcers. This may explain why successive contrast effects are more elusive than simultaneous effects. The one situation where successive contrast is usually observed is when rats are shifted from a large to a small reward: it is probable that generalization decrement is partly responsible for this result.

CHAPTER 8

Extinction

I. THEORIES OF EXTINCTION

When a CS is followed by a UCS, or an instrumental response by a reinforcing event, the probability of the CR or instrumental response usually increases. If all else is held constant, the omission of the UCS or reinforcer defines the operation of experimental extinction and usually results in a decrease in the probability of responding. The term "extinction" is used (as is the term "reinforcement") interchangeably for operation and result: the omission of reinforcement is the operation of extinction; the result of this operation is the extinction of responding.

That classical or instrumental responses, initially established by appropriate contingencies of reinforcement, should decline when reinforcement is omitted is a fact so unsurprising that it takes an effort of the imagination to see that it posed substantial problems for early S–R theories of learning. Neither Pavlovian reinforcement theory nor the law of effect, taken by themselves, contain any set of statements sufficient to account for extinction: although both attribute increases in response probability to the presentation of a reinforcing event, it is not clear how either can explain, without further assumptions, why the omission of reinforcement should result in a decrease in the probability of responding. Hence, apparently, the need for theories of extinction.

This conclusion may, however, be too strong. One major theory of extinction, most systematically analysed by Capaldi (1967a), argues that the sole effect of the omission of reinforcement is to change the stimulus situation affecting the subject: if reinforcement is omitted during extinction, responding will decline because it was not established under the conditions prevailing during extinction. The effects of extinction are attributed to generalization decrement. Two major additional theories can be discerned. According to the first, the response reinforced during acquisition disappears during extinction because of competition from other responses whose probability increases during the course of extinction. According to the second, the decline in response probability is attributed to inhibition produced by nonreinforcement. The distinction between interference and inhibition is not always clear or easy to draw (see discussion below). A final theory of extinction is that proposed by Hull (1943) which attributed decreases in responding to a process of "reactive inhibition"—a process which needs to be distinguished sharply from inhibition in the sense in which that term has been used up to this point.

The chapter begins with a preliminary discussion of some of the merits and failings of these several theories, of the sort of evidence which points to the operation of one process rather than another, and with an attempt to clarify the distinctions between these theories; after this, the main variables affecting extinction will be discussed and an attempt made to interpret them within the framework of these theoretical positions.

A. Generalization Decrement Theory

Some form of inhibitory or interference analysis of extinction will, in the end, prove necessary; that is to say, there are data which clearly require the postulation of a specific, response-decrementing, learning process that produces appropriate changes in behaviour when the reinforcement contingencies which prevailed in acquisition are changed. It is, however, a matter of some interest to press an analysis of extinction which does not invoke any such new process.

Several earlier discussions have suggested that the delivery of a reinforcer can be regarded as an event which may serve as a cue to control an animal's subsequent behaviour. In particular, the observation of reliable patterning when rats are trained on an alternating schedule of reinforced (R) and nonreinforced (N) trials, even when the interval between trials is several minutes, requires the assumption that rats remember the outcome of one trial and can use the memory of that outcome as a cue to predict the outcome of the next (p. 162). This observation may be generalized to suggest that animals remember the outcomes of preceding trials, and that the memory of these events can be regarded as forming part of the set of events or stimuli available for association with reinforcement on a subsequent trial.

If a rat is trained to run in an alley for consistent reinforcement, one part of the set of events associated with reinforcement on each trial will be the memory of preceding R trials. The omission of reinforcement in extinction will ensure that the memory of preceding N trials will start to replace the memory of preceding R trials and will therefore result in a change in the set of events from that which was associated with reinforcement during acquisition. A response which was established under one set of conditions will, by generalization decrement, be less likely to occur when those conditions are changed. Furthermore, the *progressive* decline in performance observed as extinction continues can be attributed to a progressive change in these conditions. At the end of consistently reinforced acquisition, all preceding trials have been reinforced and the set of events associated with reinforcement will include the memory of only R trials. After a small number of nonreinforced extinction trials, this set will now contain the memory of some N trials as well as the R memories from acquisition. As extinction continues, the proportion of R elements must decrease and the proportion of N elements increase. Thus the set will become progressively less similar to that which prevailed during acquisition and there will be a progressive extinction of the instrumental response, with the rate of this extinction depending upon the relative importance of external stimuli and memories of preceding trials as determinants of instrumental responding. This argument has been presented in a slightly different form by Capaldi (1967a) who assumes that a sequence of N trials generates a set of after-effects proportional to its length: the outcome of a single N trial is represented as N_1, while the outcome of 10 N trials in a row is N_{10}, and N_1 is assumed to be more similar to R than is N_{10}. A reformulation of Capaldi's theory in terms of stimulus sampling theory by Koteskey (1972) provides an analysis similar to, but more elegant than, that suggested here.

Even if generalization decrement cannot be seriously regarded as a complete account of extinction, a theory which appeals to no new principles has much to recommend it as a first approximation. It is worth seeing how many phenomena occurring in extinction can reasonably be attributed to generalization decrement before turning to other theories. There are three general kinds of result which suggest that generalization decrement contributes to the change in response probability observed in extinction.

Typical extinction procedures omit the delivery of reinforcement, but maintain other conditions that held in acquisition relatively constant. If any of these other conditions are changed, however, the chances are high that performance in extinction will decline more rapidly: resistance to extinction is usually decreased by changes from the conditions under which responding was initially established. Among the changes that have been reported to decrease resistance to extinction are changes in intertrial, interval or ITI (Sheffield, 1950; Teichner, 1952; Capaldi and Minkoff 1966), the length of time the subject is detained in the goal box of an alley (Hulse, 1958; Capaldi, 1966b; Tombaugh, 1966); the wavelength of light projected onto a pigeon's response key (Azrin and Holz, 1966); and the subject's drive level (Barry, 1958). There are, it need hardly be said, many exceptions. If there was none, it would imply that the subject's behaviour in acquisition had come under the control of every possible set of stimuli in the experimental situation. As we have seen (p. 46), and shall discuss in even greater detail in Chapters 9 and 10, not all stimuli present at the moment of reinforcement enter into associations with reinforcement. Ferster (1951), for example, found that the performance in extinction of rats trained to press a lever in the presence of a buzzer and light was unaffected by the removal of these stimuli. Several studies varying the effort required to make the instrumental response have found that changing the effort requirement from acquisition to extinction does not affect resistance to extinction (e.g., Aiken, 1957; Johnson and Viney, 1970). Changes in drive from acquisition to extinction have not always decreased resistance to extinction (e.g., Lewis and Cotton, 1957). But the general run of the evidence is consistent with the main contention of generalization decrement theory: performance in extinction declines as the conditions of extinction change from those prevailing in acquisition.

That other changes in conditions affect performance in extinction does not, of course, necessarily imply that the change produced by the omission of reinforcement affects behaviour in the same way. The second general source of support for generalization decrement theory is the observation that the rate of extinction is decreased if acquisition has been so scheduled as to ensure reinforcement of the instrumental response after the subject is exposed to N outcomes similar to those experienced in extinction. The partial reinforcement effect (PRE) is the reference result: rats trained to run in an alley, but reinforced on only some proportion (often 50 per cent) of trials, show much greater resistance to extinction than animals consistently reinforced on all trials. The PRE is a much investigated phenomenon, for which a host of competing explanations have been offered. As we shall argue below, however, there is overwhelming evidence that much of the effect is attributable to the fact that PR subjects have been reinforced for making the instrumental response in a situation similar to that encountered in extinction.

Finally, if rate of extinction can be decreased by ensuring that the instrumental response is reinforced after N outcomes similar to those prevailing in extinction, it should also be decreased by presenting reinforcement during extinction. The suggestion does not, as might at first sight appear, rest upon any paradox: if acquisition in classical and instrumental experiments is defined as the presentation of a reinforcing event contingent upon either a stimulus or a response, extinction need not be defined as the complete omission of the reinforcer, but may include the case where the contingency or contiguity between stimulus or response and reinforcer is degraded. There is considerable evidence that the continued presentation of a reinforcer during extinction, under conditions which are quite insufficient to establish responding in acquisition, will result in surprisingly effective maintenance of an already established response. By comparison with a condition in which the reinforcer is omitted, such a procedure ensures a marked decrease in rate of extinction. Extinction of classically conditioned eyelid responses in human subjects is retarded if the UCS is presented either at long delays or during the ITI (Spence, 1966). Although random presentations of a CS and shock will not normally result in the acquisition of significant conditioned suppression (p. 28), suppression already established to a tone by appropriate pairings with shock may show little evidence of extinction when subjects are shifted from contingent to random shocks (Ayres and DeCosta, 1971). Several studies have shown that the removal of the contingency between responding and shock is a less effective procedure for extinguishing avoidance responses than is the complete removal of shock (p. 285). In appetitive instrumental experiments, Rescorla and Skucy (1969) and Boakes (1973) have shown that response-independent delivery of reinforcement may retard the extinction of lever pressing in rats or key pecking in pigeons. Uhl and Garcia (1969) reported a similar finding in rats, even when the presentation of reinforcement in extinction was contingent upon not pressing the lever for a specified interval. They went on to provide evidence that the basis for this effect was the maintenance of certain aspects of the conditions prevailing during initial reinforcement of lever pressing: in the free-operant situation they employed, responding in acquisition would normally immediately follow the delivery of reinforcement; they showed that the prevention of responding for 30 sec following each reinforcement in extinction abolished the effects of delivering that reinforcement.

It is, however, important to note that the noncontingent presentation of reinforcement in these experiments did not indefinitely maintain responding at the level attained during acquisition. Reliable extinction was observed in nearly all cases. Since the continued presentation of reinforcement might be expected to have prevented the occurrence of generalization decrement, the decline in responding observed in these cases suggests that additional assumptions must be made in order to provide a complete explanation of extinction.

B. Inhibition and Interference Theories

According to a generalization decrement analysis, at least part of the decline in responding in extinction does not reflect any new learning process operating as a consequence of nonreinforcement. In view of the emphasis placed in earlier chapters on inhibitory conditioning and inhibition in instrumental learning, reliance on such a theory of extinction may seem paradoxical. But it must be remembered that the conditions of simple extinction are not those which earlier discussions have suggested are optimal for inhibitory learning. Inhibitory conditioning was shown to be markedly influenced by the concurrent presentation of reinforcement in the absence of the stimulus that was to be turned into a conditioned inhibitor (p. 36). In a simple study of extinction following consistently reinforced acquisition, nonreinforcement occurs in the absence of any further reinforcement. Furthermore, the discussion of negative contrast effects (p. 397) suggested that while an inhibitory analysis was required to account for simultaneous negative contrast, in which animals are exposed to a less favourable condition of reinforcement at the same time as they concurrently experience more favourable conditions of reinforcement in a different situation, successive negative contrast (when animals are shifted from a more to a less favourable condition of reinforcement without continued exposure to the more favourable condition) is a less general phenomenon and may be largely attributed to generalization decrement.

These considerations provide additional reasons for taking generalization decrement seriously as a partial cause of simple extinction. But it is not difficult to use these same arguments to show that many extinction situations should be such as to involve inhibition. In partial reinforcement experiments, N trials are interspersed with R trials during acquisition and nonreinforcement should therefore have inhibitory effects. In experiments on successive acquisitions and extinctions, in which subjects are given alternate blocks or sessions of R and N trials, a similar degree of interspersing is in effect. Here, indeed, it is perfectly clear that some additional learning process must occur on N trials, for animals typically show an increase in rate of extinction over such a series of repeated acquisitions and extinctions and such a change in rate implies that they learn something on N trials that decreases the probability of responding.*

1. *Distinction between Inhibition and Interference*

The nature of this learning process remains a matter of some contention. We shall define interference theories of extinction as those which assert that extinction establishes some set of responses whose occurrence competes with, and eventually prevents the appearance of, the originally reinforced response. An inhibition theory of extinction is one which asserts that nonreinforcement in extinction is a sufficient condition for subjects to learn that an expected reinforcer is no longer contingent upon

---

* It is noteworthy that the evidence for the occurrence of attack and aggressive behaviour in extinction (presumably an index of inhibition or frustration) comes from studies of repeated acquisitions and extinctions, rather than a single extinction session following consistent reinforcement (p. 263).

a particular stimulus or response, and that such learning results in the suppression of the originally reinforced response. The distinction between these two positions is essentially the same as that between interference and suppression theories of punishment (p. 278). Just as it is possible to regard the decline in the probability of a punished response as a consequence of the learning of the relationship between that response and an aversive reinforcer, so inhibition theory, as defined here, regards a decline in the probability of responding in extinction as a direct consequence of the learning of a relationship between responding and nonreinforcement. Interference theories of extinction and punishment, on the other hand, attribute a decline in the probability of extinguished or punished responses to an increase in the probability of some other, competing response.

Detailed analyses of behaviour during extinction after classical conditioning have often revealed, just as in studies of punishment, that a decline in the original CR is accompanied by an increase in other behaviour (e.g., Hilgard and Marquis, 1935; Zener, 1937); studies of instrumental extinction have also observed such effects (e.g., Wendt, 1936; Bindra, 1961; Jones and Bridges, 1966). But, as we noted in the discussion of punishment, the emergence of a new response, even if it is closely correlated with the disappearance of an old one, does not prove that the new response competes with and suppresses the old. Its appearance may be as much a consequence as a cause of the extinction of the originally reinforced response.

Some studies by McFarland (1969; McFarland and L'Angellier, 1967) have illustrated how, in a somewhat different context, these two possibilities may be distinguished. When animals are placed in a situation which generates conflict between two or more activities (e.g., between attack and flight, or eating and flight, or eating and drinking), they often engage in apparently irrelevant "displacement activities" (McFarland, 1966b). Sticklebacks, for example, start digging in the sand, and various birds start preening when in conflict between attack and flight. Such displacement activities also occur when a dominant response tendency is blocked, or when its motivation declines. If hungry doves are trained to peck a key for food, they may occasionally stop pecking if, for example, responding is made more effortful or they become satiated, or during a sequence of nonreinforced responses. During these pauses, they may be observed to engage in irrelevant and apparently interfering activities, such as preening or drinking. The question, then, is whether preening or drinking are true competing responses whose occurrence causes the pause in pecking for food, or whether the pause appears for other reasons and then permits the appearance of these other activities. McFarland has argued that one can distinguish between these two suggestions by varying the motivation for preening or drinking. If they are competing responses, then an increase in their motivational strength (by increasing thirst or disturbing the birds' feathers) should produce earlier, more frequent, and longer pauses in pecking for food. In a series of experiments, however, the only effect of increasing the strength of a particular irrelevant activity was sometimes to increase the length of each pause in ongoing pecking and always to increase the probability of that particular activity occurring during such pauses; in no case, however, did the frequency of pauses increase, or the amount of time elapsing before the first pause decrease. McFarland concluded that this ruled out any interpretation of preening or drinking as competing responses: pecking for food must have been inhibited by some other mechanism and when inhibited must have permitted the appearance of these irrelevant activities.

In at least one situation, therefore, responses that appear at first sight to satisfy appropriate criteria for interfering or competing responses, appear on further investigation not to be causally responsible for the suppression of an ongoing pattern of instrumental behaviour. While it would be absurd to generalize too widely from a single situation, it is reasonable to conclude that the mere observation of other responses emerging during the course of extinction is not sufficient grounds for accepting an interference theory of extinction. The sort of evidence required is a demonstration that the independent manipulation of those responses affects extinction performance appropriately: the prevention of an interfering response, for example, should ensure the reinstatement of the extinguished response that it displaced. It is not too dogmatic to assert that in this sense there is no evidence whatever to support an interference theory of extinction.

2. *The Nature of Potential Interfering Responses*: *Frustration Theory*

Interference theory must not only show that competing responses are the cause of the suppression of the originally reinforced response, it must also show why any new response should ever become sufficiently strengthened during extinction to compete with the original response. It is not at all clear that the theories advanced by Guthrie (1935) and Wendt (1936) succeed in doing this, and Guthrie (1935, pp. 69–70) was willing to accept that the original response might have to be first inhibited before the competing response could be securely established. Even then it is not clear what process is responsible for the later establishment of the competing response, unless it is simply a dominant response in the experimental situation that was suppressed by the appearance of the CR or instrumental response during initial acquisition (e.g., Hilgard and Marquis, 1935).

Zener (1937) appears to have been the first to suggest that interfering responses developing in extinction were specifically caused by nonreinforcement, i.e. that they were emotional responses generated by frustration at the omission of an expected reinforcer. Skinner (1938, pp. 76–8; 1950) also attributed extinction to an increase in emotional responses elicited by nonreinforcement. Frustration theory, as developed by Amsel (1958, 1962, 1967, 1972) and Spence (1960), is by now the only seriously accepted version of a competing response theory of extinction. Although it has not presented evidence satisfying the stringent criterion suggested above for distinguishing between inhibition and interference theories of response suppression, it has solved some of the other problems typically encountered by interference theories. The omission of an expected reinforcer does indeed elicit a variety of responses and patterns of behaviour (such as aggressive attacks, biting, turning away from the CS or discriminative stimulus) and these responses may be intuitively regarded as indicators of frustration (p. 263); there is, therefore, no problem in accounting for the origin of the required competing responses. Many of these patterns of behaviour also look as if they were indeed incompatible with the originally reinforced response: they are, for example, withdrawal rather than approach responses, or may be directed at other aspects of the experimental situation. In this sense, therefore, frustration theory does seem to provide a viable account of extinction.

By describing his earliest account of frustration theory as "an addition to Hull's two-factor theory" of extinction, Amsel specifically recognized that if it is simply assumed that frustration reactions cause the disappearance of the originally established response by interference or competition, then frustration theory cannot be, and is not intended to be, a complete theory of extinction. This is not particularly because it could apply only to the extinction of appetitively reinforced behaviour, for there is logically nothing to stand in the way of a completely analogous theory of aversive extinction (such as the relaxation theory proposed by Denny, 1971). More important is the evidence which suggests that frustration is only a temporary phenomenon, produced by the initially unexpected omission of reinforcement: continued omission of reinforcement results in the eventual disappearance of frustration reactions. Aggressive behaviour tends to be confined to the beginning of extinction periods (p. 263); escape from a stimulus signalling nonreinforcement occurs at the onset of that stimulus and declines with continued discrimination training (p. 264); the double-alleyway frustration effect and other transient contrast effects decline with continued exposure to nonreinforcement (p. 361). If responses elicited by frustration were genuine competing responses, and were the sole cause of the suppression of a reinforced response in extinction, their gradual disappearance should be accompanied by the reappearance of the original instrumental response. In order to account for the more permanent disappearance of responding after the withdrawal of reinforcement, it is obvious that one must appeal to some additional process underlying extinction, specifically to some process responsible for the extinction of the expectation of reinforcement. It is precisely such a process that is envisaged in what is here being called an inhibitory theory of extinction.

On grounds of parsimony, therefore, one might well wonder whether an appeal to the interfering effects of frustration is a necessary part of any theory of extinction. Even if the omission of an expected reinforcer produces a set of frustration reactions, it does not follow that the occurrence of those reactions is responsible for the extinction of the original response. Both the appearance of the frustration responses and the disappearance of the trained response might be a consequence of learning that a situation previously associated with reinforcement is no longer so associated.

C. Hull's Theory of Reactive Inhibition

The most venerable version of an inhibitory theory of extinction in American psychology has been the theory of reactive inhibition, attributed by Hull to Mowrer (Mowrer and Jones, 1943) and Miller (Miller and Dollard, 1941, p. 40), but more usually associated with Hull's name (Hull, 1943, Chapter 16). According to Hull, two factors, one transitory and one permanent, are responsible for the decline in responding observed in extinction. The basic assumption is that the occurrence of a response, whether reinforced or not, generates a transient state of reactive inhibition, which temporarily reduces the probability of repeating that response. In order to account for the permanent decline in responding typically observed in extinction, it is further assumed that the transient state of reactive inhibition will also give rise to a permanent state of conditioned (reactive) inhibition.* The decay of the transient state then serves to explain such phenomena as spontaneous recovery. In order to distinguish Hull's theory from other inhibitory theories of extinction, the term "reactive inhibition" will be used, in what follows, to refer both to the transient and to the conditioned states postulated by Hull.

The first point to note about Hull's reactive inhibition is that it has little to do with inhibition in the sense in which the term has been used here. Although it is a response-decrementing process, it is not a function of nonreinforcement as such. Hull's account of extinction does not assume that any new learning process is brought into play by the transition from

---

* The specific assumption is that reactive inhibition is a motivational state, akin to fatigue, inducing a cessation of responding. If the organism does stop responding, this drive state will be reduced, and the response of stopping responding will be reinforced by the consequent reduction in drive. The problems arising from this line of reasoning have been excellently discussed by Gleitman *et al.* (1954).

acquisition to extinction; it does not assume that anything new happens in extinction at all. Reactive inhibition is generated every time a response occurs and the only reason why the probability of the appropriate instrumental response tends to increase in acquisition and decrease in extinction is because the effects of reinforcement are sufficient to outweigh the effects of reactive inhibition during acquisition; the removal of reinforcement permits the effects of reactive inhibition to be displayed undisguised.

Unconditioned reactive inhibition is simply a drive factor produced by a response that reduces the probability of repeating that response. It is most easily understood as a process analogous to fatigue and was, indeed, postulated to be a function of the effortfulness of a response. Hull termed the unit of reactive inhibition the *Pav*, but the implication of Pavlovian parentage is somewhat misleading. It is true that Pavlov's theory of internal inhibition contains elements that suggest some affinity to reactive inhibition (e.g., Pavlov, 1927, p. 234), but Pavlovian internal inhibition is also a process attributable to nonreinforcement, rather than a process analogous to fatigue occurring in both acquisition and extinction (Pavlov, 1927, p. 60).

The second point to note about the theory of reactive inhibition is that even if it were the case that the occurrence of a response automatically produced a tendency not to repeat that response, such a process would not necessarily constitute an important part of the explanation of extinction. The evidence thought to support the theory of reactive inhibition does nothing to dispute this judgement. For even if it were universally true that more effortful responses extinguished more rapidly than less effortful responses, and that the massing of trials depressed performance in acquisition and accelerated the process of extinction (to take two examples of predictions derivable from the theory of reactive inhibition), this would do no more than suggest that reactive inhibition is a process whose effects must be taken into consideration when attempting to predict and understand any decrease in responding during extinction. It would not show that reactive inhibition was a fundamental cause of that decline.

In fact, the evidence bearing on these two issues is singularly confused. Studies of effort and extinction have been beset by methodological problems, and have produced conflicting results. The most popular procedure has been to vary the force required to depress a lever; but when this is varied between groups, there is the possibility that animals required to exert more pressure may make numerous incomplete responses which fail to meet the experimenter's criterion: in acquisition, therefore, these subjects may be on a partial reinforcement schedule, while in extinction only a fraction of their behaviour may be recorded. A procedure which, it has been claimed, may eliminate some of these problems is to vary the effort required to run in an alley by varying the slope of the alley; but this may result in variations in delay of reinforcement.

Lever-pressing studies have often found that a higher force requirement in extinction leads to fewer responses in extinction (e.g., Mowrer and Jones, 1943; Applezweig, 1951; Capehart et al., 1958), but there are numerous exceptions to this generalization (e.g., Maatsch et al., 1954; Aiken, 1957; Quartermain, 1965; Young, 1966) and in none of these studies has a group of animals, both trained and extinguished on a more effortful response, extinguished significantly faster than a group trained and extinguished on a less effortful response. In alleyway studies, Lawrence and Festinger (1962, Experiment 14) found that animals trained and extinguished with a steeply inclined alley were significantly more resistant to extinction than those trained and extinguished in the less sloping alley; Johnson and Viney (1970) reported exactly the opposite result. The most notable difference between the two studies is that the former used spaced trials, and the latter used massed trials; it is certainly true that if the difference in their outcome were attributable to this difference in procedure, such a result would be entirely consistent with a theory of reactive inhibition. But the conclusion most forcibly suggested by this rather cursory review is that effort is simply not the decisive determinant of rate of extinction that is required by a theory which regards extinction as a consequence of response-produced inhibition.

The second prediction derivable from the theory of reactive inhibition is that extinction should be affected by the distribution of trials. As will be seen below, the effects of this variable on extinction are only marginally more consistent than those of effort. Furthermore, there are other theories equally able to predict that massed extinction should proceed more rapidly than spaced extinction.

Even if none of this were true, the conclusion would stand that reactive inhibition cannot be regarded as a major cause of extinction. There are several reasons for this. For one thing, responses can be extinguished without being elicited. The first line of attack on Hull's theory of extinction was launched by Seward and Levy (1949), who showed that the response of rats trained to run in an alley for food in the goal box could be extinguished simply by placing them in the goal box without food. Although this demonstration of "latent extinction" was immediately disputed by Bugelski et al. (1952), subsequent studies have shown that the result is an entirely reliable one (e.g., Robinson and Capaldi, 1958; Dyal, 1962, 1964; Clifford, 1964).* The term "latent extinction" suggests the parallel with studies of

---

* The most notable feature of the procedure of Bugelski et al. was that their subjects might well have found it difficult to discriminate between being placed in the empty goal box and being placed in the start box at the beginning of each normal trial. Several

latent learning, and the parallel is indeed close. The fact that rats, placed in an empty goal box (in which they have previously found food), run down the alley slowly on their first test trial indicates that speed of running must be determined by anticipation of reward. This anticipation can be both established and extinguished by direct placements, and extinction therefore must be partly a matter of the disconfirmation of previously established expectancies (cf. Gonzalez and Shepp, 1965).

Other studies have shown that extinction can occur without the instrumental response occurring. Avoidance responding, indeed, may be extinguished most effectively by procedures which prevent the overt occurrence of the response: Black (1958), for example, showed that extinction of an avoidance response in dogs was substantially facilitated if the first 50 extinction trials were given while the animals were immobilized with curare.

The most striking feature of Hull's theory of extinction is that the only effect it attributes to the omission of reinforcement is that of preventing further increases in response strength. The only difference between acquisition and extinction, therefore, would appear to be that continued reinforcement in acquisition will ensure the continued growth of response strength. But even this is not true, for habit strength is assumed to have an asymptote and once asymptote is reached, further reinforcements should have no further effect and the response must spontaneously extinguish.

Pavlov (1927, pp. 234–6) observed such a phenomenon, terming it "inhibition with reinforcement". Ascribing it to the progressive development of inhibition, he said that "the function of the unconditioned reflex after the conditioned reflex has become established is merely to retard the development of inhibition (p. 234)". In spite of his arguments to the contrary, however, several features of his results, as Konorski (1948, p. 144) and Sheffield (1965) have noted, suggest a much stronger resemblance to "inhibition of delay" (p. 61). A few intrepid experimenters have looked for the spontaneous extinction of instrumental running in an alley by rats— Kendrick (1958) with some success and Fuchs (1960), in an attempt to replicate Kendrick's results, with no success at all. Kendrick's positive results should be interpreted with some caution: even if rats that are required to run a 10-foot alley 30 times per day with no interval between successive trials do eventually (after nearly 1,000 trials) stop running, this does not really provide particularly strong evidence for the importance of reactive inhibition under more normal circumstances. To maintain perspective, two points should be remembered: first, Hull's account of extinction predicts that eventually the continued delivery of reinforcement will have no effect at all on the maintenance of responding; secondly, in more

studies have shown, what is hardly surprising, that this discrimination is essential (Denny and Ratner, 1959; Hughes, et al., 1960).

typical experimental situations, to which, after all, a theory of extinction should apply, animals continue to respond for a very long time without any apparent progressive decline in performance. Pigeons trained in free-operant situations will maintain a steady rate of pecking of between 20 and 100 pecks per minute for hundreds of sessions when trained on VI schedules that deliver reinforcement no more often than once every few minutes. It is not enough to make the evident observation that pecking a key is less effortful than running a 10-foot alley, for key pecking also extinguishes when reinforcement is omitted. It seems clear that reactive inhibition can have little to do with the extinction of responses such as these.

D. Conclusions

Reactive inhibition cannot seriously be maintained as an important determinant of extinction. Traditional frustration theory provides only a partial explanation of extinction, since frustration is a transient state generated at the outset of extinction when reinforcement is still expected. Finally, generalization decrement theory, while undoubtedly pointing to an important determinant of the decline of responding during extinction, is not a theory about the changes in associations produced by extinction. Traditional S–R theory is thus left with no account of the learning processes underlying extinction. A complete explanation of extinction needs to take seriously the idea of inhibitory learning, defined as the disconfirmation of expectancies: if subjects learn to associate stimuli or responses with reinforcing events, when appropriate contingencies are arranged between these events in acquisition, then they must also be supposed to learn about the new relations between stimuli, responses and the omission of reinforcement that are scheduled in extinction. It is tempting to suppose that this is, indeed, the major associative change occurring during extinction, just as the learning of the relationship between a response and the presentation of an aversive reinforcer is the major associative change underlying punishment. This is not to deny the importance of some of these other factors. An understanding of the effects of punishment seemed to require appeal to the discriminative properties of punishing stimuli (an explanation analogous to generalization decrement), the role of responses elicited by punishment (an explanation analogous to frustration theory), as well as a direct, response-suppressing, associative process. The facts of extinction are no less complex.

II. VARIABLES AFFECTING EXTINCTION: CONSTANT ACQUISITION CONDITIONS

Theories of extinction must not only explain the fact that the probability of responding declines during extinction, they must also account for differences in resistance to extinction produced by different treatments. Not surprisingly, performance in extinction is affected by variations in the conditions of extinction; it is the effects of different acquisition treatments that are both more interesting and pose more of a theoretical challenge. Somewhat arbitrarily, the present section discusses the effects of several constant factors on performance in extinction, while the effects of acquisition treatments in which conditions are varied from trial to trial are discussed in the following section. Most of the data considered are derived from studies of instrumental extinction: this is partly because we have already briefly touched on some features of the extinction of classical conditioning (Chapter 2), but largely because the experimental and theoretical analysis of extinction has tended to concentrate on instrumental experiments.

A. Distribution of Trials

1. *Data*

The received opinion is that the massing of extinction trials increases the rate of extinction. Pavlov (1927, p. 53) reported that a salivary CR was extinguished in five trials when extinction trials were given at an intertrial interval (ITI) of 4 min, but had not completely extinguished after eight trials when the ITI was 16 min. Hilgard and Marquis (1940, p. 133) and Kimble (1961, p. 283), although somewhat erroneously stating that Pavlov's data showed no effect of ITI on extinction, concluded that there was sufficient evidence from other experiments to support the generalization. Although there are important exceptions, the generalization does seem to have some validity.

Several experiments studying extinction of running in an alley have shown that rate of extinction is decreased by the use of a relatively long ITI. Krane and Ison (1971) reported a regular decrease in rate of extinction as the ITI increased from 1 to 4 to 10 min. Mackintosh (1970a) found that a 40 min ITI resulted in slower extinction than a 30 sec ITI. Hill and Spear (1962) and Birch (1965) showed that a 24 h ITI resulted in much slower extinction than a 20 or 60 sec ITI. To these data may be added the results of two studies of discrete-trial lever pressing. Rohrer (1949) compared 10 sec and 90 sec ITIs, and Teichner (1952) varied the ITI between

15 and 90 sec. In both experiments rate of extinction decreased as the ITI increased.

There are two exceptions to the conclusions suggested by these data. First, as noted earlier (p. 407), a change in ITI from acquisition to extinction may increase rate of extinction, whether the change is to a longer or shorter ITI. Teichner (1952), for example, found that rats, trained to press a lever with a 45 sec ITI in acquisition, extinguished more slowly with a 45 sec ITI in extinction than with either a 60 sec or a 30 sec ITI. The second exception is provided by a group of studies which have found that rats trained to run in an alley may extinguish more slowly at very short ITIs than at somewhat longer ITIs. Both Sheffield (1949, 1950) and

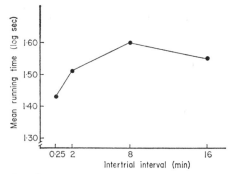

FIG. 8.1. Mean running times in an alley over 30 extinction trials as a function of the intertrial interval in acquisition and extinction. (*After* Cotton and Lewis, 1957.)

Wilson *et al.* (1955) reported slower extinction with a 15 sec ITI than with a 15 min ITI. The apparent implication is that the extreme massing of extinction trials in an alley may result in slower extinction than an ITI of a few minutes, but that further increases in ITI ensure a decline in the rate of extinction. Unfortunately, there is only one study that has examined rate of extinction across a sufficiently wide range of ITIs to provide evidence of both increases and decreases in rate of extinction (Cotton and Lewis, 1957). The remaining evidence for this conclusion depends upon comparisons across experiments which undoubtedly differ in numerous other respects.* However, Cotton and Lewis' data, which are illustrated in Fig. 8.1, clearly suggest that as the ITI is varied between 15 sec and 16 min there is first an increase and later a decrease in rate of extinction. It should,

---

* It is notable that the experiments which have shown that extreme massing of trials may decrease rate of extinction have all given only a single session of extinction, while the experiments which have shown that moderate massing of trials may increase rate of extinction have usually studied extinction over several sessions. This difference may well be critically important.

however, be noted that the evidence for a decrease in rate of extinction with very short ITIs is derived exclusively from studies employing rats in an alley. The data of Rohrer (1949) and Teichner (1952), described above, provide no evidence of such an effect when rats are trained in a lever box.

2. *Theory*

The available data suggest the following conclusions. The main effect of the distribution of trials is that increases in ITI decrease the rate of extinction. Superimposed on this tendency, however, are two other factors: by generalization decrement, a change in ITI from acquisition to extinction may increase the rate of extinction; and in alleyway studies, some factor may temporarily increase speed of running in extinction at very short ITIs.

Sheffield (1949), the first investigator to observe this last result, suggested that it might be due to frustration or rebound from inhibition. It is, indeed, tempting to see the parallel between this result and the double-alley frustration effect (p. 266): in both cases, speed of running along an alley is temporarily enhanced immediately following the omission of an expected reinforcer. Interestingly enough, Rohrer (1949) also appealed to the notion of frustration to explain the entirely different results of his experiment: he suggested that the frustration produced by omission of reinforcement produced competing responses which disrupted the required instrumental response. It is easy to deride a theoretical notion that can be used to explain apparently diametrically opposed results with equal facility. However, in the present instance, the argument may not be altogether absurd. We have seen that positive contrast effects are not readily observed when latency of response initiation is measured (p. 386); it is, therefore, conceivable that the momentarily facilitative effect of omission of reinforcement would act to enhance speed of running in an alley, but not to decrease the latency with which a rat pressed a lever. An experiment by Stanley (1952) shows directly that the effects of ITI on extinction depend upon the measure of responding used. Rats were trained in a T-maze and extinguished with a 15 sec or 15 min ITI. Although speed of running in extinction was increased by the massing of extinction trials, choice of the former positive side extinguished more rapidly when trials were massed.

It remains to consider the major effect of the distribution of trials on extinction, the finding that rate of extinction declines with increases in ITI. It is tempting to see a parallel between this result and the phenomenon of spontaneous recovery. If a partially extinguished response shows some recovery after a long interval, it is hardly surprising that long intervals between successive extinction trials should result in only a gradual disappearance of conditioned responses. It is difficult to believe that both sets of results could not be explained by related processes.

The most widely accepted account of both findings seems to have been in terms of reactive inhibition: if the occurrence of a response results in a transient decrease in the probability of repeating that response, it is obvious that the massing of extinction trials will lead to a more rapid disappearance of the conditioned response. It is possible, however, that the effects of distribution of trials may be understood in terms of generalization decrement rather than reactive inhibition. According to generalization decrement theory, a major cause of the decrease in strength of an instrumental response in extinction is the change in the set of events stored in the subject's memory. In acquisition, this set consists of traces of immediately preceding reinforced outcomes (R traces); in extinction, the set starts including traces of nonreinforced outcomes (N traces). If we assume that memory traces decay over time, then a long ITI will ensure that there is a less salient trace of the outcome of one trial present at the beginning of the next. But if rate of extinction is determined by the change in these traces from acquisition to extinction, any condition which reduces their salience must decrease the rate of extinction.

Generalization decrement theory has traditionally (Sheffield, 1949; Hull, 1952, pp. 120–21) talked of stimulus traces, such as the traces of food in the mouth, rather than memory traces, and in this form it is obvious enough that such traces will form an important part of the stimulus complex controlling responding only when trials are relatively massed. But it is rather natural to assume that memory traces also fade or decay with the passage of time. If talk of trace decay is found objectionable, it is equally possible to assume that it is interference between traces which is responsible for the decline in their salience at long ITIs. If R traces and N traces interfere with one another, then prior exposure to R outcomes in acquisition will proactively interfere with the retention of N outcomes in extinction; now since proactive interference causes a progressive increase in forgetting with the passage of time (p. 479), memory for N outcomes in extinction will be reduced by long ITIs. This leads to exactly the same prediction as before: resistance to extinction will be increased at long ITIs, because the traces of recent N outcomes will be less salient.

It will later be argued that the concepts of proactive interference provide the simplest analysis of spontaneous recovery (p. 472). It has already been argued that results showing interactions between ITIs in acquisition and extinction are readily interpreted as instances of generalization decrement. It is reasonable, then, to suggest that generalization decrement theory explains much of the effect of distribution of trials on extinction.

B. Number of Acquisition Trials

Most traditional theories of learning took resistance to extinction as one of several interchangeable measures of response strength. As successive reinforcements add successive increments to response strength during acquisition, so they should progressively retard the subsequent rate of extinction. It is now clear that resistance to extinction cannot be regarded as providing anything like a direct measure of response strength; several variables, all of which undoubtedly increase the speed or rate of responding in acquisition, have often been found to increase rather than decrease the subsequent rate of extinction. These variables include the number of trials received in acquisition, and the magnitude and delay of reward.

A serious problem, often complicating the assessment of the effect of these variables on performance in extinction, is that since they typically have marked effects on performance at the end of acquisition, it may be difficult to decide whether differences in the overall level of performance observed in extinction are simply a consequence of the maintenance of the differences established in acquisition, and equally whether differences in the *rate* of extinction are simply a consequence of differences in the amount of room for a change in performance. The only valid way to show that an increase in the number of acquisition trials or in the magnitude of reward increases the rate of extinction is to show that the extinction curves of different groups actually cross over during the course of extinction.

1. *Data*

The data on the relationship between resistance to extinction and amount of acquisition training are somewhat confused. Early studies of lever pressing in rats, on which Hull (1943), for example, relied for the function relating the two, reported that over the range of 5 to 90 reinforcements, the level of performance maintained in extinction was an increasing function of the number of reinforcements received in acquisition (Williams, 1938; Perin, 1942). This particular result seems relatively secure: several subsequent studies of both discrete-trial and free-operant lever pressing have found no evidence that an increase in the number of reinforcements, even up to 1,000, will increase the subsequent rate of extinction (e.g., Harris and Nygaard, 1961; D'Amato *et al.*, 1962; Dyal and Holland, 1963; Uhl and Young, 1967). There are, however, two exceptions. Tombaugh (1967), and Barnes and Tombaugh (1970) trained rats to press a lever in a discrete-trial situation for 120, 360 or 720 trials and found that 720 acquisition trials resulted in a more rapid rate of extinction than 360 trials. Unfortunately, these results are at variance with those reported by Uhl and Young (1967); in this study, rats were trained for 180, 360, or 720 trials in a discrete-trial situation and the rate of extinction showed no sign of increasing with increases in the number of acquisition trials.

The results of alleyway studies are equally confusing, although in this case the more usual result has been that overtraining increases the subsequent rate of extinction. North and Stimmel (1960) are usually credited with the discovery of this effect, having found that subjects receiving 45 acquisition trials ran faster in extinction than subjects receiving 90 or 135 trials. Similar results have been reported by Ison (1962), Siegel and Wagner (1963), Wagner (1963b), Clifford (1964), Ison and Cook (1964), Madison (1964), Theios and Brelsford (1964), Likely and Schnitzer (1968), and Porter *et al.* (1971). Fig. 8.2 shows the results of three of the groups

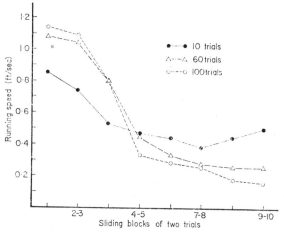

FIG. 8.2. Speed of running during the course of extinction as a function of the number of acquisition trials. (*After* Ison, 1962.)

run in Ison's experiment. It can be seen that although overtraining led to faster running at the outset of extinction, subjects given only 10 acquisition trials showed a notably less precipitous decline of running in extinction. There have, however, been several studies reporting different outcomes: Wagner (1961), Hill and Spear (1963a), Hill and Wallace (1967), and Ison and Rosen (1968), for example, found no evidence of an increase in the rate of extinction with an increase in the number of acquisition trials.

It is difficult to reconcile all of these conflicting results. One popular suggestion, derivable from frustration theory, is that overtraining may increase the rate of extinction only when a large reward is used. In general, studies using lever boxes have tended to employ small rewards, while those using alleys have tended to employ larger rewards. The results of two alley studies, moreover, have been thought to provide direct evidence of the importance of this factor (Ison and Cook, 1964; Traupmann, 1972). Nevertheless, there are many results that remained unaccounted for by this suggestion. Uhl and Young (1967) and Barnes and Tombaugh (1970) varied the concentration of sucrose used as a reward in their studies of lever pressing. Although obtaining conflicting data on the effects of over-training on extinction, neither experiment found any interaction between the level of training and the concentration of sucrose. Similarly, Wagner (1961) observed no interaction between the level of training and the size of reward in his study of extinction in an alley, and two experiments have found that overtraining will increase the rate of extinction in an alley even when a very small reward is used in acquisition (Madison, 1964; Likely and Schnitzer, 1968).

One final result should be mentioned. Miller (1960) and Karsh (1962) have reported that if rats are trained to run in an alley for food, over-training may increase the extent to which responding is disrupted by the delivery of shock in the goal box, *even when food is still presented*. The implication is that overtraining does not increase the rate of extinction as such, but rather increases the probability that the running response will be disrupted by any adverse change.

2. Theory

The theoretical interpretation of these data is no more coherent than the results themselves. It is tempting to see an analogy between these results and those such as Kendrick's (1958), in which continued training led to extinction, without the necessity of omitting reinforcement. If one assumed that pressing a lever was a less effortful response than running in an alley, the general difference between lever box and alleyway studies might be attributable to differences in the accumulation of reactive inhibition. This explanation, however, is contradicted by Likely and Schnitzer's (1968) results: they found that the inverse relationship between level of training and resistance to extinction was abolished if animals were required to climb over a hurdle placed in the alley. Since this can hardly have failed to increase the effort involved in reaching the goal box, it should, according to reactive inhibition theory, have accentuated the effects of overtraining. Lawrence and Festinger (1962, Experiment 5) also found no effect of over-training on resistance to extinction in an alley containing several hurdles.

The most popular explanation of the so-called overtraining extinction effect has been in terms of frustration theory (Ison, 1962; Siegel and Wagner, 1963; Theios and Brelsford, 1964). The argument, in its simplest form, is that the degree of frustration produced by the omission of rein-forcement is a direct function of the level of expectation of reinforcement or incentive motivation, and that rate of extinction is a direct function of level of frustration. The explanation thus assimilates the effects of an increase in training to those of an increase in size of reward: both variables increase incentive motivation and both should therefore increase frustration.

There are several problems with frustration theory's analysis of the overtraining extinction effect. As has already been noted, the effect of overtraining on the rate of extinction does not necessarily depend upon the size of reward in the manner required by the theory: the theory also predicts that the effect should not occur if subjects receive partial reinforcement during acquisition. Wagner (1963b) observed no difference in the effect of overtraining as a function of schedule of reinforcement in acquisition. In accordance with frustration theory, on the other hand, Theios and Brelsford (1964) found that rate of extinction increased with an increase in the number of times rats were placed and fed in the goal box of an alley, in the absence of any increase in the number of trials they had been trained to run in the alley: this was interpreted to mean that the result depends not on practice in running but on the strength of anticipation of reinforcement. Marx (1967a), however, found no evidence of such an effect in an analogous lever box experiment, and Porter et al. (1971) were also unable to replicate Theios and Brelsford's results when animals were placed directly over the food cup in the goal box. The most damaging finding from the point of view of frustration theory, however, is that reported by Miller (1960) and Karsh (1962): if overtraining increases the extent to which a rewarded running response is disrupted by punishment, this suggests that the omission of reinforcement is not the critical factor responsible for increasing the rate of extinction following overtraining.

A second possible interpretation, which would account for these results, has been suggested by Mackintosh (1965b) and Logan (1970, p. 117): one commonly suggested effect of overtraining in alleys and mazes is a decrease in the variability of responding and an increase in the extent to which running comes under the control of proprioceptive rather than external stimuli (Restle, 1957; Mackintosh, 1965b, see p. 554). Since there is usually a greater variety of potentially controlling external stimuli, this shift from external to internal control may be accompanied by a reduction in the number of stimuli controlling behaviour. Since, moreover, there is evidence that resistance to extinction is directly related to the number of stimuli controlling behaviour (p. 434), this reduction in control might be accompanied by a reduction in resistance to extinction. A possible virtue of this analysis is that it stresses the importance of the type of response that the animal is required to make, thus helping to explain why the overtraining extinction effect may be confined largely to runways rather than lever boxes. It is also consistent with the finding of Likely and Schnitzer (1968) (mentioned above) that the effect is abolished by the presence of a hurdle in the alley: animals required to climb over a hurdle might well find it difficult to learn to make a smooth, proprioceptively chained, series of responses.

C. Magnitude of Reinforcement

The second instance cited above of an inverse relationship between response strength and resistance to extinction has been provided by studies of size of reward and extinction. Although there are certain inconsistencies between different experiments, the conclusion that large rewards in acquisition can increase the subsequent rate of extinction is securely established; many of the discrepancies fall into a particular pattern and the theoretical analysis of the effect has made reasonable progress.

1. Data

Experiments in which rats have been trained to run in an alley for consistent food reward have almost invariably found that rate of extinction is directly related to the size of the reward.* This result was originally reported by Hulse (1958), Armus (1959) and Wagner (1961), and has since been confirmed in more than half a dozen other studies (e.g., Ison and Cook, 1964; Zaretsky, 1965; Marx, 1967b; Gonzalez *et al.*, 1967b; Gonzalez and Bitterman, 1969; Roberts, 1969; Capaldi and Sparling, 1971a; Likely *et al.*, 1971; Ratliff and Ratliff, 1971; Traupmann, 1972). The result appears to depend upon the subjects' level of motivation (Marx, 1967b); as noted by Traupmann (1972), this may explain why Hill and Wallace (1967) found no effect of size of reward on performance in extinction in a large-scale study that provided the only serious exception to the generalization suggested by this array of experiments.

In situations other than the alley, with reinforcement other than food and in species other than the rat, however, there is virtually no evidence of an inverse relationship between magnitude of reinforcement and resistance to extinction. One study of discrete-trial lever pressing in rats for different amounts of food reported faster extinction following a larger reward (Marx, 1967b), but when different concentrations of sucrose have been used as rewards in such a situation, performance in extinction has been an increasing rather than decreasing function of concentration (Uhl and Young, 1967; Barnes and Tombaugh, 1970); a similar result has been reported when rats are trained to run an alley for sucrose rewards (Ison and Rosen, 1968; Likely *et al.*, 1971).

* It is important to note that this conclusion depends on the use of a consistent schedule of reinforcement in acquisition. With partial reinforcement, the use of a large reward usually ensures the maintenance of a higher level of performance throughout extinction (p. 455).

Experiments in which goldfish have been trained either to push a paddle (Gonzalez *et al.*, 1967b; Gonzalez and Bitterman, 1967), or swim down an alleyway (Mackintosh, 1971b; Gonzalez *et al.*, 1972) have uniformly found that rate of extinction is decreased after training with a larger reward. A similar relationship occurs in turtles (Pert and Bitterman, 1970). Finally, there is evidence that classically conditioned salivary responses may extinguish more slowly in dogs after conditioning with six pellets of food as the UCS than after a single-pellet UCS (Wagner *et al.*, 1964). There is, of course, considerable evidence that increases in shock intensity increase resistance to extinction following aversive conditioning (p. 70).

2. *Theory*

Frustration theory, as we have noted above, explains the effects of over-training and of size of reward on extinction in similar terms: both factors increase incentive and both, therefore, increase the level of frustration produced by the omission of reinforcement. It is worth remembering, however, that the evidence relating the magnitude of the frustration effect to the size of the omitted reward was, at best, somewhat tenuous (p. 364). A possibly more serious problem for frustration theory is the finding, reported by Capaldi and Sparling (1971a), that the effect of large reward on resistance to extinction survives the administration of amobarbital in extinction. As we have seen (p. 265), there is evidence which suggests that this drug attenuates the inhibitory or interfering effect of frustration.

Capaldi (1967a) has argued that the effect of reward size on extinction may be interpreted in terms of generalization decrement. He assumes that trial outcomes may be ordered along a continuum, as is shown in Fig. 8.3:

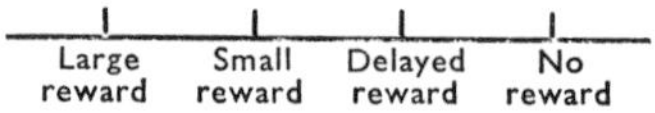

FIG. 8.3. Hypothetical continuum of trial outcomes, ranging from large, immediate reward, to no reward. Although the distances between neighbouring events are here shown arbitrarily, it is assumed that the continuum represents an ordinal scale, with a small reward being more similar to no reward than is a large reward.

small rewards differ from large rewards in the direction of nonreinforcement. Thus an instrumental response reinforced in the presence of traces of a small reward (R_S) will, other things being equal, show greater persistence in the presence of N traces, than will a response reinforced in the presence of traces of a large reward (R_L). Other things will not always be equal, of course, since the consistent presentation of a large reward will establish a stronger instrumental response than will consistent presentation of small rewards; but this difference in the reinforcing effects of large and small rewards may not be sufficient to outweigh the greater generalization decrement produced by the change from R_L to N, by comparison with the change from R_S to N.

An experiment by Leonard (1969) provides relatively detailed support for Capaldi's analysis. Leonard trained four groups of rats to run an alley, giving two trials a day for 15 days of acquisition, followed by a series of massed extinction trials. During acquisition, animals received either a large (L) or a small (S) reward on each trial, in the same sequence each day, with the sequence depending on which of the four groups they had been assigned to. The four sequences were: LL, SS, SL, and LS.

Leonard argued that in any experiment such as this, the reward received on Trial 1 of each day in acquisition, although reinforcing running on Trial 1, will have weaker reinforcing effects on later trials in a session, since the stimulus complex controlling running on Trial 1 differs from that present on later trials (among other things, the ITI preceding Trial 1 differs from that preceding later trials, and we have seen that the length of an ITI can form part of the stimulus complex controlling responding). The Trial 1 reward will, however, set up memory traces in whose presence responding on Trial 2 will be reinforced. Conversely, the reward received on Trial 2 will be more important as a reinforcer of responding on trials after the first in any session, but will not provide traces in whose presence responding is later reinforced. Two predictions follow from this argument. If subjects are given a series of massed extinction trials, rate of extinction will be a direct function of the size of reward received on Trial 1 of each day in acquisition: the increase in generalization decrement resulting from the omission of the larger reward will outweigh the difference in reinforcing effects. On the other hand, rate of extinction will be an inverse function of the size of reward received on Trial 2 of each day of acquisition, since the reinforcing effect of this reward should be more important than the generalization decrement produced by its omission.

The results of Leonard's experiments confirmed these predictions. The performance of each of his four groups in extinction is shown in Fig. 8.4. Group SS extinguished more slowly than Group LL, thus replicating the general finding that consistent large rewards increase the rate of extinction. The important groups, of course, are LS and SL. Group LS extinguished more rapidly than any other group, while Group SL was the slowest to extinguish. Thus resistance to extinction was greater for groups receiving a small reward on Trial 1 (SL and SS vs LL and LS), but less for groups receiving a small reward on Trial 2 (LS and SS vs LL and SL).* These results have been confirmed, under somewhat different conditions, by

---

* The theory predicts, and Leonard et al. (1969) have confirmed, that this result depends upon the massing of extinction trials. When extinction is given at the rate of one trial per day, the outcome of the experiment is quite different.

Bowen (1968) and by Capaldi and Capaldi (1970a). In this latter experiment, Groups SL and LS, in addition to receiving the appropriate sequence of large and small rewards, were also exposed, on some days, to an LL sequence. This was designed to prevent their learning to anticipate the occurrence of large and small rewards (since there is evidence, discussed below, that anticipated reductions in reward may have different effects on extinction than unanticipated reductions). Group SL was still significantly more resistant to extinction than Group LS.

Although frustration theory must, at best, remain silent about the outcome of these experiments, and thus cannot be said to provide a complete account of the effects of reward size on extinction, it is often asserted (e.g., by Amsel, 1967, and Gonzalez and Bitterman, 1969) that Capaldi's

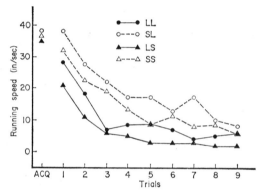

FIG. 8.4. Speed of responding at the end of acquisition (Acq.), and during the course of extinction, as a function of the sequence of large and small rewards received during acquisition. (*After* Leonard, 1969.)

explanation in terms of after-effects or memory traces of previous outcomes can apply only to experiments employing massed trials. Some other explanation, it is argued, must be found for the results of those studies in which large rewards increase the rate of extinction when subjects receive only one trial per day (e.g., Hulse, 1958; Wagner, 1961). This argument is discussed in greater detail below (p. 458); for the moment, it is only necessary to suggest that memory traces for regular, repeated events may reasonably be regarded as persisting for more than a few minutes, even in rats.

The final problem for any explanation of the effects of reward size on extinction is to show why the use of sucrose concentration rather than amount of food, or of turtles or goldfish rather than rats, produces an entirely different pattern of results. It can hardly have escaped attention that the results of experiments on extinction are precisely paralleled by those on successive contrast effects (pp. 390–395). A shift from a large to a small food reward produces a negative contrast effect in rats, but not in turtles or goldfish; and rats do not usually show contrast effects when concentration of sucrose is shifted. It should be possible to find a single explanation for both sets of results.

It was argued that successive negative contrast is probably more a matter of generalization decrement than of inhibition or frustration. This is largely because the absence of successive contrast is not correlated with the absence of simultaneous negative contrast, but does correlate with failures of alternation learning: if animals are unable to respond appropriately on a regularly alternating schedule of reinforcement, this implies that the different reinforcers used do not provide memory traces sufficient to serve as cues to predict the outcome of the following trial. If this is accepted, then generalization decrement theory may be regarded as the better explanation of the extinction data also. The use of a large reward will increase the rate of extinction only when its greater effectiveness as a reinforcer is outweighed by greater generalization decrement in extinction. The traces of different concentrations of sucrose do not appear to control instrumental responding very readily (Likely, 1970) and the behaviour of goldfish shows little evidence of control by memory traces of preceding outcomes (Mackintosh, 1971b).

D. Delay of Reinforcement

A third variable conforms to the pattern, suggested by the effects of overtraining and size of reward, that factors which appear to increase response strength in acquisition, may increase rather than decrease the subsequent rate of extinction. Studies of constant delay of reinforcement also conform to the by now familiar pattern of being somewhat inconsistent in their outcome.

The first experiment which suggested that a constant delay of reinforcement in acquisition might increase resistance to extinction, in spite of depressing performance in acquisition, was a study by Fehrer (1956). Similar results have been reported by Schoonard and Lawrence (1962), Sgro and Weinstock (1963), Capaldi and Bowen (1964), Marx *et al.* (1965), Tombaugh (1966), and McCain and Bowen (1967). Several studies, however, have reported that delay of reward may have little or no effect on subsequent extinction (Renner, 1963; Sgro *et al.* 1967; and Tombaugh, 1970).

Since delay of reinforcement in, for example, an alley involves detaining subjects in a goal box without food, and since this is precisely what happens on an extinction trial, it is not surprising that the positive results have been explained in terms of generalization decrement (Capaldi, 1967a). Equally, since it is reasonable to suppose that delay of reinforcement is only one step less frustrating than the complete omission of reinforcement, it is possible to argue that frustration reactions are elicited whenever reinforcement is delayed but that, since reinforcement is eventually presented, these reactions are counterconditioned (or adapted out) in such a way that they no longer interfere with the appropriate instrumental response. Hence, while the omission of reinforcement in extinction may elicit frustration, animals earlier exposed to delay of reinforcement will continue to run.

There is evidence that constant delay of reward may increase resistance to extinction, even when animals are rapidly removed from the goal box on extinction trials (e.g., Sgro and Weinstock, 1963; Capaldi and Bowen, 1964; Tombaugh, 1966).* These results imply that the effects of delay of reinforcement on extinction cannot be attributed solely to a reduction in generalization decrement: acquisition delay must have some further effect (e.g., that of producing frustration) which mediates its effects on performance in extinction.

On the other hand, there are two results which suggest that delay-produced frustration is neither necessary nor sufficient to ensure increased resistance to extinction. Fehrer (1956), and McCain and Bowen (1967) found that *post*-reinforcement delay (i.e., detention in the goal box after consumption of the reward) could increase resistance to extinction. It is hard to see why such detention should be frustrating and yet it is clearly similar to the detention experienced in extinction. Conversely, Wike and McWilliams (1967), found that delay of reinforcement did not increase resistance to extinction if the delay was imposed in a detention box preceding the goal box. Such detention should presumably be frustrating, but dissimilar to the detention experienced in extinction. The implication is that both frustration and the similarity of acquisition delay to extinction detention are responsible for the results under review.

Why, then are the effects of delay of reinforcement so unreliable? The most reasonable suggestion is that delay of reward, just like size of reward, may have two, opposed, effects on resistance to extinction. A delayed reward is a less effective reinforcer than an immediate reward (p. 155), but it sets up traces or after-effects relatively similar to those encountered in extinction (see Fig. 8.3). A less effective reinforcer establishes behaviour that will extinguish more rapidly, but this may be outweighed by differ-

---

* In one of these experiments (Tombaugh, 1966), it was found that if animals were actually confined on extinction trials for the same length of time as they were delayed on acquisition trials, the effect of delay on resistance to extinction disappeared. This is presumably because, as Capaldi and Bowen (1964) showed, resistance to extinction is inversely related to length of detention in the goal box on extinction trials.

ences in generalization decrement. The evidence for this argument comes from an experiment on delay of reward analogous to that of Leonard's (1969) on size of reward. Capaldi *et al.* (1968a) trained rats to run in an alley, giving each group two trials per day with either immediate (I) or delayed (D) reward. Group II received two immediately rewarded trials each day; Group ID received an immediate reward followed by a delayed reward each day; while Group DI, received the delayed reward before the immediate reward. Group DI, corresponding to Group SL of Leonard's experiment, was significantly slower to extinguish than Group II, while Group ID, corresponding to Group LS, extinguished significantly more rapidly. Similar results were obtained by Wike *et al.* (1959) and by Capaldi and Poynor (1966) in a somewhat more complicated experiment varying delay of reinforcement.

E. Conclusions

The experiments discussed in this section have provided a substantial body of information about the conditions determining the rate of extinction. Although there are conflicts and exceptions, the single generalization that summarizes the major portion of these data is that rate of extinction is determined by the extent to which the transfer from acquisition to extinction causes a change from the set of events associated with reinforcement during acquisition.

An analysis of extinction in terms of generalization decrement is obviously supported by data showing that a change in ITI from acquisition to extinction increases the rate of extinction. But it is also possible that the reason why the massing of trials, in either acquisition or extinction, increases rate of extinction, is because at short ITIs the association between the outcome of one trial and the outcome of the next forms a more important part of the set of events controlling behaviour. Moreover, many of the effects of differences in the conditions of reinforcement on resistance to extinction seem to follow from this analysis. The omission, in extinction, of a large, immediate, reward received in acquisition, constitutes a greater change than the omission of a small, delayed, reward, and this may be sufficient to explain why the use of a large or immediate reward in acquisition usually increases the rate of extinction.

None of these data, however, provide much information about the associative processes underlying extinction. The point has been emphasized by Capaldi (1967a): "the usual extinction investigation cannot be used to make inferences about either the process of extinction, or the extent to which the response decrement observed is actually due to extinction" (p. 138). If the decline in responding typically observed during extinction

is due partly to generalization decrement caused by the omission of reinforcement, then investigators who wished to study the associative processes
involved in extinction might be better advised to programme extinction
without omitting reinforcement. The removal of the contingency between
responding and reinforcement has been shown to be sufficient to extinguish
responding in a variety of situations. Since extinction cannot, in these
circumstances, be attributed to generalization decrement, such preparations would be more suitable for studying the changes in associations
produced by extinction.

III. EXTINCTION AFTER VARIABLE CONDITIONS OF REINFORCEMENT

The most extensively studied determinant of performance in extinction
(possibly the most extensively studied aspect of instrumental learning) has
been the schedule of reinforcement employed in acquisition. The partial
reinforcement effect (PRE) has been the subject of several lengthy reviews
(Jenkins and Stanley, 1950; Lewis, 1960; Lawrence and Festinger, 1962;
Robbins, 1971; Sutherland and Mackintosh, 1971) and has been seen as a
major problem for learning theory. Before discussing the phenomenon and
its interpretations, it is worth attempting to set it into a particular context
by reviewing other cases in which variable conditions of acquisition have
been shown to affect performance in extinction.

A. Stimulus and Response Variability

McClelland and McGown (1953) first suggested that habits trained and
extinguished under constant, well-controlled, experimental conditions
might extinguish much more rapidly than those established under more
variable conditions. McNamara and Wike (1958) confirmed this result:
subjects given acquisition training in three different alleys, and then
extinguished in one, showed greater resistance to extinction than subjects
trained and extinguished in a single alley. This particular result was not
confirmed in an elaborate, factorial experiment by Brown and Bass (1958),
but the differences obtained were in this direction.

Brown and Bass also varied extinction conditions for some groups and
kept them constant for others They found, as have Long et al. (1965),
and May and Beauchamp (1969), that variability increased resistance to
extinction.

The most plausible explanation of the effects of acquisition variability
is to say that constant conditions during acquisition result in the formation
of few associations, while variable conditions produce a multiplicity of

associations. Rate of extinction then depends upon the number of associations which must be unlearned—provided that sufficient training on each has been given to ensure a certain minimal strength. A possible cause of Brown and Bass' failure to find a significant effect of acquisition variability was that their subjects were trained for only 24 trials; McNamara and Wike reported a positive effect after giving subjects 36 acquisition trials.

After acquisition training under constant conditions, one would expect that any variation in conditions in extinction might lead to more rapid extinction because of generalization decrement. Both Brown and Bass, and Pavlik *et al.* (1963) reported just such an initial disruption of performance. With continued extinction trials, however, animals shifted from constant to variable conditions showed a significantly more gradual decline in performance than those extinguished under constant conditions. An interpretation of this finding is suggested by some results reported by Long *et al.* (1965). After variable acquisition in three separate alleys, they divided their subjects into three groups: one group was extinguished in a single, constant alley, one group received extinction trials in each alley in a random sequence and the third group received a block of extinction trials in one alley, followed by another block in the second alley, followed by a final block in the third. The animals in this third group produced three consecutive extinction curves: after showing a decline in performance during the first block of trials in one alley, their speed of running increased at the beginning of the second block and then declined to show a second extinction curve, with a similar increase followed by a decrease in the third block of trials. When the results for the second group, extinguished in each alley in a random sequence, were regrouped into blocks of trials in each alley, their performance was essentially similar to that of the third group. The implication is that variable extinction conditions necessitate the extinction of separate associations: exposure to nonreinforcement in one alley does not extinguish the associations formed in another alley. Thus following acquisition in a single, constant, alley, extinction trials in other alleys do not succeed in extinguishing the associations formed in the first.

B. Variability of Reinforcement: Theoretical Interpretations of PRE

The effect of partial reinforcement on extinction is well established, usually substantial and occurs over a relatively wide range of conditions. The PRE has been observed in classical and instrumental experiments, in alleyways and lever boxes, with discrete-trial or free-operant procedures, with massed or spaced trials, few or many acquisition trials, large or small rewards, and in animals other than rats. Although nearly all of these variables have some influence on the magnitude of the PRE, and some combinations of these variables have very substantial effects, relatively few abolish the PRE entirely.

1. *Response-unit and Discrimination Hypotheses*

The number of experimental studies, and the proliferation of theoretical explanations (Lewis, 1960, listed seven) make it impossible to provide a comprehensive survey of the PRE in a limited space. Several of the theories discussed by Lewis have died a natural death, and a more recent review by Robbins (1971) reduced the number of viable contenders to four. Two explanations which do not seem to have survived were originally proposed by Mowrer and Jones (1945). They reported that rats trained to press a lever on free-operant schedules showed greater resistance to extinction after variable-ratio training than after consistent reinforcement. They suggested two possible explanations. The response-unit hypothesis stated that on any intermittent schedule the response reinforced is not a single depression of the lever, but a chain of lever presses which together constitute a single unit. It is this unit, consisting of many component responses, that must then be extinguished. The analysis is not implausible for free-operant situations and may contribute to the PRE observed in such situations. It is less easily applied to a PRE obtained in a discrete-trial experiment with one trial per day. Their second suggestion was labelled the "discrimination hypothesis": the suggestion here was that any PR schedule retards extinction because it increases the difficulty of discriminating the transition from acquisition to extinction. In this version the hypothesis is probably incorrect: numerous discrete-trial experiments have shown that a significant PRE is still obtained if subjects, given initial inconsistent reinforcement, receive a long block of consistently reinforced trials before being extinguished (Jenkins, 1962; Theios, 1962; Sutherland *et al.*, 1965; Leung and Jensen, 1968; Rashotte and Surridge, 1969; Amsel *et al.*, 1971). If it is the *transition* between acquisition and extinction that is critical, then consistent reinforcement immediately before the transition should severely attenuate, if not totally abolish, the PRE. The discrimination hypothesis, of course, is easily confused with an interpretation of the PRE in terms of generalization decrement. The two are not, however, the same: according to the discrimination hypothesis, partially reinforced subjects continue to respond in extinction because they do not know they are in extinction; according to a generalization decrement analysis, they continue responding because they have been reinforced for responding in a situation similar to that encountered in extinction. These are different statements: the second, for example, does not predict that the PRE will disappear if subjects are given a series of consistently reinforced trials after exposure to partial reinforcement. As we shall see below, the analysis in terms of generalization decrement has much to recommend it.

2. *Conditioned Reinforcement and Cognitive Dissonance Hypotheses*

From one point of view, the problem posed by the PRE is the same as that posed, for example, by the effects of size of reward on extinction. Either consistent reinforcement, or a large reward, should increase response strength more than partial reinforcement or a small reward; and yet in each case, the supposedly weaker response extinguishes more slowly. In the case of the PRE, one bold solution has been to insist that partial reinforcement does establish a stronger response, because other reinforcement processes are brought into play. An early version of this idea was proposed by Denny (1946). The argument appealed to the conditioned reinforcing properties of stimuli associated with reinforcement (e.g., the stimuli of the goal box); since these stimuli are also present on N trials throughout acquisition and during extinction, they may serve to reinforce responding in a partially reinforced group. As Denny noted, this hypothesis could account for the PRE only if partially reinforced subjects received the same number of R trials as consistently reinforced subjects (i.e., if they received a greater total number of training trials); since the PRE is routinely obtained when both partially and consistently reinforced subjects receive the same total number of acquisition trials (with partially reinforced subjects, therefore, receiving fewer reinforcements) the explanation cannot be accepted. To account for the PRE in these circumstances it would be necessary to make the extremely implausible assumption that goal-box stimuli acquired more conditioned reinforcing properties on N trials for partially reinforced subjects than on R trials for consistently reinforced subjects.

In fact, even when the total number of reinforcements received on a partial and on a consistent schedule is equated, there is good evidence that animals prefer a stimulus correlated with the consistent schedule to one correlated with the partial schedule (e.g., Mason, 1957; D'Amato *et al.*, 1958; vom Saal, 1972). This finding seems to rule out any attempt to attribute the PRE to an increase in the total reinforcing value of a partial reinforcement schedule. The argument applies equally strongly to the cognitive dissonance hypothesis proposed by Lawrence and Festinger (1962). They suggested that partially reinforced rats, trained to run down an alley to a goal box in which they only occasionally find food, find "extra attractions" in the goal box to compensate for their disappointment at the omission of food. These extra attractions are then said to be sufficient to establish a stronger response tendency in partially reinforced subjects than in consistently reinforced subjects. But if this were true, it would follow

that animals should prefer a goal box correlated with 50 per cent reinforcement of one correlated with 100 per cent reinforcement. In attempting
to account for the findings of D'Amato *et al.*, Lawrence and Festinger argued
that initial choices might be determined by differences in expectation of
reward, and that only later in testing would one expect the effects of the
extra attractions to become apparent. This accords ill with their account
of the fact that partial reinforcement may increase speed of running in an
alley (p. 160): here the extra attractions are assumed to be sufficient to
outweigh the less certain expectation of food. More importantly, the
explanation assumes, without justification, that extra attractions remain
after the expectation of food (which was, of course, stronger following
consistent reinforcement) has extinguished. Most important of all, the
prediction is simply not confirmed: D'Amato *et al.* reported that over a
series of nonreinforced test trials the preference for the consistently reinforced goal box declined to chance, but that their subjects never showed
any evidence of reversing their preference. A similar conclusion, supported by a much longer test series, has been reached by vom Saal
(1972).

3. *The PRE as an Instance of Multiple Associations*

There is good reason to believe that both the statement of the problem
posed by the PRE and the solution to that problem offered by Lawrence
and Festinger (1962) are basically mistaken. The problem posed by the
PRE is not to explain how a supposedly weaker response should extinguish
more rapidly. Resistance to extinction is not a particularly good measure of
the strength of the associations formed in acquisition; it is a much better
measure of the *nature* of those associations. Partial reinforcement increases
resistance to extinction because it establishes associations that are appropriate for maintaining performance during the conditions encountered
during extinction. The most important interpretation of this general
argument is to attribute the PRE to a decrease in generalization decrement;
and versions of this theory are dealt with in the next section. There is,
however, another way in which partial reinforcement may affect the nature
of the associations formed in acquisition. D'Amato and D'Amato (1962)
and Logan (1970, p. 115) have argued that partial reinforcement may result
in the establishment of a larger number of associations than consistent
reinforcement, and thus that its effects on resistance to extinction will be
similar to those of any other variation in the conditions of reinforcement.
Sutherland (1966) has proposed a similar suggestion, arguing that any
inconsistency in reinforcement will increase the range of stimuli attended
to during acquisition, and will therefore increase the number of stimuli
associated with responding and reinforcement.

There is reasonable evidence that partial reinforcement does indeed increase breadth of learning in the way suggested. When response topographies are closely examined, it can be shown that both the force with which rats depress a lever (Herrick and Bromberger, 1965) and the location of a pigeon's key peck (Ferraro and Branch, 1968) are more variable on partial reinforcement schedules. In discrimination learning experiments, where the number of relevant stimuli can be controlled and their contribution to responding more exactly assessed, there is evidence that a consistent reinforcement schedule may result in few stimuli controlling responding, while a partial reinforcement schedule may increase significantly the number of stimuli associated with reinforcement (McFarland, 1966a; Sutherland, 1966; Wagner *et al.*, 1968; Waller, 1973). There are, however, several studies in which this result has not been replicated (Cranford and Clayton, 1970; Waller, 1971); the reason for this discrepancy is not apparent.

One prediction derivable from this account of the PRE is that the effect will be detected less readily if the response being measured is one that could be controlled only by very few stimuli. A typical instrumental learning situation, such as the alleyway, permits a large number of stimuli to enter into control of responding. As Sutherland and Mackintosh (1971, p. 398) have suggested, classical conditioning experiments normally ensure a much more specific CS-reinforcer association, and this may be one reason why the reliability of the PRE in classical conditioning is open to question (p. 73). More direct evidence, however, can be brought to bear on this prediction by studying the extinction of choice behaviour. If a subject is trained on a simultaneous discrimination, then choice of one alternative over the other must depend upon control of responding by a specifiable set of relevant cues. If the normal basis for the PRE is an increase in the number of cues controlling behaviour, no PRE should occur in such a situation. The evidence, reviewed by Sutherland and Mackintosh (1971, p. 418–20), is conflicting. There are several studies (e.g., Spear and Pavlik, 1966; Pavlik and Lehr, 1967; Mackintosh and Holgate, 1968) which have found no PRE in a discrimination situation when choice of the former positive stimulus was measured; in others, however, a PRE has been observed (e.g., Pennes and Ison, 1967; Lehr, 1970). Even though the effect does sometimes occur, its unreliability, coupled with the fact that some of these studies have reported no effect of partial reinforcement on choice behaviour at the same time as a substantial effect on speed of responding, does suggest that behaviour controlled by a small number of stimuli may be less susceptible to the PRE than is speed of running in a typical alleyway.

4. Generalization Decrement and the PRE

The most plausible, and by now most widely accepted, theory of the PRE attributes the effect to differences in generalization decrement during extinction following partial and consistent reinforcement. A partial reinforcement schedule does not establish a stronger set of associations than, nor is it preferred to, a consistent schedule of reinforcement; it does, however, establish associations or responses that are less disrupted by the conditions of extinction. A typical partial reinforcement schedule ensures that subjects are reinforced for making the required instrumental response in a situation similar to that normally encountered in extinction.

It is important to see that this general explanation of the PRE can be derived from several theories of extinction. The explanations provided by these theories would differ only in detail—specifically in the nature of the critical stimuli to which partially reinforced subjects are assumed to have learned to respond. In practice, there have been three such theories of the PRE. The first, proposed by Sheffield (1949) and Hull (1952, pp. 120–21), suggested that the stimuli in question were the short-term stimulus traces of reinforcement and nonreinforcement: consistently reinforced subjects were always reinforced for responding in the presence of the stimulus traces of the preceding reinforcement (e.g., particles of food in the mouth); partially reinforced subjects were reinforced in the absence of such stimuli; since these stimuli are not present on extinction trials, partially reinforced subjects continue to respond longer in extinction. The weight of evidence against this account led Lewis to write that "at best, Sheffieldian after-effects are not important contributors to the PRE, and they probably have no effect whatsoever (Lewis, 1960, p. 16)". The second theory, proposed by Capaldi (1966a, 1967a; E. J. Capaldi, 1971) states that the "stimuli" in question are the memory traces of the outcomes of preceding trials. Nonreinforced trials both in acquisition and extinction occasion N traces, and it is these traces, subject to interference from other preceding and succeeding traces, that come to control responding in partially reinforced subjects. The third account of the PRE can be derived from frustration theory (Amsel, 1958, 1962, 1967): unexpected nonreinforcement elicits frustration reactions, which will be conditioned to stimuli preceding nonreinforcement, such as the stimuli in the alley and goal box; the stimuli produced by these conditioned frustration reactions are experienced in extinction; since they have already been associated with appropriate responding by partially reinforced subjects, these subjects will continue to respond in extinction.*

---

* Other explanations of the PRE could equally be derived from frustration theory. One could stress that frustration elicits interfering responses following consistent reinforcement, but either that these responses habituate during partially reinforced acqui-

Before examining any further differences between these different theories, and before seeing how well they explain the major facts about the PRE, it is worth first discussing briefly two sets of results which provide rather compelling support for the basic assumption common to all attempts to analyse the PRE in terms of generalization decrement. According to such an analysis, partially reinforced acquisition trials in an alleyway ensure that subjects are reinforced for running rapidly in the presence of stimuli similar to these later encountered in extinction. Hence they will continue to run rapidly in extinction. But this can be only a special case of a much more general rule: in extinction, animals should perform whatever response they have learned to make in the presence of stimuli produced on N trials. If the response in question involves rapid running, they will show substantial resistance to extinction; if, however, it involves slow running, they will not.

Rashotte and Amsel (1968a), and Amsel and Rashotte (1969) trained rats to run in an alley on a negatively correlated schedule: subjects were reinforced only if they took longer than a certain number of seconds to enter the goal box. Rats do not learn to perform perfectly on such a schedule (p. 136), and in these particular experiments they received reinforcement on only about 50 per cent of trials. They were, therefore, effectively on a partial reinforcement schedule. When, however, they were consistently reinforced and then extinguished in a second alley, they did not continue to run rapidly during extinction, as partially reinforced subjects usually do, but slowed down after only a few trials. Even more strikingly, the particular pattern of running that had been adopted during initial training in the negatively correlated alley reappeared when they were extinguished in the second alley. It is hard to resist the conclusion that animals reinforced for one pattern of responding in a situation associated with nonreinforcement, will repeat that pattern whenever they are later exposed to further nonreinforcement.

A second case where animals do not learn to respond rapidly after N trials, and therefore extinguish rapidly, is in experiments on repeated acquisitions and extinctions. It is well established that if subjects are trained on a series of acquisitions and extinctions, performance may become progressively more efficient, with rate of both re-acquisition and re-extinction increasing over the series (e.g., Bullock and Smith, 1953;

---

sition (cf. Weinstock, 1958), or that they are counterconditioned to approach (the analysis favoured by Amsel himself). While it may eventually prove possible to distinguish between these different accounts of the PRE offered by frustration theory, there is not as yet any evidence that favours the analysis in terms of interfering responses adopted by Amsel over the analysis in terms of generalization decrement suggested here (and by Wilton, 1967). It will be simpler, in the present context, to use the latter.

Gonzalez *et al.*, 1967b; Davenport, 1969). As has been often noticed (e.g., Lewis, 1960), a series of acquisitions and extinctions bears some resemblance to a sequence of partially reinforced trials. In both cases, responding is sometimes reinforced and sometimes not reinforced. But the difference between the two situations is at least equally obvious and important. With a random partial reinforcement schedule, the occurrence of one N trial does not predict whether the next trial will be reinforced or not. In a series of acquisitions and extinctions, however, one N trial usually predicts further N trials. In the former case, therefore, animals can learn only that nonreinforcement may signal either reinforcement or nonreinforcement and will run relatively rapidly after N trials; in the latter case they can learn that nonreinforcement signals further nonreinforcement and will then run slowly after N trials.

That this is the basis for the progressive increase in rate of extinction observed over a series of acquisitions and extinctions has been suggested strongly by results reported by Capaldi *et al.* (1968b) and Leonard and Capaldi (1971). Although it is generally true that one N trial usually predicts further N trials in such a situation, there is one exception to this rule. The last N trial of an extinction session is followed by an R trial at the beginning of the next acquisition session. Normally the interval between sessions is very much greater than the interval between trials within a session, and this should permit subjects to learn that a recent N trial is a reliable signal of further N trials. Capaldi *et al.* (1968b) confirmed the prediction, derivable from this argument, that rate of extinction in a series of acquisitions and extinctions depends upon the difference between the interval between trials within a session and the interval between successive sessions. Leonard and Capaldi (1971) showed that when sessions were massed, rate of extinction depended upon the length of each extinction session: with a small number of extinction trials in each extinction session, nonreinforcement did not signal further nonreinforcement reliably enough to produce rapid extinction; with a large number of extinction trials, animals were able to learn that early N trials signalled further N trials and slowed down accordingly (as would be expected, however, they speeded up towards the end of each extinction session).

The implications of these studies are straightforward. If animals are not reinforced following an N trial, then the stimuli produced by nonreinforcement, so far from controlling rapid or persistent responding, may eventually be established as signals for further nonreinforcement, and extinction may occur rapidly. In Amsel and Rashotte's experiments, animals were, in effect, reinforced for running slowly following N trials; in experiments on successive acquisitions and extinctions, responding is not reinforced following N trials. In these circumstances extinction occurs rapidly. In view of this, it seems reasonable to infer that the reason why the typical partial reinforcement schedule results in persistent responding in extinction, is that it ensures that relatively rapid responding has been reinforced following N trials.

C. Determinants of the PRE

The magnitude of the PRE is affected by numerous variables, and by numerous combinations of variables. The main interest of most of these factors is the light they throw on various versions of a generalization decrement analysis.

1. *Percentage of Reinforcement, N-length, and N–R Transitions*

a. Data. Lewis (1960), largely on the basis of a number of studies of eyelid conditioning and performance in a gambling situation in human subjects, concluded that resistance to extinction first increases, but eventually decreases, as percentage of reinforcement is decreased. Although there

TABLE 8.1

Examples of Different Schedules of Partial Reinforcement

Row	Schedule	Percentage of Reinforcement	Number of N trials*	Maximum N-length*	Number of N–R transitions
1	N N N R N N N N R R	30	7	4	2
2	N R R N N R R N R R	60	4	2	3
3	R R R R R N N N N N	50	0	0	0
4	R R N N N N N N R R	50	5	5	1
5	N R N N R N R N N R	40	6	2	4
6	R R N N N N N N R R	40	6	6	1
7	R N R R R N N R R R	70	3	2	2
8	R R N N N N R R R R	60	4	4	1

R designates reinforced trial; N designates nonreinforced trial.
* Excluding N trials occurring after the last R trial in the series.

has been little attempt to study the effects of wide variations in percentage of reinforcement in animal experiments, the available evidence does not support Lewis' suggestion of an inverted U-shaped function. Weinstock (1958) trained six groups of rats to run in an alley, with percentage of reward ranging from 16·7 per cent to 100 per cent, and found that over the entire range, increasing percentage of reward in acquisition increased the subsequent rate of extinction. An earlier study by Weinstock (1954), and a subsequent study by Bacon (1962), varied the percentage of reinforcement between somewhat narrower limits and also found that the lower the percentage the greater the resistance to extinction.

As Capaldi (1966a) has pointed out, however, such results do not in fact prove that resistance to extinction is determined by percentage of reinforcement at all: as the first and second rows of Table 8.1 show, variations in percentage of reinforcement may produce concomitant variation in several other factors, e.g., the number of N trials, the maximum or average run-length of N trials (referred to as N-length by Capaldi), and the number of N trials followed by R trials (referred to as N–R transitions). More analytic experiments have shown that these latter variables are often more important determinants of resistance to extinction than is the overall percentage of reinforcement.*

Let us first consider the variable of N–R transitions. The first and most important conclusion that can be drawn from a long series of studies of this variable, is that N trials occurring at the end of a day (i.e., not followed within that session by further R trials) have essentially no effect on resistance to extinction. This result was first reported by Grosslight and Radlow (1956, 1957) in experiments on reversal learning in rats. They compared the effects of schedules containing exactly the same number of trials and percentage of reinforced trials, as in Rows 3 and 4 of Table 8.1, but differing in that one schedule contained no N–R transitions; they found that such a schedule did not increase resistance to extinction and concluded that a necessary condition for the occurrence of a PRE was that animals be reinforced for responding in the presence of after-effects of N trials. Their results have been confirmed in several more recent alleyway studies. Spivey and Hess (1968), for example, trained rats in an alley for three days at a rate of four trials each day; one group received the sequence NNRR, and another the sequence RRNN. Group NNRR showed a significant PRE, while Group RRNN extinguished as rapidly as a consistently reinforced control group. Mackintosh and Little (1970a) compared groups receiving only two trials each day, and found that Group NR was significantly more resistant to extinction than Group RN. Capaldi and Kassover (1970) found that although resistance to extinction was increased

---

* As we shall see below, few of these variables can be manipulated independently. In particular, the percentage of R (or N) trials cannot be varied independently of the total number of N trials, unless the total number of acquisition trials is allowed to vary. Lawrence and Festinger (1962, Experiment 5) in fact performed an experiment whose results, they argued, showed that the absolute number rather than the percentage of N trials determined resistance to extinction, but since this was tantamount to showing that, for a given percentage of reinforcement, resistance to extinction increased with the number of training trials (a result also reported for consistently reinforced groups), the conclusion is necessarily open to question. In what follows, we shall not, therefore, distinguish between percentage of reinforcement and the absolute number of N trials as possible factors in determining resistance to extinction.

by an increase in the number of N trials provided that they preceded the last R trial of each session, the addition of N trials following the last R trial had no effect on resistance to extinction.

The implication is that resistance to extinction can be thought to depend on the number or percentage of N trials, only if N trials following the last R trial of a session are ignored. The evidence for this suggestion is sufficiently impressive that it will be simpler if the expression "number of N trials" is henceforth understood to refer only to those N trials that precede the last R trial of a session.

There is, however, one important restriction on this conclusion. In the studies so far considered, both acquisition and extinction trials have been given in sessions with relatively short ITIs (20 min or less). Under these conditions, resistance to extinction is determined by N–R transitions within each acquisition session. If, however, extinction trials were given at the rate of one per day, would one then expect resistance to extinction to be determined by the occurrence of N–R transitions between successive days of acquisition training? Capaldi's assumptions about the operation of after-effects imply a positive answer to this question (cf. Leonard et al., 1969). It follows, therefore, that a group receiving an N trial at the end of each day in acquisition, followed by an R trial at the beginning of the next day, should extinguish more slowly than one receiving an R trial at the end of each day. In a test of this prediction, Mackintosh and Little (1970a) trained rats in an alley for 15 days of acquisition, with one group receiving the sequence RRN each day, and the other receiving the sequence RNR. When the two groups were extinguished at the rate of one trial per day, there was no suggestion that Group RRN extinguished more slowly. The implication is that with very long ITIs in extinction, resistance to extinction is no longer determined by the occurrence of N–R transitions.

With this important exception, however, it remains true that N–R transitions do affect performance in extinction. Furthermore, there is evidence that, up to a point, the number of N–R transitions within a daily session has an effect on resistance to extinction independent of variations in other factors such as the number of N trials. Capaldi and Hart (1962) and Spivey (1967) showed that after a small number of acquisition trials, a schedule containing more N–R transitions (e.g., Row 5 in Table 8.1) resulted in greater resistance to extinction than one containing the same number of N trials but fewer N–R transitions (e.g., Row 6). In Spivey's experiment, moreover, resistance to extinction was determined by the number of N–R transitions even when the number of N trials was allowed to vary. A schedule such as that shown in Row 7 resulted in as much resistance to extinction as the schedule in Row 1, even though it contains many fewer N trials.

These results, however, occur only following a relatively small number of acquisition trials (a finding which is consistent with Capaldi's assumption that the learning that occurs on N–R transitions reaches asymptote fairly rapidly). With more extended acquisition training, resistance to extinction is determined by factors other than the number of N–R transitions. As noted earlier, both Weinstock (1954, 1958) and Bacon (1962) reported that after extended training, rate of extinction is a direct function of the overall percentage of reinforcement. Bacon trained three groups of rats, on a 30 per cent, 50 per cent or 70 per cent schedule, at a rate of 10 trials each day, with each day both beginning and ending with an R trial. In these circumstances, as has been pointed out by Capaldi (1966a), the 30 per cent schedule will have contained fewer N–R transitions than the 50 per cent or 70 per cent schedules. Since it also resulted in the slowest rate of extinction, it follows that after extended training the number of N–R transitions is not the most important determinant of resistance to extinction. A similar conclusion is suggested by the results of Capaldi and Kassover's (1970) experiment: they showed that after 10 days of training with between four and six trials per day, a schedule containing three N trials and one N–R transition per day resulted in significantly greater resistance to extinction than one containing a single N trial and N–R transition each day.

In both of these experiments, resistance to extinction was increased by schedules containing both more N trials and longer N-lengths. The next question, therefore, is which of these two variables is the more important determinant of resistance to extinction after extended training. Two experiments have suggested that N-length may be the critical variable. By using schedules such as those shown in Rows 5 and 6 of Table 8.1, Gonzalez and Bitterman (1964) held percentage of reinforcement constant and found that resistance to extinction increased with N-length; while Capaldi and Stanley (1965) opposed the two factors by using schedules such as those shown in Rows 5 and 8, and found that N-length was more important.

In both these studies, the ITI was very short (15 or 20 sec). Although Capaldi and Minkoff (1967) have shown that increases in N-length with percentage or number of N trials held constant may still increase resistance to extinction at longer ITIs (20 min or more), the effect was a relatively small one. When the effects of N-length are opposed by percentage of reinforcement at long ITIs, there is clear evidence that the percentage of reinforcement or the total number of N trials may be more important (Koteskey, 1969; Haggbloom and Williams, 1971).*

* A similar conclusion is suggested by the results of two experiments with pigeons: Roberts, *et al.* (1963) found that resistance to extinction was not affected by increases in

b. Theory. The available data suggest the following conclusions: except in the case where extinction is scheduled at the rate of one trial per day, N trials increase resistance to extinction only if they are followed in the same session by an R trial; resistance to extinction increases with the number of N–R transitions, but this effect reaches asymptote relatively rapidly; with extended training, and a short ITI, resistance to extinction is determined more by N-length than by the percentage of reinforcement or absolute number of N trials; but with a longer ITI the relative importance of these two factors is reversed.

The picture is notably more complicated than that suggested by Lewis' (1960) generalization that resistance to extinction is a function of percentage of reinforcement. Many of these studies were undertaken in order to throw light on Capaldi's sequential version of generalization decrement theory; indeed they touch on one of the main distinctions between the explanations of the PRE provided by Capaldi and by Amsel.

According to Capaldi, the sequence of N and R trials is a crucial determinant of the PRE, since the memory trace for the outcome of one trial is modified by the outcome of succeeding or preceding trials. Performance over a series of extinction trials within a single session depends upon the extent to which the N-traces encountered in extinction are similar to the traces experienced in acquisition, and also on whether responding was reinforced in the presence of those traces in acquisition. N–R transitions affect resistance to extinction because it is on R trials immediately following N trials that subjects are reinforced for responding in a situation which includes N-traces from immediately preceding trials. N-length affects resistance to extinction because the longer the sequence of N trials preceding an R trial, the more similar the traces signalling reinforcement in acquisition to those encountered in extinction (see Capaldi, 1966a, 1967a, for a discussion of these derivations). Percentage of reinforcement, or the absolute number of N trials, should have no effect on resistance to extinction. All of these predictions are confirmed by the outcome of studies using relatively massed trials. However, in Mackintosh and Little's (1970a) study, there was no evidence that resistance to extinction was affected by between-day N–R transitions; and in the experiments of Koteskey (1969) and Haggbloom and Williams (1971), employing intermediate ITIs (15 or 20 min), the percentage or number of N trials was a more important determinant of resistance to extinction than N-length.

Frustration theory, on the other hand, has considerable problems in accounting for those data which support Capaldi's analysis. According to frustration theory (although Amsel has not been specific on this point),

---

N-length when trials were spaced, while Gonzalez *et al.* (1965b) found a highly significant effect when a short ITI was used.

if nonreinforcement results in the conditioning of frustration to apparatus cues, there is little reason to see why such a process should not be cumulative. Every N trial, therefore, should increase the amount of frustration conditioned to the apparatus: since extinction consists of a large number of N trials, resistance to extinction should simply be determined by the number of unanticipated and frustrating N trials experienced in acquisition. Although increased resistance to extinction depends upon the reinforcement of responses in the presence of frustration-produced stimuli, there seems to be no reason why reinforcement would have to follow nonreinforcement within a single session, and no reason therefore why N–R transitions (in the sense employed here) should be important. The theory states that the PRE is determined by percentage of reinforcement rather than N-length or N–R transitions. The only results supporting this conclusion are those from experiments with spaced trials, and even here, Capaldi and Kassover (1970) showed that if percentage of reinforcement was varied by adding N trials after the last R trial in each session it has no effect on resistance to extinction.

The popular resolution of this conflict of data and theories has been to attribute the PRE obtained in studies using massed trials to the operation of short-term after-effects or memory traces, and that obtained in studies using spaced trials to the operation of conditioned frustration (e.g., Amsel, 1967). It will be easier to comment on the viability of this suggestion after a more detailed discussion of the effects of the distribution of trials on the PRE (see below). But it is worth reiterating that Capaldi and Kassover's results provide little support for such a suggestion.

2. *Patterning*

Tyler *et al.* (1953) trained one group of rats on a random, 50 per cent partial reinforcement schedule and a second group on a regularly alternating (NRNRNR) schedule. They found that random partial reinforcement resulted in considerably greater resistance to extinction. Although this effect does not occur when only a small number of acquisition trials are given (Capaldi, 1958; Capaldi and Hart, 1962), with sufficient training it is entirely reliable (e.g., Capaldi, 1958; Bloom and Capaldi, 1961; Capaldi and Minkoff, 1967; Rudy, 1971).

Why should single alternation training be less effective than random partial reinforcement in increasing resistance to extinction? The first possibility to note is that a single alternation schedule has a maximum N-length of one, and we have already seen that, as Capaldi's theory predicts, after sufficient training resistance to extinction is an increasing function of N-length. While this is true, and while it may account for some of the difference between the two schedules, it is difficult to resist the inference that the regularity of the single alternation schedule must be partly responsible for its effects on extinction. After sufficient training on an alternating schedule, rats learn to anticipate R and N trials, running rapidly on the former and more slowly on the latter (p. 161). Anticipated nonreinforcement is not frustrating and as would be expected from this, an alternating schedule of reinforcement in the first alley of a double alleyway may lead eventually to the disappearance of the frustration effect in the second alley (p. 364). From all of this it follows that animals given sufficient training on an alternating schedule of reinforcement will experience less frustration than animals trained on a random schedule and will not therefore learn so effectively to continue running in the face of the frustration produced in extinction.

Several results support this derivation from frustration theory. Capaldi (1958) trained rats on alternating or random schedules for either 70 or 140 trials before extinction. Rate of extinction was unaffected by level of training in animals trained on the random schedule, but was significantly increased by the additional training on the alternating schedule. The implication is that as the alternating pattern was learned, frustration decreased and performance was more disrupted by the frustration subsequently experienced in extinction. Campbell *et al.* (1971) trained rats on an alternating schedule with either a large or a small reward on R trials. As expected, the large reward produced better learning of the pattern of N and R trials. After a small number of trials, before either group had learned the pattern, the large reward increased resistance to extinction; after further training, however, the large reward decreased resistance to extinction. Since it was only the subjects given extended training with large rewards that showed significant patterning, the implication is that it is the occurrence of patterning that reduces resistance to extinction.

Rudy (1971) has provided more direct, and quite conclusive, evidence that the possibility of anticipating nonreinforcement on alternating schedules is partly responsible for the effect of such schedules on performance in extinction. He found that an alternating schedule resulted in less resistance to extinction than a second schedule in which N-length was always one, but the occurrence of N trials was unpredictable. Furthermore, the addition of a stimulus signalling the occurrence of N trials on a random schedule significantly reduced resistance to extinction (cf. also Homzie and Rudy, 1971).

3. *Effects of Reinforced and Nonreinforced Placements in the Goal Box*

As we have seen earlier, Capaldi argues that rewards serve two separable functions: when contingent upon responding they serve as reinforcers; but whether contingent on responding or not, they are events which, if remembered, may constitute one of the events associated with reinforcement on a subsequent R trial. A reward given to a rat placed directly in the goal box of an alley would tend to be a less effective reinforcer than one received contingent on running, but it should still establish a trace which could affect performance on later trials.

The implications of this argument are simple enough. Consider a rat trained on the sequence of trials shown in Row 1 of Table 8.2. The sequence contains four N–R transitions, and from the preceding discussion it is clear that it is to a large extent the rewards given on these trials that are responsible for reinforcing responding in the presence of N-traces. If, now, before each of these trials the rat were directly placed in the goal box and fed (i.e., received the sequence shown in Row 2), the effects of

TABLE 8.2

Examples of Schedules Including Intertrial Reinforcement

Row										
1	N	R	N	R	R	N	N	R	N	R
2	N (r) R		N (r) R		R	N	N (r) R		N (r) R	
3	R (r) N		N	R	R (r) N		N	R		
4	R	N (r) N		R	R	N (r) N		R		
5	R	N	N (r) R		R	N	N (r) R			

R designates reinforced trial; N designates nonreinforced trial; (r) designates reinforced placement.

these rewards would be to replace, by retroactive interference, the traces of the preceding N outcomes, but not to reinforce responding in the presence of those traces. Thus intertrial, reinforced, placements, if given between N and R trials, should significantly reduce resistance to extinction. Capaldi *et al.* (1963), Capaldi and Spivey (1963), Spence *et al.* (1965), McCain (1966), and Homzie *et al.* (1970), among others, have all confirmed this prediction. In other experiments Capaldi (1964) and Capaldi and Wilson (1968) have used intertrial reinforcements to vary functional N-length. Capaldi (1964), for example, compared the three schedules shown in Rows 3, 4 and 5 of Table 8.2. Each schedule contains the same sequence of N and R trials, and each contains two intertrial reinforcements, but they differ in the location of these placements. Resistance to extinction was an increasing function of the number of N trials following an intertrial reinforcement: it is as though N trials preceding an intertrial reinforcement (i.e., not followed by an ordinary reinforced trial) do little to increase resistance to extinction.

However, Spence *et al.* (1965), Black and Spence (1965) and Homzie *et al.* (1970) found that with extended training, intertrial reinforcements are no longer effective in reducing the magnitude of the PRE. Since Capaldi and Olivier (1967) have shown that this change is not produced simply by extended prior acquisition training, it must, presumably, be due to extended experience with the intertrial reinforcement procedure itself. Capaldi and Olivier argued that after a few such placements, animals would learn to discriminate between direct placements and ordinary acquisition trials. There is, indeed, evidence of just such a discrimination being formed (Homzie *et al.*, 1971), but at first sight it is not clear why this should affect the outcome of such placements. E. J. Capaldi (1971) has suggested one possible reason. He argues that the retrieval of traces from memory may be, to some extent, a context-specific affair: the memory for an event occurring in one context may be less efficiently retrieved in a different context. If a rat receives an N trial in an alleyway, the trace of this N outcome will be retrieved the next time the rat is placed in the same situation (i.e., the start box of the alley); an intervening reward will interfere with the retention of this N-trace only to the extent that it is received in this situation. If the reward is received in an entirely different context, it will cause little or no interference (cf. Capaldi and Spivey, 1963). In so far as the reward received when the subject is placed directly in the goal box for an intertrial reinforcement is eventually discriminated from the reward received for running in the alley, it will tend not to interfere with the retrieval of preceding N-traces on a succeeding R trial.

Similar arguments appear to apply to the case of nonreinforced placements. Numerous experiments have studied the effects of such placements on resistance to extinction, with apparently confusing results. To the extent that reinforced placements increase the rate of extinction, it should be expected that nonreinforced placements will decrease the rate of extinction. To a first approximation this appears to be true. A small number of non-reinforced placements, in the context of a small number of acquisition trials, does significantly retard extinction (Brown and Logan, 1965; McCain *et al.*, 1969; Homzie *et al.*, 1970). Since, however, nonreinforced placements can be discriminated from nonreinforced trials (Homzie and Rudy, 1971) with more extended training and a larger number of placements, no effect of nonreinforced placements on resistance to extinction can be detected (Lawrence and Festinger, 1962, Experiment 6; Theios and Polson, 1962; Brown and Logan, 1965; Fitzgerald and Teyler, 1968; Homzie *et al.*, 1970; E. D. Capaldi, 1971b).

So far the results accord well with those of studies of reinforced placements: after a moderate amount of training, subjects discriminate between nonreinforced placements and ordinary nonreinforced trials, the memory of the former is not retrieved at the start of an ordinary trial, and is not therefore established as a signal that running will be reinforced. Two further factors, however, complicate this relatively neat picture. The effects of nonreinforced placements can be increased by interspersing nonreinforced placements with reinforced placements, i.e. by scheduling partially reinforced placements (Theios and Polson, 1962; Brown and Logan, 1965; Fitzgerald and Teyler, 1968; E. D. Capaldi, 1971b). Non-reinforced placements are also more likely to increase resistance to extinction if the rat is placed at some distance from the food cup on placement trials and is thus required to run a short distance to discover whether food is available or not (Trapold and Doren, 1966; E. D. Capaldi, 1971b). It is reasonable to suppose that requiring a brief run on placement trials will increase the similarity between placement and ordinary trials, and this, of course, would then be sufficient to increase the effects of such placements on resistance to extinction.

It remains to interpret the effects of partially reinforced placements. E. D. Capaldi (1971b) has shown that such placements increase resistance to extinction only if a nonreinforced placement is followed by a reinforced placement; if nonreinforced placements are followed only by reinforced running trials, then resistance to extinction is not increased by scheduling partially reinforced placements. This is a striking result. In one way it appears entirely consistent with previous arguments about context-specific retrieval: the memory of a nonreinforced placement will be more readily retrieved on another placement trial than on an ordinary running trial. Thus if the only N trials to which a subject is exposed are nonreinforced placements, their outcomes will be associated with subsequent reinforcement only if the subject receives reinforced placements after nonreinforced placements. The problem arises when one tries to see how such N–R transitions, confined to placement trials, could affect resistance to extinction of the response of running the alley. If, during acquisition, N-traces from a placement trial are not retrieved on the next alley trial, how do they come to affect performance in the alley in extinction? The problem is not an isolated one. Amsel and his associates have shown that partial reinforcement in one alley may increase resistance to extinction not only in that alley but in another dissimilar alley in which subjects have been consistently reinforced (e.g., Amsel, et al., 1966; Rashotte and Amsel, 1968b). If N outcomes are established as signals for reinforcement only if followed by R trials given in a similar context, this implies that they may be retrieved only in a situation similar to that in which they occurred. But if partial reinforcement in one situation increases resistance to extinction in another, this implies that once N outcomes have been established as signals for reinforcement, this association may generalize to another situation and hence serve to retard the extinction of a response trained in that situation.

4. *Partial Delay of Reinforcement*

In a study which, they argued, provided strong evidence against Hull and Sheffield's explanation of the PRE in terms of stimulus traces of reinforcement, Crum *et al.* (1951) trained rats in an alley either with immediate reward on all trials, or with the reward delayed for 30 sec on a random 50 per cent of trials. They found that this partial delay (PD) schedule significantly retarded extinction. The effect seems analogous to the PRE, but since the PD schedule ensures that all trials terminate with reinforcement, the traces available for conditioning on the next trial must be those of reinforcement rather than nonreinforcement, and PD and consistent schedules should result in equivalent performance in extinction.

That the argument is not irrelevant even to Capaldi's theory will be apparent from the discussion of the preceding section: there it was argued that intertrial reinforcements attenuate the PRE if given during N–R transitions, because the delivery of reinforcement causes some loss of the N trace from the preceding N trial. A procedure which schedules intertrial reinforcements, however, seems remarkably similar to one which schedules delay of reinforcement. In each case the subject receives no reinforcement in the goal box for a certain period of time and is later reinforced: the difference is that, when given intertrial reinforcements, the subject is removed from the goal box and later replaced, while with delayed reinforcement the entire delay period is spent in the apparatus. Capaldi *et al.* (1963) and Capaldi and Poynor (1966), noting this similarity, suggested that the critical difference between the two procedures was the length of time the subject spent in the apparatus before receiving reinforcement. Partial delay of reinforcement increases resistance to extinction only when a relatively long delay is imposed. Logan *et al.* (1956) found no effect with a 9 sec delay, but a significant effect with a 30 sec delay. Knouse and Campbell (1971), varying the delay period between 8 and 56 sec for different groups, found that resistance to extinction after PD training varied directly with the length of the delay: only subjects receiving delays longer than 30 sec extinguished more slowly than an immediately reinforced control group.

These results are entirely consistent with the evidence derived from patterning experiments. Animals trained on a regularly alternating schedule of immediate and delayed reinforcement come to run faster on immediately reinforced trials than on trials when reinforcement is delayed, only when the delay of reinforcement is longer than 20 to 30 seconds (p. 401). The implication is that the trace of a short delay of reinforcement is sufficiently similar to that of an immediate reinforcement so that the rat cannot learn to use the difference to predict the outcome of the following trial. It follows that short delays of reinforcement on a PD schedule might not be sufficient to increase resistance to extinction and since in typical studies of intertrial reinforcement subjects are confined to the goal box only for 15 sec or so before removal, the subsequent reinforcement may be sufficient to reduce any PRE.

When relatively long delays of reinforcement are used on a PD schedule, however, resistance to extinction is significantly increased in a manner very similar to that occurring following PR training. The effect, like the PRE, survives a block of consistent, immediate, reinforcements (Donin et al., 1967) and is affected in much the same way by manipulations of the sequence of delay (D) and immediate (I) reinforcements, as is the PRE by the sequence of N and R trials. As has already been noted, the PDE depends upon D trials being followed by I trials (Wike et al., 1959; Capaldi et al., 1968a). After a small number of training trials, resistance to extinction is dependent on the number of D–I transitions rather than on the percentage of D trials (Capaldi and Spivey, 1965); while after a large number of training trials, resistance to extinction is determined by D-length rather than by the percentage of D trials (Capaldi, 1967b). Capaldi and Poynor (1966) showed that the effect of length of delay on resistance to extinction could be affected by sequential variables. Two groups of rats received two D trials per day, on one of which the delay was short (5 sec), while on the other it was long; only one of these D trials, however, was followed by immediate reinforcement. Resistance to extinction was increased if the long D trial was followed by immediate reinforcement; but the group which received a short D trial preceding reinforcement and a long D trial at the end of each day extinguished only marginally more slowly than a control group.

The one apparent difference between studies of partial delay and partial reinforcement has arisen from studies of patterning. Extended training on an alternating sequence of N and R trials results in relatively rapid extinction. This seems attributable to the fact that anticipated non-reinforcement is not frustrating. Wike et al. (1959), however, found no difference in resistance to extinction following an alternating or a random schedule of D and I trials, and in a subsequent study employing longer delays, Wike et al. (1964) found only marginal evidence of faster extinction following the alternating schedule. This difference is perhaps attributable to the difference in the rate at which alternating N and R trials and alternating D and I trials permit the formation of a discrimination: patterning appears to develop more slowly and with greater difficulty with delay of reinforcement. If this is true, it reinforces the earlier conclusion that the development of patterning has an important effect on resistance to extinction following training on an alternating schedule.

5. Effects of Size of Reward

Although large rewards increase the rate of extinction following consistent reinforcement, they usually retard extinction following partial reinforcement (e.g., Hulse, 1958; Wagner, 1961; Gonzalez and Bitterman, 1969; Likely et al., 1971, Ratliff and Ratliff, 1971). Although this effect has not appeared in some studies (e.g., Roberts, 1969), there is no report of the inverse relationship between size of reward and resistance to extinction typically observed after consistent reinforcement. From this it may be inferred that the size of the PRE (i.e., the difference between partially and consistently reinforced groups) is reliably increased by an increase in size of reward. In all of these studies just such an interaction has been reported. Indeed the use of a small reward, especially if combined with other factors that reduce the size of the PRE (such as a long ITI or a small number of acquisition trials), may be sufficient to abolish the PRE altogether (Hulse, 1958; Gonzalez and Bitterman, 1969). A final point is that although variations in concentration of sucrose do not have the same effect on resistance to extinction following consistent reinforcement as do variations in the amount of food given as reward (p. 427), they do have similar effects following partial reinforcement: increases in concentration of sucrose increase resistance to extinction following both consistent and partial reinforcement (Likely et al., 1971).

Capaldi's explanation of these results (e.g., Capaldi, 1967a) follows naturally from previous arguments. Large rewards may affect extinction in two ways: as strong reinforcers they retard extinction, but as outcomes extremely dissimilar to nonreinforcement they increase the rate of extinction. Since a partial reinforcement schedule contains N trials, the traces of these N outcomes will be established as signals for reinforcement whatever the size of reward used in acquisition, and the major effect of a large reward will be to produce better learning on N–R transitions. Although the traces produced by large rewards would tend to decrease resistance to extinction, this function of rewards in a partial schedule will be less important than the reinforcing function.

Frustration theory provides a relatively similar analysis (Wagner, 1961; Amsel, 1962). Consistent large rewards increase the rate of extinction because the increase in incentive is outweighed by the subsequent increase in frustration. During partially reinforced acquisition, however, animals receiving large rewards will be reinforced for responding in the presence of a greater degree of frustration than will animals receiving small rewards. Thus the fact that they will also be more frustrated in extinction will not lead to any greater disruption of performance. In effect, both large and small reward groups will have learned to respond to frustration stimuli similar to those they will experience in extinction. Resistance to extinction, therefore, will be determined by differences in the strength of such learning: a large reward, according to Amsel, will permit "the connection of S_F to continued approach to develop more strongly" (1962, p. 314).

Both Capaldi and Amsel, therefore, can explain why high concentrations of sucrose increase resistance to extinction following partial reinforcement: higher concentrations are simply more effective reinforcers. The problems for frustration theory, it should come as no surprise, are produced by the results of experiments investigating particular sequences of large, small

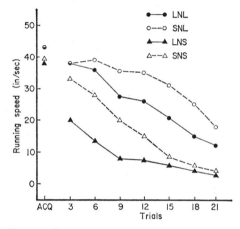

FIG. 8.5. Speed of responding at the end of acquisition (Acq.), and during the course of extinction, as a function of the sequence of large, small, and no rewards received during acquisition. (*After* Leonard, 1969.)

and no reward in partial reinforcement schedules. The simplest case is a study by Leonard (1969). Leonard trained four groups of rats to run in an alley with the same sequence of three trials on each day of acquisition. The sequences for the four groups were: LNL, SNS, LNS and SNL. The performance of these four groups in extinction is shown in Fig. 8.5. Group LNL was more resistant to extinction than Group SNS, thus replicating, for this simple and regular schedule, the standard effect of size of reward on resistance to extinction following partial reinforcement. The results for Groups LNS and SNL, however, show unquestionably that it is the large reward received following N trials that is responsible for this effect: a comparison of Group LNS with Group SNS, and of Group LNL with Group SNL, makes clear that large rewards on the first trial of each day tended to increase, rather than decrease, the rate of extinction.

Leonard employed regular and predictable schedules of reinforcement, and a relatively short ITI (3 to 4 min). Two further results make it clear that the conclusions that may be drawn from his experiment are not restricted by these conditions. Capaldi and Minkoff (1969) used a 15 min ITI and more varied schedules of reinforcement. They preserved, however, the essential feature of Leonard's SNL and LNS schedules: although all groups received both large and small rewards, for one pair of groups N trials were always followed by large rewards, while for another pair of groups N trials were always followed by small rewards. Resistance to extinction was increased by large rewards only if they followed N trials. A similar result was reported by Capaldi and Capaldi (1970b) in a study employing a 24 h ITI.

6. *Intertrial Interval*

We have already come across the suggestion that the causes of the PRE, as well as of other phenomena encountered during extinction, may depend upon the ITI employed: at short ITIs, so it is argued, traces of preceding outcomes may well form an important part of the stimulus complex controlling responding, but these traces must decay over time so that more permanent processes must be invoked to account for the PRE at long ITIs. Before considering this argument, it will be as well to review some data on the effects of variations in ITI on the PRE.

Sheffield (1949) employed ITIs of either 15 sec or 15 min, in a factorial experiment varying ITI in acquisition, ITI in extinction and schedule of reinforcement (50 per cent *vs* 100 per cent). She reported a PRE only in groups trained with the short ITI. Her experiment was repeated by Wilson *et al.* (1955) with somewhat different results: they found a PRE after both massed and spaced training, although inspection of their results shows a marked reduction in the size of the PRE for groups trained with the 15 min ITI and extinguished with the 15 sec ITI. A subsequent experiment by Lewis (1956) tended to support this conclusion: although a PRE was obtained after both massed (15 sec ITI) and spaced (2 min ITI) training, there was a suggestion that animals given spaced training and massed extinction showed a smaller PRE than other groups.

In the meantime, however, Weinstock (1954, 1958) had shown that a highly significant PRE could be obtained after extended training with an ITI of 24 h. This result clearly demolished the Hull–Sheffield after-effects theory of the PRE and also rendered untenable the strong conclusion, previously drawn from Sheffield's data, that the PRE occurs only with a short ITI. But Weinstock's results do not rule out the possibility that the magnitude of the PRE is affected by the length of ITI in acquisition or extinction and there is, in fact, good evidence that it is. Gonzalez and

Bitterman (1969) reported a significant PRE after training with both large and small rewards when trials were massed, but no PRE following spaced trials (60 min ITI) with a small reward. Mackintosh (1970a) found that groups trained and extinguished with a 30 sec ITI showed a significantly larger PRE than groups trained and extinguished with a 45 min ITI. Two recent studies have provided striking confirmation of the interaction between ITI in training and extinction suggested by the data of Wilson *et al.* (1955) and Lewis (1956). Amsel *et al.* (1971) trained rats on partial or consistent schedules with a 24 h ITI, and found essentially no evidence of a PRE when subjects were given a series of extinction trials with a 10 sec ITI. When the subjects were retrained with consistent reinforcement and then extinguished again, this time with a 24 h ITI, a significant PRE appeared. In two experiments, Capaldi *et al.* (1971) found that a shift from a 24 h ITI in acquisition to an ITI of a few minutes in extinction either greatly attenuated or completely abolished the PRE.

An additional effect of ITI has already been discussed (p. 447). With short ITIs, the features of partial schedules of reinforcement most responsible for increasing resistance to extinction are the number of N–R transitions and N-length. With extremely spaced trials, resistance to extinction is unaffected by N–R transitions and with even moderately spaced trials (15 min) there is evidence that the absolute number or percentage of N trials preceding R trials is a more important determinant of resistance to extinction than is N-length *per se*.

Several explanations have been suggested for some or all of these results. The first, and perhaps the most generally accepted, suggestion is that resistance to extinction is mediated by conditioned frustration at long ITIs, and both by conditioned frustration and by the after-effects or traces of preceding outcomes at short ITIs (e.g., Amsel, 1967; Gonzalez and Bitterman, 1969). While this suggestion certainly explains why sequential variables such as N-length and the number of N–R transitions are less important at long than at short ITIs, and can also explain why the magnitude of the PRE is affected by the distribution of trials, it cannot explain why spaced acquisition and massed extinction trials virtually abolish the PRE. Nor, in fact, is it consistent with all the schedule data. Although N-length may be less important than the proportion or number of N trials at intermediate ITIs, Capaldi and Kassover (1970) showed that even with a 20 min ITI, an increase in the number of N trials following the last R trial of each day had no effect on resistance to extinction. There is no reason why such trials should not have resulted in the conditioning of frustration to apparatus cues, which then would have been available for associating with the instrumental response on the following day.

A second possibility has been suggested by Capaldi *et al.* (1971). They also argued that the effect of ITI can be understood in terms of differences in the importance of memory traces and frustration reactions at short and long ITIs; but, rather perversely from the point of view of frustration theory, they supposed that the frustration reactions are the short-lived events which contribute to the PRE only at short ITIs. In effect, they assumed that nonreinforcement elicits a temporary, rapidly decaying, state of frustration, and that partially reinforced subjects learn to respond in the presence of the after-effects of this frustration reaction only if trained with massed trials. If acquisition is given with spaced trials, therefore, performance would be severely disrupted by massed extinction. In support of this analysis, Capaldi *et al.* reported that although the PRE was virtually abolished if spaced acquisition were followed by massed extinction, the administration of sodium amytal in extinction led to a significant increase in the magnitude of the PRE. Since there is evidence that this drug attenuates frustration reactions (p. 265), this result is consistent with the supposition that it was frustration which disrupted the performance of partially reinforced subjects that had been given acquisition training with spaced trials.

The analysis of Capaldi *et al.*, although successful in explaining the effects of trial distribution on the overall magnitude of the PRE, does not explain its effects on sequential variables. The assumption that the memory trace for the outcome of one trial persists indefinitely (or at least for 24 h) incorrectly implies that the importance of N-length should be unaffected by the length of the ITI, and is inconsistent with the observation that N–R transitions between days have no effect on resistance to extinction when extinction trials are given at a rate of one per day. Few investigators have been prepared to accept that a rat can remember on one day whether the trial it received on the previous day was reinforced or not, and there is good reason for such scepticism. Studies of alternation learning, which provide the most direct evidence of the rat's ability to remember the outcome of one trial at the beginning of the next, have consistently found that learning becomes markedly less efficient as the ITI is lengthened (e.g., Katz *et al.*, 1966; Heise *et al.*, 1969), and have usually found no evidence of learning at all with a 24 h ITI (p. 163). The implication is that rats do forget the specific outcome of the preceding trial if the ITI is long enough

It does not, however, follow from this that the PRE obtained at long ITIs must be attributed exclusively to conditioned frustration. A third account of the effects of trial distribution on the PRE has been proposed by Mackintosh (1970a), and by Haggbloom and Williams (1971). The suggestion is that some forgetting of the outcomes of preceding trials, or more particularly of the *sequence* of preceding outcomes, occurs over long ITIs. The suggestion can be interpreted in terms of an interference theory of forgetting. We have already seen that the outcome of one trial may affect performance on the next only if both trials occur in the same context. This suggests that the retrieval of memory traces is dependent on appropriate contextual cues. Capaldi's assumption that the trace of one outcome is replaced by the trace of the succeeding outcome can be regarded as an instance of retroactive interference. The assumption that some forgetting of outcomes occurs over the course of a long ITI can be regarded as an instance of proactive interference: the trace of the outcome of one trial suffers interference from the outcomes of preceding trials.

Sequential effects depend upon subjects remembering the order in which preceding outcomes have occurred. In the most obvious cases, learning to alternate depends upon remembering whether the last trial was reinforced or not and the effect of N–R transitions can be felt only if the subject remembers on the R trial that the preceding trial was not reinforced. If proactive interference causes increased forgetting over long ITIs, then the demonstration of sequential effects such as these should become increasingly difficult at longer intervals.

A plausible analysis of proactive interference, at least in studies of animal learning, is that it is less a matter of the complete loss of information about preceding outcomes and more a case of forgetting the order in which preceding outcomes occurred (p. 481). While this would be sufficient to abolish alternation learning at long ITIs, it would, presumably, do no more than attenuate the PRE itself. Animals trained on a partial reinforcement schedule with spaced trials might not remember whether the last trial was reinforced or not, but the set of traces associated with reinforcement on R trials would still be a mixture of previous N and R outcomes. Although this would differ from the set of N traces experienced in extinction, the difference would be much smaller than that between the extinction set and the set of exclusively R traces associated with reinforcement in consistently reinforced animals. Indeed, if extinction trials were also spaced, there might be some recovery of R traces from acquisition during each ITI early in extinction; at the outset of extinction there would not be too much difference between the set of traces associated with reinforcement during partially reinforced acquisition and the set encountered in extinction. Although the spacing of acquisition and extinction trials might attenuate the PRE, therefore, it would certainly not abolish it. But if extinction trials were massed, there would be no recovery of R traces during extinction ITIs and there would be a considerable difference between the set of N and R traces associated with reinforcement during spaced acquisition trials and the set of exclusively N traces experienced in extinction. The PRE, therefore, should be markedly reduced when the ITI is long in acquisition, and short in extinction.

Finally, if subjects given partial reinforcement with a long ITI cannot remember the precise sequence of trials to which they have been exposed, but do learn to respond in the presence of a mixed set of N and R traces from preceding trials, resistance to extinction would be determined less by the number of N trials in a row, or N-length (for this is a sequential variable, dependent on order information), and more by the proportion of N traces in the total set associated with reinforcement on any R trial. Thus, at long ITIs, resistance to extinction would be less influenced by N-length, and more dependent on the number or percentage of N trials.

Of these three explanations, therefore, the first—that the PRE at long ITIs is dependent exclusively on conditioned frustration—finds least support. The suggestion that short-term after-effects of frustration contribute to the PRE after massed trials is supported by some results, but cannot readily explain the decreasing significance of sequential variables as ITI increases. The suggestion that proactive interference causes some forgetting at long ITIs can handle this result, and much of the remaining data, fairly readily. Although this analysis is inconsistent with the often-voiced view that there are two PREs, one occurring with massed trials and the other with spaced trials, it does not deny the possibility that the PRE is partly determined by conditioned inhibition or frustration at all ITIs. The point is that there is little evidence for the view that the PRE after spaced trials is determined exclusively by such a process.

7. Number of Acquisition Trials

A second argument for partitioning the PRE into one effect determined largely by memory traces of preceding outcomes and another determined largely by conditioned frustration, has been advanced by Black and Spence (1965) and by Amsel et al. (1968). The basis for this distinction is the number of trials given in acquisition. The argument is *prima facie* a plausible one. Extended training on a partial reinforcement schedule appears to provide optimal conditions for the occurrence of frustration or inhibition and its conditioning to apparatus cues. Unlike the nonreinforcement of simple extinction, nonreinforcement on a partial reinforcement schedule occurs in the context of continued, concurrent reinforcement, and it would be surprising if the occasional omission of reinforcement in such a situation did not result in the conditioning of inhibition or frustration to apparatus cues. Since extinction trials may also be reasonably supposed to result in *some* inhibition or frustration, partially reinforced animals should have learned to respond under the conditions later encountered during extinction for exactly the reasons supposed by frustration theory.

All of this, however, should require some minimal number of acquisition trials. Frustration is produced only by the omission of expected reinforcement (p. 260). Frustration theory, therefore, can expect that partial reinforcement will establish persistent responding in extinction only if sufficient acquisition has been given, first to establish an expectation of reinforcement, secondly to produce frustration when reinforcement is omitted, and thirdly to reinforce responding in the presence of frustration. Amsel (1958, 1967) was quite explicit in arguing that the number of trials required in the ordinary alleyway would be of the order of 10 to 30, and there is indeed good independent evidence for such an argument: Hug (1970a), for example, has shown that the double-alleyway frustration effect appears only after at least 12 partially reinforced trials.

It follows from frustration theory, and from a theory such as Lawrence and Festinger's (1962) cognitive dissonance hypothesis, that the PRE should be observed only after a moderate number of acquisition trials. This prediction has been tested, and found wanting, in a large number of studies. Although there is some evidence that the magnitude of the PRE increases with amount of training (e.g., Bacon, 1962; Hill and Spear, 1963a; Wilson, 1964), and although a few studies have failed to find evidence of a PRE after 10 or fewer acquisition trials (e.g., Bacon, 1962; Amsel et al., 1968), a small, but reliable, PRE may be observed after as few as five or six acquisition trials (e.g., McCain, 1966, 1969; Ziff and Capaldi, 1971). It is quite possible that the effect disappears if the ITI is sufficiently long (although it is readily obtained with an ITI of 15 min) or if the reward is sufficiently small, but this is hardly surprising. These are variables that affect the magnitude of the PRE after any amount of training; that sufficiently unfavourable conditions should abolish an effect that is small in the first place is hardly reason for supposing that the effect does not exist.

Logically, it would be possible for frustration theory to argue that a minimum of three trials, in the sequence RNR, was sufficient to generate a PRE. The first R trial would establish the expectation of reinforcement, the second would generate frustration and the third would reinforce responding in the presence of frustration. Such a step would divorce the theory from independent validation by rejecting evidence that the frustration effect does not develop until a substantially larger number of trials has been given. In the event, it is clear that the step cannot save the theory: a significant PRE can occur under conditions where the N trials scheduled during acquisition cannot have produced any conditioning of frustration. The critical evidence was first provided by Spear et al. (1965) and Spear and Spitzner (1967) who showed that rats which were never fed in the experimental apparatus showed greater resistance to extinction than a consistently reinforced control group if they were given a series of initial N trials followed by a series of R trials. In both these studies animals received a relatively large number of acquisition trials 24 N and 24 R trials). Capaldi *et al.* (1970) and Capaldi and Waters (1970) have obtained entirely similar results after a very much smaller number of acquisition trials. Since partially reinforced subjects received no reward in the apparatus before their N trials, they cannot have been frustrated on these trials. Capaldi and Waters, indeed, observed no difference in resistance to extinction between partially reinforced subjects that received two R trials before their block of N trials and those that received all N trials before their first R trial. The PRE that occurs after 10 or fewer acquisition trials cannot be attributed to the conditioning of frustration to apparatus cues. Frustration theory, therefore, cannot be regarded as providing a complete explanation of the PRE.

8. *Generalization between Nonreinforcement and Punishment*

There is substantial evidence that the omission of an expected appetitive reinforcer is an aversive event, and that the frustration reactions elicited by such an event share properties in common with the reactions elicited by other aversive events such as electric shocks (p. 261). The application of this line of reasoning to the case of partial reinforcement suggests that animals trained on a standard partial reinforcement schedule have learned to respond in the presence of a set of aversive stimuli, of which the set encountered in extinction is only one example. Partial reinforcement should also, for example, increase resistance to punishment. Even more importantly, exposure to occasional punishment during the course of acquisition should increase resistance to extinction. The classic test of this hypothesis is a study by Brown and Wagner (1964). They trained rats to run in an alley under one of three conditions: one group was consistently reinforced; a second group was reinforced on only 50 per cent of trials, and the third group was reinforced on 100 per cent of trials but was shocked in the goal box on 50 per cent of trials. After 114 acquisition trials each group was divided into two: half the subjects received a series of ordinary extinction trials and the other half continued to be rewarded in the goal box but were also shocked on every trial. The results are shown in Fig. 8.6. Not only did partial reinforcement increase resistance to extinction, and partial punishment increase resistance to consistent punishment, but, to a lesser extent, each treatment increased persistence in the face of opposite "extinction" procedure.

These results, and in particular the observation that occasional punishment during acquisition increases resistance to extinction, provide relatively compelling evidence in favour of the frustration theory account of the PRE. Although they have not always been replicated (e.g., Banks and

Torney, 1969), at least two other studies have reported similar results (Uhl, 1967; Wong, 1971). With sufficient training it is reasonable to infer that the occasional nonreinforcements experienced by partially reinforced subjects produce an aversive, frustrative, state and that resistance to extinction is dependent on having learning to respond in the presence of aversive stimuli.*

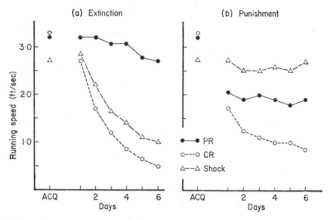

FIG. 8.6. The effect of partial reinforcement or punishment on resistance to extinction or consistent punishment. In acquisition, subjects received either consistent reinforcement (CR), partial reinforcement (PR), or consistent reinforcement plus gradually increasing intensities of shock on a partial schedule (Shock). Their performance is shown at the end of acquisition (Acq.), and (a) over blocks of extinction trials, or (b) over blocks of consistently punished trials. (*After* Brown and Wagner, 1964.)

9. *Effects of Drugs*

Barbiturates such as sodium amytal attenuate the effects of such aversive events as electric shock, and also at least some of the aversive and inhibitory effects of frustration (p. 265). It follows from frustration theory, therefore, that the administration of such a drug should abolish, or at least attenuate,

---

* Fallon (1971) has proposed an ingenious argument to counter this conclusion. He suggests that the reason why punishment on R trials increases resistance to extinction is because it reduces the functional magnitude of the reward: the combination of a large reward and a mild shock is equivalent to the receipt of a small reward. Since resistance to extinction is inversely related to the size of reward, the effects of punishment can be assimilated to those of magnitude of reward: they do not require that persistence in the face of one aversive event should generalize to persistence in the face of another. Although this is an interesting idea, it clearly runs counter to the interpretation of the effects of reward size given earlier in this chapter. Large rewards can no longer be assumed to increase resistance to extinction because they are more discriminable from nonreinforcement than are small rewards, for a large reward combined with an electric shock should be even more discriminable.

the PRE. Two experiments have tested this prediction, with results that provide only limited support for the frustration theory account of the PRE (Gray, 1969; Ison and Pennes, 1969). In both studies the PRE was markedly attenuated if animals were given acquisition training when drugged and extinguished without the drug. This implies, as frustration theory requires, that animals that receive partial reinforcement when drugged do not learn to respond under the conditions encountered in normal extinction. Again in both studies the PRE was quite severely attenuated if acquisition training was given in the undrugged state and extinction was given while animals were drugged. As would be expected, the main effect of administering the drug in extinction was to increase the resistance to extinction of consistently reinforced subjects. However, when the drug was administered both in acquisition and in extinction, the PRE seems to have been virtually unaffected.*

The strongest conclusion that can be drawn from these results, therefore, is that after quite extensive acquisition training (40 or 64 trials), extinction is an aversive event whose effects may be attenuated by a barbiturate drug and that persistence during extinction is *partly* dependent upon having been reinforced for responding in the presence of similarly aversive stimuli associated with frustration. One problem for an analysis of these results in terms of *conditioned* frustration, however, is that Capaldi and Sparling (1971b) have shown that partially reinforced animals show normal resistance to extinction if they receive the drug only on those acquisition days when N trials are not followed by R trials: the PRE is affected only by the administration of the drug on days when N–R transitions are scheduled. This suggests that the PRE is dependent on the reinforcement of responding in the presence of the immediate after-effects of unconditioned frustration rather than in the presence of conditioned frustration. Whichever interpretation proves more fruitful (and the answer could be provided by a drug study run with widely spaced trials), the finding reported by Ziff and Capaldi (1971)—that after a very small number of acquisition trials the PRE is unaffected by the drug—supports the argument of the preceding section that frustration becomes an important factor in the PRE only after a substantial number of acquisition trials.

10. *Conclusions*

In spite of the extraordinarily large number of experimental studies of the PRE, there are still some serious gaps in our knowledge. Several of the conclusions reached in the present discussion have rashly relied upon

---

* The account given here of Gray's results differs somewhat from that provided in the original report. Gray's own analysis, however, ignores the very substantial differences in terminal acquisition performance between some of his groups.

comparisons between separate experiments rather than between separate groups within a single experiment. It is not known, for example, whether the distribution of trials alters the effects of barbiturates on the PRE, nor is there a single experiment showing that the effects of these drugs depend upon the number of acquisition trials. The crucially important conclusion that the effects of the sequential variable of N-length are dependent on ITI is based on a comparison between separate studies rather than on a single study.

One should not, of course, be surprised by this. Experiments have been performed to answer numerous different questions and it is appropriate enough that new analyses should suggest new points of view that are not directly tested by existing evidence. With this caution in mind, it is worth summarizing some of the main conclusions suggested by the present discussion. Although other factors enter into the picture, the PRE seems to be basically a phenomenon caused by generalization decrement: partially reinforced subjects have been reinforced for responding under conditions similar to those encountered in extinction. The implication of this is that the N trials occurring during a partial reinforcement schedule, or the D trials occurring in a partial delay schedule, have some effect on subjects similar to the effects produced by N trials in extinction.

Perhaps the simplest assumption is that subjects just remember that they have not been reinforced and a subsequent R trial ensures that responding is reinforced at a moment when the subject still remembers this earlier nonreinforcement. This assumption is supported by the sequential data collected by Capaldi and his associates, which suggest that an interpretation in terms of memory traces is indeed a useful one. The available evidence supports three specific implications of this interpretation. First, the memory traces of preceding outcomes are subject to retroactive interference: hence the importance of N–R transitions as determinants of performance in extinction. Secondly, these traces are subject to proactive interference: hence the effects of long ITIs on the PRE itself and, in particular, on sequential determinants of the PRE. Thirdly, the retrieval of traces of preceding outcomes depends upon the reinstatement of appropriate contextual stimuli: hence the effects of reinforced and nonreinforced placements may depend on the sequence in which they are given.

It is certain, however, that nonreinforcement is not just an emotionally neutral event like the presentation of a light or a tone. There exists good, independent evidence for the assumption that unexpected nonreinforcement is inhibitory or frustrative which implies that with sufficient training partially reinforced subjects must experience frustration on N trials and may therefore learn to respond in the presence of frustration stimuli.

Studies of punishment and drugs suggest that such a factor does enter into the PRE at a later stage in training and this conclusion is strongly supported by the finding that schedules of reinforcement which make it possible to predict the occurrence of N trials during acquisition do not necessarily increase resistance to extinction.

The large majority of studies of the PRE have employed rats in alleyways. Although studies of classical conditioning in rats and dogs have sometimes revealed a PRE, the effect is notably less resilient than that observed in typical instrumental experiments. However, in the absence of more direct experimental comparisons of the different response systems in classical conditioning studies and of the classical and instrumental contingencies, it is foolish to speculate too extensively on the possible causes of this apparent discrepancy (p. 75). It may be worth noting that the discrepancy between classical and instrumental experiments also seems to extend to studies varying magnitude of reinforcement, and that both sets of results might be explicable if one supposed that relatively reflexive responses were unlikely to be controlled by memory traces of the outcomes of preceding trials.

The PRE has also been studied in instrumental experiments with animals other than the laboratory rat. It appeared at one time that fish might show no PRE in a standard situation where partially and consistently reinforced groups received equal numbers of trials in acquisition, and that the effect would occur only if subjects were given equal numbers of reinforcements in acquisition (Gonzalez et al., 1963). More recent studies, however, have demonstrated a PRE in both situations (e.g., Gonzalez and Bitterman, 1967). The effect is, in fact, controlled by the same set of variables in fish as in rats, being increased by the use of large rewards (Gonzalez and Bitterman, 1967), a short ITI (Gonzalez et al., 1965a; Schutz and Bitterman, 1969) and increases in N-length (Gonzalez et al., 1963; Zych and Wolach, 1973). It seems clear, however, that the overall magnitude of the PRE is smaller in fish than in rats: whereas a combination of unfavourable conditions (e.g., a small reward *and* a long ITI) is needed to abolish the PRE in rats, the effect disappears in fish when only one of these variables is changed. Turtles similarly show a PRE only when trained with massed trials (Murillo et al., 1961; Eskin and Bitterman, 1961). The difference between rats, turtles and fish, then, appears to be a quantitative one; like the difference in successive contrast effects and alternation learning (p. 401), and the effects of size of reward on extinction (p. 431), it may be due to differences in the extent to which the behaviour of rats and that of turtles or fish is controlled by traces of previous conditions of reinforcement (Mackintosh, 1971b).

Q

IV. CHANGES IN PERFORMANCE OVER TIME

We can account for the changes in performance that occur during acquisition and extinction by pointing to the subject's exposure to certain reinforcement contingencies in each case. But it is clear that changes in performance may also occur over an interval of time when the subject is not exposed to the experimenter's contingencies of reinforcement, and even when the subject is not exposed to the experimental situation at all. The two most common instances of such "spontaneous" changes are spontaneous recovery after extinction and forgetting. On the rare occasions when the topic of forgetting in animals is discussed at all, it is usually treated separately from spontaneous recovery. This is almost certainly a mistake.

A. Spontaneous Recovery

The phenomenon of spontaneous recovery provides one of the few instances where the extinction of classical CRs and of instrumental responses appear to produce identical outcomes. Pavlov (1927, p. 58) reported that after a series of extinction trials at a rate of one every three minutes, a salivary CR was completely extinguished, but that after an interval of two hours the renewed presentation of the CS resulted in a substantial CR. The classic study of the spontaneous recovery of an instrumental response was performed by Ellson (1938). Rats were given a single session of training on a lever-pressing response and 24 h later were extinguished to a criterion of five minutes without responding. When returned to the apparatus at intervals ranging from five minutes to three hours later, they all showed some tendency to respond again. The results of the experiment are shown in Fig. 8.7. It can be seen that the number of responses emitted in the second extinction session is an increasing function of the interval between first and second sessions. Unfortunately, no animals were tested after longer intervals, to see if spontaneous recovery continued to increase: at the longest interval studied animals responded very much less in the second extinction session than in the first.

　　Pavlov (1927, p. 60) interpreted spontaneous recovery as proof that extinction did not result in the loss of the original CR or of the associations on which it was based, but rather overlaid these excitatory associations with a new set of inhibitory associations. Inhibition, Pavlov further assumed, was more labile than excitation; since it dissipated more rapidly, the elapse of an interval of time after an extinction session would ensure that the original excitatory associations were no longer suppressed. Hull (1943, pp. 285–6) adopted essentially the same explanation of spontaneous recovery which continues to find adherents (e.g., Kimble, 1961, p. 284).

The first of Pavlov's inferences seems reasonable: at any rate, the phenomenon of spontaneous recovery is evidence that not all of the excitatory associations established during acquisition can simply have been unravelled during the course of extinction. The remaining inferences, however, are less secure. It may be possible to predict the occurrence of spontaneous recovery from a theory which explains extinction as the loss of some of the excitatory associations established during acquisition. The explanation of spontaneous recovery may not necessarily require the postulation of a special process of inhibition, and even if it does, it certainly does not require that the rate of decay of excitation and inhibition should differ.

Skinner (1950) pointed out that a series of extinction trials in a single

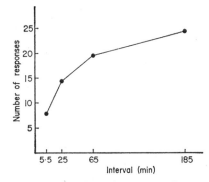

FIG. 8.7. Spontaneous recovery of an instrumental response. The number of responses made in a second session of extinction as a function of the interval between the end of the first session and the beginning of the second. (*After* Ellson, 1938.)

session could not necessarily be expected to extinguish the tendency to respond at the beginning of a new session. At its simplest the argument rests on the assumption that immediately after being placed in the apparatus, the subject is in a different stimulus situation from that which will prevail after a long series of extinction trials. As was argued earlier, the loss of responding during extinction may be partly accounted for by noting that the stimulus situation prevailing late in extinction differs from that which prevailed in acquisition, since one contains traces from a long series of N trials, while the other does not. In just the same way, the partial recovery of responding after an interval may be accounted for by noting that the stimulus situation at the beginning of the next session may contain elements which were not present late in the preceding extinction session, such as stimuli associated with being placed in the apparatus, and may not contain elements which were present then, such as the traces of recent N outcomes. Spontaneous recovery may, therefore, be partly a reinstatement effect. In support of this argument, a recent study by Burstein and Moeser (1971) showed that it was possible to increase the spontaneous recovery of an instrumental response in pigeons by providing subjects with a distinctive stimulus on the first trial of each session; while Homme (1956) found that rats given a single session of acquisition training showed significantly less spontaneous recovery than subjects given the same number of reinforcements spread over five acquisition sessions.

To the extent that the reappearance of a partially extinguished response after an interval of time is a consequence of the reinstatement of stimuli whose association with reinforcement was never extinguished, the phenomenon of spontaneous recovery does not provide unique support for Pavlov's arguments. It may be unnecessary to suppose that a series of extinction trials results in a new process of inhibitory learning which prevents the expression of the original excitatory learning; spontaneous recovery may only reflect the failure to extinguish all of these excitatory associations. In principle, therefore, it might be possible to dispense with the notion of inhibitory learning as a process fundamentally distinct from that of excitatory learning. The two terms would just refer to the increments and decrements in a single associative variable produced by reinforcement and nonreinforcement (cf. Wagner and Rescorla, 1972).

One question that may be asked about this analysis is how well it is able to account for the temporal course of spontaneous recovery. If the change in the probability of a response between the end of a session of extinction and the test for spontaneous recovery is due to a change in the set of stimuli to which the subject is exposed, and in particular to the reinstatement of stimuli dependent upon replacement in the apparatus, it is not entirely clear why the reinstatement of such stimuli should vary with the passage of time intervening between extinction and spontaneous recovery. The only attempt to provide an analysis of this problem is due to Estes (1955). Within the framework of stimulus sampling theory, Estes argued that at any given time not all stimulus elements are available for sampling by the subject and that with the passage of time elements move at random in and out of the subject's sample set. If elements fluctuate at random, then the probability of an element unavailable during extinction becoming available during the test for spontaneous recovery will be an increasing function of the interval between extinction and spontaneous recovery. Since spontaneous recovery depends upon the reappearance in the subject's sample of unextinguished elements, the magnitude of the effect should depend upon the length of this interval.

There is, unfortunately, no evidence which would show whether this is a sufficient account of spontaneous recovery. Since the point at issue is whether it is necessary to postulate a distinct process of inhibitory learning, it is surprising that the phenomenon of spontaneous recovery should have attracted so little experimental attention. Even if this question must remain unresolved, however, Pavlov's final argument may be approached more directly. According to Pavlov, spontaneous recovery implied not only that excitatory associations were simply overlaid by inhibitory associations during the course of extinction, but also that these inhibitory associations were inherently more labile than the original excitatory associations. Although it may eventually be necessary to postulate distinct processes of excitatory and inhibitory learning, there is no reason to suppose that the two processes differ in their stability. The crucial point, as Estes (1955) recognized, is that spontaneous recovery may not be the unique phenomenon it has often been thought to be. Spontaneous increases in the probability of a response from the end of one extinction session to the beginning of another may be similar to spontaneous decreases in the probability of a response from the end of one acquisition session to the beginning of another. Such spontaneous regression, although accorded less publicity than spontaneous recovery, occurs both in classical conditioning (e.g., Konorski, 1948, p. 83) and in instrumental learning (e.g., Mote and Finger, 1943; Spear et al., 1965).

The argument has been developed by Spear (1967, 1971). He has pointed to the formal similarity between spontaneous recovery and phenomena usually regarded as instances of proactive interference. In the former class of study, acquisition is followed by extinction; in the latter, the learning of list A–B is followed by the learning of list A–C. In both cases there is a conflict between the first and second task, and in both cases, if subjects are tested after a retention interval, the relative strengths of the responses appropriate to the first and second tasks appear to change. As time progresses the subjects become less likely to respond in a manner appropriate to the second task. This may not be because inhibition is intrinsically more labile than excitation, but because there is more room for changes in the strength of more recent associations than of less recent associations.

The most obvious implication of this argument is that spontaneous recovery is a more noticeable phenomenon than regression, only because the former situation automatically contains a source of proactive interference, while the latter does not. Spontaneous recovery is demonstrated by subjects given extinction after acquisition training. Chiszar and Spear (1968b) were able to show that significant recovery from suppression, produced by providing quinine-adulterated food as a reinforcer for a simple approach response, occurred only if subjects had been rewarded previously with unadulterated food in this situation. Without prior rewarded training, in other words, suppression did not disappear over time. Studies of regression, of course, have not usually given acquisition training after prior extinction. When they have done so, however, the results are very striking. Konorski and Szwejkowska (1952b) noted that a former CS — could only with great difficulty be turned into a CS+ with relatively spaced training, but extinguished very rapidly during a subsequent extinction session. Spear *et al.* (1965) and Spear and Spitzner (1967) trained rats in an alleyway, giving some groups of subjects a series of nonreinforced trials before beginning acquisition training. There was a significant increase in spontaneous regression in these groups: over several days of training they ran significantly more slowly on Trial 1 of each day than did control groups given no prior nonreinforcement. Moreover, Spear and Spitzner (1968) have shown very similar effects in rats exposed, successively, to different magnitudes of reward.

The spontaneous regression reported in Konorski's and Spear's studies implies that instances of spontaneous changes in behaviour with the passage of time are not to be explained in terms of any general difference between excitatory and inhibitory learning, but by noting that in typical experiments on spontaneous recovery, inhibitory learning occurs after excitatory learning. When the order is reversed, spontaneous regression occurs. Spontaneous recovery and regression, therefore, seem to be two sides of the same coin: they each represent a transitory failure to respond in a manner appropriate to the reinforcement contingencies most recently experienced by the subject. In this sense they are analogous to the phenomena of proactive interference studied in human verbal learning. The connection between proactive interference and spontaneous recovery is not, of course, new. Traditionally, students of verbal learning and retention have sought, not entirely successfully, to explain the former in terms of the latter (e.g., Underwood and Postman, 1960; Keppel, 1968). The implication of the present argument is that it might be better to reverse this process, to take proactive interference as fundamental and spontaneous recovery as derivative, and to explain the latter as an instance of the former. This would at least make clear that what is needed is a general theory of proactive interference.

B. Forgetting

A widely accepted generalization is that laboratory animals, like elephants, never forget (e.g., Kimble, 1961, p. 281). As with many myths, this one rests on a small core of truth surrounded by a larger mass of casually reported and uncritically accepted data. Although the same series of studies is repeatedly cited as evidence for this generalization, few of the early reports provided acceptable evidence. Liddell *et al.* (1935) claimed to have observed excellent retention of a flexion CR in sheep, but the observation is merely noted in passing, with no supporting details. Wendt (1937) is said to have shown complete retention of an avoidance response (also flexion) in dogs: here the details do not entirely bear out the claim. The single subject received 25 widely spaced sessions of acquisition training, spread out over a three-month period. The series included a four-week interval without training, which provided the first test of retention: the probability of a response declined by 10 per cent over this retention interval and the amplitude of the flexion response declined by over 50 per cent. After further training a second test for retention was given two and half years later. This test provided evidence of relatively good retention, with the subject responding on 80 per cent of trials, but it is important to note that an unspecified number of unsignalled shocks were delivered to the animal when it was replaced in the apparatus before receiving the series of test trials. On a second test for retention, given six months later with no free shocks, the subject failed to respond once in 11 trials.

A more recent study by Skinner (1950) also provided some details which do little to suggest that forgetting is negligible. Skinner tested pigeons for the retention of an appetitive instrumental response; in spite of the fact that the subjects were returned to the apparatus for several days before the test for retention, being fed on each occasion, Skinner's data reveal that the maximum amount of retention observed in four birds was of the order of 25 to 50 per cent (Gleitman, 1971).

Although, therefore, there may be some truth to the suggestion that simple CRs or instrumental responses are sometimes relatively well remembered over surprisingly long intervals, the claim merits rather closer scrutiny than it has often been accorded. The most convincing evidence of good retention comes from studies of conditioned suppression (e.g., Hoffman et al., 1963; Gleitman and Holmes, 1967). Numerous other studies have reported substantial amounts of forgetting of simple instrumental behaviour. These are reviewed briefly in the following sections.

1. Retention of Appetitive Instrumental Behaviour

A long series of studies by Gleitman and his associates has provided evidence of forgetting, over 1- or 2-month retention intervals, of a variety of patterns of behaviour (Gleitman, 1971). Rats trained to run in an alley for food, showed considerably greater loss of the habit after a 60-day retention interval than after a 1-day interval (Gleitman and Steinman, 1963); rats trained to press a lever on an FI schedule not only showed a loss of responding in the later part of each interval over a 24-day retention period, they also forgot the temporal properties of the FI schedule and responded more rapidly immediately after reinforcement than subjects tested one day after training (Gleitman and Bernheim, 1963). Part of what is forgotten in the alleyway experiment is the nature of the reward received: Gleitman and Steinman (1964) showed that a 68-day interval between pre- and post-shift training may abolish the negative contrast effect produced by a shift from large to small reward.

2. *Retention of Aversively Motivated Behaviour*

In addition to the earlier studies on retention of flexion and eyelid CRs in various animals, retention of conditioned suppression has been studied in both pigeons (Hoffman, *et al.*, 1963) and rats (Gleitman and Holmes, 1967). In the former study, retention of conditioned suppression established over several hundred trials was essentially perfect $2\frac{1}{2}$ years later. In the latter study, rats showed no forgetting of incompletely learned conditioned

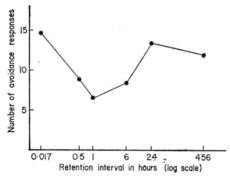

FIG. 8.8. The Kamin effect in the retention of avoidance responding in rats. The number of avoidance responses made in a second training session of 25 trials as a function of the interval between the end of the first session and the beginning of the second. (*After* Kamin, 1957d.)

suppression over a 90-day retention interval. Other studies (e.g., Campbell and Campbell, 1962) have found good retention of learned fear in adult rats, although Campbell and Spear (1972) have reported other cases where retention has been less than perfect. As might be expected in view of Gleitman and Bernheim's (1963) data on retention of FI behaviour, Hammond and Maser (1970) reported that if rats were trained on conditioned suppression with a CS long enough to produce measurable inhibition of delay, this temporal discrimination was lost over a 25-day retention interval.

Over shorter intervals, aversively motivated behaviour shows a peculiar U-shaped retention function. Kamin (1957d) trained rats for 25 trials on a signalled avoidance task in a two-way shuttle box; different groups received a second 25-trial training session after intervals ranging from 0 sec to 19 days. The results are shown in Fig. 8.8: animals that received their second training session one hour later showed a pronounced deficit. This finding, usually known as the Kamin effect, has provoked a very large body of research (Brush, 1971), much of which is outside the scope of this discussion. The effect has often been attributed to changes in motivational state. Kamin (1957d), for example, originally attributed the improvement in performance between the one-hour and one-day retention tests to an increase in fear, arguing that avoidance was facilitated by higher levels of fear. Denny and Ditchman (1962), on the other hand, attributed the decline in performance between the immediate and one-hour tests to a similar increase in fear, arguing that avoidance was disrupted by too high a level of fear. This latter explanation runs into problems from the results of studies of punishment. Both Pinel and Cooper (1966) and Singh *et al.* (1971), for example, have reported a similar U-shaped curve for the retention of punishment of appetitively reinforced responding. Any increase in fear over the first hour or two of the retention interval could lead only to better suppression of the appetitively reinforced response. Recent evidence, considered below, suggests that the effect may be caused by a failure of retrieval at intermediate retention intervals.

3. *Retention of Choice Behaviour*

Even when prior sources of interference have been introduced deliberately, many studies of choice behaviour have shown strikingly good retention over intervals of weeks or months: in the absence of such sources of interference, Gleitman and Jung (1963) found perfect retention of a spatial discrimination over a 44 day interval in rats, and Maier and Gleitman (1967) found little forgetting of a visual discrimination over a 32 day interval. Campbell and Spear (1972), however, reported some forgetting of a T-maze habit in rats over 14 days, and retention was even poorer in very young subjects. It is, of course, obvious that when choice of one alternative is consistently reinforced and choice of another is never reinforced, the probability of choosing the correct alternative will rapidly approach 1·00; thus a choice measure of performance will be relatively insensitive to small amounts of forgetting. In a probability learning situation, where choice of one alternative is reinforced on, say, 75 per cent of trials and choice of the other is reinforced on the remaining 25 per cent of trials, performance is usually less consistent (p. 191) and it seems likely that retention would be less than perfect. Indeed Shimp (1970) has shown that some forgetting may occur from one day to the next when rats are trained on a two-lever probability learning task: after more than 5,000 trials rats attained an asymptote of 94 per cent choice of the more favourable alternative, but the probability of a correct choice on Trial 1 of each day was only 0·61.

4. *Retention of the Absolute Properties of a Stimulus*

Although simultaneous discrimination problems with consistent reinforcement may show relatively slight amounts of forgetting, successive discriminations are less well retained (e.g., Campbell *et al.*, 1968), and there is evidence from a wide range of studies that animals are unable to remember the absolute characteristics of a stimulus to which they have been exposed previously. Generalization gradients become significantly flatter over the course of a long retention interval. For example, rats trained to run in an alley of a given brightness run more slowly in a different alley only if tested immediately: after a long interval speed of running is unaffected by the brightness of the alley (Perkins and Weyant, 1958; Steinman, 1967). Similar results have been obtained in studies of wavelength generalization in pigeons (e.g., Thomas and Lopez, 1962; Thomas and Burr, 1969) and in studies of the generalization of conditioned fear in rats (McAllister and McAllister, 1963).

C. Causes of Forgetting

1. *Dependence of Successful Retrieval on Contextual Stimuli*

One possible analysis of spontaneous recovery, as we have seen, is that some of the stimuli present in aquisition do not lose their association with reinforcement during the course of extinction. To the extent that extinction trials occur in a somewhat different stimulus situation from that prevailing in acquisition, the subsequent reinstatement of some of the stimuli present during acquisition, such as those associated with the beginning of a session, will result in partial recovery of the behaviour controlled by those stimuli. The change in performance from the end of an extinction session to the test for spontaneous recovery is attributed to a change in the prevailing stimuli. In a similar manner the disappearance of a response over the course of a retention interval may be attributed to a change in the set of stimuli from the final training session to the test for retention (cf. Estes, 1955). Even though the experimenter presents the same discriminative stimuli in the same apparatus to a subject maintained under the same conditions of deprivation, there are numerous other stimuli which may have changed during the course of the retention interval.

The importance of apparently incidental, contextual stimuli in the control of animal behaviour should never be underestimated. Seemingly trivial changes in the experimental situation, as Pavlov was the first to notice, can severely disrupt learned performance. A good example of this is provided by Logan (1961) who trained rats in an apparatus in which they were forced on alternate trials to take one or other of two different routes to the goal box. On one of these paths they were required to solve a brightness discrimination in order to reach the goal box. After they had mastered this problem, the discriminative stimuli were relocated at a comparable point in the other path. This change was sufficient to produce complete disruption of discriminative performance, to the point where some subjects took as many trials to relearn the discrimination as they had required to learn it in the first place.

Learned performance may also be disrupted by changes in the state of the subject. A common finding in studies utilizing drugs is that behaviour established in the drugged state may show little transfer to the undrugged state and vice-versa (e.g., Overton, 1964, 1966). Such dissociation effects suggest that subjects in one state may be unable to retrieve what they have learned in another state.*

The extension of this argument to the general study of forgetting is straightforward enough. A subject may fail to perform appropriately on a retention test because the stimulus situation prevailing differs in significant respects from that prevailing during initial training. Such changes may be hard to specify, and even harder to control. The only way to assess the validity of the argument is to show that in at least some cases the reinstatement of plausible contextual stimuli does significantly improve retention: this was the procedure employed by Irion (1949) in his classic study of "warm-up decrement", or reinstatement effects, in verbal learning.

Klein and Spear (1970a, b) have presented some results, which if substantiated (cf. Anisman and Waller, 1971; Barrett et al., 1971) suggest that the Kamin effect may represent a failure of retrieval. They argue that for some time after initial aversive training, subjects are in a temporarily different state (presumably as a consequence of stress reactions to shock: see Brush, 1971). During this stage they are unable to retrieve what they had originally learned and therefore perform poorly; but they eventually revert to their original state and are now able to retrieve the appropriate response. Klein and Spear found that poor performance at the intermediate retention interval was obtained only if subjects were retrained on exactly the same task as in original training: rats trained on an active avoidance task in a shuttle box performed very badly when tested on the same task one hour later, but learned a passive avoidance task significantly more rapidly at the one-hour test than at earlier or later tests. Similarly, if rats were tested on an active avoidance task after passive avoidance training, performance was again better one hour following original training than following shorter or longer intervals. The implication is that animals are quite capable of learning new avoidance tasks one hour after a stressful

* A principle sufficiently familiar to Wilkie Collins in 1868 to form the basis of the plot of *The Moonstone*.

experience: indeed if the new task is the reverse of the old, the absence of interference results in very rapid learning. Poor performance on the original task, therefore, should be attributed to a temporary failure to retrieve what was learned in original acquisition, rather than to any general inability to avoid as a consequence of too high or too low a level of fear.

2. Interference Theory

There have been two dominant analyses of forgetting in the history of psychology: either forgetting has been attributed to the decay of a memory trace, or it has been thought to be produced by interference from conflicting associations. The appeal of interference theory has been that it is considerably more amenable to experimental investigation. If forgetting occurs because memory traces simply fade away with the passage of time,

TABLE 8.3

Designs for Studies of Retroactive and Proactive Interference

		Train	Train	Test
Retroactive Interference	Experimental	A	B	A
	Control	A	—	A
Proactive Interference	Experimental	A	B	B
	Control	—	B	B

there is little that the student of behaviour can do about it. But if forgetting is caused by interference, then it is an easy matter to manipulate the incidence, temporal relationships and similarity of potentially interfering associations.

Interference theory has been developed largely in the context of verbal learning (Underwood and Postman, 1960; Keppel, 1968; Spear, 1970). Several aspects of the theory have suffered extensively at the hands of recalcitrant data, but the basic idea survives in one form or another, not only because of its intrinsic appeal but also because of the undoubted fact that exposing subjects to potentially interfering tasks does materially affect their performance on a retention test. In this empirical sense, interference may occur either retroactively or proactively. Table 8.3 illustrates the designs required to demonstrate the two effects. Retroactive interference (RI) occurs if an experimental subject, required to learn two tasks (A and

B) in succession, shows poorer retention of the first task (A) than a control subject trained on A alone. Proactive interference (PI) occurs if the experimental subject shows poorer retention of the *second* task (B) than a control subject trained on B alone.

In studies of retroactive interference in human subjects, the test for retention includes instructions stating that the subject is required to recall Task A. Since such instructions are not possible in experiments with animal subjects, the occurrence of retroactive interference is hardly surprising. If a rat has been trained to turn left in a T-maze, subsequent training to turn right could hardly fail to decrease the probability of a left turn on a later test trial. The observation of such an effect, therefore (e.g., Crowder, 1967), is not particularly informative. It is, however, possible to study the extent to which different kinds of intervening activity interfere with the apparent retention of an earlier learned task. Marx (1944), for example, showed that retention of a maze habit was inversely related to the number of subsequent problems the subject was exposed to during the retention interval. Chiszar and Spear (1969), training rats on a spatial discrimination and its reversal, gave reversal training for one group of subjects in a similar T-maze located in a different room. Retention of Task A was greatly improved when Task B was learned in a different context.

On the assumption that RI may be produced by the subject's extra-experimental activities, another procedure has been to manipulate the possibility of any activity during a retention interval. In a classic study Minami and Dallenbach (1946) showed that cockroaches, immobilized by being confined during the retention interval in a small box stuffed with tissue paper, showed better retention of an avoidance response than a control group, which in turn performed better than a hyperactive group forced to run along a continuously moving belt during the retention interval.

Proactive interference has also been studied extensively in choice situations. The outcome of such studies has been more variable: in some cases, notably where simple spatial discriminations have been used (e.g., Rickard, 1965; Crowder, 1967), there has been no evidence that retention of a discrimination problem is poorer following prior training on its reversal. But significant PI has been observed sufficiently often to leave little doubt that it is a genuine phenomenon (e.g., Maier and Gleitman, 1967; Maier *et al.*, 1967; Chiszar and Spear, 1968a; Cole and Hopkins, 1968; Koppenaal and Jagoda, 1968; Behrend *et al.*, 1970). As Spear (1971) has noted, the discrepancy between successful and unsuccessful studies may be due partly to differences in the difficulty of the discrimination. There is little or no evidence of a significant PI effect when rats are trained with consistent reinforcement in a simple T-maze. The majority of successful studies have used visual discriminations, or three-choice problems. None, unfortunately, has employed an inconsistent reinforcement schedule, which, as noted above, might be very much more sensitive to forgetting.

In classical interference theory, as noted above, PI is attributed to the spontaneous recovery of competing associations. Learning of Task B is assumed to involve extinction of the responses learned in Task A; during the retention interval, these extinguished responses are assumed to recover spontaneously and hence interfere with the performance of Task B in the retention test. One problem with this analysis that has emerged from studies of verbal learning is that it has often been impossible to demonstrate any recovery of the Task A associations (e.g., Koppenaal, 1963; Slamecka, 1966). Although evidence for such recovery has been detected over very short intervals (e.g., 20 minutes in a study by Postman *et al.*, 1968), this will hardly help to explain the outcome of studies where retention of Task B declines over the course of hours or days. Further evidence that PI does not necessarily depend on the recovery of extinguished responses has been provided by Maier *et al.* (1967), Koppenaal and Jagoda (1968), and Crowder *et al.* (1968), all of whom observed PI effects in discrimination learning when training on Task B involved no extinction of the responses established in Task A.

These results reinforce the conclusion suggested earlier that spontaneous recovery should be regarded as a special case of proactive interference rather than its cause. Proactive interference represents a failure to respond in a manner consistent with the subject's most recent training. Although spontaneous recovery provides one example of such a tendency, it is unnecessary to oppose excitatory and inhibitory associations in order to observe proactive interference.

It remains, therefore, to consider how proactive interference may be accounted for. One possibility that has been already noted, and which was applied by Woodworth and Schlosberg (1954, p. 561) to the case of spontaneous recovery, is to argue that more recent associations show more forgetting than less recent ones. If a subject is trained successively on two conflicting tasks, A followed by B, the learning of B will require the establishment of new associations in sufficient strength to override the now inappropriate associations of Task A. During the course of a subsequent retention interval there may be some forgetting of both sets of associations; but if curves of forgetting are negatively accelerated, the present rate of forgetting of the earlier established A associations will now be slower than that of the more recently established B associations. There will, therefore, be some change in the relative strengths of the two sets of associations: at the least, the difference between B and A will decrease and it is even possible that the A associations might end up stronger. In any event, the probability of responding in a manner appropriate to Task B will decline as the length of the retention interval increases.

A second possibility, suggested by Gleitman (1971) and Spear (1971), is to regard the development of proactive interference as the failure of a temporal discrimination. In effect, the reason why training on Task A interferes with the apparent retention of Task B is because subjects eventually forget which task they were most recently exposed to. If the subject tends to respond in a manner appropriate to the most recently experienced contingencies, correct performance must still depend upon discrimination of the relative recencies of the first and second tasks. The difference in relative recencies will decline with the passage of time. There is, moreover, unambiguous evidence from studies of generalization that animals may forget the absolute value of a discriminative stimulus, and that the temporal discriminations which form the basis of FI scallops and inhibition of delay are lost over a long retention interval. Grosser temporal discriminations may also be lost, and such forgetting may well contribute to some instances of proactive interference.

V. SUMMARY

Several different theories have been proposed to account for the decline in responding observed in extinction. The first, and in some sense the simplest, suggestion is that the probability of responding changes during extinction because the situation changes from that which prevailed during acquisition. Responding was reinforced in a context of earlier reinforcements; if the traces of those earlier reinforcements are established as signals for further reinforcement, the omission of reinforcement in extinction may disrupt performance by a process of generalization decrement.

The concept of generalization decrement may provide part of the explanation of the change in response probability observed in extinction, but it is not an account of the associative processes that might underly those changes. Traditionally, these have been explained in terms of either interference or inhibition. According to interference theory, the original CR or instrumental response disappears during the course of extinction, because it is displaced by a new, competing, response. Frustration theory is now the most widely accepted version of such a theory; previously rewarded responses are said to extinguish because the omission of an expected reward elicits a variety of emotional reactions which disrupt the execution of those responses.

According to an inhibitory analysis, the omission of reinforcement provides the occasion for inhibitory learning. Inhibitory associations are sometimes assumed to replace excitatory associations, but according to Pàvlov they simply suppressed these original associations.

It is probable that all of these factors contribute to extinction. There is independent evidence that performance is often disrupted by changes in the experimental situation; there is also evidence that the memory of the outcome of one trial can be established as a signal for reinforcement on the next. The combination of these two propositions implies that a change in the conditions of reinforcement on one trial may disrupt performance on the next trial via a process of generalization decrement. There is equally good evidence that the omission of an expected reward may elicit a variety of emotional or aggressive responses. Even if it has not been shown that such behaviour is instrumental in displacing the originally trained response, it is hard to believe that its occurrence is entirely irrelevant to an understanding of extinction. Finally, however, since frustration is only a transient response to the unexpected omission of reinforcement, the more permanent changes in performance produced by continued extinction must be attributed to more permanent associative changes: it is these that are referred to in inhibitory theories of extinction.

Of several unresolved issues in the study of extinction, two may be stressed. First, it is not clear whether competition between responses causes the disappearance of an extinguished response, or whether the appearance of new patterns of behaviour during extinction is a consequence of the disappearance of the trained response. Secondly, it is not clear whether inhibitory learning involves the replacement of excitatory associations by inhibitory associations, or merely the suppression of one by the other.

Although considerable progress has been made in elucidating the conditions which determine the rate of extinction, little of this evidence throws any light on these questions. This is because rate of extinction is often primarily determined by the extent to which the change from the conditions of acquisition to those of extinction disrupts performance by generalization decrement. Although undeniably important, this conclusion does not illuminate the nature of the associative processes specifically engaged by an extinction procedure. It may be that these would be better examined in experiments which programmed extinction not by omitting reinforcement but by decreasing the contingency between reinforcement and CS or behaviour which was responsible for the original acquisition of responding.

Rate of extinction is usually related directly to the size and immediacy of the reward received in acquisition and, in some circumstances, may be increased by increases in the amount of acquisition training. Undoubtedly the most important determinant of the rate of extinction, however, is the schedule of reinforcement received during acquisition. The partial reinforcement effect seems to provide a clear illustration of the principle that rate of extinction is determined less by the strength of the associations formed in acquisition, and more by the nature of those associations. Partial reinforcement retards the rate of extinction at least in part because it ensures that subjects have learned to respond in a situation similar to that encountered in extinction. In part, this may be because nonreinforcement during both acquisition and extinction elicits frustration. Good evidence for this is provided by the observation that if nonreinforcement is expected during acquisition, as on a regular alternation schedule, rate of extinction may be relatively rapid. Nevertheless, many of the data on the partial reinforcement effect seem satisfactorily explained by supposing that the memory of preceding nonreinforced trials is associated with reinforcement during partial reinforcement and that these associations maintain responding during extinction. Good evidence for this is provided by the data of a variety of experiments on different sequences and schedules of partial reinforcement.

One important result which should illuminate the nature of the processes involved in extinction is the phenomenon of spontaneous recovery. The reappearance of an extinguished response after an interval of time suggests that extinction involves the formation of new associations which temporarily suppress the old associations, rather than replacing them. Unfortunately this conclusion is premature, for it is possible that spontaneous recovery is partly a consequence of the reinstatement of stimuli whose association with reinforcement had never been extinguished. There is insufficient information to distinguish between these accounts. However, even if it appears that spontaneous recovery requires the assumption that inhibitory associations are added to excitatory associations, rather than replacing them, it does not follow, as Pavlov thought, that these inhibitory associations must be assumed to be inherently more labile than excitatory associations. It may simply be that, in a given period of time, more recently established associations lose more of their strength than do less recently established associations.

If acquisition and extinction are thought of as procedures for establishing incompatible excitatory and inhibitory associations, then spontaneous recovery may be regarded as a case of forgetting. Specifically, spontaneous recovery is a case where subjects fail to respond in a manner appropriate to their most recent experiences, and thus is a special case of proactive interference. The study of forgetting in animals has only recently begun to attract serious attention, and the causes of forgetting are not well understood.

CHAPTER 9

Generalization

I. GENERALIZATION AND GENERALIZATION GRADIENTS

The concept of generalization decrement has been used frequently enough in earlier chapters as an explanatory principle to account for an observed decline in performance when the stimulus situation is changed. It is time to turn to a more detailed discussion of the processes involved in those changes in behaviour that can be correlated with changes in stimulation.

Pavlov (1927, p. 113) reported that when a salivary CR had been established to one CS, other stimuli could be shown to elicit salivation. Such a result defines the empirical phenomenon of generalization: as a consequence of prior training with one stimulus, some other stimulus, never itself associated with reinforcement, may be able to elicit a response from a subject. Pavlov also reported that the effectiveness of new stimuli in eliciting a generalized CR declined in proportion to their distance from CS+. A progressive decline in generalized responding to stimuli further removed from the training stimulus, defines a sloping gradient of generalization.

Since Pavlov's early studies, generalization gradients have been plotted along a variety of stimulus dimensions, with a variety of animals serving as subjects and in both classical and instrumental experiments (see Mednick and Freedman, 1960, for a comprehensive survey). Figure 9.1 provides a few examples and illustrates wavelength gradients in pigeons trained to peck a response key illuminated with light of 550 nm, auditory frequency gradients in rabbits after eyelid conditioning in which a 1200 Hz tone served as CS, gradients of size generalization in rats trained to run in a short alley and push at a door set in the middle of a 20 cm² white circle, and finally the average results of a series of studies from Pavlov's laboratory studying visual intensity generalization in dogs given salivery conditioning with a light as the CS. With the exception of the intensity gradients, the remaining studies show orderly gradients of generalization, with the probability of a response declining progressively to stimuli further removed from S+. The intensity gradient shows a decline in responding to stimuli less intense than CS+, but an increase rather than a decrease in responding to stimuli more intense than CS+. These are not isolated results: Razran (1949) has reported that similarly shaped gradients were obtained along the dimension of auditory intensity in several Pavlovian studies;

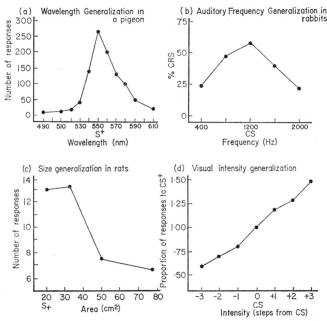

FIG. 9.1. Gradients of visual and auditory generalization in pigeons, rabbits, rats, and dogs. (a) *After* Guttman and Kalish, 1956; (b) *After* Moore, 1972; (c) *After* Grice and Saltz, 1950; (d) *After* Razran, 1949.

Hovland (1937b) observed a similar effect in a study of the generalization of the GSR; Brown (1942) found a markedly asymmetrical gradient along the dimension of visual intensity in rats; and Miller and Greene (1954) reported similar results from a study of auditory intensity generalization in rats. One possible reason why studies of intensity generalization result in such asymmetrical gradients will be discussed below (p. 536).

II. THE THEORETICAL ANALYSIS OF GENERALIZATION

A. Generalization as Irradiation or Failure of Discrimination

Pavlov (1927) assumed that the generalization of CRs to stimuli other than the training CS was a consequence of a spreading wave of excitation which travelled across the cortex from the focus of the CS: other stimuli would elicit CRs to the extent to which they shared in this irradiation of excitation. This account has been severely and justifiably criticized (e.g., by Konorski, 1948, Chapter 3). As Thompson (1965) has delicately pointed out: "concepts of spreading cortical waves of excitation and inhibition became difficult to interpret with the advent of the neuron doctrine in neurophysiology" (p. 154).

Lashley and Wade (1946) rejected not only Pavlov's physiology but also what they took to be the psychological assumptions underlying his account of generalization. Pavlov's theory, they argued, even if stripped of all references to cortical waves, would still be entirely unacceptable; what is wrong with the theory is the basic assumption that the conditioning of excitation to one stimulus automatically ensures that other stimuli will acquire excitatory properties. Excitation in Pavlovian physiological theory might refer to cortical states, but even if an excitatory stimulus was merely one which elicited responses as a consequence of prior reinforcement, the theory would still be inappropriate. Generalization of responding from a training stimulus to a test stimulus, Lashley and Wade argued, occurs only because the subject fails to discriminate the training and test stimuli. Thus was the stage set for a dispute which lingers to this day (e.g., Prokasy and Hall, 1963; Riley, 1968): is the empirical phenomenon of generalization due to a *process* of generalization, or is it a consequence of a failure of discrimination?

Interestingly enough, as Riley (1968, p. 26) has pointed out, the clearest evidence supporting the Pavlovian account of generalization, and showing that Lashley and Wade's account must be incomplete, was provided by Pavlov himself. Lashley and Wade did not suppose that generalization would occur only to the extent that the subject was incapable of detecting the difference between training and test stimuli; their point was that generalization would occur only to the extent that subjects exposed to a particular history of training in a particular apparatus, actually did fail to discriminate training from test stimuli. This is a harder position to disprove. Pavlov (1927, pp. 118–21), however, reported that on the first occasion that a test stimulus was presented, it would elicit little or no salivation, but that with successive nonreinforced presentations, the CR elicited by the test stimulus would increase. A similar result has been reported by Szwejkowska (1959). Pavlov suggested that this paradoxical result was analogous to the effects observed in studies of external inhibition: if a novel stimulus is added to the CS, it acts as an external inhibitor and inhibits salivation, but this disruptive effect rapidly wears off with repeated presentations. In just the same way, after conditioning to one tone the presentation of a novel tone initially disrupts the CR, but with repeated presentation the effect wears off and the novel tone elicits a relatively strong, generalized, CR. The important point is that the tone can be classified as novel only if it is discriminated from the training tone. If the subject starts responding to the novel tone on later trials, therefore, this cannot be due to a failure of discrimination: generalization occurs in spite of the fact that the subject has discriminated the test tone from the training tone.

This conclusion, although valid enough, should not be taken to detract from the importance of one insight contained in the arguments advanced by Lashley and Wade (1946) and by Prokasy and Hall (1963). Although subjects may respond to stimuli in a generalization test which they can, in appropriate conditions, discriminate from the training stimulus, the observed incidence of generalization may reflect the extent to which they do in fact discriminate the test stimuli from the stimulus associated with reinforcement during acquisition. The set of stimuli actually associated with reinforcement may not be the same as those manipulated by the experimenter while testing for generalization. If a pigeon, for example, is reinforced for pecking at a vertical line on a green key and continues to respond during a subsequent nonreinforced generalization test when the colour of the key remains the same but the orientation of the line is changed (Newman and Baron, 1965), this does not imply that pigeons are incapable of discriminating between lines of different orientation. A more plausible interpretation is that the main feature of the experimental situation associated with reinforcement was the colour of the key, rather than the orientation of the line projected onto the key; since the colour remained unchanged during the generalization test, responding also remained unchanged. Similarly, the fact that cats may show substantial cross-modality generalization from a light to a tone after a moderate amount of avoidance training (Thompson, 1959) does not imply that cats cannot discriminate lights from tones, but may imply that the effective stimulus, controlling the avoidance response, was any sudden increase in stimulation.

What is suggested by these observations is simply that the experimenter's specification of the CS or discriminative stimulus may not coincide with the subject's. Any stimulus situation can be regarded from different aspects, or may be represented as consisting of a large number of elements, not all of which may have been associated with reinforcement by the subject. This point of view suggests an analysis of the phenomena of stimulus generalization that may reconcile the apparent differences between Pavlov's spread of effect and Lashley and Wade's failure of discrimination. Subjects respond to stimuli in a generalization test, it could be argued, to the extent that those stimuli contain elements which overlap with those contained in the training stimulus. More precisely, the probability of a response occurring to a test stimulus is some function of the proportion of elements in the test set, which, because they formed part of the set of elements present during training, were associated with reinforcement during initial training (Bush and Mosteller, 1951; Atkinson and Estes, 1963). The notion that generalization is dependent upon overlapping sets of stimuli, which thus contain common elements, is consistent both with the Pavlovian view that generalization represents a spread of effect and with Lashley and Wade's view that generalization represents a failure of discrimination: some generalization occurs to test stimuli because there are elements common to both training and test sets; generalization is incomplete, however, because the overlap is less than perfect.

It only remains to distinguish between several classes of stimulus elements which may be present in any given experimental situation. It will be helpful to consider a concrete example. A pigeon is trained on a VI schedule to peck at a key illuminated with a white vertical line on a green surround. The key light provides the only source of illumination in the chamber. Trials are 60 sec long, separated by 10 sec intertrial intervals (ITIs) during which the chamber is completely dark. A constant masking noise is produced from a loudspeaker in the roof of the chamber. After a number of acquisition sessions, the pigeon is given a generalization test. During this test the orientation of the line on the key varies from trial to trial between 60 degrees on either side of the vertical, and responses are never reinforced; all other features of the experimental situation, however, are held constant. In spite of the fact that pigeons are capable of discriminating between lines differing in orientation by 15 degrees, several experiments similar in general design to this have shown that the pigeon generalizes almost completely to orientation 60 degrees on either side of vertical (Newman and Baron, 1965) and may show only a slight decline in rate of responding to a horizontal line (Freeman and Thomas, 1967; cited by Honig, 1970).

Since pigeons are capable of very much finer orientation discriminations, it will not do to say that lines of different orientation provide sets of elements with almost complete overlap. The obvious solution to this not very puzzling problem is to point to the numerous features of the situation that are not changed in the generalization test. Different orientations of lines may contain relatively few overlapping elements, but the proportion of elements in the total situation that is changed when the orientation of the line is changed is exceedingly small. The hue, saturation, brightness, shape and size of the response key, the auditory and visual background stimuli of the experimental chamber and the stimuli associated with pecking, all remain unchanged during the course of the generalization test. The experimenter is inclined to regard all these as irrelevant stimuli, but the only sense in which most of them are irrelevant is that the experimenter was not interested in assessing their contribution to the control of the subject's behaviour. As far as their correlation with reinforcement is concerned, most of these stimuli are as relevant as the orientation of the line on the response key. Reinforcement is available only when the key is illuminated and it is always illuminated with light of a particular hue, saturation and brightness. Responding is typically confined to the period when the key is illuminated, and response-produced stimuli, therefore, are equally well correlated with reinforcement. The only strictly irrelevant stimuli are those, such as the masking noise, present during ITIs as well as during trials.

It is not surprising, therefore, that responding should generalize relatively completely to other orientations of the line after training in such a situation. We have suggested that the slope of a generalization gradient is determined by the proportion of elements common to both training and test sets; but it is incorrect to view the test stimulus as, say, a 45-degree line—it is a 45-degree line of given length, width, and brightness, on a background of given hue, brightness, etc., in the presence of which a particular pattern of responding has been reinforced. If all these other elements have been associated with reinforcement, then the proportion of conditioned elements changed when the orientation of the line is changed is exceedingly small.

B. The Analysis of Gradients of Generalization

This analysis implies that the slope of a generalization gradient along a particular stimulus dimension will depend upon how many other aspects of the stimulus situation were permitted to become associated with reinforcement. Before consideration of the evidence bearing on this assumption, one further feature of the analysis of generalization in terms of common elements needs elaboration. Generalization gradients are typically orderly: responding declines in a regular manner as test stimuli depart further from the training stimulus (see Fig. 9.1). While the overall slope of a generalization gradient may depend on the number of controlling features unchanged during testing, the fact that the gradient shows a progressive and orderly slope implies that stimuli varying along the test dimension share progressively fewer elements in common with the training stimulus. To account for this in terms of the present analysis, it would be necessary to argue that a vertical line, or a tone of 1000 Hz, should be regarded as a set of elements and furthermore that a series of lines of different orientations, or a series of tones of different frequencies, should be regarded as further sets of partially overlapping elements. But this, surely, is entirely reasonable. Lines of particular orientations excite sets of units in the cat's striate cortex, producing maximal firing in units sensitive to that particular orientation, with graded amounts of firing in units sensitive to different orientations (Hubel and Wiesel, 1959). Thus a series of lines of different orientations will fire a series of partially overlapping cortical units. Similarly, a tone of given frequency will fire a particular set of units in the cat's auditory cortex and a series of tones will fire a series of overlapping units (Hind, 1961). In the case of mammals, therefore, common cortical units may plausibly be regarded as providing the basis for the common elements involved in generalization. As Thompson (1965) has suggested: "The amount of behavioural stimulus generalization given by an organism to a test stimulus is a monotonic (linear?) increasing function of the degree of overlap of excitation in the cerebral cortex resulting from the training and test stimuli" (p. 159).

Thompson's analysis of generalization should not be confused with Pavlov's just because it is also couched in quasi-physiological language. Pavlov's physiological account of generalization is undoubtedly open to neurophysiological objections, for the cortex is more profitably viewed as a set of interconnected units than as a medium for the transmission of waves of excitation and inhibition. It is also, however, open to crucial psychological objections. Mednick and Freedman (1960) have listed several consequences of the assumption that generalization is produced by a wave of excitation travelling across the surface of the cortex, but these are, as Thompson (1965) has shown, quite irrelevant to his hypothesis. For example, the occurrence of octave generalization, systematically analysed by Blackwell and Schlosberg (1943), is inconsistent with the regular spatial representation of frequency in the auditory cortex. As Hind (1961) has shown, however, a small proportion of cortical units respond to frequencies that are octave multiples of their optimal frequency, and an increase in generalization to octaves is therefore predicted by the hypothesis that generalization is proportional to the degree of overlap in cortical units fired by training and test stimuli.

This brief excursion into simple physiology suggests that the analysis of generalization in terms of common elements can in fact be taken to a relatively molecular level. This is a reasonably encouraging start to the task of explaining some of the more complex psychological problems posed by studies of generalization.

III. PROBLEMS OF MEASURING GENERALIZATION

The first problem is one of measurement. It is unsafe to assume that a behaviourally defined gradient of generalization will automatically be a

straightforward or direct measure of the degree of overlap between training and test stimuli. What is actually measured in a study of generalization
is the subject's responses—rate of responding in free-operant experiments,
probability or speed of responding in discrete-trial experiments, probability or amplitude of responding in classical conditioning experiments.
These response measures pose serious scaling problems (Blough, 1965).

A. Absolute and Relative Gradients

The wavelength, auditory frequency and size gradients shown in Fig. 9.1
are all based directly on the number of responses made by a group of
subjects (or different groups of subjects) to different test stimuli. Such a
measure is referred to as an absolute gradient, in contrast to a relative
gradient which plots the proportion of responses made to various test
stimuli. The intensity gradient shown in Fig. 9.1 is a relative gradient,
which expresses responding to each test stimulus during a series of test
trials as a proportion of the responses made to the training stimulus during
an equivalent series of test trials. Relative gradients may also be plotted
(as in Fig. 9.2) by expressing the number of responses to each test stimulus
as a proportion of the total number of responses made to all stimuli over
the entire test series.

Relative gradients ignore wide differences in overall rate or probability
of responding. This feature has made their use very popular in free-
operant experiments with pigeons, where the assessment of the effects of
a particular treatment is often made difficult by wide individual variations
in overall rate of responding. Fig. 9.2 shows two pairs of absolute gradients
of generalization from two experiments employing pigeons, together with
their transformation into relative gradients. It is apparent that substantial
differences in overall rate of responding in a pair of absolute gradients will
disappear when the gradients are expressed in relative terms.

A comparison of the absolute and relative gradients shown in Fig. 9.2,
however, suggests that there may be considerable problems attendant upon
attempts to interpret possible differences in the slopes of absolute gradients
when those gradients are produced by groups or individuals with different
overall levels of responding. The gradients from the two top panels were
obtained by Thomas and Switalski (1966) from two groups of pigeons
reinforced for responding to a key light of 550 nm on either a VR or a VI
schedule. The two absolute gradients give the impression of having similar
slopes, with the gradient for the VR group simply being displaced upwards.
But when the two sets of scores are converted to relative gradients, the VI
gradient appears significantly steeper than the VR gradient. Similarly, in
Hearst and Koresko's (1968) experiment, where the two groups whose data

are shown received different numbers of days of VI training in the presence of a white vertical line on an otherwise dark key, the absolute gradient for the 14 day group looks steeper, as well as higher, than the gradient for the 2 day group; and yet, when the gradients are expressed in relative terms, there is little or no difference in their slopes.

These examples were of course, deliberately chosen to illustrate the point that conclusions about the slopes of absolute and relative gradients

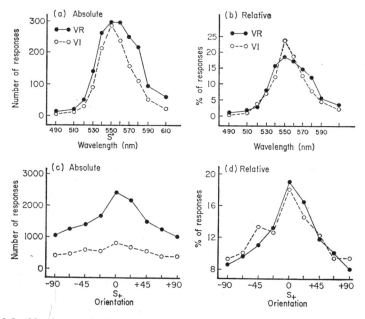

FIG. 9.2. Absolute and relative gradients of generalization. (a) and (b) Absolute and relative gradients in two groups of pigeons reinforced on either a VR or VI schedule. (*After* Thomas and Switalski, 1966.) (c) and (d) Absolute and relative gradients in two groups of pigeons that have received either 14 (•—•) or 2 (o--o) days of VI training before the test for generalization. (*After* Hearst and Koresko, 1968.)

may not necessarily coincide. In many cases there is no disagreement. If, for example, the absolute gradient for one group crosses over with the gradient of another, because one group has a higher probability of responding to S+ but a lower probability of responding to a test stimulus, then both absolute and relative gradients will necessarily differ in slope. But the occasional disagreement is by no means trivial, for it is but one symptom of the point of view that the precise slope of a generalization gradient, and therefore the assessment of differences in slopes between two gradients, is critically dependent on the response scale used. Consider again the data from Thomas and Switalski's study shown in Fig. 9.2. To say that the VI schedule produced a steeper gradient than the VR schedule is to say that the difference between the relative gradients is more meaningful than the similarity of the absolute gradients. This assumes that the difference between the absolute number of responses at two points on the VI gradient, such as 550 and 590 nm, although approximately the same as the absolute difference between these two points on the VR gradient, really reflects a greater difference in underlying response strength for the VI group. This might, indeed, be true: subjects with a very low rate of responding will necessarily show a rather flat absolute gradient, simply because of a floor effect. But the transformation to a relative scale ignores

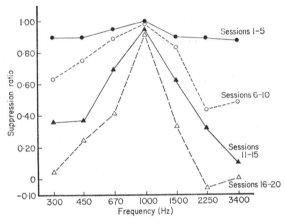

FIG. 9.3. The sharpening of a generalization gradient of conditioned suppression during the course of testing in extinction. Gradients are shown for one subject over 20 sessions of extinction. (*After* Hoffman and Fleshler, 1961.)

the possibility that the absolute gradients of subjects with a very high rate of responding may be artificially flattened by a ceiling effect: in the extreme case, where the probability of a discrete response is being measured, subjects can do no more than respond once on every test trial to S+, and if they also respond on nearly all trials to other test stimuli, they will produce a flat generalization gradient. But this may reflect no more than their high overall response strength.

The point is illustrated in Fig. 9.3. It is well established that if pigeons are given a series of test trials in extinction to a number of different stimuli, following free-operant training in the presence of one stimulus, the generalization gradient so obtained will usually become sharper during the course of testing (e.g., Jenkins and Harrison, 1960; Thomas and Barker, 1964; Friedman and Guttman, 1965). Fig. 9.3 shows a striking instance of such a sharpening of a generalization gradient during testing. The data are from a single pigeon trained to peck at a key and then given conditioned suppression training with a tone of 1000 Hz signalling shock (Hoffman and Fleshler, 1961). The gradient is thus one of conditioned suppression (to avoid confusion, the measure of conditioned suppression employed here and throughout this chapter assigns scores of 1·00 to complete suppression and 0·00 to complete lack of suppression, instead of, as previously, 0·00 for complete suppression, and 0·50 for complete lack of suppression). On the first block of test trials, the subject was virtually completely suppressed on all test trials and the gradient is, therefore, essentially flat. By the end of testing, however, the gradient has become extremely steep. Similar results have been reported for the generalization of conditioned suppression in rats (Desiderato, 1964). Although this progressive sharpening of the gradient during testing may reflect a genuine increase in control by the value of the tone used as CS (as Lashley and Wade, 1946, would have argued), it seems more reasonable to believe that it is largely caused by a ceiling effect on early test trials. The suppression ratio cannot exceed 1·0, since this represents the complete absence of responding to the stimulus; differences in the strength of suppression to the CS and to the test stimuli (i.e., suppression "below zero") cannot be recorded.

The problem does not admit of any very ready solution. It will, however, be important to bear in mind throughout the remainder of this chapter that if two gradients differ markedly in overall height at S+, then unless they actually cross over at some test stimuli, it is dangerous to accept at its face value any claim that their slopes are significantly different. This is no doubt unfortunate, but it should be recognized as part of a more pervasive difficulty encountered previously: we do not, in general, have any independently validated way of transforming our arbitrary records of speed, amplitude, probability, or rate of responding into more meaningful measures of underlying associative value, response strength, or reaction potential. Psychology will have come of age only when it has constructed some meaningful scales, but a necessary part of the ageing process is the recognition of the problems posed by the lack of such scales.

B. Within- and Between-Subject Generalization Tests

Generalization gradients may either be obtained by training a large group of subjects with a given stimulus, and then testing individual subjects at different values of the test dimension (e.g., Grice and Saltz, 1950; Hiss and Thomas, 1963), or by testing all subjects at all points along the test dimension (e.g., Pavlov, 1927; Guttman and Kalish, 1956; Hoffman and Fleshler, 1961; Hiss and Thomas, 1963). The latter procedure has the advantage of a within-subject design and is clearly less time-consuming. But it requires a training procedure which ensures that subjects will continue to respond long enough in the absence of further reinforcement to provide a meaningful measure of generalized response-strength to each test stimulus. Guttman and Kalish's technique of initially training pigeons on a VI schedule of reinforcement has been widely adopted by subsequent workers because it satisfies this requirement.

Lashley and Wade (1946), however, argued that any procedure that exposed a single subject to the complete range of test stimuli would generate sloping gradients of generalization that were partly a product of the test procedure itself. Their idea was that the subject would start comparing the training and test stimuli during the course of the test and that the resulting sloping gradient would somehow be a consequence of such a comparison. Whatever the reasoning behind this argument, it leads to the prediction that generalization should be complete upon a subject's first exposure to a test stimulus, and is thus contradicted not only by Pavlov's data cited above but also by subsequent studies (e.g., Hiss and Thomas, 1963; Newman and Grice, 1965), which have reported reliably sloping gradients of wavelength generalization in pigeons or size generalization in rats on the first test trial. The fact that gradients are often flat over the first few test trials, as in Fig. 9.3, may just reflect a ceiling effect.

There remain, however, several important ways in which repeated exposure to a series of test stimuli might affect a gradient of generalization. One possibility, suggested by Mednick and Freedman (1960), is that the extent of generalization to a given test stimulus might be determined by the number of stimuli intervening between that stimulus and S+ in the generalization test. For example, a subject reinforced for responding to a vertical (0-degree) line, might show greater generalization to a 30-degree line if the test series consisted of S+ and 30-degree, 60-degree and 90-degree lines, than if the test series consisted of S+ and 10-degree, 20-degree and 30-degree lines. Although Mednick and Freedman found some support for this "units hypothesis" in studies with human subjects (see also Thomas and Hiss, 1963), experiments with pigeons (Friedman, 1963; Hiss and Thomas, 1963) and octopuses (Muntz, 1965) have found that the amount of generalization between a training stimulus and a particular test stimulus is not necessarily affected by the number of stimuli intervening between the two in the test series.

It remains possible, however, that a repeated series of test trials with a range of stimuli may have some effect on the slope of the gradient. This would not be surprising, for such a generalization test consists essentially of a series of extinction trials to a number of different stimuli (including S+). Extinction trials provide the opportunity for learning just as do acquisition trials and the gradient produced by such a test may partly reflect the interaction of a series of inhibitory tendencies associated with each test stimulus. In at least one situation there is clear evidence from a study by Marsh (1967a) that the slope of a generalization gradient may be distorted by the use of this test procedure.

One feature of Guttman and Kalish's (1956) study of wavelength generalization in pigeons surprised them as it has surprised others. They initially trained different birds with different wavelengths as S+ and were thus able to compare the slope of gradients at different points along the wavelength dimension. They found no systematic differences: the gradient established around an S+ of 530 nm, for example, did not differ in slope from one produced by training with an S+ of 580 nm. This is, in fact, very surprising, for the best available evidence (Hamilton and Coleman, 1933) suggests that the pigeon's difference threshold for hue approximates that of human subjects: thus wavelengths in the green region of 520 to 540 nm are much less discriminable than wavelengths in the yellow–orange region of 570 to 590 nm. These differences in discriminability suggest that the difference in neural activity produced by a given difference in wavelength is not the same at all points on the spectrum. It should, therefore, be possible to detect differences in the slopes of the gradients established round these points.

In a study in which rhesus monkeys were given wavelength discrimination training before a generalization test, Ganz (1962) did obtain appropriate correlations between discriminability functions and the slope of generalization gradients. He suggested that the failure to obtain such a correlation when a multiple-stimulus test procedure followed reinforced training to S+ only, might be a consequence of an interaction between excitatory and inhibitory gradients during the course of testing. A subject trained in a region of low discriminability and tested with a long series of test stimuli would be expected to show marked generalization of extinction during the course of testing. To the extent that gradients of generalization sharpen during extinction, any increase in the overall level of extinction would tend to result in a steeper gradient. This might be sufficient to obscure an initial difference in the slope of gradients at different points on the spectrum. Whether or not this is the correct explanation, Marsh (1967a) was able to show that a good correlation between discriminability functions and observed gradients of generalization, obtained when a small number of test stimuli was used, was abolished by the use of a large number of test stimuli. Marsh found that if separate groups of pigeons were trained with an S+ of 520 or 590 nm, and then tested for generalization to a *single* test stimulus 10 nm distant, they produced a clear correlation between discriminability and the slope of the gradient. As is shown in

Fig. 9.4, the gradient between stimuli of 520 and 530 nm was significantly flatter than the gradient between stimuli of 590 and 580 nm. If, however, subjects were given a generalization test to a series of stimuli, the difference

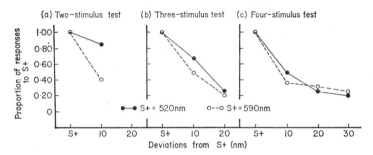

FIG. 9.4. The relationship between the slope of a generalization gradient and the discriminability of training and test stimuli may be obscured by testing subjects with several generalization stimuli. In panel (a) pigeons reinforced for responding to a wavelength of 520 nm (a region of low discriminability) show a flatter gradient to a single stimulus 10 nm from the training stimulus, than do pigeons reinforced for responding to a wavelength of 590 nm (a region of good discriminability). In panels (b) and (c) the addition of further test stimuli abolishes any difference in slope between the two gradients. (*After* Marsh, 1967a.)

in slope vanished: the gradient from 520 nm became just as steep as that from 590 nm.

A second example of test effects produced by a multiple-stimulus generalization test has been provided by Donahoe *et al.* (1970). They trained pigeons on a brightness discrimination with a bright light as S— and a dimmer light as S+. One group was given generalization tests with three

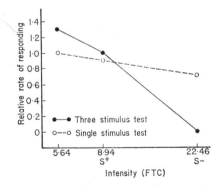

FIG. 9.5. The effect on the slope of a generalization gradient of testing subjects with all three test stimuli in a single session, or with one stimulus at a time in separate sessions. (*After* Donahoe *et al.*, 1970.)

stimuli, S-, S+ and an even dimmer stimulus, appearing in an inter-mixed, random sequence. A second group of subjects was tested with only one stimulus at a time. The results are shown in Fig. 9.5: subjects tested with all three stimuli at once responded very much less to S−, and rather more to the test stimulus, than did subjects tested with only one stimulus at a time. Donahoe *et al.* (1970) suggested that these interactions could be looked upon as a consequence of transient contrast effects (p. 357): rate of responding to S − was suppressed in the multiple-stimulus test because S − sometimes followed S +, while rate of responding to the test stimulus was elevated because presentation of the test stimulus was sometimes preceded by S −. We shall have more to say about these results later.

C. Conclusions

In the meantime, these results all point to the conclusion suggested by other data considered here. Behavioural gradients of generalization are no more a pure measure of sensory processing than any other behavioural measure. Indeed, they are quite certainly much less satisfactory measures than some. It is probable that a more sophisticated theoretical approach will be necessary if the processes underlying an observed gradient of generalization are ever to be unravelled. The theory of signal detection provides one obvious approach and some preliminary analyses have been undertaken (e.g., Blough, 1967, 1969a; Boneau and Cole, 1967). If the probability of responding in a generalization test is regarded as a function both of the subject's decision criterion or threshold for responding, and also of some measure of the discriminability of the test stimulus from the training stimulus, then the effects of certain variables on the slope of a gradient need not necessarily be interpreted in terms of changes in the stimuli controlling the subject's behaviour, but in terms of their effects on the subject's decision criterion. Perhaps gradients become sharper in extinction, as Blough (1969a) suggests, because the threshold for respond-ing is raised during the course of extinction. Similarly, the fact that an increase in drive level may flatten generalization gradients (see below) might be attributable to a lowering of the decision criterion.

IV. DISCRIMINATION TRAINING AND THE SUPPRESSION OF INCIDENTAL STIMULI

It is all too clear that the slope of a gradient of generalization can be affected by a multitude of factors—perhaps by the subject's decision criterion and by the interaction of excitatory and inhibitory tendencies during the course of testing, as well as by the overlap between the set of training and test stimuli. This overlap is itself a product of two factors. First, the physical stimuli, such as the wavelengths projected onto the pigeon's response key, may overlap with the training stimulus to a greater or lesser degree. Secondly, the proportion of the elements actually associated with reinforcement that is varied when the experimenter presents a series of test stimuli may depend to a considerable extent on the nature of the other stimuli present in acquisition and their correlation with reinforcement: the greater the proportion of unchanging stimuli controlling the subject's behaviour, the less will be the effect of the change produced by variations in the test dimension.

A. Incidental Stimuli and Overshadowing

The importance of incidental stimuli* was a central part of Lashley and Wade's (1946) account of generalization and, in a modified form, a similar argument was advanced by Hull (1952, pp. 64–9). Hull supposed that if incidental stimuli acquired significant habit strength, they would increase responding to all test stimuli and would therefore flatten the generalization gradient. There are some data that support this assumption. Pigeons trained to respond to a vertical line on a green background, it will be recalled, tend to show relatively flat generalization gradients to other line orientations. This could be partly because responding is controlled by the colour of the background as much as, or more than, by the orientation of the line on the key. If Hull's argument were correct, therefore, subjects tested with different orientations of line on a black background should show a steeper gradient of generalization. Freeman and Thomas (1967), Newman and Benefield (1968), and Thomas *et al.* (1970b) have confirmed this prediction. Interestingly enough, this result is observed only when animals are trained and tested in a chamber in which the key light provides the sole source of illumination. If the chamber is permanently illuminated with a house light, then the orientation gradient is relatively flat and either unaffected, or actually steepened, by the presence of a green background (Baron and Bresnahan, 1969; Thomas *et al.*, 1971a). This is presumably because the presence of a house light dramatically

---

* The term "incidental stimuli" will be used here to refer to any set of stimuli, other than the test set, which may come to control the subject's behaviour as a consequence of acquisition training. They are incidental, of course, only by reference to the experimenter's intentions or subsequent experimental manipulations. That is to say, such stimuli may be as well correlated with reinforcement as the supposedly relevant stimuli in which the experimenter is interested. Incidental stimuli, therefore, are not necessarily irrelevant, i.e. less well correlated with reinforcement than the relevant stimuli.

R

increases the number of other incidental visual stimuli and thus decreases the importance of the colour of the response key.

Although the results of these experiments are consistent with Hull's predictions, the reported differences were, in fact, relatively small, and are made difficult to interpret by the substantial differences in absolute levels of responding. The problem may be that incidental stimuli not only mask the "true" slope of the underlying gradient along the test dimension; they may also, as Lashley and Wade would have argued, prevent the relevant feature of the experimental situation (i.e., that aspect which the experimenter will vary in the generalization test) from becoming associated with reinforcement in the first place. Many of these incidental stimuli are as well correlated with reinforcement as the vertical line on the response key; the principle of overshadowing (p. 46) would lead one to expect that they would compete with the line and reduce the probability that it will become an effective signal for reinforcement. In the present example the implication is that animals trained and tested with a vertical line on a black background will show a substantially steeper gradient than animals trained with the green background, even if the colour was removed for testing. The results of the experiment by Thomas *et al.* (1970b) tend to confirm this prediction.

B. Neutralization of Incidental Stimuli by Discrimination Training

Both Lashley and Wade (1946) and Hull (1952) saw that the most obvious implication of their positions was that generalization gradients along a particular dimension would be made considerably sharper if subjects were initially given discrimination training instead of simply being reinforced in the presence of S+. According to Lashley and Wade, this is because discrimination training between two values of the test dimension will establish the appropriate set of comparisons; according to Hull, discrimination training will entail the random reinforcement and nonreinforcement of incidental stimuli, thus suppressing any control exercised by such stimuli and permitting control acquired by the relevant stimulus to be revealed. There are reasons for objecting to the details of both these analyses, but the initial insight has been abundantly justified. Generalization gradients can be made dramatically steeper if subjects are initially trained on a successive discrimination in which reinforced presentations of S+ are alternated with nonreinforced presentations of some other set of stimuli.

If reinforcement of responding in the presence of a vertical line produced a flat gradient of generalization to other orientations of the line, it would obviously be possible to ensure a steeper gradient by alternating

reinforced presentations of the vertical line with nonreinforced presentations of a 30-degree line. Pigeons can learn this discrimination with little
difficulty and once they have learned it, we may be reasonably confident
that they will not respond at the same rate to the vertical and 30-degree
lines in a generalization test. In other words, they will show a steeper
gradient. This is hardly surprising and it does not seem necessary to
attribute such a finding either to the opportunity for comparison provided
by discrimination training, or to the suppression of control by incidental
stimuli. It is readily explained as a consequence of the extinction of
responding to the 30-degree line: the resultant gradient would be a product
of an excitatory gradient round S+ (the vertical line) and an inhibitory
gradient round S— (30 degrees).

Jenkins and Harrison (1960) showed that discrimination training had
effects on generalization gradients that could not be attributed to a simple
interaction of excitatory and inhibitory tendencies. They first trained one
group of pigeons to peck at a white key in the presence of a 1000 Hz tone.
When tested for auditory frequency generalization to tones ranging from
300 to 3500 Hz, the subjects showed at best a very shallow gradient and
even responded at approximately the same rate when the tone was turned
off altogether. A second group of birds received, in addition to reinforced
trials with the 1000 Hz tone, interspersed nonreinforced trials with the
tone turned off. All five birds in this group showed orderly and steep
gradients of auditory frequency generalization in the subsequent test.
Since S— in discrimination training was not a specific tone, but rather the
absence of the tone, the inhibitory learning in this condition should presumably have generalized equally to all tones used in the generalization
test. The effect on the slope of the gradient produced by such orthogonal
or interdimensional training, therefore, cannot be attributed to any interaction between excitatory and inhibitory gradients. These results have been
confirmed in numerous subsequent experiments, employing a variety of
different stimulus dimensions: orientation gradients in pigeons may be
sharpened by alternating reinforced trials with a vertical line and nonreinforced trials with no line on the key (Newman and Baron, 1965; Baron and
Bresnahan, 1969; Farthing, 1972); wavelength gradients may be sharpened
by alternating reinforced trials with a coloured key and nonreinforced trials
with no colour on the key (Friedman and Guttman, 1965; Switalski et al.,
1966; Lyons and Thomas, 1967).

Jenkins and Harrison (1960) suggested that their results were consistent
with Hull's general assumption that discrimination training sharpens
generalization gradients by suppressing incidental stimuli. The point is
easily illustrated. A subject given only reinforced trials in the presence of
a tone is exposed to a large number of different stimuli, all just as well

correlated with reinforcement as the tone: there is nothing in the experimental situation that singles out the auditory stimulus as a better predictor of reinforcement than the visual stimuli from the key light or the mere fact that the pigeon is responding. But in the discrimination procedure employed by Jenkins and Harrison, all these other features remain constant on nonreinforced trials and only the tone is omitted. These incidental stimuli, therefore, are no longer simply incidental: by comparison with the tone, they are irrelevant and should be less likely to compete with the tone for control of responding.

This argument has been substantiated by several recent studies. Rudolph and Van Houten (1974) have shown that it is the presence of a key light that prevents a tone from acquiring control over responding, when pigeons are simply reinforced for pecking at an illuminated key in the presence of a tone. When they trained pigeons to peck in complete darkness in the presence of a tone, they found relatively orderly gradients of auditory frequency generalization. The other part of the argument has been directly supported by Miles *et al.* (1970). They trained pigeons on an auditory discrimination between a tone and white noise, with the key always illuminated with the same colour of light on both positive and negative trials. After different amounts of discrimination training, subjects were given generalization tests between this colour and another. Miles *et al.* found that as auditory discrimination training progressed, so the control exercised by the colour of the key declined.

We can conclude, therefore, that in the absence of discrimination training visual stimuli may prevent auditory stimuli from acquiring control over responding in pigeons, but that discrimination training between the presence and absence of a tone may be sufficient to suppress control by the now irrelevant visual stimuli and may thus enhance the slope of an auditory generalization gradient. Blough (1969b) has provided some extremely persuasive evidence of the generality of these processes. He trained pigeons on a discrete-trial compound visual and auditory discrimination. On each trial one of seven tones and seven wavelengths was presented. Reinforcement was available only when the wavelength was 582 nm and the tone was 3990 Hz; all other combinations were unreinforced. After extensive training on this problem, subjects showed steep gradients along both wavelength and frequency dimensions. At this point either the light or the tone was held constant for several sessions, while the other stimulus continued to vary, signalling reinforcement at only one value. Thus one stimulus became irrelevant, while the other remained relevant. Subsequent generalization gradients showed both an increase in control by the stimulus that remained relevant and a substantial decrease in control by the stimulus that had been temporarily irrelevant.

It remains then to consider the mechanism whereby discrimination training suppresses control by incidental stimuli. For Hull (1952) the answer was simple. Discrimination training ensures equal reinforcement and nonreinforcement of such stimuli: equal numbers of reinforced and nonreinforced trials will result in equal and opposite amounts of excitation and inhibition, leaving a net excitatory strength of zero. If incidental stimuli had zero excitatory strength, they would not increase rate of responding to all stimuli on the generalization test and the control acquired by the relevant stimuli would be revealed. As Wagner (1969a) has charitably put it, Hull's assumption was "rather gratuitous". Hull had ample evidence available to him that a 50 per cent schedule of reinforcement does not result in a net excitatory strength of zero: for example, rats run in alleys rather rapidly on a random 50 per cent schedule of reinforcement.

A series of experiments by Wagner *et al.* (1968) has directly shown that discrimination training does not suppress incidental stimuli simply by arranging that they are correlated equally often with reinforcement and nonreinforcement. Such a schedule of reinforcement is quite sufficient to produce strong control by an incidental stimulus. It is not the fact that incidental stimuli are only partially correlated with reinforcement, but rather the presence of other stimuli more highly correlated with reinforcement, that is responsible for the effects of discrimination training. Wagner *et al.* showed that if rats were given discrimination training between two auditory stimuli, T_1+ and T_2-, a light present on both reinforced and nonreinforced trials acquired little control over responding; but if reinforced and nonreinforced trials occurred exactly as before, except that T_1 and T_2 were no longer correlated with the presence and absence of reinforcement, then the light acquired strong control over responding, while T_1 and T_2 acquired little control (see p. 47 for a more detailed discussion of these experiments). It is the presence of a more reliable predictor of reinforcement that causes suppression of control by incidental stimuli. Discrimination training may not change the schedule of reinforcement in effect on a set of incidental stimuli, but, relative to other stimuli, it ensures that they become irrelevant and it is this decrease in the relative validity of such stimuli that prevents their acquiring control over responding (Wagner, 1969a, b).

The implications of these conclusions are entirely unfavourable to Hull's specific analysis. If it is not the change in reinforcement schedule that neutralizes incidental stimuli, but the decrease in their relative validity, then we seem required to assume that stimuli somehow compete for control over responding. In the absence of a better predictor of reinforcement, an incidental stimulus, imperfectly correlated with reinforcement, may gain control over responding. When other stimuli are differentially reinforced, however, they may interfere with the control exercised by the incidental stimulus. The results are similar to those on overshadowing in classical conditioning (p. 47), and imply, as we have seen, some process of stimulus selection or competition between stimuli for association with reinforcement, such as that postulated by theories of selective attention. For the moment it will be sufficient to assume that the effects of such a process can be detected in studies on generalization, without troubling to distinguish between possible causes of stimulus selection.

C. Implicit Discrimination Training

The powerful effect of interdimensional discrimination training on auditory generalization,* reported by Jenkins and Harrison (1960), has prompted the suggestion that some sort of discrimination training, whether explicitly scheduled by the experimenter, or implicit in the experimental situation, might actually be necessary for the formation of sloping gradients of generalization. From the present point of view this suggestion is not very plausible, for it would imply that the experimenter was unable to devise a set of stimuli, subsequently varied in a generalization test, which are sufficiently salient to compete with general apparatus stimuli or stimuli produced by the subject's responding. The present argument has been that during nondifferential reinforcement the specific set of stimuli which the experimenter is interested in are no more valid predictors of reinforcement than are all other incidental stimuli and they may therefore be overshadowed by these incidental stimuli. It is not suggested that nondifferential reinforcement prevents *all* stimuli from becoming associated with reinforcement.

Nevertheless, Heinemann and Rudolph (1963) pointed to a possibly crucial difference between studies in which nondifferential reinforcement was sufficient to establish a sharply sloping gradient of generalization and those in which it was not. Guttman and Kalish (1956) obtained sharp wavelength gradients in pigeons, while Jenkins and Harrison (1960) obtained barely significant gradients of auditory frequency generalization in their control subjects trained under essentially comparable conditions. While Guttman and Kalish had not explicitly scheduled differential reinforcement in the presence and absence of the stimulus subsequently varied in their generalization test, the fact that this stimulus was localized on the pigeon's response key may have been sufficient to ensure implicit differential reinforcement. Pigeons are reinforced only for pecking at the key

* The term "interdimensional" is used to refer to discrimination training given between the presence and absence of a specific stimulus, such as the tone in Jenkins and Harrison's experiment. See Fig. 9.6 for further definitions.

and thus reinforcement can occur only when the subject has just been maximally exposed to the relevant stimulus; when the pigeon is not engaged in pecking the key, the probability of reinforcement is reduced to zero and the probability of exposure to the relevant stimulus is also reduced. No such correlation will occur when a diffuse, auditory stimulus is employed, and the use of diffuse stimuli, therefore, will reduce the probability of implicit differential reinforcement being provided on an ostensibly nondifferential schedule.*

Heinemann and Rudolph (1963) supported their argument by showing that the slope of a gradient of brightness generalization in pigeons was inversely related to the area of the stimulus used in training. If pigeons were trained to peck at an illuminated response key, they showed a steep gradient of generalization to other intensities of light projected onto the key; but if the key was surrounded by a large sheet of cardboard of the same brightness as the key, rate of responding in the generalization test was essentially unaffected by changes in brightness.

We can accept one implication of these results. As Heinemann and Rudolph stated: "It would appear that some discriminative training must inevitably occur in any situation in which the stimulus is a fairly small visual area and the required response is a movement that is directed with respect to this stimulus" (p. 657). We can also accept that such implicit discrimination training increases the slope of generalization gradients in exactly the same way and for exactly the same reasons that explicit inter-dimensional training sharpens gradients. In both cases the probability of reinforcement is correlated with the effective presence or absence of one set of stimuli which will tend to neutralize other, incidental, stimuli. Furthermore, the reason why the key light in a pigeon's chamber usually overshadows auditory stimuli may be more appropriately ascribed to the differences in implicit schedules of differential reinforcement rather than to any intrinsic differences in salience. (Even if this conclusion is too strong, it is certainly possible that the effects of differences in intrinsic salience are augmented by differences in effective correlations with reinforcement.) None of this, however, entails the strong conclusion that implicit differential reinforcement is a *necessary* condition for the formation of sloping gradients of generalization. Rudolph and Van Houten (1974), for example, obtained significant gradients of auditory generalization in

* It should be stressed, as Heinemann and Rudolph noted, that these arguments do not apply to classical conditioning experiments, or to discrete-trial studies where the possibility of responding in the ITI is not excluded. These situations correspond to the case of interdimensional training in a free-operant experiment: reinforcement occurs only in the presence of a critical set of stimuli (CS+ or S+) and not in their absence (i.e., in the ITI). A similar point was made earlier in comparing the effects of stimulus intensity in classical conditioning and free-operant experiments (p.43).

pigeons, when subjects were trained in the dark with a tone on continuously throughout each training session. Hearst (1962) found significant gradients of visual intensity generalization in monkeys, after subjects had been reinforced for pulling a chain in the continuous presence of a bright overhead light. It would be surprising if pigeons did not show significant wavelength gradients after training in the presence of a diffuse overhead light of one particular wavelength.

D. Effects of Different Discrimination Procedures

The idea that prior discrimination training is necessary for the appearance of sloping gradients of generalization has usually been attributed to Lashley and Wade (1946), whose gnomic utterance that "the 'dimensions' of a stimulus series are determined by comparison of two or more stimuli and do not exist for the organism until established by differential training" (p. 74), has excited considerable controversy. As Sutherland and Mackintosh (1971, pp. 209–28) have argued, it is difficult to provide any single, unambiguous, interpretation of this remark and easy to find evidence inconsistent with any single interpretation. A more defensible position, consistent with other remarks of Lashley and Wade's and with the general argument proposed here, is that a sloping gradient of generalization along a given dimension is dependent on the subject's having associated the occurrence of reinforcement with that aspect of the experimental situation chosen by the experimenter for variation in the generalization test. While this is essentially tautologous, it focuses attention on the possibility that features other than those varied by the experimenter may be associated with reinforcement and thereby overshadow the experimenter's dimension, and provides a general framework for interpreting what are otherwise rather disorderly data on the effects of various types of discrimination training.

1. *Implicit Discrimination Training*

Most training procedures utilized in generalization studies necessarily involve some element of discrimination training. This is clearly true in classical conditioning and discrete-trial instrumental experiments. Even in free-operant experiments supposedly scheduling nondifferential reinforcement, the use of stimuli localized on the response key may ensure some implicit differential reinforcement. Where such differential reinforcement is minimized, generalization gradients may be relatively flat; but it is possible that the presence or absence of potentially more salient or more valid cues is a more important determinant of the slopes of gradients than the occurrence of implicit discrimination training.

2. *Explicit Interdimensional Discrimination Training*

Explicit discrimination training may be programmed in a variety of ways and it will be important to distinguish between the various procedures. Fig. 9.6 illustrates four possible programmes of discrimination training to which a pigeon might be exposed before being tested for generalization

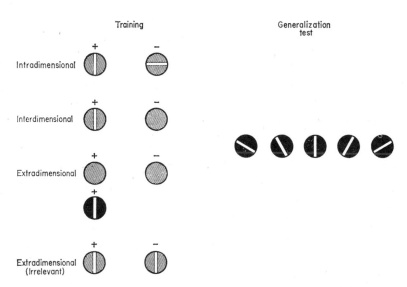

FIG. 9.6. Schematic diagram of various types of discrimination training that may be given before a test for generalization. In the present example subjects are tested for generalization along the dimension of orientation of the line, being presented with different lines on a black background. The cross-hatchings on the training stimuli represent coloured backgrounds. Thus in intradimensional and interdimensional training, the colour of the background is the same on positive and negative trials; in intradimensional training, S— is a horizontal line (another value of the test dimension); interdimensional training is given between the presence and absence of the line. In extradimensional training, the colour of the background varies from positive to negative trials; in the first type subjects are trained on a discrimination between two colours, and are also reinforced for responding to a vertical line; in extradimensional (irrelevant) training, the vertical line appears on both positive and negative trials.

along the dimension of orientation. The terminology used to refer to these procedures is that used by Honig (1970).

In interdimensional discrimination training, reinforcement is correlated with the presence or absence of a particular set of stimuli; such differential reinforcement should neutralize potentially competing incidental stimuli by making them relatively less valid predictors of reinforcement. If a pigeon is reinforced only when a vertical line is shown on a green response key and is never reinforced when the line is absent, then all stimuli except the set associated with the presence of the line (e.g., the wavelength and intensity of the background, the shape and size of the key, stimuli associated with responding, general apparatus cues) will become irrelevant.

It is important, however, to note that if interdimensional discrimination training is scheduled in such a way that important features of the situation other than the presence or absence of the test stimulus are varied on positive and negative trials, than these other stimuli remain as relevant as the test stimulus and may continue to overshadow it. For example, pigeons may be given discrimination training between the presence and absence of a vertical white line on the response key, but if the colour of the background also varies between positive and negative trials, then wavelength may continue to overshadow the line stimulus. Newman and Baron (1965), and Farthing (1972) have provided clear evidence of such an effect. They both found that interdimensional training between the presence and absence of a line on a green background resulted in a steep line-orientation gradient, but that if the colour of the background changed along with the presence or absence of the line during discrimination training, the orientation gradient was significantly flatter.

It is also important to note that the set of stimuli associated with the presence of a line on the pigeon's response key can itself be decomposed into various subsets. The line has various aspects, a particular intensity, width and length, as well as a particular orientation. It may indeed be the case that differential reinforcement correlated with the presence and absence of a particular physical stimulus virtually always increases the slope of a generalization gradient when any aspect of that stimulus object is varied in a subsequent test: by comparison with nondifferential reinforcement, such interdimensional training will sharply reduce the number of potentially relevant elements in the experimental situation and this set of relevant elements will necessarily include that aspect varied in the generalization test. However, it is still possible that this critical aspect will be overshadowed by other elements remaining in the relevant set: as Hearst (1969a) has noted, differential reinforcement between the presence and absence of a line may sharpen an orientation gradient, but it would probably sharpen a brightness gradient even more if the brightness of the line on the key were varied during generalization testing. A few studies have indeed reported that interdimensional training may do little to sharpen a generalization gradient and these failures may well be explained in terms of this argument. MacCaslin et al. (1954), for example, found that rats reinforced for responding to a card containing black and white vertical striations of a given width, and not reinforced for responding to a plain white card, generalized completely when tested with a set of stripes of different width. Boneau and Honig (1964) found that pigeons, given differential reinforcement between the presence and absence of a line, generalized completely to other orientations of the line if the differential reinforcement had been provided in the context of a complex, conditional, discrimination. Finally, Mackintosh (1965a) found that rats reinforced for responding to a white circle of a particular size and not reinforced for responding in the absence of the circle, generalized relatively completely to a circle of a different size, but showed that they had associated the brightness of the training stimulus with reinforcement.

3. *Extradimensional Discrimination Training*

It is easy to see how interdimensional discrimination training between the presence and absence of a particular stimulus will normally increase the control over responding exerted by most aspects of that stimulus. It is less easy, at first sight, to see how discrimination training between an entirely independent pair of stimuli could subsequently ensure that the gradient formed round some third stimulus is sharpened. Several studies, employing the first type of extradimensional discrimination training illustrated in Fig. 9.6, have found that gradients of generalization may be sharpened by such training procedures. Honig (1969), for example, trained pigeons on a wavelength discrimination with a blue key light signalling reinforcement and a green key light signalling no reinforcement. The birds were then reinforced for responding in the presence of three vertical lines on the key and were finally tested for generalization to other orientations of the lines. They showed reliably steeper gradients than a control group initially exposed to blue and green stimuli, but with neither stimulus consistently signalling reinforcement (an uncorrelated or "pseudo-discrimination" group). Similar results have been reported by Thomas *et al.* (1970a), and by Mackintosh and Honig (1970), in experiments in which the gradients obtained after such extradimensional discrimination training have been compared with those obtained following various control treatments. Although the effects have not always been large or even statistically significant, they have been quite consistent from one study to the next: in free-operant experiments with pigeons, extradimensional discrimination training, given independently of reinforced training in the presence of a single stimulus, usually increases the control exerted by that stimulus as measured in a subsequent generalization test.

Wagner (1969a) has pointed out how these results, although at first sight somewhat puzzling, are amenable to the general type of analysis proposed here. The situation may not, in fact, differ greatly from that assumed to hold during interdimensional discrimination teaching. Extra-dimensional training will be as effective as interdimensional training in neutralizing such incidental stimuli as those associated with the shape, size, or illumination of the response key, or with the repetitive nature of responding in a free-operant situation. The only question is whether the suppression of these incidental stimuli will be maintained when the original discriminative stimuli (the blue and green key lights in Honig's experiment) are removed and a new set of stimuli is projected onto the response key. Wagner (1969a) has reported a study of differential eyelid conditioning in rabbits which suggests that, in some circumstances at least, such an effect can occur. The design of the experiment is shown in Table 9.1. One group was given discrimination training between two visual stimuli, L_1 and L_2, interspersed with reinforced presentations of a tone T; this

TABLE 9.1

Design of Experiment by Wagner (1969a)

Groups	Stimuli and Reinforcement		
Discrimination	L_1V+	L_2V-	$TV+$
Pseudo-discrimination	$L_1V+/-$	$L_2V+/-$	$TV+$

L_1 and L_2 designate lights of different intensity; T designates tone; V designates vibratory stimulus.

group showed a steeper gradient of generalization to other tones than did a control group treated in the same way in the presence of T, but given randomly reinforced and nonreinforced trials in the presence of L_1 and L_2. This result, of course, confirms the results of the pigeon experiments under discussion. The important feature of Wagner's experiment was that all stimuli—L_1, L_2 and T—were presented throughout training in conjunction with an incidental vibratory stimulus, V. By providing a specific manipulable incidental stimulus, Wagner was able to assess its control over responding. He found that the presence or absence of V had a smaller effect on responding during test trials in the group given discrimination training between L_1 and L_2 than in the pseudo-discrimination control group. Wagner concluded that discrimination training between L_1 and L_2 had suppressed control by V, thus preventing V from overshadowing the tone during reinforced trials with the TV compound, and it was this reduction in overshadowing that enabled the tone to acquire greater control over responding.

The main problem for this analysis comes from the results of a final set of experiments which have assessed the effects of extradimensional (irrelevant) discrimination training (see Fig. 9.6). In some of these experiments, discrimination training between two stimuli differing along one dimension, but both presented in conjunction with a third constant stimulus, actually sharpens the gradient of generalization obtained round this constant stimulus (e.g., Reinhold and Perkins, 1955; Thomas et al., 1970a). In one of the experiments reported by Thomas et al., for example, pigeons given free-operant discrimination training between two wavelengths (538 and 555 nm), with a vertical line projected onto the key on both positive and negative trials, subsequently showed a steeper orientation gradient than a pseudo-discrimination control group, given randomly reinforced and nonreinforced trials in the presence of both stimulus compounds.

Such a result is not only inconsistent with the general analysis proposed here, it is, apparently, equally inconsistent with many of the results on which the analysis rests. For the vertical line in this experiment is an incidental stimulus common to both reinforced and nonreinforced trials: when reinforcement is predicted by a more valid set of stimuli, i.e. when subjects are given discrimination training between the two wavelengths, such an incidental stimulus should be neutralized. When reinforcement is not predicted by any more valid stimuli, i.e. when subjects are given pseudo-discrimination training, incidental stimuli may gain control over responding. The discrimination and pseudo-discrimination groups in the experiment of Thomas et al. (1970a) correspond exactly to the correlated and uncorrelated groups in the experiments of Wagner et al. (1968) and Wagner (1969a). Yet Thomas et al. found that discrimination training enhanced rather than suppressed control by the incidental stimulus.

Thomas et al. (1970a) followed Reinhold and Perkins (1955) in interpreting their results to mean that discrimination training increases "general attentiveness" or establishes a "set to discriminate", with the apparent consequence that all stimuli present at the time of reinforcement will increase their control over responding. It would be foolish to deny the possibility of such a process, for, as will be seen in the following chapter, there are numerous cases where discrimination training on one discrimination problem tends to facilitate the learning of another (p. 600). In the present context, however, the explanation is unlikely to be correct, for Turner and Mackintosh (1972) have shown that the difference found by Thomas et al. between the slopes of the gradients displayed by discrimination and control groups may be abolished by subsequent discrimination training. Specifically, they replicated the procedure of Thomas et al. but before giving a generalization test to lines of different orientation, they trained both groups on a second wavelength discrimination, with no line present on the key. This training sharpened the control group's orientation gradient to the point where there was no longer any difference between the two groups' gradients. In the light of this result, it is difficult to argue that the discrimination subjects trained by Thomas *et al.* showed a steeper gradient than their control subjects because they had *learned* more about the stimuli on the key during initial training.

The suggestion that discrimination training establishes a general set to discriminate is, moreover, quite inconsistent with the data reported by Wagner *et al.* (1968) with rats and rabbits, by Wagner (1969a) with rabbits, and by Blough (1969b) and Miles *et al.* (1970) with pigeons. In all these studies discrimination training decreased rather than increased control by incidental stimuli. There appears to be a straightforward conflict between the results of different experiments, with one set finding that discrimination training enhances control by a specific incidental stimulus and another set showing that discrimination training may suppress control by such a stimulus. Since both types of finding have been frequently replicated, the problem is to discover the important procedural difference between the two groups of experiments.* One suggestion is that the critical factor is the use of discrete-trial or free-operant procedures. In all studies which have found that discrimination training suppresses control by an incidental stimulus, a discrete-trial procedure has been used: Thomas *et al.* (1970a), on the other hand, employed free-operant procedures. In support of this suggestion, Turner and Mackintosh (1972) showed that discrete-trial discrimination training in pigeons significantly suppressed control by an incidental stimulus (a vertical line on the response key), while free-operant discrimination training either enhanced control by the line or had no effect. Similar results have been obtained by Gray and Mackintosh (1973).

If this suggestion is correct, it remains to explain why the two training procedures should produce such different outcomes. There are several

* It is worth interjecting at this point the observation that differences in the slopes of generalization gradients between discrimination and control groups may partly reflect differences in resistance to extinction. The typical pseudo-discrimination control group is, in effect, trained on an inconsistent reinforcement schedule, while a discrimination group receives predictable reinforcement. There is evidence that discrimination training reduces resistance to extinction by comparison with a random reinforcement schedule (e.g., Jenkins, 1961). Since there is also evidence that generalization gradients sharpen in extinction (p. 493), it may be that pseudo-discrimination training appears to flatten generalization gradients only because it increases resistance to extinction. Thomas *et al.* (1970c), for example, claimed that rats showed a steeper gradient along a light intensity dimension following discrimination training than following pseudo-discrimination training; but since their pseudo-discrimination group made over twice as many responses during the course of the generalization test as the discrimination group, this apparent difference in the slope of the relative gradient is impossible to interpret.

possibilities, but rather little evidence. One suggestion, consistent with the general argument advanced here, is that free-operant procedures contain other incidental stimuli, which are less important in discrete-trial situations: the repetitive nature of responding in free-operant situations may permit response-produced stimuli to gain control over free-operant behaviour, unless suppressed by discrimination training. Now if the main effect of free-operant discrimination training were to suppress control by such stimuli, this might work towards enhancing the relative control exercised by other incidental stimuli—including the specific incidental stimulus being manipulated by the experimenter (Turner and Mackintosh, 1972). The argument is clearly speculative, and this is an area where more detailed information is necessary.

E. Conclusions

The available data support the general suggestion that the slope of a generalization gradient is dependent on the extent to which the change from training to test stimuli changes the stimulus elements actually associated with reinforcement during initial acquisition. The reason why generalization gradients are often relatively flat is not because the subject is incapable of discriminating between training and test stimuli, but because some of the stimuli associated with reinforcement were those not varied during the generalization test. If a subject is exposed to nondifferential reinforcement in a given situation, there is no reason to suppose that the stimuli subsequently varied by the experimenter during the generalization test are any better correlated with reinforcement than are numerous other incidental stimuli.

Discriminative training nearly always sharpens gradients of generalization: the most plausible explanation of this finding is that differential reinforcement tends to neutralize such incidental stimuli. A variety of experiments have provided direct evidence of such an effect (Wagner et al., 1968; Wagner, 1969a; Miles et al., 1970); although the results of the free-operant experiments reported by Thomas et al. (1970a) clearly pose problems for such an analysis, it is hard to believe that this is not an important determinant of the effects of discriminative training on generalization. Since the suppression of control by irrelevant stimuli is not simply a consequence of the schedule of reinforcement associated with such stimuli, but depends on the presence of discriminative stimuli better correlated with reinforcement, these results imply the operation of some process of stimulus selection, such as that required to account for the phenomenon of overshadowing.

V. SOME VARIABLES AFFECTING THE SLOPE OF EXCITATORY GRADIENTS

The slope of a generalization gradient is affected by several variables in addition to the type of discriminative training experienced. Some of the more important of these variables are discussed in the present section.

A. Schedule of Reinforcement

Anrep (1923) and Pavlov (1927, pp. 113–16) noted that although a delayed conditioning procedure, where CS and UCS overlap, will typically produce relatively steep gradients of generalization, a trace conditioning procedure, where the CS is terminated before the onset of the UCS, might result in "permanent and universal generalization, involving all the analyzers" (Pavlov, 1927, p. 113). Ellison (1964), studying discriminative conditioning with both delay and trace procedures, found that when the interval between onset of CS and onset of UCS was relatively long, the discrimination between CS+ and CS− was notably poorer with the trace procedure.

Early studies of GSR conditioning in human subjects found that generalization gradients were significantly steeper after consistent reinforcement than after partial reinforcement (Humphreys, 1939; Wickens et al., 1954). Generalization of the GSR is, however, a phenomenon not easy to interpret since the response is elicited by novel stimuli and will therefore be elicited unconditionally by stimuli in a generalization test (Epstein and Burstein, 1966). The effect of schedule of reinforcement on generalization does not seem to have been studied in discrete-trial experiments with animals.

Free-operant studies, however, have examined the effects of a variety of reinforcement schedules on the slope of generalization gradients. Haber and Kalish (1963) and Hearst et al. (1964) studied wavelength and line orientation generalization in pigeons following training with VI schedules ranging from VI 15 sec to VI 4 min: in both studies gradients were significantly flatter in birds trained with the longer schedules. Hearst et al. also found that a DRL schedule of reinforcement produced a significantly flatter generalization gradient than a VI 1 min schedule; while Thomas and Switalski (1966) found that a VR schedule resulted in a flatter relative gradient of generalization than a comparable VI schedule (see Fig. 9.2).

These results are open to several possible interpretations. One of the more interesting is that different schedules of reinforcement affect the degree to which responding is controlled by particular exteroceptive stimuli (Hearst, 1965, 1969b; Donahoe, 1970). Unfortunately, other more prosaic possibilities demand consideration. One problem is that different schedules may establish different rates of responding and have different effects on resistance to extinction. Fig. 9.7 shows the absolute and relative generalization gradients from some of the groups trained by Hearst *et al.* (1964). The relative gradients present a clear and simple picture: training on a VI 1 schedule resulted in a much steeper gradient

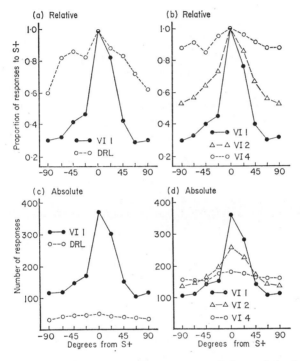

FIG. 9.7. The effect of schedule of reinforcement on relative and absolute generalization gradients. (a) and (c) Relative and absolute gradients following training on VI 1 min or DRL 6 sec schedules. (b) and (d) Relative and absolute gradients following training on VI 1 min, VI 2 min, or VI 4 min schedules. (*After* Hearst *et al.*, 1964.)

than training on either a DRL or on a VI 4 schedule. But the comparison of the VI 1 and DRL gradients is surely somewhat hazardous in view of the extreme differences in overall rate of responding: it is at least conceivable that the DRL gradient is very flat because of a floor effect. The gradients following VI 1 and VI 4 training do not seem to pose such problems. Both the absolute and relative gradients are steeper following VI 1 training, and the absolute gradients cross over. But, at the risk of seeming pedantic, it is worth remembering that a VI 4 schedule produces a lower rate of responding, accompanied by greater resistance to extinction, than a VI 1 schedule. The lower rate of responding at S+ may be a consequence of the former difference, while the higher rate of responding to the test stimuli may be a consequence of the latter.

The absolute and relative gradients from Thomas and Switalski's (1966) comparison of VR and VI schedules have already been shown in Fig. 9.2 and the problems involved in accepting the conclusion suggested by the cross-over in the relative gradients have been discussed. It is hard to decide whether the parallel slopes of the absolute gradients or the cross-over in the slopes of the relative gradients is the more valid measure of generalization. It is worth noting here, however, that Thomas and Switalski found that as the absolute rates of responding of VR and VI subjects converged during the course of extinction, so the difference in the slopes of the relative gradients tended to disappear.

A second problem arises in the comparison of the VI and the DRL gradients. As has been argued before (p. 382), the use of a DRL schedule of reinforcement makes changes in overall rate of responding at best difficult to interpret and at worst totally meaningless. If subjects have learned to pause for 5 sec before responding in the presence of a particular stimulus, it does not follow that they will pause for 10 sec before responding in the presence of a different stimulus. And yet this is what would seem to be required if a sloping gradient of overall rate of responding is to be obtained. An experiment by Gray (1974) nicely documents this point. Pigeons were trained on a DRL 8 sec schedule in the presence of a 570 nm key light and were then given generalization tests to other wavelengths. Although the gradients were relatively shallow, Gray showed that this overall measure concealed a variety of different gradients for different classes of response. Spaced responses gave orderly, quite steep excitatory gradients; but when gradients were calculated only from those responses emitted at very short IRTs, some birds actually showed inhibitory gradients, i.e., were more likely to respond to stimuli other than the training stimulus. The shallow gradient of total responses was a composite of these inhibitory and excitatory gradients produced by different classes of response.

In spite of these problems, it remains possible that some different schedules of reinforcement do have different effects on the slopes of generalization gradients and that these differences are a consequence of the extent to which cues other than the relevant stimuli manipulated by the experimenter come to control the subject's behaviour. Schedules may differ in the extent to which they suppress control by incidental stimuli; while some schedules may ensure that behaviour is controlled more by interoceptive cues than by any exteroceptive cues (including those manipulated by the experimenter).

Konorski (1961), noting that trace conditioning resulted in relatively flat generalization gradients, suggested that:

> Trace CRs . . . are formed not to the traces of the given exteroceptive stimulus itself, but rather to some of its consequences which are common for various sorts of stimuli. Since any external stimulus elicits an orientation reaction, it may be that the proprioceptive stimuli generated by this reaction form the true basis for elaboration and occurrence of trace CRs (p. 116).

Hearst *et al.* (1964), and Thomas and Switalski (1966) have advanced rather similar arguments to account for the schedule effects observed in their experiments. Behaviour on DRL schedules must be controlled partly by stimuli correlated with the passage of time, even if these stimuli are often associated with the occurrence of stereotyped patterns of behaviour which mediate the appropriate temporal discrimination (p. 173). Whatever the precise nature of the stimuli controlling spaced responses, they will not include the stimuli projected onto the response key by the experimenter. At best, therefore, a pigeon's spaced responding must be controlled jointly by temporal stimuli and by stimuli on the response key; hence variations in the latter stimuli will change only a small proportion of the stimuli controlling the subject's behaviour. Hearst *et al.* (1964) argued:

> Reinforcement contingencies (e.g., DRL and long VI schedules) which produce comparatively low response rates during training, along with a large amount of pausing and a relatively frequent occurrence of stereotyped response chains, are more likely to yield flat generalization gradients than contingencies which produce high, steady rates during training . . . Stereotyped "mediating" response chains were much more likely to develop on long VI schedules than on short VI schedules (p. 378).

Thomas and Switalski (1966) suggested that the VR gradients in their experiment were shallower than the VI gradients, because "ratio schedules make proprioceptive feedback from rapid responding a positive discriminative cue for additional responding (p. 236)".

While it is possible that both very high and very low rates of responding increase the importance of proprioceptive control in free-operant situations, the arguments advanced here seem a shade arbitrary. As was suggested in the preceding section, there is reason to believe that responding will be controlled by the occurrence of prior responses under most free-operant schedules of reinforcement: this is one possible reason why discrimination training typically sharpens generalization gradients along exteroceptive stimulus dimensions. This does not, however, necessarily imply that the differences in the slopes of gradients produced by training on different schedules can be attributed to differences in the extent to which they permit the development of such proprioceptive control. There is need here for some more explicit experimental analysis. In the meantime, one final suggestion is worth advancing. We have seen that pigeons reinforced for pecking at a key probably receive implicit differential reinforcement with respect to stimuli projected onto the key. Such differential reinforcement should tend to neutralize other, incidental, stimuli, since the stimuli on the key are relatively more valid predictors of reinforcement. Now the higher the probability of reinforcement for responding, the greater the difference in relative validities between relevant and incidental stimuli, and the more effectively, therefore, should incidental stimuli be neutralized by differential reinforcement (Wagner, 1969a, b). It follows that short VI schedules may neutralize incidental stimuli more successfully than long VI schedules and this should result in steeper gradients along relevant dimensions. The argument is similar to that advanced by Sutherland (1966) in discussing partial reinforcement and breadth of learning (p. 438).

B. Level of Training

Razran (1949), summarizing earlier Russian work on the generalization of classical CRs, concluded that generalization tended to increase after a small number of conditioning trials, but that as training continued generalization gradients became significantly steeper. The initial increase in generalization presumably represents little more than the establishment of the CR; the later decrease in the level of responding to test stimuli is more intriguing. Similar results have been reported by Hovland (1937c), for the generalization of the GSR in human subjects, and by Brown (1970) in a study of salivary conditioning in dogs. A series of experiments on signalled avoidance learning in cats (Thompson, 1958, 1959; Hoffeld, 1962) also obtained similar results. In these experiments overtraining resulted in a cross-over of the absolute gradients: after extended training, animals responded more to the training stimulus during testing but less to the remaining test stimuli.

Both classical conditioning experiments and discrete-trial signalled avoidance experiments, as has already been argued, automatically involve interdimensional discrimination training between the presence and absence of the CS or warning signal. Such explicit differential reinforcement should ensure the gradual suppression of incidental stimuli, and these results, therefore, are entirely consistent with the present general analysis.

Free-operant experiments, in which the relevant stimuli are localized on the response key, probably provide some differential reinforcement with respect to such stimuli and as would be expected, therefore, Hearst and Koresko (1968) reported that pigeons given varying amounts of training with a vertical line on the key, tended to show steeper absolute gradients after more extended training. There was, however, no cross-over in the absolute gradients: rate of responding was uniformly higher in subjects trained for 14 sessions than in those trained for fewer sessions (see Fig. 9.2). Although Hearst and Koresko found that overtraining tended to sharpen the relative gradients, the effects of overtraining were less striking than those obtained in the discrete-trial avoidance experiments. A similar result was reported by Taus and Hearst (1972) who gave pigeons discrimination training between a vertical line and either a 45-degree or a horizontal line after different numbers of sessions of VI training on the vertical line. In accordance with Razran's conclusions they found that, as measured by the proportion of responses to the vertical line during discrimination training, generalization first increased, and later decreased, as a function of prior single-stimulus training. Once again, however, there was little evidence of any cross-over in the absolute gradients.

The implication of the present argument is that overtraining will sharpen generalization gradients only if such training provides explicit or implicit differential reinforcement with respect to the stimuli varied during the generalization test. There is, unfortunately, no direct evidence bearing on this supposition. There is, however, good evidence that overtraining does not necessarily lead to the sharpening of generalization gradients. Studies of size or brightness generalization in rats trained in alleyways have found that overtraining either has no effect on, or significantly decreases the slopes of, generalization gradients on both absolute and relative measures (Margolius, 1955; Muntz, 1963; McCain and Garrett, 1964). In this situation, as was argued earlier (p. 426), overtraining may increase the probability of control by proprioceptive stimuli and will therefore decrease control by stimuli varied during any generalization test.

C. Motivational Effects

1. *Generalization as a Function of Drive Level*

The effect of variations in drive level on generalization gradients is one of those questions suggested by Hullian theory which are very much harder to answer, and possibly rather less important, than was at one time thought. Increases in drive level will increase speed or rate of responding both during training and on test trials to S+. We are, therefore, likely to be involved in the comparison of generalization gradients differing greatly in overall height, with all the attendant problems that such comparisons entail. In principle, Hull's (1943) multiplicative theory of drive predicts that absolute gradients should be higher and steeper with increases in drive level, but that the slope of relative gradients should be unaffected. In practice, differences in overall levels of responding may introduce sufficient complications from potential floor and ceiling effects as to make the predictions untestable (cf. Storms and Broen, 1966).

In fact, the predictions have received some confirmation in a number of studies. Newman and Grice (1965), studying size generalization in rats, reported that an increase in hunger increased the height and slope of the

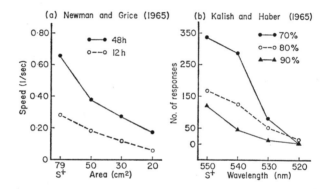

FIG. 9.8. The effect of deprivation on absolute gradients of generalization. (a) Size generalization in rats trained to respond to a stimulus of 79 cm² under 48 or 12 hours of food deprivation. (*After* Newman and Grice, 1965.) (b) Wavelength generalization in pigeons trained to peck a key illuminated with 550 nm at 70, 80, or 90 per cent of their free-feeding weights. (*After* Kalish and Haber, 1965.)

absolute gradient, leaving the slope of the relative gradient unaffected. Their results are shown in the left panel of Fig. 9.8. Healey (1965), studying size and auditory frequency generalization in rats in a similar situation, reported similar findings although, perhaps as a consequence of the smaller difference in drive levels in his study (12 *vs* 23 h food deprivation, as opposed to 12 *vs* 48 h deprivation in Newman and Grice's study), the difference in slope of the absolute gradient fell short of significance. Kalish and Haber (1965), studying wavelength generalization in pigeons reduced to 90 per cent, 80 per cent or 70 per cent of their free-feeding weights, obtained the same results as those of Newman and Grice: their results are shown in the right panel of Fig. 9.8.

There are, however, several discordant results. Brown (1942), studying intensity generalization in rats, found no evidence of an increase in the slope of the absolute gradient in animals tested under a high level of hunger; while Thomas and King (1959), although confirming Kalish and Haber's results for wavelength generalization at moderate levels of drive (90 per cent and 80 per cent of free-feeding weights), found that extreme conditions of deprivation (60 per cent of free-feeding weight) actually decreased the number of test responses to S+, and markedly flattened both absolute and relative gradients. In both these experiments, however, all subjects had been trained under one level of deprivation and tested at different levels; some of these discrepancies, therefore, may be due to the change between training and testing experienced by some groups but not by others.

Another pigeon study which suggests that relative generalization gradients may become significantly flatter with increases in deprivation (Jenkins et al., 1958) is open to a different set of objections. Jenkins et al. gave their subjects a single test session to each test stimulus, interspersed with reinforced retraining on S+. Such a procedure must have maintained a relatively high rate of responding to all test stimuli. Subjects trained and tested under conditions of high drive responded at a much higher rate to S+ than subjects under low drive; although the absolute gradient was in fact considerably steeper for highly motivated subjects, extinction may not have been permitted to proceed far enough to equalize the slopes of the relative gradients.

The trouble with this sort of special pleading is that it is equally easy to raise objections to studies reporting results consistent with Hullian predictions. That absolute gradients become steeper with increases in drive might simply reflect the fact that subjects responding at a higher rate to S+ have more room to show a greater absolute decline in rate of responding to test stimuli. Until more meaningful response scales can be devised, the effort involved in testing predictions of this general sort may not be repaid by the information gained. It is, at any rate, worth noting that if there is a discrepancy between data and theory, it is that increases in drive may lead to a flattening of relative generalization gradients. If this is the case, one solution, as suggested earlier, is that increases in drive may affect a subject's decision criterion.

2. Appetitive and Aversive Motivation

On the basis of studies of approach–avoidance conflict, Miller (1944, 1959) concluded that appetitively motivated approach responses may generalize more widely than aversively motivated avoidance responses. The argument was originally applied to spatial generalization from the goal box of an alley, in which rats were fed or shocked, to earlier points in the alley. If rats are trained to run in an alley for food and are then shocked in the goal box, when replaced in the start box, they may run part of the way down the alley until they reach a point at which they show strong signs of conflict. The assumption is that this represents the point of equilibrium of approach and avoidance tendencies which have generalized from the goal box. Since the rats do leave the start box, their approach tendencies must at that point be stronger than their avoidance tendencies. Since they stop before reaching the goal box, the avoidance tendency must increase more rapidly than the approach tendency. Hence, as is suggested in Fig. 9.9, the avoidance gradient must be steeper than the approach gradient.

If these results are due to differences in the generalization of approach

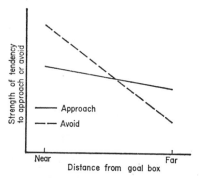

FIG. 9.9. Hypothetical gradients of approach and avoidance. The tendency of a rat to approach a goal box where it has been fed, or to avoid a goal box where it has been shocked, is assumed to vary, in the manner shown, with the distance from the goal box. The intersection of the two gradients represents the point of equilibrium at which the rat will stop. (*After* Miller, 1944.)

and avoidance tendencies along all stimulus dimensions, some simple predictions may be tested by giving animals test trials in alleys of varying degrees of similarity to the training alley. If the approach tendency generalizes more widely than the avoidance tendency, it follows that animals tested in a dissimilar alley should approach the goal box more closely before showing signs of conflict. Several experiments have confirmed this and related predictions (Miller and Kraeling, 1952; Murray and Miller, 1952; Murray and Berkun, 1955).

Why should approach behaviour generalize more widely than avoidance behaviour in these experiments? The first step towards an adequate analysis is to see that different situations may produce entirely different results. Hearst (1962) trained monkeys and rats on two concurrent tasks: pulling a chain was reinforced with food on a VI schedule while pressing a lever postponed the occurrence of unsignalled shocks on a standard free-operant avoidance schedule. When generalization tests were given with different intensities of illumination in the chamber, the avoidance response generalized much more widely than the food-reinforced response.

If approach gradients are sometimes steeper than avoidance gradients it is likely that the slope of generalization gradients is not directly affected by the nature of the subject's motivational state, but that differences in the procedures involved in these studies are responsible for variations in the slopes of exteroceptive gradients. Hearst (1965, 1969b) has pointed to a number of differences between the requirements for avoidance responding and food-reinforced responding in his experimental situation, more than one of which may have contributed to the observed differences in generalization. One factor which he has stressed is the use of a free-operant avoidance schedule, in which shocks occur in the absence of any warning signal: in such a situation, responding may be controlled largely by stimuli produced by the subject's own behaviour (p. 325). With regular response–shock intervals, animals show temporal distributions of respond-ing, consistent with the hypothesis that responding is controlled by stimuli correlated with the passage of time since the last response. To the extent that avoidance responding in such a situation is controlled by intero-ceptive stimuli, it will generalize widely when exteroceptive stimuli are varied. Since similar variations can be observed in the slopes of appetitive gradients, it is the nature of the avoidance schedule, rather than the nature of the subject's motivational state, that determines the slope of the gradient.

It is not only in Miller's conflict experiments that relatively steep gradi-ents of generalization for avoidance responses can be observed. Discrete-trial signalled avoidance procedures with a tone as the warning signal produce sharp gradients of auditory generalization in cats (e.g., Thomp-son, 1958). In an attempt to study approach and avoidance gradients under conditions made as comparable as possible, Hearst (1969b) trained rats to press a lever within 5 sec of the onset of a tone, either to avoid shock or to obtain a pellet of food. Subsequent generalization gradients along the dimension of intensity of the tone revealed no evidence of any differ-ence in the slope of the gradient for responses reinforced by food and those reinforced by avoidance of shock. Hearst concluded that differences between approach and avoidance gradients are in general a consequence of differences in procedure, type of response, schedule of reinforcement, or probability of interoceptive control; variations in any of these parameters have profound effects on the slopes of both approach and avoidance gradients and once they have been eliminated, no differences remain which could be attributed to differences in motivational state.

While this conclusion seems entirely reasonable, it does not seem suffi-cient to explain the original conflict data on which Miller (1944, 1959) and Brown (1948) based their claim that gradients of approach are flatter than gradients of avoidance. The claim may, however, be open to question, for it rests on the assumption that the strength of the approach tendency at different points in an alley is dependent on the *generalization* of approach responses to stimuli differing from the goal box. The evidence for this assumption is extremely tenuous: indeed, the available data suggest that speed of running in an alley may be inversely rather than directly related to the similarity between the goal box and early parts of the alley (Saltz *et al.*, 1963; see p. 230).* The behaviour of a rat at different points in an alley is probably influenced less by the physical similarity between the alley and the situation directly associated with reinforcement and more by the establishment of associations between these sets of stimuli. It may be a mistake, therefore, to apply the concept of stimulus generalization to such behaviour.

D. Conclusions

This sampling of variables shown to affect generalization has not been exhaustive. The data reviewed, however, tend to support the methodological and theoretical points suggested earlier. When different treatments, such as changes in the schedule of reinforcement or drive level, result in substantial differences in overall rate of responding or resistance to extinction, the assessment of the effects of those treatments on the slope of generalization gradients becomes highly problematic. Where these effects are more interpretable, however, they may often be a consequence of the extent to which incidental stimuli, not varied during the test for generalization, become associated with reinforcement. Certain schedules of either appetitive or aversive reinforcement may increase the probability that response-produced stimuli come to control further responding; others may effectively increase the difference in the relative validities of relevant and incidental stimuli. Increases in the amount of differential reinforcement almost invariably steepen subsequent gradients of generalization, but where acquisition does not explicitly involve differential reinforcement in the presence and absence of the relevant set of stimuli, variations in the amount of acquisition training may not have such marked effects on generalization.

* Even the assumption that there is a *gradient* of approach at all may be open to question. In order to show such a gradient, Brown (1948) measured approach tendencies by recording the force with which a restrained rat pulled on a harness, and reported that the strength of this tendency varied directly with proximity to the goal. His procedure, however, was to allow a harnessed rat to run down an alley and then to measure the force exerted when further progress was impeded by locking the harness to an overhead wire. That the rat pulled slightly harder when restrained at a point near the goal box than when restrained shortly after leaving the start box may not reflect differences in the strength of generalized approach tendencies, but simply differences in the speed at which the rat was running at the moment it was restrained.

VI. INHIBITORY GRADIENTS OF GENERALIZATION

A. The Measurement of Inhibitory Gradients

The discussion to this point has been confined largely to the analysis of excitatory, as opposed to inhibitory, gradients. If responding is established in the presence of S_1, other stimuli may elicit responding in proportion to their similarity to S_1, and an inverted U-shaped gradient, with its maximum at S_1, may be recorded. Similarly, it seems reasonable to suppose that if some procedure were used to decrease the probability of responding in the presence of S_2, then other stimuli might also suppress responding in proportion to their similarity to S_2 and a U-shaped gradient, with its minimum at S_2, would be recorded. It is tempting to refer to the former as an excitatory gradient and the latter as an inhibitory gradient. Excitatory and inhibitory conditioning, however, have previously been defined in terms of the correlation between stimuli and reinforcement: an excitatory stimulus is one which has been positively correlated with reinforcement; an inhibitory stimulus is one that has been negatively correlated with reinforcement (p. 36). Excitatory gradients, therefore, are those formed round a stimulus previously associated with a reinforcer, while inhibitory gradients are those formed round a stimulus previously associated with the omission of a reinforcer.

The point of these definitions is largely to distinguish between various so-called "inhibitory" gradients. A U-shaped gradient, with a minimum at the training stimulus, may be produced by a variety of training procedures. If pigeons or rats receive food for key pecking or lever pressing, then a stimulus associated with shock will suppress responding and a U-shaped gradient of responding around that stimulus may be observed: the gradients of conditioned suppression shown in Fig. 9.3 are, of course, derived from U-shaped gradients of key pecking. A stimulus signalling punishment may also generate a U-shaped gradient (Honig, 1966). On the present definitions, however, these are excitatory rather than inhibitory gradients. If the decision were taken to call them inhibitory gradients, then a further problem would arise in the description of gradients formed round a CS− for shock in a study of conditioned suppression. Such a stimulus can increase rate of responding when added to a CS+: does this mean that the inverted U-shaped gradient that might be formed round such a CS− is really an excitatory gradient? It will be far better to reserve the terms excitation and inhibition so that they refer to the direction of the correlation between stimuli and reinforcers, and to remember that a stimulus positively correlated with an aversive reinforcer may suppress appetitively reinforced responding.

For the moment, therefore, we shall use the term inhibitory gradient to refer to gradients formed round a stimulus that has previously been associated with the omission of, or a reduction in, reinforcement. In most cases, of course, such a gradient will be measured as an increase in the probability of responding to stimuli further removed from S—. The problem then becomes one of devising a procedure which will enable the experimenter to measure such an increase in rate of responding.

In early studies of inhibitory generalization, subjects were reinforced in the presence of several different values of a given stimulus dimension, to ensure an equal and high probability of responding to each, extinguished in the presence of one of these stimuli, and finally given a generalization test to the entire series of stimuli. Inhibitory gradients, in the form of a low probability of responding to the extinguished stimulus and increasing probabilities of responding to test stimuli increasingly far removed from the extinguished stimulus, have been obtained with such procedures in both classical and instrumental experiments. Pavlov (1927, pp. 153–4) reported an experiment by Krasnogorsky in which gradients of inhibition were obtained along a tactile dimension in dogs; Kling (1952) observed a gradient of inhibition along the dimension of size in rats; and Honig (1961) observed an inhibitory gradient in pigeons along the wavelength dimension.

Gradients obtained in this way must inevitably, as Honig pointed out, be a complex consequence of the interactions between excitatory and inhibitory training. A potentially simpler procedure for measuring inhibitory gradients, in which an increase in the probability of responding to test stimuli as a function of their distance from S— must be attributed to generalization of the effects of training at S—, was independently developed by Jenkins and Harrison (1962) and Honig et al. (1963). The procedure they used was to give interdimensional discrimination training, scheduling nonreinforcement in the presence of one stimulus (such as a tone or a vertical line on a pigeon's response key) and reinforcing responding in the absence of that stimulus. A generalization test can now be given to a series of stimuli, such as different frequencies of tone or different orientations of line, which differ along some dimension from S—, but which are all (hopefully) equidistant from the stimulus situation which signalled reinforcement. Thus any increase in probability of responding to test stimuli as a function of their distance from S— cannot be attributed to differential generalization of excitation from S+.

This procedure has now become the standard method of studying inhibitory gradients. Sloping gradients have been obtained along dimensions which include auditory frequency, line orientation and wavelength. Fig. 9.10 shows the inhibitory gradients obtained by Jenkins and Harrison

(1962) and Honig *et al.* (1963), along with comparable excitatory gradients obtained after interdimensional discrimination training with the tone or line serving as S+ and its absence as S−. It can be seen that the inhibitory gradients, although reliably sloping, are somewhat flatter than the excitatory gradients. Most studies of inhibitory gradients have involved free-operant discriminations with pigeons, with responses to S+ reinforced on a VI schedule and responses to S− extinguished. Weisman (1969, 1970), however, has shown that pigeons, trained on a multiple schedule with S_1 signalling a VI 1 min schedule of reinforcement, will show an inhibitory gradient round S_2 when it signals either a longer VI schedule (e.g.,

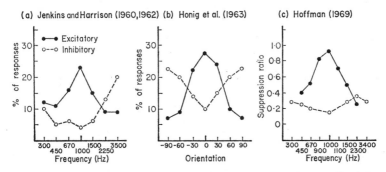

FIG. 9.10. Gradients of excitation and inhibition in pigeons. (a) Gradients formed after interdimensional training between the presence and absence of a 1000 Hz tone. (*After* Jenkins and Harrison, 1960, 1962.) (b) Gradients formed after inter-dimensional training between the presence and absence of a vertical line on the response key. (*After* Honig *et al.*, 1963.) (c) Excitatory and inhibitory gradients of conditioned suppression. The excitatory gradient was obtained following reinforcement in the presence of a CS of 1000 Hz; the inhibitory gradient was obtained following reinforcement in the presence of white noise, and nonreinforcement in the presence of a 1000 Hz tone. (*After* Hoffman, 1969.)

VI 5 min) or even a DRL or DRO schedule. In the latter case, the gradients are perhaps not "inhibitory" in the sense employed here at all.

A few studies have reported gradients of inhibition in other situations. Desiderato (1969) trained dogs on a free-operant avoidance task, and gave classical discriminative conditioning with a tone serving as either CS+ or CS−. Generalization tests were administered by superimposing tones of different frequencies upon ongoing avoidance responding. Excitatory gradients along the dimension of auditory frequency were obtained when the tone was CS+ and inhibitory gradients were obtained in animals for whom the tone was CS−. The third panel of Fig. 9.10 illustrates a gradient of inhibition obtained by Hoffman (1969) in a study of conditioned suppression in pigeons. The CS+ for shock was a white noise signal and

CS— was a 1000 Hz tone; when tones of different frequencies were super-imposed upon the noise signal, they decreased the amount of suppression displayed to the noise, with the magnitude of this effect varying as a function of the distance of the test stimulus from CS—. It should be noted that the inhibitory gradient is here measured as a *decrease* in rate of instru-mental responding to stimuli as a function of their distance from S—. Again the inhibitory gradient is somewhat flatter than the comparable excitatory gradient.

Finally, inhibitory gradients have been obtained in studies of instru-mental discrimination learning employing measures of responding other than rate. Beale and Winton (1970) trained pigeons on a concurrent schedule in which S+ (a blue key light) and S— (a vertical line) alter-nated on one key, and responses to a second key changed the stimulus in effect on the main key. Subjects were therefore able to control the amount of time spent in the presence of each stimulus on the main key and dis-crimination learning was largely evidenced as a reduction in the amount of time spent in the presence of S—. When given a generalization test to lines of different orientation, they showed steeply sloping inhibitory gradients in terms of the amount of time spent in each stimulus, but a much shallower gradient in terms of rate of responding. Robson (1970) trained pigeons on a discrete-trial simultaneous discrimination with a red key light as S+ and a 45-degree line as S—, and then gave a series of test trials on which subjects were required to choose between all possible combinations of five line orientations. This choice measure gave a steep inhibitory gradient around the 45-degree line.

B. Inhibitory Gradients and Conditioned Inhibition

The definition of inhibitory gradients as those generated round stimuli correlated with the omission of reinforcement implies a close connection between inhibitory gradients and conditioned inhibition. A condition inhibitor is usually defined as a stimulus which, by virtue of a negative correlation with reinforcement, may reduce the probability of responding controlled by excitatory stimuli. Inhibitory gradients, therefore, seem to be measuring the extent to which such conditioned inhibition generalizes to neighbouring stimuli.

It does not, however, follow that the observation of a significant inhibi-tory gradient round S— necessarily implies that S— will satisfy the strong requirement of actually suppressing the responding maintained by an excitatory stimulus. There is nothing surprising about this: nonreinforce-ment of S— may result in some inhibitory learning, but if S— is also a partially excitatory stimulus, either because of a past history of rein-

forcement, or by generalization from other reinforced stimuli, nonreinforcement may not produce sufficiently strong inhibitory learning to
ensure that S— will effectively suppress responding maintained by
another stimulus (p. 35). Nevertheless, if some inhibitory learning has
accrued to S—, it may generalize to other stimuli in proportion to their
similarity to S— and an inhibitory gradient will be observed.

Studies of inhibitory gradients in free-operant experiments have
routinely reinforced subjects for responding in the presence of both S+
and S—, before introducing discrimination training with S— signalling
extinction. Although such a procedure may prevent S— from suppressing
responding maintained by an excitatory stimulus, both in classical conditioning (p. 34) and free-operant discrimination learning(Lyons, 1969a,
b), it may have little or no systematic effect on the slope of the inhibitory
gradient formed round S— (Weisman and Palmer, 1969; Zentall et al.,
1971).

Terrace (1966c) has argued that " a U-shaped gradient, with a minimum
at S—, would indicate that S— was an inhibitory stimulus, while a flat
gradient would indicate the absence of any inhibitory function" (p. 1678).
It is, however, even more important to see that S— may be an entirely
effective conditioned inhibitor even if the gradient round S— is perfectly
flat (Hearst et al., 1970; Hearst, 1972). In the first place, the assessment of
inhibitory gradients via generalization tests conducted in extinction may
well run into problems posed by a floor effect. Nonreinforcement of S—
may generalize sufficiently widely to suppress responding to stimuli
similar to S—. An extinction test will ensure that few responses occur to
any test stimulus in spite of differences in the underlying strength of
an inhibitory process. A comparable procedure for assessing excitatory
gradients might be equally unsuccessful. Reinforcement at S+ might
generalize widely enough to ensure a high rate of responding to stimuli
similar to S+. If generalization were tested by reinforcing responses to
all test stimuli, a completely flat gradient (i.e., a uniformly high rate of
responding to all stimuli) might well be observed. The force of this argument may be judged by a glance at the excitatory gradients of conditioned
suppression obtained by Hoffman and Fleshler (1961) and shown in
Fig. 9.3. If generalization to the different tones had been tested by reinforcing all tones, it is clear that generalization would have been complete
and the gradient absolutely flat. If generalization is tested in extinction, a
flat inhibitory gradient, in which no responses occur to any test stimulus,
may signify no more than the initially flat excitatory gradient observed
by Hoffman and Fleshler. Before concluding that a flat gradient is not
simply a consequence of a floor effect, therefore, an appropriate test must
be provided. By analogy with the extinction test of excitatory gradients,

the appropriate test of inhibitory gradients presumably involves reinforcing responses in the presence of each stimulus (the resistance to reinforcement technique).

Even if this test procedure resulted in a flat gradient, however, one
would be entitled to conclude no more than that the aspect of S— varied
in the generalization test was not that which had become associated with
nonreinforcement during initial training. Flat gradients of excitation along
a particular dimension, obtained after extensive reinforced training in the
experimental situation, do not imply the complete absence of excitation.
They imply only that responding was controlled by some other feature or
features of the experimental situation than that which was varied in the
test for generalization. Flat excitatory gradients, as we have seen, are not
so very uncommon, but they have not prompted many investigators to
suppose that a particular S+ was not an excitatory stimulus.

There is considerable evidence to support these arguments. Yarczower
and Curto (1972) trained pigeons on a multiple schedule with a green key
light signalling reinforcement and a vertical line on a black ground signalling extinction. That S— was established as an effective conditioned
inhibitor was shown by superimposing the line on the green key and
observing a significant suppression of responding. Nevertheless, there was
no evidence of a U-shaped gradient of inhibition to other orientations of
the line.

Terrace (1966a, c; 1972) has argued that when discrimination learning
is established without errors (p. 317), no inhibitory gradient can be
detected round S— and that this therefore implies that an errorless S— is
not an inhibitory stimulus. Although there is good evidence that gradients
round such an S— may be essentially flat (Lyons, 1969a; Terrace, 1972;
Johnson and Anderson, 1972).* the conclusion does not follow. Both
Lyons (1969b) and Johnson (cited by Hearst, 1972) have shown that an
errorless S— is a strong conditioned inhibitor, which, when superimposed
on an excitatory stimulus, significantly suppresses responding. Thus the
absence of sloping inhibitory gradients following errorless discrimination
training does not imply an absence of conditioned inhibition. It may even,
as Deutsch (1967) has argued, imply that inhibitory learning is strong
enough to generalize completely to all test stimuli.

---

* An earlier study by Terrace (1966c), purporting to show that errorless discrimination
training results in flat inhibitory gradients, has been rightly criticized by Bernheim (1968)
and Hearst (1969a; Hearst et al., 1970): S+ was a vertical line and S— a key illuminated
with a light of 580 nm. No special training procedures were employed: errorless subjects
were those which happened never to respond to S— when it was first presented. But if
they did not respond to one arbitrarily chosen wavelength, there is no reason to expect
them to respond to any other. It is hardly surprising, therefore, that they showed a flat
inhibitory gradient when tested with other wavelengths.

VII. INTERACTIONS BETWEEN GRADIENTS

A. The Summation of Excitatory Gradients

Although most work on the interaction of generalization gradients has concentrated on the theoretically more interesting question of how excitatory and inhibitory gradients interact to determine discriminative performance, there have been some studies of the interaction of excitatory gradients. In these experiments responses are typically reinforced in the presence of two or more values of a particular stimulus dimension (e.g., two wavelengths) and the generalization gradient so formed is compared with that produced by prior reinforcement at single values on the dimension. For example, if pigeons are reinforced for responding both to a 540 nm and to a 560 nm key light, is it possible to predict the rate of responding to other wavelengths in a subsequent generalization test from some combination of the gradients produced by two further groups—one reinforced at 540 nm and the other at 560 nm? The particular point of interest has been whether it is possible to predict rate of responding to stimuli falling between two reinforced stimuli (e.g., 550 nm in the present example) from a combination of the two single-stimulus gradients.

The question raises the usual problem of an inadequate response scale. For example, if independent groups of pigeons reinforced either at 540 nm or at 560 nm both made 100 responses to a 550 nm stimulus in a subsequent test, we should hardly expect subjects reinforced at both 540 and 560 nm to respond 200 times to the 550 nm stimulus. Such a prediction would require not only the assumption that generalized response strengths combine additively, but also the assumption that rate of responding can be translated directly and linearly into response strength. Hull (1943, pp. 194–203) accepted that generalized habit strengths summated, but because he assumed that habit strength exponentially approaches an asymptote, two generalized habits, especially if they are both strong, cannot simply be added together to yield a total habit strength. In Hull's case, therefore, precise prediction of the amount of summation depends upon the value of this asymptote and of the rate of approach to asymptote.

Even if precise prediction is difficult or impossible, the weak assumption that excitatory gradients do in fact summate still generates some qualitative predictions about rate of responding to a stimulus falling between two reinforced stimuli. Pigeons reinforced at two relatively close wavelengths should respond more rapidly to intervening stimuli than subjects reinforced at only one of the two wavelengths. If the two wavelengths are sufficiently close together, rate of responding to an intervening stimulus

8

should be at least as high as that to either reinforced stimulus. As the reinforced stimuli are moved further apart, so the rate of responding to a stimulus half-way between the two should decline, until the point is reached when the two independently established individual gradients no longer overlap and no summation occurs. Some of these predictions have been tested by Kalish and Guttman (1957, 1959) and by Blough (1969a). Although some weak confirmation has been provided the evidence is not entirely favourable. On the one hand, summation may be less than predicted, for in no experiment has rate of responding to an intervening stimulus ever exceeded the rate maintained by the two reinforced stimuli. On the other hand, sometimes too much summation is observed: Blough found that when the two reinforced stimuli were relatively far apart, rate of responding to an intermediate stimulus was substantially greater than the combined rates of responding obtained from individual gradients. These are not encouraging data for the underlying assumption that generalization gradients simply measure generalized response strength. Blough has argued, rather persuasively, that they are more readily interpreted within the framework of signal detection theory.

B. The Interaction of Excitatory and Inhibitory Gradients

Somewhat greater success has attended efforts to predict some aspects of discriminative performance from the interaction of excitatory and inhibitory gradients. The theories of discrimination learning advanced by Spence (1936, 1937a) for simultaneous discriminations, and by Hull (1952) for successive discriminations, rested on a set of fundamental assumptions first proposed by Pavlov (1927). These assumptions are that discrimination training can be reduced to the scheduling of differential reinforcement in the presence of S+ and S—; that this differential reinforcement establishes S+ as an excitatory stimulus and S— as an inhibitory stimulus; that the excitatory and inhibitory processes established to S+ and S— generalize to other, similar, stimuli (including each other); and finally, that discriminative performance is a consequence of this interaction of excitatory and inhibitory gradients. Among the several consequences of this set of assumptions, it follows that the rate of discrimination learning will be inversely related to the degree of similarity between S+ and S—. This is, perhaps, hardly an astounding prediction, yet the effects of the similarity of S+ and S— must be explained by any comprehensive theory of discrimination learning. It is worth noting that this explanation is precisely the same as that provided by Perkins (1953) and Logan (1954) for the effects of CS intensity on the rate of conditioning (p. 42). The greater the intensity of a CS, the less the generalization of inhibition from nonreinforced background stimuli and therefore the faster the rate of learning. The analogy between the two effects may be stressed by using the same term to refer to both. Just as a more intense CS was said to be more salient, so a decrease in the similarity between S+ and S− in discrimination learning may be said to increase the salience of the relevant discriminative stimuli.

The general theory of discrimination learning proposed by Spence and Hull, often referred to as conditioning-extinction theory, has in fact derived several considerably more complex predictions from assumptions about the interactions of generalization gradients. Discussion of some of these predictions, such as those concerned with transposition and transfer along a continuum, is better postponed to the following chapter, which will take up the general question of whether a conditioning-extinction theory is sufficient to encompass the facts of discrimination learning. For the present we may be content to ask whether the empirical study of generalization gradients has usefully contributed to the understanding of simple discrimination learning. Although conditioning–extinction theories have frequently used theoretical gradients to derive particular predictions, it is only within the past 10 to 15 years that empirically obtained gradients have been substituted for imaginary gradients in these theoretical derivations. The earliest studies explicitly designed to see whether performance in a generalization test, following discriminative training between two points along some dimension, might be rationally reconstructed by a postulated set of interactions between appropriate excitatory and inhibitory gradients, were those of Hanson (1959) and Honig (1962). It will be simpler, however, to start with a logically prior, although historically later, study reported by Hearst (1968, 1969a).

Hearst trained independent groups of pigeons to discriminate between a vertical line as S+, signalling a VI schedule of reinforcement, and a line rotated either 90 degrees, 60 degrees, or 30 degrees from vertical as S−, signalling extinction. Subsequent generalization tests to lines varying in orientation from 90 degrees on one side of vertical to 90 degrees on the other side, provided a postdiscrimination gradient (or PDG). In order to obtain empirical excitatory and inhibitory gradients, Hearst trained other pigeons either with a vertical line as S+ and a plain white key as S−, or with the white key as S+ and either a 90-degree, 60-degree or 30-degree line as S−. Generalization tests given to these groups provided an excitatory gradient around the vertical line, and inhibitory gradients around the other three lines. The question was whether suitable combinations of these independently obtained excitatory and inhibitory gradients would yield PDGs similar to those generated by subjects trained to discriminate between different orientations of the line. Fig. 9.11 gives the answers.

In the top row of the figure are shown the absolute gradients of the subjects trained on the three line-orientation discriminations (labelled PDG) and the excitatory gradients obtained when three additional groups were trained with a vertical line as S+ and a white key as S—. It is all too obvious that the absolute PDG could never be reconstructed by subtracting any inhibitory gradient from the excitatory gradients shown here: at most points, particularly at S+ and particularly for the group trained

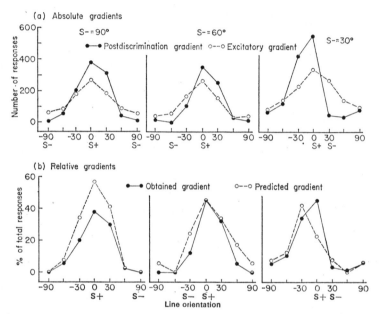

FIG. 9.11. Reconstructions of postdiscrimination gradients (PDGs) from the combination of excitatory and inhibitory gradients, in pigeons reinforced for responding to a vertical (0 degree) line, and nonreinforced in the presence of a 90, 60 or 30 degree line. (a) Absolute gradients; (b) Relative gradients. (*After* Hearst, 1968.)

on the 30 degree discrimination, the absolute level of responding in the PDG is *higher* than that of the excitatory gradient.

The problem is that free-operant discrimination training results in an increase in rate of responding to S+ and that this positive contrast effect is, within limits, an increasing function of the similarity of S+ and S— (p. 377). The generalization of inhibition from S— to S+, of course, should only decrease rate of responding to S+. Since contrast-like interactions also appear to occur during the course of the test for generalization itself (p. 498), it is not surprising that the absolute level of the PDG cannot be predicted by subtracting an inhibitory gradient from an independently established excitatory gradient. The second row of Fig. 9.11, therefore, shows Hearst's attempt to reconstruct the *relative* PDG from a combination of the relative excitatory and inhibitory gradients. Now, it can be seen, the attempt is somewhat more successful. In general, the derived PDG is of much the same general shape as the obtained gradient (with one notable exception in the gradients of the 30 degree discrimination). In the case of the two more difficult discriminations, however, there is some suggestion that the obtained gradient may be somewhat sharper than the predicted gradient. That such a discrepancy is not an isolated result is suggested by results reported by Jenkins and Harrison (1962): after training pigeons on a frequency discrimination between tones of 1000 Hz (S +) and 950 Hz (S −), they obtained extremely steep generalization gradients along the frequency dimension; independent excitatory and inhibitory gradients had been obtained from pigeons trained with a tone as S + and no tone as S − (or vice-versa), but Jenkins and Harrison concluded that the steep PDGs which they observed "cannot be reconstructed by any simple rational method of combining excitatory and inhibitory gradients" (p. 440). If these conclusions can be accepted, they reinforce a conclusion suggested earlier (p. 508): interdimensional discrimination training may be less effective than intradimensional training in establishing control exclusively by the dimension subsequently varied during testing.

Although the strategy of Hearst's study is extremely valuable, the problems created by the contrast effects inherent in free-operant discrimination procedures suggest, as Hearst himself pointed out, that the strategy would be more profitably pursued in studies of discrete-trial discrimination learning, or of discriminative classical conditioning. Relative generalization gradients are by no means easy to interpret and differences in the slopes of relative gradients are necessarily somewhat ambiguous.

C. The Peak Shift

One of the more interesting features of Hearst's derived PGDs can be seen in the gradient reconstructed for the 30-degree discrimination: the peak of this gradient does not fall at S + (the vertical line), but at the line 30 degrees on the other side of vertical from S −. Thus Hearst was able to predict a phenomenon, the peak shift, which did not in fact occur in his empirical PDG, but which was first observed by Hanson (1959) in a study of generalization following wavelength discrimination learning in pigeons.*

---

* As Hearst noted, one reason for his failure to obtain a peak shift may have been the nature of the stimulus dimension he employed. There is evidence that lines oriented at equal but opposite angles from the vertical are particularly prone to confusion (Corballis

Hanson trained various groups of pigeons with light of 550 nm as S+ and with S— for different groups ranging from 555 nm to 590 nm. Fig. 9.12 shows the results for three of these groups: as S— comes closer to S+, so there is a progressive distortion in the shape of the PDG, with a higher proportion of responses occurring to stimuli on the other side of S+, and the peak of the gradient falling at stimulus values further removed from S+. Both the occurrence of this peak shift and its dependence on the separation of S+ and S— are predicted by the derived gradients reconstructed by Hearst. An inhibitory gradient formed round S—, interacting with an excitatory gradient around S+, will produce an asymmetrical

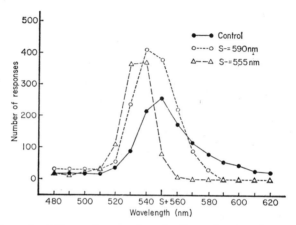

FIG. 9.12. A peak shift on the wavelength dimension in pigeons. In control subjects, trained with S+ only, the gradient is peaked at S+ (550 nm). Subjects trained on a discrimination between 550 nm as S+, and either 590 nm or 555 nm as S—, produce gradients whose peaks are displaced to 540 or 530 nm. (*After* Hanson, 1959.)

PDG and may result in a displacement of the peak of the PDG away from S+ (see Fig. 9.13).

Hanson's data have been replicated frequently: although not all subsequent studies have reported that discrimination training between two stimuli falling along a single dimension necessarily produces a shift in the peak of a PDG away from S+, the shift in the overall distribution of the gradient is entirely reliable (e.g., Honig *et al.*, 1959; Terrace, 1964; Friedman and Guttman, 1965; Bloomfield, 1967c; Ellis, 1970). One example of a peak shift which has not always been interpreted in this way is provided by the typically asymmetrical gradient of intensity generaliza-

and Beale, 1970). This mirror-image effect may have resulted in some suppression of responding to a line 30 degrees to the left of vertical, when a line 30 degrees to the right of vertical served as S—.

tion shown in Fig. 9.1. Following classical conditioning or discrete-trial instrumental training with a light or tone as S+, subjects may respond more frequently to stimuli more intense than S+, than to S+ itself. It is unnecessary to attribute this to "stimulus intensity dynamism", since the experimental procedures which result in such an asymmetrical gradient are those which involve nonreinforcement in the absence of S+, i.e., at a particular point on the intensity dimension. The Perkins–Logan account of stimulus intensity effects (p. 42), combined with an appropriate interaction between the excitatory gradient formed round S+ and the inhibitory gradient formed round the point of zero intensity, is sufficient to account for the typical shape of intensity gradients. The most obvious prediction derivable from this analysis is that a peak shift to intensities lower than S+ should be demonstrable following discrimination training with an S— more intense than S+. Thomas and Setzer (1972) have provided confirmation of this prediction in a study of intensity generalization in rats and guinea-pigs.

Just as U-shaped gradients with minima at S— may be produced by operations other than nonreinforcement at S—, and indeed by operations not properly described as inhibitory at all, so peak shifts have been observed not only when S— signals nonreinforcement, but also when it is correlated with a longer VI schedule than S+ (Guttman, 1959), noncontingent shock (Grusec, 1968), punishment (Terrace, 1968) and, rather less reliably, when S— signals a DRL schedule (Terrace, 1968).

This parallel between the operations which produce U-shaped gradients around S— following interdimensional discrimination training, and those which produce a peak shift following intradimensional discrimination training, reinforces the argument that the peak shift is a consequence of the interaction of excitatory and U-shaped gradients. One further parallel has been noted by Terrace (1964), but used by him to support a somewhat different argument from that advanced here. Terrace trained pigeons on a wavelength discrimination between 580 nm as S+ and 540 nm as S—, but employed his fading procedure to prevent the occurrence of errors to S—. The resulting PDG gave no evidence of a peak shift. This result (later confirmed by Grusec, 1968) showed, according to Terrace, that an errorless S— is not an inhibitory stimulus. We have already seen reason to question this conclusion and the absence of a peak shift does not seem sufficient to alleviate these doubts. For, as Hearst (1969a) has argued, the occurrence of a peak shift is not an automatic consequence of the interaction of an excitatory and inhibitory gradient: it depends upon the shape and slope of those gradients. Fig. 9.13 presents several imaginary excitatory and inhibitory gradients, along with the PDGs constructed by extracting the differences between the two gradients at each value along the dimension in question. It can be seen that the peak shift disappears both when the inhibitory gradient is very steep and when it is very flat. A peak shift is not a necessary consequence of an interaction between excitatory and inhibitory gradients. Since we have already had reason to argue that inhibitory gradients formed round an errorless S— are usually extremely flat, the absence of a peak shift in Terrace's experiment is easily explained. As it happens, Hoffman (1969), in his studies of the generalization of conditioned suppression in pigeons, was able to obtain a peak shift only when he used a fading procedure to introduce S— gradually during the course of initial discrimination training.

Terrace (1966b) has also shown that the peak shift may disappear if

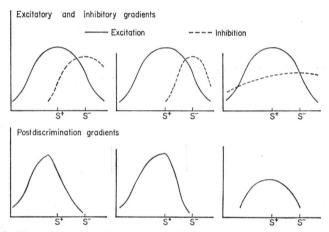

FIG. 9.13. Hypothetical gradients of excitation and inhibition, and the post-discrimination gradients resulting from their interaction. Gradients of appropriate shape and slope may produce a peak shift in the postdiscrimination gradient, but if the inhibitory gradient is too steep or too flat, no peak shift is predicted.

extensive overtraining is given on the original discrimination. Since both errorless discrimination training, and extensive overtraining were also apparently effective in abolishing positive contrast, Terrace concluded that the peak shift, like contrast, is a consequence of an emotional inhibitory process generated by the occurrence of errors to S —. However, Terrace's conclusions about the effects of errorless training and of overtraining on contrast are somewhat dubious (p. 378), and the effect of overtraining on the peak shift is also somewhat problematic. Ellis (1970), for example, reported no disappearance of peak shifts in pigeons trained on an intensity discrimination for 26 sessions. Furthermore, if the peak shift depends upon the precise slope of the inhibitory gradient round S—, it might be possible to explain its disappearance as a consequence of changes in the slope of inhibitory gradients during training. There is, indeed, evidence that inhibitory gradients may change with continued training, although the direction of this change is less certain. Farthing and Hearst (1968) reported that inhibitory gradients became steeper with continued training; Yarczower (1970) has suggested they may become flatter. Either way, of course (as Fig. 9.13 shows), it would be possible to explain the disappearance of the peak shift, but one may hope that a more consistent picture will emerge with further research.

The close connection between contrast and the peak shift postulated by Terrace, although supported by the results of some of his other experiments (e.g., Terrace, 1968), does not stand up to detailed scrutiny. Studies of classical aversive conditioning have reported evidence of a peak shift, but no evidence of positive contrast (A. Siegel, 1967; Hendry et al., 1969; Hoffman, 1969; Moore, 1972). At least one free-operant study of food-reinforced discrimination learning in pigeons has obtained positive contrast under conditions which yielded no peak shift (Ellis, 1970), while Yarczower and Curto (1972) found that S− remained an effective conditioned inhibitor long after any evidence of contrast had disappeared. Moreover, if the peak shift continues to increase in magnitude when S− and S+ are brought closer together, as Hanson's data suggest, there should come a time when the separation between S+ and S− is so small that contrast decreases and disappears (p. 378).

The more reasonable conclusion suggested by the data reviewed here is that the peak shift is a consequence of the interaction of two opposed response tendencies generalized from S+ and S−. If animals have learned to respond in the presence of S+ and not to respond, or to respond at a lower rate, in the presence of S− (whether this latter tendency is produced by an inhibitory process or not), it is possible that these two tendencies will generalize in such a way as to cause maximum responding to stimuli removed from S+ in a direction away from S−. The prediction of the peak shift and the conditions under which it occurs does not particularly require assumptions about emotionality or aversiveness. Even shorn of these additional complications, the prediction of the peak shift is a striking confirmation of the value of the conditioning–extinction approach to some of the phenomena of discrimination learning.

VIII. SUMMARY

After subjects have been reinforced in the presence of one set of stimuli, they may respond to other stimuli, more or less similar to those present during training. The occurrence of such responses defines the empirical phenomenon of generalization. The probability of responding to other stimuli may decline as the test stimuli are made progressively less similar to the training stimuli. A regular decline in response probability of this sort defines a gradient of generalization.

These definitions may suffice for some purposes. But as soon as it is desired to compare differences in the slopes of different gradients, or to assess the effects of particular treatments on the slope of generalization gradients, it becomes important to remember how these gradients are obtained. Typically, subjects receive reinforcement in the presence of one stimulus and are then tested in extinction with a repeated series of test stimuli. The gradient of generalization so produced, however, does not provide any very direct measure of the effects of changes in stimuli. It is complicated by the effects of the subject's history of reinforcement, the rate of responding produced by that history and the rate at which responding extinguishes during the test series. Differences in the slopes of gradients produced by different conditions of acquisition, are often difficult or impossible to interpret because of large differences in the rate of responding or resistance to extinction produced by those treatments.

One of the oldest issues in the study of generalization is whether the occurrence of generalization represents a "spread of effect" from training to test stimuli, or whether it is due to a failure to discriminate between the two sets of stimuli. The important aspects of both of these positions may be satisfactorily captured by the suggestion that generalization is based on common elements. Subjects respond in a test situation to the extent that it contains stimulus elements identical to those associated with reinforcement during training; they fail to respond in a test situation to the extent that it contains no elements that were originally associated with reinforcement.

The most important implication of this analysis is that the slope of a generalization gradient along a particular dimension may reflect not only the subject's sensitivity to changes along that dimension, but also the extent to which features of the training situation other than those varied during the test were originally associated with reinforcement. Pigeons reinforced for pecking at a circular key containing a white vertical line on a green background, may respond almost equally to other orientations of the line, not because they are incapable of discriminating between these orientations but because the main features associated with reinforcement were not the orientation of the line but the colour and shape of the key.

In a situation such as this, the slope of a gradient of generalization may be markedly increased by the provision of discrimination training, even if such training does not involve differential reinforcement between different orientations of the line, but only between the presence and absence of the line. It is probable that the most important reason for this is that discrimination training ensures that several constant features of the experimental situation become irrelevant. If a pigeon receives nondifferential reinforcement for pecking at a vertical line on a green key, then the colour and shape of the key, as well as innumerable other aspects of the experimental situation, are as well correlated with the availability of reinforcement as is the orientation of the line. If discrimination training is programmed with reinforced trials to the white line on the green key interspersed with nonreinforced trials on which all else is held constant but the line is omitted, then all aspects of the situation except the line become irrelevant. Direct evidence, from a number of different experiments, has shown that discrimination training can decrease the control over responding acquired by such incidental stimuli. Although there are some exceptions to this rule, which have been interpreted to mean that discrimination training may establish some more general set to attend to any change in stimulation, it can hardly be doubted that the slope of a generalization gradient is determined by the extent to which incidental stimuli have acquired control over responding, and that such control may be reduced by discrimination training.

Variables such as the amount of training and schedule of reinforcement, as well as the subject's motivational state, have been shown to affect generalization. Unfortunately, several of these variables have such marked effects on rate of responding or resistance to extinction, that it is difficult to assess the significance of apparent differences in the slopes of gradients. Some of these differences, however, may be attributed to variations in the extent to which incidental stimuli have become associated with reinforcement.

If the effects of reinforcement in the presence of one stimulus may generalize to others, so may the effects of nonreinforcement. Inhibitory gradients have been studied with a variety of preparations. In a typical situation, S−, by virtue of signalling the omission of an appetitive reinforcer, may control a low rate of responding; when tested with other stimuli differing from S−, but not necessarily more similar to S+, subjects may show an increase in rate of responding.

The omission of an appetitive reinforcer is not, of course, the only procedure for reducing responding in the presence of a particular stimulus. Responding maintained by appetitive reinforcement may be suppressed in the presence of a stimulus signalling shock. Although a U-shaped gradient of generalization might be observed round such a stimulus, it seems worth reserving the term inhibitory gradient to refer to the gradient formed round an inhibitory stimulus, i.e., one negatively correlated with reinforcement.

When pigeons are reinforced for responding to one wavelength and not reinforced for responding to another, the resulting generalization gradient may be peaked not at S+, but at a wavelength on the opposite side of S+ from S—. It is possible to derive this peak shift from an interaction between a hypothetical excitatory gradient centred round S+ and an inhibitory gradient centred round S—. Variations in the magnitude of the peak shift may be due to variations in the slopes of these underlying inferred gradients. It is important to note that a peak shift can be predicted provided only that a U-shaped gradient is formed round S—. It is not necessary that this be an inhibitory gradient in the strict sense and there is evidence that any procedure which reduces the probability of responding at S— may be sufficient to produce the peak shift.

CHAPTER 10

Discrimination Learning

I. INTRODUCTION

Studies of discrimination learning involve exposing subjects to different stimuli and arranging different schedules of reinforcement for responses to each stimulus. The discriminative stimuli may be presented either simultaneously on a single trial, or successively on consecutive trials. The simplest case of a simultaneous discrimination problem is the T-maze, where the subject is reinforced for entering the right-hand goal arm and not reinforced for choice of the left-hand goal arm. A simultaneous visual discrimination can also be arranged in a T-maze by providing black and white goal arms or doors to each arm and reinforcing only choices of the black arm, irrespective of the side it appears on. Since the position of the black and white stimuli is varied from trial to trial in some quasirandom manner, position may be said to be irrelevant. Analogous simultaneous discriminations can be arranged in free-operant experiments as concurrent schedules of reinforcement.

In successive discriminations, of the kind most frequently mentioned in earlier chapters, presentations of S+ and S− are randomly alternated and only responses to S+ are reinforced. Such discriminations may be programmed using either discrete-trial or free-operant procedures (when they are called multiple schedules). In the present context both will be referred to as successive go/no-go discriminations, in order to distinguish them from a second class of successive discrimination, which will be called sucessive conditional discriminations. In a typical successive conditional discrimination, a rat is exposed to a T-maze whose arms on successive trials are either both black or both white. When both arms are black, a response to the left is reinforced; when both are white, a response to the right is reinforced. The difference between the two types of successive discrimination is that in go/no-go problems only one response is ostensibly available to the subject, and that response is either reinforced or not reinforced, depending on the stimulus shown on that trial; in conditional problems, on the other hand, two responses are available to the subject on each trial, one of which is reinforced in the presence of one stimulus and the other in the presence of another stimulus.

Some of the issues raised by studies of discrimination learning (such as contrast effects) have already been considered in earlier chapters, and will not be further discussed here. Others, although mentioned quite extensively at earlier points, are particularly important for an understanding of the processes involved in discrimination learning and demand special treatment in this chapter. These issues include: the nature of the responses learned in simultaneous and successive discriminations; the nature of the stimuli controlling discriminative responding in different situations; the need to postulate some process of stimulus selection to account for phenomena of overshadowing and blocking; and evidence derived from studies of transfer. This suggests that a complete understanding of discrimination learning may have to appeal to more processes than are suggested by the conditioning–extinction models associated with Spence and Hull.

II. THE NATURE OF THE RESPONSE IN DISCRIMINATION LEARNING

A. Response Selection or Stimulus Approach

Traditional S–R theories have assumed that the reinforcement of a particular response in the presence of a particular stimulus situation strengthens the learned connection between that response and that situation. But how is the response learned in a typical discrimination problem to be defined? Should we speak of an animal solving a simultaneous black–white discrimination as learning to approach the black stimulus, or is it more appropriate to say that the response learned is that of turning right or left at the choice point depending on the set of stimuli present on that trial?

Table 10.1 presents a schematic layout of the stimulus configurations of simultaneous and successive conditional brightness discriminations, together with two possible descriptions of the solution of these problems. According to a "response–selection" view, subjects learn to turn left or right at the choice point (or press the left or right lever, or peck the left or right key) depending on the configuration of stimuli presented on that trial. The simultaneous discrimination is solved by going left when black is on the left and white on the right, and by going right when white is on the left and black on the right; the successive discrimination is solved by going left when both stimuli are black, and right when both stimuli are white. According to the "stimulus–approach" view, the simultaneous problem is solved by approaching black and avoiding white, while the successive discrimination is solved by approaching black on the left and white on the right, and avoiding black on the right and white on the left.

Response–selection theories of discrimination learning have been espoused (for reasons discussed by Lovejoy, 1968) by a number of mathematical learning theorists (e.g., Gulliksen and Wolfle, 1938; Bush, 1965); while the stimulus–approach view has been most ably propounded by Spence (1936; 1952). A large number of experiments (reviewed by Sutherland, 1961a; Lovejoy, 1968; and Sutherland and Mackintosh, 1971) have been designed with the intention of deciding whether animals solve simultaneous and successive discriminations by approaching and avoiding

TABLE 10.1

Descriptions of Alternative Bases for the Solution of Simultaneous and Successive Conditional Discriminations

	Stimulus Configurations		Description	
	Left	Right	Response Selection	Stimulus Approach
Simultaneous Discrimination	Black (+) White (−)	White (−) Black (+)	Turn left to B–W Turn right to W–B	Approach B, avoid W.
Successive Conditional Discrimination	Black (+) White (−)	Black (−) White (+)	Turn left to B–B Turn right to W–W	Approach B–L, avoid B–R. Approach W–R, avoid W–L.

particular discriminative stimuli, or by selecting particular responses in the presence of different stimulus configurations. As so often happens, the outcome of these experiments has not been completely unequivocal. To anticipate the conclusions suggested below: simultaneous discriminations are typically solved by subjects learning to approach particular stimuli; although successive problems may be solved by subjects learning to approach appropriate compounds, there is evidence which suggests that subjects may also solve successive problems by learning to select particular responses. A critical factor determining the basis for solution is whether the experimental situation requires the subject to make contact with the discriminative stimuli: such contact increases the probability of the solution being based on the approach and avoidance of particular stimuli, or stimulus compounds.

B. Relative Difficulty of Simultaneous and Successive Conditional Discriminations

If animals solve discrimination problems by response selection, i.e., by learning to go one way in the presence of one stimulus configuration and the other way in the presence of a different configuration, it follows that the rate of discrimination learning will depend, all else being equal, on the discriminability of the two configurations. In a successive conditional problem, the two configurations (shown in Table 10.1) are B–B and W–W. In a simultaneous discrimination, they are B–W and W–B. Essentially any assumption about the determinants of generalization and discriminability, and certainly any assumption that stresses the role of common elements, will predict that it is easier to discriminate between the two configurations of the successive discrimination than between the two configurations of the simultaneous problem. If both types of problem are learned by response selection, it follows that the successive problem should be learned more rapidly.

In the very large majority of studies employing standard T-mazes, discrimination boxes, or jumping stands, where the discriminative stimuli (e.g., black and white) are painted on the goal boxes, goal arms or doors, this prediction has been strikingly disconfirmed. Spence (1952), Bitterman and McConnell (1954), MacCaslin (1954), Wodinsky et al. (1954), Bitterman et al. (1955), and Gonzalez and Shepp (1961), for example, have all found that a simultaneous discrimination is typically learned very much faster than a successive conditional discrimination employing the same stimuli. The ease of the simultaneous discrimination implies that it is solved by the subject's learning to approach S+ and/or avoid S−. If this is the natural tendency brought by the subject to the experimental situation, the successive problem will be relatively difficult, for its solution will depend on the subject either abandoning a stimulus-approach solution in favour of response selection, or learning to approach and avoid particular *compound* stimuli (black on the right, white on the left).

The relative difficulty of simultaneous and successive problems can be altered by variations in the discriminability of the stimuli involved. Teas and Bitterman (1952), for example, trained rats on a discrimination problem involving two pairs of stimuli, a dark and a light grey, and wide and narrow vertical striations. The design of their experiment is shown in Fig. 10.1. The two grey doors were always shown in the same configuration, e.g., with the darker stimulus on the left, and the two sets of striations were also always shown in the same configuration, e.g., with the narrower striations always on the right. Responses to the dark grey and narrow striations were reinforced. These two sets of stimulus–reinforcement contingencies can be regarded as defining either two simultaneous discriminations, with dark grey and narrow striations positive, or as a single successive conditional discrimination, with the left-hand stimulus positive when two greys are shown and the right-hand stimulus positive when two sets of striations are shown. On the basis of the results of the transfer tests shown in Fig. 10.1, Teas and Bitterman concluded that their rats had learned the single successive discrimination rather than the two simultaneous discriminations: subjects showed a significant tendency to select the light grey stimulus on the left and the wide striations on the right. A subsequent study by Zeiler and Paul (1965) confirms, what one might have expected, that the outcome of such transfer tests depends upon the

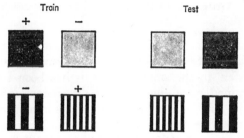

FIG. 10.1. Schematic diagram of experiment by Teas and Bitterman. Subjects are trained concurrently on two discriminations, with the reinforced darker grey stimulus always on the left, and the reinforced narrower striations always on the right. Tests are given with the same pairs of stimuli but with their positions reversed. Performance on test trials reveals whether subjects have learned to go left to two grey stimuli, and right to two sets of striations, or have learned to approach the darker of two greys, and the narrower of two sets of striations. (*After* Teas and Bitterman, 1952.)

relationship between the discriminability of the two stimuli within each pair (e.g., the dark and light greys) and the discriminability of the two pairs (the difference between the grey stimuli and the striations). The successive, configurational solution occurs only when the within-pair difference is smaller than the between-pair difference. If the stimuli within each pair are easy to discriminate, animals solve the problem as a pair of concurrent simultaneous discriminations, learning to approach one particular brightness and one particular set of striations.

A second, and more interesting, exception to the general rule that simultaneous problems are easier than successive is provided by studies which do not require the subject to make contact with the discriminative stimuli. Weise and Bitterman (1951) found that if rats were trained in a maze with small light bulbs at the entrance to each arm as the discriminative stimuli, a successive discrimination, which required the subjects to turn right when both lights were on, and left when both lights were off, was very much easier than a simultaneous discrimination. Wodinsky *et al.* (1954) and Bitterman *et al.* (1955) trained rats in a four-window jumping stand, with the subjects being required to respond to the two outer windows. If the discriminative stimuli were shown in these outer windows, a simultaneous discrimination was much easier than a successive discrimination; but if the discriminative stimuli were shown in the inner two windows, while the subjects were still required to respond to the outer windows, the relative difficulty of the two problems was completely reversed.

C. Transfer between Simultaneous and Successive Go/No-Go Discriminations

Further evidence that simultaneous discriminations are usually solved by the subject's learning to approach S+ and/or avoid S−, is provided by studies of transfer. In the simplest case, it has been shown frequently that animals trained on a simultaneous discrimination respond appropriately when S+ or S− is presented alone (as in a successive go/no-go discrimination): they respond rapidly to S+, but less readily to S− (e.g., Webb, 1950; Sutherland, 1961b, 1969; Macphail, 1972).

A number of studies have demonstrated good transfer from a simultaneous discrimination to a problem where the same discriminative stimuli are shown in a new configuration (e.g., Nissen, 1950; Bitterman *et al.*, 1955): Nissen, for example, trained chimpanzees with the discriminative stimuli side by side and found that they continued to respond accurately when one stimulus was placed above the other. Several further studies have shown good transfer when a new stimulus is substituted for either S+ or S− (Mandler, 1968; Stevens and Fechter, 1968; Derdzinski and Warren, 1969). These results, taken together with those showing that simultaneous problems are normally much easier than successive conditional problems, leave no room for doubt that solution of the standard simultaneous discrimination is achieved by the development of approach and avoidance tendencies to the discriminative stimuli.

D. The Solution of Successive Conditional Discriminations

If simultaneous discriminations are typically solved by subjects learning to approach and avoid particular stimuli, then it is reasonable to expect that successive problems are also solved by the establishment of appropriate approach and avoidance responses. In the successive case, however, subjects must approach and avoid particular compound stimuli.

Provided that subjects are required to direct their responses to the discriminative stimuli, there is good evidence that successive problems are solved in this way. Wodinsky *et al.* (1954) and Bitterman *et al.* (1955) trained rats on successive discriminations with the two different configurations shown in Fig. 10.2. Not only were the two configurations learned at the same rate, but transfer from one to the other was essentially perfect. Such transfer could not have occurred if animals had learned to go right in the presence of one pair of displays and left in the presence of another pair. These results, therefore, imply that animals learned to approach

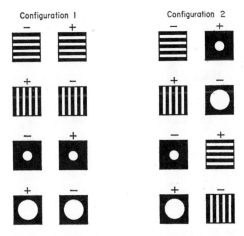

FIG. 10.2. Schematic diagram of experiments by Wodinsky *et al.*, and Bitterman *et al.* Subjects are trained on successive conditional discriminations with horizontal and vertical striations and large and small circles. In the first configuration the stimuli are presented as pairs of identical elements. In the second configuration the stimuli are recombined into different pairs, although the same stimulus is still reinforced in the same position. (*After* Wodinsky *et al.*, 1954; and Bitterman *et al.*, 1955.)

particular compounds, such as a small circle on the right or a large circle on the left.

There is evidence that visual conditional discriminations may also be solved, when the subject is required to respond directly to the discriminative stimuli, by the establishment of approach tendencies to particular compounds. A typical visual conditional discrimination is schematized in Fig. 10.3. The subject is confronted with either two black rectangles, one horizontal and the other vertical, or two white rectangles, again with one horizontal and the other vertical. When both rectangles are black, choice of horizontal is reinforced: when both are white, choice of vertical is reinforced. Such a problem could be solved by the subject's learning to approach two compounds, the black horizontal and the white vertical rectangles. Alternatively, it could be solved by the development of the conditional rule: if black choose horizontal, if white choose vertical.

North *et al.* (1958) and North and Lang (1961) showed that under the conditions described here, rats learn the compound-approach solution: when the stimuli were presented in new configurations, appropriate choice of compounds was not disturbed. For example, after solution of a conditional problem similar to that shown in Fig. 10.3, subjects were tested with a black horizontal and a white horizontal rectangle: although this configuration had not been seen before, and although the conditional rule provides no basis for choice between these two compounds, animals

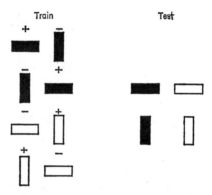

FIG. 10.3. Schematic diagram of a visual conditional discrimination.

consistently chose the previously reinforced compound and avoided the previously nonreinforced compound.

In all of these studies, animals have been required to respond directly to the discriminative stimuli. It is under these conditions, as was noted earlier, that simultaneous discriminations are easier than successive; this is consistent with the suggestion that the former problem requires the establishment of responses to component stimuli, while the latter demands the establishment of responses to compound stimuli. Where, however, the discriminative stimuli are in one location and subjects are required to respond to another, the successive discrimination may be easier than the simultaneous, and it is possible that this is because the spatial separation of stimulus and response increases the probability of a strategy of response selection. That the successive discrimination, at least, is solved in this way has been suggested by Bitterman *et al.* (1955). They trained rats in a four-window jumping stand, with the discriminative stimuli in the inner two windows, but with subjects being required to respond to the outer two windows. Under these conditions, they found relatively poor performance when rats were transferred from one of the configurations shown in Fig. 10.2 to the other. This implies that solution had been partly based on the establishment of particular responses to particular configurations of stimuli.

There is, however, reason to believe that even when rats are required to respond directly to the discriminative stimuli, the contingencies of a successive discrimination may to some extent encourage the development of a strategy of response selection. Bitterman and McConnell (1954), for example, trained rats on two consecutive discriminations. Since subjects were required to respond to the discriminative stimuli, the first problem was learned faster if it was a simultaneous discrimination rather than a successive discrimination; however, animals learning their second successive discrimination showed so much positive transfer from one problem to the next, that the difference between simultaneous and successive problems was eliminated. In a later study, Gonzalez and Shepp (1961) trained rats on a series of reversals of either a simultaneous or a successive brightness discrimination. In the case of simultaneous discrimination reversals, the subject is initially reinforced, for example, for choice of black; after criterion is reached, the reinforcement contingencies are reversed and choice of white is reinforced. In the successive problems, the initial contingencies would reinforce a response to the right when both stimuli were black and a response to the left when both stimuli were white; reversal learning would involve reinforcement of a response to the left when both stimuli were black and a response to the right when both stimuli were white. The initial problem was learned more rapidly by animals required to learn the simultaneous discrimination, but the successive animals showed more rapid reversal learning, and after a few reversals, performed very much more efficiently than the simultaneous animals.

It is rather difficult to see how these results could be explained if successive discriminations were always solved by animals' learning to approach compound stimuli, for this should always be more difficult than learning to approach a single component. But they seem consistent with the view that rats can learn to make one response in the presence of one configuration and the opposite response in the presence of another, and that once this strategy has been learned it is readily applied to new problems. A series of reversals of a successive conditional discrimination would, on this view, reduce to a series of *position* reversals: the animal would have to learn to go to the left where it had previously gone to the right. As we shall see below, spatial reversals tend to be substantially easier for the rat than visual reversals.

E. Conclusions

In situations which require contact with the discriminative stimuli, animals typically learn both simultaneous and successive conditional discriminations by learning to approach and/or avoid particular stimuli or stimulus compounds. Simultaneous discriminations are solved by learning to approach or avoid particular discriminative stimuli, while successive discriminations are solved by approaching or avoiding particular compounds. This explains why successive discriminations are usually more difficult than simultaneous discriminations.

Spence (1936) characterized this position by saying that the solution of a simultaneous discrimination does not involve "the strengthening of one response relatively to another . . . but involves, rather, the relative strengthening of the excitatory tendency of a certain component of the stimulus complex as compared with that of certain other elements" (p. 430). The implication is that, by virtue of their association with reinforcement and nonreinforcement, particular stimuli acquire attractive or aversive properties which lead to approach or avoidance. The solution of such visual discriminations, therefore, can be regarded as the outcome of a process of classical conditioning of positive and negative incentive value to S+ and S− (cf. Chapter 5).

A study by Bauer and Lawrence (1953) provides a nice illustration of this. They trained rats in a discrimination box on a brightness discrimination between black and white doors. After pushing open the door, the subjects of one group entered a goal box whose colour was the opposite of that of the door. Thus, for example, subjects were reinforced for choosing the black door rather than the white, but received reinforcement in a white goal box. The initial effect of this arrangement was a pronounced tendency for subjects to choose the white door, i.e., score significantly below chance on the discrimination problem. The association of a white goal box with reinforcement increased the attractiveness of the white door.

The importance of contact with the discriminative stimuli is readily understood in these terms. If reinforcement and nonreinforcement unconditionally elicit approach and withdrawal responses, then stimuli associated with reinforcement and nonreinforcement will become CSs for these patterns of behaviour. When animals are required to respond directly to the discriminative stimuli, then, no further learning process need be invoked. If, however, animals are required to respond to some other location, then the discriminative stimuli could serve only as conditional signals for directing approach responses to these other locations.

This brings us to the final question of the proper interpretation of what has hitherto been described as a response–selection account of discrimination learning. What is meant by saying that an animal may learn to make one response in the presence of one stimulus configuration, and another response in the presence of another? If animals normally learn to approach particular stimuli associated with reinforcement, perhaps response–selection should be interpreted as learning to approach a particular location. The question of whether spatial discrimination problems are solved by a subject's learning to approach a particular location or place, or by learning to make a particular response, was one that excited much controversy 25 years ago. The issue was described as one of place *vs* response learning, and led to a long series of experiments designed to show whether rats were "place learners" or "response learners" (e.g., Tolman *et al.*, 1946). It turns out that the question does not admit of any single unequivocal answer: whether a group of rats will solve a spatial discrimination by learning to approach a particular location, or by making a particular turn at the choice point, critically depends upon the nature of the available stimuli (Restle, 1957). If numerous intra- and extra-maze cues are available for discriminating between the two arms of a T-maze, animals will typically learn to approach the set of stimuli associated with the positive arm; in the absence of such distinguishing stimuli (i.e., in a totally homogeneous environment), they will learn to make a particular turn at the choice point. The distinction between these two possibilities may be largely a question of whether explicit instrumental contingencies, or implicit classical contingencies are affecting the subject's behaviour. Since we have earlier interpreted instrumental learning as the formation of associations between responses and reinforcers, the distinction reduces to a question of whether the animal learns that reinforcement is contingent upon a particular pattern of responding or on a particular pattern of stimulation.

III. THE NATURE OF THE EFFECTIVE STIMULI IN SPATIAL DISCRIMINATIONS

A. The Sensory Control of the Maze Habit: Place *vs* Response

The sensory control of maze learning was a topic extensively studied by early investigators of animal learning, forming one of the central questions motivating the experiments of Small (1901) and Watson (1907) on maze learning (see Munn, 1950, for a review). Watson argued, on the basis of distinctly inadequate evidence, that a multiple-unit maze was learned kinaesthetically, since he found that animals deprived of different exteroceptive sensory modalities were still able to learn the correct path through the maze. Later work by Tsang (1934, 1936) and Honzik (1936) suggested that the integrity of visual, auditory and olfactory sensory systems may be important for the solution of multiple-unit mazes. This was particularly true when elevated rather than enclosed mazes were used, for in elevated mazes a very large number of extra-maze cues may be available to the subject. In general, as Hunter (1929) and Restle (1957) have pointed out, rats are able to use a variety of cues to solve a typical maze problem, including intra- and extra-maze visual cues, auditory and olfactory cues (Honzik, 1936), and a vestibular directional sense (Douglas, 1966). Among the auditory cues used may be the possibility of echo-location (Riley and Rosenzweig, 1957). The importance of any one modality depends upon the nature of the available discriminative stimuli: as was noted above, when care is taken to train subjects in a totally homogeneous external environment and intra-maze cues are made irrelevant by randomly changing around the two goal arms, solution is necessarily based on the learning of an appropriate turn at the choice point. This may be more difficult than learning to approach a particular goal arm associated with a variety of extra-maze stimuli, but is far from impossible (e.g., Hill and Thune, 1952; Scharlock, 1955).

B. Conditions Leading to Proprioceptive Control

Two sets of circumstances seem to decrease the rat's reliance on exteroceptive stimuli in a maze, and to increase the tendency to learn a particular pattern of responses. Vincent (1915) and Carr (1917), among other early investigators of maze learning, suggested that if sufficient training were given on a maze problem, control was gradually transferred from exteroceptive to proprioceptive stimuli: "after the problem is thoroughly mastered the act is to be regarded as a kinaesthetic-motor coordination with an occasional reliance upon contact in time of emergency" (Carr, 1917, p. 259). The suggestion was reiterated by Restle (1957) who pointed out that to the extent that the subject's pattern of locomotion through the maze becomes more stereotyped with increased training, so proprioceptive stimulation varies less from trial to trial and a single pattern of such stimulation may become reliably associated with reinforcement. Although the original evidence for this hypothesis was distinctly limited (Munn, 1950, p. 198; Sutherland and Mackintosh, 1971, p. 285), two recent studies have provided considerable support. Hicks (1964) and Mackintosh (1965b) trained rats in simple, enclosed T-mazes, and gave test trials with the start arm on the opposite side of the maze. If subjects continue on test trials to enter the same goal arm as during training, they must turn in the opposite direction; conversely, if they make the same turn as during training, they will enter the opposite goal arm. Thus a choice based on approach to particular intra- or extra-maze stimuli is opposed to a choice based on a particular turn at the choice point. Both Hicks and Mackintosh found that rats given a small number of training trials made a "place" choice on test trials, but that extensive overtraining increased the probability of a "response" choice. The importance of this effect of overtraining on the behaviour of rats in alleys and mazes has already been discussed (p. 426).

A second situation which may lead to proprioceptive control of maze behaviour is the use of a multiple-unit maze rather than a single-unit T-maze, especially when some regular pattern of turns is reinforced. Under these conditions there is evidence that the stimulus determining the direction of turn in one unit is the choice made in the preceding unit. Hunter (1940) studied regular alternation learning (LRLR) in a multiple-unit elevated maze and found that learning was possible even when intra- and extra-maze cues were completely controlled by using blinded rats, changing segments of the maze and rotating the maze in the room from trial to trial.

Subsequent experiments by Dabrowska (1962; 1963a, b; 1967) have supported Hunter's conclusion that such a multiple-unit maze is solved as a proprioceptively controlled chain. Dabrowska trained rats in three- or four-unit mazes and, after they had learned, transferred them to new patterns. In one set of experiments she found that the complete reversal of the original pattern (e.g., from LRRL to RLLR) was learned more rapidly than a partial reversal (e.g., from LRRL to RLRL). In the latter case errors did not occur only at those choice points where the correct turn had been reversed, but spread to other units (Dabrowska, 1963a, b). The complete reversal of the four-unit maze, indeed, was learned with fewer than half the errors required to learn the original problem; while in a single-unit maze reversal was twice as hard as original learning (Dabrowska, 1962).

These results suggest that subjects learn the multiple-unit maze by coding turns at each point as the same as, or opposite to, the turn at the preceding choice point. Reversal learning then requires no more than reversal at the first choice point, while partial reversal requires the establishment of a new pattern. In a final experiment Dabrowska (1967) used rats with frontal lesions and observed an entirely different pattern of results. Complete reversal was difficult, but partial reversal was easy; in the latter case errors occurred only at the one unit where reversal was required. These are the results that would be expected if subjects solved the problem by learning to make a particular choice at each of the four choice points.

C. Conclusions

Perhaps the most important conclusion to be derived from maze studies is that suggested by Hunter (1929) and Restle (1957): rats will solve a discrimination problem by learning to associate any available stimuli with reinforcement. Where more than one set of stimuli is available they may all be used. Although the question of additivity of cues has not received the analytic attention in the case of spatial learning that it has in visual discrimination learning (see below), there is evidence that an increase in the number of available cues increases rate of learning (Honzik, 1936; Scharlock, 1955; Restle, 1957).

IV. THE EFFECTIVE STIMULI IN VISUAL DISCRIMINATIONS

It might be thought that the experimenter should be able to specify the effective stimuli controlling his subjects' behaviour more easily in visual discrimination problems than in a spatial discrimination in a maze, with its multiplicity of intra- and extra-maze, exteroceptive and proprioceptive stimuli. This is, however, true in only a limited sense. The experimenter may require a rat to select the window of a jumping stand containing a black card with a white vertical rectangle and not to respond to the window containing a black card with a white horizontal rectangle. Suitable randomization of the lateral arrangement of these cards from trial to trial, together with additional controls against the possibility of the rat's detecting which window is unlocked, or smelling food behind the correct window (cf. Sutherland, 1961a), make it certain that appropriate discriminative performance is based on the difference between the visual characteristics of these two cards. But there are numerous ways in which the two cards differ: one has more white in the lower half than the other; one has two black squares in the lower half separated by a patch of white; the rectangles differ in height and in width. Only extensive, and perhaps rather subtle, tests will reveal which one or more of these features is critical (Sutherland, 1961b). We do not, in fact, know which are the features of visual forms that are responded to by animals. Since the pioneering work of Kluver (1933) and Lashley (1938), a certain amount of progress has been made (e.g., Sutherland, 1961a; 1969). Since the question at issue is largely perceptual, being one concerned with the nature of visual analysis, it is perhaps not one traditionally encompassed in the field of learning. Yet ignorance can be unfortunate. The problem becomes even more acute when experimenters set up studies of transfer between different discriminations designed to test theoretical analyses of the processes involved in transfer. Since a central question in such studies is whether transfer is based more on general or specific processes, e.g., whether animals learn something from one problem sufficient to benefit performance on any subsequent discrimination or whether they learn only to attend to specific dimensions, our nearly total ignorance of the dimensions along which stimuli actually differ for a given animal, makes interpretation of some studies relatively uncertain.

Leaving on one side the problems of pattern recognition, there remain several questions that have often been asked about the nature of the effective stimulus in discrimination learning. The first has been posed in cases where subjects are trained to discriminate between two stimuli that appear to fall along some simple, physical dimension, such as size or brightness: for example, subjects might be required to discriminate between a large and a small circle, or a dark and a light grey card. In such a situation S+can be categorized either in absolute terms, as a stimulus of a particular size or brightness, or in relative terms, as the larger or brighter of the two stimuli. We can ask which of these characterizations is more appropriate, i.e., whether animals respond to the absolute or relational properties of discriminative stimuli.

The second question we shall consider is whether discriminative performance is controlled more by S+ than by S−. We have previously talked of subjects learning to approach S+ and/or avoid S−. One may ask whether one or other of these descriptions is more appropriate.

A. Absolute and Relational Views of Discrimination Learning

1. Simultaneous or Successive Presentation of Stimuli

To describe one stimulus as brighter, larger or wider than another may be natural enough when both are presented for inspection simultaneously; but this relationship may be less obvious when the two are presented successively. If relational cues are used at all, then, they should be more effective in simultaneous than in successive discriminations. From this it follows that, other things being equal, simultaneous discriminations should be easier than successive discriminations: if the only cues used by an animal are relational, then these are more obvious in the simultaneous case, for in the successive case the comparison must be made between one stimulus and the memory of the other; if animals use both absolute and relational cues, then the simultaneous discrimination provides an additional, effective set of relevant cues.

The problem with this prediction is that, as we have seen, other things are not necessarily equal when a comparison is made between simultaneous and successive discriminations. A successive conditional discrimination is usually much more difficult than the equivalent simultaneous discrimination, but the reasons for this are to be found in the differences in the correlation between stimuli, responses and reinforcement in the two situations. Only if these differences can be somehow eliminated can the comparison throw light on the importance of relational cues in learning.

Because of the complexities of the successive conditional discrimination, it has seemed to some that it would be more appropriate to compare the learning of a simultaneous discrimination with the learning of a successive, go/no-go discrimination. Here the response requirements (approach S+ and avoid S−) seem essentially equivalent. Grice (1949) found no difference in the rate at which rats learned a simultaneous and a go/no-go size discrimination, and concluded that relational cues were unimportant. The problem here, however, is one of ensuring equivalent measures of learning. Grice employed a choice criterion of performance in the simultaneous discrimination and a latency criterion in the successive discrimination. But there is good evidence that rats trained on a simultaneous discrimination may show a substantial difference in the latencies of their responses to S+ and S− while still choosing between them at a chance level of accuracy (Mahut, 1954; Turner, 1968, cited by Sutherland and Mackintosh, 1971, p. 95). If a similar effect had occurred in Grice's study, his simultaneous subjects may have reached an identical latency criterion of learning more rapidly than the successive group.

More satisfactory solutions to these problems have been provided in studies by Saldanha and Bitterman (1951), MacCaslin (1954), and Sutherland et al. (1963). Saldanha and Bitterman trained two groups of rats concurrently on two discrimination problems. The first group was trained to discriminate between a dark and a light card on some trials, and on other trials between wide and narrow vertical striations. A second group was presented with exactly the same stimuli, and exactly the same relationships between stimuli and reinforcement, but with different pairings of S+ and S−: on some trials they were shown the dark card paired with the narrow striations; on the remaining trials they discriminated between the light card and the wide striations. The former group, given the opportunity to form relational comparisons along the dimensions of brightness and width, learned both their discriminations in an average of 354 trials, with the slowest subject requiring 516 trials. Only 4 to 12 subjects in the noncomparison group learned within 516 trials. The implication is that the availability of the relational cue greatly facilitated learning.

McCaslin (1954) chose to compare the learning of a simultaneous discrimination with the learning of a successive conditional discrimination; but, by comparing discriminations differing in difficulty, he was able to show that the advantage of simultaneous over successive discriminations was greatly enhanced when the discriminative stimuli lay close together along some stimulus dimension. Both simultaneous and successive groups were trained concurrently on two problems. The easy discriminations were between horizontal and vertical striations, and between a large and a small circle. These two problems were mastered by the simultaneous group in 120 trials, and by the successive group in 144 trials. The difficult discriminations, between dark and light grey and between wide and narrow striations, were learned by the simultaneous group in 448 trials; nine of thirteen subjects in the successive group, however, failed to learn within 868 trials. MacCaslin's findings have been confirmed in a study comparing simultaneous and go/no-go discriminations in octopuses (Sutherland *et al.*, 1963): here, too, although an easy discrimination was learned equally efficiently under either condition, performance on a difficult discrimination was much less efficient when the stimuli were presented successively.

These studies strongly suggest that the direct comparison of discriminative stimuli, permitted by their simultaneous presentation, provides an additional relational cue which markedly facilitates the learning of very difficult discriminations. Although a relational view of discrimination learning is supported by these results, however, it is important to note that there is incontrovertible evidence to show that animals also learn to approach and avoid the absolute values of S+ and S— in a simultaneous discrimination. As noted earlier, animals trained on a simultaneous discrimination may show excellent transfer when shown either S+ or S— alone, or when a new stimulus is substituted for S+ or S—. Such transfer could not occur if subjects had simply learned to approach the brighter or darker, or larger or smaller, of two stimuli; it must depend upon the establishment of approach and avoidance tendencies to the appropriate absolute values of the discriminative stimuli. The implication is, therefore, that animals use absolute cues to solve both successive and simultaneous discriminations, but that simultaneous discriminations may provide additional relational cues, which assist the learning of difficult discriminations. This conclusion is entirely consistent with the results of studies on transposition, a phenomenon which provided the initial impetus for the elaboration of absolute and relational theories of discrimination.

2. *Transposition*

The clearest evidence that discrimination learning was fundamentally a matter of the learning of relations was, according to Köhler (1918), the fact of transposition. Chimpanzees and chickens, if trained on a simultaneous discrimination with the lighter of two grey cards positive, would select a still lighter grey card in preference to their original S+ in a subsequent test session. The term "transposition", technically used of transposing melodies into different keys, describes what Köhler saw as the important aspect of this result: just as a melody is a set of relationships between different notes which remains unaffected by the key in which it is played, so a brightness discrimination is based on the perception of a relationship between S+ and S— which is unaffected by transposing the two stimuli along the brightness dimension.

Spence (1937a), in the second of his two classic papers on discrimination learning, showed that the fact of transposition did not necessarily require a relational explanation. We have seen his explanation at work in the discussion of the peak shift in the preceding chapter (p. 535). As is shown below in Fig. 10.4, the interaction of gradients of excitation and inhibition around S+ and S— may be sufficient to shift the peak of a postdiscrimination gradient away from S+ towards a stimulus further removed from S—. Spence argued that transposition is a consequence of this shift in the peak of the postdiscrimination gradient. This ability to predict transposition from the interaction of excitatory and inhibitory gradients must count- as one of the more striking achievements of the analysis of discrimination learning proposed by Spence and Hull, and provides an excellent example of their general attempt to account for the phenomena of discrimination learning in terms of principles of conditioning and extinction.

The available data on the peak shift provide convincing support for this analysis of successive, go/no-go, discrimination learning: empirical gradients of excitation and inhibition can be recorded, and relatively simple assumptions about their interaction yield a displaced postdiscrimination gradient. Variables known to determine the slope of excitatory and inhibitory gradients have appropriate effects on the occurrence of the peak shift. This history of success might encourage the belief that transposition is simply the analogue, in a simultaneous discrimination, of the peak shift observed in successive discriminations.

The critical question, however, is whether the observed incidence of transposition is in fact correlated with observed shifts in the peak of postdiscrimination gradients. Although the results of a study by Honig (1962) suggested an affirmative answer to this question, subsequent experiments have shown that transposition and the peak shift do not necessarily co-vary. Honig trained pigeons on a wavelength discrimination, with one group learning a successive discrimination, and a second group learning a simultaneous discrimination. In order to measure the postdiscrimination gradient, he then gave both groups a series of single-stimulus tests; these were followed by a series of choice tests, designed to measure transposition. His results provided excellent support for Spence's analysis: transposition occurred only to those stimuli which also commanded a higher rate of responding than S+ during generalization testing. More surprisingly, however, Honig found no evidence at all of transposition, or of a peak shift, in subjects given simultaneous discrimination training.

A subsequent study by Rudolph (1967; cited by Riley, 1968, pp. 81–5) confirmed some of Honig's conclusions, but neither of these final results. If pigeons were given successive discrimination training between two wavelengths, and were then given either a choice test to measure transposition or a single-stimulus test to measure a peak shift, there was good agreement between the two measures. Following simultaneous discrimination training, however, Rudolph observed good transposition, but no evidence of a peak shift at all. Similar results have been reported by Marsh (1967b), and by Sachs (1969). In Honig's (1962) study, however, simultaneous training produced no transposition. The discrepancy between these different results has been attributed by Riley (1968, pp. 85–6) to the fact that Honig gave transposition tests to subjects after they had received a series of single-stimulus generalization tests. Both Rudolph and Marsh, on the other hand, avoided potential sequential effects in testing, by using separate groups of subjects to provide measures of transposition and of the peak shift.

The interpretation of the findings reported by Rudolph, Marsh, and Sachs seems reasonably straightforward. Successive presentation of stimuli prevents, or reduces, the utilization of relational cues: if transposition occurs on a choice test following successive training, therefore, it must be caused largely by a shift in the peak of the postdiscrimination gradient; this peak shift, as we have seen, is caused by the interaction of excitatory and inhibitory gradients, just as Spence proposed. If, on the other hand, it is assumed that the simultaneous presentation of stimuli in discrimination learning may result in relatively little inhibitory control by S − (see below, p. 568), it follows that after simultaneous discrimination training there will be little or no shift in the peak of the postdiscrimination gradient. Since, however, simultaneous presentation of stimuli does permit the use of relational cues, reliable transposition will still occur. Spence's analysis of transposition, therefore, may be less successful in explaining the phenomenon to which he originally applied it, than in explaining a phenomenon only discovered 20 years later (the peak shift).

Much of the remaining experimental evidence on transposition is consistent with this conclusion. One of the most notable and reliable features of transposition, and one which has generally been regarded as strong support for Spence's analysis, is that relational responding on test trials is dependent on the distance between the test stimuli and the stimuli used in training. Transposition occurs reliably on a "near" test, as between S_5 and S_6 in Fig. 10.4; the incidence of transposition may significantly decline, however, when a "far" test, as between S_1 and S_2, is given (e.g., Gulliksin, 1932; Kluver, 1933; Spence, 1937a; Kendler, 1950). This result has been attributed generally to the shape and extent of Spence's hypothetical gradients of excitation and inhibition (Kimble, 1961, p. 380): as can be seen from Fig. 10.4, the difference in net excitatory potential between S_1 and S_2 is negligible and there should be no consistent tendency to select one in preference to the other. There is, however, a much stronger prediction that follows from the gradients shown in Fig. 10.4: although transposition is predicted on tests with S_5 and S_6, or with S_4 and S_5, it can be seen that the net excitatory value of S_4 is greater than that of S_3, which is in turn greater than that of S_2. Tests between these stimuli,

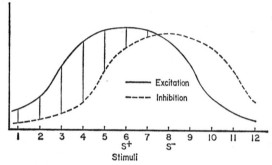

FIG. 10.4. Derivation of transposition and transposition reversal from the interaction of hypothetical gradients of excitation and inhibition. The strength of the tendency to approach each stimulus is represented by the height of the vertical line between the two gradients. After reinforcement at S_6 and nonreinforcement at S_8, therefore, subjects will select S_5 in preference to S_6, thus showing transposition, but will select S_4 in preference to S_3 or S_2, thus showing transposition reversal.

therefore, should reveal a preference for the stimulus lying closer to $S+$, i.e., a significant *absolute* choice. It follows that as one moves from near to far tests, one should observe first a decline in transposition, secondly a significant *reversal* of transposition and finally a reversion to chance responding. There is no convincing evidence to support this strong prediction. As Riley (1968) has concluded: although there is "clear evidence for a decline from strong relational responding to chance responding as stimulus values are changed from a near to far test . . . there is no evidence for a genuine reversal to absolute responding in which the non-relative stimulus is preferred" (p. 75). The one study sometimes cited as showing such a reversal (Ehrenfreund, 1952) cannot be regarded as serious evidence for Spence's position. As Riley has pointed out, the "far" test at which Ehrenfreund found absolute responding was in fact the only set of test trials that paired the original $S+$ against a new stimulus. Animals indeed chose S+ above chance on these test trials, i.e., failed to transpose, but this simply suggests that there was something about Ehrenfreund's procedure that resulted in little or no transposition (Riley, 1968, pp. 72–4, 78).

This failure to find reliable evidence of transposition reversal in choice tests following simultaneous training, contrasts sharply with the evidence provided by postdiscrimination gradients following successive discrimination training. Hanson's (1959) data, for example, showed large "reversals" in the postdiscrimination gradient (see Fig. 9.12): with an S+ of 550 nm, and an S− ranging between 555 and 590 nm, all groups showed peaks at 540 nm; in other words, they responded more rapidly to a stimulus closer to S+ (540 or 530 nm) than to several stimuli further removed from S+ (e.g., 520 or 510 nm). Once again postdiscrimination gradients following successive training provide data consistent with Spence's theory of transposition, while transposition tests do not.

The important point is that it is unnecessary to appeal to Spence's theory in order to explain the gradual decline in transposition as test stimuli move further from the training stimuli. It is true that such a decline unequivocally shows that animals have learned something about the absolute values of the training stimuli, for if they had not, a change in absolute values would go undetected; nevertheless, the observed orderly decline in transposition is most readily interpreted as a gradual breakdown of discriminative performance produced by a change in the controlling stimuli. When other features of the experimental situation are changed, well-established discriminations may be severely disrupted (Logan, 1961); a change in the absolute values of the training stimuli may disrupt performance in exactly the same way.

There is, by now, considerable evidence directly pointing to the relational basis of transposition. Spence (1942) noted that no interaction between excitatory and inhibitory gradients would predict the occurrence of transposition following training on a three-stimulus problem. If, for example, subjects are exposed to three stimuli differing in size and are reinforced for choice of the intermediate size but not for responses to the largest or smallest stimulus, Spence's account implies that they will show no tendency to select the intermediate stimulus when tested with three new stimuli differing in absolute size from the training stimuli. Although transposition does not always occur under these conditions (Spence, 1942), reliable transposition has been reported by Gonzalez et al. (1954) and by Gentry et al. (1959).

Experiments by Lawrence and DeRivera (1954) and Riley (1958) have clearly established that animals may learn about the relationship between different parts of a single-stimulus object, or between a stimulus and its background. Lawrence and DeRivera trained rats on a successive conditional

T

discrimination with stimulus cards each of which consisted of a mid-grey lower half and an upper half which was either brighter or darker. Examples of the type of stimuli and details of the reinforcement contingencies are shown in Fig. 10.5. A response to the right was reinforced when the upper halves were darker; a response to the left was reinforced when the upper halves were lighter. After they had learned, subjects were tested with new cards, in which the lower halves might now be very much brighter or very much darker than they had been during training. For example, as is shown in Fig. 10.5, they might be presented with cards

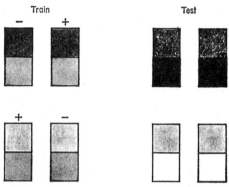

FIG. 10.5. Diagram of some of the stimuli used in training and testing by Lawrence and DeRivera. In training, subjects were exposed to stimulus cards whose lower halves were always mid-grey, and whose upper halves were either darker or lighter. On test trials the brightness of the lower halves was changed, so that the same absolute value of the initially darker top halves might now be lighter than the lower halves, and the same absolute value of the initially lighter top halves might now be darker than the lower halves. (*After* Lawrence and DeRivera, 1954.)

whose upper halves were the same shade as the dark cards used in training, but with lower halves that were even darker. Since the lower halves of the cards had always been of a fixed mid-grey throughout training, this feature of the stimuli should not have come to control any specific response tendency. If the subjects, therefore, had learned to respond in absolute terms, their responses must have been controlled by the absolute values of the top halves of the cards. Thus with the upper pair of transfer stimuli shown in Fig. 10.5 they should have responded to the right, and with the lower pair of test stimuli they should have responded to the left. If, however, they had learned to respond to the relationship between the upper and lower halves of the cards, they should have responded to the left when confronted with the top pair of test stimuli and to the right when confronted with the lower pair. The large majority of choices on test trials indicated that animals had learned the relational solution.

Riley (1958) also showed that rats are capable of responding to the relationship between a stimulus and its background. Subjects trained on a brightness discrimination and given a far transposition test showed imperfect transposition: only 70 per cent of test choices were to the relational stimulus. When the background was also changed, however, so as to preserve the same ratio between the brightness of the discriminative stimuli and that of the background, transposition was essentially perfect on the far test. It is, of course, well established that human observers judge the brightness of a stimulus by reference to the ratio of the stimulus to its background (e.g., Wallach, 1948) and there is evidence that a similar effect may operate to influence perceived size (Rock and Ebenholtz, 1959). As Riley (1968) remarked: "That it should also appear in rats is not too surprising" (p. 76). But if animals do respond to the relationship between a stimulus and its background, or between different parts of a single-stimulus object, it is hard to resist the conclusion that they may also respond to the relationship between S+ and S−. Perhaps the extent to which such relational responding occurs depends upon the spatial and temporal contiguity of the two stimuli (just as it does for human observers). Riley *et al.* (1963) have confirmed one prediction derivable from this argument: they found that the incidence of transposition decreased as the physical distance between the discriminative stimuli is increased. A second prediction derivable from a relational analysis is that transposition should be affected by the temporal contiguity of the discriminative stimuli, i.e., by whether S+ and S− are presented simultaneously or successively. This prediction has been confirmed by Baker and Lawrence (1951), and Riley *et al.* (1960), both of whom reported significantly greater amounts of transposition after simultaneous discrimination training (see also Thompson, 1955, and Sachs, 1969). The outcome of these studies brings us back to the point made earlier: the transposition that occurs following simultaneous discrimination training cannot be reduced to the interaction of gradients of excitation and inhibition, even though such an interaction does serve to explain what appears at first sight to be a similar phenomenon, i.e., the peak shift in postdiscrimination gradients. The simultaneous presentation of discriminative stimuli makes available a set of relational cues which may be used in addition to the absolute stimuli necessarily relied on in successive discriminations.

B. Relative Importance of S+ and S− in Discrimination Learning

If animals solve successive discriminations largely in terms of the absolute properties of the discriminative stimuli, and if simultaneous discriminations simply add relational cues to the absolute cues already available, then it becomes possible to ask whether the establishment of approach tendencies to S+, or of avoidance tendencies to S−, plays the more important role in discrimination learning. We have talked earlier of animals learning to approach S+ and/or avoid S−, for it is obvious enough in principle that appropriate discriminative performance could be maintained by only one of these tendencies (Jenkins, 1965). A relational cue is by definition one that is derived from a comparison of S+ and S−, and the occurrence of relational learning, therefore, shows that simultaneous discrimination learning cannot be reduced to the learning of either approach to S+ or avoidance to S− taken in isolation. For different reasons the occurrence of a peak shift following successive discrimination training is also evidence that discrimination learning depends upon control by both S+ and S−. And, in fact, the general view, incorporated into most theories of discrimination learning (e.g., Spence, 1936; Hull, 1952; Zeaman and House, 1963; Lovejoy, 1968; Sutherland and Mackintosh, 1971), has been that both approach and avoidance tendencies are involved. There have, however, been occasional exceptions (e.g., Harlow and Hicks, 1957).

1. *Tests for Control by S+ and S−*

The fact that interdimensional discrimination training results in both excitatory gradients round S+ and inhibitory gradients round S− (p. 527) provides one important line of evidence that successive discrimination training produces control by both S+ and S− (Jenkins, 1965). However, the fact that inhibitory gradients are usually shallower than excitatory gradients following free-operant discrimination training in pigeons, suggests that with this particular procedure, control by S+ may be more important than control by S− (although, as Jenkins has argued, the inference is far from rigorous).

A number of other studies, not so far considered, have employed a similar technique to assess control by S+ and S−. Following simultaneous discrimination training between two stimuli, subjects may be tested with a new stimulus substituted for either S+ or S−; alternatively, either S+ or S− may be simply removed. If change or removal of S+ disrupts performance more than change or removal of S−, then one may conclude that control by S+ was the more important. This design has been used in experiments with rats in a discrimination box by Sutherland (1961b; Sutherland *et al.*, 1962) who found good transfer in one study regardless of whether S+ or S− was removed, but in the second study observed reliable transfer only when S− was removed. Mandler (1968), training rats on a brightness discrimination in a Y-maze, found that substitution of a new stimulus for S− tended to be more disruptive when animals had only just learned the discrimination, but that following overtraining, substitution of a new stimulus for S+ was more disruptive. To add to the complications, Stevens and Fechter (1968), training rats on a horizontal-vertical discrimination in a discrimination box, found that performance was more disrupted by the change of S+ in animals trained on a shock-escape problem but that animals with water reinforcement were more disrupted by a change in S−. Finally, Derdzinski and Warren (1969) found that cats showed excellent transfer when either S+ or S− was changed.

A second procedure, used to determine the relative importance of S+ and S−, is the reverse of this. Moss and Harlow (1947) and Harlow and Hicks (1957), using monkeys in a WGTA, and Hearst (1969a, 1971a), using pigeons in a free-operant situation, trained subjects with either S+ or S− alone and then tested them on a discrimination between S+ and S−. In Harlow's experiments with monkeys, good transfer occurred following either type of training, but nonreinforced exposure to S− was significantly more effective than reinforced exposure to S+. Hearst's studies with pigeons revealed exactly the opposite outcome: subjects receiving reinforced exposure to S+ (a line of given length or orientation) learned a subsequent line-length or orientation discrimination significantly faster than subjects given comparable nonreinforced exposure to S−.

A final procedure, employed with rats by Gardner and Coate (1965) in a discrimination box and by Mandler (1970) in a Y-maze, involves training subjects on a simultaneous discrimination in which either S+ or S− varies from trial to trial. The assumption is that learning to approach or avoid a single constant stimulus is easier than learning to approach or avoid a set of variable stimuli. If, therefore, variations in S+ retard learning more than variations in S−, the implication is that learning to approach S+ is more important than learning to avoid S−. In conformity with her earlier results obtained with the substitution of novel stimuli for S+ or S−, Mandler found that variations in S− interfered with learning more than did variations in S+. Gardner and Coate, however, obtained exactly the opposite result and concluded that in their situation the establishment of control by S+ played the more important part in discrimination learning.

2. Interpretation

The impression left by this brief review of the evidence can only be one of confusion. Nearly all of these studies, however, agree on one point: in virtually all cases animals have shown learning about both S+ and S−. Where confusion enters is in the relative importance of control by S+ and S−. Here, there is no general conclusion to draw. In some types of apparatus rats have shown more learning about S+ than about S— (Sutherland *et al.*, 1962; Gardner and Coate, 1965); in others they have shown more learning about S— (Mandler, 1970). But in one and the same apparatus, the relative importance of the two may depend upon the level of training (Mandler, 1968), or the nature of the reinforcement (Stevens and Fechter, 1968).

It would be easy to dismiss such inconsistent results as unimportant vagaries of differences in procedures, subjects and apparatus. The temptation to do so might be reinforced by the thought that not all of these results necessarily imply much about control by S+ or S—. Monkeys, for example, may perform better following nonreinforced exposure to S— than following reinforced exposure to S+, not because they learn more about the first of these contingencies than the second but rather because they have a strong inclination to respond to novel stimuli. Such a tendency will increase the probability of selecting a new S+ after exposure to S—, but decrease the probability of selecting an old S+ when it is paired with a new S—.

Nevertheless, it may be possible to make some sense of some of these results. It is possible that one determinant of the control over responding acquired by the precise characteristics of S— is the extent to which the subject actually responds to S— during the course of training. Errorless discrimination training results in relatively flat inhibitory gradients, presumably because it is the occurrence of an error that ensures contact with S— (p. 530). Overtraining may reduce control by S— for similar reasons: once the subject has mastered the discrimination it no longer makes contact with S—. Furthermore, the degree of exposure to S— may vary from apparatus to apparatus. In Mandler's Y-maze, for example animals could leave the choice point to approach either S+ or S—, but at this point the stimulus not approached was no longer visible. In order to learn the discrimination, therefore, subjects had to learn to retreat from S—, retrace and approach S+. In a conventional discrimination box, on the other hand, the subject's choice is more likely to be made from a position in which both S+ and S— are visible. In these circumstances appropriate performance may depend less upon learning the absolute properties of S— and may be more controlled by the development of approach tendencies to S+.

These arguments suggest that the degree of control established by S— might also depend on whether the discriminative stimuli are presented simultaneously or successively. The availability of a reinforced alternative on all trials of a simultaneous discrimination might ensure less contact with S— than in the course of learning a successive go/no-go discrimination. Moreover, it is possible that in the simultaneous case, the appearance of S— does not provide a signal to stop responding but rather serves as a signal to respond elsewhere (Bloomfield, 1969). This would explain why simultaneous or concurrent discrimination procedures may result in relatively shallow gradients round S— when subjects are given single-stimulus tests (Honig, 1962; Beale and Winton, 1970).

There is some evidence to support these suggestions. S. Siegel (1967), for example, showed that rats, trained on a simultaneous discrimination, might learn always to look at one side first and use the appearance of S— on that side as a signal to respond to the other side, regardless of the stimulus which appeared there. Similar data have been reported by Hall (1974a) in a study of simultaneous discrimination learning by pigeons.

Two qualifications, however, need to be added. First, the use of S— as a signal to respond to another stimulus may not be confined to simultaneous discriminations. Jenkins and Sainsbury (1969) trained pigeons on successive go/no-go discriminations, in which S+ and S— were a pair of visual displays, each containing several features in common, but one of which contained a distinctive feature. For example, one display might consist of three dots, while the other consisted of two dots and a star. Jenkins and Sainsbury found that such discriminations were solved rapidly if the distinctive feature appeared on the positive display, but that they were inordinately difficult if the distinctive feature appeared on negative displays. The reason for this difficulty was that the appearance of the distinctive feature on the negative display, although clearly having an effect on the subjects' behaviour, served to direct their responses to other features of the display, rather than to prevent their responding altogether (cf. Jenkins, 1973). When the distinctive feature was on the negative display, subjects ended up by pecking at the common features on both positive and negative trials.*

* One reason why Jenkins and Sainsbury's pigeons found it difficult to withhold responses in the presence of the distinctive negative feature is probably that the spatial separation of the features made it possible to peck at a common feature without being directly stimulated by the negative feature. When the displays are compact (Sainsbury, 1971), or when the negative "feature" is a diffuse tone (Brown and Jenkins, 1967), pigeons are able to withhold responses on negative trials. The implication, confirmed by other studies of observing responses (p. 585), is that discriminative behaviour is most effectively controlled by stimuli towards which responses are actually directed. A story, of whose truth I am assured by Dr. V. Gray, illustrates the importance of adequate observation of negative stimuli. A monkey was trained on a successive discrimination in an operant chamber: lever pressing was reinforced when the chamber was dark and the illumination of an overhead light signalled nonreinforcement. Although the monkey learned the discrimination rapidly, his discriminative performance later deteriorated to the point where he responded throughout S— periods. The surprised experimenter decided to observe what was going on inside the chamber: the monkey had one hand resting on the lever in order to press it, and the other pressed tightly over his eyes in order to shut out the light.

In the second place, it is quite certain that even in simultaneous discriminations S— does not merely serve to direct responses to an alternative location. Webb (1950), Sutherland (1969) and Macphail (1972) have shown that rats, octopuses and pigeons, trained on a simultaneous discrimination, will respond to S+ and withhold responses to S— when the two stimuli are presented successively.

If there is, therefore, a difference in the role of S— in simultaneous and successive discriminations, it is one of degree and may well be understood in terms of differences in the number of nonreinforced responses actually directed to S— that occur in the two situations. It remains probable that this variable is the most important determinant of the inhibitory control exercised by S—.

V. DISCRIMINATIONS WITH REDUNDANT RELEVANT STIMULI: SELECTIVE LEARNING AND CONTROL

In most discrimination problems S+ and S— differ from each other in a variety of ways. In some cases this is a matter of deliberate experimental strategy: the experimenter explicitly arranges that S+ and S— be compound stimuli, such as a black horizontal rectangle and a white vertical rectangle, or a red triangle and a green circle. In other cases this may be an inherent aspect of the discriminative stimuli and the subject's discriminative capacities: the simultaneous presentation of a large and a small circle may provide the subject with both absolute and relational discriminative stimuli; and many, apparently simple stimuli may be analysed into numerous different elements. An important and general question in the analysis of visual discrimination learning, therefore, is whether all available stimuli contribute to discriminative performance and whether their associative strengths are all changed as a consequence of the outcome of a trial.

The question is often phrased by asking whether animals attend to all features of a discrimination problem on every trial, or whether attention is selective. Phrased thus, the question may sound deceptively simple, but it is important to see that such simplicity is achieved only at the cost of confounding several different questions. The statement that a subject attends to a particular set of discriminative stimuli may be translated by saying first that the subject's discriminative behaviour is being controlled by those stimuli and second that the subject is learning about the correlation of those stimuli with reinforcement. Thus, questions about the direction and extent of a subject's attention are questions about what stimuli control responding on a given trial and what stimuli change their associative strength on that trial.

Perhaps one reason why these distinctions have not been kept clear is that questions of selective learning and control have been bound up with a venerable theoretical controversy, with the consequence that they are not always treated in a dispassionate empirical manner. In what follows, an attempt is made to distinguish between a number of different questions. First, if an animal's performance is apparently controlled by one set of discriminative stimuli on a particular trial, do the associative strengths of other stimuli change as a consequence of the outcome of that trial? Secondly, is performance controlled by only one set of stimuli on each trial, or by all available stimuli? Thirdly, do the associative strengths of all stimuli change on each trial, or do animals learn the significance of one set of stimuli at a time? Finally, is it possible that each of these questions is posing a false dichotomy and that the truth lies somewhere between the two extreme positions implied by the phrasing of these questions?

A. Continuity and Noncontinuity in Discrimination Learning

The possible importance of selective attention in discrimination learning formed one of the central points at issue between Lashley and Krechevsky on the one hand and Spence on the other, in what came to be known as the continuity–noncontinuity controversy. Lashley (1929, p. 135) had rather casually noted that when rats were trained on a simultaneous visual discrimination problem, they often developed a position habit (i.e., consistently selected one side rather than the other even though $S+$ and $S-$ varied randomly from side to side) before solving the problem. The observation was systematically replicated by Krechevsky (1932) in his studies of "hypothesis behaviour" during discrimination learning and has been confirmed in numerous subsequent studies of simultaneous discrimination learning (e.g., Sutherland and Mackintosh, 1971, pp. 88–92). Such observations suggested to both Lashley and Krechevsky that during early trials on a visual discrimination, rats might attend predominantly to irrelevant spatial stimuli and that while their behaviour was controlled by these stimuli, they would learn little about the relevant visual stimuli. Not until they stopped responding systematically to position and started attending to the visual stimuli, so the argument ran, could they be expected to learn about the correlation between $S+$, $S-$ and reinforcement (Krechevsky, 1938). The question at issue, therefore, is the first of those considered above: do animals learn about stimuli other than those apparently controlling their behaviour?

Spence (1936) argued that the development of position habits during the course of visual discrimination learning could be explained without recourse to any assumptions about selective learning or control. In this, he was certainly correct. It remained to be determined, however, whether rats did in fact learn about the relevant visual stimuli while they were responding systematically to one position or the other. The question provoked a series of experiments, which have been the subject of frequent review (e.g., Goodrich *et al.*, 1959; Sutherland and Mackintosh, 1971, pp. 87–106).

The results of these experiments leave little doubt that animals learn something about one set of stimuli while systematically responding to another. The evidence for this is provided by three observations. First, if the reinforcement schedule associated with the relevant stimuli is reversed while the rat is still responding to position, learning of the discrimination may be significantly retarded. This experimental design was described as the study of "presolution reversal" and was the favourite of both continuity and noncontinuity theorists. Although Krechevsky (1938) reported one study in which such reversal did not retard subsequent discrimination learning, he himself found that reversal after a larger number of presolution trials did retard learning, and this result was obtained in most other experiments of this design (e.g., Spence, 1945). Secondly, if latencies of responding to S+ and S— are recorded, it is found that rats start responding significantly faster to S+ than to S— even while they are still systematically responding to position (Mahut, 1954; Turner, 1968, cited by Sutherland and Mackintosh, 1971, p. 95). Thirdly, if rats are trained on a simultaneous discrimination and develop a position habit, systematically selecting the same side for at least 20 trials in a row, it has been shown that when they break their position habit (i.e., respond to the other side), they virtually always do so by responding to S+ rather than S— (Sutherland and Mackintosh, 1971, p. 96): at the point where they stop responding to position they have already learned something about the correlation of S+ and S— with reinforcement.

All of these results clearly imply that the associative strengths of the relevant visual stimuli are changing at the same time as subjects are responding systematically to irrelevant spatial stimuli. Control of choice behaviour by one set of stimuli does not entirely preclude the possibility of concurrent learning about another set of stimuli. But it is important to see that no stronger conclusion is implied: it might be the case, as Krechevsky (1938) had suggested, that systematic responding to irrelevant stimuli, even if it does not completely prevent learning about the relevant stimuli, does interfere with such learning. Unfortunately, none of the experiments stimulated by the controversy between Lashley, Krechevsky and Spence was designed to answer this question.

That animals, systematically responding to one set of stimuli, may fail to learn about another set was suggested by the results of a different experimental design employed by Goodwin and Lawrence (1955). They trained rats on a simultaneous black–white discrimination and, after the problem had been learned, shifted subjects to a new problem in which the discriminative stimuli consisted of high and low hurdles at the entrance to each goal box, with the original black and white stimuli still present but now irrelevant. The question they asked was whether continued training on the hurdles discrimination would eventually abolish the preference for the original S+ (e.g., black) established during initial training on the brightness discrimination. Since responses to black and white were no longer differentially reinforced, the difference between the associative strengths of black and white should have declined during the course of training on the second discrimination. In order to test this assumption, Goodwin and Lawrence returned subjects to the black–white discrimination, half being retrained with black as S+ and the other half being required to reverse to white as S+. They found that relearning the original problem was significantly easier than learning the reversal; even more important, however, they found that this difference was unaffected by extensive overtraining on the intervening hurdles discrimination. The implication is that while animals were responding systematically to the relevant hurdles stimuli in the second stage of the experiment, they learned relatively little about the now irrelevant black and white stimuli: the continued nondifferential reinforcement scheduled in the presence of black and white stimuli had little or no effect on subsequent performance. Goodwin and Lawrence's results have been confirmed in different situations by Mackintosh (1963b) and by Stettner (1965), both of whom found that even if sufficient nondifferential reinforcement on the original S+ and S− were given to equalize latencies of responding to the two stimuli, a subsequent comparison of reversal and nonreversal learning would still reveal a strong preference for the former S+.

In a study of successive discrimination learning in pigeons, Jenkins (1973) has presented additional evidence to suggest that the outcome of a trial has a greater effect on the associative strength of the stimulus controlling responding on that trial than of stimuli not apparently controlling performance. The experiments discussed in this section, therefore, although certainly showing that systematic responding to one set of stimuli does not prevent changes in the associative strength of other stimuli, do imply that control by one stimulus may interfere with learning about another.

B. Additivity of Cues

The question whether discriminative behaviour may be simultaneously controlled by several sets of stimuli, and whether the associate strength of all those stimuli changes simultaneously, has prompted the study of discrimination problems in which more than one stimulus is relevant. The general aim is to see whether learning and performance are improved by the addition of redundant relevant stimuli. The assumption has been that such a finding would imply an affirmative answer to these questions.

1. *Control of Performance by Redundant Relevant Stimuli*

In order to see whether discriminative performance is simultaneously controlled by several redundant sets of stimuli, experimenters have trained animals concurrently on separate component discriminations and then given test trials with two or more of the component stimuli combined. Hara and Warren (1961) and McGonigle (1967), for example, respectively trained cats and rats on brightness, size and orientation discriminations. Subjects learned all three discriminations concurrently and the values of the discriminative stimuli were deliberately chosen to ensure that performance remained well below 100 per cent correct. After performance had stabilized, subjects were given test trials with two or more component stimuli combined. They made significantly more correct responses on these compound test trials than on component training trials: if, for example, they made 80 per cent correct responses to brightness or size stimuli alone, they responded correctly on over 90 per cent of test trials when both sets of stimuli were present. Similar additivity has been observed in studies of free-operant successive discrimination learning (e.g., Weiss, 1971), and classical conditioning (e.g., Van Houten *et al.*, 1970). These studies, however, lend themselves less readily to quantitative analysis and the following discussion will be confined to the results of experiments on simultaneous discrimination learning, such as those of Hara and Warren and of McGonigle.

Although it has been thought that results such as these imply that discriminative performance is simultaneously controlled by several relevant stimuli, the conclusion does not necessarily follow. Consider a subject in one of these experiments, who, at asymptote, has learned to select $S+$ on 80 per cent of trials when $S+$ and $S-$ differ only in brightness or only in size. If we make the simplifying assumption that subjects always respond correctly provided that responses are controlled by the relevant stimuli, so that all errors are a consequence of control by irrelevant stimuli (such as position), then it is a simple matter to calculate the proportion of trials on which responding must have been controlled by relevant stimuli. Since 50 per cent of responses controlled by irrelevant stimuli will be correct by chance, it follows that:

Pr (correct response)

$$= \text{Pr (relevant control)} + 0.5 \, [1\text{-Pr (relevant control)}]$$

Solving this gives a value of 0·6 for the probability of control by brightness or size, when subjects respond correctly on 80 per cent of trials in the presence of either set of stimuli alone:

$$0·8 = 0·6 + 0·5 (1 - 0·6)$$

In order now to predict the probability that either brightness or size stimuli will control behaviour when both sets of stimuli are presented on a compound test trial, we need only combine the individual probabilities in accordance with standard rules for the combination of probabilities:

$$0·6 + 0·6 - (0·6 \times 0·6) = 0·84$$

It follows, therefore, that the probability that responses will be controlled by one or other relevant set of stimuli on a test trial when both are present is 0·84. From this, in accordance with the general expression given above, it is easy to calculate the probability of a correct response:

$$\text{Pr (correct response)} = 0·84 + 0·5 (1 - 0·84) = 0·92$$

Thus performance on a compound test trial will be better than performance on component training trials, even if choice behaviour is controlled by only a single relevant stimulus. In fact, the prediction of 92 per cent correct provides a good approximation to Hara and Warren's and also McGonigle's data.*

In general, therefore, the observation of some additivity is consistent with the assumption that discriminative responding is not necessarily controlled by all available relevant stimuli on a single trial. Provided that subjects have learned about the significance of the relevant stimuli, discriminative performance will be more accurate when several stimuli are relevant, even if responding is controlled by only one stimulus on any one trial. In this sense, many of the results obtained in experiments on additivity are somewhat inconclusive.

There are, however, exceptions. Turner (1968; cited by Sutherland and Mackintosh, 1971, pp. 142–3) trained rats on a compound brightness and orientation discrimination, with interspersed trials on one or both component discriminations. The question of interest was whether subjects with a position habit early in discrimination training would first break their position habit to S+ on a compound trial when both relevant stimuli were present, or on a component trial when only one set of relevant stimuli was present. Since 20 out of 22 rats broke their position habit on a

---

* It should, however, be stressed that this prediction is the highest derivable from any model which assumes that behaviour is controlled by only one set of stimuli on each trial. It requires the assumption that all errors are due to control by irrelevant stimuli and that the addition of a second relevant stimulus detracts from control by these irrelevant stimuli rather than from control by the first relevant stimulus. To the extent that these assumptions are unrealistic, data such as these may be inconsistent with this class of model.

compound trial, the implication must be that on such trials both brightness and orientation stimuli contributed to the choice of S+.

A second example where it must be assumed that discriminative performance is controlled by more than one set of stimuli occurs in experiments on extradimensional shifts (Tighe and Frey, 1972; Tighe and Graf, 1972). A rat is trained, for example, on a simultaneous brightness discrimination, with black positive and white negative, and is then shifted to a new discrimination, in which the presence of chains serves as S+ and their absence as S−. During this second problem, however, brightness differences are still present, although no longer relevant: on 50 per cent of trials the positive alternative is black and the negative is white; on the remaining trials black is negative and white positive. In these circumstances, as is not entirely surprising, animals initially continue to select the black alternative on every trial. The important observation, however, is that during the course of learning the chains discrimination, they make essentially no errors on trials when S+is black: all errors are confined to trials when S+ is white and the solution of the problem depends entirely on learning to select S+ even when it is white. As Medin (1973) has shown in detail, this outcome is inconsistent with any theory which assumes that behaviour is always controlled either by brightness or by the chains alone and never by both sets of stimuli together.

Although, therefore, the occurrence of additivity is not itself sufficient to prove that several stimuli may simultaneously control discriminative performance, there is evidence that requires this assumption. Such a conclusion was, perhaps, to be expected. It is clear that the solution of successive conditional discriminations requires that responding be simultaneously controlled by both visual and spatial stimuli. It is not entirely surprising, then, that when two sets of stimuli are redundantly relevant, or even, as in experiments on extradimensional shifts, when one set is relevant and another irrelevant, both may simultaneously control discriminative performance.

2. Learning with Redundant Relevant Stimuli

The experiments on additivity discussed in the preceding section were addressed to the question whether several relevant stimuli might combine to determine discriminative performance. Additivity of cues is also observed in experiments examining the rate of learning: subjects trained on a compound discrimination with several relevant stimuli may learn significantly faster than subjects trained on the component discriminations with only a single relevant stimulus.

The effect has already been noted in studies of maze learning (p. 556). It is also observed in other situations. Eninger (1952) reported that rats learned a compound visual and auditory discrimination in 55 trials, but took 148 and 234 trials to learn when either the visual or auditory stimuli alone were available. Warren (1953) reported that monkeys learned a visual discrimination faster when both shape and size stimuli were relevant, than when either alone was relevant; Sutherland and Holgate (see Sutherland and Mackintosh, 1971, p. 138) found that rats learned a compound brightness and orientation discrimination faster than either component discrimination; Miles and Jenkins (1973) found similar results in pigeons trained with auditory and visual stimuli. Finally, Kamin (1969), studying conditioned suppression in rats, reported faster acquisition of suppression when the CS was a noise-light compound than when either noise alone or light alone served as the CS. Additivity of cues, therefore, is a general and reliable phenomenon.*

From what has been said before, however, it should be clear that an increase in the rate of learning with an increase in the number of relevant stimuli may not by itself prove that the associative strengths of all relevant stimuli are changing on all trials. Following the argument presented in the previous section, it can be shown that the probability of learning about one or other of two relevant stimuli, when both are present, may be greater than the probability of learning about either on its own. Sutherland and Holgate have suggested an additional factor which may contribute to additivity in rate of learning. If a group of subjects is trained on a compound visual and auditory discrimination, there may be some subjects with a higher probability of learning about visual than about auditory stimuli and others with a higher probability of learning about auditory than about visual stimuli. In the compound discrimination, all subjects will have a high probability of learning about one of the relevant stimuli. When only one compound is relevant, however, some subjects will have a low probability of learning about the relevant stimulus.

If animals are trained on a compound discrimination with two sets of relevant stimuli, tests with the components in isolation usually reveal that individual subjects have learned the significance of both sets of component stimuli (e.g., Sutherland and Holgate, 1966; Miles and Jenkins, 1973). Even this observation, of course, does not necessarily imply that such learning occurred simultaneously: it is always possible for a strict noncontinuity theorist to maintain that subjects learned about one component on some trials and about the second component on other trials. To provide a decisive test of this possibility, the ideal procedure would be to use a

---

* There is one set of exceptions. Warren (1953), Kamin (1969), and Miles and Jenkins (1973) all found that animals trained with compound stimuli, where one component was very salient and the other very weak, learned no faster than animals trained with the more salient component alone.

preparation which ensured that measurable amounts of learning would occur in a single trial. One such preparation is the conditioned suppression of licking: a single pairing of a CS with shock is sufficient to establish significant suppression to that CS. Using this procedure, Mackintosh (1971a) found significant levels of suppression to both components of a compound visual and auditory CS after a single conditioning trial.

If this result can be generalized to other situations, it implies that where two sets of stimuli are relevant, the associative strengths of both may change simultaneously. This conclusion is incompatible with the strict noncontinuity theory associated with Krechevsky (1938) and Lashley (1942). However, it is important to note that, just as the observation that subjects can learn about one stimulus while their behaviour is controlled by another does not rule out the possibility that the controlling stimulus causes some interference with learning about the other, so the observation that subjects may learn about two stimuli simultaneously does not rule out the possibility of some interaction between the two.

Several studies of compound discrimination learning have suggested that such an interaction may occur: a common finding is that many subjects solve compound discriminations largely in terms of one of the components. Reynolds (1961c) trained two pigeons on a successive go/no-go discrimination with a white triangle on a red background signalling reinforcement and a white circle on a green background signalling nonreinforcement. Subsequent test trials, with the component stimuli (triangle, circle, red and green) presented in isolation, revealed that one bird had learned only about the significance of the coloured background, while the other (gratifyingly enough) had learned only about the triangle and circle. Cohen et al. (1969), and Telegdy and Cohen (1971) have shown that this tendency to solve a compound discrimination largely in terms of one component may be increased by an increase in drive level in rats. And Sutherland and Holgate (1966) found that within a group of animals trained on a compound discrimination, there was often a negative correlation between the amount individual animals learn about each component. They trained rats on compound brightness and orientation discriminations (e.g., between a black horizontal and a white vertical rectangle) and tested them with the component stimuli in isolation. In general, they found that the better a subject performed on test trials with one component, the worse it performed on tests with the other. It should be noted, however, that not all the correlations reported by Sutherland and Holgate were very large, and although Warren and Warren (1969) reported the appearance of a significant negative correlation when they trained monkeys on a compound discrimination, neither Miles and Jenkins (1973), in their experiment with pigeons, nor Warren et al. (1970), in an experiment with cats, found any significant negative correlation between performance with the two components of a compound discrimination.

The implication of these results is that changes in the associative strength of one component of a compound discrimination may not proceed entirely independently of changes in the other. In particular, the negative correlations reported by Sutherland and Holgate (1966) and Warren and Warren (1969) suggest that even if changes in the associative strength of several components can occur concurrently, the magnitude of the changes accruing to one component may affect the magnitude of those accruing to the other: the more subjects learn about one set of stimuli, the less they learn about others. This conclusion is confirmed by the results of a group of experiments which have explicitly studied how the presence of one set of stimuli may affect learning about others.

C. Stimulus Selection in Discrimination Learning

That changes in the associative strength of one set of stimuli are affected by the presence of other stimuli has been documented fully by studies of overshadowing and blocking. The basic design of these experiments is

TABLE 10.2

Design of Experiments on Overshadowing and Blocking

			Train	Test
Overshadowing	Control		B	B
	Experimental		AB	B
Blocking	Control		AB	B
	Experimental	A	AB	B

In studies of classical conditioning, A and B would signify CSs, such as a light and a tone. In studies of instrumental discrimination learning, A and B would signify pairs of discriminative stimuli, such as red *vs* green, and vertical *vs* horizontal line. Note that the experimental group in an experiment on overshadowing receives the same treatment as the control group in an experiment on blocking.

shown in Table 10.2. In studies of overshadowing, subjects trained with a compound (AB) signalling reinforcement may be found to have learned less about one of the components (B) then a control group trained with B alone. In studies of blocking, the amount learned about B may be found to be yet further reduced if subjects receive preliminary training on A alone, before being trained with AB compound.

1. *Overshadowing*

The term overshadowing was used by Pavlov (1927, p. 141) to refer to the observation that the presence of an intense CS may reduce conditioning to a less intense CS. Since overshadowing has also been observed in studies of conditioned reinforcement and generalization (pp. 249, 499), it would be surprising if it did not occur in studies of discrimination learning. Several studies of both simultaneous and successive (go/no-go) discrimination learning have obtained significant overshadowing effects: D'Amato and Fazzaro (1966b), employing monkeys, and Lovejoy and Russell (1967), employing rats, found that the presence of relatively salient colour or brightness stimuli in a simultaneous discrimination significantly interfered with learning about more difficult orientation stimuli; and Ray (1967) with monkeys, and Miles and Jenkins (1973) with pigeons, have observed significant overshadowing in successive discriminations.

The results of some studies of classical conditioning seemed to confirm Pavlov's supposition that overshadowing is a unidirectional phenomenon: a salient CS may overshadow a weak one, but two salient or highly discriminable CSs do not necessarily overshadow each other (p. 46). A similar conclusion is suggested by the results of discrimination learning experiments (e.g., Sutherland and Andelman, 1967; Miles and Jenkins, 1973). Miles and Jenkins' experiment is particularly instructive. They trained several groups of pigeons on a light-intensity discrimination with L_1 (a bright light) as $S+$ for all groups, and successively dimmer lights (L_2 to L_5) as $S-$. For each light-intensity group there was a second group of subjects with the same intensity for $S-$, but with a tone added on $S+$ trials: for example, one pair of groups was trained either on TL_1 *vs* L_2 or on L_1 *vs* L_5. A final group was trained on the auditory discrimination alone, with the same intensity of light on both $S+$ and $S-$ trials (i.e., TL_1 *vs* L_1).

Miles and Jenkins examined both the extent to which the presence of the tone affected learning about the visual stimuli and the extent to which the light affected learning about the tone. The question of interest was whether these interactions would be affected by the discriminability of the visual stimuli. After receiving 12 sessions of discrimination training, all subjects received a series of test trials to the various light intensities, with or without the tone; thus the degree of control over responding acquired by the visual and auditory stimuli could be assessed by calculating the proportion of test responses emitted in the presence of L_1 (the reinforced visual stimulus) and in the presence of the tone. The groups trained without the tone (Groups L_1-L_2 etc.) all showed good control by light; but, as can be seen in Fig. 10.6, the addition of the tone resulted in a significant overshadowing of the light in Group TL_1-L_2. The remaining compound groups, however, all showed good control by the light, with little suggestion that this was affected by the difficulty of the visual discrimination. On the other hand, when control by the tone is examined, it can be seen that Group TL_1-L_2 is the only group to have learned very nearly as much about the tone as Group TL_1-L_1. In all the remaining groups the light significantly overshadowed the tone. The critical point is that there is little evidence that any one group displayed significant overshadowing both of tone by light and of light by tone. To a large extent, when the visual stimuli were relatively discriminable, they overshadowed

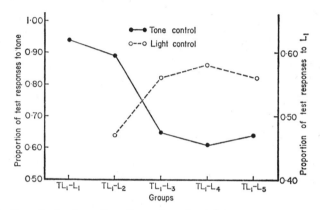

FIG. 10.6. Overshadowing of auditory by visual stimuli, and vice-versa, as a function of the discriminability of the visual stimuli. Different groups were trained with visual stimuli varying in discriminability, from TL_1-L_1, where there was no difference between the visual stimulus on positive and negative trials, to TL_1-L_5, where the visual difference was large. The data suggest that where the visual discrimination was difficult (for Group TL_1-L_2), the tone overshadowed the light, but that for all other groups the light overshadowed the tone. (*After* Miles and Jenkins, 1973.)

the tone without themselves being overshadowed; when they were relatively indiscriminable, they were overshadowed by the tone without themselves detracting from the associative strength of the tone.

There is one case where one stimulus may overshadow another, even when both are relatively salient. Differences in the relative validities of the two components of a compound discrimination may result in significant overshadowing of the less valid component. Although studies of simultaneous discrimination learning have shown that animals will learn something about stimuli partially correlated with reinforcement, even though other stimuli are perfectly correlated with reinforcement (e.g., Bitterman and Coate, 1950; Jeeves and North, 1956), these studies did not incorporate the control groups necessary to ascertain whether the presence of the more reliable predictor of reinforcement reduced the amount learned about the partially correlated stimuli. Wagner *et al.* (1968), whose studies of classical conditioning have already been described (p. 47), have established that if rats are trained on a successive go/no-go discrimination, a light signalling reinforcement with a probability of 0·5 may acquire relatively little associative strength if a set of auditory stimuli is perfectly correlated with reinforcement, whereas the same light will acquire substantial strength when the auditory stimuli are uncorrelated with reinforcement.

2. *Blocking*

In studies of overshadowing the amount learned about one stimulus may be reduced by the presence of another more salient, or more valid set of stimuli. In studies of blocking the amount learned about one component of a compound may be yet further reduced if subjects have received prior training with the other component alone signalling reinforcement. Lashley (1942) was the first to study blocking in discrimination learning, arguing that if subjects attend to only one stimulus at a time then prior training on one component of a compound discrimination should completely prevent learning about the other. He initially trained rats on a simultaneous size discrimination between a large and small circle, and after they had learned continued training with a large triangle substituted for the large circle. Although these subjects thus received training on a compound size and shape discrimination, they performed at chance on a series of test trials on which they were required to choose between a triangle and circle of equal size. Lashley's study is, however, open to several criticisms (Sutherland and Mackintosh, 1971, pp. 107–8), and subsequent studies (e.g., Bitterman and Coate, 1950) have found that complete blocking is the exception rather than the rule in studies of instrumental discrimination learning. Since, however, these studies were designed simply to answer the question of whether animals pretrained on one component learned *anything* about a second, redundant component, they were unable to answer the question whether pretraining has some effect on learning about the added component.

Several more recent studies have, indeed, reported that significant blocking does occur in instrumental discrimination learning. Miles (1970), for example, in a companion study to the Miles and Jenkins (1973) experiment on overshadowing, showed that if pigeons were initially trained on the successive visual discrimination between L_1 and L_2, and then trained on the compound discrimination between TL_1 and L_2, they showed very much less learning about the tone than did birds simply trained on TL_1 *vs* L_2 from the outset. Although pretraining on the visual discrimination did not completely block learning about the tone, a significant reduction was readily observed. Similar findings have been reported in a number of other studies of successive discrimination learning in pigeons (e.g., Chase, 1968; D. F. Johnson, 1970; vom Saal and Jenkins, 1970; Mackintosh and Honig, 1970) and rats (Seraganian and vom Saal, 1969). Similarly, significant but incomplete blocking has also been observed in an experiment on simultaneous discrimination learning in rats (Mackintosh, 1965c).

Although blocking, therefore, is clearly a reliable enough phenomenon in instrumental discrimination learning, it should be stressed that the effect has usually been a relatively small one which has not been obtained in all groups in the above experiments and, in some studies (e.g., Farthing and Hearst, 1970), has not been observed at all. This contrasts with the results of Kamin's (1969) experiments on blocking of conditioned suppression, where rats initially exposed to pairings of a noise and shock show little or no conditioning to a light CS after a large number of pairings of a noise-light compound and shock (p. 48). There has, unfortunately, been no attempt to analyse the causes of this apparent discrepancy and relatively little attempt to study variables which might affect the magnitude of blocking effects in discrimination learning. There is some evidence that blocking is increased by increases in drive level (Bruner et al., 1955) and is more readily obtained when the stimuli on which subjects are pretrained are more salient than the redundant stimuli (Mackintosh, 1965c; Chase, 1968). Both of these results fit in readily with the findings of other studies of compound learning discussed earlier (e.g., Telegdy and Cohen, 1971; Miles and Jenkins, 1973).

D. Theoretical Analysis of Stimulus Selection

The available data are consistent neither with the view identified with Lashley (1929, 1942) and Krechevsky (1938) that animals attend to only one stimulus at a time, nor with the view identified with Spence (1936, 1940) and Hull (1952) that learning proceeds independently about all available stimuli. On the one hand, there is evidence that changes in the associative strength of one set of stimuli may occur at a time when discriminative behaviour is apparently controlled by other stimuli, and even that significant changes in the associative strength of several stimuli may occur in a single trial. On the other hand, there can be little doubt that when animals are trained on a compound discrimination with both A and B relevant, changes in the associative strength of A do not depend simply on the salience of A, or the schedule of reinforcement associated with A, but are also affected by the salience of B, the reinforcement schedule associated with B and the subject's prior experience with B.

When animals are trained on a compound discrimination with both A and B relevant, it is sometimes possible to show a negative correlation between the two components' contribution to discriminative performance (Reynolds, 1961c; Sutherland and Holgate, 1966; Warren and Warren, 1969). This finding may provide the clearest illustration of the general rule underlying all instances of stimulus selection. There may be an inverse relationship between the associative strength acquired by one stimulus and that acquired by others. An increase in the salience or validity of one component of a compound discrimination will increase the associative strength of that component, but will be accompanied by a proportional decrease in the associative strength of other components.

This "inverse hypothesis" of stimulus selection can be derived from more than one theoretical account of discrimination learning. Perhaps the most obvious interpretation is in terms of theories of selective attention (Zeaman and House, 1963; Lovejoy, 1968; Sutherland and Mackintosh, 1971). Although these theories differ substantially in details, they all assume that changes in the associative strength of a stimulus depend on the probability or strength of attention to that stimulus. The crucial assumption, again common to all these theories, is that the probabilities or strengths of attention to the various stimuli of any experimental situation sum to 1·0. Hence, an increase in the probability or strength of attention to one set of stimuli must be accompanied by a decrease in attention to others and large changes in the associative strength of one stimulus will be at the expense of small changes for other stimuli.

Kamin (1969), Revusky (1971), and Rescorla and Wagner (1972) have interpreted the inverse hypothesis not in terms of a limitation on attention, but in terms of a limitation on the effectiveness of reinforcers. A given reinforcer is assumed to support only a given level of conditioning and this must be shared between all components of a compound stimulus. Thus if one component is more salient or more valid than another, it will acquire associative strength more rapidly than the other and hence reduce the associative strength available for conditioning to the other. In the limiting case, pretraining on one component of a compound may ensure asymptotic conditioning to that component and hence pre-empt all the associative strength available for conditioning with that reinforcer. In this case, pretraining on one component would completely block conditioning to the other.

There are, it should be pointed out, other plausible interpretations of some of the data on stimulus selection in discrimination learning. Wagner (1969a), for example, has noted that the presence of a more salient relevant cue (especially if the subject has already been trained with it) will reduce the number of errors made in the presence of a less salient cue, and that this change in effective reinforcement schedule may account for some overshadowing effects. While this is true of some experiments there are others where such a possibility has been effectively eliminated, without this materially affecting the degree of blocking or overshadowing observed (e.g., Mackintosh, 1965c; Mackintosh and Honig, 1970). Moreover, as Wagner argued, such a criticism cannot apply to demonstrations of overshadowing or blocking in classical conditioning where the experimenter rather than the subject controls the effective reinforcement schedule.

It is obvious that the probability of learning the significance of a given stimulus is dependent on the probability of observing that stimulus. It is possible that in many situations animals may learn various "observing responses", whose effect is to modify the stimuli to which the subject is exposed immediately before reinforcement. We have already noted that the control over responding acquired by S+ and S− in a particular discrimination problem varies considerably from apparatus to apparatus, and that these differences are most plausibly attributed to differences in the degree of exposure to, or contact with, S+ and S− imposed by the nature of the apparatus (p. 568). Moreover, the results of Jenkins and Sainsbury's (1969, 1970) experiments, in which pigeons were trained to discriminate between displays containing several discrete features, suggested that the feature at which the pigeon pecks on each trial is the most important determinant of discriminative performance on that trial. Perhaps pigeons tend to look only at stimuli at the tip of their beak. There is abundant evidence, reviewed by Meyer *et al.* (1965) and by Stollnitz (1965), which suggests that monkeys tend to look only at stimuli at the tips of their fingers. Rate of discrimination learning is critically affected by any spatial separation between discriminative stimulus and response: a gap of as little as an inch between the discriminative stimuli and the point touched by the subject when making its response may severely interfere with discrimination learning (e.g., McClearn and Harlow, 1954; Schrier and Harlow, 1957). Finally, there is some evidence that rats in a jumping stand tend to look more at the lower than the upper half of the windows in which the discriminative stimuli are displayed (Ehrenfreund, 1948).

The importance of observing responses in many discrimination learning situations should not be ignored. Nevertheless, it would surely be a mistake to interpret all cases of stimulus selection in terms of observing responses. Many studies of overshadowing and blocking have employed diffuse stimuli such as overhead lights, tones, or white noise (e.g., Kamin, 1969; Wagner *et al.*, 1968); several pigeon studies have observed stimulus selection between tones and key lights (e.g., vom Saal and Jenkins, 1970; Miles and Jenkins, 1973); Mackintosh (1965c), and Sutherland and Holgate (1966), in their studies of stimulus selection with rats in a jumping stand, employed black and white horizontal and vertical rectangles: it is rather hard to see how a rat could look at a small black horizontal rectangle in such a way as to be stimulated by its horizontality but not by its blackness.

If stimulus selection cannot be explained adequately by an appeal to observing responses or to changes in the effective schedule of reinforcement, then it seems reasonable to argue that the basis of such effects is that there will always be an inverse relationship between the associative strengths of the components of a compound stimulus. The question then reduces to one of deciding between alternative explanations of this inverse hypothesis. Is it a consequence of a limitation on attentional or on associative capacity?

There are data that raise problems for the analysis proposed by Kamin (1969) and by Rescorla and Wagner (1972). For example, if blocking occurs because sufficient pretraining with A drives the associative strength of A close to asymptote, and hence ensures that there is no room for conditioning to B when the AB compound is reinforced, there should equally be no room for further conditioning to A on these compound trials. However, Mackintosh (1971a), in a study of conditioned suppression in rats, found that even when blocking of the added component was essentially complete, the pretrained component might acquire additional associative strength over a series of compound trials.

On the other hand, there are data which make it difficult to believe that stimulus selection is always caused by a competition for attention. For example, there is excellent evidence that the critical factor responsible for blocking is that the added component should signal no change in reinforcement. In Kamin's (1969) studies of conditioned suppression, the added component B acquired little or no associative strength when AB signalled the same shock as that previously signalled by A alone. However, if AB signalled a stronger shock than that signalled by A alone, B acquired significant associative strength; while if AB signalled the omission of the shock signalled by A alone, B was established as an effective conditioned inhibitor. This latter result has been confirmed by Wagner (1971), who further showed that an increase in the amount of prior conditioning with A alone, so far from decreasing learning about B when AB signalled the omission of reinforcement, established B as a more effective conditioned inhibitor. It is difficult, therefore, to maintain that conditioning to B is blocked because prior training with A increases the strength of attention to A and there is an inverse relationship between the probabilities of attending to, or learning about, the two components of a compound stimulus. Data such as these are more easily explained by Kamin's suggestion that only unexpected events are effective reinforcers, or by Rescorla and Wagner's more formal version of such a theory.

Instead of considering further the detailed evidence favouring one of these two accounts of stimulus selection over the other, it is, perhaps, worth questioning the validity of the one assumption they hold in common. They both assume that there will, in general, be an inverse relationship between the associative strengths of the several components of a reinforced compound. Although this assumption is consistent with much of the data on overshadowing and blocking, and with the finding of a negative correlation between performance maintained by separate components following compound discrimination learning, there are data which suggest it may be mistaken. A negative correlation between the associative strengths of the components of a compound discrimination has not been observed in all studies (e.g., Warren *et al.*, 1970). More important, the results of experiments on overshadowing do not provide any very good grounds for believing that each component of a compound stimulus detracts from the associative strength of the other. A salient component may reduce conditioning to a less salient component, but there is no evidence that the latter significantly interferes with conditioning to the former (e.g., Miles and Jenkins, 1973); when animals are trained on a compound discrimination with two equally salient components, there may be no evidence of overshadowing at all (Sutherland and Andelman, 1967). While it might be possible to reconcile these results with a strict inverse hypothesis by assuming that two relevant stimuli may detract more from the associative strength of irrelevant stimuli than from each other, it is also possible that overshadowing should be regarded as a largely unidirectional phenomenon: more salient or more valid components may reduce conditioning to less salient or less valid components, but no reciprocal effects are to be expected. Finally, there remains an unexplained discrepancy in the data on blocking; in instrumental discrimination learning, pretraining on one component of a compound may have only slight effects on learning about the other component; in studies of aversive conditioning, on the other hand, the effect may be extremely large.

It is possible, then, that stimulus selection should not be regarded as an automatic consequence of any limitation on attentional or associative capacity. A different interpretation has been suggested by Mackintosh and Turner (1971) and Mackintosh (1973). Changes in the associative strength of a stimulus as a consequence of reinforcement or nonreinforcement may indeed be thought of as dependent on the strength of attention to that stimulus; but blocking and overshadowing are not a consequence of competition for a fixed pool of attention, but rather of the fact that attention is maintained only to stimuli that are informative. If a subject is reinforced in the presence of a compound stimulus, one of whose components is already established as a reliable signal for that reinforcer, then the strength of attention to the other component will decline. Blocking, therefore, occurs because the added component is redundant and thus ignored. Overshadowing, however, is largely a unidirectional phenomenon, because it depends on the more salient component being established as a signal for reinforcement first, and then causing a suppression of attention to the less salient component. While it would be possible to predict a negative correlation between learning about the components of a compound discrimination, such an effect would depend on individual differences in the salience of the two components.

There is, in fact, evidence which points to the operation of some such process as is here suggested in experiments on blocking. In a study of conditioned suppression, Mackintosh and Turner (1971) showed that after initial conditioning to A, four trials with an AB compound and unchanged reinforcement significantly interfered with subsequent excitatory or inhibitory conditioning to B, when, for separate groups, AB signalled either an increase in the intensity of shock, or the omission of shock. Thus the rate of conditioning to B, when reinforcement was changed, was significantly impaired by exposure to B in a context where it signalled no change in reinforcement. Related evidence has been obtained in studies of blocking in instrumental discrimination learning by Mackintosh (1965c) and Sutherland and Holgate (1966): although pretraining on a brightness discrimination interfered to a slight extent with learning about orientation when animals were subsequently trained on a compound brightness and orientation discrimination, blocking of learning about orientation was markedly increased by pretraining on a brightness discrimination with orientation irrelevant.

E. Conclusions

When animals are trained on discrimination problems with several relevant stimuli, the available data imply that discriminative performance is not simply determined by one set of stimuli at a time and that changes in the associative strength of several stimuli may occur in a single trial. Nevertheless, the evidence of interactions between the associative strengths of the components of a compound stimulus is sufficient to require some assumptions about stimulus selection in discrimination learning. The nature of those assumptions is a matter of dispute.

Theories of selective attention have traditionally assumed that there is an inverse relationship between the momentary probabilities of learning about different stimuli. Even if several stimuli change their associative strength on a single trial, the greater the change to one stimulus, the smaller will be the change to others. One rationale for this assumption is that there must be some limit to an organism's capacity to process incoming information. This is, of course, a reasonable assumption, but one may still wonder whether this capacity is so limited as to prevent the concurrent processing of information about two rather salient stimuli in typical experiments on conditioned suppression.

Rescorla and Wagner (1972) have argued that instances of stimulus selection should be attributed not to any limit on attentional capacity, but to limits on the effectiveness of reinforcers. A given reinforcer can maintain only a certain level of conditioning, and the greater the associative strength of one component of a compound stimulus paired with reinforcement, the less will be the associative strength available for conditioning to the other. From this it follows that blocking will depend upon the maintenance of the original conditions of reinforcement. Perhaps the most attractive feature of their theory is that it provides a symmetrical account of the conditions necessary for excitatory and inhibitory conditioning. If inhibitory conditioning depends on the unexpected omission of reinforcement, so excitatory conditioning depends on the unexpected presentation of reinforcement.

Finally, it is not difficult to see the parallels between experiments on blocking and those on latent inhibition (p. 36). In both cases subjects are exposed to a stimulus without any correlated change in the probability of reinforcement, and in both cases the effect of this procedure is to increase the difficulty of associating that stimulus with subsequent changes in reinforcement. The implication is that the strength of attention to a particular stimulus may change with the subject's experience of the validity or redundancy of that stimulus—although there is no necessary implication that changes in attention to one component of a compound are accompanied by equal and opposite changes in attention to all other components.

There is, indeed, further evidence, from some of the studies of transfer between discriminations considered in the following section, that the associability of a stimulus with reinforcement is not a fixed consequence of the salience of that stimulus, but may change with experience. It would be surprising, therefore, if such changes did not also occur in experiments on overshadowing and blocking; and if this is true, it may be a mistake to explain stimulus selection solely in terms of a limitation in the effectiveness of reinforcement.

VI. TRANSFER EFFECTS IN DISCRIMINATION LEARNING

It may be generally accepted that among the processes involved in discrimination learning are the establishment of approach and avoidance responses to S+ and S−, and their generalization to other stimuli. Few theories of discrimination learning, however, have been content with such a bald description and it is equally generally accepted that other processes must be invoked. The nature of these other processes may be illuminated by studies of the transfer of discriminative training from one problem to another.

If additional associations are formed during the course of training on one discrimination, their effects should be observed when subjects are tested on another discrimination. Systematic analysis of the relationship between training and test problems should provide important information about the nature of these additional processes. For example, if discrimination learning involves the establishment of appropriate observing responses, the extent of transfer from one problem to another might depend upon the location of the discriminative stimuli in the two problems. While if discriminative training with one set of relevant stimuli increases attention to those stimuli, or to the dimension along which they differ, transfer should depend on whether the dimension relevant in the second problem is the same as, or different from, that which was relevant in the first.

At one time or another, discrimination learning has been thought to involve a variety of processes over and above the establishment of approach and avoidance responses to the discriminative stimuli, and many of these suggestions have been encountered in earlier discussions. They include several versions of the idea that animals learn to make appropriate observing or orienting responses, ranging from the "receptor-orienting acts" of Spence (1937a) to the more complicated strategies suggested by Mandler (1966), S. Siegel (1967), and Hall (1974b). Instead of postulating overt observing responses, two-stage theories of discrimination learning have proposed that subjects learn to identify or attend to the relevant stimuli, or stimulus dimension (Zeaman and House, 1963; Lovejoy, 1968; Sutherland and Mackintosh, 1971). It is also possible to emphasize the neutralization of incidental or irrelevant stimuli as a process distinct from the strengthening of attention to relevant stimuli (e.g., Restle, 1955; Wagner, 1969a); while Thomas et al. (1970a) have argued that discrimination training establishes a general set to attend to any potentially relevant stimuli, rather than a specific tendency to attend only to the particular stimuli or stimulus dimension relevant during training. Frustration theory implies that transfer from one problem to another may depend upon what subjects have learned to do in the presence of stimuli signalling nonreinforcement or anticipatory frustration (Amsel and Ward, 1965). Finally, Restle (1958) and Levine (1965) have suggested that at least some subjects, under at least some conditions, may learn to respond not simply on the basis of the physical characteristics of the discriminative stimuli, but to the relationship between those stimuli and the outcome of the preceding trial. It might seem that such theoretical proliferation must have outstripped any accumulation of relevant data. And yet the bewildering variety of situations in which discrimination learning has been studied, to say nothing of the variety of results obtained from such studies, suggests that many, if not most, of these principles may contribute to a full understanding of discrimination learning.

The discussion in the following sections will be organized round different experimental situations in which transfer effects have been observed, and will attempt to assess the contribution of some of these processes in each situation.

A. Acquired Distinctiveness of Cues

Lawrence (1949, 1950, 1952) was one of the first investigators to use transfer tasks as a method for studying the processes involved in discrimination learning. He was specifically interested in the possibility that differential reinforcement in the presence of one set of relevant stimuli might increase the associability of those stimuli with reinforcement in subsequent problems. The possibility, of course, is one implied by attentional theories of discrimination learning. Even without the assumption of an inverse relationship between the momentary probabilities of attending to different stimuli, such a theory may still assume that attention increases to stimuli correlated with differences in reinforcement and decreases to stimuli associated with no change in reinforcement.

In order to test this idea, Lawrence (1949, 1950) initially studied transfer between simultaneous and successive conditional discriminations. The argument was seemingly straightforward: according to conditioning-extinction theory (e.g., Spence, 1952) the solution of a simultaneous black–white discrimination involves the establishment of approach tendencies to white, while the solution of a successive black–white discrimination involves the establishment of approach tendencies to the compounds black–left and white–right, and avoidance tendencies to the compounds black–right and white–left (p. 545). An animal transferred from a successive to a simultaneous problem, therefore, will on 50 per cent of trials be confronted with two previously positive compounds (black–left and white–right), and on the remaining 50 per cent of trials be confronted with two previously negative compounds (white–left and black–right). Similarly, an animal transferred from a simultaneous to a successive problem will be confronted with two positive stimuli on some trials and with two negative stimuli on others. It seems to follow that if subjects do show positive transfer between the two types of discrimination, this must be based on features that are common to the two problems. The most obvious common feature is that the relevant stimuli (black and white) remain the same. In order to assess the importance of this factor, Lawrence trained rats on simultaneous or successive discriminations with one of several sets of stimuli relevant, and then transferred some subjects to a second problem where the same stimuli remained relevant and others to a problem with different relevant stimuli. The results were unequivocal: transfer from one brightness problem to another was significantly better than transfer from, say, a texture discrimination to the brightness problem. Similar results have been reported in several other studies (e.g., Mumma and Warren, 1968; Winefield and Jeeves, 1971).

The crux of Lawrence's argument was that transfer between simultaneous and successive discriminations could not be based on the particular approach and avoidance tendencies involved, and therefore that any transfer specific to the relevant stimuli must have depended upon changes in attention to those stimuli. This argument has been challenged by S. Siegel (1967), who showed that, in some circumstances at any rate, the first assumption might be invalid. Siegel's argument was that if a rat developed a position habit during the course of simultaneous discrimination training (e.g., always went left), then instead of solving a black–white problem simply by learning to approach black and avoid white, it might learn to approach black on the left, and avoid white on the left. But such a compound solution will be recognized as very similar to the solution suggested by Spence as underlying the successive discrimination: if the contingencies of the successive discrimination require the subject to go left to two black stimuli and right to two white stimuli, then a rat that has learned to approach black on the left, and avoid white on the left, may have learned enough to show perfect transfer to the successive discrimination. The critical test of this analysis derives from its predictions concerning the behaviour of a rat which is similarly exposed to consecutive simultaneous and successive brightness discriminations, but which happens to develop a right-hand position habit during training on the simultaneous problem. Such a rat might learn to approach black on the right and avoid white on the right, but these response tendencies, instead of permitting immediate solution of the successive discrimination with the contingencies outlined above, are precisely the opposite of those required for solution. By recording position habits during simultaneous discrimination learning, and noting the contingencies of the successive problem, Siegel was able to predict which of his subjects should learn the successive discrimination rapidly and which should not. These predictions were confirmed.

As Sutherland and Mackintosh (1971, pp. 178–80) have pointed out, however, Siegel's analysis necessarily predicts that in any study of transfer

between simultaneous and successive problems, some subjects will fortuitously have learned the appropriate responses while others will not; and it
follows, therefore, that transfer scores should always be bimodally distributed. Neither in Lawrence's original data, nor in those of Mumma and
Warren (1968), is there any evidence of such a bimodal distribution. It is
only Siegel who has obtained this result. This suggests that the situation
employed by Siegel may have forced his subjects to adopt the strategy
postulated by his analysis, whereas animals in other situations do not solve
simultaneous problems in the way he proposes. Since Siegel employed a
special T-maze in which the subjects had to push open a swing door at the
entrance to each arm in order to observe the discriminative stimulus in that
arm, and thus were physically prevented from solving the simultaneous
discrimination by choosing between two simultaneously available stimuli,
this suggestion is entirely reasonable. It is unlikely, therefore, that the
results obtained in other situations, including Lawrence's original results,
can be explained away in the manner Siegel suggests.

B. Transfer along a Continuum

Lawrence (1952) also provided a definitive study of a result earlier reported
by Pavlov (1927, p. 121): performance on a very difficult discrimination,
such as that between two close shades of grey, may benefit more from
prior training on an easier version of the same problem (i.e., between black
and white), than from an equal number of trials with the test stimuli
themselves. Pavlov suggested that a discrimination too difficult to establish
by conventional training techniques, might be learned rapidly if subjects
were initially trained on a problem with discriminative stimuli that differed
more widely along the same dimension. Lawrence confirmed this suggestion
by showing that a group of rats, trained from the outset on a difficult
brightness discrimination, averaged less than 75 per cent correct responses
after 80 training trials, while groups initially trained on easier brightness
discriminations, but shifted to the difficult discrimination after 30 or 50
trials, maintained an average of between 80 per cent and 90 per cent correct
responses on the difficult problem. After an equal number of training
trials, animals that had received some of their training on easier brightness
discriminations performed more accurately on the difficult problem than
animals that had received all their training on the difficult problem.

Lawrence used the term "transfer along a continuum" to describe his
results. They have been confirmed in a number of subsequent studies,
employing a variety of subjects and procedures. Sutherland et al. (1963)
trained octopuses on a simultaneous shape discrimination; Logan (1966)
used rats and a successive (go/no-go) auditory discrimination; Marsh

(1969), and Mackintosh and Little (1970b) used pigeons and successive wavelength discriminations; finally, Haberlandt (1971) gave differential eyelid conditioning to rabbits, using an auditory discrimination.

Lawrence (1952) argued that transfer along a continuum could be most easily explained on the assumption that training on an easy discrimination enabled subjects to identify the relevant stimulus dimension and that once they had learned to attend to this dimension, they would perform accurately enough on the difficult problem. If they were trained from the outset on the difficult problem, they might never learn to attend to the relevant stimuli. Wagner (1969a) has stressed the related possibility that the establishment of efficient discriminative performance with a salient relevant cue might be more effective in neutralizing irrelevant stimuli than a comparable amount of training with a less salient relevant cue.

At first sight it would seem reasonable to suppose that transfer along a continuum could not be explained by the principles of conditioning-extinction theory. Lawrence (1955), Logan (1966) and Bitterman (1967), however, have all pointed out that the requisite prediction can be derived from an appropriate set of assumptions about the generalization of approach and avoidance responses to $S+$ and $S-$. The most important assumption is that generalization gradients should be of the shape shown in Fig. 10.7 (Lawrence pointed to certain additional requirements imposed by additional aspects of his data). Fig. 10.7 illustrates that training on an easy discrimination between S_1 and S_4 may establish a greater difference in the net associative strengths of S_2 and S_3, than is established by an equivalent amount of training directly at S_2 and S_3. Since the hypothetical gradients illustrated in Fig. 10.7 are similar in shape to the gradients required to predict the peak shift (p. 560), this seems an additional reason for acceptance of a gradient-interaction analysis of transfer along a continuum.

The analysis, however, is open to a very simple test. If efficient performance on the discrimination between S_2 and S_3 is simply a consequence of reinforcement at S_1 and nonreinforcement at S_4, it follows that animals given interdimensional discrimination training between one or other of these stimuli and an independent stimulus (S_0) would show an effect analogous to transfer along a continuum. Specifically, animals trained on the discriminations $S_1 + vs\ S_0-$, or $S_0 + vs\ S_4-$, should transfer more accurately to the discrimination between S_2+ and S_3-, then animals given comparable training on either $S_2+ vs\ S_0-$, or $S_0 + vs\ S_3-$. Mackintosh and Little (1970b) tested this prediction in an experiment on successive discrimination learning in pigeons, where S_1, S_2, S_3 and S_4 were lights of different wavelengths, ranging from 501 nm to 576 nm, projected onto the response key, and S_0 was a plain white light. Under these conditions they found that subjects trained on a discrimination between S_0 and S_2 or S_3 actually transferred rather more successfully to the discrimination between S_2 and S_3, than those trained with S_0 and S_1 or S_4. This finding makes it unlikely that differential reinforcement in the presence of S_1 or S_4 alone establishes a greater effective difference between the associative strengths of S_2 and S_3, than does a comparable amount of differential reinforcement in the presence of S_2 or S_3.

It seems to follow that transfer along a continuum can be explained only by supposing that certain associations or tendencies beneficial to subsequent discrimination learning are established more effectively by training

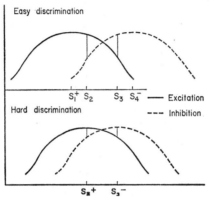

FIG. 10.7. Derivation of transfer along a continuum from the interaction of hypothetical gradients of excitation and inhibition. Training on an easy discrimination between S_1 and S_4 is assumed to result in the gradients of excitation and inhibition shown in the top panel, while training on a difficult discrimination between S_2 and S_3 results in the gradients shown below. The accuracy of performance on the discrimination between S_3 and S_4 is dependent on the difference between the net associative strengths of S_2 and S_3.

on an easy discrimination than by training on a difficult discrimination. The effect cannot, however, be attributed to any general facilitation of discriminative performance resulting from exposure to easy discriminations. Performance on a difficult discrimination benefits from prior training on an easier discrimination, only if the relevant stimuli of the easier problem differ more widely along the same dimension as the relevant stimuli of the harder problem. Marsh (1969), for example, showed that performance on a difficult wavelength discrimination between 520 and 530 nm was improved by prior training on an easier wavelength discrimination between 520 and 550 nm, but not by similar training on an easy brightness discrimination between a bright 520 nm stimulus and a dim 530 nm stimulus.

υ

A second possibility, suggested by Wagner (1969a), is that training on an easy discrimination may be a more effective procedure for suppressing control by irrelevant stimuli. Haberlandt (1971) has provided direct support for this idea by showing that if rabbits are trained on an easy auditory discrimination, they not only perform more accurately on a more difficult discrimination than subjects trained from the outset on the latter but they also respond significantly less to an irrelevant stimulus common to both positive and negative trials.

One might describe Haberlandt's data as a demonstration that the more salient the relevant stimuli, the more effectively they overshadow any irrelevant stimuli. Such an assimilation of transfer along a continuum to overshadowing suggests that both phenomena might be explained by a theory such as that of Rescorla and Wagner (1972), which makes no appeal to changes in attention at all. It may be sufficient to argue that when subjects are trained on an easy discrimination, the relevant stimuli will pre-empt the available associative strength, and that the consequent reduction in control by irrelevant stimuli will facilitate performance when subjects are transferred to a more difficult discrimination.

There is, however, one group of experiments which implies that this account is not sufficient and that a full explanation of transfer along a continuum may require the assumption that the easier the discrimination the greater the increase in the strength of attention to the relevant stimuli. The critical observation, reported by Mackintosh and Little (1970b) in a study of successive discrimination learning in pigeons, and by Logan (1971) and Sweller (1972) in studies of successive and simultaneous discrimination learning in rats, is that transfer along a continuum may still occur even if subjects are initially trained on the reverse of the test problem. Sweller, for example, trained rats on either an easy or a difficult brightness discrimination. After equivalent exposure to this problem, all subjects were transferred to a difficult brightness discrimination in which the reinforcement contingencies were the reverse of those in effect during initial training: if subjects had initially been reinforced for choice of the darker stimulus, the lighter stimulus was now S+. Although animals first trained on the easy discrimination started off by performing less accurately on the reverse problem they learned considerably more rapidly, and reached criterion in significantly fewer trials, than those first trained on the difficult discrimination.

These results cannot be explained by saying that the more salient the relevant stimuli the less will be the associative strength available for conditioning to irrelevant stimuli. When subjects are reversed on a discrimination, there must be a point at which the associative strengths of S+ and S− are equal: at this point, according to Rescorla and Wagner's (1972) model, the difficulty of the *initial* discrimination can have no further effect on learning, for the associative strength accruing to irrelevant stimuli will be the same whatever the salience of those initial stimuli. If differences in the difficulty of the initial discrimination produce differences in the rate of learning a subsequent discrimination with similar relevant stimuli, this is precisely the type of result which points to changes in the strength of attention to those stimuli.

C. Intradimensional and Extradimensional Shifts

Perhaps the best evidence that transfer between discrimination problems may be based partly on increases in attention to relevant dimensions and decreases in attention to irrelevant dimensions, is provided by an experimental design that has been employed extensively with human subjects and, in particular, with young children. The design is illustrated in Fig. 10.8. Subjects are typically trained on two consecutive simultaneous discriminations: for subjects exposed to an intradimensional shift (IDS), the relevant dimension remains the same from one problem to the next; for subjects exposed to an extradimensional shift (EDS), the relevant dimension changes from one problem to the next. The terms "intradimensional" and "extradimensional" are thus used in the same sense as was used to describe different procedures for providing discrimination training before generalization tests (p. 507). An important feature of ID and ED shifts, however, is that for both shifts the actual values of the discriminative stimuli change between the first and second problem. In the example illustrated in Fig. 10.8, subjects are trained initially either on a wavelength discrimination (red *vs* yellow) with shape (square and circle) irrelevant, or on a shape discrimination (square *vs* circle) with wavelength (red and yellow) irrelevant. In the second stage of the experiment all subjects are trained on a wavelength discrimination (blue *vs* green), with shape (diamond and triangle) irrelevant. Subjects shifted from one wavelength discrimination to another are learning an IDS problem; those shifted from the shape discrimination to the wavelength discrimination are learning an EDS problem.

With impressive regularity experiments of this design with children and adults have found that the IDS problem is learned very much faster than the EDS problem (Wolff, 1967). There is also a pleasing uniformity to the results of a number of animal studies. Rats (Shepp and Eimas, 1964; Schwartz *et al.*, 1971), pigeons (Mackintosh and Little, 1969) and monkeys (Rothblat and Wilson, 1968; Shepp and Schrier, 1969) have all shown better performance on IDS problems than on EDS problems.

Since the specific values of the relevant and irrelevant dimensions are changed from Stage 1 to Stage 2, there is no way in which these results can be explained in terms of the establishment of approach and avoidance responses to specific stimuli. Appeals to observing responses and changes in general attentiveness are equally inappropriate. The implication is that differential reinforcement correlated with one set of stimuli increases the rate at which subjects associate other stimuli differing along the same dimension with reinforcement, while if a set of stimuli is not correlated with reinforcement, subjects are less able to associate stimuli differing along that dimension with reinforcement. Attention to relevant stimulus dimensions is increased; attention to irrelevant stimulus dimensions is decreased.

In fact, it is not clear whether both increases and decreases in attention

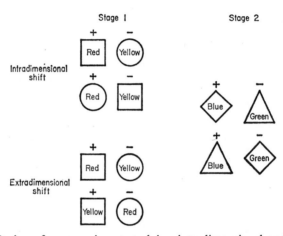

FIG. 10.8. Design of an experiment studying intradimensional and extradimensional shifts. Although all subjects are exposed to the same stimuli in Stage 1, for the intradimensional group reinforcement is correlated with differences in colour, while for the extradimensional group reinforcement is correlated with differences in shape. In Stage 2 a new set of stimuli is used, but for both groups reinforcement is correlated with differences in colour.

must be postulated to account for differences in the difficulty of ID and ED shifts. There is some evidence from studies with both young children (Kemler and Shepp, 1971) and animals (Turrisi *et al.*, 1969), that suppression of attention to irrelevant dimensions is at least as important as any increase in attention to relevant dimensions. In both these experiments the difference between IDS and EDS performance was attenuated or abolished if there were no irrelevant stimuli in Stage 1: subjects trained, for example, on a wavelength discrimination (without shape irrelevant) learned a subsequent wavelength discrimination no faster than subjects trained initially on a shape discrimination (without colour irrelevant).

D. General Transfer Effects

Studies of acquired distinctiveness and transfer along a continuum, together with comparisons of IDS and EDS learning, have all been designed to demonstrate that some dimension- or stimulus-specific transfer effects occur. The results of these experiments, taken as a whole, leave little doubt that when other factors have been controlled, there is some positive transfer attributable to changes in attention to relevant and irrelevant stimuli. It is time to see whether there is evidence of transfer effects that cannot be based on such changes. Studies of learning sets, of course, indicate quite clearly that animals trained on a long series of discrimination problems may gradually acquire information that enables them to learn any new discrimination problem very rapidly; but for the present we shall confine attention to the question whether training on a single discrimination will facilitate the learning of a second discrimination with entirely different relevant stimuli.

A variety of experimental situations has been used to study this question and the answers provided have varied as widely as the situations used. As has already been noted, rats trained in a jumping stand on successive conditional discriminations, learn a second problem considerably more rapidly than the first (p. 551): Bitterman and McConnell (1954) and Wodinsky et al. (1954) attributed this improvement to the establishment of a response-selection set. In neither of these experiments, however, did animals trained on two consecutive simultaneous discriminations show significant improvement from the first problem to the second. The actual discriminative stimuli were counterbalanced between subjects, with some subjects learning an orientation discrimination first and a size discrimination second, and others receiving the two problems in reverse order. In the experiment by Wodinsky et al. the first simultaneous discrimination was learned with an average of 12·6 errors; the second with an average of 12·2 errors. In Bitterman and McConnell's experiment, the subjects received some overtraining after reaching criterion on the first problem. They averaged 16·3 errors on the first problem and 12·3 on the second, but this difference was not, in fact, statistically significant.

The pattern of results obtained in these experiments has been confirmed in some, although not all, subsequent studies of simultaneous discrimination learning in rats. Mandler (1966), training rats on brightness and orientation discriminations in a Y-maze, with different subjects learning different problems first, found that both first and second problems were learned in exactly 79 trials. If, however, subjects were overtrained on the first problem, they learned the second in 50 trials. Waller (1970), training rats in a discrimination box, also found that subjects overtrained on the first problem might learn a second more rapidly than subjects trained only to criterion on the first problem. The effect was significant, however, only when the second discrimination was relatively easy. Although Sutherland and Mackintosh (1971, p. 313) cite a number of studies that tend to confirm Waller's conclusions, Hall (1974b) found that overtraining on an orientation discrimination tended marginally to interfere with rather than facilitate the learning of a black-white discrimination.

Mandler (1966) argued that the transfer obtained in her study might be due to the establishment of appropriate observing or orienting responses. Although this is possible, it is equally possible that positive transfer is based on the suppression of control by irrelevant stimuli. Since the positions of S+ and S− vary from trial to trial, any simultaneous discrimination contains salient irrelevant stimuli. As was noted earlier (p. 510), Wagner (1969a) has shown that when differential reinforcement succeeds in neutralizing an irrelevant stimulus, the effect can be observed as an increase in learning about another stimulus presented in conjunction with the irrelevant stimulus.

An observing response analysis could be readily tested by altering the location of the relevant stimuli from one problem to the next to see if this disrupted performance on the second. It is, however, worth emphasizing that the positive transfer from one simultaneous discrimination to another is not particularly large, and there is no evidence of such transfer if animals are trained only to criterion on the first problem. Just why it should require overtraining to establish an observing response or neutralize irrelevant stimuli has not been considered seriously. Moreover, neither of these analyses accounts for the puzzling finding that even overtraining fails to facilitate the learning of a second, difficult discrimination.

General transfer effects from one problem to another seem to occur much more reliably in successive (go/no-go) discriminations. Eck et al. (1969), for example, found that pigeons learned a free-operant successive wavelength discrimination very much more rapidly if they had previously been trained on a successive orientation discrimination, than if they had simply been reinforced for responding to a single line, or if they had received randomly reinforced and nonreinforced trials in the presence of different line orientations. A similar finding was reported by Keilitz and Frieman (1970). These results are reminiscent of those obtained by Thomas et al. (1970a), who found that free-operant discrimination training might significantly sharpen generalization gradients around an incidental stimulus common to both positive and negative trials (p. 511).

In the experiments by Eck et al. and by Keilitz and Frieman, the discriminative stimuli for both problems were located on the pigeon's response key: thus positive transfer from one problem to the second might have been due to the establishment of appropriate observing responses. Other experiments, however, have ruled out this possibility but have still observed significant transfer. Frieman and Goyette (1973) reported that pigeons showed positive transfer when shifted from a wavelength to an auditory discrimination, and Thomas et al. (1971b) reported positive transfer when rats were trained on consecutive free-operant successive discriminations— the first between two intensities of the houselight, the second between two auditory stimuli. Thomas et al. (1970a) have argued that discrimination training may increase general attentiveness and thus increase the rate at which any new discrimination is learned. One problem with this suggestion, however, is that it does not help to explain why general transfer effects should be so much more prevalent in free-operant successive discriminations than in simultaneous discriminations. A second suggestion, noted in Chapter 9 (p. 513), is that free-operant situations may contain other irrelevant cues (perhaps dependent on the repetitive nature of responding in such situations) which are neutralized by discrimination training.

A third possibility is that animals exposed to differential reinforcement may learn to stop responding in the face of nonreinforcement, whereas subjects exposed to random reinforcements and nonreinforcements may learn to persist in the face of nonreinforcement. The former habit would facilitate, and the latter would interfere with, the subsequent establishment of appropriate discriminative behaviour. Such a possibility is suggested by frustration theory (Amsel and Ward, 1965) and is supported by a number of results. Mandler (1966), for example, found that rats exposed to random reinforcements and nonreinforcements in a Y-maze in the absence of any discriminative stimuli, took significantly longer to learn a visual discrimination than did naive rats. Flaherty and Davenport (1972) reported that rats exposed to a random partial reinforcement schedule in a grey alley, subsequently learned a successive brightness discrimination between black and white alleys significantly more slowly than a group given alternation training (i.e., with trials in the sequence RNRNRN) in the grey alley. They suggested that alternation training establishes a tendency for subjects to stop running when they anticipate nonreinforcement (frustration), while random partial reinforcement establishes a tendency to persist in the presence of anticipatory frustration. It is, at least, clear that alternation training in a grey alley cannot have reinforced any observing responses appropriate for discriminating between black and white alleys.

E. Reversal Learning

Reversal training involves reversing the reinforcement contingencies in effect during an initial discrimination problem. For example, after animals have learned a black–white discrimination with black as S+ and white as S−, the contingencies may be reversed and black become S− and white S+. If discrimination learning involves nothing more than the establish ment of approach and avoidance tendencies to S+ and S−, then reversal learning will involve nothing more than the extinction of these tendencies and the establishment of new approach and avoidance responses. If, how- ever, other processes are involved in discrimination learning, then the analysis of reversal learning may become notably more complex. A reversal of reinforcement contingencies does not, for example, alter the fact that the same discriminative stimuli remain relevant, and that the same stimuli remain irrelevant. Thus if discriminative training results in changes in attention to relevant and irrelevant stimuli, or if it establishes appropriate observing responses, then there will be some sources of positive transfer from acquisition to reversal. The theoretical problem is that the reversal of reinforcement contingencies must be presumed to produce some negative transfer, for no one has seriously supposed that discrimination learning, whatever else it may involve, does not also require the establishment of appropriate response tendencies. The prediction of the outcome of any given reversal procedure is, therefore, bound to be complicated. Certain comparisons, indeed, such as that of the relative difficulty of reversal learning and EDS learning, in spite of their popularity with some investigators (e.g., Kendler and Kendler, 1962; Kendler, 1971) are singularly inappropriate for deciding whether or not animals learn any mediating, attentional responses (Sutherland and Mackintosh, 1971, pp. 305–6). In spite of these problems, two situations for the study of reversal learning have generated a large volume of research in recent years.

1. *The Overtraining Reversal Effect*

Reid (1953) trained rats on a simultaneous black–white discrimination in a Y-maze: all animals were trained to a criterion of 9 out of 10 correct responses and then divided into three groups for reversal training. One group was reversed immediately, one after 50 overtraining trials and the third after 150 trials of overtraining. Reid found that speed of reversal was directly related to the amount of overtraining on the original problem, with the groups receiving 0, 50 and 150 trials of over- training requiring respectively 138, 129, and 70 trials to reach criterion during reversal. Two replications of Reid's results were reported a few years later (Pubols, 1956; Capaldi and Stevenson, 1957), and shortly thereafter, the overtraining reversal effect (ORE) began to occupy the experimental, theoretical and polemical energies of experimental psychol- ogists in much the same way that the study of latent learning had oc- cupied an earlier generation in the late 1940s and 1950s. These experiments have been fairly exhaustively reviewed (e.g., by Paul, 1965; Sperling, 1965; Lovejoy, 1966; Denny, 1970; Sutherland and Mackintosh, 1971). All reviewers have been forcibly impressed by the inconsistency of the effect and its reluctance to submit to any simple analysis.

To take a single example: if overtrained subjects are said to reverse faster than those trained only to criterion, one might hope that there would be some agreement on the definitions of criterion training and overtraining. In practice, different experimenters have adopted quite different definitions, and also appear to have obtained quite different data. In Reid's (1953) experiment, for example, animals trained to a 9/10 criterion reversed only slightly more slowly than those given an additional 50 trials before reversal, but a further 100 trials of overtraining markedly facilitated reversal. Capaldi and Stevenson (1957) also found that reversal was facilitated by a moderate number of trials beyond a 7/8 criterion, and was substantially facilitated by more extended overtraining. Both of these results are quite inconsistent with those reported by Sperling (1970), however, who found that training to a stringent criterion of learning (21/24) facilitated reversal by comparison with training to a less stringent criterion (10/12), but that a further 180 overtraining trials had no additional effect on speed of reversal. If reversal performance is critically affected by small variations in the number of acquisition trials, as Sperling's data suggest, then it is hardly surprising that a number of studies should have employed inappropriate levels of training and in consequence failed to find a significant ORE. Sperling's results also suggest one possible reason for the difficulty of observing the effect in simple, spatial reversals: it is conceivable that the use of a strict acquisition criterion ensures that animals learning a simple problem are already overtrained (c.f., Richman and Coussens, 1970).

Several variables have been shown, more or less convincingly, to affect the incidence of the ORE. No study using a small reward has ever observed the effect, and several studies explicitly varying size of reward have reported a significant ORE in animals given a large reward, but none in animals given a small reward (Hooper, 1967; Mackintosh, 1969). In general, it has proved difficult to obtain an ORE in spatial discriminations, regardless of the size of reward used (e.g., Eimas, 1967; Kendler and Kimm, 1967; Mackintosh, 1969). The one condition where the effect seems reasonably reliable is when rats are trained with a large reward on a moderately difficult visual discrimination: these were the conditions employed by Reid (1953), Pubols (1956), and Capaldi and Stevenson (1957), as well as in several subsequent studies (Paul, 1966; Hooper, 1967; S. Siegel, 1967; Mandler, 1968; Mackintosh, 1969; Sperling, 1970).

Although there is probably more than one difference between visual and spatial problems responsible for this difference in outcome (Mackintosh, 1965b), it is almost certainly largely a consequence of differences in the difficulty of typical visual and spatial discriminations. Rats usually learn a simple spatial discrimination in a T-maze more rapidly than visual discriminations between black and white cards, or horizontal and vertical striations, and the magnitude of the ORE in visual problems is a function of the difficulty of the discrimination, whether this is manipulated by an increase in the number of irrelevant stimuli (Mackintosh, 1963a), or by a decrease in the discriminability of S+ and S− (Mackintosh, 1969). Indeed, if the visual discrimination is exceptionally easy, such as that between black and white painted goal boxes, there is evidence that the ORE may disappear altogether (Eimas, 1967; Mackintosh, 1969; Denny and Tortora, 1971). Finally, if an apparatus is so designed as to ensure that rats learn a simultaneous brightness discrimination (with position irrelevant) more rapidly than a spatial discrimination (with brightness irrelevant), an ORE may be observed in the spatial discrimination but not in the visual discrimination (Richman et al., 1972).

It seems obvious enough that the ORE poses something of a problem for a simple conditioning-extinction theory of discrimination learning: other things being equal one would expect that further reinforcement of responses to S+, and further nonreinforcement of responses to S−, would increase the difficulty of adjusting to the contingencies of reversal training. One suggestion consonant with conditioning-extinction theory has been advanced by Birch et al. (1960) and Theios and Blosser (1965): overtraining may facilitate reversal for the same reason that it facilitates extinction in an alleyway (p. 424). One immediate problem with this suggestion, however, is that there is reason to believe that overtraining may facilitate the extinction of only certain classes of behavioural chains (p. 426). A second problem is that no explanation is provided for the finding that the incidence of the ORE is affected by the difficulty of the discrimination. The major objection to this analysis, however, is that there is unequivocal evidence that overtraining facilitates reversal, not because it facilitates extinction of the initially established discriminative responses, but in spite of the fact that it usually increases resistance to extinction. Fig. 10.9 illustrates that the ORE may occur even though overtrained subjects persist in their selection of the former S+ for a considerably greater number of trials than subjects trained only to criterion. The reason why they reach criterion sooner is because later in reversal they respond less persistently at a chance level of accuracy. The data shown in Fig. 10.9 appear to be entirely typical of those reported in other successful studies of the ORE (Sutherland and Mackintosh, 1971, pp. 258–61): the one important feature of reversal performance not shown there is that during the period of chance performance animals are usually responding systematically to one or other position. Thus overtraining facilitates reversal, not by decreasing resistance to extinction, but by reducing the incidence of position habits.

It remains possible that the effects of overtraining on extinction contribute to the ORE. One plausible suggestion is that the reason why the ORE depends on the use of a large reward is because size of reward determines

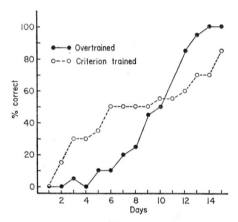

FIG. 10.9. An example of the overtraining reversal effect in rats. The median percentage of correct responses on each day of reversal training for groups trained either to criterion or overtrained on the original discrimination. The problem was a brightness discrimination between black and white rectangles, with both the position and the orientation of the stimuli irrelevant. (*After* Mackintosh, 1963a.)

the effect of overtraining on extinction (Theios and Blosser, 1965). The evidence, however, is conflicting. Although Furstenau and Schaeffer (1970) obtained results consistent with the suggestion, other analyses of the effect of size of reward on the ORE have reported no interaction between size of reward and overtraining on the persistence of responding to the former S+ (Mandler and Hooper, 1967; Mackintosh, 1969).

These considerations suggest that the explanation of the ORE must be found largely in the effects of overtraining on changes in rate of learning about the discriminative stimuli. Although overtraining may increase the excitatory strength of S+ and inhibitory strength of S−, and therefore increase the number of initial errors at the outset of reversal, it must also increase the rate at which changes in the associative strengths of the relevant stimuli are brought about by the new contingencies of reinforcement. Different versions of this suggestion have been proposed by a number of investigators. Reid (1953), S. Siegel (1967), Mandler (1968), and Hall (1974b) have argued that overtraining may increase the strength of appropriate observing responses or orienting strategies; Sutherland (1959) and Lovejoy (1966) have assumed that overtraining increases the probability or strength of attention to the relevant stimuli; while Denny (1970) has suggested that overtraining may decrease generalization between S+ and S−.*

Lovejoy (1966) has shown that a simple two-stage model of discrimination learning, which assumes that differential reinforcement increases both the probability of attending to or observing the relevant stimuli, and the probability of approaching S+ and avoiding S−, can formally predict the ORE. Because overtraining increases the probability of attending to the relevant stimuli, it will both increase errors at the outset of reversal and decrease the probability that responding will be controlled by irrelevant stimuli such as position. This class of model, therefore, is capable of providing an accurate description of the data shown in Fig. 10.9. Lovejoy also showed that the model predicted no ORE if the probability of attending to the relevant stimuli at the outset of initial training was relatively high: this is consistent with the observation that the ORE does not occur if the discrimination is an extremely easy one.

It remains to consider whether overtraining should be regarded as increasing the strength of observing responses or orienting strategies that might be appropriate to any new discrimination, or whether overtraining increases learning only about the relevant stimuli of the discrimination problem, or stimuli differing along the same dimension, in the manner envisaged by attentional theories. The most obvious way of deciding between these alternatives is to examine the effects of overtraining on transfer to other discriminations. As noted in the previous section, there is some evidence that overtraining on one simultaneous visual discrimination may facilitate the learning of another, but the effect is neither large nor consistent. If overtraining facilitates reversal in spite of increasing the strength of inappropriate response tendencies, then an analysis stressing the importance of observing responses would seem bound to predict that the effect on transfer to a new problem should be even larger and more reliable. Moreover, in experiments on extradimensional shifts, where the stimuli relevant in Stage 1 become irrelevant in Stage 2, and those

* It is worth remarking that however different the language used by Denny from that used by Sutherland and Lovejoy, the explanations they propose may not differ in substance. In both cases animals are assumed to learn about the new correlation between the relevant stimuli and reinforcement more rapidly after overtraining, and in both cases this change in rate of learning is confined to stimuli differing along the same dimension as S+ and S−. The differences in language may represent different theoretical predilections rather than differences in the formal properties of the theories.

irrelevant in Stage 1 becomes relevant in Stage 2, there is virtually no evidence that overtraining facilitates EDS learning. Some studies have reported no effect of over training(e.g., S. Siegel, 1967); in others, over-training has retarded EDS learning (Goodwin and Lawrence, 1955; Mackintosh, 1963b).

At best, these results provide extremely equivocal support for an analysis in terms of observing responses. A recent study by Hall (1974b) complicates the picture yet further. Hall first trained rats, either to criterion or with additional overtraining, on an orientation discrimination, then transferred them to a brightness discrimination and finally gave them reversal training on the brightness problem. Although overtraining on the orientation discrimination tended to retard rather than facilitate the initial learning of the brightness discrimination, it reliably facilitated its reversal. The implication is that overtraining may establish some response strategy that, although not assisting the learning of new discriminations, somehow benefits the learning of any reversal problem. What this strategy might be is not entirely clear.

There is, however, one set of results which is difficult to reconcile with the suggestion that overtraining facilitates reversal by increasing the strength of any observing response or strategy. As was noted above, the magnitude of the ORE depends on the salience of the relevant stimuli and on the number of irrelevant stimuli. While this finding can be represented in a formal two-stage model of discrimination learning by varying the initial probability of the hypothetical attentional or observing response, it is extremely doubtful whether the procedures that have actually been employed in these experiments to vary the difficulty of the discrimination could be said to have affected the probability that subjects would in fact observe the relevant stimuli or make appropriate orienting strategies. It is hard to see how a change in the colour of two goal boxes from black and white to dark grey and light grey could decrease the probability that sub-jects would observe these stimuli; it is even harder to see how the use of extremely distinctive black and white goal boxes could decrease the probability of observing their spatial location. Yet just such an increase in the discriminability of brightness stimuli reduces the probability that overtraining will facilitate brightness reversal (Mackintosh, 1969) and increases the probability that it will facilitate spatial reversal (Richman et al., 1972). Results such as these, in conjunction with the general finding that overtraining is more likely to facilitate reversal than EDS learning, seem to point to changes in attention as an important determinant of the ORE.

2. Serial Reversal Learning

If a rat is trained on a simultaneous spatial or visual discrimination, and the reinforcement contingencies are then repeatedly reversed, performance is initially relatively inefficient, with a large number of errors being made on each reversal. With continued reversal training, however, performance improves and each new reversal may be learned with fewer errors than were required to learn the original discrimination. Eventually, indeed, each new reversal may be learned with only a single error. A large number of studies has provided evidence of these effects. They include: Dufort *et al.* (1954); Bitterman *et al.* (1958); North (1959); Pubols (1962b); Stretch *et al.* (1964); Mackintosh *et al.* (1968); and Mackintosh and Holgate (1969).

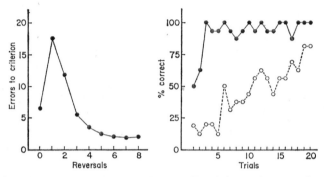

FIG. 10.10. Serial reversal learning in rats. The left panel shows the number of errors required to learn the initial black-white discrimination (Reversal 0) and each of eight reversals. The right panel shows the within-problem learning curves from Reversals 1 and 2 (O–––O) and Reversals 7 and 8 (●——●). (*After* Mackintosh *et al.*, 1968.)

Fig. 10.10 shows data from an experiment by Mackintosh *et al.* (1968). The left panel shows the number of errors made on each of eight consecutive reversals of a black–white discrimination: as can be seen, the rats made many errors over the first two reversals, but eventually learned each new reversal with an average of less than two errors. The right panel provides more detailed information about this overall reduction in errors, showing the within-problem learning curve from early reversals and from late reversals.

Unlike the ORE, the improvement in reversal learning with serial reversal training is an entirely reliable finding and has been observed, regardless of the difficulty of the discrimination and the size of reward, in both simultaneous discriminations, successive conditional discriminations (Gonzalez and Shepp, 1961), and successive go/no-go discriminations (Khavari and Heise, 1967). Improvement has also been observed in a variety of animals other than the rat: the most extensively studied subjects have been pigeons, who show improvement in both simultaneous problems (e.g., Gonzalez et al., 1966b) and successive go/no-go problems (e.g., Beale, 1970), and one or two species of fish, which although often showing no significant evidence of improvement (e.g., Behrend et al., 1965), have subsequently been found to be capable of some slight improvement in both simultaneous (Setterington and Bishop, 1967) and successive problems (Woodard et al., 1971). Several more exotic mammalian and avian species have also been shown to improve over a series of reversals (e.g., Gossette, 1966; Gossette et al., 1966).

There is reason to believe that several factors contribute to this reduction in errors with repeated reversal training. The within-problem learning curves illustrated in Fig. 10.10 show at least two differences between performance on early and late reversals. First, the probability of an error on Trial 1 of each reversal declines from nearly 1·0 to about 0·5; secondly, there is a large change in the subsequent rate of error reduction from early to late reversals. These data show one similarity and one difference between the effects of overtraining and those of serial reversal training on reversal learning. Both procedures increase the rate of within-problem learning; but while overtraining increases the probability of errors at the outset of reversal, serial reversal training decreases the probability of these initial errors. This presumably accounts for the difference in the reliability of the two phenomena.

The observation that the probability of an error on Trial 1 of a new reversal may decline to a chance level over the course of a series of reversals was first reported by Stretch et al. (1964), and confirmed by Gonzalez et al. (1966b). Gonzalez et al. (1967a) and Mackintosh et al. (1968) suggested that this might be due to forgetting. Repeated daily reversals of a discrimination problem should provide optimal conditions for the development of a proactive interference (p. 478): at the beginning of each day the strength of the associations established on the preceding day may be no greater than those established the day before that. Subjects may, in other words, forget from one day to the next which alternative was most recently correct. Gonzalez et al. (1967a) and Mackintosh et al. (1968) confirmed this suggestion by showing that when pigeons or rats were trained on a series of reversals, they still performed at chance at the beginning of each day, even if they were not being required to reverse at that point. Mackintosh et al. reversed one group of rats only in the middle of each daily session; the reinforcement contingencies at the beginning of one day were always those in effect at the end of the preceding day. In spite of this, animals still responded at chance on Trial 1 of each day.

The development of proactive interference, therefore, may account for some of the decline in negative transfer with repeated reversal training. However, the increase in the slope of the within-problem learning curve from early to late reversals, as shown in Fig. 10.10, requires additional explanation. One possibility is that animals learn to maintain attention to the set of stimuli which remains relevant from one reversal to the next. There is, indeed, some evidence that the effects of reversal training may be specific to the stimuli relevant during reversal. Rats trained on a series of spatial reversals show negative rather than positive transfer when shifted to visual reversals (Bitterman *et al.*, 1958; Mackintosh *et al.*, 1968); conversely, rats trained on a series of brightness reversals, while becoming much more proficient at brightness reversal learning, show no comparable improvement when shifted to a spatial problem (Mackintosh and Holgate, 1969). There are, however, grounds for supposing that other factors must be involved. For example, it is difficult to see how an increase in the strength of attention to the relevant stimuli could be responsible for the consistent one-trial reversal performance of the well-trained rat. Moreover, there is evidence that pigeons may concurrently improve their performance on the reversal of more than one problem (Gonzalez and Bitterman, 1968): this would seem inconsistent with any theory which postulates an inverse relationship between the probabilities of attending to different stimuli. It is, of course, possible that some of the observing or orienting strategies proposed to account for the effects of overtraining on reversal may also develop during the course of repeated reversals. However, the finding that serial reversal performance is markedly affected by the spacing of trials (North, 1959; Stretch *et al.*, 1964; Williams, 1971) has prompted a number of investigators to suggest that animals may eventually learn to use the outcome of one trial as a cue to control choice behaviour on the next trial (Hull, 1952, pp. 114–16). In other terms, they may develop win–stay, lose–shift strategies (Restle, 1958; Levine, 1965), dependent upon the memory of the outcome of the preceding trial. If such a process were involved, it would certainly explain why animals such as fish, whose behaviour in other situations shows little control by traces of previous outcomes (p. 467), are also relatively inefficient at reversal learning. The possibility of such a strategy being acquired during extended reversal training in other animals, moreover, suggests important parallels between the processes involved in serial reversal learning and in learning-set tasks.

F. Learning Sets

If monkeys are trained on a long series of independent discrimination problems, they may become progressively more proficient at solving each new problem. In his classic study of learning sets, for example, Harlow

(1949) showed that rhesus macaques trained on several hundred visual discriminations, for a limited number of trials on each, eventually solved each new problem in a single trial, with performance jumping from a chance level (50 per cent correct) on Trial 1, to over 90 per cent correct on Trial 2. These and other results have been reviewed by Miles (1965).

Perhaps the most important finding of learning-set studies has been that not all animals are capable of the level of performance attained by macaques and apes (Warren, 1965). Fig. 10.11 shows the performance of a variety of animals in standard learning-set experiments, and illustrates the

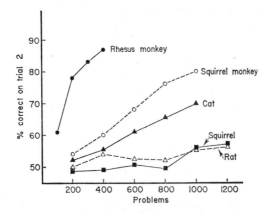

FIG. 10.11. Performance of different animals trained on a series of visual discrimination problems. As in most experiments on discrimination learning sets, the establishment of the learning set is measured by the probability of a correct response on Trial 2 of each new problem. (*After* Warren, 1965.)

striking superiority of rhesus monkeys to squirrel monkeys, and of both to members of several other mammalian orders. Warren was sufficiently cautious to wonder how long this tidy picture would last. His caution was justified. Although Shell and Riopelle (1958), for example, had presented data showing that the performance of such new world primates as cebus and spider monkeys was less proficient than that of macaques, subsequent experiments have not substantiated this difference (Devine, 1970). Herman and Arbeit (1973) have shown that a bottlenose dolphin is capable of performing at a level of 86 per cent to 100 per cent correct on Trial 2 of each problem of an auditory learning-set task. Although pigeons have shown relatively poor performance on a visual learning-set, some passerine birds such as mynas and bluejays have been found to attain a level of about 70 per cent correct on Trial 2 after about 600 problems (Kamil and Hunter, 1970; Hunter and Kamil, 1971).

As with any comparison of the performance of different animal groups trained under supposedly comparable conditions on supposedly comparable tasks, it is difficult to rule out the possibility that differences in performance are due simply to differences in conditions of testing, or at least to differences in sensory or motor capacities. It hardly needs emphasizing that the visual capacities of the rat, for example, are greatly inferior to those of the majority of primates. Several specific examples of the importance of such differences may be cited. Devine (1970) showed that differences in the learning-set performance of rhesus and cebus monkeys were largely a consequence of differences in the salience of colour as a cue for the two species. Although dolphins perform extremely well on an auditory learning-set task, with visual stimuli their performance is much less impressive (Herman *et al.*, 1969).

In spite of all these qualifications, however, there is good reason to believe that the performance of an adult, sophisticated, rhesus monkey, trained on several hundred visual discrimination problems, depends upon a number of associative processes developed only imperfectly, if at all, in animals such as rats and pigeons. If this is so, then the attempt by Reese (1964) to explain the rhesus monkey's learning-set performance in terms of a conditioning-extinction theory of discrimination learning, may be fundamentally misguided. What is needed is a theoretical analysis which takes as its point of departure the observation that many other animals simply do not show a similar level of performance.

A series of simultaneous visual discriminations contains a number of incidental stimuli that remain irrelevant from one problem to the next. It is possible, therefore, to attribute some improvement to the suppression of control by such irrelevant stimuli. Harlow (1959) has gone further than this and has attempted to explain the performance of the sophisticated rhesus monkey exclusively in terms of the suppression of what he terms "error factors". By a careful analysis of the performance of monkeys on a series of problems, Harlow was able to identify four error factors, or systematic, but incorrect, response tendencies. These included: position habits; a tendency to select the last rewarded position; a tendency to choose the stimulus *not* chosen on the preceding trial even though that choice had been rewarded; and a tendency to perseverate with choice of the stimulus chosen on Trial 1 even though that choice was not rewarded. During early problems monkeys would make numerous errors on each problem and the majority of these errors fell into one or other of these patterns. After a few hundred problems they made no more than one error per problem, and necessarily did not display any of these systematically erroneous response tendencies.

One may still ask, however, whether the elimination of error factors is a cause rather than a consequence of the improvement in performance. It can hardly be the sole cause, for the standard six-trial problem typically presented in a learning-set experiment permits 2^5, or 32 possible sequences of choices of the two discriminative stimuli after Trial 1, and the elimination of a small subset of these sequences could not possibly guarantee the emergence of the single correct sequence of choices. An animal that has simply learned not to respond in terms of spatial cues, for example, can still choose at random between the visual stimuli from one trial to the next. There is, in fact, evidence that the elimination of one error factor does not necessarily decrease the overall probability of errors, but results only in a redistribution of the same total number of errors (Bessemer and Stollnitz, 1971, p. 37). Although, therefore, some improvement over a series of visual discriminations may depend upon learning to ignore spatial cues and learning to attend to visual cues, and although this may indeed comprise an important part of the explanation of the relatively limited amount of improvement shown by some animals, it is surely necessary to suppose that the sophisticated rhesus monkey learns what to do in order to solve new visual discrimination problems, rather than simply what not to do.

A series of experiments by Schusterman (1962, 1964), Schrier (1966) and Warren (1966) has demonstrated that chimpanzees and macaques will perform efficiently in a learning-set situation after they have been trained simply on a series of discrimination reversals. In Warren's experiment, moreover, rhesus monkeys trained on a series of spatial reversals also showed excellent transfer to a visual learning-set task. These results provide convincing evidence that efficient learning-set performance cannot be based solely on the elimination of error factors. Only one of Harlow's four error factors, the overall tendency to select a single position, could be eliminated by training on a series of spatial reversals. It is apparent, therefore, that during the course of reversal training, primates must learn some strategy which transfers to any new two-choice simultaneous visual discrimination. It is not hard to see that this must be a generalized version of the win–stay, lose–shift strategy proposed by Hull (1952) to account for serial reversal performance, and by Restle (1958) and Levine (1965) to account for learning-set performance. In confirmation of this suggestion, Schusterman (1962) showed that chimpanzees trained on an object-alternation discrimination, which requires a win–shift, lose–stay strategy, showed extremely poor performance when tested on a learning-set task.

The generally accepted assumption has been that an animal following a win–stay, lose–shift strategy is utilizing traces of the outcome of one trial as a cue to control discriminative performance on the next (Bessemer and Stollnitz, 1971). The obvious implication, confirmed in a series of studies reported by Bessemer and Stollnitz, is that learning-set perform-ance should deteriorate as the intertrial interval (ITI) is lengthened. Deets *et al.* (1970) have also shown the importance of this factor: they found that rhesus monkeys, given a series of two-trial visual discriminations, per-formed significantly more accurately on Trial 2 of each problem with a 5 sec ITI than with a 20 sec ITI.

The effect of the ITI on the learning-set performance of primates is, of course, the same as its effect on serial reversal learning in rats and pigeons (p. 610); and Kamil *et al.* (1973) have reported a similar result when bluejays are tested in a series of learning-set problems. What, then, is responsible for the wide variations in learning-set proficiency between different animal groups? If rats learn reversals with a single error by utilizing the outcome of one trial as a cue to control behaviour on the next, why is their learning-set performance so much less efficient than that of rhesus monkeys and dolphins? It may, of course, be that the mechanisms for the development of a win–stay, lose–shift strategy are less well developed in some animals than in others, but it seems probable that an important factor determining learning-set performance is the ability to generalize this strategy to new discriminative stimuli. Whereas the rhesus monkeys in Warren's (1966) experiment showed excellent transfer from spatial reversal training to visual discrimination learning, rats show nega-tive rather than positive transfer when trained on spatial reversals and tested on visual reversals (p. 610). Warren (1966) found no evidence of positive transfer when he shifted cats from either visual or spatial re-versal training to a series of visual discriminations; and Ricciardi and Treichler (1970) found only marginal evidence of positive transfer in squirrel monkeys, one of the least proficient of primates at learning-set tasks. It is probable that an understanding of the processes involved in the ability to generalize rules across changes in specific stimuli will throw considerable light on the cognitive capacities of more advanced primates.

G. Conclusions

The observation of positive transfer between discrimination problems, when either the specific stimuli or the contingencies of reinforcement are changed from one problem to the next, creates some difficulties for any simple, conditioning-extinction theory of discrimination learning. The apparent implication of such transfer is that additional processes, other than those envisaged in these theories, must be involved in discrimination learning. In the final analysis this implication must be accepted, but it should be noted that conditioning-extinction theories are surprisingly resilient: for example, it is, in principle possible to predict the phenomenon of transfer along a continuum in terms of the generalization of approach and avoidance responses; moreover, Rescorla and Wagner (1972) are able to predict overshadowing without appeal to additional processes, and it may be possible to explain certain instances of transfer in terms of the overshadowing of irrelevant stimuli.

At the moment, however, there seem to be clear limits to the success of this approach. Neither an analysis in terms of generalization gradients, nor one in terms of differential overshadowing of irrelevant stimuli, is able to explain how pretraining on an easy discrimination may facilitate the learning of a more difficult discrimination between stimuli differing along the same dimension, even when the contingencies of reinforcement are reversed from one problem to the next. Similarly, Lawrence's experiments on transfer between simultaneous and successive conditional discriminations, and, even more importantly, the comparison of ID and ED shifts, clearly show that the effects of exposure to one discrimination on performance in a second problem depend upon the relationship between the relevant stimuli in each problem. The implication is that exposure to particular correlations between certain stimuli and reinforcement not only changes the current associative strength of those stimuli, but also affects their associability with subsequent changes in reinforcement.

Whether such changes in the associability of stimuli with reinforcement should be described as changes in attention, or whether they may be ascribed to changes in observing responses or strategies of orientation, is not always easy to decide. It is beyond question that the use of discrete, localized, discriminative stimuli may require the subject to learn appropriate observing behaviour, and that spatial contiguity between stimulus and response may have marked effects on discriminative performance. Nevertheless, it is doubtful whether changes in observing behaviour will explain why ID shifts are easier to learn than ED shifts, nor why overtraining may facilitate reversal more reliably than the learning of an entirely new discrimination. There is, of course, unequivocal evidence from studies of latent inhibition, that the repeated presentation of a salient stimulus (whose observation cannot seriously be thought to require elaborate orientation of the subject's receptors), in the absence of any correlated change in reinforcement, may significantly decrease the associability of that stimulus with future changes in reinforcement. It is reasonable to suggest that, in a similar manner, the presentation of a rather less salient stimulus, correlated with significant changes in reinforcement, might increase the associability of that stimulus with further changes in reinforcement.

Differential reinforcement in the presence of one set of stimuli, as was noted in Chapter 9, does undoubtedly neutralize incidental or irrelevant stimuli that might otherwise gain control over behaviour. Thus, it is not surprising that, in at least some situations, positive transfer may be observed from one discrimination to another, even when there is no relationship between the relevant stimuli of the two problems, and where any appeal to observing responses is clearly inappropriate.

Such general transfer has also been explained by supposing that discrimination training establishes a "set to discriminate". However unsatisfactory and vague such an explanation may appear, the formation of discrimination learning-sets by primates has encouraged the search for more respectable versions of such an account. The finding that the performance of a learning-set experienced monkey is adversely affected by increases in the intertrial interval has suggested that the basis for this set may lie in the utilization of the memory traces of one trial as a cue for appropriate choices on the next. There is some plausibility to this suggestion. However, one may question the implication of a sharp distinction between reliance on short-term memory traces and long-term storage of information. It is true that monkeys may show relatively rapid forgetting of the outcome of the first trial or two of a new visual discrimination, but by the time they have received six trials their retention is surprisingly good (e.g., Stollnitz and Schrier, 1968; Balogh and Zimmerman, 1971). Furthermore, such forgetting as does occur, is almost certainly a consequence of proactive interference from preceding problems (Conner and Meyer, 1971).

VII. SUMMARY

There has been a certain progression to the questions considered in this chapter. Beginning with a consideration of the proper definition of discriminative learning, of the associations formed during such learning, and the nature of the stimuli entering into those associations, the discussion proceeded with the question whether there is any selection between stimuli for control of discriminative performance and for association with reinforcement, and finally considered whether discriminative learning could be analysed solely in terms of changes in the current associative strengths of the discriminative stimuli and their generalization to other stimuli.

In a successive go/no-go discrimination, a particular response is reinforced in the presence of one stimulus and not in the presence of another. In successive conditional discriminations two responses are available to the subject (a choice between left and right): one response is reinforced in the presence of one configuration of stimuli, and the other response is reinforced in the presence of another configuration. Even a simultaneous visual discrimination may be described in these terms: a left turn, for example, may be reinforced when black is on the left, and a right turn when black is on the right. This set of descriptions of discrimination learning, although natural enough from the point of view of S–R theory, does seem very largely incorrect. The response tendencies usually established by discriminative training are those of approaching and avoiding particular stimuli or stimulus compounds, and the basis for these tendencies is presumably the establishment of associations between those stimuli and the presentation and omission of reinforcement. In most situations both approach and avoidance responses seem to be established, although the relative importance of the two may vary from situation to situation, in a manner partly determined by the physical arrangement of the discriminative stimuli and their spatial relationship to the location of the subject's responses.

Since experiments on discrimination learning necessarily require the subject to respond differently in the presence of different stimuli, they obviously raise questions about the proper description of the stimuli controlling the subject's behaviour. Although students of animal learning have typically ignored most questions related to perception, they have considered one particular problem: whether animals characterize S+ and S− as stimuli having certain absolute properties, such as a particular size or brightness, or whether they learn their relational characteristics, such as the fact that S+ is brighter or smaller than S−. Although the question has traditionally been posed as a dichotomy, there can hardly be doubt that discrimination learning may involve both processes. The importance of relational learning depends on the similarity of the discriminative stimuli and on the opportunity provided for their simultaneous comparison. Although the occurrence of transposition can, in principle, be accounted for in absolute terms, by postulating an appropriate interaction between excitatory and inhibitory gradients, and although this analysis provides a convincing explanation of shifts in the peak of a postdiscrimination gradient following successive discrimination training, transposition following simultaneous discrimination training is not reducible to such a peak shift. Nevertheless, the decline in transposition as subjects are tested with stimuli more and more dissimilar to those used during training, can only mean that discriminative performance was based partly on a characterization of those stimuli in absolute terms.

The data on transposition thus imply that animals may utilize all available cues, both absolute and relational, when solving visual discriminations; and studies of maze learning have also implied that rats may utilize available stimuli from any modality in order to learn the correct path through a maze. Nevertheless, a persistent theme in the theoretical analysis of discrimination learning has been the suggestion that discriminative performance is selectively controlled by a subset of the available stimuli and that not all stimuli change their associative strengths on every trial. In one version of a theory of stimulus selection, and one which remains popular for its formal simplicity, subjects are said to attend to only one set of stimuli on each trial; only those stimuli control performance on that trial, and enter into new associations on that trial.

Although there are good grounds for rejecting such an account of discrimination learning, there is a considerable body of evidence which appears to support a closely related position: studies of discrimination learning with compound relevant stimuli, and in particular studies of overshadowing and blocking, suggest that where several stimuli are simultaneously relevant, an increase in control by one stimulus is accompanied by a decrease in control by the remaining stimuli. This generalization would obviously be consistent with a less rigid theory of selective attention. Other explanations, however, have been suggested: such interactions may be a consequence not of any limitation on a subject's attentional capacity, but on the level of conditioning supportable by a given reinforcer; alternatively, when several stimuli are correlated with reinforcement, animals may learn that reinforcement is signalled by one component of the compound, and treat other components as redundant or uninformative.

Whatever may be the correct interpretation of stimulus selection, there is good evidence that the associability of a stimulus with reinforcement is not a fixed consequence of the salience of that stimulus and its temporal relationship with reinforcement, but may also be affected by the subject's previous experience with that stimulus. If experiments on latent inhibition suggest that prior exposure to a stimulus uncorrelated with changes in reinforcement may reduce the associability of that stimulus with reinforcement, experiments on the acquired distinctiveness of cues, transfer along a continuum, intradimensional and extradimensional shifts, and reversal learning, are all consistent with the proposition that prior exposure to a set of stimuli correlated with changes in reinforcement may increase the associability of those stimuli with further changes in reinforcement.

This is not to imply that the diversity of results obtained in studies of transfer of discrimination learning should all be explained by appeal to this single principle. It is obvious enough that instrumental discrimination learning may involve the establishment of appropriate observing responses, and that transfer between two problems may be mediated by such overt patterns of behaviour. Moreover, there are well-documented examples of positive transfer between successive discriminations, where the relevant stimuli change modality from one problem to the next and are both diffuse and unlocalized. General transfer effects such as these are obviously not to be explained in terms of changes in attention to the relevant stimuli, nor in terms of the transfer of observing responses. They may be partly a consequence of the suppression of control by irrelevant stimuli common to both problems; but a further possibility is that discrimination training may establish patterns of behaviour, such as learning to withhold responses in the presence of stimuli signalling nonreinforcement, that will benefit the learning of any new problem.

The most striking example of transfer between discrimination problems remains that observed in studies of learning-set formation. A rhesus monkey, trained on a series of brief, visual discrimination problems, soon learns to solve each new problem in a single trial. Similar performance has been observed in several other primate species and, in a study of auditory discrimination learning, in a dolphin; in general, however, although some species of passerine birds have shown surprisingly efficient performance, one-trial learning of new discriminations appears to be beyond the capacity of most other animals so far tested. Although some of the transfer from one problem to another typically shown by primates may depend on the neutralization of irrelevant stimuli, it is hard to see how this could be sufficient to produce one-trial solution of any new problem. It has, accordingly, been argued that the formation of a learning-set must involve the development of a generalized strategy of repeating rewarded choices and shifting from unrewarded choices.

REFERENCES AND AUTHOR INDEX

Numbers in italic list the pages on which the references are referred to.

Adamec, R. and Melzack, R. (1970). The role of motivation and orientation in sensory preconditioning. *Can. J. Psychol.* **24**, 230–9. [*86*]

Adelman, H. M. and Maatsch, J. L. (1956). Learning and extinction based upon frustration, food reward and exploratory tendency. *J. Exp. Psychol.* **52**, 311–315. [*264*]

Adrian, E. D. (1928). *The Basis of Sensation: The Action of the Sense Organs.* London: Christophers. [*63*].

Aiken, E. G. (1957). The effort variable in the acquisition, extinction and spontaneous recovery of an instrumental response. *J. Exp. Psychol.* **53**, 47–51. [*408, 416*]

Allison, J., Larson, D., and Jensen, D. D. (1967). Acquired fear, brightness preference, and one-way shuttlebox performance. *Psychon. Sci.* **8**, 269–70. [*312*]

Amsel, A. (1950). The combination of a primary appetitional need with primary and secondary emotionally derived needs. *J. Exp. Psychol.* **40**, 1–14. [*352*]

Amsel, A. (1958). The role of frustrative nonreward in noncontinuous reward situations. *Psychol. Bull.* **55**, 102–19. [*260, 349, 413, 440, 462*].

Amsel, A. (1962). Frustrative nonreward in partial reinforcement and discrimination learning. *Psychol. Rev.* **69**, 306–28. [*260, 349, 413, 440, 455, 456*]

Amsel, A. (1967). Partial reinforcement effects on vigor and persistence. In K. W. Spence and J. T. Spence (Eds.), *The Psychology of Learning and Motivation.* Vol. 1. New York: Academic Press. Pp. 1–65. [*160, 349, 398, 430, 440, 448, 458, 462*]

Amsel, A. (1971). Positive induction, behavioral contrast, and generalization of inhibition in discrimination learning. In H. H. Kendler and J. T. Spence (Eds.), *Essays in Neobehaviorism: A Memorial Volume to Kenneth W. Spence.* New York: Appleton-Century-Crofts. Pp. 217–36. [*349, 365, 367, 385, 398*]

Amsel, A. (1972). Inhibition and mediation in classical, Pavlovian and instrumental conditioning. In R. A. Boakes and M. S. Halliday (Eds.), *Inhibition and Learning.* London: Academic Press. Pp. 275–99. [*413*]

Amsel, A. and Cole, K. F. (1953). Generalization of fear-motivated interference with water intake. *J. Exp. Psychol.* **46**, 243–7. [*265, 352*]

Amsel, A., Hug, J. J., and Surridge, C. T. (1968). Number of food pellets, goal approaches, and the partial reinforcement effect after minimal acquisition. *J. Exp. Psychol.* **77**, 530–4. [*461, 462*]

Amsel, A., Hug, J. J., and Surridge, C. T. (1969). Subject-to-subject trial sequence, odor trails, and patterning at 24-h ITI. *Psychon. Sci.* **15**, 119–20. [*162, 163*]

Amsel, A. and Maltzman, I. (1950). The effect upon generalized drive strength of emotionality as inferred from the level of consummatory response. *J. Exp. Psychol.* **40**, 563–9. [*265*]

Amsel, A. and Rashotte, M. E. (1969). Transfer of experimenter-imposed slow-response patterns to extinction of a continuously rewarded response. *J. Comp. Physiol. Psychol.* **69**, 185–9. [*136, 441*]

Amsel, A., Rashotte, M. E., and MacKinnon, J. R. (1966). Partial reinforcement effects within subject and between subjects. *Psychol. Monogr.* **80** (20, Whole No. 628). [*160, 452*]

Amsel, A. and Roussel, J. (1952). Motivational properties of frustration: I. Effect on a running response of the addition of frustration to the motivational complex. *J. Exp. Psychol.* **43**, 363–8. [*266, 361*]

Amsel, A. and Ward, J. S. (1965). Frustration and persistence: resistance to discrimination following prior experience with the discriminanda. *Psychol. Monogr.* **79** (4, Whole No. 597). [*360, 362, 364, 365, 590, 601*]

Amsel, A., Wong, P. T., and Traupmann, K. L. (1971). Short-term and long-term factors in extinction and durable persistence. *J. Exp. Psychol.* **90**, 90–5. [*436, 458*]

Anderson, A. C. (1932). Time discrimination in the white rat. *J. Comp. Psychol.* **13**, 27–55. [*186, 187, 188*]

Anderson, D. C., O'Farrell, T., Formica, R. and Caponigri, V. (1969). Preconditioning CS exposure: Variation in place of conditioning and of presentation. *Psychon. Sci.* **15**, 54–5. [*37*]

Anger, D. (1963). The role of temporal discriminations in the reinforcement of Sidman avoidance behavior. *J. Exp. Anal. Behav.* **6**, 477–506. [*322, 324*]

Anisman, H. and Waller, T. G. (1971). Effects of conflicting response requirements and shock-compartment confinement on the Kamin effect in rats. *J. Comp. Physiol. Psychol.* **77**, 240–4. [*477*]

Annau, Z. and Kamin, L. J. (1961). The conditioned emotional response as a function of intensity of the US. *J. Comp. Physiol. Psychol.* **54**, 428–32. [*9, 10, 14, 15, 17, 70, 71, 333, 352*]

Anrep, G. V. (1920). Pitch discrimination in the dog. *J. Physiol.* **53**, 367–85. [*9*]

Anrep, G. V. (1923). The irradiation of conditioned reflexes. *Proc. Roy. Soc.* Series B, **94**, 404–26. [*514*]

Appel, J. B. (1963). Punishment and shock intensity. *Science, N.Y.* **141**, 528–9. [*274*]

Applezweig, M. H. (1951). Response potential as a function of effort. *J. Comp. Physiol. Psychol.* **44**, 225–35. [*416*]

Armus, H. L. (1959). Effect of magnitude of reinforcement on acquisition and extinction of a running response. *J. Exp. Psychol.* **58**, 61–3. [*151, 427*]

Armus, H. L. and Sniadowski-Dolinsky, D. (1966). Startle decrement and secondary reinforcement stimulation. *Psychon. Sci.* **4**, 175–6. [*224*]

Aronfreed, J. (1969). The problem of imitation. In L. P. Lipsitt and H. W. Reese (Eds.), *Advances in Child Development and Behavior*, Vol. 4, New York: Academic Press. Pp. 209–319. [*201*]

Asratyan, E. A. (1965). *Conditioned Reflex and Compensatory Mechanism*. Oxford: Pergamon Press. [*57, 60, 79*]

Atkinson, R. C. and Estes, W. K. (1963). Stimulus sampling theory. In R. D. Luce, R. R. Bush, and E. Galanter (Eds.), *Handbook of Mathematical Psychology*, Vol. 2. New York: Wiley. Pp. 121–268. [*191, 488*]

Autor, S. M. (1969). The strength of conditioned reinforcers as a function of frequency and probability of reinforcement. In D. P. Hendry (Ed.), *Conditioned Reinforcement*. Homewood, Illinois: The Dorsey Press. Pp. 127–62. [*243, 245, 382*]

Ayres, J. J. B. (1966). Conditioned suppression and the information hypothesis. *J. Comp. Physiol. Psychol.* **62**, 21–5. [*248*]

Ayres, J. J. B. and DeCosta, M. J. (1971). The truly random control as an extinction procedure. *Psychon. Sci.* **24**, 31–3. [*409*]

Azrin, N. H. (1956). Some effects of two intermittent schedules of immediate and nonimmediate punishment. *J. Psychol.* **42**, 3–21. [*274, 275*]

Azrin, N. H. (1960). Effects of punishment intensity during variable-interval reinforcement. *J. Exp. Anal. Behav.* **3**, 123–42. [*274*]

Azrin. N. H. (1961). Time-out from positive reinforcement. *Science, N.Y.* **133**, 382–3. [*264*]

Azrin, N. H. (1970). Punishment of elicited aggression. *J. Exp. Anal. Behav.* **14**, 7–10. [*288*]

Azrin, N. H. and Hake, D. F. (1969). Positive conditioned suppression: Conditioned suppression using positive reinforcers as the unconditioned stimuli. *J. Exp. Anal. Behav.* **12**, 167–73. [*225, 226*]

Azrin, N. H. and Holz, W. C. (1966). Punishment. In W. K. Honig (Ed.), *Operant Behavior: Areas of Research and Application*. New York: Appleton-Century-Crofts. Pp. 380–447. [*274, 292, 407*]

Azrin, N. H., Hutchinson, R. R., and Hake, D. F. (1966). Extinction-induced aggression. *J. Exp. Anal. Behav.* **9**, 191–204. [*263*]

Azrin, N. H., Hutchinson, R. R., and Hake, D. F. (1967). Attack, avoidance, and escape reactions to aversive shock. *J. Exp. Anal. Behav.* **10**, 131–48. [*108*]

Azrin, N. H., Hutchinson, R. R., and Sallery, R. D. (1964). Pain-aggression toward inanimate objects. *J. Exp. Anal. Behav.* 223–8. [*305*]

Bacon, W. E. (1962). Partial-reinforcement extinction effect following different amounts of training. *J. Comp. Physiol. Psychol.* **55**, 998–1003. [*160, 443, 446, 462*]

Bailey, C. J. and Miller, N. E. (1952). The effect of sodium amytal on an approach-avoidance conflict in cats. *J. Comp. Physiol. Psychol.* **45**, 205–8. [*265*]

Baker, R. A. and Lawrence, D. H. (1951). The differential effects of simultaneous and successive stimuli presentation on transposition. *J. Comp. Physiol. Psychol.* **44**, 378–82. [*565*]

Baker, T. W. (1968). Properties of compound conditioned stimuli and their components. *Psychol. Bull.* **70**, 611–25. [*47*]

Balogh, B. A. and Zimmermann, R. R. (1971). Short-term retention of object discriminations in experienced and naive rhesus monkeys. *Percept. Mot. Skills* **33**, 543–9. [*616*]

Banks, R. K. and Torney, D. (1969). Generalization of persistence: The transfer of approach behaviour to differing aversive stimuli. *Can. J. Psychol.* **23**, 268–73. [*464*]

Barlow, J. A. (1952). Secondary motivation through classical conditioning: One trial nonmotor learning in the white rat. *Am. Psychol.* **7**, 273. [*112*]

Barnes, G. W. (1956). Conditioned stimulus intensity and temporal factors in spaced-trial classical conditioning. *J. Exp. Psychol.* **51**, 192–8. [*41*]

Barnes, W. and Tombaugh, T. N. (1970). Effects of sucrose rewards on the overtraining extinction effect. *J. Exp. Psychol.* **86**, 355–9. [*423, 425, 427*]

Barnett, S. A. (1963). *The Rat: A Study in Behaviour*. London: Methuen. [*53*]

Baron, A., Kaufman, A., and Fazzini, D. (1969). Density and delay of punishment of free operant avoidance. *J. Exp. Anal. Behav.* **12**, 1029–37. [*274, 286, 328*]

Baron, M. R. (1965). The stimulus, stimulus control, and stimulus generalization.

In D. I. Mostofsky (Ed.), *Stimulus Generalization*. Stanford: Stanford University Press. Pp. 62–71. [*42*]

Baron, M. R. and Bresnahan, E. L. (1969). The effect of chromatic surround upon generalization along an angularity dimension in pigeons. *Psychon. Sci.* **15**, 9–10. [*449, 501*]

Barrett, R. J., Leith, N. J., and Ray, O. S. (1971). Kamin effect in rats: Index of memory or shock-induced inhibition? *J. Comp. Physiol. Psychol.* **77**, 234–9. [*477*]

Barrett, R. J., Peyser, C. S., and McHose, J. H. (1965). Effects of complete and incomplete reward reduction on a subsequent response. *Psychon. Sci.* **3**, 277–8. [*363*]

Barry, H., III. (1958). Effects of strength of drive on learning and extinction. *J. Exp. Psychol.* **55**, 473–81. [*151, 407*]

Barry, H., III and Miller, N. E. (1965). Comparison of drug effects on approach, avoidance and escape motivation. *J. Comp. Physiol. Psychol.* **59**, 18–24. [*265*]

Barry, H., III, Miller, N. E., and Tidd, G. E. (1962a). Control for stimulus change while testing effects of amobarbital on conflict. *J. Comp. Physiol. Psychol.* **55**, 1071–4. (a) [*265*]

Barry, H., III, Wagner, A. R. and Miller, N. E. (1962b). Effects of alcohol and amobarbital on performance inhibited by experimental extinction. *J. Comp. Physiol. Psychol.* **55**, 464–8 (b) [*265*]

Bauer, F. J. and Lawrence, D. H. (1953). Influence of similarity of choice-point and goal cues on discrimination learning. *J. Comp. Physiol. Psychol.* **46**, 241–52. [*552*]

Baum, M. (1966). Rapid extinction of an avoidance response following a period of response prevention in the avoidance apparatus. *Psychol. Rep.* **18**, 59–64. [*109, 341*]

Baum, M. (1969). Paradoxical effect of alcohol on the resistance to extinction of an avoidance response. *J. Comp. Physiol. Psychol.* **69**, 238–40. [*332*]

Baum, M. (1970). Extinction of avoidance responding through response prevention (Flooding). *Psychol. Bull.* **74**, 276–84. [*336*]

Baum, M. and Gleitman, H. (1967). "Conditioned anticipation" with an extinction base-line: the need for a disinhibition control group. *Psychon. Sci.* **8**, 95–6. [*226*]

Baum, W. M. and Rachlin, H. C. (1969). Choice as time allocation. *J. Exp. Anal. Behav.* **12**, 861–74. [*183*]

Beach, F. A. (1950). The snark was a boojum. *Am. Psychol.* **5**, 115–24. [*2*]

Beale, I. L. (1970). The effects of amount of training per reversal on successive reversals of a color discrimination. *J. Exp. Anal. Behav.* **14**, 345–52. [*609*]

Beale, I. L. and Winton, A. S. W. (1970). Inhibitory control in concurrent schedules. *J. Exp. Anal. Behav.* **14**, 133–7. [*528, 569*]

Beck, E. C. and Doty, R. W. (1957). Conditioned flexion reflexes acquired during combined catalepsy and de-efferentation. *J. Comp. Physiol. Psychol.* **50**, 211–16. [*80*]

Beecroft, R. S. (1967). Near-goal punishment of avoidance running. *Psychon. Sci.* **8**, 109–10. [*332, 341*]

Beer, T., Bethe, A., and Von Uexküll, J. (1899). Vorschläge zu einer objektivieren-der Nomenklatur in der Physiologie der Nervensystems. *Z. Physiol.* **13**, 137–41. [*2*]

Beery, R. G. (1968). A negative contrast effect of reward delay in differential conditioning. *J. Exp. Psychol.*, **77**, 429–34. [*393*]

Behrend, E. R. and Bitterman, M. E. (1961). Probability-matching in the fish. *Am. J. Psychol.* **74**, 542–51. [*192*]

Behrend, E. R. and Bitterman, M. E. (1966). Probability-matching in the goldfish. *Psychon. Sci.* **6**, 327–8. [*192*]

Behrend, E. R., Domesick, V. B., and Bitterman, M. E. (1965). Habit reversal in the fish. *J. Comp. Physiol. Psychol.* **60**, 407–11. [*609*]

Behrend, E. R., Powers, A. S., and Bitterman, M. E. (1970). Interference and forgetting in birds and fish. *Science, N.Y.* **167**, 389–90. [*479*]

Bekhterev, V. M. (1932). *General Principles of Human Reflexology.* New York: International Press. [*17*]

Bendig, A. W. (1952). Latent learning in a water maze. *J. Exp. Psychol.* **43**, 134–7. [*212*]

Benedict, J. O. and Ayres, J. J. B. (1972). Factors affecting conditioning in the truly random control procedure in the rat. *J. Comp. Physiol. Psychol.* **78**, 323–30. [*30*]

Beninger, R. J. (1972). Positive behavioral contrast with qualitatively different reinforcing stimuli. *Psychon. Sci.* **29**, 307–8. [*376*]

Berger, B. D., Yarczower, M. and Bitterman, M. E. (1965). Effect of partial reinforcement on the extinction of a classically conditioned response in the goldfish. *J. Comp. Physiol. Psychol.* **59**, 399–405. [*72, 74*]

Berlyne, D. E. (1960). *Conflict, Arousal, and Curiosity.* New York: McGraw-Hill. [*251*]

Bernheim, J. W. (1968). Comment. *Psychon. Sci.*, **11**, 327. [*530*]

Bernheim, J. W. and Williams, D. R. (1967). Time-dependent contrast effects in a multiple schedule of food reinforcement. *J. Exp. Anal. Behav.* **10**, 243–9. [*206, 357, 359, 366, 378*]

Bersh, P. J. (1951). The influence of two variables upon the establishment of a secondary reinforcer for operant responses. *J. Exp. Psychol.* **41**, 62–73. [*242, 244, 245*]

Bersh, P. J. and Paynter, W., Jr. (1972). Pavlovian extinction in rats during avoidance response prevention. *J. Comp. Physiol. Psychol.* **78**, 255–9. [*336*]

Bertsch, G. J. and Leitenberg, H. (1970). A "frustration effect" following electric shock. *Learning and Motivation*, **1**, 150–6. [*365*]

Bessemer, D. W. and Stollnitz, F. (1971). Retention of discriminations and an analysis of learning set. In A. M. Schrier and F. Stollnitz (Eds.), *Behavior of Nonhuman Primates*, Vol 4. New York: Academic Press. Pp. 1–58. [*613*]

Bevan, W. and Adamson, R. (1960). Reinforcers and reinforcement: their relation to performance. *J. Exp. Psychol.* **59**, 226–32. [*370, 390*]

Biederman, G. B., D'Amato, M. R., and Keller, D. M. (1964). Facilitation of discriminated avoidance learning by dissociation of CS and manipulandum. *Psychon. Sci.* **1**, 229–30. [*340, 342, 343*]

Bindra, D. (1961). Components of general activity and the analysis of behavior. *Psychol. Rev.* **68**, 205–15. [*411*]

Bindra, D. (1969). The interrelated mechanisms of reinforcement and motivation, and the nature of their influence on response. In W. J. Arnold and D. Levine (Eds.), *Nebraska Symposium on Motivation*, Vol. 17. Lincoln: University of Nebraska Press. Pp. 1–33. [*199*]

Bindra, D. (1972). A unified account of classical conditioning and operant training. In A. H. Black and W. F. Prokasy (Eds.), *Classical Conditioning II: Current Research and Theory*. New York: Appleton-Century-Crofts. Pp. 453–81. [*79, 143, 199, 205, 222*]

Birch, D. (1965). Extended training extinction effect under massed and spaced extinction trials. *J. Exp. Psychol.* **70**, 315–22. [*419*]

Birch, D., Ison, J. R., and Sperling, S. E. (1960). Reversal learning under single stimulus presentation. *J. Exp. Psychol.* **60**, 36–40. [*604*]

Birch, H. A. and Bitterman, M. E. (1949). Reinforcement and learning: The process of sensory integration. *Psychol. Rev.* **56**, 292–308. [*79*]

Bitterman, M. E. (1957). Review of K. W. Spence's *Behavior Theory and Conditioning*. *Am. J. Psychol.* **70**, 141–5. [*211*]

Bitterman, M. E. (1964). Classical conditioning in the goldfish as a function of the CS-US interval. *J. Comp. Physiol. Psychol.* **58**, 359–66. [*57, 61, 65, 350*]

Bitterman, M. E. (1965). The CS-US interval in classical and avoidance conditioning. In W. F. Prokasy (Ed.), *Classical Conditioning: A Symposium.* New York: Appleton-Century-Crofts. Pp. 1–19. [*10*]

Bitterman, M. E. (1967). Learning in animals. In H. Helson and W. Bevan (Eds.), *Contemporary Approaches to Psychology.* Princeton: Van Nostrand. Pp. 139–79. [*1, 594*]

Bitterman, M. E. (1971). Visual probability learning in the rat. *Psychon. Sci.* **22**, 191–2. [*192*]

Bitterman, M. E. and Coate, W. B. (1950). Some new experiments on the nature of discrimination learning in the rat. *J. Comp. Physiol. Psychol.* **43**, 198–210. [*581, 582*]

Bitterman, M. E., Fedderson, W. E., and Tyler, D. W. (1953). Secondary re inforcement and the discrimination hypothesis. *Am. J. Psychol.* **66**, 456–64. [*235*]

Bitterman, M. E. and McConnell, J. V. (1954). The role of set in successive discrimination. *Am. J. Psychol.* **67**, 129–32. [*546, 551, 599*]

Bitterman, M. E., Reed, P. C., and Krauskopf, J. (1952). The effect of the duration of the unconditioned stimulus upon conditioning and extinction. *Am. J. Psychol.* **65**, 256–62. [*112*]

Bitterman, M. E., Tyler, D. W., and Elam, C. B. (1955). Simultaneous and successive discrimination under identical stimulating conditions. *Am. J. Psychol.* **68**, 237–48. [*546, 548, 549, 550*]

Bitterman, M. E., Wodinsky, J., and Candland, D. K. (1958). Some comparative psychology. *Am. J. Psychol.* **71**, 94–110. [*191, 192, 608, 610*]

Black, A. H. (1958). The extinction of avoidance responses under curare. *J. Comp. Physiol. Psychol.* **51**, 519–24. [*305, 336, 417*]

Black, A. H. (1965). Cardiac conditioning in curarized dogs: The relationship between heart rate and skeletal behavior. In W. F. Prokasy (Ed.), *Classical Conditioning: A Symposium.* New York: Appleton-Century-Crofts. Pp. 20–47. [*17, 80, 104*]

Black, A. H. (1971). Autonomic aversive conditioning in infrahuman subjects. In F. R. Brush (Ed.), *Aversive Conditioning and Learning.* New York: Academic Press. Pp. 3–104. [*17, 78, 100, 103, 130, 131, 132*]

Black, A. H. and de Toledo, L. (1972). The relationship among classically conditioned responses: Heart rate and skeletal behavior. In A. H. Black and W. F. Prokasy (Eds.), *Classical Conditioning II: Current Research and Theory.* New York: Appleton-Century-Crofts. Pp. 290–311. [*103*]

Black, A. H. and Morse, P. (1961). Avoidance learning in dogs without a warning stimulus. *J. Exp. Anal. Behav.* **4**, 17–23. [*286*]

Black, A. H. and Young, G. A. (1972). Constraints on the operant conditioning drinking. In R. M. Gilbert and J. R. Millenson (Eds.), *Reinforcement: Behavioral Analyses*. New York: Academic Press. Pp. 35–50. [*305*]

Black, R. W. (1968). Shifts in magnitude of reward and contrast effects in instrumental and selective learning. *Psychol. Rev.* **75**, 114–26. [*215, 396, 398*]

Black, R. W. (1969). Incentive motivation and the parameters of reward in instrumental conditioning. In W. J. Arnold and D. Levine (Eds.), *Nebraska Symposium on Motivation*, Vol. 17. Lincoln: University of Nebraska Press. Pp. 85–137. [*153*]

Black, R. W. and Black, P. E. (1967). Heart rate conditioning as a function of interstimulus interval in rats. *Psychon. Sci.* **8**, 219–20. [*65, 103*]

Black, R. W. and Spence, K. W. (1965). Effects of intertrial reinforcement on resistance to extinction following extended training. *J. Exp. Psychol.* **70**, 559–63. [*451, 461*]

Blackman, D. (1968). Conditioned suppression or facilitation as a function of the behavioral baseline. *J. Exp. Anal. Behav.* **11**, 53–61. [*82, 228*]

Blackman, D. E. (1970). Conditioned suppression of avoidance behavior in rats. *Q. Jl Exp. Psychol.* **22**, 547–53. [*83*]

Blackwell, H. R. and Schlosberg, H. (1943). Octave generalization, pitch discrimination, and loudness thresholds in the white rat. *J. Exp. Psychol.* **33**, 407–19. [*490*]

Blanchard, R. J. and Blanchard, D. C. (1969a). Crouching as an index of fear. *J. Comp. Physiol. Psychol.* **67**, 370–5. (a) [*82*]

Blanchard, R. J. and Blanchard, D. C. (1969b). Passive and active reactions to fear-eliciting stimuli. *J. Comp. Physiol. Psychol.* **68**, 129–35. (b) [*109*]

Blanchard, R. J. and Blanchard, D. C. (1970). Dual mechanisms in passive avoidance: I. *Psychon. Sci.* **19**, 1–2. [*109*]

Blodgett, H. C. (1929). The effect of the introduction of reward upon the maze performance of rats. *University of California Publications in Psychology*, **4**, 113–34. [*207*]

Bloom, J. M. (1967). Early acquisition responding on trials following different rewards and nonrewards. *Psychon. Sci.* **7**, 37–8. [*161*]

Bloom, J. M. and Capaldi, E. J. (1961). The behavior of rats in relation to complex patterns of partial reinforcement. *J. Comp. Physiol. Psychol.* **54**, 261–5. [*163, 448*]

Bloom, J. M. and Malone, P. (1968). Single alternation patterning without a trace for blame. *Psychon. Sci.* **11**, 335–6. [*162*]

Bloomfield, T. M. (1966). Two types of behavioral contrast in discrimination learning. *J. Exp. Anal. Behav.* **9**, 155–61 [*372*].

Bloomfield, T. M. (1967a). Behavioral contrast and relative reinforcement frequency in two multiple schedules. *J. Exp. Anal. Behav.* **10**, 151–8. (a) [*380*]

Bloomfield, T. M. (1967b). Some temporal properties of behavioral contrast. *J. Exp. Anal. Behav.* **10**, 159–64. (b) [*369, 372, 373, 374*]

Bloomfield, T. M. (1967c). A peak shift on a line tilt continuum. *J. Exp. Anal. Behav.* **10**, 361–6. (c) [*536*]

Bloomfield, T. M. (1969). Behavioral contrast and the peak shift. In R. M. Gilbert and N. S. Sutherland (Eds.), *Animal Discrimination Learning*. London: Academic Press. Pp. 215–41. [*382, 569*]

Bloomfield, T. M. (1972). Contrast and inhibition in discrimination learning by the pigeon: Analysis through drug effects. *Learning and Motivation*, **3**, 162–78. [*381*]

x

Blough, D. S. (1965). Definitions and measurement in generalization research. In D. I. Mostofsky (Ed.), *Stimulus Generalization*. Stanford: Stanford University Press. Pp. 30–37. [*491*]

Blough, D. S. (1966). The reinforcement of least-frequent interresponse times. *J. Exp. Anal. Behav.* **9**, 581–91. [*172, 174*]

Blough, D. S. (1967). Stimulus generalization as signal detection in pigeons. *Science, N.Y.* **158**, 940–1. [*498*]

Blough, D. S. (1969a). Generalization gradient shape and summation in steady-state tests. *J. Exp. Anal. Behav.* **12**, 91–104. (a) [*498, 532*]

Blough, D. S. (1969b). Attention shifts in a maintained discrimination. *Science, N.Y.* **166**, 125–6. [*502, 512*]

Boakes, R. A. (1973). Response decrements produced by extinction and by response-independent reinforcement. *J. Exp. Anal. Behav.* **19**, 293–302. [*409*]

Boe, E. E. and Church, R. M. (1967). Permanent effects of punishment during extinction. *J. Comp. Physiol. Psychol.* **63**, 486–92. [*274, 275, 291, 292, 352*]

Boice, R. and Denny, M. R. (1965). The conditioned licking response in rats as a function of the CS-UCS interval. *Psychon. Sci.* **3**, 93–4. [*17, 65*]

Bolles, R. C. (1967). *Theory of Motivation*. New York: Harper & Row. [*224, 349*]

Bolles, R. C. (1969). Avoidance and escape learning: Simultaneous acquisition of different responses. *J. Comp. Physiol. Psychol.* **68**, 355–8. [*302*]

Bolles, R. C. (1970). Species-specific defense reactions and avoidance learning. *Psychol. Rev.* **77**, 32–48. [*137, 303, 340, 342*]

Bolles, R. C. (1971). Species-specific defense reactions. In F. R. Brush (Ed.), *Aversive Conditioning and Learning*. New York: Academic Press. Pp. 183–233. [*137, 302, 303, 314, 340, 342, 343*]

Bolles, R. C. (1972). Reinforcement, expectancy, and learning. *Psychol. Rev.* **79**, 394–409. [*231*]

Bolles, R. C. and Grossen, N. E. (1969). Effects of an informational stimulus on the acquisition of avoidance behavior in rats. *J. Comp. Physiol. Psychol.* **68**, 90–9. [*319*]

Bolles, R. C. and Grossen, N. E. (1970). The noncontingent manipulation of incentive motivation. In J. H. Reynierse (Ed.), *Current Issues in Animal Learning*. Lincoln: University of Nebraska Press. Pp. 143–74. [*226*]

Bolles, R. C., Moot, S. A., and Grossen, N. E. (1971). The extinction of shuttlebox avoidance. *Learning and Motivation*, **2**, 324–33. [*286, 333, 338, 339*]

Bolles, R. C. and Popp, R. J., Jr. (1964). Parameters affecting the acquisition of Sidman avoidance. *J. Exp. Anal. Behav.* **7**, 315–21. [*328*]

Bolles, R. C. and Seelbach, S. E. (1964). Punishing and reinforcing effects of noise onset and termination for different responses. *J. Comp. Physiol. Psychol.* **58**, 127–31. [*219*]

Bolles, R. C., Stokes, L. W., and Younger, M. S. (1966). Does CS termination reinforce avoidance behavior? *J. Comp. Physiol. Psychol.* **62**, 201–7. [*116, 206, 302, 304, 305, 318, 321*]

Boneau, C. A. (1958). The interstimulus interval and the latency of the conditioned eyelid response. *J. Exp. Psychol.* **56**, 464–72. [*62*]

Boneau, C. A. and Axelrod, S. (1962). Work decrement and reminiscence in pigeon operant responding. *J. Exp. Psychol.* **64**, 352–4. [*357, 359*]

Boneau, C. A. and Cole, J. L. (1967). Decision theory, the pigeon, and the psychophysical function. *Psychol. Rev.* **74**, 123–35. [*498*]

Boneau, C. A. and Honig, W. K. (1964). Opposed generalization gradients based upon conditional discrimination training. *J. Exp. Psychol.* **66**, 89–93. *[509]*

Booth, J. H. and Hammond, L. J. (1971). Configural conditioning: greater fear in rats to compound than component through overtraining of the compound. *J. Exp. Psychol.* **87**, 255–62. *[47]*

Borgealt, A. J., Donahoe, J. W., and Weinstein, A. (1972). Effects of delayed and trace components of a compound CS on conditioned suppression and heart rate. *Psychon. Sci.* **26**, 13–15. *[103, 249]*

Bowen, J. (1968). Effect of ascending, descending and irregular order of varied reward magnitude. *Psychon. Sci.* **12**, 209–10. *[430]*

Bower, G. H. (1961). A contrast effect in differential conditioning. *J. Exp. Psychol.* **62**, 196–9. *[385, 388]*

Bower, G. H. (1962). The influence of graded reductions in reward and prior frustrating events upon the magnitude of the frustration effect. *J. Comp. Physiol. Psychol.* **55**, 582–7. *[362]*

Bower, G. H., Fowler, H., and Trapold, M. A. (1959). Escape learning as a function of amount of shock reduction. *J. Exp. Psychol.* **58**, 482–4. *[152]*

Bower, G. and Grusec, T. (1964). Effect of prior Pavlovian discrimination training upon learning an operant discrimination. *J. Exp. Anal. Behav.* **7**, 401–4. *[203, 225]*

Bower, G., McLean, J., and Meacham, J. (1966). Value of knowing when reinforcement is due. *J. Comp. Physiol. Psychol.* **62**, 184–92. *[252, 253, 254]*

Bower, G., Starr, R., and Lazarovitz, L. (1965). Amount of response-produced change in the CS and avoidance learning. *J. Comp. Physiol. Psychol.* **59**, 13–17. *[319, 320]*

Bower, G. H. and Trapold, M. A. (1959). Reward magnitude and learning in a single-presentation discrimination. *J. Comp. Physiol. Psychol.* **52**, 727–9. *[151, 384, 385]*

Brahlek, J. A. (1968). Conditioned suppression as a function of the number of stimuli that precede shock. *Psychon. Sci.* **12**, 189–90. *[62]*

Brandauer, C. M. (1953). A confirmation of Webb's data concerning the action of irrelevant drives. *J. Exp. Psychol.* **45**, 150–2. *[350]*

Breland, K. and Breland, M. (1961). The misbehavior of organisms. *Am. Psychol.* **16**, 661–4. *[137, 219]*

Breland, K. and Breland, M. (1966). *Animal Behavior.* New York: The Macmillan Company. *[108, 137, 219]*

Brelsford, J., Jr. and Theios, J. (1965). Single session conditioning of the nictitating membrane in the rabbit: Effect of intertrial interval. *Psychon. Sci.* **2**, 81–2. *[11, 17]*

Brener, J. and Goesling, W. J. (1970). Avoidance conditioning of activity and immobility in rats. *J. Comp. Physiol. Psychol.* **70**, 276–80. *[343]*

Brener, J. and Hothersall, D. (1966). Heart rate control under conditions of augmented sensory feedback. *Psychophysiology* **3**, 23–8. *[133]*

Brethower, D. M. and Reynolds, G. S. (1962). A facilitative effect of punishment on unpunished behavior. *J. Exp. Anal. Behav.* **5**, 191–9. *[380]*

Brimer, C. J. (1970). Disinhibition of an operant response. *Learning and Motivation* **1**, 346–71. *[16, 32, 44]*

Brimer, C. J. and Dockrill, F. J. (1966). Partial reinforcement and the CER. *Psychon. Sci.* **5**, 185–6. *[24, 73, 74]*

Broadhurst, P. L. (1957). Emotionality and the Yerkes-Dodson law. *J. Exp. Psychol.* **54**, 345–52. [*184*]

Brogden, W. J. (1939a). Unconditioned stimulus-substitution in the conditioning process. *Am. J. Psychol.* **52**, 46–55. (a) [*110, 111*]

Brogden, W. J. (1939b). The effect of frequency of reinforcement upon the level of conditioning. *J. Exp. Psychol.* **24**, 419–31. (b) [*17, 24, 72, 73, 117*]

Brogden, W. J. (1939c). Sensory pre-conditioning. *J. Exp. Psychol.* **25**, 323–32. (c) [*20*]

Brogden, W. J. and Gantt, W. H. (1937). Cerebellar conditioned reflexes. *Am. J. Physiol.* **119**, 277–8. [*19, 94*]

Brogden, W. J., Lipman, E. A. and Culler, E. (1938). The role of incentive in conditioning and extinction. *Am. J. Psychol.* **51**, 109–17. [*116, 135, 304, 305, 308*]

Bronstein, P. and Spear, N. E. (1972). Acquisition of a spatial discrimination by rats as a function of age. *J. Comp. Physiol. Psychol.* **78**, 208–12. [*183*]

Brower, L. P. (1969). Ecological chemistry. *Sci. Am.* **220**, 22–9. [*55*]

Brown, B. L. (1970). Stimulus generalization in salivary conditioning. *J. Comp. Physiol. Psychol.* **71**, 467–77. [*518*]

Brown, J. S. (1942). The generalization of approach responses as a function of stimulus intensity. *J. Comp. Psychol.* **33**, 209–26. [*485, 520*]

Brown, J. S. (1948). Gradients of approach and avoidance responses and their relation to level of motivation. *J. Comp. Physiol. Psychol.* **41**, 450–65. [*523, 524*]

Brown, J. S. (1969). Factors affecting self-punitive locomotor behavior. In B. A. Campbell and R. M. Church (Eds.), *Punishment and Aversive Behavior*. New York: Appleton-Century-Crofts. Pp. 467–514. [*286*]

Brown, J. S. and Bass, B. (1958). The acquisition and extinction of an instrumental response under constant and variable stimulus conditions. *J. Comp. Physiol. Psychol.* **51**, 499–504. [*434*]

Brown, J. S. and Jacobs, A. (1949). The role of fear in the motivation and acquisition of responses. *J. Exp. Psychol.* **39**, 747–59. [*311*]

Brown, J. S., Kalish, H. I., and Farber, I. E. (1951). Conditioned fear as revealed by magnitude of startle response to an auditory stimulus. *J. Exp. Psychol.* **41**, 317–28. [*310*]

Brown, P. L. and Jenkins, H. M. (1967). Conditioned inhibition and excitation in operant discrimination learning. *J. Exp. Psychol.* **75**, 255–66. [*33, 569*]

Brown, P. L. and Jenkins, H. M. (1968). Auto-shaping of the pigeon's key-peck. *J. Exp. Anal. Behav.* **11**, 1-8. [*17, 199*]

Brown, R. T. and Logan, F. A. (1965). Generalised partial reinforcement effect. *J. Comp. Physiol. Psychol.* **60**, 64–9. [*451, 452*]

Brown, R. T. and Wagner, A. R. (1964). Resistance to punishment and extinction following training with shock or nonreinforcement. *J. Exp. Psychol.* **68**, 503–7. [*160, 463, 464*]

Brownlee, A. and Bitterman, M. E. (1968). Differential reward conditioning in the pigeon. *Psychon. Sci.* **12**, 345–6. [*394*]

Brownstein, A. J. (1971). Concurrent schedules of response-independent reinforcement: duration of a reinforcing stimulus. *J. Exp. Anal. Behav.* **15**, 211–14. [*186*]

Brownstein, A. J. and Hughes, R. G. (1970). The role of response suppression in behavioral contrast: signalled reinforcement. *Psychon. Sci.* **18**, 50–1. [*381*]

Bruner, A. (1965). UCS properties in classical conditioning of the albino rabbit's nictating membrane response. *J. Exp. Psychol.* **69**, 186–92. [*92, 96, 97*]

Bruner, A. (1969). Reinforcement strength in classical conditioning of leg flexion, freezing and heart rate in cats. *Conditional Reflex* **4**, 24–31. [*22, 71, 92, 96*]

Bruner, J. S., Matter, J., and Papanek, M. L. (1955). Breadth of learning as a function of drive level and mechanization. *Psychol. Rev.* **62**, 1–10. [*583*]

Brush, F. R. (1957). The effects of shock intensity on the acquisition and extinction of an avoidance response in dogs. *J. Comp. Physiol. Psychol.* **50**, 547–52. [*332, 333, 344*]

Brush, F. R. (1962). The effects of intertrial interval on avoidance learning in the rat. *J. Comp. Physiol. Psychol.* **55**, 888–92. [*320*]

Brush, F. R. (1971). Retention of aversively motivated behavior. In F. R. Brush (Ed.), *Aversive Conditioning and Learning*. New York: Academic Press. Pp. 401–65. [*475, 477*]

Bryant, R. C. (1972). Conditioned suppression of free-operant avoidance. *J. Exp. Anal. Behav.* **17**, 257–60. [*83*]

Bugelski, B. R. (1938). Extinction with and without sub-goal reinforcement. *J. Comp. Psychol.* **26**, 121–34. [*234*]

Bugelski, B. R., Coyer, R. A., and Rogers, W. A. (1952). A criticism of pre-acquisition and pre-extinction of expectancies. *J. Exp. Psychol.* **44**, 27–30. [*416*]

Bull, J. A. III (1970). An interaction between appetitive Pavlovian CSs and instrumental avoidance responding. *Learning and Motivation* **1**, 18–26. [*18, 224*]

Bull, J. A. III and Overmier, J. B. (1968). Additive and subtractive properties of excitation and inhibition. *J. Comp. Physiol. Psychol.* **66**, 511–14. [*33, 204, 311*]

Bullock, D. H. and Bitterman, M. E. (1962). Probability-matching in the pigeon. *Am. J. Psychol.* **75**, 634–9. [*192*]

Bullock, D. H. and Smith, W. C. (1953). An effect of repeated conditioning-extinction upon operant strength. *J. Exp. Psychol.* **46**, 349–52. [*441*]

Burstein, K. R. (1965). The influence of UCS intensity upon the acquisition of the conditioned eyelid response. *Psychon. Sci.* **2**, 303–4. [*71*]

Burstein, K. R. and Moeser, S. (1971). The informational value of a distinctive stimulus associated with the initiation of acquisition trials. *Learning and Motivation* **2**, 228–34. [*470*]

Burt, D. E. and Wike, E. L. (1963). Effects of alternating partial reinforcement and alternating delay of reinforcement on a runway response. *Psychol. Rep.* **13**, 439–42. [*401*]

Bush, R. R. (1965). Identification learning. In R. D. Luce, R. R. Bush, and E. Galanter (Eds.), *Handbook of Mathematical Psychology*, Vol. 3. New York: Wiley. Pp. 161–203. [*545*]

Bush, R. R. and Mosteller, F. (1951). A mathematical model for simple learning. *Psychol. Rev.* **58**, 313–23. [*11, 488*]

Butter, C. M. and Thomas, D. R. (1958). Secondary reinforcement as a function of the amount of primary reinforcement. *J. Comp. Physiol. Psychol.* **51**, 346–8. [*242*]

Bykov, K. M. (1957). *The Cerebral Cortex and the Internal Organs*. New York: Chemical Publishing Co. [*19*]

Byrd, L. D. and Marr, M. J. (1969). Relations between patterns of responding and the presentation of stimuli under second-order schedules. *J. Exp. Anal. Behav.* **12**, 713–22. [*241*]

Camp, D. S., Raymond, G. A., and Church, R. M. (1967). Temporal relationship between response and punishment. *J. Exp. Psychol.* **74**, 114–23. [*274, 275, 282, 352*]

Campbell, B. A. and Campbell, E. H. (1962). Retention and extinction of learned fear in infant and adult rats. *J. Comp. Physiol. Psychol.* **55**, 1–8. [*474*]

Campbell, B. A. and Cicala, G. A. (1962). Studies of water deprivation in rats as a function of age. *J. Comp. Physiol. Psychol.* **55**, 763–8. [*350*]

Campbell, B. A., Jaynes, J., and Misanin, J. R. (1968). Retention of a light-dark discrimination in rats of different ages. *J. Comp. Physiol. Psychol.* **66**, 467–72. [*476*]

Campbell, B. A. and Kraeling, D. (1953). Response strength as a function of drive level and amount of drive reduction. *J. Exp. Psychol.* **45**, 97–101. [*150, 152*]

Campbell, B. A. and Sheffield, F. D. (1953). Relation of random activity to food deprivation. *J. Comp. Physiol. Psychol.* **46**, 320–2. [*350*]

Campbell, B. A., Smith, N. F. S., and Misanin, J. R. (1966a). Effects of punishment on extinction of avoidance behavior: Avoidance-avoidance conflict or vicious circle behavior? *J. Comp. Physiol. Psychol.* **62**, 495–8. (a) [*286, 289*]

Campbell, B. A., Smith, N. F., Misanin, J. R., and Jaynes, J. (1966b). Species differences in activity during hunger and thirst. *J. Comp. Physiol. Psychol.* **61**, 123–7. (b) [*350*]

Campbell, B. A. and Spear, N. E. (1972). Ontogeny of memory. *Psychol. Rev.* **79**, 215–36. [*474, 475*]

Campbell, P. E., Batsche, C. J., and Batsche, G. M. (1972). Spaced-trials reward magnitude effects in the rat: single versus multiple food pellets. *J. Comp. Physiol. Psychol.* **81**, 360–4. [*153*]

Campbell, P. E., Crumbaugh, C. M., Knouse, S. B., and Snodgrass, M. E. (1970). A test of the "ceiling effect" hypothesis of positive contrast. *Psychon. Sci.* **20**, 17–18. [*389*]

Campbell, P. E., Crumbaugh, C. M., Rhodus, D. M., and Knouse, S. B. (1971). Magnitude of partial reward and amount of training in the rat: An hypothesis of sequential effects. *J. Comp. Physiol. Psychol.* **75**, 120–8. [*449*]

Capaldi, E. D. (1971a). Simultaneous shifts in reward magnitude and level of food deprivation. *Psychon. Sci.* **23**, 357–9. (a) [*151, 389, 390*]

Capaldi, E. D. (1971b). Effect of nonrewarded and partially rewarded placements on resistance to extinction in the rat. *J. Comp. Physiol. Psychol.* **76**, 483–90. (b) [*451, 452*]

Capaldi, E. J. (1958). The effect of different amounts of training on the resistance to extinction of different patterns of partially reinforced responses. *J. Comp. Physiol. Psychol.* **51**, 367–71. [*161, 162, 448, 449*]

Capaldi, E. J. (1964). Effect of N-length, number of different N-lengths, and number of reinforcements on resistance to extinction. *J. Exp. Psychol.* **68**, 230–9. [*450*]

Capaldi, E. J. (1966a). Partial reinforcement: a hypothesis of sequential effects. *Psychol. Rev.* **73**, 459–77. (a) [*440, 444, 446, 447*]

Capaldi, E. J. (1966b). Stimulus specificity: nonreward. *J. Exp. Psychol.* **72**, 410–14. (b) [*407*]

Capaldi, E. J. (1967a). A sequential hypothesis of instrumental learning. In K. W. Spence and J. T. Spence (Eds.), *The Psychology of Learning and Motivation*, Vol. 1. New York: Academic Press. Pp. 67–156. (a) [*153, 162, 398, 405, 407, 428, 432, 433, 440, 447, 455*]

Capaldi, E. J. (1967b). Sequential versus nonsequential variables in partial delay of reward. *J. Exp. Psychol.* **74**, 161–6. (b) [*454*]

Capaldi, E. J. (1971). Memory and learning: a sequential viewpoint. In W. K. Honig and P. H. R. James (Eds.), *Animal Memory*. New York: Academic Press. Pp. 111–54. [*163, 440, 451*]

Capaldi, E. J. (1972). Successive negative contrast effect: intertrial interval, type of shift, and four sources of generalization decrement. *J. Exp. Psychol.* **96**, 433–8. [*391, 392, 400*]

Capaldi, E. J., Berg, R. F., and Sparling, D. L. (1971). Trial spacing and emotionality in the rat. *J. Comp. Physiol. Psychol.* **76**, 290–9. [*458*]

Capaldi, E. J. and Bowen, J. N. (1964). Delay of reward and goal box confinement time in extinction. *Psychon. Sci.* **1**, 141–2. [*431, 432*]

Capaldi, E. J. and Capaldi, E. D. (1970a). A discrepancy between anticipated reward and obtained reward with no increase in resistance to extinction. *Psychon. Sci.* **18**, 19–20. (a) [*430*]

Capaldi, E. J. and Capaldi, E. D. (1970b). Magnitude of partial reward, irregular reward schedules, and a 24-hour ITI: a test of several hypotheses. *J. Comp. Physiol. Psychol.* **72**, 203–9. (b) [*457*]

Capaldi, E. J. and Cogan, D. (1963). Magnitude of reward and differential stimulus consequences. *Psychol. Rep.* **13**, 85–6. [*401*]

Capaldi, E. J., Godbout, R. C., and Ksir, C. (1968a). A comparison of two delay of reward procedures, pre-reinforcement delay vs. post-reinforcement delay. *Psychon. Sci.* **13**, 279–80. (a) [*433, 454*]

Capaldi, E. J. and Hart, D. (1962). Influence of a small number of partial reinforcement training trials on resistance to extinction. *J. Exp. Psychol.* **64**, 166–71. [*445, 448*]

Capaldi, E. J., Hart, D., and Stanley, L. R. (1963). Effect of intertrial reinforcement on the aftereffect of nonreinforcement and resistance to extinction. *J. Exp. Psychol.* **65**, 70–4. [*450, 453*]

Capaldi, E. J. and Kassover, K. (1970). Sequence, number of nonrewards, anticipation, and intertrial interval in extinction. *J. Exp. Psychol.* **84**, 470–6. [*444, 446, 448, 458*]

Capaldi, E. J., Leonard, D. W., and Ksir, C. (1968b). A reexamination of extinction rate in successive acquisitions and extinctions. *J. Comp. Physiol. Psychol.* **66**, 128–32. (b) [*442*]

Capaldi, E. J. and Lynch, D. (1967). Repeated shifts in reward magnitude: Evidence in favor of an associational and absolute (noncontextual) interpretation. *J. Exp. Psychol.* **75**, 226–35. [*389, 398*]

Capaldi, E. J. and Minkoff, R. (1966). Change in the stimulus produced by nonreward as a function of time. *Psychon. Sci.* **6**, 321–2. [*407*]

Capaldi, E. J. & Minkoff, R. (1967). Reward schedule effects at a relatively long intertrial interval. *Psychon. Sci.* **9**, 169–70. [*162, 446, 448*]

Capaldi, E. J. and Minkoff, R. (1969). Influence of order of occurrence of nonreward and large and small reward on acquisition and extinction. *J. Exp. Psychol.* **81**, 156–60. [*457*]

Capaldi, E. J. and Olivier, W. P. (1967). Effect of intertrial reinforcement following a substantial number of consistently rewarded trials. *J. Exp. Psychol.* **75**, 135–8. [*451*]

Capaldi, E. J. and Poynor, H. (1966). Aftereffects and delay of reward. *J. Exp. Psychol.* **71**, 80–8. [*433, 453*]

Capaldi, E. J. and Senko, M. G. (1962). Acquisition and transfer in partial reinforcement. *J. Exp. Psychol.* **63**, 155–9. [*163*]

Capaldi, E. J. and Sparling, D. L. (1971a). Amobarbital vs. saline extinction following different magnitudes of consistent reinforcement. *Psychon. Sci.* **23**, 215–17. (a). [*427, 428*]

Capaldi, E. J. and Sparling, D. L. (1971b). Amobarbital and the partial reinforcement effect in rats: isolating frustrative control over instrumental responding. *J. Comp. Physiol. Psychol.* **74**, 467–77. (b). [*465*]

Capaldi, E. J. and Spivey, J. E. (1963). Effect of goal-box similarity on the aftereffect of nonreinforcement and resistance to extinction. *J. Exp. Psychol.* **66**, 461–5. [*450, 451*]

Capaldi, E. J. and Spivey, J. E. (1964). Stimulus consequences of reinforcement and nonreinforcement: Stimulus traces or memory. *Psychon. Sci.* **1**, 403–4. [*162*]

Capaldi, E. J. and Spivey, J. E. (1965). Schedule of partial delay of reinforcement and resistance to extinction. *J. Comp. Physiol. Psychol.* **60**, 274–6. [*454*]

Capaldi, E. J. and Stanley, L. R. (1963). Temporal properties of reinforcement aftereffects. *J. Exp. Psychol.* **65**, 169–75. [*162*]

Capaldi, E. J. and Stanley, L. R. (1965). Percentage of reward vs. N-length in the runway. *Psychon. Sci.* **3**, 263–4. [*446*]

Capaldi, E. J. and Stevenson, H. W. (1957). Response reversal following different amounts of training. *J. Comp. Physiol. Psychol.* **50**, 195–8. [*602, 603*]

Capaldi, E. J. and Waters, R. W. (1970). Conditioning and nonconditioning interpretations of small-trial phenomena. *J. Exp. Psychol.* **84**, 518–22. [*463*]

Capaldi, E. J. and Wilson, N. B. (1968). Intertrial reinforcement: A test of several hypotheses. *Psychon. Sci.* **13**, 169–70. [*450*]

Capaldi, E. J. and Ziff, D. R. (1969). Schedule of partial reward and the negative contrast effect. *J. Comp. Physiol. Psychol.* **68**, 593–6. [*400*]

Capaldi, E. J., Ziff, D. R., and Godbout, R. C. (1970). Extinction and the necessity or non-necessity of anticipating reward on nonrewarded trials. *Psychon. Sci.* **18**, 61–3. [*463*]

Capehart, J., Viney, W. and Hulicka, I. M. (1958). The effect of effort upon extinction. *J. Comp. Physiol. Psychol.* **51**, 505–7. [*416*]

Capretta, P. J. (1961). An experimental modification of food preferences in chickens. *J. Comp. Physiol. Psychol.* **54**, 238–42. [*55*]

Carlson, J. G. and Wielkiewicz, R. M. (1972). Delay of reinforcement in instrumental discrimination learning of rats. *J. Comp. Physiol. Psychol.* **81**, 365–70. [*186*]

Carlson, N. J. and Black, A. H. (1960). Traumatic avoidance learning: The effect of preventing escape responses. *Can. J. Psychol.* **14**, 21–8. [*302*]

Carlton, P. L. (1961). The interacting effects of deprivation and reinforcement schedule. *J. Exp. Anal. Behav.* **4**, 379–81. [*151*]

Carlton, P. L. and Marks, R. A. (1958). Cold exposure and heat reinforced operant behavior. *Science, N.Y.* **128**, 1344. [*247*]

Carlton, P. L. and Vogel, J. R. (1967). Habituation and conditioning. *J. Comp. Physiol. Psychol.* **63**, 348–51. [*37, 39*]

Carr, H. A. (1917). Maze studies with the white rat. I. Normal animals. *J. Anim. Behav.* **7**, 259–75. [*554*]

Catania, A. C. (1963a). Concurrent performances: Reinforcement interaction and response independence. *J. Exp. Anal. Behav.* **6**, 253–63. (a) [*193*]

Catania, A. C. (1963b). Concurrent performances: A baseline for the study of reinforcement magnitude. *J. Exp. Anal. Behav.* **6**, 299–300. (b) [*152, 186*]

Catania, A. C. (1966). Concurrent operants. In W. K. Honig (Ed.), *Operant Behavior: Areas of Application and Research.* New York: Appleton-Century-Crofts. Pp. 213–70. [*183, 193*]

Catania, A. C. (1971a). Elicitation, reinforcement, and stimulus control. In R. Glaser (Ed.). *The Nature of Reinforcement.* New York: Academic Press. Pp. 196–220. (a) [*125*]

Catania, A. C. (1971b). Reinforcement schedules: the role of responses preceding the one that produces the reinforcer. *J. Exp. Anal. Behav.* **15**, 271–87. [*178*]

Catania, A. C. and Reynolds, G. S. (1968). A quantitative analysis of the responding maintained by interval schedules of reinforcement. *J. Exp. Anal. Behav.* **11**, 327–83. [*167, 169, 178*]

Chacto, C. and Lubow, R. E. (1967). Classical conditioning and latent inhibition in the white rat. *Psychon. Sci.* **9**, 135–6. [*37*]

Champion, R. A. and Jones, J. E. (1961). Forward, backward, and pseudoconditioning of the GSR. *J. Exp. Psychol.* **62**, 58–61. [*58, 59*]

Chase, S. (1968). Selectivity in multidimensional stimulus control. *J. Comp. Physiol. Psychol.* **66**, 787–92. [*583*]

Chechile, R. and Fowler, H. (1973). Primary and secondary negative incentive contrast in differential conditioning. *J. Exp. Psychol.* **97**, 189–97. [*393, 396*]

Chiszar, D. A. and Spear, N. E. (1968a). Proactive interference in a T-maze brightness-discrimination task. *Psychon. Sci.* **11**, 107–8. (a) [*479*]

Chiszar, D. A. and Spear, N. E. (1968b). Proactive interference in retention of nondiscriminative learning. *Psychon. Sci.* **12**, 87–8. (b) [*471*]

Chiszar, D. A. and Spear, N. E. (1969). Stimulus change, reversal learning, and retention in the rat. *J. Comp. Physiol. Psychol.* **69**, 190–95. [*479*]

Chung, S. H. and Herrnstein, R. J. (1967). Choice and delay of reinforcement. *J. Exp. Anal. Behav.* **10**, 67–74. [*159, 186*]

Church, R. M. (1963). The varied effects of punishment on behavior. *Psychol. Rev.* **70**, 369–402. [*274, 291*]

Church, R. M. (1964). Systematic effect of random error in the yoked control design. *Psychol. Bull.* **62**, 122–31. [*120, 130*]

Church, R. M. (1969). Response suppression. In B. A. Campbell and R. M. Church (Eds.), *Punishment and Aversive Behavior.* New York: Appleton-Century-Crofts. Pp. 111–56. [*274, 284, 352*]

Church, R. M. and Black, A. H. (1958). Latency of the conditioned heart rate as a function of the CS-US interval. *J. Comp. Physiol. Psychol.* **51**, 478–82. [*65*]

Church, R. M., Raymond, G. A., and Beauchamp, R. D. (1967). Response suppression as a function of intensity and duration of a punishment. *J. Comp. Physiol. Psychol.* **63**, 39–44. [*274, 282*]

Church, R. M., Wooten, C. L., and Matthews, T. J. (1970). Discriminative punishment and the conditional emotional response. *Learning and Motivation* **1**, 1–17. [*275, 276*]

Cicala, G. A. (1961). Running speed in rats as a function of drive level and presence or absence of competing response trials. *J. Exp. Psychol.* **62**, 329–34. [*150*]

Clark, F. C. (1958). The effect of deprivation and frequency of reinforcement on variable-interval responding. *J. Exp. Anal. Behav.* **1**, 221–7. [*151, 167*]

Clayton, K. N. (1964). T-maze learning as a joint function of the reward magnitudes for the alternatives. *J. Comp. Physiol. Psychol.* **58**, 333–8. [*184, 186*]

Clayton, K. N. (1969). Reward and reinforcement in selective learning: Considerations with respect to a mathematical model of learning. In J. T Tapp (Ed.), *Reinforcement and Behavior*. New York: Academic Press. Pp. 96–119. [*148, 149, 184*]

Clayton, K. N. and Koplin, S. T. (1964). T-maze learning as a joint function of probability and magnitude of reward. *Psychon. Sci.* **1**, 381–2. [*184, 190*]

Clifford, T. (1964). Extinction following continuous reward and latent extinction. *J. Exp. Psychol.* **68**, 456–65. [*416, 424*]

Coate, W. B. (1956). Weakening of conditioned bar-pressing by prior extinction of its subsequent discriminated operant. *J. Comp. Physiol. Psychol.* **49**, 135–8. [*90, 242*]

Cochrane, T., Fallon, D., and Scobie, S. (1973). Negative contrast in goldfish (*Carassius auratus*). *Bull. Psychon. Soc.* **1**, 411–13. [*394, 395*]

Cogan, D. and Capaldi, E. J. (1961). Relative effects of delayed reinforcement and partial reinforcement on acquisition and extinction. *Psychol. Rep.* **9**, 7–13. [*401*]

Cohen, J. S., Stettner, L. J., and Michael, D. J. (1969). Effect of deprivation level on span of attention in a multi-dimension discrimination task. *Psychon. Sci.* **15**, 31–2. [*578*]

Cohen, P. S. (1968). Punishment: The interactive effects of delay and intensity of shock. *J. Exp. Anal. Behav.* **11**, 789–99. [*274, 335*]

Colavita, F. B. (1965). Dual function of the US in classical salivary conditioning. *J. Comp. Physiol. Psychol.* **60**, 218–22. [*92, 93*]

Cole, M. and Hopkins, D. (1968). Proactive interference for maze habits in the rat. *Psychon. Sci.* **10**, 365–6. [*479*]

Collier, G. and Marx, M. H. (1959). Changes in performance as a function of shifts in the magnitude of reinforcement. *J. Exp. Psychol.* **57**, 305–9. [*370, 389*]

Conger, J. J. (1951). The effects of alcohol on conflict behavior in the albino rat. *Q. J. Stud. Alcohol* **12**, 1–29. [*265*]

Conner, J. B. and Meyer, D. R. (1971). Assessment of the role of transfer suppression in learning-set formation in monkeys. *J. Comp. Physiol. Psychol.* **75**, 141–5. [*616*]

Coons, E. E., Anderson, N. H., and Myers, A. K. (1960). Disappearance of avoidance responding during continued training. *J. Comp. Physiol. Psychol.* **53**, 290–2. [*333*]

Corballis, M. C. and Beale, I. L. (1970). Bilateral symmetry and behavior. *Psychol. Rev.* **77**, 451–64. [*535*]

Cornell, J. M. and Strub, H. M. (1965). A technique for demonstrating the inhibitory function of S^{Δ}. *Psychon. Sci.* **3**, 25–6. [*33*]

Cornish, E. R. and Mrosovsky, N. (1965). Activity during food deprivation and satiation of six species of rodent. *Anim. Behav.* **13**, 242–8. [*350*]

Cotton, J. W. (1953). Running time as a function of amount of food deprivation. *J. Exp. Psychol.* **46**, 188–98. [*150, 151*]

Cotton, J. W. and Lewis, D. J. (1957). Effect of intertrial interval on acquisition and extinction of a running response. *J. Exp. Psychol.* **54**, 15–20. [*420*]

Coulson, G., Coulson, V., and Gardner, L. (1970). The effect of two extinction procedures after acquisition on a Sidman avoidance contingency. *Psychon. Sci.* **18**, 309–10. [*284, 285, 333*]

Cowles, J. T. (1937). Food-tokens as incentives for learning by chimpanzees. *Comp. Psychol. Monogr.* **14** (5, Serial No. 71). [*237, 239, 241, 244*]

Cowles, J. T. and Nissen, H. W. (1937). Reward-expectancy in delayed-responses of chimpanzees. *J. Comp. Psychol.* **24**, 345–58. [*214, 215*]

Cranford, J. L. and Clayton, K. N. (1970). Effects of percentage of reward and amount of stimulus exposure on compound-cue discrimination learning by rats. *J. Comp. Physiol. Psychol.* **71**, 497–502. [*439*]

Crespi, L. P. (1952). Quantitative variation of incentive and performance in the white rat. *Am. J. Psychol.* **55**, 467–517. [*151, 153, 214, 368, 369, 389*]

Crisler, G. (1930). Salivation is unnecessary for the establishment of the salivary conditioned reflex induced by morphine. *Am. J. Physiol.* **94**, 553–6. [*18, 80*]

Crowder, R. G. (1967). Proactive and retroactive inhibition in the retention of a T-maze habit in rats. *J. Exp. Psychol.* **74**, 167–71. [*479*]

Crowder, R. G., Cole, M., and Boucher, R. (1968). Extinction and response competition in original and interpolated learning of a visual discrimination. *J. Exp. Psychol.* **77**, 422–8. [*480*]

Crowder, W. F., Gill, K., Jr., Hodge, C. C., and Nash, F. A. (1959). Secondary reinforcement. II. Response acquisition. *J. Psychol.* **48**, 303–6. [*236*]

Crowell, C. R. and Anderson, D. C. (1972). Variations in intensity, interstimulus interval, and interval between preconditioning CS exposures and conditioning with rats. *J. Comp. Physiol. Psychol.* **79**, 291–8. [*39*]

Crum. J., Brown, W. L., and Bitterman, M. E. (1951). The effect of partial and delayed reinforcement on resistance to extinction. *Am. J. Psychol.* **64**, 228–37. [*453*]

Culler, E. (1938). Recent advances in some concepts of conditioning. *Psychol. Rev.* **45**, 134–53. [*100, 113*]

Culler, E., Finch, G., Girden, E., and Brogden, W. J. (1935). Measurements of acuity by the conditioned-response technique. *J. Gen. Psychol.* **12**, 223–7. [*101*]

Cumming, W. W. and Schoenfeld, W. N. (1958). Behavior under extended exposure to a high-value fixed interval reinforcement schedule. *J. Exp. Anal. Behav.* **1**, 245–63. [*176*]

Dabrowska, J. (1962). An analysis of reversal learning in relation to the complexity of task in white rats. *Acta Biol. Exp.* **22**, 139–45. [*555*]

Dabrowska, J. (1963a). An analysis of reversal learning in relation to the pattern of reversal in rats. *Acta Biol. Exp.* **23**, 11–24. (a) [*208, 555*]

Dabrowska, J. (1963b). Reversal learning in relation to the pattern of reversal in a three-unit double-choice apparatus. *Acta Biol. Exp.* **23**, 263–6. (b) [*208, 555*]

Dabrowska, J. (1967). Reversal learning in relation to the pattern maze alterations in frontal rats. *Acta Biol. Exp.* **27**, 421–8. [*555*]

Daly, H. B. (1968). Excitatory and inhibitory effects of complete and incomplete reward reduction in the double runway. *J. Exp. Psychol.* **76**, 430–8. [*363*]

Daly, H. B. (1969). Learning of a hurdle-jump response to escape cues paired with reduced reward or frustrative nonreward. *J. Exp. Psychol.* **79**, 146–57. [*264*]

Daly, H. B. (1972). Learning to escape cues paired with reward reductions following single- or multiple-pellet rewards. *Psychon. Sci.* **26**, 49–52. [*152, 364, 397*]

D'Amato, M. R. (1955). Secondary reinforcement and magnitude of primary reinforcement. *J. Comp. Physiol. Psychol.* **48**, 378–80. [*242*]

D'Amato, M. R. and D'Amato, M. F. (1962). The partial reinforcement extinction effect following conventional and placed training trials. *J. Gen. Psychol.* **66**, 17–23. [*438*]

D'Amato, M. R., Etkin, M., and Fazzaro, J. (1968a). Effects of shock type and intensity on anticipatory responses. *J. Comp. Physiol. Psychol.* **66**, 527–9. (a) [344]

D'Amato, M. R. and Fazzaro, J. (1966a). Discriminated lever-press avoidance learning as a function of type and intensity of shock. *J. Comp. Physiol. Psychol.* **61**, 313–15. (a) [*343*]

D'Amato, M. R. and Fazzaro, J. (1966b). Attention and cue-producing behavior in the monkey. *J. Exp. Anal. Behav.* **9**, 469–73. (b) [*580*]

D'Amato, M. R., Fazzaro, J., and Etkin, M. (1967). Discriminated bar-press avoidance maintenance and extinction in rats as a function of shock intensity. *J. Comp. Physiol. Psychol.* **63**, 351–4. [*344*]

D'Amato, M. R., Fazzaro, J., and Etkin, M. (1968b). Anticipatory responding and avoidance discrimination as factors in avoidance conditioning. *J. Exp. Psychol.* **77**, 41–7. (b) [*319*]

D'Amato, M. R., Keller, D., and DiCara, L. (1964). Facilitation of discriminated avoidance learning by discontinuous shock. *J. Comp. Physiol. Psychol.* **58**, 344–9. [*302*]

D'Amato, M. R., Lachman, R. and Kivy, P. (1958). Secondary reinforcement as affected by reward schedule and the testing situation. *J. Comp. Physiol. Psychol.* **51**, 734–41. [*242, 245, 437*]

D'Amato, M. R., Schiff, D., and Jagoda, H. (1962). Resistance to extinction after varying amounts of discriminative or nondiscriminative instrumental training. *J. Exp. Psychol.* **64**, 526–32. [*423*]

Darby, C. L. and Riopelle, A. J. (1959). Observational learning in the rhesus monkey. *J. Comp. Physiol. Psychol.* **52**, 94–8. [*202*]

Davenport, D. G. and Olson, R. D. (1968). A reinterpretation of extinction in discriminated avoidance. *Psychon. Sci.* **13**, 5–6. [*338, 339*]

Davenport, J. W. (1962). The interaction of magnitude and delay of reinforcement in spatial discrimination. *J. Comp. Physiol. Psychol.* **55**, 267–73. [*186*]

Davenport, J. W. (1966). Higher-order conditioning of fear (CER). *Psychon. Sci.* **4**, 27–8. [*20*]

Davenport, J. W. (1969). Successive acquisitions and extinctions of discrete bar-pressing in monkeys and rats. *Psychon. Sci.* **16**, 242–4. [*442*]

Davis, H. and Kreuter, C. (1972). Conditioned suppression of an avoidance response by a stimulus paired with food. *J. Exp. Anal. Behav.* **17**, 277–85. [*84*]

Davison, M. C. (1969a). Preference of mixed-interval versus fixed-interval schedules. *J. Exp. Anal. Behav.* **12**, 247–52. (a) [*188, 256*]

Davison, M. C. (1969b). Successive interresponse times in fixed-ratio and second-order fixed-ratio performances. *J. Exp. Anal. Behav.* **12**, 385–9. (b) [*168*]

Davitz, J. R. (1955). Reinforcement of fear at the beginning and end of shock. *J. Comp. Physiol. Psychol.* **48**, 152–5. [*112*]

Davitz, J. R. and Mason, D. J. (1955). Socially facilitated reduction of a fear response in rats. *J. Comp. Physiol. Psychol.* **48**, 149–51. [*201*]

Davydova, E. K. (1967). Changes in excitability of motor cortex during formation of conditioned reflex in response to its stimulation. *Neuroscience Translations*, **1**, 53–6. [*110*]

Deaux, E. B. and Patten, R. L. (1964). Measurement of the anticipatory goal response in instrumental runway conditioning. *Psychon. Sci.* **1**, 357–8. [*223, 230*]

DeBold, R. C., Miller, N. E., and Jensen, D. D. (1965). Effect of strength of drive determined by a new technique for appetitive classical conditioning of rats. *J. Comp. Physiol. Psychol.* **59**, 102–8. [*17, 91*]

Deets, A. C., Harlow, H. F., and Blomquist, A. J. (1970). Effects of intertrial interval and Trial 1 reward during acquisition of an object-discrimination learning set in monkeys. *J. Comp. Physiol. Psychol.* **73**, 501–5. [*614*]

de Lorge, J. (1967). Fixed-interval behavior maintained by conditioned reinforcement. *J. Exp. Anal. Behav.* **10**, 271–6. [*241, 242*]

Dember, W. N. and Fowler, H. (1958). Spontaneous alternation behavior. *Psychol. Bull.* **55**, 412–28. [*149*]

Denny, M. R. (1946). The role of secondary reinforcement in a partial reinforcement learning situation. *J. Exp. Psychol.* **36**, 373–89. [*437*]

Denny, M. R. (1970). Elicitation theory applied to an analysis of the overlearning reversal effect. In J. H. Reynierse (Ed.), *Current Issues in Animal Learning.* Lincoln: University of Nebraska Press. Pp. 175–94. [*603, 606*]

Denny, M. R. (1971). Relaxation theory and experiments. In F. R. Brush (Ed.), *Aversive Conditioning and Learning.* New York: Academic Press. Pp. 235–95. [*113, 307, 413*]

Denny, M. R. and Davis, R. H. (1951). A test of latent learning for a nongoal significate. *J. Comp. Physiol. Psychol.* **44**, 590–95. [*210*]

Denny, M. R. and Ditchman, R. E. (1962). The locus of maximal "Kamin effect" in rats. *J. Comp. Physiol. Psychol.* **55**, 1069–70. [*475*]

Denny, M. R. and Ratner, S. C. (1959). Distal cues and latent extinction. *Psychol. Rec.* **9**, 33–5. [*417*]

Denny, M. R. and Tortora, D. F. (1971). A stimulus trace interpretation of the overlearning reversal effect in a black-white discrimination. *Learning and Motivation* **2**, 371–5. [*604*]

Denny, M. R. and Weisman, R. G. (1964). Avoidance behavior as a function of length of non-shock confinement. *J. Comp. Physiol. Psychol.* **58**, 252–7. [*320*]

Derdzinski, D. and Warren, J. M. (1969). Perimeter, complexity, and form discrimination learning by cats. *J. Comp. Physiol. Psychol.* **68**, 407–11. [*548, 567*]

Desiderato, O. (1964). Generalization of conditioned suppression. *J. Comp. Physiol. Psychol.* **57**, 434–7. [*494*]

Desiderato, O. (1969). Generalization of excitation and inhibition in control of avoidance responding by Pavlovian CSs in dogs. *J. Comp. Physiol. Psychol.* **68**, 611–16. [*527*]

deToledo, L. and Black, A. H. (1966). Heart rate: Changes during conditioned suppression in rats. *Science, N.Y.* **152**, 1404–6. [*22, 103*]

Deutsch, J. A. (1960). *The Structural Basis of Behavior.* Cambridge: Cambridge University Press. [*187, 199, 213, 222*]

Deutsch, J. A. (1967). Discrimination learning and inhibition. *Science, N.Y.* **156**, 988. [*530*]

Devine, J. V. (1970). Stimulus attributes and training procedures in learning-set formation of rhesus and cebus monkeys. *J. Comp. Physiol. Psychol.* **73**, 62–7. [*611, 612*]

Dews, P. B. (1962). The effect of multiple S^Δ periods on responding on a fixed-interval schedule. *J. Exp. Anal. Behav.* **5**, 369–74. [*170, 173, 178*]

Dews, P. B. (1966). The effect of multiple S$^\Delta$ periods on responding on a fixed-interval schedule. V. Effect of periods of complete darkness and of occasional omissions of food presentations. *J. Exp. Anal. Behav.* **9**, 573–8. [*173*]

Dews, P. B. (1969). Studies on responding under fixed-interval schedules of reinforcement: the effects on the pattern of responding of changes in requirement at reinforcement. *J. Exp. Anal. Behav.* **12**, 191–9. [*177*]

DiCara, L. V. and Miller, N. E. (1968a). Changes in heart rate instrumentally learned by curarized rats as avoidance responses. *J. Comp. Physiol. Psychol.* **65**, 8–12. (a) [*131*]

DiCara, L. V. and Miller, N. E. (1968b). Instrumental learning of systolic blood pressure by curarized rats: Dissociation of cardiac and vascular changes. *Psychosom. Med.* **30**, 489–94. (b) [*132*]

DiCara, L. V. and Miller, N. E. (1968c). Instrumental learning of vasomotor responses by rats: Learning to respond differentially in the two ears. *Science, N.Y.* **159**, 1485–6. (c) [*132*]

DiCara, L. V. and Miller, N. E. (1969a). Heart-rate learning in the noncurarized state, transfer to the curarized state, and subsequent retraining in the noncurarized state. *Physiol. Behav.* **4**, 621–4. [*131*]

DiCara, L. V. and Miller, N. E. (1969b). Transfer of instrumentally learned heart-rate changes from curarized to noncurarized state: Implications for a mediational hypothesis. *J. Comp. Physiol. Psychol.* **68**, 159–62. (b) [*132*]

Dickinson, A. (1972). Septal damage and response output under frustrative non-reward. In R. A. Boakes and M. S. Halliday (Eds.), *Inhibition and Learning.* London: Academic Press. Pp. 461–6. [*366, 369, 376, 381*]

DiLollo, V. (1964). Runway performance in relation to runway-goal-box similarity and changes in incentive amount. *J. Comp. Physiol. Psychol.* **58**, 327–9. [*389*]

DiLollo, V. and Beez, V. (1966). Negative contrast effects as a function of magnitude of reward decrements. *Psychon. Sci.* **5**, 99–100. [*390*]

Dinsmoor, J. A. (1954). Punishment: I. The avoidance hypothesis. *Psychol. Rev.* **61**, 34–46. [*280*]

Dinsmoor, J. A. (1958). Pulse duration and food deprivation in escape-from-shock training. *Psychol. Rep.* **4**, 531–4. [*352*]

Dinsmoor, J. A., Browne, M. P., and Lawrence, C. E. (1972). A test of the negative discriminative stimulus as a reinforcer of observing. *J. Exp. Anal. Behav.* **18**, 79–85. [*259*]

Dinsmoor, J. A., Flint, G. A., Smith, R. F., and Viemeister, N. F. (1969). Differential reinforcing effects of stimuli associated with the presence or absence of a schedule of punishment. In D. P. Hendry (Ed.), *Conditioned Reinforcement.* Homewood, Illinois: The Dorsey Press. Pp. 357–84. [*258*]

Dodwell, P. C. and Bessant, D. E. (1960). Learning without swimming in a water maze. *J. Comp. Physiol. Psychol.* **53**, 422–5. [*200*]

Dollard, J. and Miller, N. E. (1950). *Personality and Psychotherapy.* New York: McGraw-Hill. [*265, 280*]

Domjan, M. and Wilson, N. E. (1972). Contribution of ingestive behaviors to taste-aversion learning in the rat. *J. Comp. Physiol. Psychol.* **80**, 403–12. [*53*]

Donahoe, J. W. (1970). Stimulus control within response sequences. In J. H. Reynierse (Ed.), *Current Issues in Animal Learning.* Lincoln: University of Nebraska Press. Pp. 233–93. [*514*]

Donahoe, J. W., McCroskery, J. H., and Richardson, W. K. (1970). Effects of context on the post discrimination gradient of stimulus generalization. *J. Exp. Psychol.* **84**, 58–63. [*497*]

Donahoe, J. W., Schulte, V. G., and Moulton, A. E. (1968). Stimulus control of approach behavior. *J. Exp. Psychol.* **78**, 21–30. [*230*]

Donin, J. A., Surridge, C. T., and Amsel, A. (1967). Extinction following partial delay of reward with immediate continuous reward interpolated, at 24-hour intertrial intervals. *J. Exp. Psychol.* **74**, 50–3. [*454*]

Doty, R. W. (1969). Electrical stimulation of the brain in behavioral context. *A. Rev. of Psychology*, **20**, 289–320. [*18, 94*]

Doty, R. W. and Giurgea, C. (1961). Conditioned reflexes established by coupling electrical excitation of two cortical areas. In J. F. Delafresnaye (Ed.), *Brain Mechanisms and Learning*. Oxford: Blackwells Scientific Publications. Pp. 133–51. [*94, 95*]

Douglas, R. J. (1966). Cues for spontaneous alternation. *J. Comp. Physiol. Psychol.* **62**, 171–83. [*554*]

Dufort, R. M., Guttman, N., and Kimble, G. A. (1954). One-trial discrimination reversal in the white rat. *J. Comp. Physiol. Psychol.* **47**, 248–9. [*608*]

Duncan, B. and Fantino, E. (1972). The psychological distance to reward. *J. Exp. Anal. Behav.* **18**, 23–34. [*187*]

Dunham, P. J. (1968). Contrasted conditions of reinforcement: A selective critique. *Psychol. Bull.* **69**, 295–315. [*215, 385*]

Dunham, P. J. (1971). Punishment: Method and theory. *Psychol. Rev.* **78**, 58–70. [*289*]

Dunham, P. J. (1972). Some effects of punishment upon unpunished responding. *J. Exp. Anal. Behav.* **17**, 443–50. [*290*]

Dunham, P. J. and Kilps, B. (1969). Shifts in magnitude of reinforcement: Confounded factors or contrast effects? *J. Exp. Psychol.* **79**, 373–4. [*389*]

Dweck, C. S. and Wagner, A. R. (1970). Situational cues and correlation between CS and US as determinants of the conditioned emotional response. *Psychon. Sci.* **18**, 145–7. [*30*]

Dyal, J. A. (1962). Latent extinction as a function of number and duration of pre-extinction exposures. *J. Exp. Psychol.* **63**, 98–104. [*416*]

Dyal, J. A. (1964). Latent extinction as a function of placement-test interval and irrelevant drive. *J. Exp. Psychol.* **68**, 486–91. [*416*]

Dyal, J. A. and Goodman, E. (1966). Fear conditioning as a function of CS duration during acquisition and suppression tests. *Psychon. Sci.* **4**, 249–50. [*65*]

Dyal, J. A. and Holland, T. A. (1963). Resistance to extinction as a function of number of reinforcements. *Am. J. Psychol.* **76**, 332–3. [*423*]

Ebel, H. C. and Prokasy, W. F. (1963). Classical eyelid conditioning as a function of sustained and shifted interstimulus intervals. *J. Exp. Psychol.* **65**, 52–8. [*62*]

Eck, K. O., Noel, R. C., and Thomas, D. R. (1969). Discrimination learning as a function of prior discrimination and nondifferential training. *J. Exp. Psychol.* **82**, 156–62. [*600*]

Egger, M. D. and Miller, N. E. (1962). Secondary reinforcement in rats as a function of information value and reliability of the stimulus. *J. Exp. Psychol.* **64**, 97–104. [*220, 245, 248, 249, 250*]

Egger, M. D. and Miller, N. E. (1963). When is a reward reinforcing? An experimental study of the information hypothesis. *J. Comp. Physiol. Psychol.* **56**, 132–7. [*248, 249*]

Ehrenfreund, D. (1948). An experimental test of the continuity theory of discrimination learning with pattern vision. *J. Comp. Physiol. Psychol.* **41**, 408–22. [*585*]

Ehrenfreund, D. (1952). A study of the transposition gradient. *J. Exp. Psychol.* **43**, 81–7. [*562*]

Eimas, P. D. (1967). Overtraining and reversal discrimination learning in rats. *Psychol. Rec.* **17**, 239–48. [*184, 603, 604*]

Eisman, E., Asimow, A., and Maltzman, I. (1956). Habit strength as a function of drive in a brightness discrimination problem. *J. Exp. Psychol.* **51**, 58–64. [*183*]

Elam, C. B., Tyler, D. W., and Bitterman, M. E. (1954). A further study of secondary reinforcement and the discrimination hypothesis. *J. Comp. Physiol. Psychol.* **47**, 381–4. [*235*]

Elliott, M. H. (1928). The effect of change of reward on the maze performance of rats. *University of California Publicatons in Psychology* **4**, 19–30. [*214, 368, 391*]

Ellis, W. R., III (1970). Role of stimulus sequences in stimulus discrimination and stimulus generalization. *J. Exp. Psychol.* **83**, 155–63. [*369, 374, 536, 538*]

Ellison, G. D. (1964). Differential salivary conditioning to traces. *J. Comp. Physiol. Psychol.* **57**, 373–80. [*58, 61, 514*]

Ellison, G. D. and Konorski, J. (1964). Separation of the salivary and motor responses in instrumental conditioning. *Science, N.Y.* **146**, 1071–2. [*230, 231*]

Ellson, D. G. (1938). Quantitative studies of the interaction of simple habits. I. Recovery from specific and generalized effects of extinction. *J. Exp. Psychol.* **23**, 339–58. [*468, 469*]

Eninger, M. U. (1952). Habit summation in a selective learning problem. *J. Comp. Physiol. Psychol.* **45**, 604–8. [*576*]

Epstein, S. and Burstein, K. R. (1966). A replication of Hovland's study of stimulus generalization to frequencies of tone. *J. Exp. Psychol.* **72**, 782–4. [*514*]

Eskin, R. M. and Bitterman, M. E. (1961). Partial reinforcement in the turtle. *Q. Jl. Exp. Psychol.* **13**, 112–16. [*467*]

Estes, W. K. (1943). Discriminative conditioning. I. A discriminative property of conditioned anticipation. *J. Exp. Psychol.* **32**, 150–55. [*226*]

Estes, W. K. (1944). An experimental study of punishment. *Psychol. Monogr.* **57**, (3, Whole No. 263). [*273, 274, 275, 278, 281, 291, 293, 294*]

Estes, W. K. (1950). Toward a statistical theory of learning. *Psychol. Rev.* **57**, 94–107. [*11*]

Estes, W. K. (1955). Statistical theory of spontaneous recovery and regression. *Psychol. Rev.* **62**, 145–54. [*32, 470, 471, 476*]

Estes, W. K. (1958). Stimulus-response theory of drive. In M. R. Jones (Ed.), *Nebraska Symposium on Motivation*, Vol. 6. Lincoln: University of Nebraska Press, Pp. 35–68. [*151, 349*]

Estes, W. K. (1959). The statistical approach to learning theory. In S. Koch (Ed.), *Psychology: A Study of a Science*, Vol. 2. New York: McGraw-Hill, Pp. 380–491. [*13, 147, 148, 191*]

Estes, W. K. (1969a). New perspectives on some old issues in association theory. In N. J. Mackintosh and W. K. Honig (Eds.), *Fundamental Issues in Associative Learning*. Halifax: Dalhousie University Press. Pp. 162–89. (a) [*132, 198, 216, 222, 231*]

Estes, W. K. (1969b). Outline of a theory of punishment. In B. A. Campbell and R. M. Church (Eds.), *Punishment and Aversive Behavior*. New York: Appleton-Century-Crofts. Pp. 57–82. (b) [*84, 132, 216, 277, 278, 294*]

Estes, W. K. and Skinner, B. F. (1941). Some quantitative properties of anxiety. *J. Exp. Psychol.* **29**, 390–400. [*10, 17, 278*]

Evans, S. M. (1966a). Non-associative avoidance learning in Nereid polychaetes. *Anim. Behav.* **14**, 102–6. (a) [*27*]

Evans, S. M. (1966b). Non-associative behavioural modifications in the polychaete *Nereis diversicolor. Anim. Behav.* **14**, 107–19. (b) [*27*]

Fallon, D. (1971). Increased resistance to extinction following punishment and reward: High frustration tolerance or low frustration magnitude? *J. Comp. Physiol. Psychol.* **77**, 245–55. [*464*]

Fantino, E. (1967). Preference for mixed- versus fixed-ratio schedules. *J. Exp. Anal. Behav.* **10**, 35–43. [*256*]

Fantino, E. (1968). Effects of required rates of responding upon choice *J. Exp. Anal. Behav.* **11**, 15–22. [*187, 382*]

Fantino, E. and Duncan, B. (1972). Some effects of interreinforcement time upon choice. *J. Exp. Anal. Behav.* **17**, 3–14. [*194*]

Fantino, E., Squires, N., Delbruck, N., and Peterson, C. (1972a). Choice behavior and the accessibility of the reinforcer. *J. Exp. Anal. Behav.* **18**, 35–43. (a) [*186, 193*]

Fantino, E., Weigele, S., and Lancy, D. (1972b). Aggressive display in the Siamese fighting fish (*Betta splendens*). *Learning and Motivation* **3**, 457–68. (b) [*288*]

Farris, H. E. (1967). Classical conditioning of courting behavior in the Japanese quail, *Coturnix coturnix japonica. J. Exp. Anal. Behav.* **10**, 213–17. [*17*]

Farthing, G. W. (1971). Effect of a signal previously paired with free food on operant response rate in pigeons. *Psychon. Sci.* **23**, 343–4. [*226*]

Farthing, G. W. (1972). Overshadowing in the discrimination of successive compound stimuli. *Psychon. Sci.* **28**, 29–32. [*501, 508*]

Farthing, G. W. and Hearst, E. (1968). Generalization gradients of inhibition after different amounts of training. *J. Exp. Anal. Behav.* **11**, 743–52. [*539, 583*]

Farthing, G. W. and Hearst, E. (1970). Attention in the pigeon: testing with compounds or elements. *Learning and Motivation* **1**, 65–78. [*583*]

Fechter, L. D. and Ison, J. R. (1972). The inhibition of the acoustic startle reaction in rats by food and water deprivation. *Learning and Motivation* **3**, 109–24. [*353*]

Fehrer, E. (1956). Effects of amount of reinforcement and of pre- and postreinforcement delays on learning and extinction. *J. Exp. Psychol.* **52**, 167–76. [*431, 432*]

Felton, M. and Lyon, D. O. (1966). The post-reinforcement pause. *J. Exp. Anal. Behav.* **9**, 131–4. [*168*]

Ferraro, D. P. and Branch, K. H. (1968). Variability of response location during regular and partial reinforcement. *Psychol. Rep.* **23**, 1023–31. [*439*]

Ferster, C. B. (1951). The effect on extinction responding of stimuli continuously present during conditioning. *J. Exp. Psychol.* **42**, 443–9. [*247, 408*]

Ferster, C. B. (1958). Control of behavior in chimpanzees and pigeons by time-out from positive reinforcement. *Psychol. Monogr.* **72**, (8, Whole No. 461). [*262*]

Ferster, C. B. and Skinner, B. F. (1957). *Schedules of Reinforcement.* New York: Appleton-Century-Crofts. [*136, 151, 166, 167, 168, 170, 176, 177, 241*]

Finch, G. (1938a). Hunger as a determinant of conditional and unconditional salivary response magnitude. *Am. J. Physiol.* **123**, 379–82. (a) [*71*]

Finch, G. (1938b). Salivary conditioning in atropinized dogs. *Am. J. Physiol.* **124**, 136–41. (b) [*80*]

Finch, G. (1938c). Pilocarpine conditioning. *Am. J. Physiol.* **124**, 679–82. (c) [*94*]

Fitzgerald, R. D. (1963). Effects of partial reinforcement with acid on the classically conditioned salivary response in dogs. *J. Comp. Physiol. Psychol.* **56**, 1056–60. [*11, 24, 72, 73*]

Fitzgerald, R. D. and Teyler, T. J. (1968). Extinction of a runway response following noncontingent partial reinforcement and nonreinforcement in the goal box. *J. Comp. Physiol. Psychol.* **65**, 542–4. [*451, 452*]

Fitzgerald, R. D., Vardaris, R. M., and Brown, J. S. (1966a). Classical conditioning of heart rate deceleration in the rat with continuous and partial reinforcement. *Psychon. Sci.* **6**, 437–8. (a) [*103*]

Fitzgerald, R. D., Vardaris, R. M., and Teyler, T. J. (1966b). Effects of partial reinforcement followed by continuous reinforcement on classically conditioned heart-rate in the dog. *J. Comp. Physiol. Psychol.* **62**, 483–6. (b) [*72, 74*]

Fitzwater, M. E. and Reisman, M. N. (1952). Comparison of forward, simultaneous, backward, and pseudo-conditioning. *J. Exp. Psychol.* **44**, 211–14. [*59*]

Flaherty, C. F. and Davenport, J. W. (1972). Successive brightness discrimination in rats following regular versus random intermittent reinforcement. *J. Exp. Psychol.* **96**, 1–9. [*162, 163, 601*]

Flaherty, C. F., Riley, E. P., and Spear, N. E. (1973). Effects of sucrose concentration and goal units on runway behavior in the rat. *Learning and Motivation* **4**, 163–75. [*392, 393*]

Flory, R. (1969). Attack behavior as a function of minimum inter-food interval. *J. Exp. Anal. Behav.* **12**, 825–8. [*263*]

Foree, D. D. and LoLordo, V. M. (1970). Signalled and unsignalled free-operant avoidance in the pigeon. *J. Exp. Anal. Behav.* **13**, 283–90. [*340*]

Foree, D. D. and LoLordo, V. M. (1973). Attention in the pigeon: the differential effects of food-getting vs. shock-avoidance procedures. *J. Comp. Physiol. Psychol.* **85**, 551–8. [*55, 56*]

Fowler, H. (1971). Suppression and facilitation by response contingent shock. In F. R. Brush (Ed.), *Aversive Conditioning and Learning.* New York: Academic Press. Pp. 537–604. [*283, 285, 289*]

Fowler, H., Goldman, L., and Wischner, G. J. (1968). Sodium amytal and the shock-right intensity function for visual discrimination learning. *J. Comp. Physiol. Psychol.* **65**, 155–9. [*283*]

Fowler, H. and Miller, N. E. (1963). Facilitation and inhibition of runway performance by hind- and forepaw shock of various intensities. *J. Comp. Physiol. Psychol.* **56**, 801–5. [*287, 288*]

Fowler, H., Spelt, P. F., and Wischner, G. J. (1967). Discrimination performance as affected by training procedure, problem difficulty and shock for the correct response. *J. Exp. Psychol.* **75**, 432–6. [*283*]

Fowler, H. and Trapold, M. A. (1962). Escape performance as a function of delay of reinforcement. *J. Exp. Psychol.* **63**, 464–7. [*156*]

Fowler, H. and Wischner, G. J. (1965). Discrimination performance as affected by problem difficulty and shock for either the correct or incorrect response. *J. Exp. Psychol.* **69**, 413–18. [*283*]

Fowler, H. and Wischner, G. J. (1969). The varied functions of punishment in discrimination learning. In B. A. Campbell and R. M. Church (Eds.), *Punishment and Aversive Behavior.* New York: Appleton-Century-Crofts. Pp. 375–420. [*283*]

Fowler, R. L. and Kimmel, H. D. (1962). Operant conditioning of the GSR. *J. Exp. Psychol.* **63**, 536–7. [*130*]

Fox, P. A., Calef, R. S., Gavelek, J. R., and McHose, J. H. (1970). Synthesis of differential conditioning and double alley data: Performance to S+ as a function of intertrial interval and antedating reward events. *Psychon. Sci.* **18**, 141–2. [*362*]

Franchina, J. J. and Brown, T. S. (1971). Reward magnitude shift effects in rats with hippocampal lesions. *J. Comp. Physiol Psychol.* **76**, 365–70. [*388, 389*]

Freedman, P. E. and Rosen, A. J. (1969). The effects of psychotropic drugs on the double alley frustration effect. *Psychopharmacologia* **15**, 39–47, [*353*]

Freeman, F. and Thomas, D. R. (1967). Attention vs. cue utilization in generalization testing. Paper presented at Midwestern Psychological Association, Chicago. [*488, 499*]

Frey, P. W. (1969). Within- and between-session CS intensity performance effects in rabbit eyelid conditioning. *Psychon. Sci.* **17**, 1–2. [*45*]

Frey, P. W. and Ross, L. E. (1967). Differential conditioning of the rabbit's eyelid response with an examination of Pavlov's induction hypothesis. *J. Comp. Physiol. Psychol.* **64**, 277–83. [*357, 387*]

Frey, P. W. and Sheldon, D. R. (1970). Effect of food deprivation on differential eyelid conditioning in New Zealand white and American Dutch rabbits. *Psychon. Sci.* **20**, 23–5. [*352*]

Friedes, D. (1957). Goal-box cues and pattern of reinforcement. *J. Exp. Psychol.* **53**, 361–71. [*235*]

Friedman, H. (1963) Wave-length generalization as a function of spacing of test stimuli. *J. Exp. Psychol.* **65**, 334–8. [*495*]

Friedman, H. and Guttman, N. (1965). A further analysis of effects of discrimination training on stimulus generalization gradients. In D. I. Mostofsky (Ed.), *Stimulus Generalization*. Stanford: Stanford University Press. Pp. 255–67. [*493,501,536*]

Frieman, J. and Goyette, C. H. (1973). Transfer of training across stimulus modality and response class. *J. Exp. Psychol.* **97**, 235–41. [*601*]

Fromer, R. (1963). Conditioned vasomotor responses in the rabbit. *J. Comp. Physiol. Psychol.* **56**, 1050–5. [*17, 92*]

Fuchs, S. S. (1960). An attempt to obtain inhibition with reinforcement. *J. Exp. Psychol.* **59**, 343–4. [*417*]

Furchtgott, E. and Rubin, R. D. (1953). The effect of magnitude of reward on maze learning in the white rat. *J. Comp. Physiol. Psychol.* **46**, 9–12. [*185*]

Furedy, J. J. and Champion, A. A. (1963). Cognitive and S–R interpretations of incentive-motivational phenomena. *Am. J. Psychol.* **76**, 616–23. [*211*]

Furstenau, P. and Schaeffer, B. H. (1970). Overtraining, extinction and reversal in the rat. *J. Comp. Physiol. Psychol.* **71**, 292–7. [*605*]

Galef, B. G., Jr. (1970). Target novelty elicits and directs shock-associated aggression in wild rats. *J. Comp. Physiol. Psychol.* **71**, 87–91. [*108*]

Gamzu, E. and Schwartz, B. (1973). The maintenance of key pecking by stimulus-contingent and response-independent food presentation. *J. Exp. Anal. Behav.* **19**, 65–72. [*375*]

Gamzu, E. and Williams, D. R. (1971). Classical conditioning of a complex skeletal response. *Science, N.Y.* **171**, 923–5. [*25, 29, 40, 375*]

Ganz, L. (1962). Hue generalization and hue discriminability in *Macaca mulatta*. *J. Exp. Psychol.* **64**, 142–50. [*496*]

Garcia, J. and Ervin, F. R. (1968). A neuropsychological approach to appropriateness of signals and specificity of reinforcers. *Commun. Behav. Biol.* **1**, Part A, 389–415. [*53, 69*]

Garcia, J., Ervin, F. R., and Koelling, R. A. (1966). Learning with prolonged delay of reinforcement. *Psychon. Sci.* **5,** 121–2. [*54, 67, 68*]

Garcia, J., Kimmeldorf, D. J., and Koelling, R. A. (1955). Conditioned aversion to saccharin resulting from exposure to gamma radiation. *Science, N.Y.* **122,** 157–8. [*67*]

Garcia, J. and Koelling, R. A. (1966). Relation of cue to consequence in avoidance learning. *Psychon. Sci.* **4,** 123–4. [*52*]

Garcia, J., Kovner, R., and Green, K. F. (1970). Cue properties vs. palatability of flavours in avoidance learning. *Psychon. Sci.* **20,** 313–14. [*53*]

Garcia, J., McGowan, B. K., Ervin, F. R., and Koelling, R. A. (1968). Cues: their effectiveness as a function of the reinforcer. *Science, N.Y.* **160,** 794–5. [*53*]

Gardner, E. L. and Engel, D. R. (1971). Imitational and social facilitatory aspects of observational learning in the laboratory rat. *Psychon. Sci.* **25,** 5–6. [*201*]

Gardner, R. A. and Coate, W. B. (1965). Reward versus nonreward in a simultaneous discrimination. *J. Exp. Psychol.* **69,** 579–82. [*567, 568*]

Gentry, G. V., Overall, J. E., and Brown, W. L. (1959). Transpositional responses of rhesus monkeys to stimulus objects of intermediate size. *Am. J. Psychol.* **72,** 453–5. [*563*]

Gentry, W. D. (1968). Fixed-ratio schedule-induced aggression. *J. Exp. Anal. Behav.* **11,** 813–17. [*263*]

Gerall, A. A. and Obrist, P. A. (1962). Classical conditioning of the pupillary dilation response of normal and curarized cats. *J. Comp. Physiol. Psychol.* **55,** 486–91. [*17, 92, 96*]

Gerall, A. A., Sampson, P. B., and Boslov, G. L. (1957). Classical conditioning of human pupillary dilation. *J. Exp. Psychol.* **54,** 467–74. [*92*]

Ginsberg, N. (1957). Matching in pigeons. *J. Comp. Physiol. Psychol.* **50,** 261–3. [*184*]

Girden, E. (1938). Conditioning and problem-solving behavior. *Am. J. Psychol.* **51,** 677–86. [*135*]

Gleitman, H. (1955). Place learning without prior reinforcement. *J. Comp. Physiol. Psychol.* **48,** 77–9. [*200*]

Gleitman, H. (1971). Forgetting of long-term memories in animals. In W. K. Honig and P. H. R. James (Eds.), *Animal Memory.* New York: Academic Press. Pp. 1–44. [*473, 481*]

Gleitman, H. and Bernheim, J. W. (1963). Retention of fixed-interval performance in rats. *J. Comp. Physiol. Psychol.* **56,** 839–41. [*473, 474*]

Gleitman, H. and Holmes, P. A. (1967). Retention of incompletely learned CER in rats. *Psychon. Sci.* **7,** 19–20. [*473, 474*]

Gleitman, H. and Jung, L. (1963). Retention in rats: the effect of proactive interference. *Science, N.Y.* **142,** 1683–4. [*475*]

Gleitman, H., Nachmias, J., and Neisser, U. (1954). The S–R reinforcement theory of extinction. *Psychol. Rev.* **61,** 23–33. [*414*]

Gleitman, H. and Steinman, F. (1963). Retention of runway performance as a function of proactive interference. *J. Comp. Physiol Psychol.* **56,** 834–8. [*473*]

Gleitman, H. and Steinman, F. (1964). Depression effect as a function of retention interval before and after shift in reward magnitude. *J. Comp. Physiol. Psychol.* **57,** 158–60. [*474*]

Goesling, W. J. and Brener, J. (1972). Effects of activity and immobility conditioning upon subsequent heart-rate conditioning in curarized rats. *J. Comp. Physiol. Psychol.* **81,** 311–17. [*131*]

Gonzalez, R. C. (1971). A comparative analysis of contrast phenomena. Paper presented at Canadian Psychological Association, St. John's. [394]

Gonzalez, R. C. (1972). Patterning in goldfish as a function of magnitude of reinforcement. Psychon. Sci. 28, 53–5. [401]

Gonzalez, R. C., Bainbridge, P., and Bitterman, M. E. (1966a). Discrete-trials lever pressing in the rat as a function of pattern of reinforcement, effortfulness of response and amount of reward. J. Comp. Physiol. Psychol. 61, 110–22 (a) [162]

Gonzalez, R. C., Behrend, E. R. and Bitterman, M. E. (1965a). Partial reinforcement in the fish: experiments with spaced trials and partial delay. Am. J. Psychol. 78, 198–207. (a) [467]

Gonzalez, R. C., Behrend, E. R., and Bitterman, M. E. (1967a). Reversal learning and forgetting in bird and fish. Science, N.Y. 158, 519–21. (a) [609]

Gonzalez, R. C., Berger, B. D., and Bitterman, M. E. (1966b). Improvement in habit-reversal as a function of amount of training per reversal and other variables. Am. J. Psychol. 79, 517–30. (b) [609]

Gonzalez, R. C. and Bitterman, M. E. (1964). Resistance to extinction in the rat as a function of percentage and distribution of reinforcement. J. Comp. Physiol. Psychol. 58, 258–63. [446]

Gonzalez, R. C. and Bitterman, M. E. (1967). Partial reinforcement effect in the goldfish as a function of amount of reward. J. Comp. Physiol. Psychol. 64, 163–7. [428, 467]

Gonzalez, R. C. and Bitterman, M. E. (1968). Two-dimensional discriminative learning in the pigeon. J. Comp. Physiol. Psychol. 65, 427–32. [610]

Gonzalez, R. C. and Bitterman, M. E. (1969). Spaced-trials partial reinforcement effect as a function of contrast. J. Comp. Physiol. Psychol. 67, 94–103. [152, 160, 427, 430, 455, 458]

Gonzalez, R. C. and Diamond, L. (1960). A test of Spence's theory of incentive motivation. Am. J. Psychol. 73, 396–403. [136, 199, 211, 231, 232]

Gonzalez, R. C., Eskin, R. M., and Bitterman, M. E. (1961). Alternating and random partial reinforcement in the fish with some observations on asymptotic resistance to extinction. Am. J. Psychol. 74, 561–8. [401]

Gonzalez, R. C., Eskin, R. M., and Bitterman, M. E. (1963). Further experiments on partial reinforcement in the fish. Am. J. Psychol. 76, 366–75. [467]

Gonzalez, R. C., Gentry, G. V., and Bitterman, M. E. (1954). Relational discrimination of intermediate size in the chimpanzee. J. Comp. Physiol. Psychol. 47, 385–8. [563]

Gonzalez, R. C., Gleitman, H., and Bitterman, M. E. (1962). Some observations on the depression effect. J. Comp. Physiol. Psychol. 55, 578–81. [390]

Gonzalez, R. C., Graf, V., and Bitterman, M. E. (1965b). Resistance to extinction in the pigeon as a function of secondary reinforcement and pattern of partial reinforcement. Am. J. Psychol. 78, 278–84. (b) [447]

Gonzalez, R. C., Holmes, N. K., and Bitterman, M. E. (1967b). Asymptotic resistance to extinction in fish and rat as a function of interpolated retraining. J. Comp. Physiol. Psychol. 63, 342–4. (b) [427, 428, 442]

Gonzalez, R. C., Potts, A., Pitcoff, K., and Bitterman, M. E. (1972). Runway performance of goldfish as a function of complete and incomplete reduction in amount of reward. Psychon. Sci. 27, 305–7. [395, 428]

Gonzalez, R. C. and Powers, A. S. (1973). Simultaneous contrast in goldfish. Animal Learning and Behavior 1, 96–8. [394]

Gonzalez, R. C. and Shepp, B. E. (1961). Simultaneous and successive discrimination-reversal in the rat. *Am. J. Psychol.* **74**, 584–9. [*546, 551, 608*]

Gonzalez, R. C. and Shepp, B. (1965). The effects of endbox-placement on subsequent performance in the runway with competing responses controlled. *Am. J. Psychol.* **78**, 441–7. [*90, 211, 417*]

Goodrich, K. P. (1959). Performance in different segments of an instrumental response chain as a function of reinforcement schedule. *J. Exp. Psychol.* **57**, 57–63. [*75, 160*]

Goodrich, K. P. (1960). Running speed and drinking rate as functions of sucrose concentration and amount of consummatory activity. *J. Comp. Physiol. Psychol.* **53**, 245–50. [*152*]

Goodrich, K. P. (1966). Experimental analysis of response slope and latency as criteria for characterizing voluntary and nonvoluntary responses in eyeblink conditioning. *Psychol. Monogr.* **80** (4, Whole No. 622). [*102*]

Goodrich, K. P., Ross, L. E., and Wagner, A. R. (1959). An examination of selected aspects of the continuity and noncontinuity positions in discrimination learning. *Psychol. Rec.* **11**, 105–17. [*572*]

Goodwin, W. R. and Lawrence, D. H. (1955). The functional independence of two discrimination habits associated with a constant stimulus situation. *J. Comp. Physiol. Psychol.* **48**, 437–43. [*573, 607*]

Gormezano, I. (1965). Yoked comparisons of classical and instrumental conditioning of the eyelid response; and an addendum on "voluntary responders". In W. F. Prokasy (Ed.), *Classical Conditioning: A Symposium.* New York: Appleton-Century-Crofts. Pp. 48–70. [*102, 119, 120, 122, 304*]

Gormezano, I. (1966). Classical conditioning. In J. B. Sidowski (Ed.), *Experimental Methods and Instrumentation in Psychology.* New York: McGraw-Hill. Pp. 385–420. [*9, 26*]

Gormezano, I. (1972). Investigations of defense and reward conditioning in the rabbit. In A. H. Black and W. F. Prokasy (Eds.), *Classical Conditioning II: Current Research and Theory.* New York: Appleton-Century-Crofts. Pp. 151–181. [*17, 44, 45, 63, 65, 66*]

Gormezano, I. and Coleman, S. R. (1973). The law of effect and CR contingent modification of the UCS. *Conditional Reflex* **8**, 41–56. [*119*]

Gormezano, I. and Hiller, G. W. (1972). Omission training of the jaw-movement response of the rabbit to a water US. *Psychon. Sci.* **29**, 276–8. [*120, 121, 122*]

Gossette, R. L. (1966). Comparisons of successive discrimination reversal (SDR) performances across fourteen different avian and mammalian species. *Am. Zool.* **6**, 545. [*609*]

Gossette, R. L., Gossette, M. F., and Riddell, W. (1966). Comparisons of successive discrimination reversal performances among closely and remotely related avian species. *Anim. Behav.* **14**, 560–64. [*609*]

Gottwald, P. (1967). The role of punishment in the development of conditioned suppression. *Physiol. Behav.* **2**, 283–6. [*81*]

Graf, V., Bullock, D. H., and Bitterman, M. E. (1964). Further experiments on probability matching in the pigeon. *J. Exp. Anal. Behav.* **7**, 151–7. [*192*]

Gray, J. A. (1965a). Relation between stimulus intensity and operant response rate as a function of discrimination training and drive. *J. Exp. Psychol.* **69**, 9–24. (a) [*43*]

Gray, J. A. (1965b). Stimulus intensity dynamism. *Psychol. Bull.* **63**, 180–96. (b) [*41*]

Gray, J. A. (1969). Sodium amobarbital and effects of frustrative nonreward. *J. Comp. Physiol. Psychol.* **69**, 55–64. [*265, 353, 465*]

Gray, V. A. (1974). The effects of DRL training on stimulus control during maintained generalization. *J. Exp. Anal. Behav.* (in press). [*516*]

Gray, V. A. and Mackintosh, N. J. (1973). Control by an irrelevant stimulus in discrete-trial discrimination learning by pigeons. *Bull. Psychon. Soc.* **1**, 193–5. [*512*]

Green, D. M. and Swets, J. A. (1966). *Signal Detection Theory and Psychophysics.* New York: Wiley. [*45*]

Greene, W. A. (1966). Operant conditioning of the GSR using partial reinforcement. *Psychol. Rep.* **19**, 571–8. [*130*]

Grether, W. F. (1938). Pseudo-conditioning without paired stimulation encountered in attempted backward conditioning. *J. Comp. Psychol.* **25**, 91–6. [*27, 58*]

Grice, G. R. (1942). An experimental study of the gradient of reinforcement in maze learning. *J. Exp. Psychol.* **30**, 475–89. [*187*]

Grice, G. R. (1948a). The relation of secondary reinforcement to delayed reward in visual discrimination learning. *J. Exp. Psychol.* **38**, 1–16. (a) [*157, 158*]

Grice, G. R. (1948b). An experimental test of the expectation theory of learning. *J. Comp. Physiol. Psychol.* **41**, 137–43. (b) [*212*]

Grice, G. R. (1949). Visual discrimination learning with simultaneous and successive presentation of stimuli. *J. Comp. Physiol Psychol.* **42**, 365–73. [*558*]

Grice, G. R. (1968). Stimulus intensity and response evocation. *Psychol. Rev.* **75**, 359–73. [*45*]

Grice, G. R. and Davis, J. D. (1957). Effect of irrelevant thirst motivation on a response learned with food reward. *J. Exp. Psychol.* **53**, 347–52. [*351*]

Grice, G. R. and Hunter, J. J. (1964). Stimulus intensity effects depend upon the type of experimental design. *Psychol. Rev.* **71**, 247–56. [*45*]

Grice, G. R. and Saltz, E. (1950). The generalization of an instrumental response to stimuli varying in the size dimension. *J. Exp. Psychol.* **40**, 702–8. [*485, 494*]

Grindley, G. C. (1929). Experiments on the influence of amount of reward on learning in young chickens. *Br. J. Psychol.* **20**, 173–80. [*235*]

Groesbeck, R. W. and Duerfeldt, P. H. (1971). Some relevant variables in observational learning of the rat. *Psychon. Sci.* **22**, 41–3. [*202*]

Grossen, N. E. and Bolles, R. C. (1968). Effects of a classical conditioned "fear signal" and "safety signal" on nondiscriminated avoidance behavior. *Psychon. Sci.* **11**, 321–2. [*83, 311*]

Grossen, N. E. and Kelley, M. J. (1972). Species-specific behavior and acquisition of avoidance behavior in rats. *J. Comp. Physiol. Psychol.* **81**, 307–10. [*343*]

Grossen, N. E., Kostansek, D. J., and Bolles, R. C. (1969). Effects of appetitive discriminative stimuli on avoidance behavior. *J. Exp. Psychol.* **81**, 340–43. [*21, 84, 224*]

Grosslight, J. H., Hall, J. F., and Scott, W. (1954). Reinforcement schedules in habit reversal—a confirmation. *J. Exp. Psychol.* **48**, 173–4. [*190*]

Grosslight, J. H. and Radlow, R. (1956). Patterning effect of the nonreinforcement-reinforcement sequence in a discrimination situation. *J. Comp. Physiol. Psychol.* **49**, 542–6. [*190, 444*]

Grosslight, J. H. and Radlow, R. (1957). Patterning of nonreinforcement-reinforcement sequence involving a single nonreinforced trial. *J. Comp. Physiol. Psychol.* **50**, 23–5. [*444*]

Grossman, S. P. and Miller, N. E. (1961). Control for stimulus-change in the evaluation of alcohol and chlorpromazine as fear-reducing drugs. *Psychopharmacologia* 2, 342–51. [*265*]

Groves, P. M. and Thompson, R. F. (1970). Habituation: A dual-process theory. *Psychol. Rev.* 77, 419–50. [*16, 39*]

Grusec, T. (1968). The peak-shift in stimulus generalization: equivalent effects of errors and non-contingent shock. *J. Exp. Anal. Behav.* 11, 39–49. [*537*]

Gulliksen, H. (1932). Studies of transfer of response: I. Relative versus absolute factors in the discrimination of size by the white rat. *J. Genet. Psychol.* 40, 37–51. [*562*]

Gulliksen, H. and Wolfle, H. L. (1938). A theory of learning and transfer: I. *Psychometrika* 3, 127–49. [*545*]

Guthrie, E. R. (1935). *The Psychology of Learning.* New York: Harper. [*7, 79, 91, 260, 279, 292, 412*]

Guttman, N. (1953). Operant conditioning, extinction, and periodic reinforcement in relation to concentration of sucrose used as reinforcing agent. *J. Exp. Psychol.* 46, 213–24. [*152*]

Guttman, N. (1959). Generalization gradients around stimuli associated with different reinforcement schedules. *J. Exp. Psychol.* 58, 335–40. [*537*]

Guttman, N. and Kalish, H. I. (1956). Discriminability and stimulus generalization. *J. Exp. Psychol.* 51, 79–88. [*485, 494, 496, 504*]

Gwinn, G. T. (1949). The effects of punishment on acts motivated by fear. *J. Exp. Psychol.* 39, 260–9. [*286*]

Gynther, M. D. (1957). Differential eyelid conditioning as a function of stimulus similarity and strength of response to the CS. *J. Exp. Psychol.* 53, 408–16. [*387*]

Haber, A. and Kalish, H. I. (1963). Prediction of discrimination from generalization after variations in schedule of reinforcement. *Science, N.Y.* 142, 412–13. [*514*]

Haberlandt, K. (1971). Transfer along a continuum in classical conditioning. *Learning and Motivation* 2, 164–72. [*594, 596*]

Haggard, D. F. (1959). Acquisition of a simple running response as a function of partial and continuous schedules of reinforcement. *Psychol. Rec.* 9, 11–18. [*75, 160*]

Haggbloom, S. J. and Williams, D. T. (1971). Increased resistance to extinction following partial reinforcement: A function of N-length or percentage of reinforcement. *Psychon. Sci.* 24, 16–18. [*446, 447, 459*]

Hake, D. F. and Powell, T. (1970). Positive reinforcement and suppression from the same occurrence of the unconditioned stimulus in a positive conditioned suppression procedure. *J. Exp. Anal. Behav.* 14, 247–57. [*226*]

Hall, G. (1974a). Strategies of simultaneous discrimination learning in the pigeon. *Q. J. Exp. Psychol.* in press. (a) [*569*]

Hall, G. (1974b). Transfer effects produced by overtraining in the rat. *J. Comp. Physiol. Psychol.* in press. (b) [*590, 600, 606, 607*]

Hall, J. F. (1951). Studies in secondary reinforcement: I. Secondary reinforcement as a function of the frequency of primary reinforcement. *J. Comp. Physiol. Psychol.* 44, 246–51. [*241*]

Hall, K. R. L. (1963). Observational learning in monkeys and apes. *Br. J. Psychol.* 54, 201–26. [*200, 201*]

Halliday, M. S. and Boakes, R. A. (1972). Discrimination involving response-independent reinforcement: implications for behavioural contrast. In R. A. Boakes and M. S. Halliday (Eds.), *Inhibition and Learning*. London: Academic Press. Pp. 73–97. [*381, 382*]

Halliday, M. S. and Boakes, R. A. (1974). Behavioral contrast without response rate reduction. *J. Exp. Anal. Behav.* in press. [*378, 381*]

Hamilton, W. F. and Coleman, T. B. (1933). Trichromatic vision in the pigeon as illustrated by the spectral hue discrimination curve. *J. Comp. Psychol.* **15**, 183–91. [*496*]

Hamm, H. D. (1967). Perseveration and summation of the frustration effect. *J. Exp. Psychol.* **73**, 196–203. [*363*]

Hammond, L. J. (1967). A traditional demonstration of the active properties of Pavlovian inhibition using differential CER. *Psychon. Sci.* **9**, 65–6. [*33*]

Hammond, L. J. (1968). Retardation of fear acquisition by a previously inhibitory CS. *J. Comp. Physiol. Psychol.* **66**, 756–9. [*32*]

Hammond, L. J. and Daniel, R. (1970). Negative contingency discrimination: Differentiation by rats between safe and random stimuli. *J. Comp. Physiol. Psychol.* **72**, 486–91. [*34*]

Hammond, L. J. and Maser, J. (1970). Forgetting and conditioned suppression: role of a temporal discrimination. *J. Exp. Anal. Behav.* **13**, 333–8. [*474*]

Hanson, H. M. (1959). Effects of discrimination training on stimulus generalization. *J. Exp. Psychol.* **58**, 321–34. [*533, 535, 536, 563*]

Hara, K. and Warren, J. M. (1961). Stimulus additivity and dominance in discrimination performance by cats. *J. Comp. Physiol. Psychol.* **54**, 86–90. [*574*]

Harker, G. S. (1956). Delay of reward and performance of an instrumental response. *J. Exp. Psychol.* **51**, 303–10. [*156, 393*]

Harlow, H. F. (1937). Experimental analysis of the role of the original stimulus in conditioned responses in monkeys. *Psychol. Rec.* **1**, 62–8. [*87, 89*]

Harlow, H. F. (1939). Forward conditioning, backward conditioning and pseudo-conditioning in the goldfish. *J. Genet. Psychol.* **55**, 49–58. [*59*]

Harlow, H. F. (1949). The formation of learning sets. *Psychol. Rev.* **56**, 51–65. [*202, 611*]

Harlow, H. F. (1959). Learning set and error factor theory. In S. Koch (Ed.), *Psychology: a Study of a Science*, Vol. 2. New York: McGraw-Hill. Pp. 492–537. [*612*]

Harlow, H. F. and Hicks, L. H. (1957). Discrimination learning theory: uniprocess vs. dualprocess. *Psychol. Rev.* **64**, 104–9. [*566, 567*]

Harris, D. R., Collerain, I., Wolf, J. C., and Ludvigson, H. W. (1970). Negative S– contrast with minimally contingent large reward as a function of trial initiation procedure. *Psychon. Sci.* **19**, 189–90. [*391, 396*]

Harris, J. D. (1943a). Studies of nonassociative factors inherent in conditioning. *Comp. Psychol. Monogr.* **18** (1, Serial No. 93). (a) [*27*]

Harris, J. D. (1943b). Habituatory response decrement in the intact organism. *Psychol. Bull.* **40**, 385–422. (b) [*39*]

Harris, J. H. and Thomas, G. J. (1966). Learning single alternation of running speeds in a runway without handling between trials. *Psychon. Sci.* **6**, 329–30. [*162*]

Harris, P. and Nygaard, J. E. (1961). Resistance to extinction and number of reinforcements. *Psychol. Rep.* **8**, 233–4. [*423*]

Hartman, T. F. and Ross, L. E. (1961). An alternative criterion for the elimination of "voluntary" responses in eyelid conditioning. *J. Exp. Psychol.* **61**, 334–8. [*102*]

Healey, A. F. (1965). Compound stimuli, drive strength and primary stimulus generalization. *J. Exp. Psychol.* **69**, 536–8. [*520*]

Hearst, E. (1962). Concurrent generalization gradients for food-controlled and shock-controlled behavior. *J. Exp. Anal. Behav.* **5**, 19–31. [*506, 522*]

Hearst, E. (1965). Approach, avoidance and stimulus generalization. In D. I. Mostofsky (Ed.), *Stimulus Generalization*. Stanford: Stanford University Press. Pp. 331–55. [*514, 523*]

Hearst, E. (1968). Discrimination learning as the summation of excitation and inhibition. *Science, N.Y.* **162**, 1303–6. [*533, 534*]

Hearst, E. (1969a). Excitation, inhibition and discrimination learning. In N. J. Mackintosh and W. K. Honig (Eds.), *Fundamental Issues in Associative Learning*. Halifax: Dalhousie University Press. Pp. 1–41. (a) [*377, 378, 508, 530, 533, 537, 567*]

Hearst, E. (1969b). Aversive conditioning and external stimulus control. In B. A. Campbell and R. M. Church (Eds.), *Punishment and Aversive Behavior*. New York: Appleton-Century-Crofts. Pp. 235–77. (b) [*523*]

Hearst, E. (1971a). Differential transfer of excitatory versus inhibitory pretraining to intradimensional discrimination learning in pigeons. *J. Comp. Physiol. Psychol.* **75**, 206–15. (a) [*567*]

Hearst, E. (1971b). Contrast and stimulus generalization following prolonged discrimination training. *J. Exp. Anal. Behav.* **15**, 355–63. (b) [*373*]

Hearst, E. (1972). Some persistent problems in the analysis of conditioned inhibition. In R. A. Boakes and M. S. Halliday (Eds.), *Inhibition and Learning*. London: Academic Press. Pp. 5–39. [*32, 38, 529, 530*]

Hearst, E., Besley, S., and Farthing, G. W. (1970). Inhibition and the stimulus control of operant behavior. *J. Exp. Anal. Behav.* **14**, 373–409. [*529, 530*]

Hearst, E. and Koresko, M. B. (1968). Stimulus generalization and amount of prior training on variable-interval reinforcement. *J. Comp. Physiol. Psychol.* **66**, 133–8. [*491, 492, 519*]

Hearst, E., Koresko, M. B., and Poppen, R. (1964). Stimulus generalization and the response-reinforcement contingency. *J. Exp. Anal. Behav.* **7**, 369–80. [*514, 515, 517*]

Heathcote, M. J. and Champion, R. A. (1963). A test of the relative-valence theory of instrumental-reward learning. *Am. J. Psychol.* **76**, 679–82. [*211*]

Hebb, D. O. (1955). Drives and the C.N.S. (Conceptual Nervous System). *Psychol. Rev.* **62**, 243–54. [*151*]

Heinemann, E. G. and Rudolph, R. L. (1963). The effect of discriminative training on the gradient of stimulus generalization. *Am. J. Psychol.* **76**, 653–8. [*504, 505*]

Heise, G. A., Keller, C., Khavari, K., and Laughlin, N. (1969). Discrete-trial alternation in the rat. *J. Exp. Anal. Behav.* **12**, 609–22. [*162, 163, 459*]

Hemmes, N. S. (1970). DRL efficiency depends upon the operant. Paper presented at Psychonomic Society, San Antonio. [*180*]

Hemmes, N. S. (1973). Behavioral contrast in pigeons depends upon the operant. *J. Comp. Physiol. Psychol.* **85**, 171–8. [*376*]

Hemmes, N. S. and Eckerman, D. A. (1972). Positive interaction (induction) in multiple variable-interval, differential-reinforcement-of-high-rate schedules. *J. Exp. Anal. Behav.* **17**, 51–7. [*380*]

Henderson, K. (1966). Within-subjects partial-reinforcement effects in acquisition and in later discrimination learning. *J. Exp. Psychol.* **72**, 704–13. [*385*]

Hendry, D. P. (1969a). D. P. Hendry (Ed.), *Conditioned Reinforcement*. Homewood, Illinois: The Dorsey Press. (a) [*233, 238*]

Hendry, D. P. (1969b). Reinforcing value of information: fixed-ratio schedules. In D. P. Hendry (Ed.), *Conditioned Reinforcement*. Homewood, Illinois: The Dorsey Press. Pp. 300–41. (b) [*251, 252, 253, 256*]

Hendry, D. P., Switalski, R., and Yarczower, M. (1969). Generalization of conditioned suppression after differential training. *J. Exp. Anal. Behav.* **12,** 799–806. [*539*]

Herb, F. H. (1940). Latent learning—non-reward followed by food in blinds. *J. Comp. Psychol.* **29,** 247–55. [*209*]

Herendeen, D. and Anderson, D. C. (1968). Dual effects of a second-order conditioned stimulus: Excitation and inhibition. *Psychon. Sci.* **13,** 15–16. [*20*]

Herman, L. M. and Arbeit, W. R. (1973). Stimulus control and auditory discrimination learning sets in the bottlenose dolphin. *J. Exp. Anal. Behav.* **19,** 379–94. [*611*]

Herman, L. M., Beach, F. A., III, Pepper, R. L., and Stalling, R. B. (1969). Learning set performance in the bottlenose dolphin. *Psychon. Sci.* **14,** 98–9. [*612*]

Herrick, R. M. and Bromberger, R. A. (1965). Lever displacement under a variable ratio schedule and subsequent extinction. *J. Comp. Physiol. Psychol.* **59,** 392–8. [*439*]

Herrick, R. M., Myers, J. L., and Korotkin, A. L. (1959). Changes in S^D and S^Δ rates during the development of an operant discrimination. *J. Comp. Physiol. Psychol.* **52,** 359–63. [*376*]

Herrnstein, R. J. (1958). Some factors influencing behavior in a two-response situation. *Trans. N.Y. Acad. Sci.* **21,** 35–45. [*194*]

Herrnstein, R. J. (1961). Relative and absolute strength of response as a function of frequency of reinforcement. *J. Exp. Anal. Behav.* **4,** 267–72. [*167, 192*]

Herrnstein, R. J. (1964a). Secondary reinforcement and the rate of primary reinforcement. *J. Exp. Anal. Behav.* **7,** 27–36. (a) [*243, 245, 382*]

Herrnstein, R. J. (1964b). Aperiodicity as a factor in choice. *J. Exp. Anal. Behav.* **7,** 179–82. (b) [*256*]

Herrnstein, R. J. (1969). Method and theory in the study of avoidance. *Psychol. Rev.* **76,** 49–69. [*299, 309, 312, 313, 323, 326, 329, 330*]

Herrnstein, R. J. (1970). On the law of effect. *J. Exp. Anal. Behav.* **13,** 243–66. [*178, 179, 191, 193*]

Herrnstein, R. J. and Hineline, P. N. (1966). Negative reinforcement as shock-frequency reduction. *J. Exp. Anal. Behav.* **9,** 421–30. [*301, 323, 329*]

Herrnstein, R. J. and Loveland, D. H. (1972). Food-avoidance in hungry pigeons and other perplexities. *J. Exp. Anal. Behav.* **18,** 369–83. [*121*]

Herrnstein, R. J. and Morse, W. H. (1958). A conjunctive schedule of reinforcement. *J. Exp. Anal. Behav.* **1,** 15–24. [*174, 175*]

Herrnstein, R. J. and Sidman, M. (1958). Avoidance conditioning as a factor in the effects of unavoidable shocks on food-reinforced behavior. *J. Comp. Physiol. Psychol.* **51,** 380–5. [*83*]

Hick, W. E. (1952). On the rate of gain of information. *Q. J. Exp. Psychol.* **4,** 11–26. [*386*]

Hicks, L. H. (1964). Effects of overtraining on acquisition and reversal of place and response learning. *Psychol. Rep.* **15,** 459–62. [*554*]

Hilgard, E. R. (1931). Conditioned eyelid reactions to a light stimulus based on the reflex wink to sound. *Psychol. Monogr.* **41** (1, Whole No. 184). [*17*]

Hilgard, E. R. (1936). The nature of the conditioned response. I. The case for and against stimulus substitution. *Psychol. Rev.* **43**, 366–85. [*97, 100, 101, 102, 105*]

Hilgard, E. R. and Allen, M. K. (1938). An attempt to condition finger reactions based on motor point stimulation. *J. Gen. Psychol.* **18**, 203–7. [*94*]

Hilgard, E. R. and Campbell, A. A. (1936). The course of acquisition and retention of conditioned eyelid responses in man. *J. Exp. Psychol.* **19**, 227–47. [*101, 102*]

Hilgard, E. R. and Marquis, D. G. (1935). Acquisition, extinction, and retention of conditioned lid responses to light in dogs. *J. Comp. Psychol.* **19**, 29–58. [*411, 412*]

Hilgard, E. R. and Marquis, D. G. (1940). *Conditioning and Learning.* New York: Appleton-Century-Crofts. [*16, 100, 102, 110, 308, 419*]

Hill, C. W. and Thune, L. E. (1952). Place and response learning in the white rat under simplified and mutually isolated conditions. *J. Exp. Psychol.* **43**, 289–97. [*554*]

Hill, W. F., Cotton, J. W. and Clayton, K. N. (1962). Effect of reward magnitude, percentage of reinforcement, and training method on acquisition and reversal in a T-maze. *J. Exp. Psychol.* **64**, 81–6. [*190*]

Hill, W. F. and Spear, N. E. (1962). Resistance to extinction as a joint function of reward magnitude and the spacing of extinction trials. *J. Exp. Psychol.* **64**, 636–9. [*419*]

Hill, W. F. and Spear, N. E. (1963a). Extinction in a runway as a function of acquisition level and reinforcement percentage. *J. Exp. Psychol.* **65**, 495–500. (a) [*424, 462*]

Hill, W. F. and Spear, N. E. (1963b). Choice between magnitudes of reward in a T-maze. *J. Comp. Physiol. Psychol.* **56**, 723–6. (b) [*186*]

Hill, W. F. and Wallace, W. P. (1967). Reward magnitude and number of training trials as joint factors in extinction. *Psychon. Sci.* **7**, 267–8. [*152, 424, 427*]

Hillman, B., Hunter, W. S., and Kimble, G. A. (1953). The effect of drive level on the maze performance of the white rat. *J. Comp. Physiol. Psychol.* **46**, 87–9. [*150, 183, 184*]

Hilton, A. (1969). Partial reinforcement of a conditioned emotional response in rats. *J. Comp. Physiol. Psychol.* **69**, 253–60. [*74, 81*]

Hind, J. E. (1961). Unit activity in the auditory cortex. In G. L. Rasmussen and W. F. Windle (Eds.), *Neural Mechanisms of the Auditory and Vestibular Systems.* Springfield: Thomas. Pp. 201–10 [*490*]

Hinde, R. A. (1970). *Animal Behaviour. A Synthesis of Ethology and Comparative Psychology.* New York: McGraw-Hill. [*6, 39, 224, 350*]

Hinde, R. A. and Tinbergen, N. (1958). The comparative study of species-specific behavior. In A. Roe and G. G. Simpson (Eds.), *Behavior and Evolution.* New Haven: Yale University Press. Pp. 251–68. [*2*]

Hineline, P. N. (1970). Negative reinforcement without shock reduction. *J. Exp. Anal. Behav.* **14**, 259–68. [*328*]

Hineline, P. N. and Herrnstein, R. J. (1970). Timing in free-operant and discrete-trial avoidance. *J. Exp. Anal. Behav.* **13**, 113–26. [*325*]

Hineline, P. N. and Rachlin, H. (1969). Escape and avoidance of shock by pigeons pecking a key. *J. Exp. Anal. Behav.* **12**, 533–8. [*340, 343*]

Hiss, R. H. and Thomas, D. R. (1963). Stimulus generalization as a function of testing procedure and response measure. *J. Exp. Psychol.* **65**, 587–92. [*494, 495*]

Hoffeld, D. R. (1962). Primary stimulus generalization and secondary extinction as a function of strength of conditioning. *J. Comp. Physiol. Psychol.* **55**, 27–31. [*518*]

Hoffeld, D. R., Kendall, S. B., Thompson, R. F., and Brogden, W. J. (1960). Effect of amount of preconditioning training upon the magnitude of sensory preconditioning. *J. Exp. Psychol.* **59**, 198–204. [*21, 96*]

Hoffeld, D. R., Thompson, R. F., and Brogden, W. J. (1958). Effect of stimuli-time relations during preconditioning training upon the magnitude of sensory preconditioning. *J. Exp. Psychol.* **56**, 437–42. [*21*]

Hoffman, H. S. (1969). Stimulus factors in conditioned suppression. In B. A. Campbell and R. M. Church (Eds.), *Punishment and Aversive Behaviour.* New York: Appleton-Century-Crofts. Pp. 185–234. [*527, 538, 539*]

Hoffman, H. S. and Barrett, J. (1971). Overt activity during conditioned suppression: a search for punishment artifacts. *J. Exp. Anal. Behav.* **16**, 343–8. [*10, 81*]

Hoffman, H. S. and Fleshler, M. (1959). Aversive control with the pigeon. *J. Exp. Anal. Behav.* **2**, 213–18. [*137*]

Hoffman, H. S. and Fleshler, M. (1961). Stimulus factors in aversive controls: the generalization of conditioned suppression. *J. Exp. Anal. Behav.* **4**, 371–81 [*493, 494, 529*]

Hoffman, H. S. and Fleshler, M. (1965). Stimulus aspects of aversive controls: The effects of response contingent shock. *J. Exp. Anal. Behav.* **8**, 89–96. [*275*]

Hoffman, H. S., Fleshler, M., and Abplanalp, P. L. (1964). Startle reaction to electrical shock in the rat. *J. Comp. Physiol. Psychol.* **58**, 132–9. [*82*]

Hoffman, H. S., Fleshler, M., and Jensen, P. (1963). Stimulus aspects of aversive controls: the retention of conditioned suppression. *J. Exp. Anal. Behav.* **6**, 575–83. [*473, 474*]

Holman, G. L. (1969). Intragastric reinforcement effect. *J. Comp. Physiol. Psychol.* **69**, 432–41. [*93*]

Holmes, P. W. (1972). Conditioned suppression with extinction as the signalled stimulus. *J. Exp. Anal. Behav.* **18**, 129–32. [*262*]

Holz, W. C. and Azrin, N. H. (1961). Discriminative properties of punishment. *J. Exp. Anal. Behav.* **4**, 225–32. [*282, 292*]

Holz, W. C. and Azrin, N. H. (1962). Interactions between the discriminative and aversive properties of punishment. *J. Exp. Anal. Behav.* **5**, 229–34. [*283*]

Homme, L. E. (1956). Spontaneous recovery and statistical learning theory. *J. Exp. Psychol.* **51**, 205–12. [*470*]

Homzie, M. J., Gohmann, T., and Hall, S. W. Jr. (1971). Runway performance in rats as determined by the predictive value of intertrial reinforcements. *J. Comp. Physiol. Psychol.* **74**, 90–5. [*451*]

Homzie, M. J. and Ross, L. E. (1962). Runway performance following a reduction in the concentration of a liquid reward. *J. Comp. Physiol. Psychol.* **55**, 1029–33. [*391*]

Homzie, M. J. and Rudy, J. W. (1971). Effect on runway performance of reinforcement contingencies established to empty goal-box placements. *Learning and Motivation* **2**, 95–101. [*449, 51*]

Homzie, M. J., Rudy, J. W., and Carter, E. N. (1970). Runway performance in rats as a function of goal-box placements and goal-event sequence. *J. Comp. Physiol. Psychol.* **71**, 283–91. [*450, 451*]

Honig, W. K. (1961). Generalization of extinction on the spectral continuum. *Psychol. Rec.* **11**, 269–78. [*526*]

Honig, W. K. (1962). Prediction of preference, transposition, and transposition-reversal from the generalization gradient. *J. Exp. Psychol.* **64**, 239–48. [*533, 560, 561, 569*]

Honig, W. K. (1966). The role of discrimination training in the generalization of punishment. *J. Exp. Anal. Behav.* **9**, 377–84. [*525*]

Honig, W. K. (1969). Attentional factors governing the slope of the generalization gradient. In R. M. Gilbert and N. S. Sutherland (Eds.), *Animal Discrimination Learning*. London: Academic Press. Pp. 35–62. [*509*]

Honig, W. K. (1970). Attention and the modulation of stimulus control. In D. Mostofsky (Ed.), *Attention: Contemporary Theory and Analysis*. New York: Appleton-Century-Crofts. Pp. 193–238. [*488, 507*]

Honig, W. K., Boneau, C. A., Burstein, K. R., and Pennypacker, H. S. (1963). Positive and negative generalization gradients obtained after equivalent training conditions. *J. Comp. Physiol. Psychol.* **56**, 111–16. [*526, 527*]

Honig, W. K., Thomas, D. R., and Guttman, N. (1959). Differential effects of continuous extinction and discrimination training on the generalization gradient. *J. Exp. Psychol.* **58**, 145–52. [*536*]

Honzik, C. H. (1936). The sensory basis of maze learning in rats. *Comp. Psychol. Monogr.* **13** (4, Serial No. 64). [*553, 554, 556*]

Hooper, R. (1967). Variables controlling the overlearning reversal effect (ORE). *J. Exp. Psychol.* **73**, 612–19. [*184, 603*]

Hovland, C. I. (1937a). The generalization of conditioned responses. I. The sensory generalization of conditioned response with varying frequencies of tone. *J. Gen. Psychol.* **17**, 125–48. (a) [*14*]

Hovland, C. I. (1937b). The generalization of conditioned responses. II. The sensory generalization of conditioned responses with varying intensities of tone. *J. Genet. Psychol.* **51**, 279–91. (b) [*14, 485*]

Hovland, C. I. (1937c). The generalization of conditioned responses. IV. The effect of varying amounts of reinforcement upon the degree of generalization of conditioned responses. *J. Exp. Psychol.* **21**, 261–76. (c) [*518*]

Howard, I. P. (1966). *Motor System*. In J. A. Deutsch and D. Deutsch, *Physiological Psychology*. Homewood, Illinois: The Dorsey Press. Pp. 386–441. [*133*]

Huang, I. (1969). Successive contrast effects as a function of type and magnitude of reward. *J. Exp. Psychol.* **82**, 64–9. [*399*]

Hubel, D. H. and Wiesel, T. N. (1959). Receptive fields of single neurones in the cat's striate cortex. *J. Physiol.* **148**, 574–91. [*490*]

Hug, J. J. (1970a). Number of food pellets and the development of the frustration effect. *Psychon. Sci.* **21**, 59–60. (a) [*462*]

Hug, J. J. (1970b). Frustration effect after development of patterned responding to single-alternation reinforcement. *Psychon. Sci.* **21**, 61–2. (b) [*364*]

Hughes, D., Davis, J. D., and Grice, G. R. (1960). Goal box and alley similarity as a factor in latent extinction. *J. Comp. Physiol. Psychol.* **53**, 612–14. [*417*]

Hughes, L. F. and Dachowski, L. (1973). The role of reinforcement and non-reinforcement in an operant frustration effect. *Animal Learning and Behavior* **1**, 68–72. [*366*]

Hughes, R. N. (1969). Social facilitation of locomotion and exploration in rats. *Br. J. Psychol.* **60**, 385–8. [*201*]

Hull, C. L. (1929). A functional interpretation of the conditioned reflex. *Psychol. Rev.* **36**, 495–511. [*99, 301*]

Hull, C. L. (1934). Learning: II. The factor of the conditioned reflex. In C. Murchison (Ed.), *A Handbook of General Experimental Psychology.* Worcester: Clark University Press. Pp. 382–455. [*16, 27, 100, 105*]

Hull, C. L. (1943). *Principles of Behavior.* New York: Appleton-Century-Crofts. [*7, 11, 49, 61, 79, 98, 111, 151, 153, 156, 188, 198, 222, 233, 260, 301, 406, 414, 423, 468, 520, 531*]

Hull, C. L. (1947). Reactively heterogeneous compound trial-and-error learning with distributed trials and terminal reinforcement. *J. Exp. Psychol.* **37**, 118–35. [*189*]

Hull, C. L. (1949). Stimulus intensity dynamism (V) and stimulus generalization. *Psychol. Rev.* **56**, 67–76. [*41, 43*]

Hull, C. L. (1952). *A Behaviour System.* New Haven: Yale University Press. [*154, 188, 189, 208, 213, 216, 222, 223, 246, 422, 440, 499, 500, 503, 532, 566, 583, 610, 613*]

Hull, C. L. and Spence, K. W. (1938). "Correction" vs. "non-correction" method of trial-and-error learning in rats. *J. Comp. Psychol.* **25**, 127–45. [*185*]

Hulse, S. H., Jr. (1958). Amount and percentage of reinforcement and duration of goal confinement in conditioning and extinction. *J. Exp. Psychol.* **56**, 48–57. [*407, 427, 430, 455*]

Humphreys, L. G. (1939). Generalization as a function of method of reinforcement. *J. Exp. Psychol.* **25**, 361–72. [*514*]

Hunt, H. F. and Brady, J. V. (1955). Some effects of punishment and intercurrent "anxiety" on a simple operant. *J. Comp. Physiol. Psychol.* **48**, 305–10. [*274*]

Hunter, M. W., III and Kamil, A. C. (1971). Object discrimination learning set and hypothesis behavior in the northern bluejay (*Cyanocitta cristata*). *Psychon. Sci.* **22**, 271–3 [*611*]

Hunter, W. S. (1929). The sensory control of the maze habit in the white rat. *J. Genet. Psychol.* **36**, 505–37. [*554, 556*]

Hunter, W. S. (1935). Conditioning and extinction in the rat. *Br. J. Psychol.* **26**, 135–48. [*116*]

Hunter, W. S. (1940). A kinaesthetically controlled maze habit in the rat. *Science, N.Y.* **91**, 267–9. [*555*]

Hurwitz, H. M. B. and Roberts, A. E. (1971). Conditioned suppression of an avoidance response. *J. Exp. Anal. Behav.* **16**, 278–81. [*83*]

Hutchinson, R. R., Azrin, N. H., and Hunt, G. M. (1968). Attack produced by intermittent reinforcement of a concurrent operant response. *J. Exp. Anal. Behav.* **11**, 489–95. [*263*]

Hutchinson, R. R., Renfrew, J. W., and Young, G. A. (1971). Effects of long term shock and associated stimuli on aggressive and manual responses. *J. Exp. Anal. Behav.* **15**, 141–66. [*18*]

Hutt, P. J. (1954). Rate of bar pressing as a function of quality and quantity of food reward. *J. Comp. Physiol. Psychol.* **47**, 235–9. [*152*]

Imanishi, K. (1957). Social behaviour in Japanese monkeys, *Macaca fuscata. Psychologia* **1**, 47–54. [*200*]

Irion, A. L. (1949). Retention and warming-up effects in paired-associate learning. *J. Exp. Psychol.* **39**, 669–75. [*477*]

Ison, J. R. (1962). Experimental extinction as a function of number of reinforcements. *J. Exp. Psychol.* **64**, 314–17. [*424, 425*]

Ison, J. R. (1964). Acquisition and reversal of a spatial response as a function of sucrose concentration. *J. Exp. Psychol.* **67**, 495–6. [*184*]

Ison, J. R. and Cook, P. E. (1964). Extinction performance as a function of incentive magnitude and number of acquisition trials. *Psychon. Sci.* **1**, 245–6. [*424, 425, 427*]

Ison, J. R., Daly, H. B., and Glass, D. H. (1967). Amobarbital sodium and the effects of reward and nonreward in the Amsel double runway. *Psychol. Rep.* **20**, 491–6. [*353*]

Ison, J. R. and Glass, D. H. (1969). Effects of concurrent exposure to different food and sucrose rewards in differential conditioning. *Psychon. Sci.* **15**, 149–50. [*392*]

Ison, J. R. and Krane, R. V. (1969). Induction in differential instrumental conditioning. *J. Exp. Psychol.* **80**, 183–5. [*357, 385*]

Ison, J. R. and Pennes, E. S. (1969). Interaction of amobarbital sodium and reinforcement schedule in determining resistance to extinction of an instrumental running response. *J. Comp. Physiol. Psychol.* **68**, 215–19. [*265, 465*]

Ison, J. R. and Rosen, A. J. (1967). The effects of amobarbital sodium on differential instrumental conditioning and subsequent extinction. *Psychopharmacologia* **10**, 417–25. [*265*]

Ison, J. R. and Rosen, A. J. (1968). Extinction and reacquisition performance as a function of sucrose-solution rewards and number of acquisition trials. *Psychol. Rep.* **22**, 375–9. [*424, 427*]

James, J. P. (1971a). Acquisition, extinction, and spontaneous recovery of conditioned suppression of licking. *Psychon. Sci.* **22**, 156–8. (a) [*59*]

James, J. P. (1971b). Latent inhibition and the preconditioning-conditioning interval. *Psychon. Sci.* **24**, 97–8. (b) [*37*]

Jeeves, M. A. and North, A. J. (1956). Irrelevant or partially correlated stimuli in discrimination learning. *J. Exp. Psychol.* **52**, 90–4. [*581*]

Jenkins, H. M. (1961). The effect of discrimination training on extinction. *J. Exp. Psychol.* **61**, 111–21. [*386, 512*]

Jenkins, H. M. (1962). Resistance to extinction when partial reinforcement is followed by regular reinforcement. *J. Exp. Psychol.* **64**, 441–50. [*436*]

Jenkins, H. M. (1965). Generalization gradients and the concept of inhibition. In D. I. Mostofsky (Ed.), *Stimulus Generalization.* Stanford: Stanford University Press. Pp. 55–61. [*566*]

Jenkins, H. M. (1970). Sequential organization in schedules of reinforcement. In W. N. Schoenfeld (Ed.), *The Theory of Reinforcement Schedules.* New York: Appleton-Century-Crofts. Pp. 63–109. [*181, 182*]

Jenkins, H. M. (1973). Noticing and responding in a discrimination based on a distinguishing element. *Learning and Motivation* **4**, 115–37. [*121, 569, 573*]

Jenkins, H. M. and Harrison, R.H. (1960). Effect of discrimination training on auditory generalization. *J. Exp. Psychol.* **59**, 246–53. [*493, 501, 504*]

Jenkins, H. M. and Harrison, R. H. (1962). Generalization gradients of inhibition following auditory discrimination learning. *J. Exp. Anal. Behav.* **5**, [*526, 527, 535*] 435–41.

Jenkins, H. M. and Moore, B. R. (1973). The form of the auto-shaped response with food or water reinforcers. *J. Exp. Anal. Behav.* **20**, 163–81. [*17, 101, 206*]

Jenkins, H. M. and Sainsbury, R. S. (1969). The development of stimulus control through differential reinforcement. In N. J. Mackintosh and W. K. Honig

(Eds.), *Fundamental Issues in Associative Learning*. Halifax: Dalhousie University Press. Pp. 123–61. [*15, 569, 585*]

Jenkins, H. M. and Sainsbury, R. S. (1970). Discrimination learning with the distinctive feature on positive or negative trials. In D. Mostofsky (Ed.), *Attention: Contemporary Theory and Analysis*. New York: Appleton-Century-Crofts. Pp. 239–73. [*15, 585*]

Jenkins, W. O. (1950). A temporal gradient of derived reinforcement. *Am. J. Psychol.* **63**, 237–43. [*242, 244, 245*]

Jenkins, W. O., Pascal, G. R., and Walker, R. W., Jr. (1958). Deprivation and generalization. *J. Exp. Psychol.* **56**, 274–7. [*521*]

Jenkins, W. O. and Stanley, J. C. Jr. (1950). Partial reinforcement: a review and critique. *Psychol. Bull.* **47**, 193–204. [*434*]

John, E. R., Chesler, P., Bartlett, F., and Victor, I. (1968). Observation learning in cats. *Science, N.Y.* **159**, 1489–91. [*201, 204*]

Johnson, D. F. (1970). Determiners of selective stimulus control in the pigeon. *J. Comp. Physiol. Psychol.* **70**, 298–307. [*583*]

Johnson, D. F. and Anderson, W. H. (1970). Generalization gradients around S^Δ following errorless discrimination learning. *Psychon. Sci.* **21**, 298–300. [*530*]

Johnson, E. E. (1952). The role of motivational strength in latent learning. *J. Comp. Physiol. Psychol.* **45**, 526–30. [*212*]

Johnson, N. and Viney, W. (1970). Resistance to extinction as a function of effort. *J. Comp. Physiol. Psychol.* **71**, 171–4. [*408, 416*]

Johnson, R. N. (1970). Spatial probability learning and brain stimulation in rats. *Psychon. Sci.* **18**, 33. [*191*]

Jones, E. C. and Bridges, C. C. (1966). Competing responses and the partial reinforcement effect. *Psychon. Sci.* **6**, 483–4. [*411*]

Kalat, J. W. and Rozin, P. (1970). "Salience": a factor which can override temporal contiguity in taste-aversion learning. *J. Comp. Physiol. Psychol.* **71**, 192–7. [*42, 54*]

Kalat, J. W. and Rozin, P. (1971). Role of interference in taste-aversion learning. *J. Comp. Physiol. Psychol.* **77**, 53–8. [*70*]

Kalish, D. (1946). The non-correction method and the delayed response problem of Blodgett and McCutchan. *J. Comp. Psychol.* **39**, 91–107. [*185*]

Kalish, H. I. and Guttman, N. (1957). Stimulus generalization after equal training on two stimuli. *J. Exp. Psychol.* **53**, 139–44. [*532*]

Kalish, H. I. and Guttman, N. (1959). Stimulus generalization after training on three stimuli: a test of the summation hypothesis. *J. Exp. Psychol.* **57**, 268–72. [*532*]

Kalish, H. I. and Haber, A. (1965). Prediction of discrimination from generalization following variations in deprivation level. *J. Comp. Physiol. Psychol.* **60**, 125–8. [*520*]

Kamano, D. K. (1970). Types of Pavlovian conditioning procedures used in establishing CS + and their effect upon avoidance behavior. *Psychon. Sci.* **18**, 63–4. [*83, 311*]

Kamil, A. C. (1969). Some parameters of the second-order conditioning of fear in rats. *J. Comp. Physiol. Psychol.* **67**, 364–9. [*20*]

Kamil, A. C. and Hunter, M. W. III (1970). Performance on object-discrimination learning set by the greater hill myna (*Gracula religiosa*). *J. Comp. Physiol. Psychol.* **73**, 68–73. [*611*]

Y

Kamil, A. C., Lougee, M., and Shulman, R. J. (1973). Learning-set behavior in the learning-set experienced bluejay (*Cyanocitta cristata*). *J. Comp. Physiol. Psychol.* **82**, 394–405. [*614*]

Kamin, L. J. (1954). Traumatic avoidance learning: The effects of CS-US interval with a trace-conditioning procedure. *J. Comp. Physiol. Psychol.* **47**, 65–72. [*321*]

Kamin, L. J. (1956). The effects of termination of the CS and avoidance of the US on avoidance learning. *J. Comp. Physiol. Psychol.* **49**, 420–4. [*116,304,316*]

Kamin, L. J. (1957a). The effects of termination of the CS and avoidance of the US on avoidance learning: An extension. *Can. J. Psychol.* **11**, 48–56. (a) [*116, 304, 316, 317*]

Kamin, L. J. (1957b). The gradient of delay of secondary reward in avoidance learning. *J. Comp. Physiol. Psychol.* **50**, 445–9. (b) [*315*]

Kamin, L. J. (1957c). The delay of secondary reward gradient in avoidance learning tested on avoidance trials only. *J. Comp. Physiol. Psychol.* **50**, 450–6. (c) [*315*]

Kamin, L. J. (1957d). The retention of an incompletely learned avoidance response. *J. Comp. Physiol. Psychol.* **50**, 457–60. (d) [*474, 475*]

Kamin, L. J. (1959). The delay-of-punishment gradient. *J. Comp. Physiol. Psychol.* **52**, 434–7. [*274, 286, 332*]

Kamin, L. J. (1963). Backward conditioning and the conditioned emotional response. *J. Comp. Physiol. Psychol.* **56**, 517–19. [*59*]

Kamin, L. J. (1965). Temporal and intensity characteristics of the conditioned stimulus. In W. F. Prokasy (Ed.), *Classical Conditioning: A Symposium*. New York: Appleton-Century-Crofts. Pp. 118–47. [*41, 43, 44, 58, 62, 65, 159*]

Kamin, L. J. (1968). 'Attention-like' processes in classical conditioning. In M. R. Jones (Ed.), *Miami Symposium on the Prediction of Behavior: Aversive Stimulation*. Miami: University of Miami Press. Pp. 9–33. [*48, 49*]

Kamin, L. J. (1969). Predictability, surprise, attention and conditioning. In B. A. Campbell and R. M. Church (Eds.), *Punishment and Aversive Behavior*. New York: Appleton-Century-Crofts. Pp. 279–96. [*46, 48, 49, 50, 335, 577, 583, 584, 585, 586*]

Kamin, L. J. and Brimer, C. J. (1963). The effects of intensity of conditioned and unconditioned stimuli on a conditioned emotional response. *Can. J. Psychol.* **17**, 194–8. [*41, 70*]

Kamin, L. J., Brimer, C. J., and Black, A. H. (1963). Conditioned suppression as a monitor of fear of the CS in the course of avoidance training. *J. Comp. Physiol. Psychol.* **56**, 497–501. [*81, 333*]

Kamin, L. J. and Schaub, R. E. (1963). Effects of conditioned stimulus intensity on the conditioned emotional response. *J. Comp. Physiol. Psychol.* **56**, 502–7. [*41*]

Kappauf, W. E. and Schlosberg, H. (1937). Conditioned responses in the white rat. III. Conditioning as a function of the length of the period of delay. *J. Genet. Psychol.* **50**, 27–45. [*61*]

Karsh, E. B. (1962). Effects of number of rewarded trials and intensity of punishment on running speed. *J. Comp Physiol. Psychol.* **55**, 44–51. [*274, 425, 426*]

Karsh, E. B. (1963). Changes in intensity of punishment: Effect on runway behavior of rats. *Science, N.Y.* **140**, 1084–5. [*284*]

Katkin, E. S. and Murray, E. N. (1968). Instrumental conditioning of autonomically mediated behavior: Theoretical and methodological issues. *Psychol. Bull.* **70**, 52–68. [*130*]

Katz, D. (1937). *Animals and Men*. London: Longmans, Green. [*201*]

Katz, S., Woods, G. T., and Carrithers, J. H. (1966). Reinforcement aftereffects and intertrial interval. *J. Exp. Psychol.* **72**, 624–6 [*162*]

Katzev, R. (1967). Extinguishing avoidance responses as a function of delayed warning signal termination. *J. Exp. Psychol.* **75**, 339–44 [*337*]

Kaufman, A. (1969). Response suppression in the CER paradigm with extinction as the aversive event. *Psychon. Sci.* **15**, 15–16. [*262*]

Keehn, J. D. (1959). The effect of a warning signal on unrestricted avoidance behaviour. *Br. J. Psychol.* **50**, 125–35. [*312*]

Keesey, R. E. and Kling, J. W. (1961). Amount of reinforcement and free-operant responding. *J. Exp. Anal. Behav.* **4**, 125–32. [*152*]

Keilitz, I. and Frieman, J. (1970). Transfer of training following errorless discrimination learning. *J. Exp. Psychol.* **85**, 293–9 [*600*]

Kelleher, R. T. (1957). Conditioned reinforcement in chimpanzees. *J. Comp. Physiol. Psychol.* **49**, 571–5. [*237*]

Kelleher, R. T. (1958a). Stimulus-producing responses and attention in the chimpanzee. *J. Exp. Anal. Behav.* **1**, 87–102. (a) [*252*]

Kelleher, R. T. (1958b). Fixed-ratio schedules of conditioned reinforcement with chimpanzees. *J. Exp. Anal. Behav.* **1**, 281–9. [*237, 244*]

Kelleher, R. T. (1961). Schedules of conditioned reinforcement during experimental extinction. *J. Exp. Anal. Behav.* **4**, 1–5. [*236, 239*]

Kelleher, R. T. (1966a). Chaining and conditioned reinforcement. In W. K. Honig (Ed.), *Operant Behavior: Areas of Research and Application*. New York: Appleton-Century-Crofts. Pp. 160–212. (a) [*146, 238*]

Kelleher, R. T. (1966b). Conditioned reinforcement in second-order schedules. *J. Exp. Anal. Behav.* **9**, 475–85. (b) [*240, 241*]

Kelleher, R. T. and Gollub, L. R. (1962). A review of positive conditioned reinforcement. *J. Exp. Anal. Behav.* **5**, 543–97. [*233, 238, 239*]

Kelleher, R. T. and Morse, W. H. (1969). Schedules using noxious stimuli. IV. An interlocking shock-postponement schedule in the squirrel monkey. *J. Exp. Anal. Behav.* **12**, 1063–79. [*286*]

Kelleher, R. T., Riddle, W. C., and Cook, L. (1963). Persistent behavior maintained by unavoidable shocks. *J. Exp. Anal. Behav.* **6**, 507–17. [*285*]

Keller, F. S. and Schoenfeld, W. N. (1950). *Principles of Psychology*. New York: Appleton-Century-Crofts. [*246*]

Kello, J. E. (1972). The reinforcement-omission effect on fixed-interval schedules: frustration or inhibition? *Learning and Motivation* **3**, 138–47. [*170, 361*]

Kellogg, W. N. (1938). Evidence for both stimulus-substitution and original anticipatory responses in the conditioning of dogs. *J. Exp. Psychol.* **22**, 186–92. [*100, 101, 117, 206*]

Kemler, D. G. and Shepp, B. E. (1971). Learning and transfer of dimensional relevance and irrelevance in children. *J. Exp. Psychol.* **90**, 120–7. [*598*]

Kendall, S. B. (1965). An observing response analysis of fixed-ratio discrimination. *Psychon. Sci.* **3**, 281–2. [*252*]

Kendall, S. B. and Gibson, D. A. (1965). Effects of discriminative stimulus removal on observing behavior. *Psychol. Rec.* **15**, 545–51. [*258*]

Kendall, S. B. and Thompson, R. F. (1960). Effect of stimulus similarity on sensory preconditioning within a single stimulus dimension. *J. Comp. Physiol. Psychol.* **53**, 439–42. [*87*]

Kendler, H. H. and Kendler, T. S. (1962). Vertical and horizontal processes in problem solving. *Psychol. Rev.* **69**, 1–16. [*602*]

Kendler, H. H. and Kimm, J. (1967). Reversal learning as a function of the size of reward during acquisition and reversal. *J. Exp. Psychol.* **73**, 66–71. [*603*]

Kendler, H. H. and Lachman, R. (1958). Habit reversal as a function of schedule of reinforcement and drive strength. *J. Exp. Psychol.* **55**, 584–91. [*190*]

Kendler, H. H. and Levine, S. (1953). A more sensitive test of irrelevant-incentive learning under conditions of satiation. *J. Comp. Physiol. Psychol.* **46**, 271–3. [*212*]

Kendler, T. S. (1950). An experimental investigation of transposition as a function of the difference between training and test stimuli. *J. Exp. Psychol.* **40**, 552–62. [*562*]

Kendler, T. S. (1971). Continuity theory and cue-dominance. In H. H. Kendler and J. T. Spence (Eds.), *Essays in Neobehaviorism: A Memorial Volume to Kenneth W. Spence.* New York: Appleton-Century-Crofts. pp. 237–64. [*42, 602*]

Kendrick, D. C. (1958). Inhibition with reinforcement (conditioned inhibition). *J. Exp. Psychol.* **56**, 313–18. [*417, 425*]

Kennedy, J. S. (1965). Co-ordination of successive activities in an aphid. Reciprocal effects of settling on flight. *J. Exp. Biol.* **43**, 489–509. [*349*]

Keppel, G. (1968). Retroactive and proactive inhibition. In T. R. Dixon and D. L. Horton (Eds.), *Verbal Behavior and General Behavior Theory.* Englewood Cliffs, New Jersey: Prentice Hall. Pp. 172–213. [*472, 478*]

Khavari, K. A. and Eisman, E. H. (1971). Some parameters of latent learning and generalized drives. *J. Comp. Physiol. Psychol.* **77**, 463–9. [*213*]

Khavari, K. A. and Heise, G. A. (1967). Analysis of discrimination reversal in the rat. *Psychon. Sci.* **9**, 271–2. [*608*]

Kierylowicz, H., Soltysik, S., and Divac, I. (1968). Conditioned reflexes reinforced by direct and indirect food presentation. *Acta Biol. Exp.* **28**, 1–10. [*9, 108*]

Killeen, P. (1968). On the measurement of reinforcement frequency in the study of preference. *J. Exp. Anal. Behav.* **11**, 263–9. [*178, 186, 188, 256*]

Killeen, P. (1969). Reinforcement frequency and contingency as factors in fixed-ratio behavior. *J. Exp. Anal. Behav.* **12**, 391–5. [*171*]

Killeen, P. (1970). Preference for fixed-interval schedules of reinforcement. *J. Exp. Anal. Behav.* **14**, 127–31. [*186, 194, 243*]

Kimball, R. C., Kimball, L. T., and Weaver, H. E. (1953). Latent learning as a function of the number of differential cues. *J. Comp. Physiol. Psychol.* **46**, 274–80. [*209*]

Kimble, G. A. (1955). Shock intensity and avoidance learning. *J. Comp. Physiol. Psychol.* **48**, 281–4 [*82*]

Kimble, G. A. (1961) *Hilgard and Marquis' Conditioning and Learning.* 2nd Ed. New York: Appleton-Century-Crofts. [*63, 72, 86, 97, 100, 102, 130, 244, 315, 419, 468, 472, 562*]

Kimble, G. A. (1964). Categories of learning and the problem of definition: Comments on Professor Grant's paper. In A. W. Melton (Ed.,), *Categories of Human Learning.* New York: Academic Press. Pp. 32–45. [*101, 102*]

Kimble, G. A., Mann, L. I., and Dufort, R. H. (1955). Classical and instrumental eyelid conditioning. *J. Exp. Psychol.* **49**, 407–17. [*119*]

Kimmel, H. D. and Baxter, R. (1964). Avoidance conditioning of the GSR. *J. Exp. Psychol.* **68**, 482–5. [*131*]

Kimmel, H. D. and Hill, F. A. (1960). Operant conditioning of the GSR. *Psychol. Rep.* **7**, 555–62. *[130]*

Kimmel, H. D. and Sternthal, H. S. (1967). Replication of GSR avoidance conditioning with concomitant EMG measurement and subjects matched in responsivity and conditionability. *J. Exp. Psychol.* **74**, 144–6. *[130]*

Kintsch, W. (1962). Runway performance as a function of drive strength and magnitude of reinforcement. *J. Comp. Physiol. Psychol.* **55**, 882–7. *[150, 152, 154]*

Kintsch, W. and Witte, R. S. (1962). Concurrent conditioning of bar press and salivation responses. *J. Comp. Physiol. Psychol.* **55**, 963–8. *[223, 227]*

Kintz, B. L. and Bruning, J. L. (1967). Punishment and compulsive avoidance behavior. *J. Comp. Physiol. Psychol.* **63**, 323–6 *[286]*

Kish, G. B. (1966). Studies of sensory reinforcement. In W. K. Honig (Ed.), *Operant Behavior: Areas of Research and Application.* New York: Appleton-Century-Crofts. Pp. 109–59. *[95]*

Klein, B. (1969). Counterconditioning and fear reduction in the rat. *Psychon. Sci.* **17**, 150–1. *[18]*

Klein, R. M. (1959). Intermittent primary reinforcement as a parameter of secondary reinforcement. *J. Exp. Psychol.* **58**, 423–7. *[242, 246]*

Klein, S. B. and Spear, N. E. (1970a). Foregetting by the rat after intermediate retention intervals ("Kamin Effect") as retrieval failure. *J. Comp. Physiol. Psychol.* **71**, 165–70. (a) *[477]*

Klein, S. B. and Spear, N. E. (1970b). Reactivation of avoidance-learning memory in the rat after intermediate retention intervals. *J. Comp. Physiol. Psychol.* **72**, 498–504. (b) *[477]*

Kleitman, N. (1927). The influence of starvation on the rate of secretion of saliva elicited by pilocarpine, and its bearing on conditioned salivation. *Am. J. Physiol.* **82**, 686–92. *[94]*

Kleitman, N. and Crisler, G. (1927). A quantitative study of a salivary conditioned reflex. *Am. J. Physiol.* **79**, 571–614. *[101]*

Kling, J. W. (1952). Generalization of extinction of an instrumental response to stimuli varying in the size dimension. *J. Exp. Psychol.* **44**, 339–46. *[526]*

Kluver, H. (1933). *Behavior Mechanisms in Monkeys.* Chicago: University of Chicago Press. *[556, 562]*

Knapp, R. K. (1965). Acquisition and extinction of avoidance with similar and different shock and escape situations. *J. Comp. Physiol. Psychol.* **60**, 272–3. *[320, 341]*

Knott, P. D. and Clayton, K. N. (1966). Durable secondary reinforcement using brain stimulation as the primary reinforcer. *J. Comp. Physiol. Psychol.* **61**, 151–3. *[247]*

Knouse, S. B. and Campbell, P. E. (1971). Partially delayed reward in the rat: a parametric study of delay duration. *J. Comp. Physiol. Psychol.* **75**, 116–19. *[453]*

Koch, S. and Daniel, W. J. (1945). The effect of satiation on the behavior mediated by a habit of maximum strength. *J. Exp. Psychol.* **35**, 167–87. *[350]*

Köhler, W. (1918). Nachweis einfacher Strukturfunkitonen beim Schimpansen und beim Haushuhn: Uber eine neue Methode zur Untersuchung des bunten Farbensystems. *Abh. Preuss. Akad. Wiss.* **2**, 1–101. **Translated as:** Simple structural functions in the chimpanzee and in the chicken. In W. D. Ellis (Ed.) *A Source Book of Gestalt Psychology.* New York: Harcourt, Brace, 1939 P. 217–27. *[559]*

Kohn, B. and Dennis, M. (1972). Observation and discrimination learning in the rat: Specific and nonspecific effects. *J. Comp. Physiol. Psychol.* **78**, 292–6. [*202, 203*]

Konorski, J. (1948). *Conditioned Reflexes and Neuron Organization.* Cambridge: Cambridge University Press. [*33, 36, 59, 65, 79, 88, 206, 219, 417, 471, 486*]

Konorski, J. (1961). The physiological approach to the problem of recent memory. In J. F. Delafresnaye (Ed.), *Brain Mechanisms and Learning.* Oxford: Blackwell Scientific Publications. Pp. 115–30. [*517*]

Konorski, J. (1967). *Integrative Activity of the Brain.* Chicago: University of Chicago Press. [*13, 18, 21, 22, 24, 79, 84, 92, 100, 107, 125, 219, 225, 231, 307, 348*]

Konorski, J. and Miller, S. (1937). On two types of conditioned reflex. *J. Gen. Psychol.* **16**, 264–72. [*110, 124, 132*]

Konorski, J. and Szwejkowska, G. (1950). Chronic extinction and restoration of conditioned reflexes. I. Extinction against the excitatory background. *Acta Biol. Exp.* **15**, 155–70. [*34*]

Konorski, J. and Szwejkowska, G. (1952a). Chronic extinction and restoration of conditioned reflexes. III. Defensive motor reflexes. *Acta Biol. Exp.* **16**, 91–4. (a) [*34*]

Konorski, J. and Szwejkowska, G. (1952b). Chronic extinction and restoration of conditioned reflexes: IV. The dependence of the course of extinction and restoration of conditioned reflexes on the "history" of the conditioned stimulus (the principle of the primacy of first training). *Acta Biol. Exp.* **16**, 95–113. (b) [*32, 34, 35, 472*]

Konorski, J. and Szwejkowska, G. (1956). Reciprocal transformations of heterogeneous conditioned reflexes. *Acta Biol. Exp.* **17**, 141–65. [*18, 23, 24, 88, 111*]

Koppenaal, R. J. (1963). Time changes in the strengths of A–B, A–C lists: spontaneous recovery? *J. Verbal Learn. Verbal Behav.* **2**, 310–19. [*480*]

Koppenaal, R. J. and Jagoda, E. (1968). Proactive interference of a maze position habit. *J. Exp. Psychol.* **76**, 664–8. [*479, 480*]

Kotesky, R. L. (1969). The effect of unreinforced–reinforced sequences on resistance to extinction following partial reinforcement. *Psychon. Sci.* **14**, 34–6. [*446, 447*]

Koteskey, R. L. (1972). A stimulus-sampling model of the partial reinforcement effect. *Psychol. Rev.* **79**, 161–71. [*407*]

Kraeling, D. (1961). Analysis of amount of reward as a variable in learning. *J. Comp. Physiol. Psychol.* **54**, 560–5. [*152*]

Kramer, T. J. and Rilling, M. (1969). Effects of lowering the magnitude of reinforcement on the response rate from a baseline during a successive discrimination. *Psychon. Sci.* **16**, 249–50. [*379*]

Kramer, T. J. and Rilling, M. (1970). Differential reinforcement of low rates: A selective critique. *Psychol. Bull.* **74**, 225–54. [*172, 179, 228*]

Kramer, T. J. and Rodriguez, M. (1971). The effect of different operants on spaced responding. *Psychon. Sci.* **25**, 177–8. [*180*]

Krane, R. V. and Ison, J. R. (1971). Positive induction in differential instrumental conditioning: effect of the inter-stimulus interval. *J. Comp. Physiol. Psychol.* **75**, 129–35. [*357, 385, 419*]

Krechevsky, I. (1932). Hypotheses in rats. *Psychol. Rev.* **39**, 516–32. [*571*]

Krechevsky, I. (1938). A study of the continuity of the problem-solving process. *Psychol. Rev.* **45**, 107–33. [*571, 572, 578, 583*]

Kremer, E. F. (1971). Truly random and traditional control procedures in CER conditioning in the rat. *J. Comp. Physiol. Psychol.* 76, 441–8. [*12, 29, 37, 40*]

Kremer, E. F. and Kamin, L. J. (1971). The truly random control procedure: Associative or nonassociative effects in rats. *J. Comp. Physiol. Psychol.* 74, 203–10. [*29, 30*]

Krieckhaus, E. E. and Wolf, G. (1968). Acquisition of sodium by rats: Interaction of innate mechanisms and latent learning. *J. Comp. Physiol. Psychol.* 65, 197–201. [*212*]

Kurtz, K. H. and Siegel, A. (1966). Conditioned fear and magnitude of startle response: A replication and extension. *J. Comp. Physiol. Psychol.* 62, 8–14. [*310*]

Lachman, R. (1961). The influence of thirst and schedules of reinforcement-nonreinforcement ratios upon brightness discrimination. *J. Exp. Psychol.* 62, 80–87. [*183*]

Lashley, K. S. (1929). *Brain Mechanisms and Intelligence: A Quantitative Study of Injuries to the Brain.* Chicago: University of Chicago Press. [*149, 571, 583*]

Lashley, K. S. (1938). The mechanism of vision: XV. Preliminary studies of the rat's capacity for detail vision. *J. Gen. Psychol.* 18, 123–93. [*556*]

Lashley, K. S. (1942). An examination of the continuity theory as applied to discriminative learning. *J. Gen. Psychol.* 26, 241–65. [*578, 582, 583*]

Lashley, K. S. and Ball, J. (1929). Spinal conduction and kinesthetic sensitivity in the maze habit. *J. Comp. Psychol.* 9, 71–105. [*199*]

Lashley, K. S. and Wade, M. (1946). The Pavlovian theory of generalization. *Psychol. Rev.* 53, 72–87. [*486, 487, 494, 499, 500, 506*]

Laties, V. G., Weiss, B., Clark, R. L., and Reynolds, M. D. (1965). Overt "mediating" behavior during temporally spaced responding. *J. Exp. Anal. Behav.* 8, 107–16. [*173*]

Laties, V. G., Weiss, B., and Weiss, A. B. (1969). Further observations on overt "mediating" behavior and the discrimination of time. *J. Exp. Anal. Behav.* 12, 43–57. [*173*]

Lawrence, D. H. (1949). Acquired distinctiveness of cues: I. Transfer between discriminations on the basis of familiarity with the stimulus. *J. Exp. Psychol.* 39, 770–84. [*591*]

Lawrence, D. H. (1950). Acquired distinctiveness of cues: II. Selective association in a constant stimulus situation. *J. Exp. Psychol.* 40, 175–88. [*591*]

Lawrence, D. H. (1952). The transfer of a discrimination along a continuum. *J. Comp. Physiol. Psychol.* 45, 511–16. [*591, 593, 594*]

Lawrence, D. H. (1955). The applicability of generalization gradients to the transfer of a discrimination along a continuum. *J. Gen. Psychol.* 52, 37–48. [*594*]

Lawrence, D. H. and DeRivera, J. (1954). Evidence for relational transposition. *J. Comp. Physiol. Psychol.* 47, 465–71. [*563, 564*]

Lawrence, D. H. and Festinger, L. (1962). *Deterrents and Reinforcement.* Stanford: Stanford University Press. [*416, 425, 434, 437, 438, 444, 451, 462*]

Lawrence, D. H. and Hommel, L. (1961). The influence of differential goal boxes on discrimination learning involving delay of reinforcement. *J. Comp. Physiol. Psychol.* 54, 552–5. [*158*]

Lawson, R. (1957). Brightness discrimination performance and secondary reward strength as a function of primary reward amount. *J. Comp. Physiol. Psychol.* 50, 35–9. [*152, 185*]

Leach, D. A. (1971). Rats' extinction performance as a function of deprivation level during training and partial reinforcement. *J. Comp. Physiol. Psychol.* **75**, 317–23. [*151*]

Leaf, R. and Muller, S. (1965). Simple method for CER conditioning and measurement. *Psychol. Rep.* **17**, 211–15. [*10*]

Leander, J. D. (1973). Effects of food deprivation on free-operant avoidance behavior.. *J. Exp. Anal. Behav.* **19**, 17–24. [*352*]

Lehr, R. (1970). Partial reinforcement and variable magnitude of reward effects in rats in a T-maze. *J. Comp. Physiol. Psychol.* **70**, 286–93. [*190, 439*]

Leitenberg, H. (1965). Is time-out from positive reinforcement an aversive event? A review of the experimental evidence. *Psychol. Bull.* **64**, 428–41. [*262*]

Leitenberg, H. (1966). Conditioned acceleration and conditioned suppression in pigeons. *J. Exp. Anal. Behav.* **9**, 205–12. [*262*]

Leitenberg, H., Bertsch, G. J., and Coughlin, R. C., Jr. (1968). "Time-out from positive reinforcement" as the UCS in a CER paradigm with rats. *Psychon. Sci.* **13**, 3–4. [*262*]

Leonard, D. W. (1969). Amount and sequence of reward in partial and continuous reinforcement. *J. Comp. Physiol. Psychol.* **67**, 204–11. [*429, 430, 433, 456*]

Leonard, D. W., Albin, R., and Leibowitz, M. (1969). Performance under massed or spaced extinction following different sequences of varied reward training. *Psychon. Sci.* **16**, 130–32. [*429, 445*]

Leonard, D. W. and Capaldi, E. J. (1971). Successive acquisitions and extinctions in the rat as a function of number of nonrewards in each extinction session. *J. Comp. Physiol. Psychol.* **74**, 102–7. [*442*]

Leonard, D. W. and Monteau, J. E. (1971). Does CS intensity determine CR amplitude? *Psychon. Sci.* **23**, 369–71. [*45*]

Leonard, D. W. and Theios, J. (1967a). Effect of CS-US interval shift on classical conditioning of the nictitating membrane in the rabbit. *J. Comp. Physiol. Psychol.* **63**, 355–8. (a) [*61*]

Leonard, D. W. and Theios, J. (1967b). Classical eyelid conditioning in rabbits under prolonged single alternation conditions of reinforcement. *J. Comp. Physiol. Psychol.* **64**, 273–6. (b) [*24, 73*]

Leonard, D. W., Weimer, J., and Albin, R. (1968). An examination of Pavlovian induction phenomena in differential instrumental conditioning. *Psychon. Sci.* **12**, 89–90. [*357*]

Lett, B. T. (1973). Delayed reward learning: disproof of the traditional theory. *Learning and Motivation*, **3**, 237–46. [*158*]

Leung, C. M. & Jensen, G. D. (1968). Shifts in percentage of reinforcement viewed as changes in incentive. *J. Exp. Psychol.* **76**, 291–6. [*436*]

Leventhal, A. M., Morrell, R. F., Morgan, E. F., Jr., and Perkins, C. C., Jr. (1959). The relation between mean reward and mean reinforcement. *J. Exp. Psychol.* **57**, 284–7. [*255*]

Levine, M. (1965). Hypothesis behavior. In A. M. Schrier, H. F. Harlow, and F. Stollnitz (Eds.), *Behavior of Nonhuman Primates*, Vol. 1. New York: Academic Press. Pp. 97–127. [*590, 610, 613*]

Levine, S. (1966). UCS intensity and avoidance learning. *J. Exp. Psychol.* **71**, 163–4. [*343*]

Lewis, D. J. (1956). Acquisition, extinction, and spontaneous recovery as a function of percentage of reinforcement and intertrial intervals. *J. Exp. Psychol.* **51**, 45–53. [*457, 458*]

Lewis, D. J. (1960). Partial reinforcement: a selective review of the literature since 1950. *Psychol. Bull.* **57**, 1–28. [*434, 436, 440, 442, 443, 447*]

Lewis, D. J. and Cotton, J. W. (1957). Learning and performance as a function of drive strength during acquisition and extinction. *J. Comp. Physiol. Psychol.* **50**, 189–94. [*408*]

Ley, R. (1965). Effects of food and water deprivation on the performance of a response motivated by acquired fear. *J. Exp. Psychol.* **69**, 583–9. [*352*]

Liddell, H. S., James, W. T., and Anderson, O.D. (1935). The comparative physiology of the conditioned motor reflex. *Comp. Psychol. Monogr.* **11** (1, Serial No. 51). [*22, 118, 472*]

Lieberman, D. A. (1972). Secondary reinforcement and information as determinants of observing behavior in monkeys (*Macaca mulatta*). *Learning and Motivation*, **3**, 341–58. [*259*]

Light, J. S. and Gantt, W. H. (1936). Essential part of reflex arc for establishment of conditioned reflex. Formation of conditioned reflex after exclusion of motor peripheral end. *J. Comp. Psychol.* **21**, 19–36 [*80*]

Likely, D. (1970). Patterning of instrumental responding to sequences of varied food and sucrose rewards. Paper presented at Eastern Psychological Association, Atlantic City. [*152, 163, 401, 431*]

Likely, D., Little, L., and Mackintosh, N. J. (1971). Extinction as a function of magnitude and percentage of food or sucrose reward. *Can. J. Psychol.* **25**, 130–7. [*427, 455*]

Likely, D. and Schnitzer, S. B. (1968). Dependence of the overtraining extinction effect on attention to runway cues. *Q. Jl. Exp. Psychol.* **20**, 193–6. [*424, 425, 426*]

Linden, D. R. (1969). Attenuation and reestablishment of the CER by discriminated avoidance conditioning in rats. *J. Comp. Physiol Psychol.* **69**, 573–8. [*334*]

Lockard, J. S. (1963). Choice of a warning signal or no warning signal in an unavoidable shock situation. *J. Comp. Physiol. Psychol.* **56**, 526–30. [*253*]

Lockard, R. B. (1971). Reflections on the fall of comparative psychology: is there a message for us all? *Am. Psychol.* **26**, 168–79. [*2*]

Loeb, J. (1900). *Comparative Physiology of the Brain and Comparative Psychology.* New York: Putnam. [*2*]

Logan, F. A. (1951). A comparison of avoidance and non-avoidance eyelid conditioning. *J. Exp. Psychol.* **42**, 390–3. [*119*]

Logan, F. A. (1954). A note on stimulus intensity dynamism (V). *Psychol. Rev.* **61**, 77–80. [*42, 44, 532*]

Logan, F. A. (1960). *Incentive.* New Haven: Yale University Press. [*136, 151, 155*]

Logan, F. A. (1961). Specificity of discrimination learning to the original context. *Science, N.Y.* **133**, 1355–6. [*476, 563*]

Logan, F. A. (1965). Decision making by rats: Delay versus amount of reward. *J. Comp. Physiol. Psychol.* **59**, 1–12. [*159, 186*]

Logan, F. A. (1966). Transfer of discrimination. *J. Exp. Psychol.* **71**, 616–18. [*593, 594*]

Logan, F. A. (1969). The negative incentive value of punishment. In B. A. Campbell and R. M. Church (Eds.), *Punishment and Aversive Behavior.* New York: Appleton-Century-Crofts. Pp. 43–54. [*278*]

Logan, F. A. (1970). *Fundamentals of Learning and Motivation.* Dubuque, Iowa: W. C. Brown & Co. [*426, 438*]

Logan, F. A. (1971). Essentials of a theory of discrimination learning. In H. H. Kendler and J. T. Spence (Eds.), *Essays in Neobehaviorism: A Memorial Volume to Kenneth W. Spence*. New York: Appleton-Century-Crofts. Pp. 265–82. [*596*]

Logan, F. A., Beier, E. M., and Kincaid, W. D. (1956). Extinction following partial and varied reinforcement. *J. Exp. Psychol.* **52**, 65–70. [*453*]

Logan, F. A. and Boice, R. (1969). Aggressive behaviors of paired rodents in an avoidance context. *Behaviour* **34**, 161–83. [*342*]

Logan, F. A. and Spanier, D. (1970). Relative effect of delay of food and water reward. *J. Comp. Physiol. Psychol.* **72**, 102–4. [*155*]

Logan, F. A. and Wagner, A. R. (1962). Supplementary report: Direction of change in CS in eyelid conditioning. *J. Exp. Psychol.* **64**, 325–6. [*43*]

Logan, F. A. and Wagner, A. R. (1965). *Reward and Punishment*. Boston: Allyn and Bacon. [*216*]

LoLordo, V. M. (1967). Similarity of conditioned fear responses based upon different aversive events. *J. Comp. Physiol. Psychol.* **64**, 154–8. [*33, 34, 83, 85, 296, 305, 311*]

LoLordo, V. M. (1971). Facilitation of food-reinforced responding by a signal for response-independent food. *J. Exp. Anal. Behav.* **15**, 49–55 [*226*]

LoLordo, V. M., McMillan, J. C. and Riley, A. L. (1974). The effects upon food-reinforced pecking and treadle-pressing of auditory and visual signals for response-independent food. *Learning and Motivation* **5**, 24–41. [*226*]

Long, J. B., McNamara, H. J., and Gardner, J. O. (1965). Resistance to extinction after variable training as a function of multiple associations. *J. Comp. Physiol. Psychol.* **60**, 252–5. [*434, 435*]

Lorenz, K. (1935). Der Kumpan in der Umwelt des Vogels. *J. Orn. Lpz.* **83**, 137–213, 289–413. **Translated as:** Companions as factors in the bird's environment. In K. Lorenz *Studies in Animal and Human Behaviour*, Vol. 1, London: Methuen 1970. Pp. 101–258. [*6*]

Loucks, R. B. (1935). The experimental delimitation of neural structures essential for learning: The attempt to condition striped muscle responses with faradization of the sygmoid gyri. *J. Psychol.* **1**, 5–44. [*18, 94*]

Lovejoy, E. (1966). Analysis of the overlearning reversal effect. *Psychol. Rev.* **73**, 87–103. [*603, 606*]

Lovejoy, E. (1968). *Attention in Discrimination Learning*. San Francisco: Holden-Day. [*42, 545, 566, 584, 590*]

Lovejoy, E. and Russell, D. G. (1967). Suppression of learning about a hard cue by the presence of an easy cue. *Psychon. Sci.* **8**, 365–6. [*580*]

Low, L. A. and Low, H. I. (1962). Effects of CS-US interval length upon avoidance responding. *J. Comp. Physiol. Psychol.* **55**, 1059–61. [*325*]

Lowes, G. and Bitterman, M. E. (1967). Reward and learning in the goldfish. *Science, N.Y.* **157**, 455–7. [*395*]

Lubow, R. E. (1965). Latent inhibition: Effects of frequency of nonreinforced preexposure of the CS. *J. Comp. Physiol. Psychol.* **60**, 454–9. [*37*]

Lubow, R. E., Markman, R. E., and Allen, J. (1968). Latent inhibition and classical conditioning of the rabbit pinna response. *J. Comp. Physiol. Psychol.* **66**, 688–94. [*36, 37*]

Lubow, R. E. and Moore, A. U. (1959). Latent inhibition: The effect of nonreinforced preexposure to the conditioned stimulus. *J. Comp. Physiol. Psychol.* **52**, 415–19. [*36, 38*]

Lubow, R. E. and Siebert, L. (1969). Latent inhibition within the CER paradigm. *J. Comp. Physiol. Psychol.* **68**, 136–8. [*37*]

Ludvigson, H. W. (1967). A preliminary investigation of the effects of sodium amytal, prior reward in G_1, and activity level on the FE. *Psychon. Sci.* **8**, 115–16. [*353, 363*]

Ludvigson, H. W. and Gay, S. E. (1966). Differential reward conditioning: S-contrast as a function of the magnitude of S +. *Psychon. Sci.* **5**, 289–90. [*390*]

Ludvigson, H. W. and Gay, R. A. (1967). An investigation of conditions determining contrast effects in differential reward conditioning. *J. Exp. Psychol.* **75**, 37–42. [*394, 396*]

Ludvigson, H. W. and Sytsma, D. (1967). The sweet smell of success: Apparent double alternation in the rat. *Psychon. Sci.* **9**, 283–4. [*162*]

Lynch, J. J. (1966). Overtraining in classical conditioning: A comparison of motor and cardiac systems in dogs. *Conditional Reflex* **1**, 266–79. [*22*]

Lynn, R. (1966). *Attention, Arousal and the Orientation Reaction.* Oxford: Pergamon Press. [*16*]

Lyon, D. O. and Ozolins, D. (1970). Pavlovian conditioning of shock elicited aggression: a discrimination procedure. *J. Exp. Anal. Behav.* **13**, 325–31. [*18*]

Lyons, J. (1969a). Stimulus generalization as a function of discrimination learning with and without errors. *Science, N.Y.* **163**, 490–1. (a) [*529, 530*]

Lyons, J. (1969b). Stimulus generalization along the dimension of S + as a function of discrimination learning with and without error. *J. Exp. Psychol.* **81**, 95–100. (b) [*33, 529, 530*]

Lyons, J. and Thomas, D. R. (1967). Effects of interdimensional training on stimulus generalization: II. Within subjects design. *J. Exp. Psychol.* **75**, 572–4. [*501*]

Maatsch, J. L., Adelman, H. M., and Denny, M. R. (1954). Effort and resistance to extinction of the bar-pressing response. *J. Comp. Physiol. Psychol.* **47**, 47–50. [*416*]

McAdam, D., Knott, J. R., and Chiorini, J. (1965). Classical conditioning in the cat as a function of the CS-UCS interval. *Psychon. Sci.* **3**, 89–90. [*64*]

McAllister, W. R. (1952). The spatial relation of irrelevant and relevant goal objects as a factor in simple selective learning. *J. Comp. Physiol. Psychol.* **45**, 531–7. [*212*]

McAllister, W. R. and McAllister, D. E. (1962a). Postconditioning delay and intensity of shock as factors in the measurement of acquired fear. *J. Exp. Psychol.* **64**, 110–16. (a) [*109, 311*]

McAllister, W. R. and McAllister, D. E. (1962b). Role of the CS and of apparatus cues in measurement of acquired fear. *Psychol. Rep.* **11**, 749–56. (b) [*109, 311*]

McAllister, W. R. and McAllister, D. E. (1963). Increase over time in the stimulus generalization of acquired fear. *J. Exp. Psychol.* **65**, 576–82. [*476*]

McAllister, W. R., McAllister, D. E., and Keith, W. (1971). The inverse relationship between shock intensity and shuttle-box avoidance learning in rats: A reinforcement explanation. *J. Comp. Physiol. Psychol.* **74**, 426–33. [*344*]

McCain, G. (1966). Partial reinforcement effects following a small number of acquisition trials. *Psychon. Monogr. Suppl.* **1**, 251–70. [*161, 450, 462*]

McCain, G. (1969). The partial reinforcement effect after minimal acquisition: single pellet reward, spaced trials. *Psychon. Sci.* **15**, 146. [*462*]

McCain, G., Baerwaldt, J., and Brown, E. R. (1969). Extinction following a small number of goal-box placements. *Can. J. Psychol.* **23**, 274–84. [*451*]

McCain, G. and Bowen, J. (1967). Pre- and postreinforcement delay with a small number of acquisition trials. *Psychon. Sci.* **7**, 121–2. [*431, 432*]

McCain, G., Dyleski, K., and McElvain, G. (1971). Reward magnitude and instrumental responses: Consistent reward. *Psychon. Monogr. Suppl.* **3**, 249–256. [*153*]

McCain, G. and Garrett, B. L. (1964). Generalization to stimuli of different brightness in three straight alley studies. *Psychol. Rep.* **15**, 368–70. [*519*]

MacCaslin, E. F. (1954). Successive and simultaneous discrimination as a function of stimulus similarity. *Am. J. Psychol.* **67**, 308–14. [*546, 558*]

MacCaslin, E. F., Wodinsky, J., and Bitterman, M. E. (1952). Stimulus-generalization as a function of prior training. *Am. J. Psychol.* **65**, 1–15. [*508*]

McClearn, G. E. and Harlow, H. F. (1954). The effect of spatial contiguity on discrimination learning by rhesus monkeys. *J. Comp. Physiol. Psychol.* **47**, 391–4. [*585*]

McClelland, D. C. and McGown, D. R. (1953). The effect of variable food reinforcement on the strength of a secondary reward. *J. Comp. Physiol. Psychol.* **46**, 80–6. [*434*]

MacCorquodale, K. and Meehl, P. E. (1951). On the elimination of cul entries without obvious reinforcement. *J. Comp. Physiol. Psychol.* **44**, 367–71. [*209*]

MacCorquodale, K. and Meehl, P. E. (1954). Edward C. Tolman. In W. K. Estes *et al. Modern Learning Theory*. New York: Appleton-Century-Crofts. Pp. 177–266. [*212*]

McDiarmid, C. G. and Rilling, M. E. (1965). Reinforcement delay and reinforcement rate as determinants of schedule preference. *Psychon. Sci.* **2**, 195–6. [*159*]

McEwen, D. (1972). The effects of terminal-link fixed-interval and variable-interval schedules on responding under concurrent chained schedules. *J. Exp. Anal. Behav.* **18**, 253–61. [*186, 188, 194*]

McFarland, D. J. (1964). Interaction of hunger and thirst in the Barbary dove. *J. Comp. Physiol. Psychol.* **58**, 174–9. [*351*]

McFarland, D. J. (1966a). The role of attention in the disinhibition of displacement activities. *Q. Jl Exp. Psychol.* **18**, 19–30. (a) [*190, 439*]

McFarland, D. J. (1966b). On the causal and functional significance of displacement activities. *Z. Tierpsychol.* **23**, 217–35. (b) [*349, 352, 411*]

McFarland, D. J. (1969). Mechanisms of behavioral disinhibition. *Anim. Behav.* **17**, 238–42. [*411*]

McFarland, D. J. and L'Angellier, A. B. (1966). Disinhibition of drinking during satiation of feeding behaviour in the Barbary dove. *Anim. Behav.* **14**, 463–7. [*411*]

MacFarlane, D. A. (1930). The role of kinesthesis in maze learning. *University of California Publications in Psychology* **4**, 277–305. [*199*]

McGill, W. J. (1963). Stochastic latency mechanisms. In R. D. Luce, R. R. Bush, and E. Galanter (Eds.), *Handbook of Mathematical Psychology*, Vol. 1. New York: Wiley. Pp. 309–60. [*45*]

McGonigle, B. (1967). Stimulus additivity and dominance in visual discrimination performance by rats. *J. Comp. Physiol. Psychol.* **64**, 110–13. [*574*]

McGuigan, F. J. and Crockett, F. (1958). Evidence that the secondary reinforcing stimulus must be discriminated. *J. Exp. Psychol.* **55**, 184–7. [*247*]

McHose, J. H. (1963). Effect of continued nonreinforcement on the frustration effect. *J. Exp. Psychol.* **65**, 444–50. [*360, 361, 363*]

McHose, J. H. (1970). Relative reinforcement effects: S_1/S_2 and S_1/S_1 paradigms in instrumental conditioning. *Psychol. Rev.* **77**, 135–46. [*363*]

McHose, J. H. and Gavelek, J. R. (1969). The frustration effect as a function of training magnitude: Within- and between-Ss designs. *Psychon. Sci.* **17**, 261–2. [*364*]

McHose, J. H. and Ludvigson, H. W. (1965). Role of reward magnitude and incomplete reduction of reward magnitude in the frustration effect. *J. Exp. Psychol.* **70**, 490–5. [*363*]

McHose, J. H. and Tauber, L. (1972). Changes in delay of reinforcement in simple instrumental conditioning. *Psychon. Sci.* **27**, 291–2. [*393*]

McKearney, J. W. (1969). Fixed-interval schedules of electric shock presentation: Extinction and recovery of performance under different shock intensities and fixed-interval durations. *J. Exp. Anal. Behav.* **12**, 301–13. [*276, 285*]

MacKinnon, J. R. (1967). Interactive effects of the two rewards in a differential magnitude of reward discrimination. *J. Exp. Psychol.* **75**, 329–38. [*384, 385*]

MacKinnon, J. R. and Amsel, A. (1964). Magnitude of the frustration effect as a function of confinement and detention in the frustrating situation. *J. Exp. Psychol.* **67**, 468–74. [*360*]

Mackintosh, N. J. (1963a). The effect of irrelevant cues on reversal learning in the rat. *Br. J. Psychol.* **54**, 127–34. (a) [*604, 605*]

Mackintosh, N. J. (1963b). Extinction of a discrimination habit as a function of overtraining. *J. Comp. Physiol. Psychol.* **56**, 842–7. (b) [*573, 607*]

Mackintosh, N. J. (1965a). Transposition after single-stimulus pretraining. *Am. J. Psychol.* **78**, 116–19. (a) [*509*]

Mackintosh, N. J. (1965b). Overtraining, transfer to proprioceptive control, and position reversal. *Q. Jl Exp. Psychol.* **17**, 26–36. (b) [*426, 554, 604*]

Mackintosh, N. J. (1965c). Incidental cue learning in rats. *Q. Jl Exp. Psychol.* **17**, 292–300. (c) [*583, 585, 588*]

Mackintosh, N. J. (1969). Further analysis of the overtraining reversal effect. *J. Comp. Physiol. Psychol. Monograph* **67**, No. 2, Part 2. [*148, 184, 185, 603, 604, 605, 607*]

Mackintosh, N. J. (1970a). Distribution of trials and the partial reinforcement effect in the rat. *J. Comp. Physiol. Psychol.* **73**, 341–8. (a) [*160, 419, 458, 459*]

Mackintosh, N. J. (1970b). Attention and probability learning. In D. Mostofsky (Ed.), *Attention: Contemporary Theory and Analysis*. New York: Appleton-Century-Crofts. Pp. 173–91. (b) [*192*]

Mackintosh, N. J (1971a) An analysis of overshadowing and blocking *Q. Jl Exp. Psychol.* **23**, 118–25. (a) [*46, 578, 586*]

Mackintosh, N. J. (1971b). Reward and aftereffects of reward in the learning of goldfish. *J. Comp. Physiol. Psychol.* **76**, 225–32. (b) [*163, 395, 401, 402, 428, 431, 467*]

Mackintosh, N. J. (1973). Stimulus selection: learning to ignore stimuli that predict no change in reinforcement. In R. A. Hinde and J. Stevenson-Hinde (Eds.), *Constraints on Learning*. London: Academic Press. Pp. 75–96. [*40, 51, 56, 587*]

Mackintosh, N. J. (1974). A search for contrast effects in discrete-trial discrimination learning by pigeons. *Learning and Motivation* (in press). [*386*]

Mackintosh, N. J. and Holgate, V. (1968). Effects of inconsistent reinforcement on reversal and nonreversal shifts. *J. Exp. Psychol.* **76**, 154–9. [*439*]

Mackintosh, N. J. and Holgate, V. (1969). Serial reversal training and nonreversal shift learning. *J. Comp. Physiol. Psychol.* **67,** 89–93. [*608, 610*]

Mackintosh, N. J. and Honig, W. K. (1970). Blocking and attentional enhancement in pigeons. *J. Comp. Physiol. Psychol.* **73,** 78–85. [*509, 583, 585*]

Mackintosh, N. J. and Little, L. (1969). Intradimensional and extradimensional shift learning by pigeons. *Psychon. Sci.* **14,** 5–6. [*597*]

Mackintosh, N. J. and Little, L. (1970a). Effects of different patterns of reinforcement on performance under massed or spaced extinction. *Psychon. Sci.* **20,** 1–2. (a) [*444, 445, 447*]

Mackintosh, N. J. and Little, L. (1970b). An analysis of transfer along a continuum. *Can. J. Psychol.* **24,** 362–9. (b) [*594, 596*]

Mackintosh, N. J., Little, L., and Lord, J. (1972). Some determinants of behavioral contrast in pigeons and rats. *Learning and Motivation* **3,** 148–61. [*359, 366, 369, 373, 374, 375, 376, 378, 379*]

Mackintosh, N. J. and Lord, J. (1973). Simultaneous and successive contrast with delay of reward. *Animal Learning and Behavior,* **1,** 283–6. [*147, 393*]

Mackintosh, N. J., Lord, J., and Little, L. (1971) Visual and spatial probability learning in pigeons and goldfish. *Psychon. Sci.* **24,** 221–3. [*192*]

Mackintosh, N. J., McGonigle, B., Holgate, V., and Vanderver, V. (1968). Factors underlying improvement in serial reversal learning. *Can. J. Psychol.* **22,** 85–95. [*608, 609, 610*]

Mackintosh, N. J. and Turner, C. (1971). Blocking as a function of novelty of CS and predictability of UCS. *Q. Jl Exp. Psychol.* **23,** 359–66. [*51, 587, 588*]

McNamara, H. J., Long, J. B., and Wike, E. L. (1956). Learning without response under two conditions of external cues. *J. Comp. Physiol. Psychol.* **49,** 477–80. [*200*]

McNamara, H. J. and Wike, E. L. (1958). The effects of irregular learning conditions upon the rate and permanence of learning. *J. Comp. Physiol. Psychol.* **51,** 363–6. [*434*]

Macphail, E. M. (1968). Avoidance responding in pigeons. *J. Exp. Anal. Behav.* **11,** 629–32. [*340*]

Macphail, E. M. (1971). Hyperstriatal lesions in pigeons: Effects on response inhibition, behavioral contrast, and reversal learning. *J. Comp. Physiol. Psychol.* **75,** 500–7. [*381*]

Macphail, E. M. (1972). Inhibition in the acquisition and reversal of simultaneous discriminations. In R. A. Boakes and M. S. Halliday (Eds.), *Inhibition and Learning.* London: Academic Press. Pp. 121–51. [*548, 570*]

Madison, H. L. (1964). Experimental extinction as a function of number of reinforcements. *Psychol. Rep.* **14,** 647–50. [*424, 425*]

Maher, W. B. and Wickens, D. D. (1954). Effect of differential quantity of reward on acquisition and performance of a maze habit. *J. Comp. Physiol. Psychol.* **47,** 44–6. [*185*]

Mahut, H. (1954). The effect of stimulus position on visual discrimination by the rat. *Can. J. Psychol.* **8,** 130–8. [*558, 572*]

Maier, S. F. (1970). Failure to escape traumatic electric shock: Incompatible skeletal-motor responses or learned helplessness? *Learning and Motivation* **1,** 157–69. [*218*]

Maier, S. F., Allaway, T. A., and Gleitman, H. (1967). Proactive inhibition in rats after prior partial reversal: A critique of the spontaneous recovery hypothesis. *Psychon. Sci.* **9,** 63–4. [*479, 480*]

Maier, S. F. and Gleitman, H. (1967). Proactive interference in rats. *Psychon. Sci.* **7,** 25–6. [*475, 479*]

Maier, S. F., Seligman, M. E. P. and Solomon, R. L. (1969). Pavlovian fear conditioning and learned helplessness: Effects on escape and avoidance behavior of (a) the CS-US contingency and (b) the independence of the US and voluntary responding. In B. A. Campbell and R. M. Church (Eds.), *Punishment and Aversive Behavior.* New York: Appleton-Century-Crofts. Pp. 299–342. [*218*]

Malagodi, E. F. (1967a). Acquisition of the token-reward habit in the rat. *Psychol. Rep.* **20,** 1335–42. (a) [*237*]

Malagodi, E. F. (1967b). Fixed-ratio schedules of token reinforcement. *Psychon. Sci.* **8,** 469–70. (b) [*237*]

Mandler, J. M. (1966). Behavior changes during overtraining and their effects on reversal and transfer. *Psychon. Monogr. Suppl.* **1,** 187–202. [*590, 599, 600, 601*]

Mandler, J. M. (1968). The effect of overtraining on the use of positive and negative stimuli in reversal and transfer. *J. Comp. Physiol. Psychol.* **66,** 110–15. [*548, 566, 568, 603, 606*]

Mandler, J. M. (1970). Two-choice discrimination learning using multiple stimuli. *Learning and Motivation* **1,** 261–6. [*567, 568*]

Mandler, J. M. and Hooper, W. R. (1967). Overtraining and goal approach strategies in discrimination reversal. *Q. J. Exp. Psychol.* **19,** 142–9. [*605*]

Mandler, G., Preven, D. W., and Kuhlman, C. K. (1962). Effects of operant reinforcement on the GSR. *J. Exp. Anal. Behav.* **5,** 317–21. [*130*]

Marchant, R. G., III, Mis, F. W., and Moore, J. W. (1972). Conditioned inhibition of the rabbit's nictitating membrane response. *J. Exp. Psychol.* **95,** 408–11. [*32, 33, 34*]

Margolius, G. (1955). Stimulus generalization of an instrumental response as a function of the number of reinforced trials. *J. Exp. Psychol.* **49,** 105–11. [*519*]

Marler, P. (1970). A comparative approach to vocal learning: song-development in white-crowned sparrows. *J. Comp. Physiol. Psychol. Monograph* **71,** No. 2, Part 2. [*6*]

Marsh, G. (1967a). Inverse relationship between discriminability and stimulus generalization as a function of number of test stimuli. *J. Comp. Physiol. Psychol.* **64,** 284–9. (a) [*496, 497*]

Marsh, G. (1967b). Relational learning in the pigeon. *J. Comp. Physiol. Psychol.* **64,** 519–21. [*561*]

Marsh, G. (1969). An evaluation of three explanations for the transfer of discrimination effect. *J. Comp. Physiol. Psychol.* **68,** 268–75. [*594, 595*]

Martin, R. C. and Melvin, K. B. (1964). Vicious circle behavior as a function of delay of punishment. *Psychon. Sci.* **1,** 415–16. [*286*]

Marx, M. H. (1944). The effects of cumulative training upon retroactive inhibition and transfer. *Comp. Psychol. Monogr.* **18,** (2, Serial No. 94). [*479*]

Marx, M. H. (1967a). Nonreinforced response vigor as a function of number of training rewards. *Psychol. Rep.* **21,** 197–204. (a) [*426*]

Marx, M. H. (1967b). Interaction of drive and reward as a determiner of resistance to extinction. *J. Comp. Physiol. Psychol.* **64,** 488–9. (b) [*427*]

Marx, M. H. (1969). Positive contrast in instrumental learning from qualitative shift in incentive. *Psychon. Sci.* **16,** 254–5. [*390*]

Marx, M. H., McCoy, D. F., Jr., and Tombaugh, J. W. (1965). Resistance to extinction as a function of constant delay of reinforcement. *Psychon. Sci.* **2,** 333–4. [*431*]

Mason, D. J. (1957). The relation of secondary reinforcement to partial reinforcement. *J. Comp. Physiol. Psychol.* **50**, 264–8. [*242, 245, 437*]

Masterson, F. A. (1970). Is termination of a warning signal an effective reward for the rat? *J. Comp. Physiol. Psychol.* **72**, 471–5. [*341*]

Matsumoto, R. T. (1969). Relative reward effects in differential conditioning. *J. Comp. Physiol. Psychol.* **68**, 489–592. [*390*]

Maxwell, F. R., Jr., Meyer, P. A., Calef, R. S., and McHewitt, E. R. (1969). Discrimination contrast: Speeds to small reward as a function of locus and amount of interpolated reinforcement. *Psychon. Sci.* **14**, 35–6. [*391*]

May, R. B. and Beauchamp, K. L. (1969). Stimulus change, previous experience and extinction. *J. Comp. Physiol. Psychol.* **68**, 607–10. [*434*]

May, R. B., Tolman, C. W., and Schoenfeldt, M. G. (1967). Effects of pre-training exposure to the CS on conditioned suppression. *Psychon. Sci.* **9**, 61–2. [*37*]

Medin, D. L. (1973). Subproblem analysis of discrimination shift learning. *Behavior Research Methods and Instrumentation*, **5**, 332–6. [*576*]

Mednick, S. A. (1964). *Learning.* Englewood Cliffs, New Jersey: Prentice-Hall. [*130*]

Mednick, S. A. and Freedman, J. L. (1960). Stimulus generalization. *Psychol. Bull.* **57**, 169–200. [*484, 490, 495*]

Melching, W. H. (1954). The acquired reward value of an intermittently presented neutral stimulus. *J. Comp. Physiol. Psychol.* **47**, 370–3. [*234, 235, 281*]

Mellgren, R. L. (1971a). Shift in magnitude of reward after minimal acquisition. *Psychon. Sci.* **23**, 243–4. (a) [*389*]

Mellgren, R. L. (1971b). Positive contrast in the rat as a function of number of preshift trials in the runway. *J. Comp. Physiol. Psychol.* **77**, 329–36. (b) [*390*]

Mellgren, R. L. and Ost, J. W. P. (1969). Transfer of Pavlovian differential conditioning to an operant discrimination. *J. Comp. Physiol. Psychol.* **67**, 390–4. [*203, 225*]

Mellgren, R. L., Wrather, D. M., and Dyck, D. G. (1972). Differential conditioning and contrast effects in rats. *J. Comp. Physiol. Psychol.* **80**, 478–83. [*386*]

Meltzer, D. and Brahlek, J. A. (1968). Quantity of reinforcement and fixed-interval performance. *Psychon. Sci.* **12**, 207–8. [*152*]

Meltzer, D. and Brahlek, J. A. (1970). Conditioned suppression and conditioned enhancement with the same positive UCS: An effect of CS duration. *J. Exp. Anal. Behav.* **13**, 67–73. [*225, 226*]

Melvin, K. B. and Anson, J. E. (1969). Facilitative effect of punishment on aggressive behavior in the Siamese fighting fish. *Psychon. Sci.* **14**, 89–90. [*288*]

Melvin, K. B. and Martin, R. C. (1966). Facilitative effects of two modes of punishment on resistance to extinction. *J. Comp. Physiol. Psychol.* **62**, 491–4. [*296*]

Meredith, A. L. and Schneiderman, N. (1967). Heart rate and nictitating membrane classical discrimination conditioning in rabbits under delay versus trace procedures. *Psychon. Sci.* **9**, 139–40. [*65*]

Meyer, D. R., Cho, C., and Weseman, A. F. (1960). On problems of conditioned discriminated lever-press avoidance responses. *Psychol. Rev.* **67**, 224–8. [*137, 340*]

Meyer, D. R., Treichler, F. R., and Meyer, P. M. (1965). Discrete-trial training techniques and stimulus variables. In A. M. Schrier, H. F. Harlow, and F. Stollnitz (Eds.), *Behavior of Nonhuman Primates*, Vol. 1. New York: Academic Press. Pp. 1–49. [*585*]

Meyer, M. E., Adams, W. A., and Worthen, V. K. (1969). Deprivation and escape conditioning with various intensities of shock. *Psychon. Sci.* **14**, 212–14. [*352*]

Miczek, K. A. and Grossman, S. P. (1971). Positive conditioned suppression: Effects of CS duration. *J. Exp. Anal. Behav.* **15**, 243–7. [*225, 226*]

Mikulka, P. J., Lehr, R., and Pavlik, W. B. (1967). Effect of reinforcement schedules on reward shifts. *J. Exp. Psychol.* **74**, 57–61. [*391*]

Mikulka, P. J. and Pavlik, W. B. (1966). Deprivation level, competing responses and the PREE. *Psychol. Rep.* **18**, 95–102. [*160*]

Miles, C. G. (1970). Blocking the acquisition of control by an auditory stimulus with pretraining on brightness. *Psychon. Sci.* **19**, 133–4. [*582*]

Miles, C. G. and Jenkins, H. M. (1973). Overshadowing in operant conditioning as a function of discriminability. *Learning and Motivation* **4**, 11–27. [*46, 577, 578, 580, 581, 582, 583, 585, 587*]

Miles, C. G., Mackintosh, N. J., and Westbrook, R. F. (1970). Redistributing control between the elements of a compound stimulus. *Q. J. Exp. Psychol.* **22**, 478–83. [*502, 512, 513*]

Miles, R. C. (1959). Discrimination in the squirrel monkey as a function of deprivation and problem difficulty. *J. Exp. Psychol.* **57**, 15–19. [*184*]

Miles, R. C. (1965). Discrimination-learning sets. In A. M. Schrier, H. F. Harlow, and F. Stollnitz (Eds.), *Behavior of Nonhuman Primates*, Vol. 1. New York: Academic Press. Pp. 51–95. [*611*]

Millenson, J. R. (1966). Probability of response and probability of reinforcement in a response-defined analogue of an interval schedule. *J. Exp. Anal. Behav.* **9**, 87–94. [*178*]

Miller, N. E. (1944). Experimental studies of conflict. In J. McV. Hunt (Ed.), *Personality and the Behavior Disorders*. New York: Ronald Press. Pp. 431–65. [*521, 522, 523*]

Miller, N. E. (1948). Studies of fear as an acquirable drive. *J. Exp. Psychol.* **38**, 89–101. [*271, 306, 311*]

Miller, N. E. (1959). Liberalization of basic S-R concepts: extensions to conflict behavior, motivation, and social learning. In S. Koch (Ed.), *Psychology: A Study of a Science*, Vol. 2. New York: McGraw-Hill, Pp. 196–293. [*521, 523*]

Miller, N. E. (1960). Learning resistance to pain and fear: Effects of overlearning, exposure and rewarded exposure in context. *J. Exp. Psychol.* **60**, 137–45. [*283, 425, 426*]

Miller, N. E. (1961). Some recent studies of conflict behavior and drugs. *Am. Psychol.* **16**, 12–24. [*265*]

Miller, N. E. (1963). Some reflections on the law of effect produce a new alternative to drive reduction. In M. R. Jones (Ed.), *Nebraska Symposium on Motivation*, Vol. 11. Lincoln: University of Nebraska Press. Pp. 65–112. [*216*]

Miller, N. E. (1969). Learning of visceral and glandular responses. *Science, N.Y.* **163**, 434–45. [*130*]

Miller, N. E. and Banuazizi, A. (1968). Instrumental learning by curarized rats of a specific visceral response, intestinal or cardiac. *J. Comp. Physiol. Psychol.* **65**, 1–7. [*131*]

Miller, N. E. and Carmona, A. (1967). Modification of a visceral response, salivation in thirsty dogs, by instrumental training with water reward. *J. Comp. Physiol. Psychol.* **63**, 1–6. [*135*]

Miller, N. E. and DeBold, R. C. (1965). Classically conditioned tongue-licking and operant bar pressing recorded simultaneously in the rat. *J. Comp. Physiol. Psychol.* **59**, 109–11. [*223, 227*]

Miller, N. E. and DiCara, L. V. (1968). Instrumental learning of urine formation by rats; changes in renal blood flow. *Am. J. Physiol.* **215**, 677–83. [*131*]

Miller, N. E. and Dollard, J. C. (1941). *Social Learning and Imitation.* New Haven: Yale University Press. [*414*]

Miller, N. E. and Dworkin, B. R. (1974). Visceral learning: recent difficulties with curarized rats and significant problems for human research. In P. A. Obrist, A. H. Black, J. Brener, and L. V. DiCara (Eds.), *Cardiovascular Psychophysiology: Current Issues in Response Mechanisms, Biofeedback, and Methodology.* Chicago: Aldine Press. [*132*]

Miller, N. E. and Kraeling, D. (1952). Displacement: greater generalization of approach than avoidance in generalized approach-avoidance conflict. *J. Exp. Psychol.* **43**, 217–21. [*522*]

Miller, S. and Konorski, J. (1928). Sur une forme particulière des reflexes conditionnels. *C. R. Séanc. Soc. Biol.* **99**, 1155–7. [*124, 128, 132*]

Miller, W. C. and Greene, J. E. (1954). Generalization of an avoidance response to varying intensities of sound. *J. Comp. Physiol. Psychol.* **47**, 136–9. [*485*]

Minami, H. and Dallenbach, K. M. (1946). The effect of activity upon learning and retention in the cockroach, *Periplaneta americana. Am. J. Psychol.* **59**, 1–58. [*479*]

Minturn, L. (1954). A test for sign-Gestalt expectancies under conditions of negative motivation. *J. Exp. Psychol.* **48**, 98–100. [*210*]

Misanin, J. R. and Campbell, B. A. (1969). Effects of hunger and thirst on sensitivity and reactivity to shock. *J. Comp. Physiol. Psychol.* **69**, 207–13. [*352*]

Misanin, J. R., Campbell, B. A., and Smith, N. F. (1966). Duration of punishment and the delay of punishment gradient. *Can. J. Psychol.* **20**, 407–12. [*286, 289*]

Miyadi, D. (1964). Social life of Japanese monkeys. *Science, N.Y.* **143**, 783–6. [*200*]

Moore, B. R. (1971). On Directed Respondents. Doctoral thesis, Stanford University. Ann Arbor, Michigan: University Microfilms, Number 72–11, 623. [*106*]

Moore, B. R. (1973). The role of directed Pavlovian reactions in simple instrumental learning in the pigeon. In R. A. Hinde and J. Stevenson-Hinde (Eds.), *Constraints on Learning.* London: Academic Press. Pp. 159–86. [*17, 80, 121, 122, 126, 136, 147, 205, 342, 343*]

Moore, J. W. (1972). Stimulus control: studies of auditory generalization in rabbits. In A. H. Black and W. F. Prokasy (Eds.), *Classical Conditioning II: Current Research and Theory.* New York: Appleton-Century-Crofts. Pp. 206–30. [*14, 387, 485, 539*]

Moore, J. W. and Gormezano, I. (1961). Yoked comparisons of instrumental and classical eyelid conditioning. *J. Exp. Psychol.* **62**, 552–9. [*119*]

Moore, M. J. and Capretta, P. J. (1968). Changes in colored or flavored food preferences in chicks as a function of shock. *Psychon. Sci.* **12**, 195–6. [*55*]

Morgan, C. L. (1894). *An Introduction to Comparative Psychology.* London: Scott. [*1*]

Morse, W. H. (1966). Intermittent reinforcement. In W. K. Honig (Ed.), *Operant Behavior: Areas of Research and Application.* New York: Appleton-Century-Crofts. Pp. 52–108. [*170, 171*]

Morse, W. H., Mead, R. N., and Kelleher, R. T. (1967). Modulation of elicited behavior by a fixed-interval schedule of electric shock presentation. *Science, N.Y.* **157**, 215–17. [*276, 287, 288*]

Moscovitch, A. and LoLordo, V. M. (1968). Role of safety in the Pavlovian backward fear conditioning procedure. *J. Comp. Physiol. Psychol.* **66**, 673–8. [*59, 60*]

Moss, E. M. and Harlow, H. F. (1947). The role of reward in discrimination learning in monkeys. *J. Comp. Physiol. Psychol.* **40**, 333–42. [*567*]

Mote, F. A., Jr. and Finger, F. W. (1943). The retention of a simple running response after varying amounts of reinforcement. *J. Exp. Psychol.* **33**, 317–22. [*471*]

Mowrer, O. H. (1947). On the dual nature of learning—a reinterpretation of "conditioning" and "problem-solving". *Harvard Educational Review* **17**, 102–48. [*112, 125, 129, 271, 280, 306, 308, 309, 330*]

Mowrer, O. H. (1950). *Learning Theory and Personality Dynamics.* New York: The Ronald Press Co. [*112, 125*]

Mowrer, O. H. (1956). Two-factor learning theory reconsidered, with special reference to secondary reinforcement and the concept of habit. *Psychol. Rev.* **63**, 114–28. [*278*]

Mowrer, O. H. (1960a) *Learning Theory and Behavior.* New York: Wiley. (a) [*21, 84, 112, 113, 132, 156, 158, 216, 231, 280*]

Mowrer, O. H. (1960b). *Learning Theory and the Symbolic Processes.* New York: Wiley. (b) [*254*]

Mowrer, O. H. and Aiken, E. G. (1954). Contiguity vs. drive-reduction in conditioned fear: Temporal variations in conditioned and unconditioned stimulus. *Am. J. Psychol.* **67**, 26–38. [*112, 247*]

Mowrer, O. H. and Jones, H. M. (1943). Extinction and behavior variability as functions of effortfulness of task. *J. Exp. Psychol.* **33**, 369–86. [*414, 416*]

Mowrer, O. H. and Jones, H. M. (1945). Habit strength as a function of the pattern of reinforcement. *J. Exp. Psychol.* **35**, 293–311. [*436*]

Mowrer, O. H. and Lamoreaux, R. R. (1942). Avoidance conditioning and signal duration—a study of secondary motivation and reward. *Psychol. Mongr.* **54**, (5, Whole No. 247). [*315*]

Mowrer, O. H. and Lamoreaux, R. R. (1946). Fear as an intervening variable in avoidance conditioning. *J. Comp. Psychol.* **39**, 29–50. [*302*]

Mowrer, O. H. and Lamoreaux, R. R. (1951). Conditioning and conditionality (discrimination). *Psychol. Rev.* **58**, 196–212. [*321*]

Mowrer, O. H. and Solomon, L. N. (1954). Contiguity vs. drive reduction in conditioned fear: The proximity and abruptness of drive reduction. *Am. J. Psychol.* **67**, 15–25. [*112*]

Moyer, K. E. (1965). Effect of experience with emotion provoking stimuli on water consumption in the rat. *Psychon. Sci.* **2**, 251–2. [*265*]

Moyer, K. E. and Korn, J. H. (1964). Effect of UCS intensity on the acquisition and extinction of an avoidance response. *J. Exp. Psychol.* **67**, 352–9. [*343*]

Moyer, K. E. and Korn, J. H. (1966). Effect of UCS intensity on the acquisition and extinction of a one-way avoidance response. *Psychon. Sci.* **4**, 121–2. [*343*]

Mrosovsky, N. (1971). *Hibernation and the Hypothalamus.* New York: Appleton-Century-Crofts. [*350*]

Muenzinger, K. F. (1934). Motivation in learning: I. Electric shock for correct responses in the visual discrimination habit. *J. Comp. Psychol.* **17**, 267–77. [*283*]

Muenzinger, K. F., Bernstone, A. H., and Richards, L. (1938). Motivation in learning. VIII. Equivalent amounts of electric shock for right and wrong responses in a visual discrimination habit. *J. Comp. Psychol.* **26**, 177–86. [*283*]

Muenzinger, K. F. and Conrad, D. G. (1954). Latent learning observed through negative transfer. *J. Comp. Physiol. Psychol.* **46**, 1–8. [*207*]

Mullin, A. D. and Mogenson, G. J. (1963). Effects of fear conditioning on avoidance learning. *Psychol. Rep.* **13**, 707–10. [*218*]

Mumma, R. and Warren, J. M. (1968). Two-cue discriminatory learning by cats. *J. Comp. Physiol. Psychol.* **66**, 116–22. [*592, 593*]

Munn, N. L. (1950). *Handbook of Psychological Research on the Rat.* Boston: Houghton Mifflin Co. [*187, 188, 553, 554*]

Muntz, W. R. A. (1963). Stimulus generalization following monocular training in the rat. *J. Comp. Physiol. Psychol.* **56**, 1003–6. [*519*]

Muntz, W. R. A. (1965). Stimulus generalization and the 'units hypothesis' in *Octopus*. *J. Comp. Physiol. Psychol.* **59**, 144–6. [*495*]

Murillo, N. R., Diercks, J. K., and Capaldi, E. J. (1961). Performance of the turtle, *Pseudemys scripta troostii*, in a partial reinforcement situation. *J. Comp. Physiol. Psychol.* **54**, 204–6. [*467*]

Murray, E. J. and Berkun, M. M. (1955). Displacement as a function of conflict. *J. Abnorm. Social Psychol.* **51**, 47–56. [*522*]

Murray, E. J. and Miller, N. E. (1952). Displacement: steeper gradient of generalization of approach than avoidance in generalized approach-avoidance conflict. *J. Exp. Psychol.* **43**, 222–6. [*522*]

Myers, W. A. (1970). Observational learning in monkeys. *J. Exp. Anal. Behav.* **14**, 225–35. [*202*]

Nachman, M. (1970). Learned taste and temperature aversions due to lithium chloride sickness after temporal delays. *J. Comp. Physiol. Psychol.* **73**, 22–30. [*54, 69*]

Neuringer, A. J. (1967). Effects of reinforcement magnitude on choice and rate of responding. *J. Exp. Anal. Behav.* **10**, 417–24. [*152, 243, 382*]

Neuringer, A. and Schneider, B. A. (1968). Separating the effects of interreinforcement time and number of interreinforcement responses. *J. Exp. Anal. Behav.* **11**, 661–7. [*171*]

Nevin, J. A. (1968). Differential reinforcement and stimulus control of not responding. *J. Exp. Anal. Behav.* **11**, 715–26. [*380*]

Nevin, J. A. and Shettleworth, S. J. (1966). An analysis of contrast effects in multiple schedules. *J. Exp. Anal. Behav.* **9**, 305–15. [*349, 357, 358, 360, 365, 373, 380*]

Newman, F. L. and Baron, M. R. (1965). Stimulus generalization along the dimension of angularity. *J. Comp. Physiol. Psychol.* **60**, 59–63. [*487, 488, 501, 508*]

Newman, F. L. and Benefield, R. L. (1968). Stimulus control, cue utilization, and attention. Effects of discrimination training. *J. Comp. Physiol. Psychol.* **66**, 101–4. [*499*]

Newman, J. R. and Grice, G. R. (1965). Stimulus generalization as a function of drive level, and the relation between two measures of response strength. *J. Exp. Psychol.* **69**, 357–62. [*495, 520*]

Newton, J. E. O. and Gantt, W. H. (1966). One-trial cardiac conditioning in dogs. *Conditional Reflex* **1**, 251–65. [*22*]

Nissen, H. W. (1950). Description of the learned response in discrimination behavior. *Psychol. Rev.* **59**, 121–37. [*548*]

Nissen, H. W. and Elder, J. H. (1935). The influence of amount of incentive on delayed response performances of chimpanzees. *J. Genet. Psychol.* **47**, 49–72. [*214, 215*]

Norman, M. F. (1964). A two-phase model and an application to verbal discrimination learning. In R. C. Atkinson (Ed.), *Studies in Mathematical Psychology*. Stanford: Stanford University Press. Pp. 173–87. [*11, 71*]

North, A. J. (1959). Discrimination reversal with spaced trials and distinctive cues. *J. Comp. Physiol. Psychol.* **52**, 426–9. [*608, 610*]

North, A. J. and Lang, P. (1961). Conditional discrimination in rats. *J. Genet. Psychol.* **98**, 113–18. [*550*]

North, A. J., Maller, O., and Hughes, C. L. (1958). Conditional discrimination and stimulus patterning. *J. Comp. Physiol. Psychol.* **51**, 711–15. [*550*]

North, A. J. and Stimmel, D. T. (1960). Extinction of an instrumental response following a large number of reinforcements. *Psychol. Rep.* **6**, 227–34. [*424*]

Norton, F. T. M. and Kenshalo, D. R. (1954). Incidental learning under conditions of unrewarded irrelevant motivation. *J. Comp. Physiol. Psychol.* **47**, 375–7. [*212*]

Notterman, J. M. (1951). A study of some relations among aperiodic reinforcement, discrimination training, and secondary reinforcement. *J. Exp. Psychol.* **41**, 161–9. [*247*]

Osgood, C. E. (1953). *Method and Theory in Experimental Psychology*. New York: Oxford University Press. [*85, 100, 110*]

Ost, J. W. P. and Lauer, D. W. (1965). Some investigations of classical salivary conditioning in the dog. In W. F. Prokasy (Ed.), *Classical Conditioning: A Symposium*. New York: Appleton-Century-Crofts. Pp. 192–207. [*17, 65, 70*]

Overmier, J. B., Bull, J. A., III, and Pack, K. (1971a). On instrumental response interaction as explaining the influences of Pavlovian CSs upon avoidance behavior. *Learning and Motivation*, **2**, 103–12. (a) [*84*]

Overmier, J. B., Bull, J. A. III, and Trapold, M. A. (1971b). Discriminative cue properties of different fears and their role in response selection in dogs. *J. Comp. Physiol. Psychol.* **76**, 478–82. (b) [*85, 296*]

Overton, D. A. (1964). State-dependent or "dissociated" learning produced with pentobarbital. *J. Comp. Physiol. Psychol.* **57**, 3–12. [*477*]

Overton, D. A. (1966). State-dependent learning produced by depressant and atropine-like drugs. *Psychopharmacologia*, **10**, 6–31. [*477*]

Padilla, A. M., Padilla, C., Ketterer, T., and Giacolone, D. (1970). Inescapable shocks and subsequent escape/avoidance conditioning in goldfish, *Carassius auratus*. *Psychon. Sci.* **20**, 295–6. [*218*]

Page, H. A. (1955). The facilitation of experimental extinction by response prevention as a function of the acquisition of a new response. *J. Comp. Physiol. Psychol.* **48**, 14–16. [*332, 336, 341*]

Paige, A. B. and McNamara, H. J. (1963). Secondary reinforcement and the discrimination hypothesis: The role of discrimination training. *Psychol. Rep.* **13**, 679–86. [*235*]

Parrish, J. (1967). Classical discrimination conditioning of heart rate and bar press suppression in the rat. *Psychon. Sci.* **9**, 267–8. [*22, 61, 103*]

Passey, G. E. (1957). Net discriminatory reaction potential as a function of stimulus separation along an intensive stimulus continuum. *J. Gen. Psychol.* **56**, 59–66. [*385*]

Patten, R. L. and Deaux, E. B. (1966). Classical conditioning and extinction of the licking response in rats. *Psychon. Sci.* **4**, 21–2. [*17*]

Patten, R. L. and Rudy, J. W. (1967). The Sheffield omission training procedure applied to the conditioning of the licking response in rats. *Psychon. Sci.* **8**, 463–4. [*120*]

Patterson, M. M. (1970). Classical conditioning of the rabbit's (*Oryctolagus cuniculus*) nictitating membrane response with fluctuating ISI and intracranial CS. *J. Comp. Physiol. Psychol.* **72**, 193–202. [*64*]

Paul, C. (1965). Effects of overlearning upon single habit reversal in rats. *Psychol. Bull.* **63**, 65–72. [*603*]

Paul, C. (1966). Effects of overtraining and two non-correction training procedures upon a brightness discrimination reversal. *Psychon. Sci.* **5**, 423–4. [*603*]

Pavlik, W. B. and Lehr, R. (1967). Strength of alternative responses and subsequent choices. *J. Exp. Psychol.* **74**, 562–73. [*439*]

Pavlik, W. B. and Reynolds, W. F. (1963). Effects of deprivation schedule and reward magnitude on acquisition and extinction performance. *J. Comp. Physiol. Psychol.* **56**, 452–5. [*154*]

Pavlik, W. B., Thompson, R. R., and Reynolds, W. F. (1963). Extinction under changed stimulus conditions. *Percept. Mot. Skills* **16**, 171–7. [*435*]

Pavlov, I. P. (1927). *Conditioned Reflexes*. Oxford: Oxford University Press. [*2, 13, 14, 15, 16, 17, 18, 19, 32, 33, 38, 41, 46, 58, 61, 62, 67, 70, 71, 79, 108, 110, 222, 281, 292, 348, 354, 356, 366, 415, 417, 419, 468, 484, 485, 486, 494, 514, 526, 532, 580, 593*]

Pavlov, I. P. (1928). *Lectures on Conditioned Reflexes*. New York: International Publishers. [*59, 60*]

Pavlov, I. P. (1941). *Conditioned Reflexes and Psychiatry*. New York: International Publishers. [*106, 107*]

Pear, J. J., Moody, J. E. and Persinger, M. A. (1972). Lever attacking by rats during free-operant avoidance. *J. Exp. Anal. Behav.* **18**, 517–23. [*305, 324*]

Pear, J. J. and Wilkie, D. M. (1971). Contrast and induction in rats on multiple schedules. *J. Exp. Anal. Behav.* **15**, 289–96. [*376*]

Peckham, R. H. and Amsel, A. (1967). Within-subject demonstration of a relationship between frustration and magnitude of reward in a differential magnitude of reward discrimination. *J. Exp. Psychol.* **73**, 187–95. [*364*]

Pennes, E. S. and Ison, J. R. (1967). Effects of partial reinforcement on discrimination learning and subsequent reversal or extinction. *J. Exp. Psychol.* **74**, 219–24. [*439*]

Perin, C. T. (1942). Behavior potentiality as a joint function of the amount of training and degree of hunger at the time of extinction. *J. Exp. Psychol.* **30**, 93–113. [*222, 350, 423*]

Perin, C. T. (1943). A quantitative investigation of the delay-of-reinforcement gradient. *J. Exp. Psychol.* **32**, 37–51. [*156*]

Perkins, C. C., Jr. (1947). The relation of secondary reward to gradients of reinforcement. *J. Exp. Psychol.* **37**, 377–92. [*157*]

Perkins, C. C., Jr. (1953). The relation between conditioned stimulus insensity and response strength. *J. Exp. Psychol.* **46**, 225–31. [*42, 43, 44, 532*]

Perkins, C. C., Jr. (1968). An analysis of the concept of reinforcement. *Psychol. Rev.* **75**, 155–72. [*113, 253*]

Perkins, C. C., Jr., Seymann, R. C., Levis, D. J., & Spencer, H. R., Jr. (1966). Factors affecting preference for signal-shock over shock-signal. *J. Exp. Psychol.* **72**, 190–6. [*253*]

Perkins, C. C., Jr. and Weyant, R. G. (1958). The interval between training and test trials as determiner of the slope of generalization gradients. *J. Comp. Physiol. Psychol.* **51**, 596–600. [*476*]

Pert, A. and Bitterman, M. E. (1970). Reward and learning in the turtle. *Learning and Motivation* **1**, 121–8. [*394, 428*]

Peterson, G. B., Ackil, J. E., Frommer, G. P., and Hearst, E. (1972). Conditioned approach and contact behavior for food or brain stimulation reinforcement. *Science, N.Y.* **177**, 1009–11. [*136*]

Pierrel, R. and Blue, S. (1967). Antecedent reinforcement contingencies in the stimulus control of an auditory discrimination. *J. Exp. Anal. Behav.* **10**, 545–50. [*378*]

Pinel, J. P. J. and Cooper, R. M. (1966). Demonstration of the Kamin effect after one-trial avoidance learning. *Psychon. Sci.* **4**, 17–18. [*475*]

Platt, J. R. and Senkowski, P. C. (1970a). Response-correlated stimulus functioning in homogeneous behavior chains. In J. H. Reynierse (Ed.), *Current Issues in Animal Learning.* Lincoln: University of Nebraska Press. Pp. 195–231. (a) [*168*]

Platt, J. R. and Senkowski, P. C. (1970b). Effects of discrete-trials reinforcement frequency and changes in reinforcement frequency on preceding and subsequent fixed-ratio performance. *J. Exp. Psychol.* **85**, 95–104. (b) [*367*]

Pliskoff, S. S. and Goldiamond, I. (1966). Some discriminative properties of fixed-ratio performance in the pigeon. *J. Exp. Anal. Behav.* **9**, 1–9. [*173*]

Porter, J. J., Madison, H. L., and Swatek, A. J. (1971). Incentive and frustration effect of direct goal placements. *Psychon. Sci.* **22**, 314–16. [*424, 426*]

Postman, L., Stark, K., and Fraser, J. (1968). Temporal changes in interference. *J. Verbal Learn. Verbal Behav.* **7**, 672–94. [*480*]

Premack, D. (1969). On some boundary conditions of contrast. In J. Tapp (Ed.), *Reinforcement and Behavior.* New York: Academic Press. Pp. 120–45. [*382*]

Prewitt, E. P. (1967). Number of preconditioning trials in sensory preconditioning using CER training. *J. Comp. Physiol. Psychol.* **64**, 360–2. [*21*]

Prokasy, W. F. (1956). The acquisition of observing responses in the absence of differential external reinforcement. *J. Comp. Physiol. Psychol.* **49**, 131–4. [*251, 254*]

Prokasy, W. F. (1965). Classical eyelid conditioning: Experimenter operations, task demands, and response shaping. In W. F. Prokasy (Ed.), *Classical Conditioning: A Symposium.* New York: Appleton-Century-Crofts. Pp. 208–25. [*28, 62, 66, 113*]

Prokasy, W. F. and Hall, J. F. (1963). Primary stimulus generalization. *Psychol. Rev.* **70**, 310–22. [*486, 487*]

Prokasy, W. F., Hall, J. F., and Fawcett, J. T. (1962). Adaptation, sensitization, forward and backward conditioning, and pseudo-conditioning of the GSR. *Psychol. Rep.* **10**, 103–6. [*59*]

Prokasy, W. F. and Harsanyi, M. A. (1968). Two-phase model for human classical conditioning. *J. Exp. Psychol.* **78**, 359–68. [*71*]

Prokasy, W. F. and Papsdorf, J. D. (1965). Effects of increasing the interstimulus interval during classical conditioning of the albino rabbit. *J. Comp. Physiol. Psychol.* **60**, 249–52. [*61*]

Pschirrer, M. (1972). Goal events as discriminative stimuli over extended intertrial intervals. *J. Exp. Psychol.* **96**, 425–32. [*162, 163*]

Pubols, B. H., Jr. (1956). The facilitation of visual and spatial discrimination reversal by overlearning. *J. Comp. Physiol. Psychol.* **49**, 243–8. [*602, 603*]

Pubols, B. H., Jr. (1960). Incentive magnitude, learning and performance in animals. *Psychol. Bull.* **57**, 89–115. [*151, 184*]

Pubols, B. H., Jr. (1962a). Constant versus variable delay of reinforcement. *J. Comp. Physiol. Psychol.* **55**, 52–6. (a) [*255*]

Pubols, B. H., Jr. (1962b). Serial reversal learning as a function of the number of trials per reversal. *J Comp. Physiol. Psychol.* **55**, 66–8. (b) [*608*]

Quartermain, D. (1965). Effect of effort on resistance to extinction of the bar-pressing response. *Q. Jl Exp. Psychol.* **17**, 63–4. [*416*]

Quinsey, V. L. (1971). Conditioned suppression with no CS-US contingency in the rat. *Can. J. Psychol.* **25**, 69–82. [*29*]

Rachlin, H. (1966). Recovery of responses during mild punishment. *J. Exp. Anal. Behav.* **9**, 251–63. [*293*]

Rachlin, H. (1973). Contrast and matching. *Psychol. Rev.* **80**, 217–34. [*375*]

Rachlin, H. and Herrnstein, R. J. (1969). Hedonism revisited: On the negative law of effect. In B. A. Campbell and R. M. Church (Eds.), *Punishment and Aversive Behavior.* New York: Appleton-Century-Crofts. Pp. 83–109. [*274, 291*]

Rackham, D. W. (1971). Conditioning of the pigeon's courtship and aggressive display. Unpublished M. A. thesis, Dalhousie University. [*17*]

Ramond, C. K. (1954). Performance in selective learning as a function of hunger. *J. Exp. Psychol.* **48**, 265–70. [*150*]

Rashotte, M. E., Adelman, L. and Dove, L. D. (1972). Influence of percentage-reinforcement on runway running of rats. *Learning and Motivation* **3**, 194–208. [*160, 161*]

Rashotte, M. E. and Amsel, A. (1968a). Transfer of slow-response rituals to extinction of a continuously rewarded response. *J. Comp. Physiol. Psychol.* **66**, 432–43. (a) [*136, 441*]

Rashotte, M. E. and Amsel, A. (1968b). The generalized PRE: within-S PRF and CRF training in different runways, at different times of day, by different experimenters. *Psychon. Sci.* **11**, 315–16. (b) [*452*]

Rashotte, M. E. and Surridge, C. T. (1969). Partial reinforcement and partial delay of reinforcement effects with 72-hour intertrial intervals and interpolated continuous reinforcement. *Q. Jl. Exp. Psychol.* **21**, 156–61. [*436*]

Ratliff, R. G. and Ratliff, A. R. (1971). Runway acquisition and extinction as a joint function of magnitude of reward and percentage of rewarded acquisition trials. *Learning and Motivation* **2**, 289–95. [*427, 455*]

Ray, B. A. (1967). The course of acquisition of a line-tilt discrimination by rhesus monkeys. *J. Exp. Anal. Behav.* **10**, 17–33. [*580*]

Raymond, B., Aderman, M., and Wolach, A. H. (1972). Incentive shifts in the goldfish. *J. Comp. Physiol. Psychol.* **78**, 10–13. [*395*]

Razran, G. (1949). Stimulus generalization of conditioned responses. *Psychol. Bull.* **46**, 337–65. [*485, 518*]

Razran, G. (1956). Backward conditioning. *Psychol. Bull.* **53**, 55–69 [*60*]

Razran, G. (1957). The dominance-contiguity theory of the acquisition of classical conditioning. *Psychol. Bull.* **54**, 1–46. [*41*]

Razran, G. (1961). The observable unconscious and the inferable conscious in current Soviet psychophysiology: Interoceptive conditioning, semantic conditioning, and the orienting reflex. *Psychol. Rev.* **68**, 81–147. [*19, 20*]

Razran, G. (1965). Russian physiologists' psychology and American experimental psychology. *Psychol. Bull.* **63**, 42–64. [*47*]

Razran, G. (1971). *Mind in Evolution: An East-West Synthesis of Learned Behavior and Cognition.* Boston: Houghton Mifflin. [*26, 27, 42, 60, 106*]

Reberg, D. (1970). Differential conditioning and extinction as inhibitory training procedures. Paper presented at Eastern Psychological Association, Atlantic City. [*35, 36*]

Reberg, D. (1972). Compound tests for excitation in early acquisition and after prolonged extinction of conditioned suppression. *Learning and Motivation* **3,** 246–58 [*36, 97*]

Reberg, D. and Black, A. H. (1969). Compound testing of individually conditioned stimuli as an index of excitatory and inhibitory properties. *Psychon. Sci.* **17,** 30–1. [*33*]

Reese, H. W. (1964). Discrimination learning set in rhesus monkeys. *Psychol. Bull.* **61,** 321–40. [*612*]

Reid, L. S. (1953). The development of noncontinuity behavior through continuity learning. *J. Exp. Psychol.* **46,** 107–12. [*602, 603, 606*]

Reinhold, D. B. and Perkins, C. C., Jr. (1955). Stimulus generalization following different methods of training. *J. Exp. Psychol.* **49,** 423–7. [*511*]

Reiss, S. and Wagner, A. R. (1972). CS habituation produces a "latent inhibition effect" but no active "conditioned inhibition". *Learning and Motivation* **3,** 237–45. [*38*]

Renner, K. E. (1963). Influence of deprivation and availability of goal box cues on the temporal gradient of reinforcement. *J. Comp. Physiol. Psychol.* **56,** 101–4 [*155, 431*]

Rescorla, R. A. (1966). Predictability and number of pairings in Pavlovian fear conditioning. *Psychon. Sci.* **4,** 383–4. [*29, 33, 83*]

Rescorla, R. A. (1967a). Pavlovian conditioning and its proper control procedures. *Psychol. Rev.* **74,** 71–80. (a) [*28*]

Rescorla, R. A. (1967b). Inhibition of delay in Pavlovian fear conditioning. *J. Comp. Physiol. Psychol.* **64,** 114–20. (b) [*62, 311*]

Rescorla, R. A. (1968a). Pavlovian conditioned fear in Sidman avoidance learning. *J. Comp. Physiol. Psychol.* **65,** 55–60. (a) [*325*]

Rescorla, R. A. (1968b). Probability of shock in the presence and absence of CS in fear conditioning. *J. Comp. Physiol. Psychol.* **66,** 1–5. (b) [*25, 28, 81, 295, 329*]

Rescorla, R. A. (1969a). Pavlovian conditioned inhibition. *Psychol. Bull.* **72,** 77–94. (a) [*32, 35*]

Rescorla, R. A. (1969b). Conditioned inhibition of fear resulting from negative CS–US contingencies. *J. Comp. Physiol. Psychol.* **67,** 504–9. (b) [*28, 32, 33, 34, 35*]

Rescorla, R. A. (1971). Summation and retardation tests of latent inhibition. *J. Comp. Physiol. Psychol.* **75,** 77–81. [*38*]

Rescorla, R. A. (1973). Effect of US habituation following conditioning. *J. Comp. Physiol. Psychol.* **82,** 137–43. [*88, 89*]

Rescorla, R. A. and LoLordo, V. M. (1965). Inhibition of avoidance behavior. *J. Comp. Physiol. Psychol.* **59,** 406–12. [*10, 33, 34, 83, 311*]

Rescorla, R. A. and Skucy, J. C. (1969). Effect of response-independent reinforcers during extinction. *J. Comp. Physiol. Psychol.* **67,** 381–9. [*409*]

Rescorla, R. A. and Solomon, R. L. (1967). Two-process learning theory: Relationships between Pavlovian conditioning and instrumental learning. *Psychol. Rev.* **74,** 151–82. [*81, 84, 124, 125, 223, 225*]

Rescorla, R. A. and Wagner, A. R. (1972). A theory of Pavlovian conditioning: Variations in the effectiveness of reinforcement and nonreinforcement. In

A. H. Black and W. F. Prokasy (Eds.), *Classical Conditioning II: Current Research and Theory*. New York: Appleton-Century-Crofts. Pp. 64–99. [*11, 25, 31, 42, 50, 153, 584, 586, 589, 596, 615*]

Restle, F. (1955). A theory of discrimination learning. *Psychol. Rev.* **62**, 11–19. [*590*]

Restle, F. (1957). Discrimination of cues in mazes: a resolution of the "place-vs.-response" question. *Psychol. Rev.* **64**, 217–28. [*426, 553, 554, 556*]

Restle, F. (1958). Toward a quantitative description of learning set data. *Psychol. Rev.* **65**, 77–91. [*590, 610, 613*]

Revusky, S. H. (1962). Mathematical analysis of the durations of reinforced inter-response times during variable interval reinforcement. *Psychometrika* **27**, 307–14. [*177*]

Revusky, S. H. (1968). Aversion to sucrose produced by contingent X-irradiation: temporal and dosage parameters. *J. Comp. Physiol. Psychol.* **65**, 17–22. [*54, 67*]

Revusky, S. (1971). The role of interference in association over a delay. In W. K. Honig and P. H. R. James (Eds.), *Animal Memory*. New York: Academic Press. Pp. 155–213. [*50, 54, 69, 70, 159, 217, 584*]

Revusky, S. (1973). Long-delay learning in rats: a black-white discrimination. *Psychol. Rep.* (In press). [*159*]

Revusky, S. H. and Bedarf, E. W. (1967). Association of illness with prior ingestion of novel foods. *Science, N.Y.* **155**, 219–20. [*54, 69*]

Revusky, S. and Garcia, J. (1970). Learned associations over long delays. In G. H. Bower (Ed.), *The Psychology of Learning and Motivation*, Vol. 4. New York: Academic Press. Pp. 1–84. [*53, 67, 156*]

Reynierse, J. H., Scavio, M. J., Jr. and Ulness, J. D. (1970). An ethological analysis of classically conditioned fear. In J. H. Reynierse (Ed.), *Current Issues in Animal Learning*. Lincoln: University of Nebraska Press. Pp. 33–54. [*82*]

Reynolds, B. (1949). The acquisition of a black-white discrimination habit under two levels of reinforcement. *J. Exp. Psychol.* **39**, 760–9. [*185*]

Reynolds, B. (1950). Acquisition of a simple spatial discrimination as a function of the amount of reinforcement. *J. Exp. Psychol.* **40**, 152–60. [*184, 185*]

Reynolds, G. S. (1961a). Behavioral contrast. *J. Exp. Anal. Behav.* **4**, 57–71. (a) [*369, 371, 375, 380*]

Reynolds, G. S. (1961b). Relativity of response rate and reinforcement frequency in a multiple schedule. *J. Exp. Anal. Behav.* **4**, 179–84. (b) [*380*]

Reynolds, G. S. (1961c). Attention in the pigeon. *J. Exp. Anal. Behav.* **4**, 203–8. (c) [*578, 584*]

Reynolds, G. S. (1966). Discrimination and emission of temporal intervals by pigeons. *J. Exp. Anal. Behav.* **9**, 65–8 [*172, 179*]

Reynolds, G. S. (1968). *Primer of Operant Conditioning*. Glenview, Illinois: Scott, Foresman. [*177, 182*]

Reynolds, G. S. and Limpo, A. J. (1968). On some causes of behavioral contrast. *J. Exp. Anal. Behav.* **11**, 543–7. [*382*]

Reynolds, G. S. and McLeod, A. (1970). On the theory of interresponse-time reinforcement. In G. H. Bower (Ed.), *The Psychology of Learning and Motivation*, Vol. 4. New York: Academic Press. Pp. 85–107. [*174*]

Reynolds, W. F., Anderson, J. E., and Besch, N. F. (1963a). Secondary reinforcement effects as a function of method of testing. *J. Exp. Psychol.* **66**, 53–6. [*247*]

Reynolds, W. F. and Pavlik, W. B. (1960). Running speed as a function of depriva-tion period and reward magnitude. *J. Comp. Physiol. Psychol.* 53, 615–18. [*154*]

Reynolds, W. F., Pavlik, W. B., Schwartz, M. M., and Besch, N. F. (1963b). Maze learning by secondary reinforcement without discrimination training. *Psychol. Rep.* 12, 775–81. [*247*]

Ricciardi, A. M. and Treichler, F. R. (1970). Prior training influences on transfer to learning set by squirrel monkeys. *J. Comp. Physiol. Psychol.* 73, 314–19. [*614*]

Richman, C. L. and Coussens, W. (1970). Undertraining reversal effect in rats. *J. Exp. Psychol.* 86, 340–2. [*603*]

Richman, C. L., Knoblock, K., and Coussens, W. (1972). The overtraining reversal effect in rats: a function of task difficulty. *Q. Jl Exp. Psychol.* 24, 291–8. [*604, 607*]

Rickard, S. (1965). Proactive inhibition involving maze habits. *Psychon. Sci.* 3, 401–2. [*479*]

Riess, D. (1971). Shuttleboxes, Skinner boxes, and Sidman avoidance in rats: Acquisition and terminal performance as a function of response topography. *Psychon. Sci.* 25, 283–6. [*340*]

Riess, D. and Farrar, C. H. (1972). Unsignalled avoidance in a shuttlebox: a rapid acquisition, high-efficiency paradigm. *J. Exp. Anal. Behav.* 18, 169–78. [*324*]

Riley, D. A. (1958). The nature of the effective stimulus in animal discrimination learning: transposition reconsidered. *Psychol. Rev.* 65, 1–7, [*563, 565*]

Riley, D. A. (1968). *Discrimination Learning*. Boston: Allyn and Bacon. [*486, 561, 562, 563, 565*]

Riley, D. A., Goggin, J. P., and Wright, D. C. (1963). Training level and cue separation as determiners of transposition and retention in rats. *J. Comp. Physiol. Psychol.* 56, 1044–9 [*565*]

Riley, D. A., Ring, K., and Thomas, J. (1960). The effect of stimulus comparison on discrimination learning and transposition. *J. Comp. Physiol. Psychol.* 53, 415–21. [*565*]

Riley, D. A. and Rosenzweig, M. R. (1957). Echolocation in rats. *J. Comp. Physiol. Psychol.* 50, 323–8. [*554*]

Rilling, M. (1967). Number of responses as a stimulus in fixed interval and fixed ratio schedules. *J. Comp. Physiol. Psychol.* 63, 60–5. [*173*]

Rilling, M. E., Askew, H. R., Ahlskog, J. E., and Kramer, T. J. (1969). Aversive properties of the negative stimulus in a successive discrimination. *J. Exp. Anal. Behav.* 12, 917–32. [*258, 264, 373*]

Rilling, M. and McDiarmid, C. (1965). Signal detection in fixed-ratio schedules. *Science, N.Y.* 148, 526–7. [*173*]

Riopelle, A. J. (1960). Observational learning of a position habit by monkeys. *J. Comp. Physiol. Psychol.* 53, 426–8. [*202*]

Rizley, R. C. and Rescorla, R. A. (1972). Associations in second-order conditioning and sensory preconditioning. *J. Comp. Physiol. Psychol.* 81, 1–11. [*20, 88, 89*]

Robbins, D. (1971). Partial reinforcement: a selective review of the alleyway literature since 1960. *Psychol. Bull.* 76, 415–31. [*434, 436*]

Roberts, A. E. and Hurwitz, H. M. B. (1970). The effect of pre-shock signal on a free-operant avoidance response. *J. Exp. Anal. Behav.* 14, 331–40. [*83*]

Roberts, W. A. (1966). Learning and motivation in the immature rat. *Am. J. Psychol.* 79, 3–23. [*191, 192*]

Roberts, W. A. (1969). Resistance to extinction following partial and consistent reinforcement with varying magnitudes of reward. *J. Comp. Physiol. Psychol.* **67**, 395–400. [*152, 427, 455*]

Roberts, W. A., Bullock, D. H., and Bitterman, M. E. (1963). Resistance to extinction in the pigeon after partially reinforced instrumental training under discrete-trials conditions. *Am. J. Psychol.* **76**, 353–65. [*446*]

Robinson, A. W. and Clayton, K. N. (1963). Effect of duration of confinement in a nonbaited goal box on the "apparent frustration effect". *J. Exp. Psychol.* **66**, 613–14. [*360*]

Robinson, D. E. and Capaldi, E. J. (1958). Spontaneous recovery following non-response extinction. *J. Comp. Physiol. Psychol.* **51**, 644–6. [*416*]

Robson, C. (1970). Paired comparison technique for measuring stimulus control functions in the pigeon after training with discrete trials. *Nature, Lond.* **228**, 1112–13. [*528*]

Rock, I. and Ebenholtz, S. (1959). The relational determination of perceived size. *Psychol. Rev.* **66**, 387–401. [*565*]

Rohrer, J. H. (1949). A motivational state resulting from non-reward. *J. Comp. Physiol. Psychol.* **42**, 476–85. [*419, 421*]

Rollin, R. R. (1958). Irrelevant-incentive learning. A repetition and a revision. *J. Comp. Physiol. Psychol.* **51**, 721–4. [*212*]

Romanes, G. J. (1882). *Animal Intelligence.* London: Kegan Paul. [*1*]

Rosen, A. J. (1966). Incentive-shift performance as a function of magnitude and number of sucrose rewards. *J. Comp. Physiol. Psychol.* **62**, 487–90. [*152, 391*]

Rosen, A. J., Glass, D. H., and Ison, J. R. (1967). Amobarbital sodium and instrumental performance changes following reward reduction. *Psychon. Sci.* **9**, 129–30. [*397*]

Rosen, A. J. and Ison, J. R. (1965). Runway performance following changes in sucrose rewards. *Psychon. Sci.* **2**, 335–6. [*391*]

Ross, L. E. and Hartman, T. F. (1965). Human-eyelid conditioning: The recent experimental literature. *Genet. Psychol. Monogr.* **71**, 177–220. [*17, 73*]

Rossman, B. B. and Homzie, M. J. (1967). Contrast effects in instrumental differential conditioning with a non-nutritive liquid reinforcement. *Psychon. Sci.* **9**, 173–4. [*392*]

Rothblat, L. A. and Wilson, W. A., Jr. (1968). Intradimensional and extradimensional shifts in the monkey within and across sensory modalities. *J. Comp. Physiol. Psychol.* **66**, 549–53. [*597*]

Rozin, P. (1969). Central or peripheral mediation of learning with long CS-UCS intervals in the feeding system. *J. Comp. Physiol. Psychol.* **67**, 421–9. [*53, 69*]

Rozin, P. and Kalat, J. W. (1971). Specific hungers and poisoning as adaptive specializations of learning. *Psychol. Rev.* **78**, 459–86. [*53, 54, 67*]

Rudolph, R. L. (1967). Transposition and the post-discrimination gradient: absolute and relational learning during hue discrimination. Paper presented at Eastern Psychological Association, Boston. [*561*]

Rudolph, R. L. and Van Houten, R. (1974). Auditory stimulus control in pigeons: Jenkins and Harrison (1960) revisited. *J. Exp. Anal. Behav.* in press. [*502, 505*]

Rudy, J. W. (1971). Sequential variables as determiners of the rat's discrimination of reinforcement events: effects on extinction performance. *J. Comp. Physiol. Psychol.* **77**, 476–81. [*448, 449*]

Runquist, W. N. and Spence, K. W. (1959). Performance in eyelid conditioning as a function of UCS duration. *J. Exp. Psychol.* **57**, 249–52. [*112*]

Rzoska, J. (1953). Bait shyness, a study in rat behaviour. *Br. J. Anim. Behav.* **1**, 128–35. [*53*]

Sachs, L. B. (1969). Effects of stimulus comparison during discrimination training on subsequent transposition and generalization gradients. *Psychon. Sci.* **14**, 247–8. [*561, 565*]

Sadler, E. W. (1968). A within- and between-subjects comparison of partial reinforcement in classical salivary conditioning. *J. Comp. Physiol. Psychol.* **66**, 695–8. [*72*]

Sainsbury, R. S. (1971). Effect of proximity of elements on the feature-positive effect. *J. Exp. Anal. Behav.* **16**, 315–25. [*569*]

St. Claire-Smith, R. (1970). Blocking of punishment. Paper presented at Eastern Psychological Association, Atlantic City. [*220*]

Saldanha, E. L. and Bitterman, M. E. (1951). Relational learning in the rat. *Am. J. Psychol.* **64**, 37–53. [*558*]

Saltz, E., Whitman, R. N., and Paul, C. (1963). Performance in the runway as a function of stimulus-differentiation. *Am. J. Psychol.* **76**, 124–7. [*230, 524*]

Saltzman, I. J. (1949). Maze learning in the absence of primary reinforcement: A study of secondary reinforcement. *J. Comp. Physiol. Psychol.* **42**, 161–73. [*236, 239, 242, 246, 247*]

Santos, J. F. (1960). The influence of amount and kind of training on the acquisition and extinction of escape and avoidance responses. *J. Comp. Physiol. Psychol.* **53**, 284–9. [*332*]

Scharlock, D. P. (1955). The role of extramaze cues in place and response learning. *J. Exp. Psychol.* **50**, 249–54. [*206, 554, 556*]

Schaub, R. E. (1969). Response-cue contingency and cue effectiveness. In D. P. Hendry (Ed.), *Conditioned Reinforcement.* Homewood, Illinois: The Dorsey Press. Pp. 342–56. [*259*]

Scheuer, C. (1969). Resistance to extinction of the CER as a function of shock-reinforcement training schedules. *Psychon. Sci.* **17**, 181–2. [*74*]

Schlosberg, H. (1934). Conditioned responses in the white rat. *J. Genet. Psychol.* **45**, 303–5. [*114, 115, 124, 304*]

Schlosberg, H. (1936). Conditioned responses in the white rat: II. Conditioned responses based upon shock to the foreleg. *J. Genet. Psychol.* **49**, 107–38. [*114, 115, 124, 304*]

Schlosberg, H. (1937). The relationship between success and the laws of conditioning. *Psychol. Rev.* **44**, 379–94. [*79, 124, 125, 129*]

Schneider, B. A. (1969). A two-state analysis of fixed-interval responding in the pigeon. *J. Exp. Anal. Behav.* **12**, 677–87. [*167*]

Schneiderman, N. (1966). Interstimulus interval function of the nictitating membrane response of the rabbit under delay versus trace conditioning. *J. Comp. Physiol. Psychol.* **62**, 397–402. [*58, 61, 63*]

Schneiderman, N. (1972). Response system divergencies in aversive classical conditioning. In A. H. Black and W. F. Prokasy (Eds.), *Classical Conditioning II: Current Research and Theory.* New York: Appleton-Century-Crofts. Pp. 341–76. [*65, 66*]

Schneiderman, N., Fuentes, I., and Gormezano, I (1962). Acquisition and extinction of the classically conditioned eyelid response in the albino rabbit. *Science, N.Y.* **136**, 650–2. [*9, 14, 17, 28*]

Schneiderman, N. and Gormezano, I. (1964). Conditioning of the nictitating membrane of the rabbit as a function of CS-US interval. *J. Comp. Physiol. Psychol.* **57**, 188–95. [*17, 28, 61, 63*]

Schoenfeld, W. N. (1950). An experimental approach to anxiety, escape and avoidance behavior. In P. H. Hock and J. Zubin (Eds.), *Anxiety.* New York: Grune and Stratton. Pp. 70–99. [*306, 322*]

Schoenfeld, W. N., Antonitis, J. J., and Bersh, P. J. (1950). A preliminary study of training conditions necessary for secondary reinforcement. *J. Exp. Psychol.* **40**, 40–5. [*246*]

Schoonard, J. and Lawrence, D. H. (1962). Resistance to extinction as a function of the number of delay of reward trials. *Psychol. Rep.* **11**, 275–8. [*431*]

Schrier, A. M. (1966). Transfer by macaque monkeys between learning-set and repeated-reversal tasks. *Percept. Mot. Skills* **23**, 787–92. [*613*]

Schrier, A. M. (1967). Effects of an upward shift in amount of reinforcer on runway performance of rats. *J. Comp. Physiol. Psychol.* **64**, 490–2. [*389*]

Schrier, A. M. and Harlow, H. F. (1957). Direct manipulation of the relevant cue and difficulty of discrimination. *J. Comp. Physiol. Psychol.* **50**, 576–80. [*585*]

Schuster, R. and Rachlin, H. (1968). Indifference between punishment and free shock: Evidence for the negative law of effect. *J. Exp. Anal. Behav.* **11**, 777–86. [*275, 382*]

Schusterman, R. J. (1962). Transfer effects of successive discrimination reversal training in chimpanzees. *Science, N.Y.* **137**, 422–3. [*613*]

Schusterman, R. J. (1964). Successive discrimination-reversal training and multiple discrimination training in one-trial learning by chimpanzees. *J. Comp. Physiol. Psychol.* **58**, 153–6. [*613*]

Schutz, S. L. and Bitterman, M. E. (1969). Spaced-trials partial reinforcement and resistance to extinction in the goldfish. *J. Comp. Physiol. Psychol.* **68**, 126–128. [*467*]

Schwartz, B. and Williams, D. R. (1971). Discrete-trials spaced responding in the pigeon: the dependence of efficient performance on the availability of a stimulus for collateral pecking. *J. Exp. Anal. Behav.* **16**, 155–60. [*180*]

Schwartz, B. and Williams, D. R. (1972a). The role of the response-reinforcer contingency in negative automaintanance. *J. Exp. Anal. Behav.* **17**, 351–7. (a) [*121, 122*]

Schwartz, B. and Williams, D. R. (1972b). Two different kinds of key peck in the pigeon: some properties of responses maintained by negative and positive response-reinforcer contingencies. *J. Exp. Anal. Behav.* **18**, 201–16. (b) [*121, 122*]

Schwartz, R. M., Schwartz, M., and Tees, R. C. (1971). Optional intradimensional and extradimensional shifts in the rat. *J. Comp. Physiol. Psychol.* **77**, 470–5. [*597*]

Scobie, S. R. (1972). Interaction of an aversive Pavlovian conditional stimulus with aversively and appetitively motivated operants in rats. *J. Comp. Physiol. Psychol.* **79**, 171–88. [*83*]

Scull, J., Davies, K., and Amsel, A. (1970). Behavioral contrast and frustration effect in multiple and mixed fixed-interval schedules in the rat. *J. Comp. Physiol. Psychol.* **71**, 478–83. [*362, 367*]

Sechenov, I. M. (1863). *Refleksy Golovnogo Mozga.* St. Petersburg. **Translated as:** *Reflexes of the Brain.* Cambridge: The M.I.T. Press, 1965. [*2*]

Segundo, J. P., Galeano, C., Sommer-Smith, J. A., and Roig, J. A. (1961). Behavioural and EEG effects of tones 'reinforced' by cessation of painful stimuli. In J. F. Delafresnaye (Ed.), *Brain Mechanisms and Learning*. Oxford: Blackwells Scientific Publications. Pp. 265–91. [*59, 113*]

Seidel, R. J. (1959). A review of sensory preconditioning. *Psychol. Bull.* **56,** 58–73. [*21*]

Seligman, M. E. P. (1966). CS redundancy and secondary punishment. *J. Exp. Psychol.* **72,** 546–50. [*247, 248*]

Seligman, M. E. P. (1968). Chronic fear produced by unpredictable electric shock. *J. Comp. Physiol. Psychol.* **66,** 402–11. [*254*]

Seligman, M. E. P. and Campbell, B. A. (1965). Effect of intensity and duration of punishment on extinction of an avoidance response. *J. Comp. Physiol. Psychol.* **59,** 295–7. [*286, 333*]

Seligman, M. E. P. and Maier, S. F. (1967). Failure to escape traumatic shock. *J. Exp. Psychol.* **74,** 1–9. [*217, 218, 312*]

Seligman, M. E. P., Maier, S. F., and Solomon, R. L. (1971). Unpredictable and uncontrollable aversive events. In F. R. Brush (Ed.), *Aversive Conditioning and Learning*. New York: Academic Press. Pp. 347–400. [*217, 218, 312*]

Senf, G. M. and Miller, N. E. (1967). Evidence for positive induction in discrimination learning. *J. Comp. Physiol. Psychol.* **64,** 121–7. [*357*]

Seraganian, P. and vom Saal, W. (1969). Blocking the development of stimulus control when stimuli indicate periods of nonreinforcement. *J. Exp. Anal. Behav.* **12,** 767–72. [*583*]

Setterington, R. G. and Bishop, H. E. (1967). Habit reversal improvement in the fish. *Psychon. Sci.* **7,** 41–2. [*609*]

Sevenster, P. (1968). Motivation and learning in sticklebacks. In D. Ingle (Ed.), *The Central Nervous System and Fish Behavior*. Chicago: University of Chicago Press. Pp. 233–45. [*137*]

Sevenster, P. (1973). Incompatibility of response and reward. In R. A. Hinde and J. Stevenson-Hinde (Eds.), *Constraints on Learning*. London: Academic Press. Pp. 265–83. [*137*]

Seward, J. P. (1949). An experimental analysis of latent learning. *J. Exp. Psychol.* **39,** 177–86. [*209, 210*]

Seward, J. P., Datel, W. E., and Levy, N. (1952). Tests of two hypotheses of latent learning. *J. Exp. Psychol.* **43,** 274–80. [*210*]

Seward, J. P. and Levy, H. (1949). Latent extinction: Sign learning as a factor in extinction. *J. Exp. Psychol.* **39,** 660–8. [*90, 416*]

Seward, J. P., Levy, N., and Handlon, J. H., Jr. (1950). Incidental learning in the rat. *J. Comp. Physiol. Psychol.* **43,** 240–51. [*212*]

Seward, J. P., Pereboom, A. C., Butler, B., and Jones, R. B. (1957). The role of prefeeding in an apparent frustration effect. *J. Exp. Psychol.* **54,** 445–50. [*362*]

Seward, J. P., Shea, R. A., and Davenport, R. H. (1960). Further evidence for the interaction of drive and reward. *Am. J. Psychol.* **73,** 370–9. [*154*]

Seward, J. P., Shea, R. A., and Elkind, D. (1958). Evidence for the interaction of drive and reward. *Am. J. Psychol.* **71,** 404–7. [*154*]

Sgro, J. A., Dyal, J. A., and Anastasio, E. J. (1967). Effects of constant delay of reinforcement on acquisition asymptote and resistance to extinction. *J. Exp. Psychol.* **73,** 634–6. [*155, 431*]

Sgro, J. A. and Weinstock, S. (1963). Effects of delay on subsequent running under immediate reinforcement. *J. Exp. Psychol.* **66,** 260–3. [*389, 431, 432*]

Shanab, M. E. (1971). Positive transfer between nonreward and delay. *J. Exp. Psychol.* **91**, 98–102. [*393*]

Shanab, M. E. and McCuiston, S. (1970). Effects of shifts in magnitude and delay of reward upon runway performance in the rat. *Psychon. Sci.* **21**, 264–6. [*390, 393*]

Shanab, M. E., Sanders, R., and Premack, D. (1969). Positive contrast in the runway obtained with delay of reward. *Science, N.Y.* **164**, 724–5. [*389*]

Shapiro, M. M., Miller, T. M., and Bresnahan, J. L. (1966). Dummy trials, novel stimuli and Pavlovian-trained stimuli: Their effect upon instrumental and consummatory response relationships. *J. Comp. Physiol. Psychol.* **61**, 480–3. [*223, 228*]

Shapiro, M. M., Sadler, E. W., and Mugg, G. J. (1971). Compound stimulus effects during higher order salivary conditioning in dogs. *J. Comp. Physiol. Psychol.* **74**, 222–6. [*20*]

Shearn, D. (1961). Does the heart learn? *Psychol. Bull.* **58**, 452–8. [*104*]

Sheffield, F. D. (1948). Avoidance training and the contiguity principle. *J. Comp. Physiol. Psychol.* **41**, 165–77. [*109, 117, 127, 135, 302*]

Sheffield, F. D. (1965). Relation between classical conditioning and instrumental learning. In W. F. Prokasy (Ed.), *Classical Conditioning: A Symposium.* New York: Appleton-Century-Crofts. Pp. 302–22. [*15, 114, 120, 123, 134, 417*]

Sheffield, F. D. (1966). A drive-induction theory of reinforcement. In R. N. Haber (Ed.), *Current Research and Theory in Motivation.* New York: Holt, Rinehart, & Winston. Pp. 98–111. [*216*]

Sheffield, F. D. and Campbell, B. A. (1954). The role of experience in the "spontaneous" activity of hungry rats. *J. Comp. Physiol. Psychol.* **47**, 97–100. [*224*]

Sheffield, F. D. and Temmer, H. W. (1950). Relative resistance to extinction of escape training and avoidance training. *J. Exp. Psychol.* **40**, 287–98. [*332*]

Sheffield, V. F. (1949). Extinction as a function of partial reinforcement and distribution of practice. *J. Exp. Psychol.* **39**, 511–26. [*367, 419, 421, 422, 440, 457*]

Sheffield, V. F. (1950). Resistance to extinction as a function of the distribution of extinction trials. *J. Exp. Psychol.* **40**, 305–13. [*367, 407, 419*]

Shell, W. F. and Riopelle, A. J. (1958). Progressive discrimination learning in platyrrhine monkeys. *J. Comp. Physiol. Psychol.* **51**, 467–70. [*611*]

Shepp, B. E. and Eimas, P. D. (1964). Intradimensional and extradimensional shifts in the rat. *J. Comp. Physiol. Psychol.* **57**, 357–61. [*597*]

Shepp, B. E. and Schrier, A. M. (1969). Consecutive intradimensional and extradimensional shifts in monkeys. *J. Comp. Physiol. Psychol.* **67**, 199–203. [*597*]

Sherrington, C. S. (1906). *The Integrative Action of the Nervous System.* New Haven: Yale University Press. [*2, 349*]

Shettleworth, S. J. (1972a). Stimulus relevance in the control of drinking and conditioned fear responses in domestic chicks (*Gallus gallus*). *J. Comp. Physiol. Psychol.* **80**, 175–98. (a) [*55*]

Shettleworth, S. J. (1972b). Constraints on learning. In D. S. Lehrman, R. A. Hinde, and E. Shaw (Eds.), *Advances in the Study of Behavior.* New York: Academic Press. Pp. 1–68. (b) [*137*]

Shettleworth, S. J. (1973). Food reinforcement and the organization of behaviour in golden hamsters. In R. A. Hinde and J. Stevenson-Hinde (Eds.), *Constraints on Learning.* London: Academic Press. Pp. 143–263. [*219*]

Shettleworth, S. and Nevin, J. A. (1965). Relative rate of response and relative magnitude of reinforcement in multiple schedules. *J. Exp. Anal. Behav.* **8**, 199–202. [*152, 379*]

Shimp, C. P. (1966). Probabilistically reinforced choice behavior in pigeons. *J. Exp. Anal. Behav.* **9**, 443–55. [*192, 194*]

Shimp, C. P. (1967). The reinforcement of short interresponse times. *J. Exp. Anal. Behav.* **10**, 425–34. [*172*]

Shimp, C. P. (1970). A within-session effect after prolonged training in probability learning by rats. *Psychon. Sci.* **18**, 152–3. [*475*]

Shimp, C. P. and Wheatley, K. L. (1971). Matching to relative reinforcement frequency in multiple schedules with a short component duration. *J. Exp. Anal. Behav.* **15**, 205–10. [*152*]

Shipley, R. H., Mock, L. A., and Levis, D. J. (1971). Effects of several response prevention procedures on activity, avoidance responding, and conditioned fear in rats. *J. Comp. Physiol. Psychol.* **77**, 256–70. [*336, 337*]

Shnidman, S. R. (1968). Extinction of Sidman avoidance behavior. *J. Exp. Anal. Behav.* **11**, 153–6. [*333*]

Shull, R. L. and Brownstein, A. J. (1968). Effects of prefeeding in a fixed-interval reinforcement schedule. *Psychon. Sci.* **11**, 89–90. [*151*]

Shull, R. L. and Pliskoff, S. S. (1967). Changeover delay and concurrent schedules: Some effects on relative performance measures. *J. Exp. Anal. Behav.* **10**, 517–27. [*193*]

Sidman, M. (1953a). Avoidance conditioning with brief shock and no exteroceptive warning signal. *Science, N.Y.* **118**, 157–8. (a) [*300, 322*]

Sidman, M. (1953b). Two temporal parameters of the maintenance of avoidance behavior by the white rat. *J. Comp. Physiol. Psychol.* **46**, 253–61. (b) [*322, 323, 324*]

Sidman, M. (1955a). On the persistence of avoidance behavior. *J. Abnorm. Social Psychol.* **50**, 217–20. (a) [*333*]

Sidman, M. (1955b). Some properties of the warning stimulus in avoidance behavior. *J. Comp. Physiol. Psychol.* **48**, 444–50. (b) [*312*]

Sidman, M. (1962a). Classical avoidance without a warning stimulus. *J. Exp. Anal. Behav.* **5**, 97–104. (a) [*301, 326, 327*]

Sidman, M. (1962b). Reduction of shock frequency as reinforcement for avoidance behavior. *J. Exp. Anal. Behav.* **5**, 247–57. (b) [*309, 326*]

Sidman, M. (1966). Avoidance behavior. In W. K. Honig (Ed.), *Operant Behavior: Areas of Research and Application*. New York: Appleton-Century-Crofts. Pp. 448–98. [*324, 328*]

Sidman, M. and Boren, J. J. (1957). A comparison of two types of warning stimulus in an avoidance situation. *J. Comp. Physiol. Psychol.* **50**, 282–7. [*312*]

Sidman, M., Herrnstein, R. J., and Conrad, D. G. (1957). Maintenance of avoidance behavior by unavoidable shocks. *J. Comp. Physiol. Psychol.* **50**, 553–7. [*10, 21, 83, 225*]

Sidman, M. and Stebbins, W. C. (1954). Satiation effects under fixed-ratio schedules of reinforcement. *J. Comp. Physiol. Psychol.* **47**, 114–16. [*151*]

Sidman, M. and Stoddard, L. T. (1966). Programming perception and learning for retarded children. In N. R. Ellis (Ed.), *International Review of Research in Mental Retardation*, Vol. 2. New York: Academic Press. Pp. 151–208. [*146*]

Siegel, A. (1967). Stimulus generalization of a classically conditioned response along a temporal dimension. *J. Comp. Physiol. Psychol.* **64**, 461–6. [*539*]

z

Siegel, S. (1967). Overtraining and transfer processes. *J. Comp. Physiol. Psychol.* **64,** 471–7. [*569, 590, 592, 603, 606, 607*]

Siegel, S. (1969a). Effect of CS habituation on eyelid conditioning. *J. Comp. Physiol. Psychol.* **68,** 245–8. (a) [*37*]

Siegel, S. (1969b). Generalization of latent inhibition. *J. Comp. Physiol. Psychol.* **69,** 157–9. (b) [*37*]

Siegel, S. (1970). Retention of latent inhibition. *Psychon. Sci.* **20,** 161–2. [*39*]

Siegel, S. and Domjan, M. (1971). Backward conditioning as an inhibitory procedure. *Learning and Motivation* **2,** 1–11. [*59*]

Siegel, S., Hearst, E., George, N., and O'Neal, E. (1968). Generalization gradients obtained from individual subjects following classical conditioning. *J. Exp. Psychol.* **78,** 171–4. [*14*]

Siegel, S. and Wagner, A. R. (1963). Extended acquisition training and resistance to extinction. *J. Exp. Psychol.* **66,** 308–10. [*424, 425*]

Singer, B. F. (1969). Role of delay in magnitude of reward effects in discrimination learning in the rat. *J. Comp. Physiol. Psychol.* **69,** 692–8. [*184*]

Singh, P. J., Sakellaris, P. C., and Brush, F. R. (1971). Retention of active and passive avoidance responses tested in extinction. *Learning and Motivation* **2,** 305–23. [*475*]

Skinner, B. F. (1935). Two types of conditioned reflex and a pseudo type. *J. Gen. Psychol.* **12,** 66–77. [*124*]

Skinner, B. F. (1937). Two types of conditioned reflex: A reply to Konorski and Miller. *J. Gen. Psychol.* **16,** 272–9. [*124*]

Skinner, B. F. (1938). *The Behavior of Organisms.* New York: Appleton-Century-Crofts. [*124, 125, 129, 130, 131, 155, 176, 198, 233, 235, 246, 260, 272, 281, 291, 292, 354, 413*]

Skinner, B. F. (1948). Superstition in the pigeon. *J. Exp. Psychol.* **38,** 168–72. [*98, 99, 111*]

Skinner, B. F. (1950). Are theories of learning necessary? *Psychol. Rev.* **57,** 193–216. [*32, 261, 292, 413, 469, 473*]

Skinner, B. F. (1953). *Science and Human Behavior.* New York: Macmillan. [*273*]

Skinner, B. F. (1966). Operant behavior. In W. K. Honig (Ed.), *Operant Behavior: Areas of Research and Application.* New York: Appleton-Century-Crofts Pp. 12–32. [*144*]

Slamecka, N. J. (1966). A search for spontaneous recovery of verbal associations. *J. Verbal Learn. Verbal Behav.* **5,** 205–7. [*480*]

Slivka, R. M. and Bitterman, M. E. (1966). Classical appetitive conditioning in the pigeon: Partial reinforcement. *Psychon. Sci.* **4,** 181–2. [*17, 73, 74, 224, 350*]

Small, W. S. (1901). Experimental study of the mental processes of the rat. *Am. J. Psychol.* **12,** 206–39. [*553*]

Smith, J. C. and Roll, D. L. (1967). Trace conditioning with X-rays as the aversive stimulus. *Psychon. Sci.* **9,** 11–12. [*54, 67, 68*]

Smith, K. (1954). Conditioning as an artifact. *Psychol. Rev.* **61,** 217–25. [*104*]

Smith, M. C. (1968). CS-US interval and US intensity in classical conditioning of the rabbit's nictitating membrane response. *J. Comp. Physiol. Psychol.* **66,** 679–87. [*17, 61, 62, 63, 66, 71*]

Smith, M. C., Coleman, S. R., and Gormezano, I. (1969). Classical conditioning of the rabbit's nictitating membrane response at backward, simultaneous, and forward CS-US intervals. *J. Comp. Physiol. Psychol.* **69,** 226–31. [*57, 58, 61, 63, 64*]

Smith, M. C., DiLollo, V. D., and Gormezano, I. (1966). Conditioned jaw movement in the rabbit. *J. Comp. Physiol. Psychol.* **62**, 479–83. [*17*]

Smith, M. H., Jr. and Hoy, W. J. (1954). Rate of response during operant discrimination. *J. Exp. Psychol.* **48**, 259–64. [*369, 376*]

Smith, N. F., Misanin, J. R., and Campbell, B. A. (1966). Effect of punishment on extinction of an avoidance response: Facilitation or inhibition? *Psychon. Sci.* **4**, 271–2. [*286*]

Smith, R. F. and Keller, F. R. (1970). Free-operant avoidance in the pigeon using a treadle response. *J. Exp. Anal. Behav.* **13**, 211–14. [*340, 343*]

Snyder, H. L. (1962). Saccharine concentration and deprivation as determinants of instrumental and consummatory response strengths. *J. Exp. Psychol.* **63**, 610–15. [*152*]

Sokolov, Y. N. (1963). *Perception and the Conditioned Reflex.* Oxford: Pergamon Press. [*16, 39*]

Solomon, R. L. (1964). Punishment. *Am. Psychol.* **19**, 239–53. [*280*]

Solomon, R. L. and Brush, E. S. (1956). Experimentally deprived conceptions of anxiety and aversion. In M. R. Jones (Ed.), *Nebraska Symposium on Motivation*, Vol. 4. Lincoln: University of Nebraska Press. Pp. 212–305. [*4*]

Solomon, R. L., Kamin, L. J., and Wynne, L. C. (1953). Traumatic avoidance learning: The outcomes of several extinction procedures with dogs. *J. Abnorm. Social Psychol.* **48**, 291–302. [*89, 286, 331, 333, 336*]

Solomon, R. L. and Turner, L. H. (1962). Discriminative classical conditioning in dogs paralyzed by curare can later control discriminative avoidance responses in the normal state. *Psychol. Rev.* **69**, 202–19. [*80*]

Solomon, R. L. and Wynne, L. C. (1953). Traumatic avoidance learning: Acquisition in normal dogs. *Psychol. Monogr.* **67**, (4, Whole No. 354). [*300, 344*]

Soltysik, S. (1960). Studies on the avoidance conditioning: 3. Alimentary conditioned reflex model of the avoidance reflex. *Acta Biol. Exp.* **20**, 183–92. [*335*]

Soltysik, S. (1963). Inhibitory feedback in avoidance conditioning. *Boln. Inst. Estud. Méd. Biol.* **21**, 433–49. [*307, 335*]

Soltysik, S. (1971). The effect of satiation upon conditioned and unconditioned salivary responses. *Acta Biol. Exp.* **31**, 59–63. [*71*]

Soltysik, S. and Jaworska, K. (1962). Studies on the aversive classical conditioning. 2. On the reinforcing role of shock in the classical leg flexion conditioning. *Acta Biol. Exp.* **22**, 181–91. [*17, 118*]

Spear, N. E. (1967). Retention of reinforcer magnitude. *Psychol. Rev.* **64**, 216–34. [*471*]

Spear, N. E. (1970). Verbal learning and retention. In M. R. D'Amato, *Experimental Psychology. Methodology, Psychophysics, and Learning.* New York: McGraw-Hill. Pp. 543–638. [*478*]

Spear, N. E. (1971). Forgetting as retrieval failure. In W. K. Honig and P. H. R. James (Eds.), *Animal Memory.* New York: Academic Press. Pp. 45–109. [*471, 479, 481*]

Spear, N. E., Hill, W. F., and O'Sullivan, D. J. (1965). Acquisition and extinction after initial trials without reward. *J. Exp. Psychol.* **69**, 25–9. [*462, 471, 472*]

Spear, N. E. and Pavlik, W. B. (1966). Percentage of reinforcement and reward magnitude effects in a T-maze: between and within subjects. *J. Exp. Psychol.* **71**, 521–8. [*439*]

Spear, N. E. and Spitzner, J. H. (1966). Simultaneous and successive contrast effects of reward magnitude in selective learning. *Psychol. Monogr.*, **80**, (10, Whole No. 618). [*399*]

Spear, N. E. and Spitzner, J. H. (1967). Effect of initial nonrewarded trials: Factors responsible for increased resistance to extinction. *J. Exp. Psychol.* **74**, 525–37. [*462, 472*]

Spear, N. E. and Spitzner, J. H. (1968). Residual effects of reinforcer magnitude. *J. Exp. Psychol.* **77**, 135–49. [*153, 472*]

Spear, N. E. and Spitzner, J. H. (1969). Influence of degree of training and prior reinforcer magnitude on contrast effects and resistance to extinction within S. *J. Comp. Physiol. Psychol.* **68**, 427–33. [*399*]

Spence, K. W. (1932). The order of eliminating blinds in maze learning by the rat. *J. Comp. Psychol.* **14**, 9–27. [*188*]

Spence, K. W. (1936). The nature of discrimination learning in animals. *Psychol. Rev.* **43**, 427–49. [*11, 149, 532, 545, 552, 566, 571, 583*]

Spence, K. W. (1937a). The differential response in animals to stimuli varying within a single dimension. *Psychol. Rev.* **44**, 430–44. (a) [*532, 560, 562, 590*]

Spence, K. W. (1937b). Experimental studies of learning and the higher mental processes in infra-human primates. *Psychol. Bull.* **34**, 806–50. (b) [*200, 201*]

Spence, K. W. (1940). Continuous versus non-continuous interpretations of discrimination learning. *Psychol. Rev.* **47**, 271–88. [*583*]

Spence, K. W. (1942). The basis of solution by chimpanzees of the intermediate size problem. *J. Exp. Psychol.* **31**, 257–71. [*563*]

Spence, K. W. (1945). An experimental test of continuity and non-continuity theories of discrimination learning. *J. Exp. Psychol.* **35**, 253–66. [*572*]

Spence, K. W. (1947). The role of secondary reinforcement in delayed reward learning. *Psychol. Rev.* **54**, 1–8. [*157*]

Spence, K. W. (1951). Theoretical interpretations of learning. In C. P. Stone (Ed.), *Comparative Psychology*. Englewood Cliffs, New Jersey: Prentice-Hall. Pp. 239–91. [*94, 95, 213, 223*]

Spence, K. W. (1952). The nature of the response in discrimination learning. *Psychol. Rev.* **59**, 89–93. [*545, 546, 591*]

Spence, K. W. (1956). *Behavior Theory and Conditioning*. New Haven: Yale University Press. [*11, 71, 149, 154, 156, 159, 211, 216, 223, 389, 393*]

Spence, K. W. (1960). *Behavior Theory and Learning*. Englewood Cliffs, New Jersey: Prentice-Hall. [*160, 413*]

Spence, K. W. (1966). Cognitive and drive factors in the extinction of the conditioned eye-blink in human subjects. *Psychol. Rev.* **73**, 445–58. [*409*]

Spence, K. W., Bergmann, G., and Lippitt, R. A. (1950). A study of simple learning under irrelevant motivational-reward conditions. *J. Exp. Psychol.* **40**, 539–51. [*212*]

Spence, K. W., Goodrich, K. P., and Ross, L. E. (1959). Performance in differential conditioning and discrimination learning as a function of hunger and relative response frequency. *J. Exp. Psychol.* **58**, 8–16. [*183*]

Spence, K. W., Haggard, D. F., and Ross, L. E. (1958a). UCS intensity and the associative (habit) strength of the eyelid CR. *J. Exp. Psychol.* **55**, 404–11. (a) [*71*]

Spence, K. W., Haggard, D. F., and Ross, L. E. (1958b). Intrasubject conditioning as a function of the intensity of the unconditioned stimulus. *Science, N.Y.* **128**, 774–5. (b) [*71*]

Spence, K. W. and Lippitt, R. (1946). An experimental test of the sign-Gestalt theory of trial-and-error learning. *J. Exp. Psychol.* **36**, 491–502. [*212, 213*]

Spence, K. W. and Platt, J. R. (1966). UCS intensity and performance in eyelid conditioning. *Psychol. Bull.* **65**, 1–10. [*71*]

Spence, K. W., Platt, J. R., and Matsumoto, R. (1965). Intertrial reinforcement and the partial reinforcement effect as a function of number of training trials. *Psychon. Sci.* **3**, 205–6. [*450, 451*]

Spence, K. W. and Ross, L. E. (1959). A methodological study of the form and latency of eyelid responses in conditioning. *J. Exp. Psychol.* **58**, 376–85. [*102*]

Sperling, S. E. (1965). Reversal learning and resistance to extinction: a review of the rat literature. *Psychol. Bull.* **63**, 281–97. [*603*]

Sperling, S. E. (1970). The ORE in simultaneous and differential reversal: the acquisition task, the acquisition criterion and the reversal task. *J. Exp. Psychol.* **84**, 349–60. [*603*]

Spivey, J. E. (1967). Resistance to extinction as a function of number of N-R transitions and percentage of reinforcement. *J. Exp. Psychol.* **75**, 43–8. [*445*]

Spivey, J. E. and Hess, D. T. (1968). Effect of partial reinforcement trial sequences on extinction performance. *Psychon. Sci.* **10**, 375–6. [*444*]

Spragg, S. D. S. (1940). Morphine addiction in chimpanzees. *Comp. Psychol. Monogr.* **15** (7, Serial No. 79). [*18, 67*]

Sprow, A. J. (1947). Reactively homogeneous compound trial-and-error learning with distributed trials and terminal reinforcement. *J. Exp. Psychol.* **37**, 197–213. [*189*]

Stabler, J. R. (1962). Performance in instrumental conditioning as a joint function of time of deprivation and sucrose concentration. *J. Exp. Psychol.* **63**, 248–53. [*150, 154*]

Staddon, J. E. R. (1965). Some properties of spaced responding in pigeons. *J. Exp. Anal. Behav.* **8**, 19–28. [*179*]

Staddon, J. E. R. (1970). Temporal effects of reinforcement: A negative "frustration" effect. *Learning and Motivation* **1**, 227–47. [*361*]

Staddon, J. E. R. and Innis, N. K. (1969). Reinforcement omission on fixed-interval schedules. *J. Exp. Anal. Behav.* **12**, 689–700. [*170, 361*]

Staddon, J. E. R. and Simmelhag, V. L. (1971). The "superstition" experiment: A reexamination of its implications for the principles of adaptive behavior. *Psychol. Rev.* **78**, 3–43. [*106, 107, 111*]

Stanley, W. C. (1952). Extinction as a function of the spacing of extinction trials. *J. Exp. Psychol.* **43**, 249–60. [*421*]

Stebbins, W. C., Mead, P. B., and Martin, J. M. (1959). The relation of amount of reinforcement to performance under a fixed-interval schedule. *J. Exp. Anal. Behav.* **2**, 351–6. [*152*]

Stein, L. (1958). Secondary reinforcement established with subcortical stimulation. *Science, N.Y.* **127**, 466–7. [*246*]

Steiner, J. (1967). Observing responses and uncertainty reduction. *Q. Jl Exp. Psychol.* **19**, 18–29. [*253*]

Steinman, F. (1967). Retention of alley brightness in the rat. *J. Comp. Physiol. Psychol.* **64**, 105–9. [*476*]

Stern, R. M. (1967). Operant conditioning of spontaneous GSRs: Negative results. *J. Exp. Psychol.* **75**, 128–30. [*130*]

Sterritt, G. M. and Smith, M. P. (1965). Reinforcing effects of specific components of feeding in young leghorn chicks. *J. Comp. Physiol. Psychol.* **59**, 171–5. [*93*]

Stettner, L. J. (1965). Effect of prior reversal and elimination of inhibition on the persistence of a discrimination despite subsequent equal reinforcement of the discriminanda. *J. Comp. Physiol. Psychol.* **60**, 262–4. [*573*]

Stevens, D. A. and Fechter, L. D. (1968). Relative strengths of approach and avoidance tendencies in discrimination learning of rats trained under two types of reinforcement. *J. Exp. Psychol.* **76**, 489–91. [*548, 567, 568*]

Stimmel, D. T. and Adams, P. C. (1969). The magnitude of the frustration effect as a function of the number of previously reinforced trials. *Psychon. Sci.* **16**, 31–2. [*365*]

Stollnitz, F. (1965). Spatial variables, observing responses, and discrimination learning sets. *Psychol. Rev.* **72**, 247–61. [*585*]

Stollnitz, F. and Schrier, A. M. (1968). Learning set without transfer suppression. *J. Comp. Physiol. Psychol.* **66**, 780–83. [*616*]

Stolz, S. B. (1965). Vasomotor response in human subjects: Conditioning and pseudo-conditioning. *Psychon. Sci.* **2**, 181–2. [*92*]

Storms, L. H. and Broen, W. E., Jr. (1966). Drive theories and stimulus generalization. *Psychol. Rev.* **73**, 113–27. [*520*]

Stretch, R. G., McGonigle, B., and Morton, A. (1964). Serial position-reversal learning in the rat: trials/problem and intertrial interval. *J. Comp. Physiol. Psychol.* **57**, 461–3. [*608, 609, 610*]

Stretch, R., Orloff, E. R., and Dalrymple, S. D. (1968). Maintenance of responding by a fixed-interval schedule of electric shock presentation in squirrel monkeys. *Science, N.Y.* **162**, 583–5. [*285*]

Strong, P. N., Jr. (1957). Activity in the white rat as a function of apparatus and hunger. *J. Comp. Physiol. Psychol.* **50**, 596–600. [*350*]

Strongman, K. T. (1965). The effect of anxiety on food intake in the rat. *Q. Jl Exp. Psychol.* **17**, 255–60. [*265*]

Strouthes, A. (1965). Effect of CS-onset UCS-termination delay, UCS duration, CS-onset UCS onset interval, and number of CS-UCS pairings on conditioned fear response. *J. Exp. Psychol.* **69**, 287–91. [*112*]

Stubbs, A. (1969). Contiguity of briefly presented stimuli with food reinforcement. *J. Exp. Anal. Behav.* **12**, 271–8. [*242*]

Stubbs, D. A. and Cohen, S. L. (1972). Second-order schedules: comparison of different procedures for scheduling paired and nonpaired brief stimuli. *J. Exp. Anal. Behav.* **18**, 403–13. [*241*]

Stubbs, D. A. and Pliskoff, S. S. (1969). Concurrent responding with fixed relative rate of reinforcement. *J. Exp. Anal. Behav.* **12**, 887–95. [*183*]

Suiter, R. D. and LoLordo, V. M. (1971). Blocking of inhibitory Pavlovian conditioning in the conditioned emotional response procedure. *J. Comp. Physiol. Psychol.* **76**, 137–44. [*335*]

Surridge, C. T. and Amsel, A. (1966). Acquisition and extinction under single alternation and random partial-reinforcement conditions with a 24-hour intertrial interval. *J. Exp. Psychol.* **72**, 361–8. [*163*]

Surridge, C. T. and Amsel, A. (1968). Confinement duration on rewarded and nonrewarded trials and patterning at 24-hour ITI. *Psychon. Sci.* **10**, 107–8. [*163*]

Sutherland, N. S. (1959). Stimulus analyzing mechanisms. In *Proceedings of a Symposium on the Mechanization of Thought Processes*, Vol. 2. London: Her Majesty's Stationery Office. Pp. 575–609. [*606*]

Sutherland, N. S. (1961a). The methods and findings of experiments on the visual discrimination of shape by animals. *Q. Jl Exp. Psychol. Monogr.* **1**, 1–68. (a) [*545, 556*]

Sutherland, N. S. (1961b). Visual discrimination of horizontal and vertical rectangles by rats in a new discrimination training apparatus. *Q. Jl Exp. Psychol.* **13**, 117–21. (b) [*548, 556, 566*]

Sutherland, N. S. (1966). Partial reinforcement and breadth of learning. *Q. Jl Exp. Psychol.* **18**, 289–302. [*190, 438, 439, 518*]

Sutherland, N. S. (1969). Shape discrimination in rat, octopus, and goldfish: a comparative study. *J. Comp. Physiol. Psychol.* **67**, 160–76. [*548, 556, 570*]

Sutherland, N. S. and Andelman, L. (1967). Learning with one and two cues. *Psychon. Sci.* **15**, 253–4. [*580, 587*]

Sutherland, N. S., Carr, A. E., and Mackintosh, J. A. (1962). Visual discrimination of open and closed shapes by rats. I. Training. *Q. Jl Exp. Psychol.* **14**, 129–39. [*566, 568*]

Sutherland, N. S. and Holgate, V. (1966). Two cue discrimination learning in rats. *J. Comp. Physiol. Psychol.* **61**, 198–207. [*577, 578, 579, 584, 585, 588*]

Sutherland, N. S. and Mackintosh, N. J. (1971). *Mechanisms of Animal Discrimination Learning.* New York: Academic Press. [*42, 51, 149, 192, 213, 434, 439, 506, 545, 554, 558, 566, 571, 572, 575, 577, 582, 584, 590, 592, 600, 602, 603, 605*]

Sutherland, N. S., Mackintosh, N. J., and Mackintosh, J. (1963). Simultaneous discrimination training of *Octopus* and transfer of discrimination along a continuum. *J. Comp. Physiol. Psychol.* **56**, 150–6. [*558, 559, 593*]

Sutherland, N. S., Mackintosh, N. J., and Wolfe, J. B. (1965). Extinction as a function of the order of partial and consistent reinforcement. *J. Exp. Psychol.* **69**, 56–9. [*74, 436*]

Sutterer, J. R. and Obrist, P. A. (1972). Heart rate and general activity alterations of dogs during several aversive conditioning procedures. *J. Comp. Physiol. Psychol.* **80**, 314–26. [*103, 104*]

Sweller, J. (1972). A test between selective attention and stimulus generalization interpretations of the easy-to-hard effect. *Q. Jl Exp. Psychol.* **24**, 352–5. [*596*]

Switalski, R. W., Lyons, J., and Thomas, D. R. (1966). Effects of interdimensional training on stimulus generalization. *J. Exp. Psychol.* **72**, 661–6. [*501*]

Switzer, S. A. (1930). Backward conditioning of the lid reflex. *J. Exp. Psychol.* **13**, 76–97. [*58*]

Szwejkowska, G. (1959). The transformation of differentiated inhibitory stimuli into positive conditioned stimuli. *Acta Biol. Exp.* **19**, 151–9. [*32, 34, 486*]

Szwejkowska, G. and Konorski, J. (1959). The influence of the primary inhibitory stimulus upon the salivary effect of excitatory conditioned stimulus. *Acta Biol. Exp.* **19**, 162–74. [*33, 34*]

Tait, R. W., Marquis, H. A., Williams, R., Weinstein, L., and Suboski, M. D. (1969). Extinction of sensory preconditioning using CER training. *J. Comp. Physiol. Psychol.* **69**, 170–2. [*21*]

Taub, E. and Berman, A. J. (1968). Movement and learning in the absence of sensory feedback. In S. J. Freedman (Ed.), *The Neuropsychology of Spatially Oriented Behavior.* Homewood, Illinois: The Dorsey Press. Pp. 173–92. [*133, 206*]

Taus, S. E. and Hearst, E. (1970). Effects of intertrial (blackout) duration on response rate to a positive stimulus. *Psychon. Sci.* **19**, 265–6. [*375*]

Taus, S. E. and Hearst, E. (1972). Operant discrimination learning after different amounts of reinforced pretraining to the positive stimulus. *J. Exp. Psychol.* **94**, 33–40. [*519*]

Teas, R. C. and Bitterman, M. E. (1952). Perceptual organization in the rat. *Psychol Rev.* **59**, 130–40. [*546, 547*]

Teel, K. S. (1952). Habit strength as a function of motivation during learning. *J. Comp. Physiol. Psychol.* **45**, 188–91. [*183*]

Teghtsoonian, R. and Campbell, B. A. (1960). Random activity of the rat during food deprivation as a function of environmental conditions. *J. Comp. Physiol. Psychol.* **53**, 242–4. [*350*]

Teichner, W. H. (1952). Experimental extinction as a function of the intertrial intervals during conditioning and extinction. *J. Exp. Psychol.* **44**, 170–8. [*407, 419, 420, 421*]

Telegdy, G. A. and Cohen, J. S. (1971). Cue utilization and drive level in albino rats. *J. Comp. Physiol. Psychol.* **75**, 248–53. [*578, 583*]

Terrace, H. S. (1963). Discrimination learning with and without "errors". *J. Exp. Anal. Behav.* **6**, 1–27. [*377, 386*]

Terrace, H. S. (1964). Wavelength generalization after discrimination learning with and without errors. *Science, N.Y.* **144**, 78–80. [*536, 537*]

Terrace, H. S. (1966a). Stimulus control. In W. K. Honig (Ed.), *Operant Behavior: Areas of Research and Application.* New York: Appleton-Century-Crofts. Pp. 271–344. (a) [*373, 377, 530*]

Terrace, H. S. (1966b). Behavioral contrast and the peak shift: Effects of extended discrimination training. *J. Exp. Anal. Behav.* **9**, 613–17. (b) [*372, 538*]

Terrace, H. S. (1966c). Discrimination learning and inhibition. *Science, N.Y.* **154**, 1677–80. (c) [*529, 530*]

Terrace, H. S. (1968). Discrimination learning, the peak shift and behavioral contrast. *J. Exp. Anal. Behav.* **11**, 727–41. [*380, 394, 537, 539*]

Terrace, H. S. (1971). Escape from S-. *Learning and Motivation* **2**, 148–63. [*259, 264*]

Terrace, H. S. (1972). By-products of discrimination learning. In G. H. Bower (Ed.), *The Psychology of Learning and Motivation*, Vol. 5. New York: Academic Press. Pp. 195–265. [*264, 530*]

Theios, J. (1962). The partial reinforcement effect sustained through blocks of continuous reinforcement. *J. Exp. Psychol.* **64**, 1–6. [*436*]

Theios, J. (1963a). Drive stimulus generalization increments. *J. Comp. Physiol. Psychol.* **56**, 691–5. (a) [*151*]

Theios, J. (1963b). Simple conditioning as two-stage all-or-none learning. *Psychol. Rev.* **70**, 403–17. (b) [*340*]

Theios, J. and Blosser, D. (1965). The overlearning reversal effect and magnitude of reward. *J. Comp. Physiol. Psychol.* **59**, 252–7. [*184, 604, 605*]

Theios, J. and Brelsford, J. (1964.) Overlearning-extinction effect as an incentive phenomenon. *J. Exp. Psychol.* **67**, 463–7. [*424, 425, 426*]

Theios, J. and Brelsford, J., Jr. (1966a). Theoretical interpretations of a Markov model for avoidance conditioning. *J. Math. Psychol.* **3**, 140–62. (a) [*302*]

Theios, J. and Brelsford, J. W., Jr. (1966b). A Markov model for classical conditioning: Application to eye-blink conditioning in rabbits. *Psychol. Rev.* **73**, 393–408. (b) [*12, 71*]

Theios, J., Lynch, A. D., and Lowe, W. F., Jr. (1966). Differential effects of shock intensity on one-way and shuttle avoidance conditioning. *J. Exp. Psychol.* **72**, 294–9. [*341, 343*]

Theios, J. and Polson, P. (1962). Instrumental and goal responses in nonresponse partial reinforcement. *J. Comp. Physiol. Psychol.* **55**, 987–91. [*451, 452*]

Thistlethwaite, D. (1951). A critical review of latent learning and related experiments. *Psychol. Bull.* **48**, 97–129. [*212*]

Thistlethwaite, D. (1952). Conditions of irrelevant-incentive learning. *J. Comp. Physiol. Psychol.* **45**, 517–25. [*212*]

Thomas, D. R. and Barker, E. G. (1964). The effects of extinction and "central tendency" on stimulus generalization in pigeons. *Psychon. Sci.* **1**, 119–20. [*493*]

Thomas, D. R., Berman, D. L., Serednesky, G. E., and Lyons, J. (1968). Information value and stimulus configuring as factors in conditioned reinforcement. *J. Exp. Psychol.* **76**, 181–9. [*47, 248, 249*]

Thomas, D. R. and Burr, D. E. S. (1969). Stimulus generalization as a function of the delay between training and testing procedures: a reevaluation. *J. Exp. Anal. Behav.* **12**, 105–9. [*476*]

Thomas, D. R., Burr, D. E. S., and Eck, K. O. (1970c). Stimulus selection in animal discrimination learning: an alternative interpretation. *J. Exp. Psychol.* **86**, 53–62. (c) [*512*]

Thomas, D. R., Ernst, A. J., and Andry, D. K. (1971a). More on masking of stimulus control during generalization testing. *Psychon. Sci.* **23**, 85–6. (a) [*499*]

Thomas, D. R., Freeman, F., Svinicki, J. G., Burr, D. E. S., and Lyons, J. (1970a). Effects of extradimensional training on stimulus generalization. *J. Exp. Psychol. Monogr.* **83**, No. 1, Part 2. (a) [*509, 511, 512, 513, 590, 600, 601*]

Thomas, D. R. and Hiss, R. H. (1963). A test of the "units hypothesis" employing wave-length generalization in human subjects. *J. Exp. Psychol.* **65**, 59–62. [*495*]

Thomas, D. R. and King, R. A. (1959). Stimulus generalization as a function of level of motivation. *J Exp Psychol* **57**, 323–8. [*521*]

Thomas, D. R. and Lopez, L. J. (1962). The effects of delayed testing on generalization slope. *J. Comp. Physiol. Psychol.* **55**, 541–4. [*476*]

Thomas, D. R., Miller, J. T., and Svinicki, J. G. (1971b). Nonspecific transfer effects of discrimination training in the rat. *J. Comp. Physiol. Psychol.* **74**, 96–101. (b) [*601*]

Thomas, D. R. and Setzer, J. (1972). Stimulus generalization gradients for auditory intensity in rats and guinea pigs. *Psychon. Sci.* **28**, 22–4. [*537*]

Thomas, D. R., Svinicki, M. D., and Svinicki, J. G. (1970b). Masking of stimulus control during generalization testing. *J. Exp. Psychol.* **84**, 479–82. (b) [*499, 500*]

Thomas, D. R. and Switalski, R. W. (1966). Comparison of stimulus generalization following variable-ratio and variable-interval training. *J. Exp. Psychol.* **71**, 236–40. [*176, 491, 492, 515, 516, 517*]

Thomas, E. (1971). Role of postural adjustments in conditioning of dogs with electrical stimulation of the motor cortex as the unconditioned stimulus. *J. Comp. Physiol. Psychol.* **76**. 187–98. [*33, 94, 97*]

Thomas, E. and Basbaum, C. (1972). Excitatory and inhibitory processes in hypothalamic conditioning in cats: role of the history of the negative stimulus. *J. Comp. Physiol. Psychol.* **79**, 419–24. [*33, 35*]

Thomas, E. and Wagner, A. R. (1964). Partial reinforcement of the classically conditioned eyelid response in the rabbit. *J. Comp. Physiol. Psychol.* **58**, 157–8. [*24, 73*]

Thompson, M. E. (1944). Learning as a function of absolute and relative amounts of work. *J. Exp. Psychol.* **34**, 506–15. [*187*]

Thompson, R. (1955). Transposition in the white rat as a function of stimulus comparison. *J. Exp. Psychol.* **50**, 185–90. [*565*]

Thompson, R. F. (1958). Primary stimulus generalization as a function of acquisition level in the cat. *J. Comp. Physiol. Psychol.* **51**, 601–6. [*518, 528*]

Thompson, R. F. (1959). Effect of acquisition level upon the magnitude of stimulus generalization across sensory modality. *J. Comp. Physiol. Psychol.* **52**, 183–5. [*487, 518*]

Thompson, R. F. (1965). The neural basis of stimulus generalization. In D. I. Mostofsky (Ed.), *Stimulus Generalization.* Stanford: Stanford University Press. Pp. 154–78. [*486, 490*]

Thompson, R. F. and Spencer, W. A. (1966). Habituation: A model phenomenon for the study of neuronal substrates of behavior. *Psychol. Rev.* **173**, 16–43. [*6, 16, 39*]

Thompson, T. I. (1964). Visual reinforcement in fighting cocks. *J. Exp. Anal. Behav.* **7**, 45–9. [*264*]

Thompson, T. and Bloom, W. (1966). Aggressive behavior and extinction-induced response-rate increase. *Psychon. Sci.* **5**, 335–6. [*263*]

Thompson, T. and Sturm, T. (1965). Classical conditioning of aggressive display in Siamese fighting fish. *J. Exp. Anal. Behav.* **8**, 397–403. [*18*]

Thorndike, E. L. (1898). Animal intelligence: an experimental study of the associative processes in animals. *Psychol. Monogr.* **2** (4, Whole No. 8). [*1*]

Thorndike, E. L. (1911). *Animal Intelligence: Experimental Studies.* New York: Macmillan. [*1, 5, 198, 200, 219*]

Thorndike, E. L. (1913). *Educational Psychology. Vol. II. The Psychology of Learning.* New York: Teachers College, Columbia University. [*272, 291*]

Thorndike, E. L. (1931). *Human Learning.* New York: Appleton-Century-Crofts. [*272, 273, 277, 291*]

Thorndike, E. L. (1932a). *Fundamentals of Learning.* New York: Teachers College. (a) [*272, 273*]

Thorndike, E. L. (1932b). Reward and punishment in animal learning. *Comp. Psychol. Monogr.* **8** (4, Serial No. 39). (b) [*273*]

Thorpe, W. H. (1961). *Bird Song. The Biology of Vocal Communication and Expression in Birds.* Cambridge: Cambridge University Press. [*6*]

Thorpe, W. H. (1963). *Learning and Instinct in Animals.* London: Methuen. [*201*]

Tighe, T. J. and Frey, K. (1972). Subproblem analysis of discrimination shift learning in the rat. *Psychon. Sci.* **28**, 129–33. [*576*]

Tighe, T. J. and Graf, V. (1972). Subproblem analysis of discrimination shift learning in the pigeon. *Psychon. Sci.* **29**, 139–41. [*576*]

Tinklepaugh, O. L. (1928). An experimental study of representative factors in monkeys. *J. Comp. Psychol.* **8**, 197–236. [*214, 215*]

Todorov, J. C. (1972). Component duration and relative response rates in multiple schedules. *J. Exp. Anal. Behav.* **17**, 45–9. [*152*]

Tolman, E. C. (1932). *Purposive Behavior in Animals and Men.* New York: Century. [*7, 100, 105, 199, 205, 207, 214, 215*]

Tolman, E. C. (1934). Theories of learning. In F. A. Moss (Ed.), *Comparative Psychology.* New York: Prentice-Hall. Pp. 367–408. [*79*]

Tolman, E. C. (1948). Cognitive maps in rats and men. *Psychol. Rev.* **55**, 189–208. [*213*]

Tolman, E. C. (1949). There is more than one kind of learning. *Psychol. Rev.* **56**, 144–55. [*7, 125*]

Tolman, E. C. and Gleitman, H. (1949). Studies in learning and motivation: I. Equal reinforcements in both end-boxes, followed by shock in one end-box. *J. Exp. Psychol.* **39**, 810–19. [*210*]

Tolman, E. C. and Honzik, C. H. (1930). Introduction and removal of reward, and maze performance in rats. *University of California Publications in Psychology* **4**, 257–75. [*207, 208, 209*]

Tolman, E. C., Ritchie, B. F., and Kalish, D. (1946). Studies in spatial learning. I. Orientation and the short-cut. *J. Exp. Psychol.* **36**, 13–24. [*553*]

Tombaugh, T. N. (1966). Resistance to extinction as a function of the interaction between training and extinction delays. *Psychol. Rep.* **19**, 791–8. [*407, 431, 432*]

Tombaugh, T. N. (1967). The overtraining extinction effect with a discrete-trial bar-press procedure. *J. Exp. Psychol.* **73**, 632–4. [*423*]

Tombaugh, T. N. (1970). A comparison of the effects of immediate reinforcement, constant delay of reinforcement, and partial delay of reinforcement on performance. *Can. J. Psychol.* **24**, 276–88. [*431*]

Towart, E. M. and Boe, E. E. (1965). Comparison of the correction and the rerun noncorrection methods in maze learning. *Psychol. Rep.* **16**, 407–15. [*185*]

Trapold, M. A. (1962). The effect of incentive motivation on an unrelated reflex response. *J. Comp. Physiol. Psychol.* **55**, 1034–9. [*224*]

Trapold, M. A. (1970). Are expectancies based upon different positive reinforcing events discriminably different? *Learning and Motivation* **1**, 129–40. [*227*]

Trapold, M. A., Carlson, J. G., and Myers, W. A. (1965). The effect of noncontingent fixed- and variable-interval reinforcement upon subsequent acquisition of the fixed-interval scallop. *Psychon. Sci.* **2**, 261–2. [*170*]

Trapold, M. A. and Doren, D. G. (1966). Effect of noncontingent partial reinforcement on the resistance to extinction of a runway response. *J. Exp. Psychol.* **71**, 429–31. [*452*]

Trapold, M. A., Homzie, M., and Rutledge, E. (1964). Backward conditioning and UCR latency. *J. Exp. Psychol.* **67**, 387–91. [*58, 59*]

Trapold, M. A., Lawton, G. W., Dick, R. A., and Goss, D. M. (1968). Transfer of training from differential classical to differential instrumental conditioning. *J. Exp. Psychol.* **76**, 568–73. [*203, 225*]

Trapold, M. A. and Overmier, J. B. (1972). The second learning process in instrumental learning. In A. H. Black and W. F. Prokasy (Eds.), *Classical Conditioning II: Current Research and Theory.* New York: Appleton-Century-Crofts. Pp. 427–52. [*225, 227, 231*]

Trapold, M. A. and Winokur, S. (1967). Transfer from classical conditioning and extinction to acquisition, extinction, and stimulus generalization of a positively reinforced instrumental response. *J. Exp. Psychol.* **73**, 517–25. [*204*]

Traupmann, K. L. (1971). Acquisition and extinction of an instrumental running response with single- or multiple-pellet reward. *Psychon. Sci.* **22**, 61–3. [*152*]

Traupmann, K. L. (1972). Drive, reward, and training parameters, and the overlearning–extinction effect (OEE). *Learning and Motivation* **3**, 359–68. [*425, 427*]

Treichler, F. R., Graham, M. M., and Schweikert, G. E. (1971). Social facilitation of the rat's responding in extinction. *Psychon. Sci.* **22**, 291–3. [*201*]

Treichler, F. R. and Hall, J. R. (1962). The relationship between deprivation weight loss and several measures of activity. *J. Comp. Physiol. Psychol.* **55**, 346–9. [*350*]

Tsang, Y-C. (1934). The functions of the visual areas of the cerebral cortex of the rat in the learning and retention of the maze. I. *Comp. Psychol. Monogr.* **10** (4, Serial No. 50). [*553*]

Tsang, Y-C. (1936). The function of the visual areas of the cerebral cortex of the rat in the learning and retention of the maze. II. *Comp. Psychol. Monogr.* **12** (2, Serial No. 57). [*553*]

Turner, C. (1968). Models of Discrimination Learning. Doctoral thesis, Oxford University. [*558, 573, 575*]

Turner, C. and Mackintosh, N. J. (1972). Stimulus selection and irrelevant stimuli in discrimination learning by pigeons. *J. Comp. Physiol. Psychol.* **78**, 1–9. [*511, 512*]

Turner, L. H. and Solomon, R. L. (1962). Human traumatic avoidance learning: theory and experiments on the operant-respondent distinction and failure to learn. *Psychol. Monogr.* **76** (40, Whole No. 559). [*133, 134, 304*]

Turrisi, F. D., Shepp, B. E., and Eimas, P. D. (1969). Intra- and extra-dimensional shifts with constant- and variable-irrelevant dimensions in the rat. *Psychon. Sci.* **14**, 19–20. [*598*]

Tyler, D. W., Wortz, E. C., and Bitterman, M. E. (1953). The effect of random and alternating partial reinforcement on resistance to extinction in the rat. *Am. J. Psychol.* **66**, 57–65. [*161, 162, 448*]

Uhl, C. N. (1963). Two-choice probability learning in the rat as a function of incentive, probability of reinforcement, and training procedure. *J. Exp. Psychol.* **66**, 443–9. [*191*]

Uhl, C. N. (1967). Persistence in punishment and extinction testing as a function of percentages of punishment and reward in training. *Psychon. Sci.* **8**, 193–4. [*464*]

Uhl, C. N. and Garcia, E. E. (1969). Comparison of omission with extinction in response elimination in rats. *J. Comp. Physiol. Psychol.* **69**, 554–62. [*409*]

Uhl, C. N. and Young, A. G. (1967). Resistance to extinction as a function of incentive, percentage of reinforcement, and number of nonreinforced trials *J. Exp. Psychol.* **73**, 556–64. [*423, 425, 427*]

Underwood, B. J. and Postman, L. (1960). Extraexperimental sources of interference in forgetting. *Psychol. Rev.* **67**, 73–95. [*472, 478*]

Upton, M. (1929). The auditory sensitivity of guinea pigs. *Am. J. Psychol.* **41**, 412–21. [*105*]

Vandercar, D. H. and Schneiderman, N. (1967). Interstimulus interval functions in different response systems during classical discrimination conditioning of rabbits. *Psychon. Sci.* **9**, 9–10 [*65*]

Vanderwolf, C. H. (1971). Limbic-diencephalic mechanisms of voluntary movements. *Psychol. Rev.* **78**, 83–113. [*104*]

Van Dyne, G. C. (1971). Conditioned suppression with a positive US in the rat. *J. Comp. Physiol. Psychol.* **77**, 131–5. [*226*]

Van Houten, R., O'Leary, K. D., and Weiss, S. J. (1970). Summation of conditioned suppression. *J. Exp. Anal. Behav.* **13**, 75–81. [*574*]

Vardaris, R. M. and Fitzgerald, R. R. (1969). Effects of partial reinforcement on a classically conditioned eyeblink response in dogs. *J. Comp. Physiol. Psychol.* **67**, 531–4. [*73*]

Verplanck, W. S. and Hayes, J. R. (1953). Eating and drinking as a function of maintenance schedule. *J. Comp. Physiol. Psychol.* **46**, 327–33. [*212, 351*]

Vieth, A. and Rilling, M. (1972). Comparison of time-out and extinction as determinants of behavioral contrast: an analysis of sequential effects. *Psychon. Sci.* **27**, 281–2. [*375*]

Vincent, S. B. (1915). The white rat and the maze problem. II. The introduction of an olfactory control. *J. Anim. Behav.* **5**, 140–57. [*554*]

Vogel, J. R., Mikulka, P. J., and Spear, N. E. (1966). Effect of interpolated extinction and level of training on the "depression" effect. *J. Exp. Psychol.* **72**, 51–60. [*390*]

Vogel, J. R., Mikulka, P. J., and Spear, N. E. (1968). Effects of shifts in sucrose and saccharine concentrations on licking behavior in the rat. *J. Comp. Physiol. Psychol.* **66**, 661–6. [*391*]

vom Saal, W. (1972). Choice between stimuli previously presented separately. *Learning and Motivation* **3**, 209–222. [*437, 438*]

vom Saal, W. and Jenkins, H. M. (1970). Blocking the development of stimulus control. *Learning and Motivation* **1**, 52–64. [*583, 585*]

Wagner, A. R. (1959). The role of reinforcement and nonreinforcement in an "apparent frustration effect". *J. Exp. Psychol.* **57**, 130–6. [*266, 362*]

Wagner, A. R. (1961). Effects of amount and percentage of reinforcement and number of acquisition trials on conditioning and extinction. *J. Exp. Psychol.* **62**, 234–42. [*75, 151, 160, 424, 425, 427, 430, 455*]

Wagner, A. R. (1963a). Conditioned frustration as a learned drive. *J. Exp. Psychol.* **66**, 142–8. (a) [*263, 264, 311*]

Wagner, A. R. (1963b). Overtraining and frustration. *Psychol. Rep.* **13**, 717–18. (b) [*424, 426*]

Wagner, A. R. (1966). Frustration and punishment. In R. N. Haber (Ed.), *Current Research in Motivation*. New York: Holt, Rinehart, & Winston, Pp. 229–39. [*265*]

Wagner, A. R. (1969a). Incidental stimuli and discrimination learning. In R. M. Gilbert and N. S. Sutherland (Eds.), *Animal Discrimination Learning*. London: Academic Press. Pp. 83–111. (a) [*503, 509, 510, 511, 512, 513, 518, 584, 590, 594, 596, 600*]

Wagner, A. R. (1969b). Stimulus validity and stimulus selection in associative learning. In N. J. Mackintosh and W. K. Honig (Eds.), *Fundamental Issues in Associative Learning*. Halifax: Dalhousie University Press. Pp. 90–122. (b) [*47, 48, 249, 503, 518*]

Wagner, A. R. (1969c). Frustrative nonreward: A variety of punishment. In B. A. Campbell and R. M. Church (Eds.), *Punishment and Aversive Behavior*. New York: Appleton-Century-Crofts. Pp. 157–81. (c) [*260, 262*]

Wagner, A. R. (1971). Elementary associations. In H. H. Kendler and J. T. Spence (Eds.), *Essays in Neobehaviorism: A Memorial Volume to Kenneth W. Spence*. New York: Appleton-Century-Crofts. Pp. 187–213. [*34, 35, 49, 586*]

Wagner, A. R., Logan, F. A., Haberlandt, K., and Price, T. (1968). Stimulus selection in animal discrimination learning. *J. Exp. Psychol.* **76**, 171–80. [*47, 48, 249, 439, 503, 511, 512, 513, 582, 585*]

Wagner, A. R. and Rescorla, R. A. (1972). Inhibition in Pavlovian conditioning: application of a theory. In R. A. Boakes and M. S. Halliday (Eds.), *Inhibition and Learning*. London: Academic Press, Pp. 301–36. [*35, 36, 50, 51, 470*]

Wagner, A. R., Siegel, L. S. and Fein, G. G. (1967a). Extinction of conditioned fear as a function of percentage of reinforcement. *J. Comp. Physiol. Psychol.* **63**. 160–4. (a) [*24, 72, 74*]

Wagner, A. R., Siegel, S., Thomas, E., and Ellison, G. D. (1964). Reinforcement history and the extinction of a conditioned salivary response. *J. Comp. Physiol. Psychol.* **58**, 354–8. [*11, 14, 24, 70, 72, 73, 428*]

Wagner, A. R., Thomas, E., and Norton, T. (1967b). Conditioning with electrical stimulation of motor cortex: Evidence of a possible source of motivation. *J. Comp. Physiol. Psychol.* **64**, 191–9. (b) [*33, 94*]

Wahlsten, D. L. and Cole, M. (1972). Classical and avoidance training of leg flexion in the dog. In A. H. Black and W. F. Prokasy (Eds.), *Classical Conditioning II: Current Research and Theory.* New York: Appleton-Century-Crofts. Pp. 379–408. [*17, 116, 117, 118, 127, 135, 206, 305, 325, 333*]

Walker, E. L., Knotter, M. C., and DeValois, R. L. (1950). Drive specificity and learning: The acquisition of a spatial response to food under conditions of water deprivation and food satiation. *J. Exp. Psychol.* **40**, 161–8. [*212*]

Wallach, H. (1948). Brightness constancy and the nature of achromatic colors. *J. Exp. Psychol.* **38**, 310–24. [*565*]

Waller, T. G. (1968). Effects of magnitude of reward in spatial and brightness discrimination tasks. *J. Comp. Physiol. Psychol.* **66**, 122–7. [*184*]

Waller, T. G. (1970). Facilitation of an extradimensional shift with overtraining in rats. *Psychon. Sci.* **20**, 172–4. [*599*]

Waller, T. G. (1971). The effect of percentage of reward on compound-cue discrimination learning by rats. *Learning and Motivation* **2**, 376–85. [*439*]

Waller, T. G. (1973). Effect of consistency of reward during runway training on subsequent discrimination performance in rats. *J. Comp. Physiol. Psychol.* **83**, 120–3. [*439*]

Walters, G. C. and Glazer, R. D. (1971). Punishment of instinctive behavior in the Mongolian gerbil. *J. Comp. Physiol. Psychol.* **75**, 331–40. [*276, 287, 288*]

Warden, C. J. (1927). The historical development of comparative psychology. *Psychol. Rev.* **34**, 135–68. [*1*]

Warden, C. J. and Aylesworth, M. (1927). The relative value of reward and punishment in the formation of a visual discrimination habit in the white rat. *J. Comp. Psychol.* **7**, 117–27. [*273*]

Warden, C. J. and Jackson, T. A. (1935). Imitative behavior in the rhesus monkey. *J. Genet. Psychol.* **46**, 103–5. [*201*]

Warner, L. H. (1932). The association span of the white rat. *J. Genet. Psychol.* **41**, 57–90. [*321*]

Warren, J. M. (1953). Additivity of cues in visual pattern discriminations by monkeys. *J. Comp. Physiol. Psychol.* **46**, 484–6. [*577*]

Warren, J. M. (1965). Primate learning in comparative perspective. In A. M. Schrier, H. F. Harlow, and F. Stollnitz (Eds.), *Behavior of Nonhuman Primates*, Vol. 1. New York: Academic Press. Pp. 249–81. [*611*]

Warren, J. M. (1966). Reversal learning and the formation of learning sets by cats and rhesus monkeys. *J. Comp. Physiol. Psychol.* **61**, 421–8. [*613, 614*]

Warren, J. M., Derdzinski, D., Hirayoshi, I., and Mumma, R. (1970). Some tests of attention theory with cats. In D. Mostofsky (Ed.), *Attention: Contemporary Theory and Analysis.* New York: Appleton-Century-Crofts. Pp. 275–99. [*578, 587*]

Warren, J. M. and Warren, H. B. (1969). Two-cue discrimination learning by rhesus monkeys. *J. Comp. Physiol. Psychol.* **69**, 688–91. [*578, 579, 584*]

Wasserman, E. A. (1973). The effect of redundant contextual stimuli on auto-shaping the pigeon's keypeck. *Animal Learning and Behavior* **1**, 198–206. [*107*]

Watson, J. B. (1907). Kinaesthetic and organic sensations: their role in the reactions of the white rat to the maze. *Psychol. Monogr.* **8** (2, Whole No. 33). [*553*]

Watson, J. B. (1914). *Behavior: An Introduction to Comparative Psychology*. New York: Holt. [*79*]

Webb, W. B. (1949). The motivational aspect of an irrelevant drive in the behavior of the white rat. *J. Exp. Psychol.* **39**, 1–14. [*350*]

Webb, W. B. (1950). A test of "relational" vs. "specific stimulus" learning in discrimination problems. *J. Comp. Physiol. Psychol.* **43**, 70–2. [*548, 570*]

Webb, W. B. and Nolan, C. Y. (1953). Cues for discrimination as secondary reinforcing agents: A confirmation. *J. Comp. Physiol. Psychol.* **46**, 180–1. [*247*]

Wegner, N. and Zeaman, D. (1958). Strength of cardiac conditioned responses with varying unconditioned stimulus durations. *Psychol. Rev.* **65**, 238–41. [*112*]

Weinstein, L. (1970). Negative incentive contrast with sucrose. *Psychon. Sci.* **19**, 13–14. [*391*]

Weinstock, S. (1954). Resistance to extinction of a running response following partial reinforcement under widely spaced trials. *J. Comp. Physiol. Psychol.* **47**, 318–22. [*443, 446, 457*]

Weinstock, S. (1958). Acquisition and extinction of a partially reinforced running response at a 24-hour intertrial interval. *J. Exp. Psychol.* **46**, 151–8. [*72, 160, 441, 443, 446, 457*]

Weise, P. and Bitterman, M. E. (1951). Response selection in discrimination learning. *Psychol. Rev.* **58**, 185–95. [*547*]

Weisman, R. G. (1965). Experimental comparison of classical and instrumental appetitive conditioning. *Am. J. Psychol.* **78**, 423–31. [*143*]

Weisman, R. G. (1969). Some determinants of inhibitory stimulus control. *J. Exp. Anal. Behav.* **12**, 443–50. [*378, 380, 394, 527*]

Weisman, R. G. (1970). Factors influencing inhibitory stimulus control: differential reinforcement of other behavior during discrimination training. *J. Exp. Anal. Behav.* **14**, 87–91. [*527*]

Weisman, R. G. and Litner, J. S. (1969). Positive conditioned reinforcement of Sidman avoidance behavior in rats. *J. Comp. Physiol. Psychol.* **68**, 597–603. [*29, 33, 83, 307, 311, 319, 320*]

Weisman, R. G. and Litner, J. S. (1971). Role of the intertrial interval in Pavlovian differential conditioning of fear in rats. *J. Comp. Physiol. Psychol.* **74**, 211–18. [*307, 320, 323*]

Weisman, R. G. and Palmer, J. A. (1969). Factors influencing inhibitory stimulus control: discrimination training and prior non-differential reinforcement. *J. Exp. Anal. Behav.* **12**, 229–37. [*529*]

Weisman, R. G. and Ramsden, M. (1973). Discrimination of a response-independent component in a multiple schedule. *J. Exp. Anal. Behav.* **19**, 55–64. [*381*]

Weisner, M. H., Finger, F. W., and Reid, L. S. (1960). Activity changes under food deprivation as a function of recording device. *J. Comp. Physiol. Psychol.* **53**, 470–4. [*350*]

Weiss, J. M., Krieckhaus, E. E., and Conte, R. (1968). Effects of fear conditioning on subsequent avoidance behavior and movement. *J. Comp. Physiol. Psychol.* **65**, 413–21. [*218*]

Weiss, R. F. (1960). Deprivation and reward magnitude effects on speed through-out the goal gradient. *J. Exp. Psychol.* **60**, 384–90. [*154*]

Weiss, S. J. (1971). Discrimination training and stimulus compounding: consideration of non-reinforcement and response differentiation consequences of S△. *J. Exp. Anal. Behav.* **15**, 387–402. [*574*]

Weissman, A. (1962). Nondiscriminated avoidance behavior in a large sample of rats. *Psychol. Rep.* **10**, 591–600. [*301*]

Wells, M. (1968). *Lower Animals.* London: Weidenfeld and Nicolson. [*27*]

Wendt, G. R. (1936). An interpretation of inhibition of conditioned reflexes as competition between reaction systems. *Psychol. Rev.* **43**, 258–81. [*91, 411, 412*]

Wendt, G. R. (1937). Two and one-half year retention of a conditioned response. *J. Gen. Psychol.* **17**, 178–80. [*473*]

Westbrook, R. F. (1973). Failure to obtain positive contrast when pigeons press a bar. *J. Exp. Anal. Behav.* **20**, 499–510 [*376*]

Wever, E. G. (1930). The upper limit of hearing in the cat. *J. Comp. Psychol.* **10**, 221–33. [*105*]

Whatmore, G. B., Morgan, E. A., and Kleitman, N. (1946). The influence of avoidance conditioning on the course of nonavoidance conditioning in dogs. *Am. J. Physiol.* **145**, 432–5. [*117*]

White, C. T. and Schlosberg, H. (1952). Degree of conditioning of the GSR as a function of the period of delay. *J. Exp. Psychol.* **43**, 357–62. [*17*]

Wickens, D. D., Born, D. G., and Wickens, C. D. (1963). Response strength to a compound conditioned stimulus as a function of the element interstimulus interval. *J. Comp. Physiol. Psychol.* **56**, 727–31. [*45, 46, 245*]

Wickens, D. D., Schroder, H. M. and Snide, J. D. (1954). Primary stimulus generalization of the GSR under two conditions. *J. Exp. Psychol.* **47**, 52–6. [*514*]

Wickens, D. D. and Wickens, C. D. (1942). Some factors related to pseudocondi-tioning. *J. Exp. Psychol.* **31**, 518–26. [*27, 109*]

Wike, E. L., Blocher, O., and Knowles, J. M. (1963). Effect of drive level and turn-ing preference on selective learning and habit reversal. *J. Comp. Physiol. Psychol.* **56**, 696–9. [*183*]

Wike, E. L. and Chen, J. S. (1971). Runway performance and reward magnitude. *Psychon. Sci.* **21**, 139–40. [*152*]

Wike, E. L., Kintsch, W., and Gutekunst, R. (1959). Patterning effects in partially delayed reinforcement. *J. Comp. Physiol. Psychol.* **52**, 411–14. [*433, 454*]

Wike, E. L. and McWilliams, J. (1967). Duration of delay, delay-box confine-ment, and runway performance. *Psychol. Rep.* **21**, 865–70. [*432*]

Wike, E. L., Platt, J. R., and Knowles, J. M. (1962). The reward value of getting out of a starting box: further extensions of Zimmerman's work. *Psychol. Rec.* **12**, 397–400. [*242*]

Wike, E. L., Platt, J. R., Wicker, A., and Tesar, V. (1964). Effects of random vs. alternating delay of reinforcement in training upon extinction performance. *Psychol. Rep.* **14**, 826. [*454*]

Wilcoxon, H. C., Dragoin, W. B., and Kral, P. A. (1971). Illness-induced aversions in rat and quail: relative salience of visual and gustatory cues. *Science, N.Y.* **171**, 826–8. [*55, 69*]

Williams, B. A. (1971). The effects of intertrial interval on discrimination reversal learning in the pigeon. *Psychon. Sci.* **23**, 241–3. [*610*]

Williams, B. A. (1972). Probability learning as a function of momentary reinforcement probability. *J. Exp. Anal. Behav.* **17,** 363–8. [*194*]

Williams, D. R. (1965). Classical conditioning and incentive motivation. In W. F. Prokasy (Ed.), *Classical Conditioning: A Symposium.* New York: Appleton-Century-Crofts. Pp. 340–57. [*61, 62, 206, 223, 229, 232, 245, 349, 357, 358, 359*]

Williams, D. R. (1968). The structure of response rate. *J. Exp. Anal. Behav.* **11,** 251–8. [*177*]

Williams, D. R. and Williams, H. (1969). Auto-maintenance in the pigeon: Sustained pecking despite contingent non-reinforcement. *J. Exp. Anal. Behav.* **12,** 511–20. [*17, 106, 121, 136*]

Williams, S. B. (1938). Resistance to extinction as a function of the number of reinforcements. *J. Exp. Psychol.* **23,** 506–21. [*423*]

Willis, R. D. (1969). The partial reinforcement of conditioned suppression. *J. Comp. Physiol. Psychol.* **68,** 289–95. [*73*]

Wilson, G. T. (1973). Counterconditioning versus forced exposure in extinction of avoidance responding and conditioned fear in rats. *J. Comp. Physiol. Psychol.* **82,** 105–14. [*333, 336*]

Wilson, J. J. (1964). Level of training and goal-box movements as parameters of the partial reinforcement effect. *J. Comp. Physiol. Psychol.* **57,** 211–13. [*462*]

Wilson, M. P. and Keller, F. S. (1953). On the selective reinforcement of spaced responses. *J. Comp. Physiol. Psychol.* **46,** 190–3. [*173*]

Wilson, W., Weiss, E. J., and Amsel, A. (1955). Two tests of the Sheffield hypothesis concerning resistance to extinction, partial reinforcement, and distribution of practice. *J. Exp. Psychol.* **50,** 51–60. [*420, 457*]

Wilton, R. N. (1967). On frustration and the PRE. *Psychol. Rev.* **74,** 149–50. [*441*]

Wilton, R. N. (1972). The role of information in the emission of observing responses and partial reinforcement acquisition phenomena. *Learning and Motivation* **3,** 479–99. [*161*]

Wilton, R. N. and Clements, R. O. (1971a). Observing responses and informative stimuli. *J. Exp. Anal. Behav.* **15,** 199–204. (a) [*256*]

Wilton, R. N. and Clements, R. O. (1971b). Behavioral contrast as a function of the duration of an immediately preceding period of extinction. *J. Exp. Anal. Behav.* **16,** 425–8. (b) [*359, 374*]

Wilton, R. N. and Clements, R. O. (1972). A failure to demonstrate behavioral contrast when the S+ and S− components of a discrimination schedule are separated by about 23 hours. *Psychon. Sci.* **28,** 137–9. [*374*]

Winefield, A. H. and Jeeves, M. A. (1971). The effect of overtraining on transfer between tasks involving the same stimulus dimension. *Q. Jl Exp. Psychol.* **23,** 234–42. [*592*]

Wittlin, W. A. and Brookshire, K. H. (1968). Apomorphine-induced conditioned aversion to a novel food. *Psychon. Sci.* **12,** 217–18. [*54, 69*]

Wodinsky, J., Varley, M. A., and Bitterman, M. E. (1954). Situational determinants of the relative difficulty of simultaneous and successive discrimination. *J. Comp. Physiol. Psychol.* **47,** 337–40. [*546, 548, 549, 599*]

Wolach, A. H., Sayeed, H., and Foster, K. (1972). Pattern running to differential reinforcement. *Learning and Motivation* **3,** 500–508. [*401*]

Wolfe, J. B. (1934). The effect of delayed reward upon learning in the white rat. *J. Comp. Psychol.* **17,** 1–21. [*157*]

Wolfe, J. B. (1936). Effectiveness of token-rewards for chimpanzees. *Comp. Psychol. Monogr.* **12** (5, Serial No. 60). [*237, 239, 241, 242*]

Wolff, J. L. (1967). Concept-shift and discrimination-reversal learning in humans. *Psychol. Bull.* **68**, 369–408. [*597*]

Wolfle, H. M. (1930). Time factors in conditioning finger withdrawal. *J. Gen. Psychol.* **4**, 372–8. [*58*]

Wong, P. T. P. (1971). Coerced approach to shock, punishment of competing responses and resistance to extinction in the rat. *J. Comp. Physiol. Psychol* **76**, 275–81. [*464*]

Woodard, W. T. and Bitterman, M. E. (1973). Pavlovian analysis of avoidance conditioning in the goldfish (*Carassius auratus*). *J. Comp. Physiol. Psychol.* **82**, 123–9. [*304*]

Woodard, W. T., Schoel, W. M., and Bitterman, M. E. (1971). Reversal learning with singly presented stimuli in pigeons and goldfish. *J. Comp. Physiol. Psychol.* **76**, 460–7. [*609*]

Woodworth, R. S. and Schlosberg, H. (1954). *Experimental Psychology*. New York: Holt. [*480*]

Wyckoff, L. B., Jr. (1952). The role of observing responses in discrimination learning. Part I. *Psychol. Rev.* **59**, 431–42. [*252, 253*]

Wyckoff, L. B. (1959). Toward a quantitative theory of secondary reinforcement. *Psychol. Rev.* **66**, 68–78 [*254*]

Wyckoff, L. B., Sidowski, J., and Chambliss, D. J. (1958). An experimental study of the relationship between secondary reinforcing and cue effects of a stimulus. *J. Comp. Physiol. Psychol.* **51**, 103–9. [*236*]

Wynne, J. D. and Brogden, W. J. (1962). Effect upon sensory preconditioning of backward, forward and trace preconditioning training. *J. Exp. Psychol.* **64**, 422–3. [*21*]

Yarczower, M. (1970). Behavioral contrast and inhibitive stimulus control. *Psychon. Sci.* **18**, 1–3. [*369, 539*]

Yarczower, M. and Curto, K. (1972). Stimulus control in pigeons after extended discriminative training. *J. Comp. Physiol. Psychol.* **80**, 484–9. [*530, 539*]

Yehle, A. L. (1968). Divergences among rabbit response systems during three-tone classical discrimination conditioning. *J. Exp. Psychol.* **77**, 468–73. [*22*]

Yehle, A., Dauth, G., and Schneiderman, N. (1967). Correlates of heart-rate classical conditioning in curarized rabbits. *J. Comp. Physiol. Psychol.* **64**, 98–104. [*80*]

Yelen, D. (1969). Magnitude of the frustration effect and number of training trials. *Psychon. Sci.* **15**, 137–8. [*365*]

Yerkes, R. M. and Dodson, J. D. (1908). The relation of strength of stimulus to rapidity of habit-formation. *J. Comp. Neurol. Psychol.* **18**, 458–82. [*184*]

Yoshioka, J. G. (1929). Weber's law in the discrimination of maze distance by the white rat. *University of California Publications in Psychology* **4**, 155–84. [*187*]

Young, A. G. (1966). Resistance to extinction as a function of number of non-reinforced trials and effortfulness of response. *J. Exp. Psychol.* **72**, 610–13. [*416*]

Young, F. A. (1965). Classical conditioning of autonomic functions. In W. F. Prokasy (Ed.), *Classical Conditioning: A Symposium*. New York: Appleton-Century-Crofts. Pp. 358–77. [*97*]

Young, J. Z. (1960). Unit processes in the formation of representations in the memory of *Octopus*. *Proc. Roy. Soc.* Series B, **153**, 1–17. [*27*]

Zamble, E. (1967). Classical conditioning of excitement anticipatory to food reward. *J. Comp. Physiol. Psychol.* **63**, 526–9. [*17, 224*]

Zamble, E. (1969). Conditioned motivational patterns in instrumental responding of rats. *J. Comp. Physiol. Psychol.* **69**, 536–43. [*170, 224*]

Zaretsky, H. H. (1965). Runway performance during extinction as a function of drive and incentive. *J. Comp. Physiol. Psychol.* **60**, 463–4. [*427*]

Zaretsky, H. H. (1966). Learning and performance in the runway as a function of the shift in drive and incentive. *J. Comp. Physiol. Psychol.* **62**, 218–21. [*151*]

Zbrozyna, A. W. (1958). On the conditioned reflex of the cessation of the act of eating. I. Establishment of the conditioned cessation reflex. *Acta Biol. Exp.* **18**, 137–62. [*59, 262*]

Zeaman, D. (1949). Response latency as a function of the amount of reinforcement. *J. Exp. Psychol.* **39**, 446–83. [*151, 153, 369, 389*]

Zeaman, D. and House, B. J. (1963). The role of attention in retardate discrimination learning. In N. R. Ellis (Ed.), *Handbook of Mental Deficiency: Psychological Theory and Research*. New York: McGraw-Hill. Pp. 159–223. [*42, 150, 566, 584, 590*]

Zeaman, D. and Smith, R. W. (1965). Review of some recent findings in human cardiac conditioning. In W. F. Prokasy (Ed.), *Classical Conditioning: A Symposium*. New York: Appleton-Century-Crofts. Pp. 378–418. [*103*]

Zeiler, M. D. and Paul, B. J. (1965). Intra-pair similarity as a determinant of component and configuration discrimination. *Am. J. Psychol.* **78**, 476–80. [*547*]

Zeiner, A. and Grings, W. W. (1968). Backward conditioning: A replication with emphasis on conceptualizations by the subject. *J. Exp. Psychol.* **76**, 232–5. [*59*]

Zener, K. (1937). The significance of behavior accompanying conditioned salivary secretion for theories of the conditioned response. *Am. J. Psychol.* **50**, 384–403. [*100, 105, 106, 107, 123, 411, 412*]

Zentall, T., Collins, N., and Hearst, E. (1971). Generalization gradients around a formerly positive S−. *Psychon. Sci.* **22**, 257–9. [*529*]

Zielinski, K. (1966). "Inhibition of delay" as a mechanism of the gradual weakening of the conditioned emotional response. *Acta Biol. Exp.* **26**, 407–18. [*61*]

Ziff, D. R. and Capaldi, E. J. (1971). Amytal and the small trial partial reinforcement effect: Stimulus properties of early trial nonrewards. *J. Exp. Psychol.* **87**, 263–9. [*462, 465*]

Zimmerman, D. W. (1959). Sustained performance in rats based on secondary reinforcement. *J. Comp. Physiol. Psychol.* **52**, 353–8. [*242*]

Zimmerman, D. W. (1969). Concurrent schedules of primary and conditioned reinforcement in rats. *J. Exp. Anal. Behav.* **12**, 261–8. [*239, 241, 242*]

Zych, K. A. and Wolach, A. H. (1973). Resistance to extinction in the goldfish (*Carassius auratus*). *J. Comp. Physiol. Psychol.* **82**, 115–22. [*401, 467*]

SUBJECT INDEX

A

Absolute cues in discrimination learning, 557–9, 563, 565

Acquired distinctiveness of cues, 591–3, 615

Acquisition,
of classical CRs, 9–13
of choice behaviour, 148–9
of instrumental responses, 147–50

Acquisition–extinction series, 410, 441–442

Acquisition level (*see* Overtraining)

Activity,
as measure of general drive, 224, 350

Adaptation level theory,
and contrast effects, 370–1, 390

Adaptation to shock, 283–4

Additivity of cues (*see also* Summation),
in control of choice behaviour, 573–576
in learning, 573, 576–7

After effects (*see* Alternation learning, Generalization decrement, Memory traces, Sequential variables and resistance to extinction, Stimulus traces)

Aggression,
as conditioned response, 18
as evidence of frustration, 263

Alcohol, *see* Drugs

Alternation learning,
cues used in, 161–4
effect of ITI on, 162–4, 459
and frustration effect, 364
in multiple-unit mazes, 555
parallel with FI schedules, 166
and resistance to extinction, 448–9

and successive negative contrast, 401
and transfer to brightness discrimination, 601

Amobarbital (*see* Drugs)

Anxiety (*see* Fear)

Approach-avoidance conflict,
effect of drugs, 265
generalization of, 521–4

Associative competition theory,
and blocking, 50–1, 584, 586–9
and inhibitory conditioning, 51, 589
and overshadowing, 50–1, 584, 586–589
and transfer along a continuum, 596–597

Associative learning (*see also* Expectancy theory, Stimulus-response theory),
historical background to, 1–3
differences between classical and instrumental associations, 125–6, 138–9
nature of associations in classical conditioning, 79–91, 138–9
nature of associations in instrumental learning, 138–9, 198–222
and non associative learning, 6
parallels between classical and instrumental associations, 216–22
theories of, 7

Attention (*see also* Stimulus selection),
and acquired distinctiveness of cues, 593–4
and blocking and overshadowing, 51, 584, 586, 588–9
and discrimination learning with redundant stimuli, 570–89

V

Variability,
  of acquisition and extinction and
    resistance to extinction, 434–5
Variable interval (VI) schedules,
  definition, 164–5
  characteristics of behaviour main-
    tained by, 167
  and momentary probability of rein-
    forcement, 169
  yoked to VR schedule, 176–9
Variable ratio (VR) schedules,
  definition, 164–5
  characteristics of behaviour main-
    tained by, 168
  rate of responding compared to VI
    schedules, 176–7
Vicious-circle effect (*see* Punishment of
  avoidance and escape responses)
Voluntary responses,
  in eyelid conditioning, 101–3

W

Warning signals,

as CSs eliciting fear, 299, 312–13
as discriminative stimuli for avoidance, 312–13
loss of aversive properties during extinction, 333–5
preference for signalled *vs* unsignalled
  shock, 253–5
protection from extinction, 335–7
termination of as reinforcement for
  avoidance, 315–20
trace signals and avoidance, 321
Weber's law, 187
Win–stay, lose–shift (*see* Response
  strategies)

Y

Yerkes–Dodson Law, 183–4
Yoked control,
  definition, 119
  in omission schedules, 119–22
  problems of bias in, 120, 130